SIGNAL THEORY AND PROCESSING

The Artech House Communication and Electronic Defense Library

Principles of Secure Communication Systems by Don J. Torrieri
Introduction to Electronic Warfare by D. Curtis Schleher
Electronic Intelligence: The Analysis of Radar Signals by Richard G. Wiley
Electronic Intelligence: The Interception of Radar Signals by Richard G. Wiley
Signal Theory and Random Processes by Harry Urkowitz
Signal Theory and Processing by Frederic de Coulon
Digital Signal Processing by Murat Kunt
Analysis and Synthesis of Logic Systems by Daniel Mange
Mathematical Methods of Information Transmission by Kurt Arbenz and Jean-Claude Martin
Advanced Mathematics for Practicing Engineers by Kurt Arbenz and Alfred Wohlhauser
Codes for Error-Control and Synchronization by Djimitri Wiggert
Machine Cryptography and Modern Cryptanalysis by Cipher A. Deavours and Louis Kruh
Microcomputer Tools for Communication Engineering by S.T. Li, J.C. Logan, J.W. Rockway, and D.W.S. Tam

SIGNAL THEORY AND PROCESSING

Frederic de Coulon

ARTECH HOUSE, INC.
610 Washington Street
Dedham, MA 02026

International Standard Book Number: 0-89006-185-8
Library of Congress Catalog Card Number: 85-073435

Translation of *Theorie et Traitement des Signaux*, originally published in French as volume VI of the *Traité d'Électricité* by the Presses Polytechniques Romandes, Lausanne, Switzerland. © 1984.

Introduction

Overview

The main tasks of *signal processing systems* are to generate, detect, and interpret signals carrying valuable information. Electronics and computer science are the usual required tools. Applications are in the areas where perception, transmission, and processing of information are important: telecommunications, scientific instrumentation, industrial automation, biomedical engineering, *et cetera.* This field is also related to image processing, pattern recognition, robotics, and artificial intelligence.

Signal theory is the basic tool for analyzing and designing signal processing systems; it is a set of concepts and mathematical models based on functional analysis, linear algebra, and probability theory.

Its starting point is the series expansion of orthogonal functions, the Fourier model being the most fruitful one. It induces the rich concepts of time-frequency duality and frequency spectrum, which, combined with the notion of correlation and adequate statistical models, can also be applied to the analysis of analog or sampled, deterministic or random, signals. The analytic signal and complex envelope theory is an extension of the concept of phasors, introduced in electrical engineering. This yields a convenient representation of bandpass signals and facilitates the development of modulation theory.

Signal processing systems are best described by functional schematics: assembly of building blocks, each one performing an elementary task. The models for these blocks are the functional relationships between their input and output signals. This systemic approach greatly simplifies the search for and the performance evaluation of efficient signal conversion, detection, and measurement *systems.*

This book provides the theoretical background needed for an understanding and the efficient use of signal theory and processing.

Organization

This book consists of two parts. Chapters 1 to 7 and chapters 14 and 15 are a general introduction to signal theory, while chapters 8 to 13 are devoted to the modeling of the fundamental operations encountered in signal processing.

The first chapters deal with a general introduction on the nature of signals, the evolution of processing *methods,* and signal classification. Mathematical representations of deterministic signals and, especially, of their spectral representations are introduced in chapters 3 and 4. Random signal models are presented in chapter 5. Chapter 6 deals with noise theory. The analytic signal and complex envelope concepts are described in chapter 7.

Chapter 8 attempts to develop a systematic analysis of the main functional operators encountered in signal processing. Chapters 9 and 10 are dedicated to the study of sampling and digital representation conditions.

A modulation theory, based on the complex envelope model, is sketched out in chapter 11. The principles of experimental spectral analysis are described in chapter 12. Finally, signal detection and estimation methods, the basis of signal processing, are depicted in chapter 13.

Two appendices, a review of probability theory (chap. 14) and a collection of reference tables (chap. 15), conclude this book.

TEACHING REQUIREMENTS

On the order of 80 to 100 hours are normally needed to teach a course based on the content of this book. With an adequate selection, it can be taught in a somewhat shorter time, if the objective is to provide only the fundamentals. Various examples and problems are offered to facilitate self-training. Additional practical works (labs, projects, *et cetera*) are required to obtain a good understanding of the relationship between the abstract concepts of signal theory and the applications of signal processing.

Conventions

This book was originally published in French as part of the "Traite d'Electricite" by Presses Polytechniques Romandes. The "Traite d'Electricite" is composed of volumes (vol.), designated by a roman numeral (vol. VI). Other volumes in the series are cited in the text by volume number, and a complete list is given in the Select Bibliography at the end of the book.

This book is divided into chapters (chap.) referred to by number (chap. 2). Each chapter is divided into sections (sect.) referred to by 2 numbers separated by a period (sect. 2.3). Each section is subdivided into sub-sections (sub-sect.) referred to by 3 numbers and 2 periods (sub-sect. 2.3.11).

Bibliographic references are numbered sequentially throughout the text and referred to by a single number in brackets [33]. The first time a term is defined, it is printed in medium italics. An important sentence is highlighted by composing it in boldface italics. Equations are numbered sequentially by chapter and referred to by 2 numbers separated by a period placed in parentheses (3.14). Boldface equation numbers emphasize important results. Figures and tables are consecutively numbered in a common sequence by chapter and referred to by 2 numbers after Fig. (Fig. 4.12) or Tab. (Tab. 4.13).

Contents

Chapter 1

Signals and Information

1.1 SIGNAL AND INFORMATION THEORY

1.1.1 Importance of signal theory and processing to electrical engineering and science in general

Electrical engineering applications can be grouped into two broad fields, with strong interdependence:

- Power engineering
- Information processing

Signal theory and processing belong to the second field, providing it with a fundamental theoretical basis and special methods.

Nonetheless, they also have influenced power engineering, where various phenomena occur (network load fluctuations, engine vibrations, electromagnetic disturbances, *et cetera*), which can be analyzed with the same set of theoretical and experimental tools.

In fact, signal theory and processing are useful in all the fields where information is perceived through experimental observations of measurable variables.

The two key words, ***perception*** and ***processing***, explain why this discipline has an evolution linked to the applications of electrical engineering and, more precisely, metrology (perception) and telecommunications and computer engineering (data transmission and processing). Metrology deals with sensors, translating almost any physical phenomenon into an electrical variable easily amplified, filtered, conditioned, encoded, *et cetera*, by appropriate electronic devices. Telecommunication circuits carry the electrical signal to its destination. Data processing, thanks to its huge computation capabilities, can execute very sophisticated manipulations and interpretations of the information carried by the signal.

The universality of signal theory and processing is proved by the diversity of its application fields: industrial, scientific, biomedical, military, space, *et cetera*.

1.1.2 Historical survey [1]

The word ***signal*** comes from sign (***signum*** in Latin), which denotes an object, a mark, a language element, or an agreed-upon symbol that serves as an information vector. Sign history can be traced back to prehistorical times.

However, it was only during the 19th century that electrical signals appear with the invention of the telegraph (Morse, Cooke, Wheatstone, 1830–1840). This invention was closely followed by the telephone (Bell, 1876) and by the first radiowave links (Popov, Marconi, 1895–1896). The birth of electronics (Fleming, Lee de Forest, 1904–1907) at the dawn of the 20th century enabled the detection and amplification of weak signals. This was the true starting point of signal processing.

The first contributors to the mathematical study of electrical current fluctuations tried to adapt the analytical methodology of Fourier (1822) developed in his investigation of heat propagation. The first important works extending this methodology to random signals and phenomena were published during the early 1930s by Wiener and Khinchin [2,3,4].

The improvement of telecommunications and radar (mainly during World War II) provided the incentive for developing information and signal theory. During the 1920s, Nyquist and Hartley already tried to evaluate the quantity of information transmitted on a telegraphic line and observed that the maximum transmission rate is proportional to the available frequency bandwidth. However, Shannon's fundamental works on the mathematical theory of communication [5,6] and Wiener's books [6,7] on cybernetics and the optimal processing of signals or data affected by noise were only published in 1948–1949. The novelty was to take into account the statistical character of the studied phenomena.

Other researchers contributed to the initial development of this theory, especially Kupfmuller [9], Gabor [10], Woodward [11], Kolmogorov [12], Kotelnikov [13], Rice [14], Goldman [15], Lawson and Uhlenbeck [16], Ville [17], Blanc-Lapierre and Fortet [18], and Brillouin [19].

The 1950s have been a maturing period, directly followed by a flow of books and articles of a tutorial nature [20–42]. Simultaneously, the invention of the transistor, in 1948, followed ten years later by the introduction of integrated circuit technology, enabled the achievement of complex processing systems and the widespread pervasiveness of their applications.

Nowadays, signal processing is an established field of research and application, its scope broadening to a wide variety of areas such as pattern recognition, robotics, and artificial intelligence. It is complementary to electronics and computer engineering on which it is based.

A good introduction to the modern concepts of signal analysis and processing has been published by Lynn [43]. He provides in-depth coverage of the current tendency toward digital methods [44–48]. The technological evolution, which yields low cost specialized processors, opens a bright future to this field.

1.1.3 Signal definition

A *signal* is the physical representation of the information that it carries from its source to its destination.

Although we will consider here only electrical variables (generally voltage or current), the theory presented in the following chapters can be applied to any kind of signal, whatever its physical nature.

1.1.4 Noise definition

We call *noise* any phenomenon (interference, random distortion, *et cetera*) disturbing signal perception or interpretation, after the acoustical effects bearing the same name.

1.1.5 Signal-to-noise ratio

The *signal-to-noise ratio* (*S/N* or *SNR*) is a measure of the extent of signal contamination by noise. It can be expressed as the ratio ξ of the signal and noise powers P_s and P_n

$$\xi = P_s/P_n \tag{1.1}$$

Generally, the signal-to-noise ratio is given on a logarithmic scale measured in *decibels* (dB).

$$\xi_{dB} = 10 \log_{10}\xi \tag{1.2}$$

1.1.6 Signal-noise dichotomy

The difference between signal and noise is purely artificial and depends only on the user's criteria. Some electromagnetical phenomena of galactic origin sensed by antennas are considered as noise by telecommunication engineers and as signals of the utmost interest by radio astronomers. What makes the difference between signal and noise is only the observer's focus of interest.

A disturbed signal remains a signal and the same models are applied to the description of the useful signal and of its disturbances. Thus, signal theory must encompass noise theory.

1.1.7 Signal theory

The mathematical description of signals is the fundamental goal of *signal theory* [49,50].

Complementary to circuit theory and electromagnetic wave propagation, signal theory provides the means to emphasize, in a convenient mathematical form, the primary signal features : its spectral energy distribution or its statistical amplitude distribution, for example. It also provides the methods to analyze the nature of the modifications imposed on the signal while it proceeds through electric and electronic functional blocks. Thus, it gives essential information needed by the designer (preliminary specifications) or the user (operating instructions) of those systems. With this approach, we can, for example, establish the key rules to go from an analog signal to a digital one (chaps. 9 and 10). It also enables us to define and take into account the operating limits imposed by the existence of random distortions (chap.6).

The basic tool of signal theory is the series expansion of orthogonal functions (sect. 3.3), the Fourier expansion being the most interesting case. Its generalized form is known as the *Fourier transform,* the main properties of which are summarized in chapter 4. With the usual signal processing notations (sub-sect. 4.1.3), the Fourier transform of a time signal $x(t)$ is a function of the frequency f defined by the integral relation:

$$X(f) = \int_{-\infty}^{\infty} x(t)\exp(-\mathrm{j}2\pi ft)\mathrm{d}t \tag{1.3}$$

It introduces the fruitful concept of time-domain and frequency-domain duality. This yields the notion of spectrum, which plots a signal's essential characteristics (amplitude, energy, power) *versus* frequency. The spectral analysis method (chap. 12) is its direct practical application.

This very useful concept, also applicable to the analysis of random signals (chap. 5), thanks to the development of appropriate statistical models, gives way to a high level of abstraction for the study of complex signal processing methods.

The introduction of analytic signal and complex envelope models (chap. 7) enables the representation of bandpass signals and the development of a modulation theory (chap. 11). Detection methods (chap. 13) are based on the statistical decision and estimation theory. Their natural extension is pattern recognition (fig. 1.1).

1.1.8 Information and coding theory

Information is closely linked to the concept of communications: transfer of a message from its source to its destination.

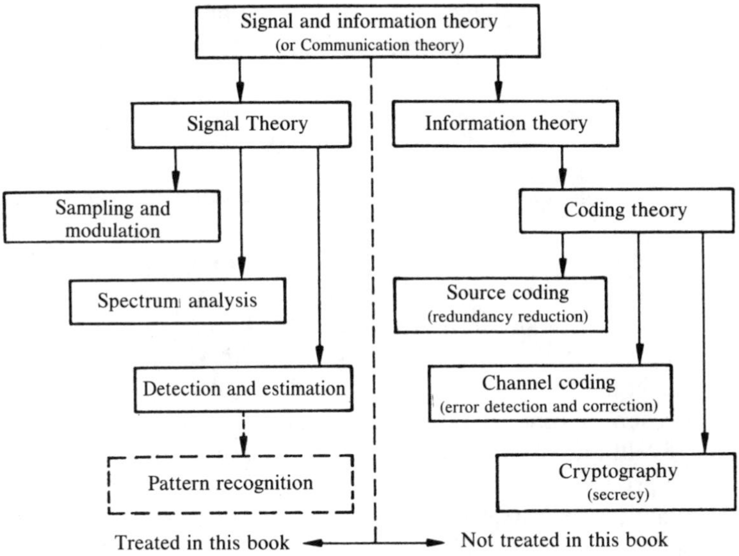

Fig. 1.1

Information (or *communication* [5]) *theory* is a probabilistic theory of messages, which takes into account their statistical properties. It provides a set of concepts allowing the evaluation of the performances of information transfer systems, especially when the message signal is disturbed by noise.

This leads to the study of information *coding* methods. Coding techniques have three seemingly conflicting goals. The first one is to increase the compactness of signals, or information vectors, by eliminating useless redundancy (source coding). The second one is to increase the transmission reliability in spite of noise by including some redundancy, cleverly structured to allow the detection, and even the correction, of the main errors (channel coding). Finally, the third one is to ensure the secrecy of communication (cryptography). These topics [51–54] are closely related to signal theory, but are not treated in this book.

1.1.9 Importance of statistical models and methods

By nature, information is random: only what cannot be predicted can carry messages. The signal information vectors are, therefore, also random. Except in some forms of interference generated by an industrial environment (influence of the power network, *et cetera*), the noise must also be considered as a random phenomenon.

It is not surprising that signal theory and signal processing methods widely relate to statistical concepts (probabilistic theory, stochastic processes, *et cetera*).

1.1.10 Signal models and measurements : functions and functionals

Mathematically, a function is defined as a correspondence rule (mapping) between two sets of real or complex values.

The mathematical model of a signal is a function of one, sometimes two, and even three variables: $s(t)$, $i(x,y)$, $i(x,y,t)$. Figure 1.2 displays illustrations. The first case is the most usual : the symbol t is usually set for time (but it can also represent another entity—distance, for example). The function represents the evolution of an electrical variable, possibly translated into this form by an appropriate sensor (microphone: acoustical signal; television camera: video signal; accelerometer: vibration signal, *et cetera*).

The second case describes bidimensional signals. These are generally functions of spatial coordinates x and y, and are thus more usually referred to as images.

The last case corresponds, for example, to a sequence of television or movie images where time appears as a third variable.

The input and output signals of a system (fig. 1.3) are often denoted, by convention, respectively $x(t)$ and $y(t)$. For instance, $y(t) = x^2(t)$ indicates the output of a quadratic nonlinear system the characteristic of which is defined by $y = x^2$.

A functional is a correspondence rule between a set of functions and a set of real or complex values. In other words, a functional is a function of functions. Processed signals or some of their parameters are often represented by functional relations. For example:

- weighted integral value [weighting function $g(t)$; see figure 1.4]

$$f_1(x) = \int_{-\infty}^{\infty} x(t)\, g(t)\, \mathrm{d}t \tag{1.4}$$

- integral weighted square value

$$f_2(x) = \int_{-\infty}^{\infty} x^2(t) g(t) \mathrm{d}t \tag{1.5}$$

- convolution product (fig. 1.5)

$$y(t) = x(t) * g(t) = \int_{-\infty}^{\infty} x(\tau) g(t - \tau) \mathrm{d}\tau \tag{1.6}$$

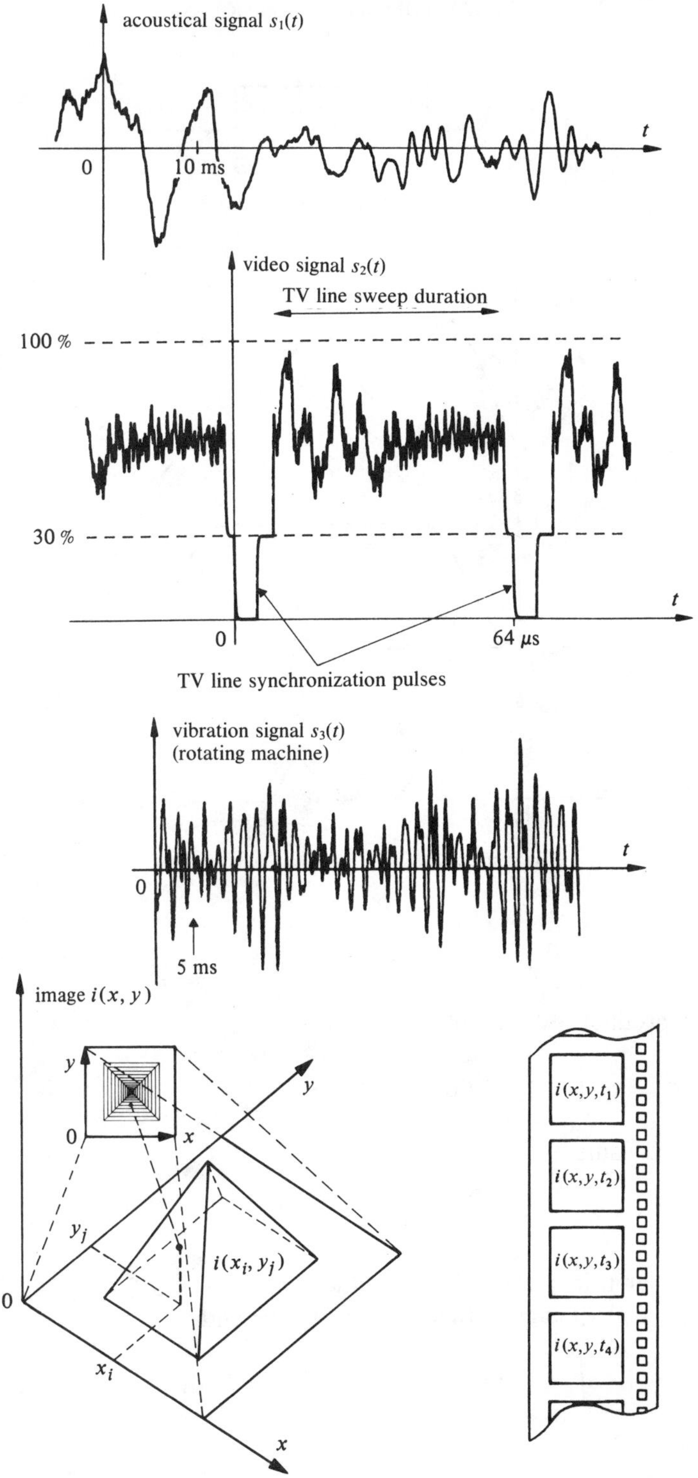
acoustical signal $s_1(t)$
t
0
10 ms
video signal $s_2(t)$
TV line sweep duration
100 %
30 %
t
0
64 μs
TV line synchronization pulses
vibration signal $s_3(t)$
(rotating machine)
t
0
5 ms
image $i(x, y)$
y
0
x
y
y_j
0
$i(x_i, y_j)$
x_i
x
$i(x,y,t_1)$
$i(x,y,t_2)$
$i(x,y,t_3)$
$i(x,y,t_4)$

Fig. 1.2

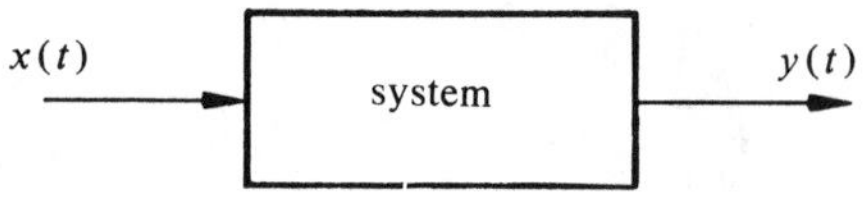

Fig. 1.3

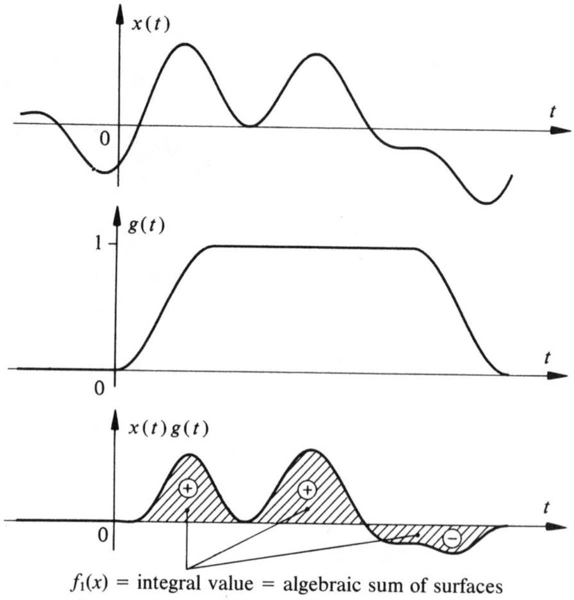

Fig. 1.4

- inner product (evaluated on interval T)

$$< x,y^* > = \int_T x(t)y^*(t)\mathrm{d}t \tag{1.7}$$

- sampled value

$$x(t_0) = < x,\delta_{t0} > = \int_{-\infty}^{\infty} x(t)\delta(t - t_0)\mathrm{d}t \tag{1.8}$$

The Fourier transform (1.3) is another example of a functional. For bidimensional signals (images), this transform is given by

$$I(u,v) = \int_{-\infty}^{\infty} \int_{-\infty}^{\infty} i(x,y) \exp[-\mathrm{j}\ 2\pi(ux + vy)]\mathrm{d}x\ \mathrm{d}y \tag{1.9}$$

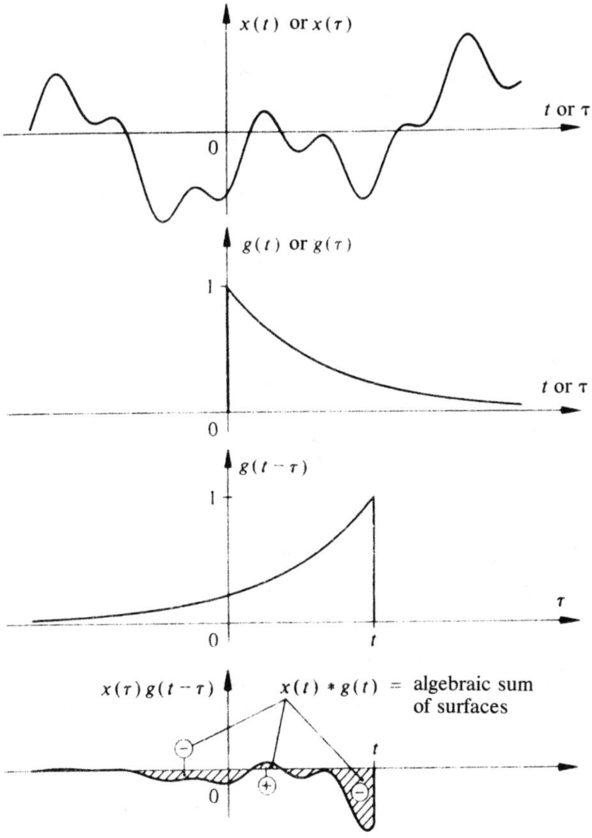

Fig. 1.5

where u and v represent ***spatial frequencies,*** measured in m^{-1} if the position variables x and y are expressed in meters.

1.2 SIGNAL PROCESSING

1.2.1 Definition

The mathematical description—or modeling—of signals is the purpose of signal theory, as we saw in sub-section 1.1.7.

Signal processing is the technical discipline which, based on the methods of signal and information theory, deals with the elaboration or interpretation of signals carrying information with the resources of electronics, computer engineering, and applied physics. It finds application in all fields concerned with perception, transmission, or interpretation of such information (fig. 1.6).

Some authors give a more restrictive meaning to signal processing by limiting its scope to those methods that allow extraction of a signal embedded in noise.

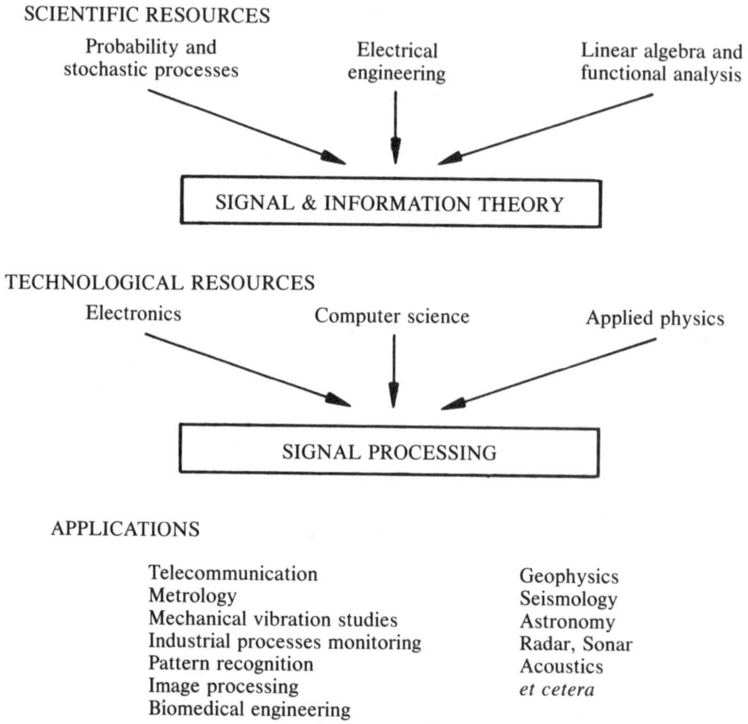

Fig. 1.6

1.2.2 Description

The relationship of man to his natural environment or to the systems he builds are characterized by high information exchanges. The observation (measure) of physical phenomena or the dialogue (communication) between men, between man and machine, or among machines themselves, are made through signals (time functions) or visual perceptions (images), the nature of which is complex and can be masked by undesirable distortion (background noise, atmospheric effects, interference).

Extracting useful information embedded in these signals (by analysis, filtering, regeneration, measurement, detection, and identification) and displaying the corresponding results in appropriate form to man or to machine is one of the main tasks of signal processing (fig. 1.7). To this can be added the generation of signals, which allow the study of physical system behavior, or are useful for information transmission or storage (synthesis, modulation and frequency translation, and coding to reduce the noise effect or the redundancy).

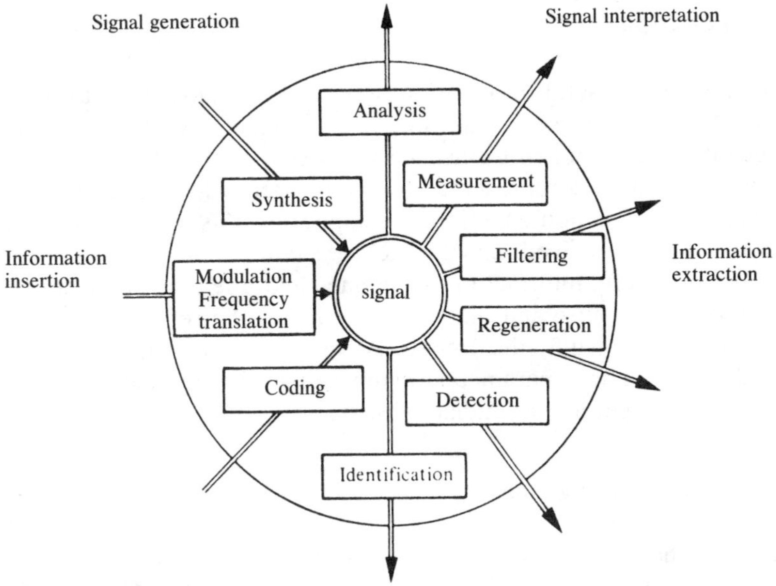

Fig. 1.7 Basic signal processing operations.

Through *analysis,* we attempt to isolate the main components of a signal of complex shape in order to better understand its nature and origin. To *measure* a signal, and especially a random one, we try to estimate the value of a characteristic variable, which is associated with it, with a given confidence level. *Filtering* is a well known function, which consists of taking some undesired components out of the signal. *Regeneration* is the operation by which we attempt to return the signal to its initial form after having undergone various distortions. With a *detection* method, we attempt to extract a useful signal from the background noise which is superimposed on it. *Identification* is often a complementary process, which allows us to classify the observed signal. *Correlation* techniques are often used for this purpose.

Synthesis, which is the opposite operation of analysis, consists of creating a signal with an appropriate shape by combining, for example, a set of elementary signals. *Coding,* besides its function of translation into digital language, is used either to minimize the effects of noise, or to attempt to conserve bandwidth or computer memory by way of a reduction of signal redundancy (sub-sect. 1.1.8). *Modulation* and *frequency translation* are primarily ways of adapting a signal to the characteristics of a transmission line, an analyzing filter, or a recording medium.

1.2.3 Observations

The concept of useful information previously mentioned is closely related to context. In a phone communication, it is essentially associated with the intelligibility of the spoken exchanged messages. In the case of radio-astronomy observations, it is represented by the frequency and the amplitude of an electromagnetic radiation emission. In geophysics, this is rather the statistical parameters of a perceived signal which are interpretable. In Doppler radar technique, the useful information is, on one hand, the time delay between the emission of a sine pulse and the reception of its echo reflected by the target, and, on the other hand, the frequency variation between the emitted and received waves. These parameters estimate the distance between the transmitter and the target and its radial velocity.

1.2.4 Signal processing language

At the higher level, the signal processing language is that of the block diagram, also familiar to specialists in control and in system theory in general, both of which are related to signal processing. A block diagram is the graphical symbolic connection of functional blocks, normally independent, realizing a given function. Figure 1.8 illustrates the principle of a swept-frequency spectrum analyzer (described in sect. 12.3).

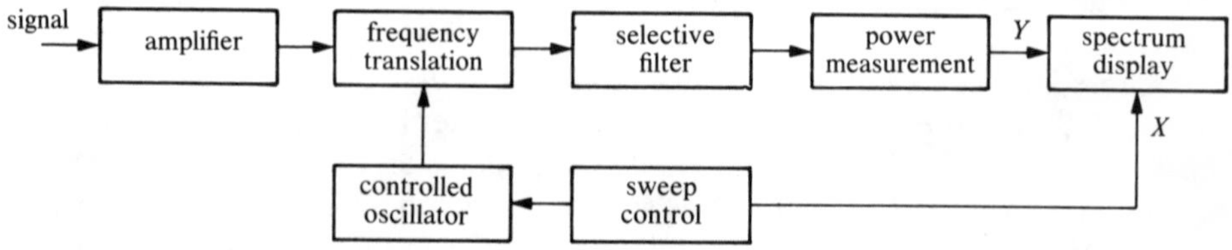

Fig. 1.8

The theoretical behavior of each block can be described by one or several mathematical relations. The functional operators developed in chapter 8 are models of blocks, which produce an output signal, depending on one or several input excitations.

1.2.5 Example and definitions : contribution of linear systems theory

It can be shown that the output signal $y(t)$ of a linear causal time-invariant system is given by the *convolution* (1.6) of the input signal $x(t)$ with a function $g(t)$, called *impulse response* of the system:

$$y(t) = x(t)*g(t) = \int_{-\infty}^{t} x(\tau)g(t - \tau)\mathrm{d}\tau \qquad \textbf{(1.10)}$$

This is the most basic, and probably the most familiar, operation of signal theory. It shows, as can be seen from fig.1.5, that ***the value of the output signal at time t is the weighted sum (integral = continuous summation) of passed values of the input signal x(t) [for a causal system g(t) = 0 for t < 0]***. The weighting function is precisely the system impulse response $g(t)$ (or $h(t)$, as it is often denoted).

The easiest example is when $g(t) = 1/T$ for $0 < t < T$, and $g(t) = 0$ elsewhere. The output signal $y(t)$ given by (1.10) then corresponds to $\overline{x}(t,T)$, *running average* of input signal $x(t)$, evaluated on the duration interval T (fig. 1.9).

$$y(t) = \overline{x}(t,T) = \frac{1}{T}\int_{t-T}^{t} x(\tau)\mathrm{d}\tau \qquad (1.11)$$

Another straightforward example is the lowpass filter built with a first order RC integrator circuit, the impulse response of which is given by $g(t) = (RC)^{-1}\exp[-t/(RC)]$:

$$y(t) = \frac{1}{RC}\int_{-\infty}^{t} x(\tau)\exp[-(t - \tau)/(RC)]\mathrm{d}\tau \qquad (1.12)$$

where $y(t)$ represents the weighted average of prior and present values of the $x(t)$ signal. A ***progressive loss*** of memory is introduced by the circuit (fig.1.5 gives an illustration of this).

The operation which corresponds, in the frequency domain, to the convolution (1.10) in the time domain, is simply a product of the Fourier transforms of the input signal $x(t)$ and of the impulse response $g(t)$. The Fourier transform of $g(t)$, often referred to as the *transfer function*, is the frequency response of the system

$$Y(f) = X(f) \cdot G(f) \qquad \textbf{(1.13)}$$

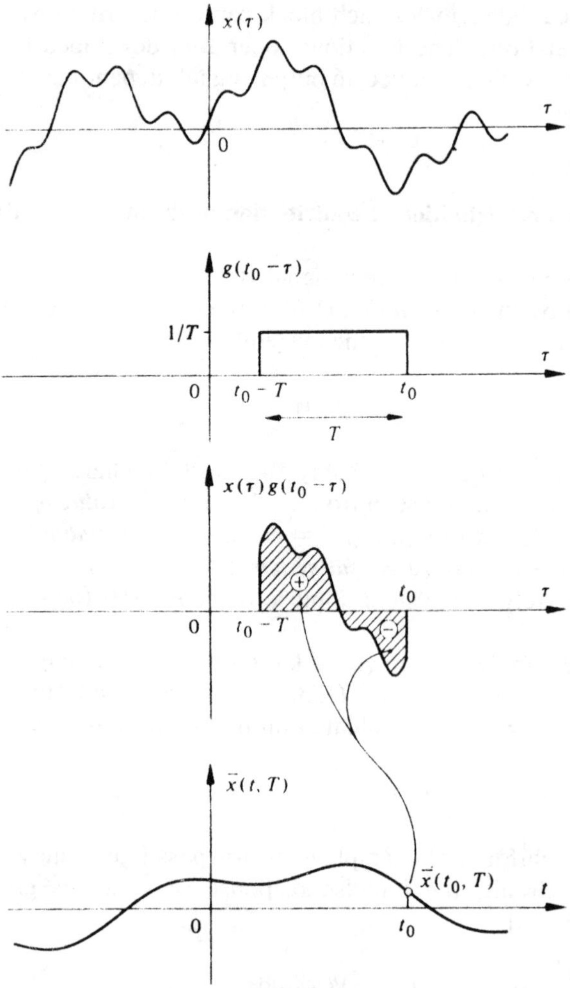

Fig. 1.9

It can be easily deduced that convolution is a ***commutative, associative, and distributive operation***:

$$x(t) * g(t) = g(t) * x(t) \tag{1.14}$$

$$[x_1(t) + x_2(t)] * g(t) = [x_1(t) * g(t)] + [x_2(t) * g(t)] \tag{1.15}$$

$$[x(t) * g_1(t)] * g_2(t) = x(t) * [g_1(t) * g_2(t)] \tag{1.16}$$

These results remain valid (with some new *graphic* notation) in the case of bidimensional signals and systems. So, the relation :

$$i_2(x,y) = i_1(x,y) ** g(x,y) \tag{1.17}$$

where the double asterisk indicates a bidimensional convolution, represents the transformation of an image $i_1(x,y)$ by a linear bidimensional system of impulse response $g(x,y)$. In optics, this relation gives the image of an object observed through an instrument (lens), the behavior of which is depicted by $g(x,y)$, often referred to as the *point-spread function* (response of the instrument to a bright spot of light).

In the spatial frequency domain, thanks to Fourier transform properties, relation (1.17) becomes

$$I_2(u,v) = I_1(u,v) \cdot G(u,v) \tag{1.18}$$

where $G(u,v)$ is the *bidimensional frequency response* of the corresponding system.

1.2.6 Technological tools

When a complex system is designed, each block of its complete block diagram becomes a hardware or software ***module,*** depending on the design's trade-offs:

- analog electronics
- digital electronics (specialized processors)
- programmed digital electronics (universal processors)
- other techniques

Technological evolution (microelectronics, microacoustics, or optics) enables the development of a wide range of specialized (analog or digital) devices, able to process rapidly and economically a growing volume of information. This trend, supported by new needs, yields to an ever-growing number of applications of signal processing.

Although the development of integrated circuits greatly helped analog signal processing, it has been far more beneficial to digital signal processing, thanks to the simultaneous introduction of powerful algorithms (such as the fast Fourier transform (FFT)), that give a clear advantage to digital processing (except at very high frequencies).

In addition to classical electronic circuits and programmable devices or systems, applied physics provides signal processing with less universal but more efficient tools dedicated to specific applications; for instance, charge coupled devices (CCD) or switched capacitor devices, built of integrated capacitors and switches, or surface acoustic wave (SAW) devices, using the limited propagation speed of elastic surface waves in special piezoelectric media are such signal processing devices. The principle of a SAW device is depicted in fig. 1.10. SAW devices are mainly used in radar and TV broadcasting applications.

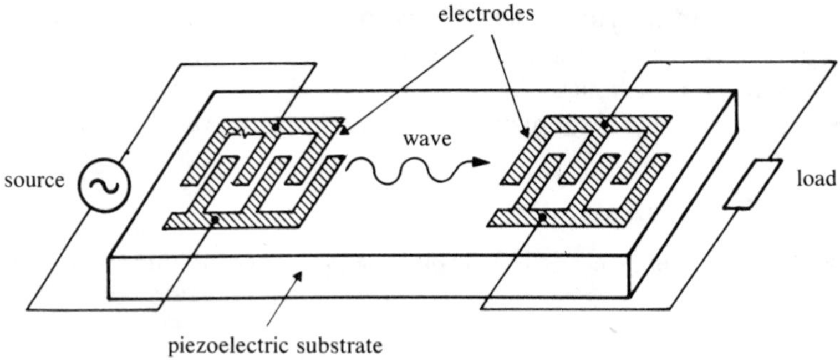

Fig. 1.10 Principle of an interdigital transducer (IDT): electrodes on a piezoelectric substrate translate the signal into an elastic wave that travels with defined delays to a network of electrodes with different dimensions, which accomplishes a weighted sum (convolution).

Optical processing systems offer an analog-parallel computation method with unmatchable speeds, but with limited accuracy and flexibility. The basic operation (fig. 1.11) is the bidimensional Fourier transform straightforwardly achieved by an optical lens excited by coherent light (laser). These methods can be best applied to the processing of information contained in an image or hologram. Nonetheless, computers remain the favorite image processing tools.

Integrated optics is a new field of optoelectronics offering interesting opportunities to build signal processing devices, especially in the fields of telecommunications and metrology.

1.3 SPECIAL NOTATIONS

1.3.1 Introduction

To simplify the mathematical representation of some signals, functions, or operators frequently encountered in signal theory, it is convenient to refer to them in an easy and concise way.

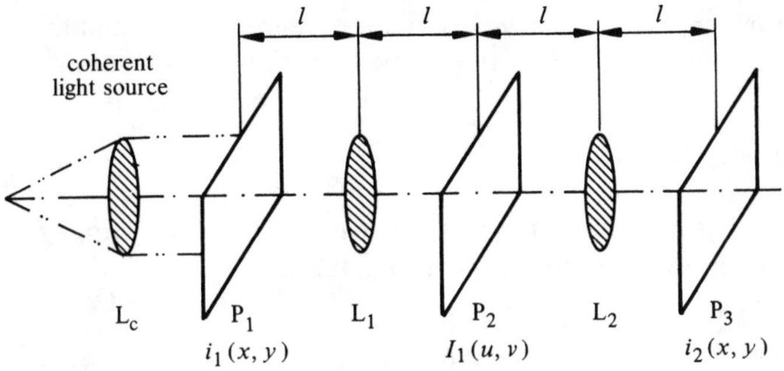

Fig. 1.11 Optical processing principle: if an image with a variable transmittance $i_1(x,y)$ is placed in the frontal focal plane P_1 of lens L_1 and illuminated by a coherent light, the light distribution in focal plane P_2 is proportional to the image Fourier transform $I_1(u,v)$. By placing an optical filter in P_2 with a transfer function $G(u,v)$, the convolution $i_2(x,y)$ of the initial image with the filter impulse response $g(x,y)$ appears in P_3.

Some of these shorthand notations are common in mathematics, but others are less conventional [11,23] and are subject to diverse notations depending on the authors.

1.3.2 Signum function

The *signum function* is defined as follows (fig. 1.12):

$$\text{sgn}(t) = \begin{cases} -1 & t < 0 \\ 1 & t > 0 \end{cases}$$

$$= \frac{t}{|t|} \text{ for } t \neq 0 \tag{1.19}$$

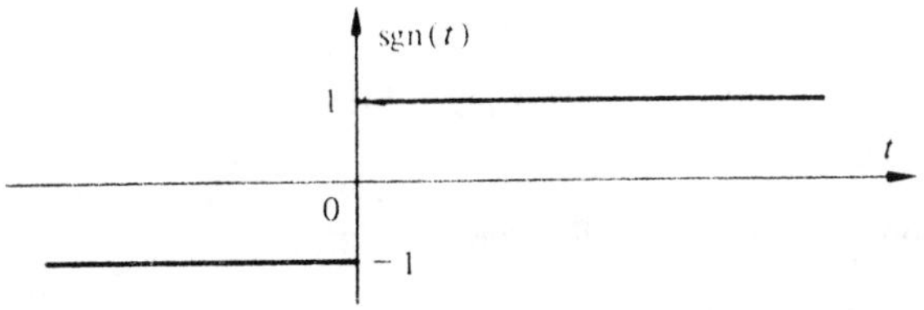

Fig. 1.12

Its value for $t = 0$ is arbitrary, between $+1$ and -1. To maintain symmetry, we will admit, in general, that this value is zero by convention.

1.3.3 Unit-step function

The *unit-step function* (also well known as *Heaviside's unit function*) can be defined from the signum function (fig. 1.13):

$$\epsilon(t) = \frac{1}{2} + \frac{1}{2}\,\text{sgn}\,(t) = \begin{cases} 0 & t < 0 \\ 1 & t > 0 \end{cases} \tag{1.20}$$

Its value for $t = 0$ is arbitrarily located between 0 and $+1$. It is fixed by convention at 1/2. In some applications, it is preferable to assign it the value 1.

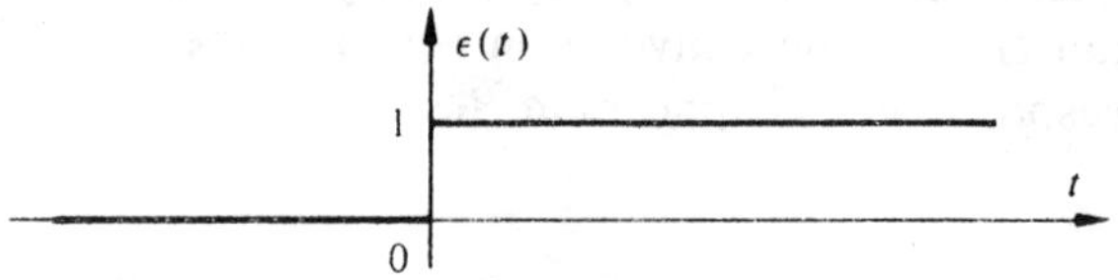

Fig. 1.13

1.3.4 Ramp function

The *ramp function* can be defined from the unit function (fig. 1.14) as

$$r(t) = \int_{-\infty}^{t} \epsilon(\tau)\,d\tau = t \cdot \epsilon(t) \tag{1.21}$$

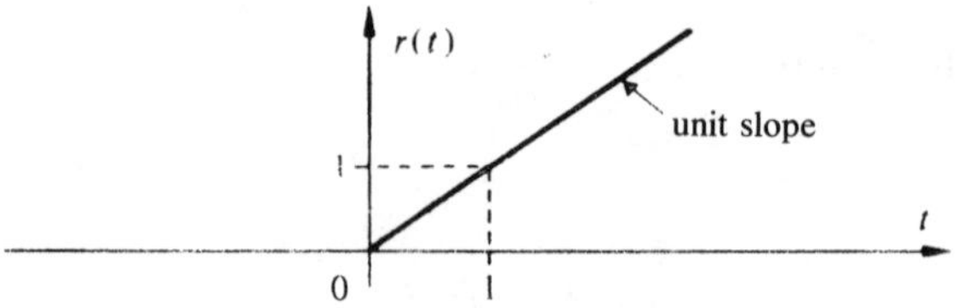

Fig. 1.14

Conversely, the unit-step function can be defined as

$$\epsilon(t) = dr(t)/dt \text{ for } t \neq 0 \tag{1.22}$$

Some authors [55] define a limited growth ramp function as the integral of the rectangular function introduced in the following sub-section.

1.3.5 Rectangular function

The *normalized rectangular function* (unit integral), sometimes also referred to in mathematics as the *gating function,* is noted and defined as follows (fig. 1.15):

$$\mathrm{rect}(t') = \epsilon(t' + 1/2) - \epsilon(t' - 1/2) = \begin{cases} 1 & |t'| < 1/2 \\ 0 & |t'| > 1/2 \end{cases} \tag{1.23}$$

where the prime indicates a dimensionless ***variable***. The conventional value of this function assigned to $t' = \pm 1/2$ is 1/2.

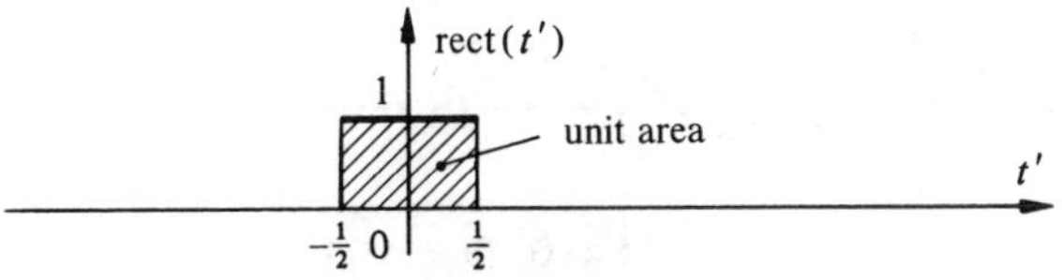

Fig. 1.15

Introducing the variable transformation $t' = t/T$, we can generalize this notation (fig. 1.16) for a rectangular pulse centered on $t = T$ with an amplitude A and a duration T:

$$x(t) = A\mathrm{rect}[(t - \tau)/T] \tag{1.24}$$

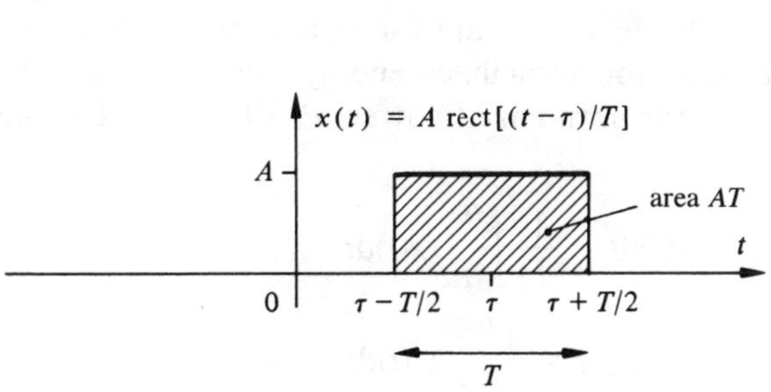

Fig. 1.16

1.3.6 Application

In addition to its use in representing rectangular shaped signals, the rectangular function is often used as a "window," or multiplicative factor, to observe a segment of duration T of any signal (fig. 1.17); e.g.,

$$x(t,T) = x(t) \cdot \text{rect}(t/T) \tag{1.25}$$

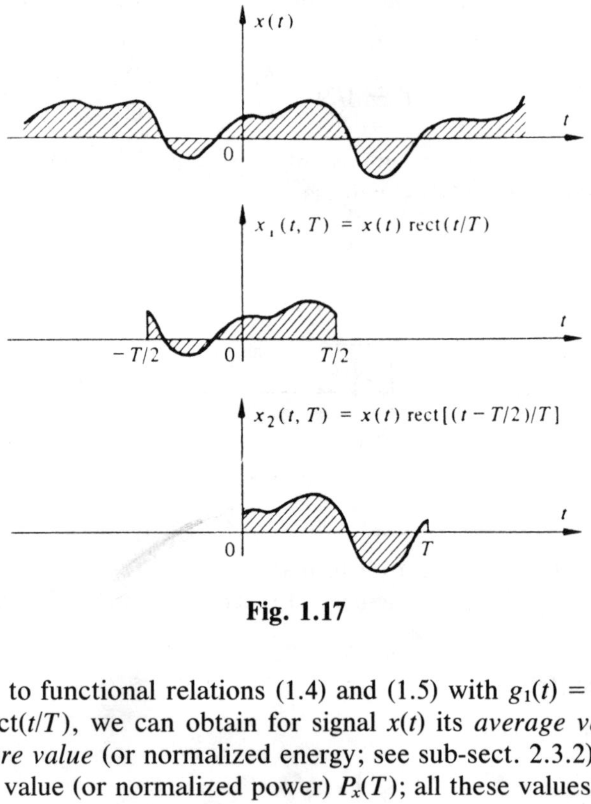

Fig. 1.17

According to functional relations (1.4) and (1.5) with $g_1(t) = T^{-1}\,\text{rect}(t/T)$ or $g_2(t) = \text{rect}(t/T)$, we can obtain for signal $x(t)$ its *average value* $\overline{x}(T)$, its *integral square value* (or normalized energy; see sub-sect. 2.3.2) $W_x(T)$, or its mean square value (or normalized power) $P_x(T)$; all these values are ***evaluated on interval*** T:

$$\overline{x}(T) = \int_{-\infty}^{\infty} x(t)g_1(t)\mathrm{d}t = \frac{1}{T}\int_{-T/2}^{T/2} x(t)\mathrm{d}t \tag{1.26}$$

$$W_x(T) = \int_{-\infty}^{\infty} x^2(t)g_2(t)\mathrm{d}t = \int_{-T/2}^{T/2} x^2(t)\mathrm{d}t \tag{1.27}$$

$$P_x(T) = \int_{-\infty}^{\infty} x^2(t)g_1(t)\mathrm{d}t = \frac{1}{T}\int_{-T/2}^{T/2} x^2(t)\mathrm{d}t \tag{1.28}$$

Similarly, according to functional relation (1.6) with $g_3(\tau) = T^{-1}$ rect[$(\tau - T/2)/T$], we obtain the running average, defined by relation (1.11) and illustrated by fig. 1.9.

$$\bar{x}(t,T) = x(t)*g_3(t) = \frac{1}{T}\int_{t-T}^{t} x(\tau)\,d\tau \tag{1.29}$$

The average value, $\bar{x}$, of the signal, measured on the complete real axis, is the limit of (1.26):

$$\bar{x} = \lim_{T\to\infty} \frac{1}{T}\int_{-T/2}^{T/2} x(t)\,dt \tag{1.30}$$

The square root of (1.28) is by definition the *root mean square* value (rms) of the signal on the interval T:

$$x_{\text{eff}}(T) = \sqrt{P_x(T)} \tag{1.31}$$

1.3.7 Triangular function

The *normalized triangular function* (unit integral and dimensionless variable t') is noted and defined as follows (fig. 1.18):

$$\text{tri}(t') = \begin{cases} 1 - |t'| & |t'| \leqslant 1 \\ 0 & |t'| > 1 \end{cases} \tag{1.32}$$

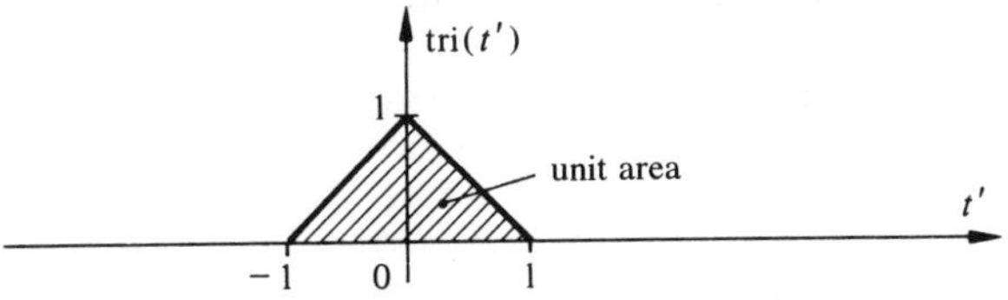

Fig. 1.18

This function also corresponds to the convolution:

$$\text{tri}(t') = \text{rect}(t')*\text{rect}(t') \tag{1.33}$$

It is noted $\Lambda\,(t')$ by some authors [23,56].

More generally, by introducing the variable transformation $t' = t/T$, a triangular shaped pulse with a maximum amplitude A and with a base $2T$, centered in $t = \tau$, will be noted (fig. 1.19) in the form

$$x(t) = A\ \text{tri}[(t - \tau)/T] \tag{1.34}$$

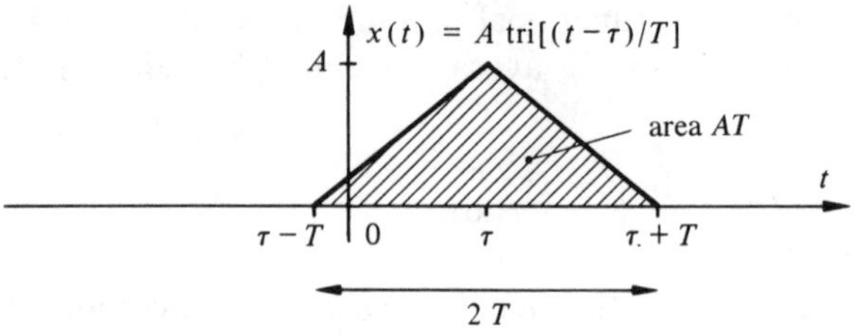

Fig. 1.19

1.3.8 Delta function

The *delta function* or *unit impulse function* $\delta(t)$ is theoretically defined by the inner product (1.8):

$$x(0) = < x, \delta > = \int_{-\infty}^{\infty} x(t)\delta(t)\mathrm{d}t \tag{1.35}$$

In other words, the delta function $\delta(t)$ is a ***sampling operator*** which gives the value $x(0)$ of a function $x(t)$ continuous in $t = 0$. Its dimension is consequently the inverse of that of the integration variable. In a more general way, for every function $x(t)$, continuous in $t = t_0$, we have

$$x(t_0) = \int_{-\infty}^{\infty} x(t)\delta(t - t_0)\mathrm{d}t \tag{1.36}$$

Particularly, by writing $x(t) = 1$, we obtain

$$\int_{-\infty}^{\infty} \delta(t)\mathrm{d}t = 1 \tag{1.37}$$

with

$$\int_{-\infty}^{t} \delta(\tau)\mathrm{d}\tau = \begin{cases} 0 & t < 0 \\ 1 & t > 0 \end{cases}$$

$$= \epsilon(t) \tag{1.38}$$

The value of the integral for $t = 0$ is generally fixed at 1/2, so that we can also admit the equivalence:

$$\delta(t) = \mathrm{d}\epsilon(t)/\mathrm{d}t \tag{1.39}$$

By taking (1.23) into account, the rectangular function derivative can be written as

$$\frac{\mathrm{d}}{\mathrm{d}t}\operatorname{rect}(t) = \delta(t + 1/2) - \delta(t - 1/2) \tag{1.40}$$

1.3.9 Interpretation

The expression (1.36) corresponds to the limit, when $T \to 0$, of the local average value $\bar{x}(t_0,T)$ of $x(t)$, measured on interval T centered in $t = t_0$:

$$x(t_0) = \lim_{T \to 0} \bar{x}(t_0,T) \tag{1.41}$$

with (fig. 1.20)

$$\bar{x}(t_0,T) = \int_{-\infty}^{\infty} x(t)g(t)\mathrm{d}t = \frac{1}{T}\int_{t_0-T/2}^{t_0+T/2} x(t)\mathrm{d}t \tag{1.42}$$

where

$$g(t) = T^{-1}\mathrm{rect}[(t - t_0)/T] \tag{1.43}$$

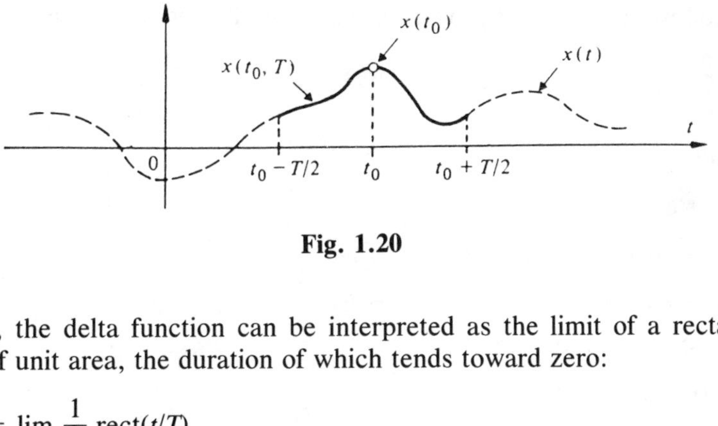

Fig. 1.20

Thus, the delta function can be interpreted as the limit of a rectangular pulse, of unit area, the duration of which tends toward zero:

$$\delta(t) = \lim_{T \to 0} \frac{1}{T}\,\mathrm{rect}(t/T) \tag{1.44}$$

In a similar way, the delta function corresponds also to the limit taken by many functions of unit area, such as $T^{-1}\,\mathrm{tri}(t/T)$. Some other cases are mentioned in sub-sections 1.3.15 and 1.3.16.

1.3.10 Product of a continuous function by a delta function

Let $x(t)$ be a continuous function in $t = 0$ or $t = t_0$. Equations (1.35), (1.36), and (1.37) yield the identities:

$$x(t)\,\delta(t) = x(0)\,\delta(t) \tag{1.45}$$

$$x(t)\,\delta(t - t_0) = x(t_0)\,\delta(t - t_0) \tag{1.46}$$

The conventional graphic representation of a delta function $c \cdot \delta(t - t_0)$ is a vertical arrow positionned at $t = t_0$, the length of which is proportional to the weighting factor c (fig. 1.21)

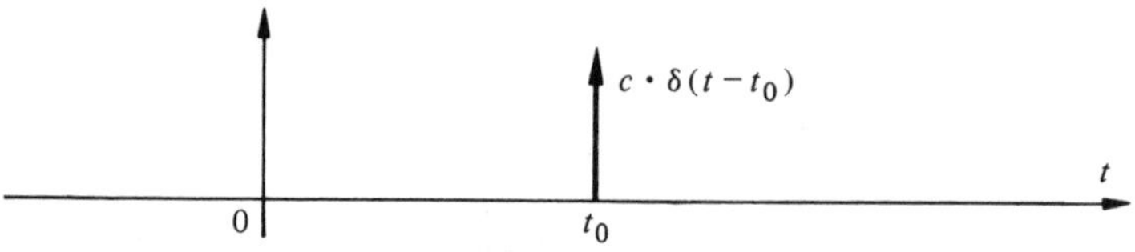

Fig. 1.21

1.3.11 Other properties

The following important properties can be easily deduced from the above:

- identity

$$x(t) * \delta(t) = x(t) \quad \textbf{(1.47)}$$

- translation

$$x(t) * \delta(t - t_0) = x(t - t_0) \quad \textbf{(1.48)}$$

$$x(t - t_1) * \delta(t - t_2) = x(t - t_1 - t_2) \quad \textbf{(1.49)}$$

$$\delta(t - t_1) * \delta(t - t_2) = \delta(t - t_1 - t_2) \quad \textbf{(1.50)}$$

- transformation of variable

$$\delta(at) = |a|^{-1}\, \delta(t) \quad \textbf{(1.51)}$$

with the particular case, when $\omega = 2\pi f$, where

$$\delta(\omega) = \frac{1}{2\pi}\, \delta(f) \quad (1.52)$$

1.3.12 Impulse response and unit-step response

A linear system *impulse response* $g(t)$, already mentioned in sub-section 1.2.5, is the response to a theoretical excitation that has a delta-function shape. When $x(t)$ is substituted by $\delta(t)$ in equation (1.10), we obtain, due to (1.47), the identity: $y(t) = g(t)$.

From the property of (1.46), it can be deduced that the general convolution equation (1.10) is an application of the superposition principle: the response to an excitation of any shape is the sum (integral) of the partial responses to a continuous sequence of delta functions—shifted in time—whose weighting factors form an image of the excitation signal.

A linear system *unit-step response* $\gamma(t)$ is the response to a unit-step excitation (1.20). According to (1.39), the impulse and unit-step responses relate to each other as follows:

$$\gamma(t) = \int_{-\infty}^{t} g(\tau)\mathrm{d}\tau \tag{1.53}$$

1.3.13 Periodic sequence of delta-functions

A sequence of delta functions separated by a fixed time interval or period T (fig. 1.22) will be denoted by

$$\delta_T(t) = \sum_{k=-\infty}^{\infty} \delta(t - kT) \tag{1.54}$$

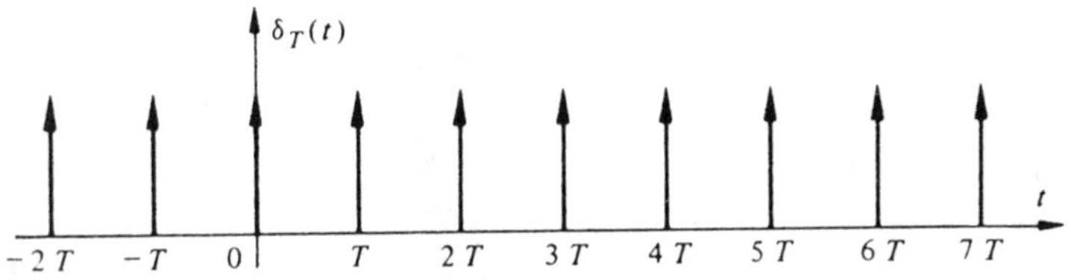

Fig. 1.22

Such a sequence is often referred to as the *sampling function* or the *comb-function*.

According to (1.46), we have

$$x(t)\delta_T(t) = \sum_{k=-\infty}^{\infty} x(kT)\delta(t - kT) \tag{1.55}$$

This expression enables the representation (chap. 9) of a periodic sampling of signal $x(t)$ with a sampling rate $f_e = 1/T$.

1.3.14 Repetition operator

The *repetition operator* $\text{rep}_T\,\{x(t)\}$ is a convenient notation for the description of periodic signals.

$$\text{rep}_T\{x(t)\} = \sum_{k=-\infty}^{\infty} x(t - kT) \tag{1.56}$$

According to (1.48), it is equivalent to the convolution:

$$\text{rep}_T\{x(t)\} = x(t) * \delta_T(t) \tag{1.57}$$

A graphic description is given in Fig. 1.23.

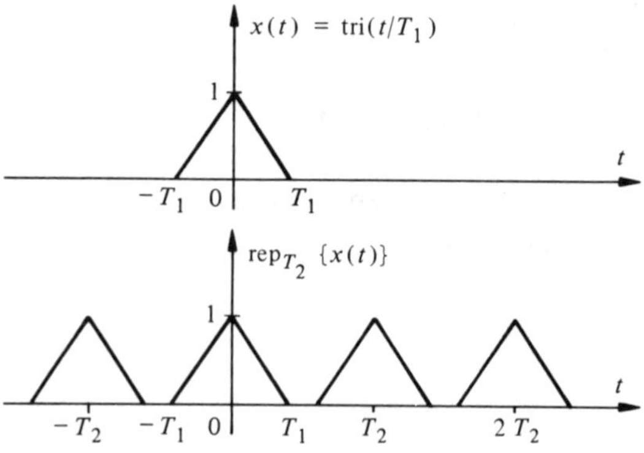

Fig. 1.23

1.3.15 Sinc function

The ratio of a sine function to its argument plays a very important role in signal theory. It is named cardinal sine. Its normalized form (unit integral and dimensionless variable α) is noted and defined by (fig. 1.24):

$$\text{sinc}(\alpha) = \frac{\sin(\pi\alpha)}{\pi\alpha} \tag{1.58}$$

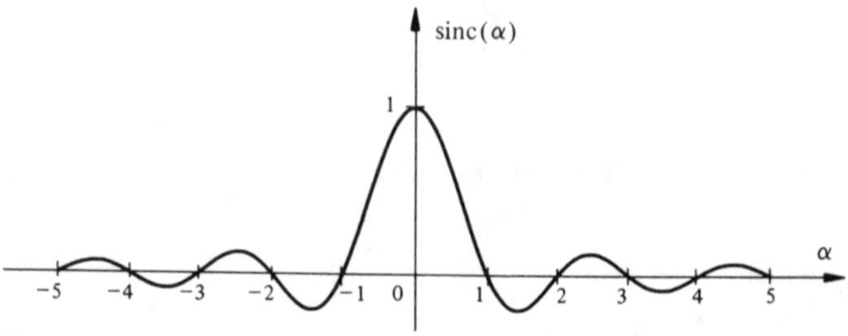

Fig. 1.24

Its value at $\alpha = 0$ is 1. It is an even function and its zeros correspond to the non-zero integer values of α. Tables of sinc(α) and $\text{sinc}^2(\alpha)$ are given at the end of chapter 15.

Taking into account the series expansion of the sine function, the sinc function can be expressed as

$$\text{sinc}(\alpha) = 1 - (\pi\alpha)^2/3! + (\pi\alpha)^4/5! - (\pi\alpha)^6/7! + \ldots \tag{1.59}$$

The rect(t) and sinc(f) functions are a fundamental pair of Fourier transforms (sub-sect. 4.2.4), as do tri(t) and $\text{sinc}^2(f)$. (sub-sect. 4.2.6).

The following properties are a consequence of (1.58):

$$\int_{-\infty}^{\infty} \text{sinc}(\alpha)\, \mathrm{d}\alpha = 1 \tag{1.60}$$

$$\int_{-\infty}^{\infty} \text{sinc}^2(\alpha)\, \mathrm{d}\alpha = 1 \tag{1.61}$$

and with $\alpha = Tf$ (i.e., positioning the function zeros at integer multiples of $f = 1/T$):

$$\int_{-\infty}^{\infty} T\, \text{sinc}(Tf)\, \mathrm{d}f = 1 \tag{1.62}$$

$$\int_{-\infty}^{\infty} T\, \text{sinc}^2(Tf)\, \mathrm{d}f = 1 \tag{1.63}$$

Hence, by similarity with (1.44), we have

$$\lim_{T\to\infty} T\text{sinc}(Tf) = \lim_{T\to\infty} T\text{sinc}^2(Tf) = \delta(f) \tag{1.64}$$

The cardinal sine integral (fig. 1.25) is related to the usual sine integral Si(u) by

$$\text{Si}(u) = \int_0^u \frac{\sin x}{x}\, \mathrm{d}x = \pi \int_0^{u'=u/\pi} \text{sinc}(\alpha)\, \mathrm{d}\alpha \tag{1.65}$$

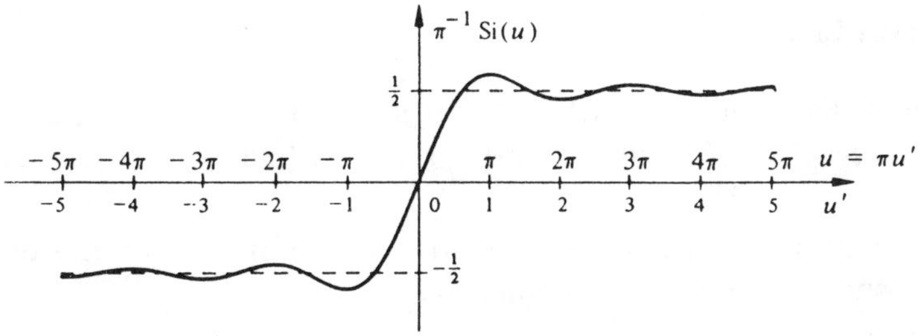

Fig. 1.25

1.3.16 Gaussian pulse

The Gaussian function (or normal distribution) is often referred to in statistical problems, but its implications in signal theory are much larger.

A *Gaussian pulse* (fig. 1.26) in its normalized form (unit integral and dimensionless variable t') is defined by

$$\mathrm{ig}(t') = \exp(-\pi t'^2) \tag{1.66}$$

With $t' = t/T$, where T is a measure of the pulsewidth and is related to the *standard deviation* σ_t by $T = \sqrt{2\pi}\,\sigma_t$, we obtain

$$\int_{-\infty}^{\infty} \mathrm{ig}(t')\,\mathrm{d}t' = T^{-1}\int_{-\infty}^{\infty} \mathrm{ig}(t/T)\,\mathrm{d}t = 1 \tag{1.67}$$

hence,

$$\lim_{T\to 0} T^{-1} \cdot \mathrm{ig}(t/T) = \delta(t) \tag{1.68}$$

One of the most interesting properties of the Gaussian pulse is that ig(t) and ig(f) form a pair of Fourier transforms [22].

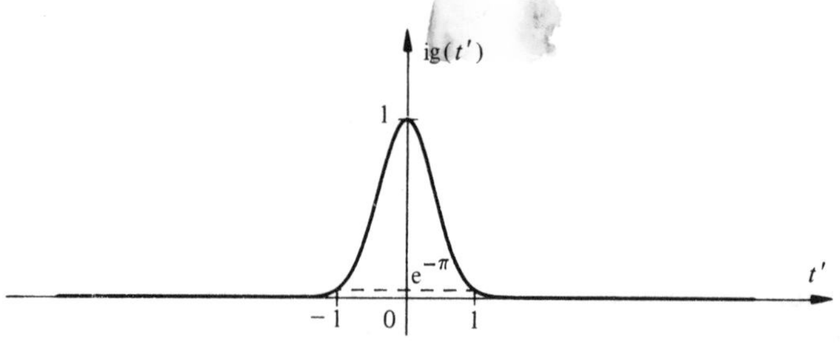

Fig. 1.26

1.4 PROBLEMS

1.4.1 Find the integral of $x_1(t) = A\ \mathrm{rect}(t/\Delta)$ and $y_1(t) = A\ \mathrm{tri}(t/\Delta)$, and the average values $\bar{x}$ and $\bar{y}$ of signals $x_2(t) = \mathrm{rep}_T\{x_1(t)\}$ and $y_2(t) = \mathrm{rep}_T\{y_1(t)\}$, for $-\infty < t < \infty$.

1.4.2 Express the signal $x(t) = A\ \mathrm{rect}[(t - t_0 - T/2)T]$ as a series of signum functions only. Plot the corresponding graph.

1.4.3 Find and plot for $t_0 < t_1$ and $t_0 > t_1$ the convolution product $z_i(t) = x_i(t) * y_i(t)$, given:

- $x_1(t) = A[\delta(t + t_0) + \delta(t - t_0)]$ and

  $y_1(t) = B\delta(t) + \frac{1}{2}B[\delta(t + t_1) + \delta(t - t_1)]$;

- $x_2(t) = \cos(\pi t/T)\,\mathrm{rect}(t/T)$ and

  $y_2(t) = A\delta_T(t)$.

1.4.4 Analytically and graphically check relation (1.33).

1.4.5 Determine the running average $\overline{x}(t,T_1)$, when $x(t) = A\sin(2\pi f_0 t)$, and evaluate the results for $T_1 = T_0/2$ and $T_1 = kT_0$, where k is an integer and $T_0 = 1/f_0$.

1.4.6 Find and plot the convolution product of

$$x(t) = \sum_{i=0}^{\infty} a_i\,\delta(t - iT) \quad \text{and} \quad y(t) = \sum_{j=0}^{2} b_j\,\delta(t - jT).$$

where $a_i = \exp(-i)$ and $b_0 = b_2 = -b_1 = 1$.

1.4.7 Find the average value (1.26), the integral square value (1.27), the mean square value (1.28), and the root mean square value (1.31) of $x(t) = A\,\mathrm{tri}(t/T)$ on the interval$T_1 = [-T,T]$.

Chapter 2

Signal Classification

2.1 PHYSICAL SIGNALS AND THEORETICAL MODELS

2.1.1 Experimental constraints

An experimental signal is the image of a physical process and, thus, must be physically achievable. It is subject to a series of constraints:

- its energy must be finite
- its amplitude is necessarily bounded
- this amplitude is a continuous function because the source system inertia prohibits any discontinuity
- the signal spectrum is also necessarily bounded and must tend towards zero when the frequency tends towards infinity

2.1.2 Theoretical models

The theoretical model of a signal (sub-sect. 1.1.10) is a real or complex function, or a functional, depending on time t, for instance, as a variable. It is convenient to classify each model in a specific subset encompassing signals with similar properties. We can often simplify the representation used by appropriate model selection, but this will not necessarily comply with the previously stated constraints.

This is why signal models with theoretically unlimited energy are widely used, with unbounded or discontinuous amplitude, which can be described by their distribution.

The model quality finally depends on the merit and the convenience of the selected approximation.

2.1.3 Examples

A sinusoidal signal (or sine wave) is represented by a function defined on the complete real axis : its theoretical energy is therefore infinite.

The usual model for disturbing signals, called noise (chap. 6), assumes the possibility, although with a probability tending toward zero, of infinite amplitudes.

The state transitions of binary-logic signals are generally represented by ordinary discontinuities.

Percussion-like excitation is symbolized by the delta function $\delta(t)$.

2.1.4 Classification methods

Different classification methods of the signal models can be considered. Among the principal ones, we can mention:

- phenomenological classification (sect.2.2) based on the evolution type of the signal, its predefined character, or its random behavior
- energy classification (sect.2.3) defining two classes of signals: those with a finite energy and those with a finite mean power but an infinite energy
- morphological classification (sect.2.4) based on the continuous or discrete character of the amplitude or of the independent variable of the signal
- dimensional classification, which is based on the number of independent variables, in the signal model
- spectral classification, which is based on the shape of the frequency distribution of the signal spectrum

2.2 DETERMINISTIC OR RANDOM SIGNALS

2.2.1 Definitions

The first classification (table 2.1) is made by considering the nature of the signal evolution along the time t axis. Two fundamental types can be defined:

- *deterministic signals* (or nonrandom [57] signals), which have an evolution that is perfectly predictable by an appropriate mathematical model
- *random signals,* which have unpredictable behavior and can generally be described only through statistical observations

2.2.2 Remarks

Deterministic signals are the most comfortable ones in theory because they can be easily depicted mathematically with handy formulas. Unfortunately, they hardly represent observable signals. They are mainly used in laboratories as test signals. They are also encountered in the production of electrical energy by electromechanical generators driven by a turbine.

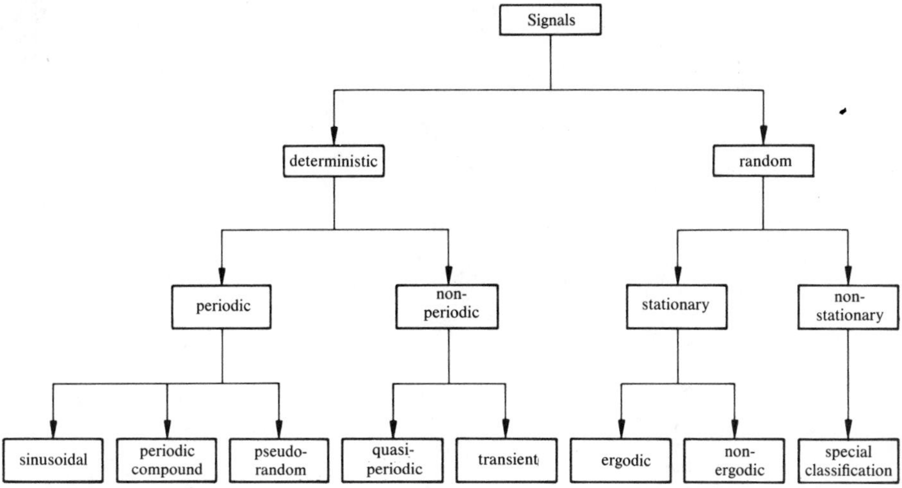

Table 2.1

A signal of known shape and whose position along the time axis is unknown (e.g., a sine wave with unknown initial phase) is already a random signal!

2.2.3 Subdivisions of deterministic signals. Definitions

Among the deterministic signals, there are:

- the *periodic signals*, complying with the relation:

$$x(t) = x(t + kT) \tag{2.1}$$

  i.e., which obey a regular cyclical repetition law with a period T
- the *nonperiodic signals*, which do not enjoy this property.

Sinusoidal signals (fig. 2.2), from the general equation:

$$\begin{aligned} x(t) &= A \sin\left(\frac{2\pi}{T} t + \alpha\right) \\ &= A \sin\left[\frac{2\pi}{T}(t + \tau)\right] \end{aligned} \tag{2.2}$$

form the most familiar group of periodic signals.

Pseudorandom signals (fig. 2.3) form a particular category of periodic signals with quasirandom behavior (sect.5.10). Among nonperiodic signals, we can distinguish between *quasiperiodic signals* (fig. 2.4), resulting from the sum

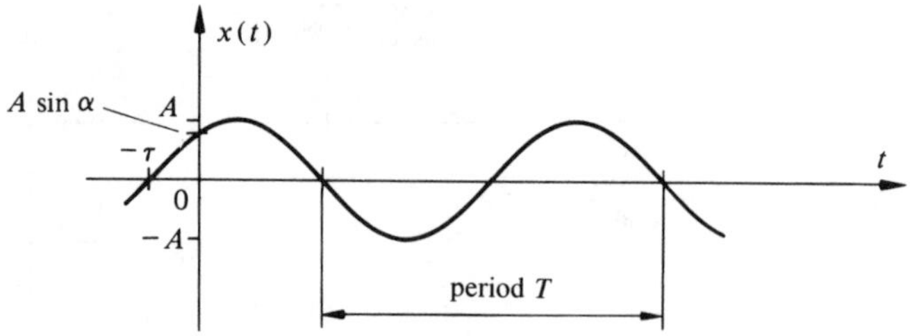

Fig. 2.2

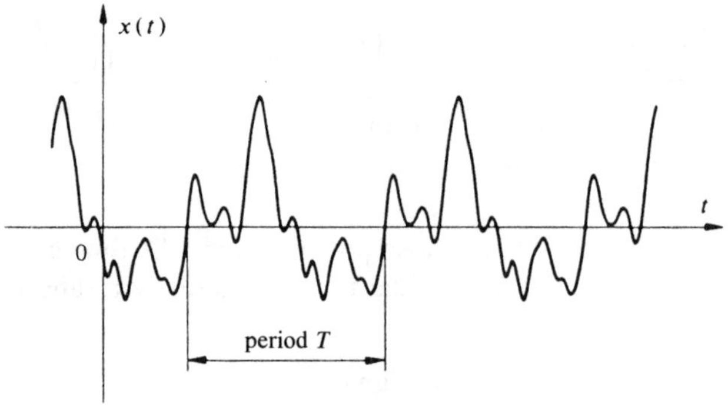

Fig. 2.3 Pseudorandom signal.

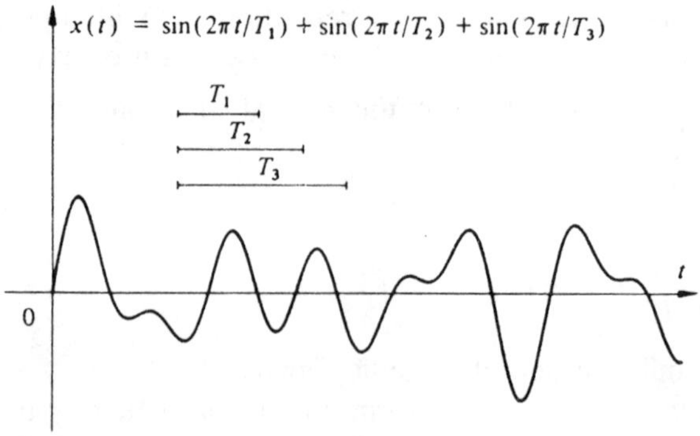

Fig. 2.4 Quasiperiodic signal.

of sinewaves with uncommensurable periods, and *transient signals* that are short-lived (fig. 2.5).

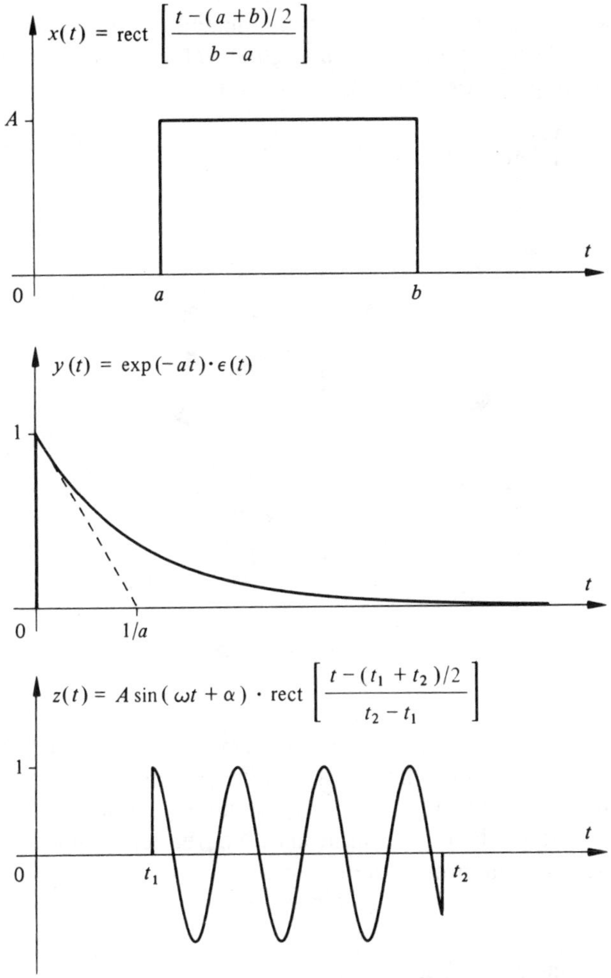

Fig. 2.5 Transient signals: $x(t)$ = rectangular pulse; $y(t)$ = decaying exponential pulse; $z(t)$ = sinusoidal pulse.

2.2.4 Complex notation of sine waves and the negative frequency concept

We often represent a sine function by the imaginary part, or the real part for a cosine, of a complex exponential:

$$A \sin\left(\frac{2\pi}{T} t + \alpha\right) = \mathrm{Im}\left\{ A \exp\left[\mathrm{j}\left(\frac{2\pi}{T} t + \alpha\right)\right]\right\} \tag{2.3}$$

A generalization of this process is possible by considering the sine wave (or cosine) as the vectorial sum of two conjugated phasors with amplitude $A/2$ turning in opposite directions with an angular velocity $\pm\omega = \pm 2\pi/T$ (fig. 2.6). This is a direct application of the Euler formula:

$$\mathrm{j}A \sin(\omega t) = \frac{A}{2} \exp(\mathrm{j}\omega t) - \frac{A}{2} \exp(-\mathrm{j}\omega t) \tag{2.4}$$

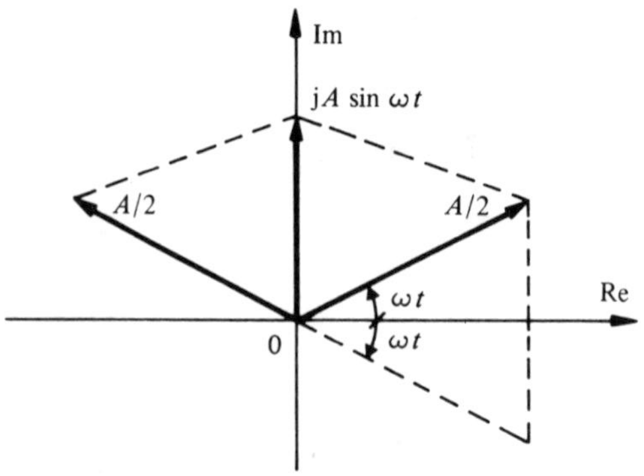

Fig. 2.6

When the angular velocity is counterclockwise, we say that the frequency is positive ($+\omega = +2\pi f$); when it is clockwise, the frequency is said to be negative ($-\omega = -2\pi f$). This negative frequency concept has no physical meaning. It is used for the representation of the frequency functions (spectrum, frequency response function), where $-\infty < f < \infty$.

2.2.5 Subdivisions of random signals

Random signals can be classified into two large categories:

- the *stationary random signals,* the statistical characteristics of which are invariant in time (fig. 2.7)
- the *nonstationary random signals,* which do not have this property (fig. 2.8)

If the statistical average values, or moments, of a stationary signal are equal to its time average values, it is said to be *ergodic* (sect. 5.1).

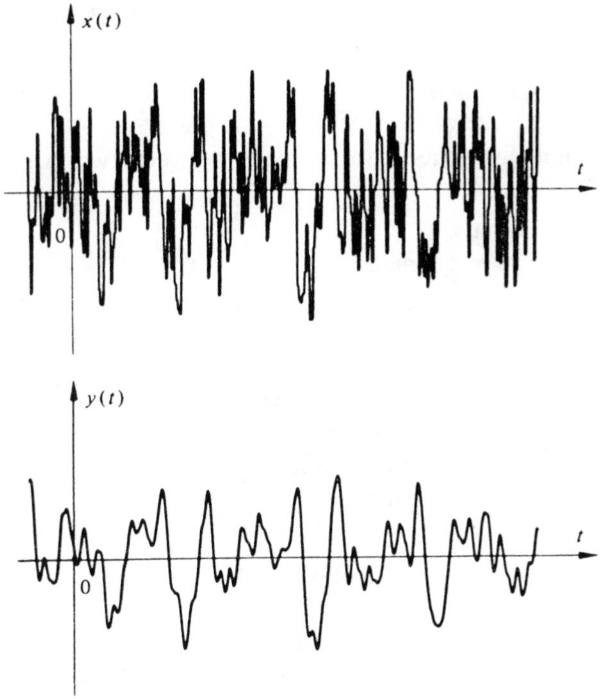

Fig. 2.7 Stationary random signal: $x(t)$ = wideband signal (white noise); $y(t)$ = lowpass filtered signal.

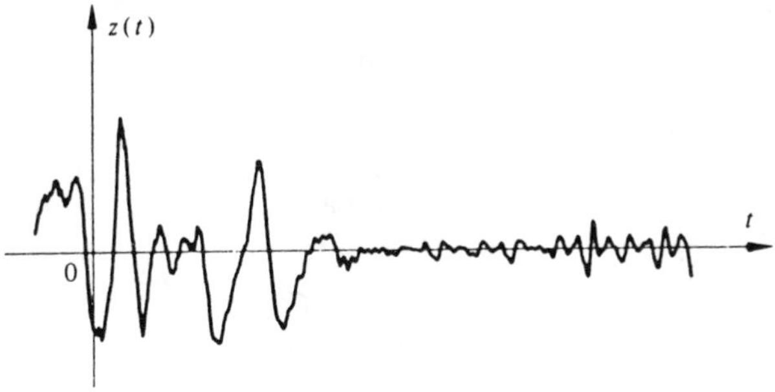

Fig. 2.8 Nonstationary random signal.

2.2.6 Observation

A random signal with a transient behavior is nonstationary.

The stationarity concept, as the permanent characteristic associated with periodic signals, is a comfortable abstraction since we often consider, in practice, a signal as being stationary ***during the observation span***.

2.3 FINITE ENERGY OR FINITE AVERAGE POWER SIGNALS

2.3.1 Energy classification

A fundamental distinction can be made between two large categories of signals:

- finite energy signals
- finite average power signals

The first category includes all the transient deterministic or random signals. The second category includes nearly all the periodic or quasiperiodic signals and the permanent random signals.

Some theoretical signals do not belong to either category: this is the case, for example, of $x(t) = \exp(at)$ for $-\infty < t < \infty$. The delta function, a convenient mathematical abstraction, or the periodic sequence of delta functions, also cannot be classified in this manner.

2.3.2 Energy and average power of a signal

In electrical engineering, the instantaneous power supplied to a bipole is defined as the product of the instantaneous values of the voltage $u(t)$ and of the current $i(t)$

$$p(t) = u(t) \cdot i(t) \quad \mathrm{W} = \mathrm{V} \cdot \mathrm{A} \tag{2.5}$$

In the case of a linear resistor R, respectively of a linear conductance G, we have

$$p(t) = Ri^2(t) = \frac{1}{R}u^2(t) = Gu^2(t) \quad \mathrm{W} \tag{2.6}$$

The energy dissipated during the interval $[t_1,t_2]$, with $t_2 > t_1$, is the integral of this instantaneous power. It is measured in joules. Thus,

$$W(t_1,t_2) = \int_{t_1}^{t_2} p(t)\mathrm{d}t = R\int_{t_1}^{t_2} i^2(t)\mathrm{d}t = G\int_{t_1}^{t_2} u^2(t)\mathrm{d}t \quad \mathrm{J} \tag{2.7}$$

By dividing this energy by the interval duration, we obtain an average power, measured in watts:

$$P(t_1,t_2) = \frac{1}{t_2 - t_1}\int_{t_1}^{t_2} p(t)\mathrm{d}t = \frac{R}{t_2 - t_1}\int_{t_1}^{t_2} i^2(t)\mathrm{d}t = \frac{G}{t_2 - t_1}\int_{t_1}^{t_2} u^2(t)\mathrm{d}t \tag{2.8}$$

Similarly, the *(normalized) energy* and *(normalized) average power* of a real signal $x(t)$, calculated on an interval $[t_1,t_2]$, are defined in signal theory by the following integral square (1.27) and mean square values (1.28):

$$W_x(t_1,t_2) = \int_{t_1}^{t_2} x^2(t)\mathrm{d}t \tag{2.9}$$

$$P_x(t_1,t_2) = \frac{1}{t_2 - t_1} \int_{t_1}^{t_2} x^2(t)\mathrm{d}t \tag{2.10}$$

The square root of (2.10) is the root mean square value (1.31). This is the same definition as for periodic signals, but is extended here to signals of any kind.

The normalized average power has the same dimension as the square of $x(t)$. Multiplying by the time dimension, we get the dimension of the normalized energy. If $x(t)$ is a voltage or an electrical current, (2.9) and (2.10) correspond to the dissipated energy and the dissipated power in a one-ohm resistor.

The *total energy* and the *total average power* of a signal are obtained by integrating time over the complete real axis. Relations (2.9) and (2.10) are then modified as follows:

$$W_x = \int_{-\infty}^{\infty} x^2(t)\mathrm{d}t \tag{2.11}$$

$$P_x = \lim_{T\to\infty} \frac{1}{T} \int_{-T/2}^{T/2} x^2(t)\mathrm{d}t \tag{2.12}$$

The total average power is defined as a Cauchy principal value. For periodic signals, the total average power (2.12) is equal to the average power of one period.

When the signal is represented as a complex function of a real variable t, we substitute $x^2(t)$ in (2.11) and (2.12) by $|x(t)|^2$.

2.3.3 Finite energy signals

Finite energy signals are those for which the integral (2.11) is finite. These signals, also called *square integrable signals*, comply with the condition:

$$\int_{-\infty}^{\infty} |x(t)|^2\mathrm{d}t < \infty \tag{2.13}$$

Their average power is zero.

2.3.4 Finite average power signals

Signals with finite average power (non-zero) are those which comply with the condition:

$$0 < \lim_{T\to\infty} \frac{1}{T} \int_{-T/2}^{T/2} |x(t)|^2 \mathrm{d}t < \infty \tag{2.14}$$

2.3.5 Observations

The function $x^2(t)$ corresponds to a time distribution of the signal energy (fig. 2.9). The average power $P_x(T)$ is, in other words, the energy mean distribution over the chosen interval T.

The study of conditions (2.13) and (2.14) clearly points out that a signal with non-zero finite average power has an unlimited energy and a signal with finite energy has a zero average power. Obviously, only the last one is physically achievable.

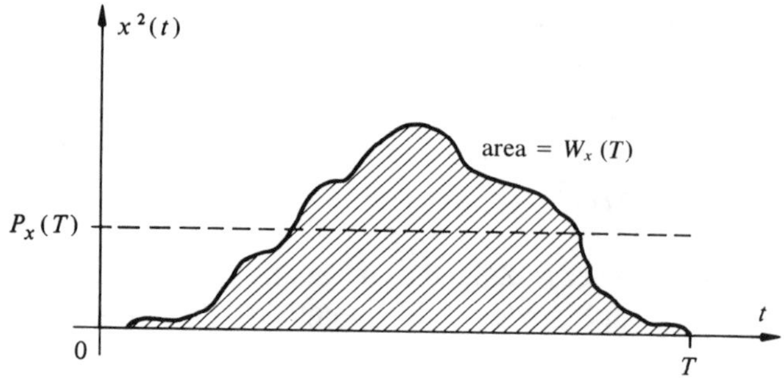

Fig. 2.9

2.4 DISCRETE AND CONTINUOUS VARIABLES

2.4.1 Morphology classification

A signal can be described in many ways, depending on whether its amplitude is a continuous or discrete variable and the argument t (here considered as time) is itself continuous or discrete (fig. 2.10). Therefore, four types of signals can be distinguished:

- signals with continuous amplitude and continuous time, usually called *analog signals*
- signals with discrete amplitude and continuous time, called *quantized signals*

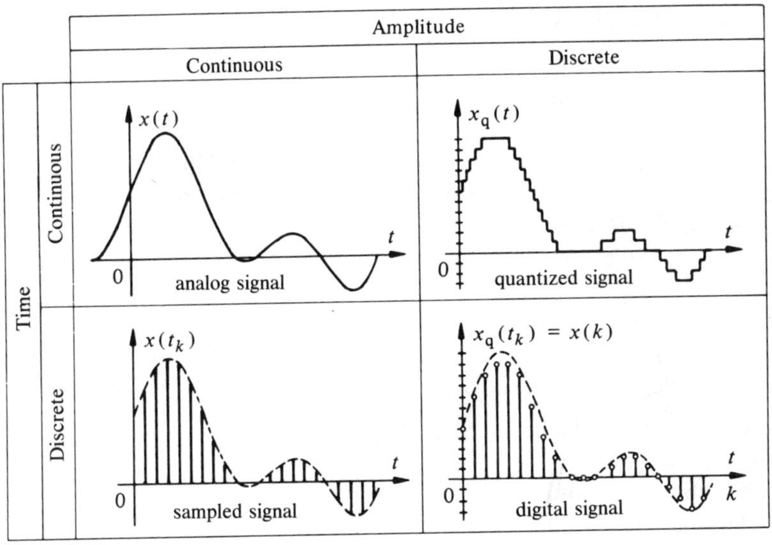

Fig. 2.10

- signals with continuous amplitude and discrete time, called *sampled signals*
- signals with discrete amplitude and discrete time, called *digital signals* because they can be represented as a sequence of numbers (*digits*).

2.4.2 Processing system classification

Signal processing systems are also classified according to the nature of the signals on which they act. Thus, we have:

- *analog systems:* amplifiers, classical filters, multipliers, signals modulators, *et cetera*
- *sampled data systems:* charge coupled devices, switched capacitor filters, *et cetera*
- *digital systems:* digital filters, correlators, Fourier transformers, and other specialized processors.

There are also hybrid structures, such as analog-digital converters.

In the sampled data systems, the effectively used signal usually corresponds to an intermediate case between the analog signal and the sampled signal; it is produced by a sampling process (chap. 9), which keeps the amplitude at the level of the last value captured between two samples.

2.5 OTHER IMPORTANT CLASSES

2.5.1 Spectral classification

The spectral analysis of a signal leads to a classification based on the frequency distribution spectrum $\Phi_x(f)$ of its energy or of its power.

The *bandwidth B* of a signal is the primary range of the frequencies (positive or negative) occupied by its spectrum. It is defined by the relation:

$$B = f_2 - f_1 \tag{2.15}$$

with $0 \leqslant f_1 < f_2$, where f_1 and f_2 are the characteristic low and high cut-off frequencies. We usually speak of

- low frequency signals (fig. 2.11)
- high frequency signals (fig. 2.12)
- narrowband signals (fig. 2.13)
- wideband signals (fig. 2.14)

These designations are unclear and depend on the context (see also subsect. 8.2.23).

A signal whose spectrum is zero except in a specified frequency bandwidth B:

$$\Phi_x(f) = 0 \quad \forall |f| \notin B \tag{2.16}$$

is said to be a *band-limited signal*.

2.5.2 Duration-limited signals

Signals whose amplitude is zero outside a specified time interval T:

$$x(t) = 0 \quad t \notin T \tag{2.17}$$

are called *duration-limited signals*.

2.5.3 Bounded signals

This is the case for all physical signals whose amplitude cannot exceed certain limit, often enforced by some electronic processing devices.

In this case, we can write

$$|x(t)| \leqslant K \text{ for } -\infty < t < \infty \tag{2.18}$$

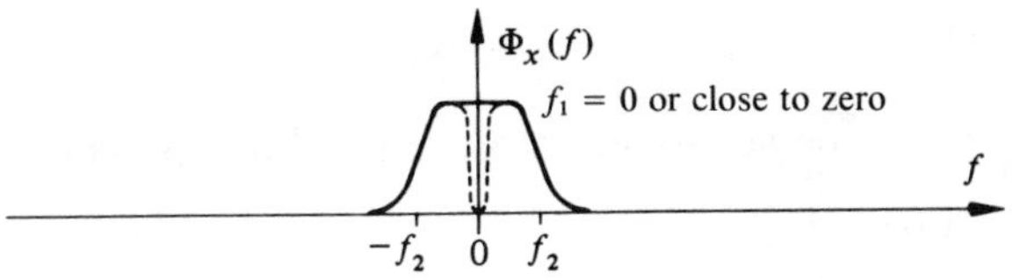

Fig. 2.11

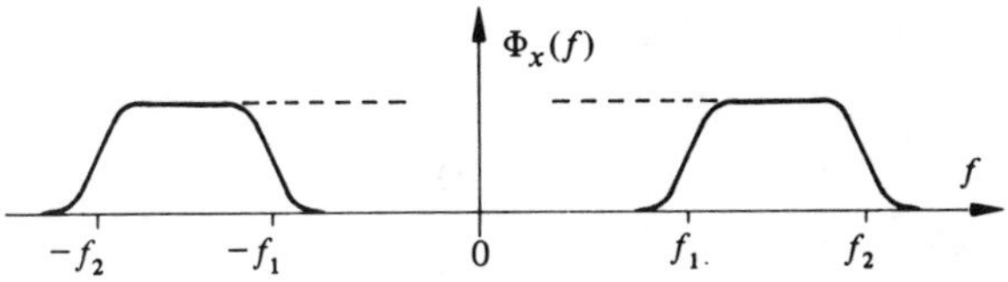

Fig. 2.12

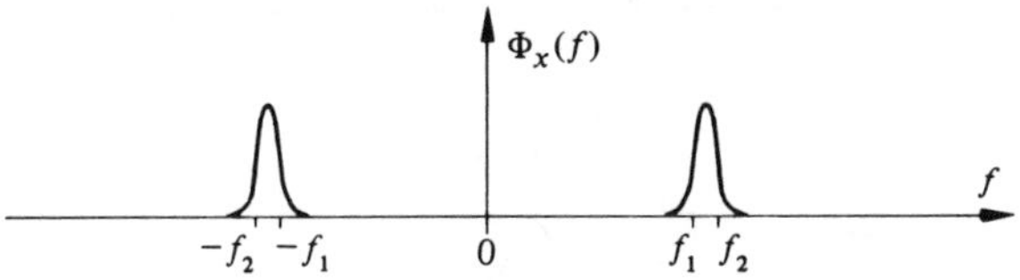

Fig. 2.13

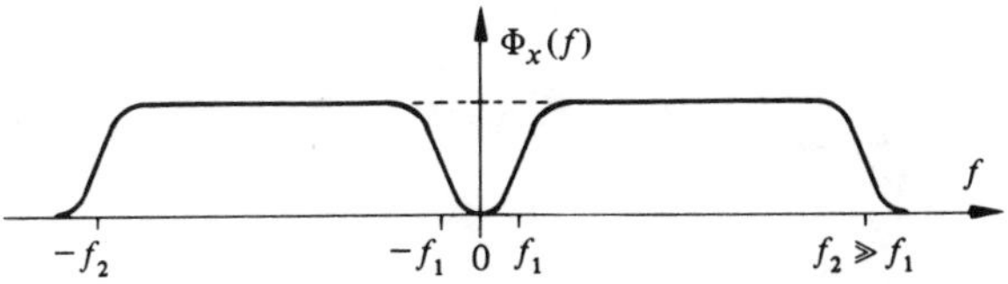

Fig. 2.14

2.5.4 Odd and even signals

A signal is said to be even if

$$x(t) = x(-t) \tag{2.19}$$

A signal is said to be odd if

$$x(t) = -x(-t) \tag{2.20}$$

2.5.5 Application

Any real signal can be separated into even and odd parts (fig. 2.15):

$$x(t) = x_p(t) + x_i(t) \tag{2.21}$$

with

$$x_p(t) = \tfrac{1}{2}[x(t) + x(-t)] \tag{2.22}$$

$$x_i(t) = \tfrac{1}{2}[x(t) - x(-t)] \tag{2.23}$$

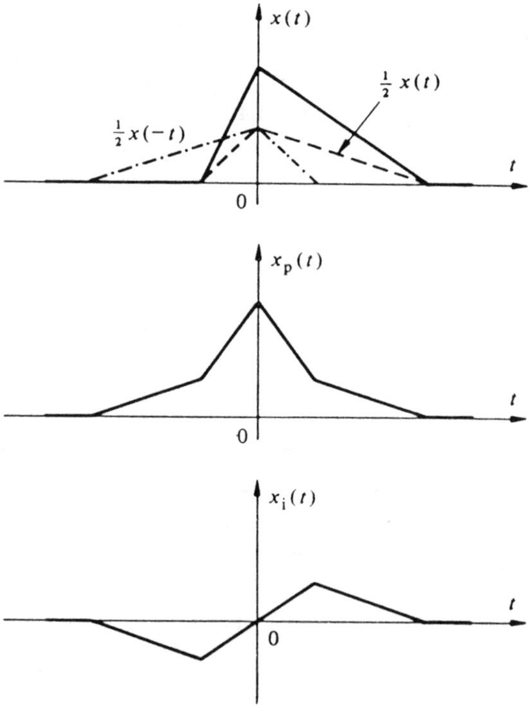

Fig. 2.15

2.5.6 Causal signals

A signal is said to be causal if, for any negative time value, its amplitude is zero:

$$x(t) \equiv 0 \quad t < 0 \tag{2.24}$$

By taking (2.21) into consideration, we see that a causal, real signal can be expressed as (fig. 2.16):

$$x_i(t) = x_p(t) \cdot \text{sgn}(t) \tag{2.25}$$

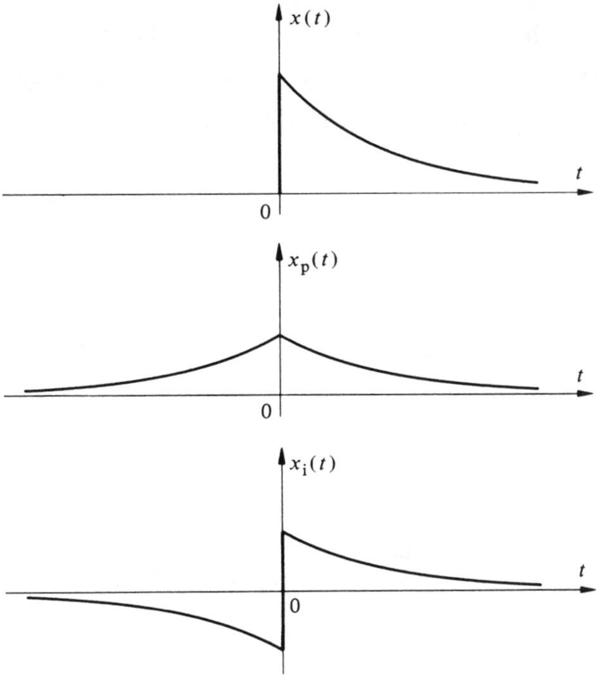

Fig. 2.16

2.5.7 Observation

Experimentally, all signals are causal, i.e., they begin at instant $t = 0$. For theoretical convenience, we usually define signals as existing on the total time axis.

A linear system is said to be causal if its impulse response is a causal signal.

2.6 PROBLEMS

2.6.1 Are the following signals of finite energy, finite average power, or of neither kind? Calculate in each case the total energy and the total average power ($a > 0$).

$A\,\mathrm{rect}(t/T)$; $A \sin \omega t$; $A \sin \omega t \cdot \epsilon(t)$; $\epsilon(t)$; $t \cdot \epsilon(t)$; $A \exp(-at) \cdot \epsilon(t)$; $A \exp(-at)$; $A\,\mathrm{tri}(t/T)$.

2.6.2 Write the average power equation of $A \sin(2\pi t/T_0)$ in terms of the measurement interval T, and demonstrate that the total average power as defined by formula (2.12) is exactly the same as the one calculated on a period T_0. For which other value of the interval do we obtain the same result?

2.6.3 Define the odd and even parts of $x(t) = A \sin(\omega t - \alpha)$.

2.6.4 Demonstrate that the average value of the odd part of a real signal is always zero.

Chapter 3

Vector Representation of Signals

3.1 SIGNAL SPACE

3.1.1 Discrete representation of signals

A discrete representation of a signal $x(t)$ is based on its development as a linear combination of known functions $\psi_k(t)$; $k = 1, 2, \ldots, n$:

$$x(t) = \sum_{k=1}^{n} \alpha_k \psi_k(t) \tag{3.1}$$

The n coefficients α_k form a ***discrete representation*** of the signal that depends on the chosen functions set. This constitutes the fundamental concept of signal analysis.

Our interest in such a description is threefold:

- an adequate choice of the $\psi_k(t)$ functions greatly facilitates the extraction of its key characteristics and the study of the transformations it is undergoing during its propagation through a given physical system, especially a linear one;
- the image of a vector in an n-dimensional (possibly infinite) space is naturally associated with this discrete representation; this brings geometrical interpretation to otherwise abstract ideas, such as distance, inner product, orthogonalization, crosscorrelation of two signals, *et cetera*;
- discrete transformation is the only way to approach digital signal processing.

3.1.2 The concept of a functional vector space

A *vector space* is a set of elements complying with the following axioms: the sum of two elements and the product of any of them by a (real or complex) scalar number also belong to this set; an n-dimensional linear vector space is spanned by a *basis,* that is, a subset of n linearly independent vectors, i.e., any vector $\boldsymbol{x}$ in this space can be expressed as a unique linear combination of those basis vectors.

A vector space is said to be *normal* if we can associate with any vector $\boldsymbol{x}$, a real, positive number (equal to zero if $\boldsymbol{x}$ is the origin), called the *norm* $\|\boldsymbol{x}\|$.

This can be seen as a generalization of the fundamental concept of length. Such a normal space is said to be *metric* when we can associate with any pair of elements $(\boldsymbol{x},\boldsymbol{y})$, a real, positive number (equal to zero when $\boldsymbol{x} = \boldsymbol{y}$) called the *distance* $d(\boldsymbol{x},\boldsymbol{y})$ between those elements. The usual metric is $d(\boldsymbol{x},\boldsymbol{y}) = \|\boldsymbol{x} - \boldsymbol{y}\|$.

An infinite sequence of elements $\{\boldsymbol{x}_n\}$ in a metric space *converges* towards an element, $\boldsymbol{x}$, when

$$\lim_{n\to\infty} d(\boldsymbol{x}_n,\boldsymbol{x}) = 0$$

Various possible metrics correspond to the various convergence modes. A space in which any sequence converges is said to be *complete*.

These abstract concepts, introduced in linear algebra [58], can be extended to the case of functions belonging to a given family. Any member of this family, assimilable to a vector, can be expressed as a linear combination of a particular set of functions of this family, forming a basis of the considered vector space (*functional space*). This basis can be composed of an infinity of elements: the space is then of infinite dimension.

3.1.3 Vector representation of a signal

A signal is usually represented as a function which belongs to a family having a common property (for example, finite energy, finite average power, *et cetera*). It is then possible to consider a signal $x(t)$ as a ***vector*** in an adequate metric space, called the *signal space* [49].

Let us consider a set $\{\psi_k(t)\}$ of n linearly independent functions forming a basis of the signal space. A unique linear combination of the functions $\psi_k(t)$ corresponds to any member of this space, as given in (3.1). The ordered sequence of the $\{\alpha_k\}$ coefficients forms a n-tuple defining a point (fig. 3.1) of coordinates $(\alpha_1,\alpha_2, \ldots, \alpha_n)$ in this n-dimensional space with respect to the $\{\psi_k(t)\}$ basis.

Thus, there is a one-to-one correspondence between vectors of the arbitrary signal space and the space of the n-tuples, often denoted R^n (real coefficients) or C^n (complex coefficients).

We say that the n-tuple $\boldsymbol{\alpha} = \{\alpha_k\}$ is a representation (in R^n or C^n) for $x(t)$ relative to the basis $\{\psi_k(t)\}$.

To each basis corresponds a particular vector representation $\boldsymbol{\alpha}$ for $x(t)$. This provides us with a large variety of signal analysis possibilities. The simplicity, efficiency, and usefulness of such an analysis depend on the appropriate choice of the basis.

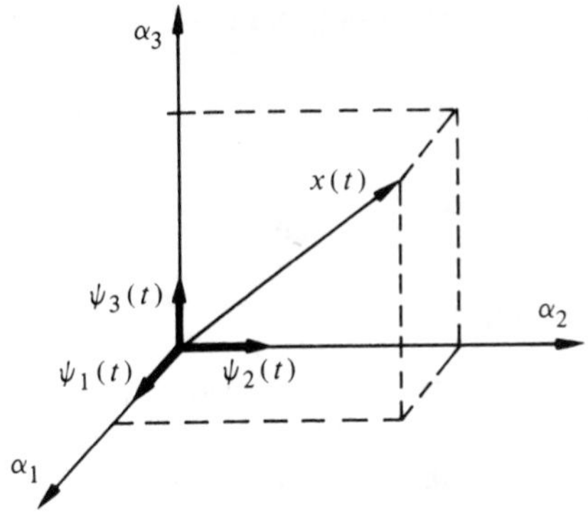

Fig. 3.1 Vector representation of a signal (for $n = 3$).

3.1.4 Distance between two signals

The *distance* $d(x,y)$ between two signals $x(t)$ and $y(t)$ is a measure of their dissimilarity. It is equal to zero if the signals are identical.

This concept plays an important role in signal theory where it is used for signal comparison. In signal detection and in pattern recognition, we calculate the distances of a signal or of a pattern to be identified with a group of possible candidates. The preferred candidate is generally the one corresponding to the smallest distance (maximum likelihood principle [20,59]).

A *filter* is a device that minimizes a certain distance $d(x,y)$ between an input signal $x(t)$, contaminated with undesired interference, and an output signal $y(t)$, presenting desired properties.

For vectors $\boldsymbol{x} = (x_1, x_2, \ldots, x_n)$ and $\boldsymbol{y} = (y_1, y_2, \ldots, y_n)$ the standard Euclidian distance is

$$d(\boldsymbol{x},\boldsymbol{y}) = \left(\sum_{i=1}^{n} |x_i - y_i|^2 \right)^{1/2} \tag{3.2}$$

The *Euclidian distance between two signals* $x(t)$ and $y(t)$, defined on a time interval T, is by analogy

$$d_1(x,y) = \left(K \int_T |x(t) - y(t)|^2 \mathrm{d}t \right)^{1/2} \tag{3.3}$$

We also call this the *mean square distance*. The coefficient K is either equal to 1, or equal to $1/T$. The definition (3.3) is the most familiar and the most useful of the distance measurements. However, other definitions are sometimes used, either because they are more adapted to a given context, or simply because they imply a lower computational load. Let us mention, for example,

$$d_2(x,y) = K\int_T |x(t) - y(t)|\, \mathrm{d}t \tag{3.4}$$

$$d_3(x,y) = K\int_T |\, \mathrm{sgn}\,\{ x(t) - a \} - \mathrm{sgn}\,\{ y(t) - b \}\,|\, \mathrm{d}t \tag{3.5}$$

$$d_4(x,y) = \sup\,\{ |x(t) - y(t)| \ ;\ t \in T\} \tag{3.6}$$

In (3.5), a and b are constants, often chosen equal to the average values of the corresponding signals on the definition interval. The notation sup $\{z(t); t \in T\}$ of relation (3.6) indicates the maximum value of $z(t)$ in the definition interval T.

To protect the signals carrying sequences of binary information (words) against interference present on the transmission path, detecting or error-correcting codes have been developed. In the study of such codes, the *Hamming distance* [59]:

$$d_5(x,y) = \sum_{i=1}^{n} [c_i \oplus c_i'] \tag{3.7}$$

is used to compare a received word $C' = (c_1', c_2', \ldots, c_n)$ to a possible candidate $C = (c_1, c_2, \ldots, c_n)$. The c_i and c_i' here denote binary symbols 0 or 1 (bits) and the sign $\oplus$ denotes the modulo-2 addition (exclusive—OR function). This special distance is equal to the number of bits by which the two words differ.

To illustrate the concept of distance between two signals and show the influence of their definitions, we must consider some examples.

3.1.5 Example 1

Assume that we want to compare two signals $x(t) = A \cos \omega_0 t$ and $y(t) = x(t - \tau) = A \cos \omega_0(t - \tau)$ in order to determine their distance according to the delay parameter τ (or the phase shift $\theta = \omega_0\tau$). Such a situation can occur, for example, in synchronization problems.

Let us make the comparison on a period $T = 2\pi/\omega_0$ for $-T/2 \leqslant T \leqslant T/2$ by simultaneously considering the distances (3.3), (3.4), (3.5), and (3.6), where $a = b = 0$ and $K = 1/T$.

We have

$$\begin{aligned} |x(t) - y(t)| &= A|\cos(2\pi t/T) - \cos[2\pi(t - \tau)/T]\,| \\ &= 2A|\sin\,(\pi\tau/T)| \cdot |\sin\,[\pi(2t - \tau)/T]\,| \end{aligned}$$

$$d_1(x,y) = 2\,A|\sin(\pi\tau/T| \cdot \left\{T^{-1} \int_0^T \sin^2[\pi(2t - \tau)/T]\,\mathrm{d}t\right\}^{1/2}$$
$$= \sqrt{2}A|\sin\,(\pi\tau/T)|$$

$$d_2(x,y) = 2\,A|\sin(\pi\tau/T)| \cdot \frac{1}{T}\int_0^T |\sin\,[\pi(2t - \tau)/T]\,|\,\mathrm{d}t = \frac{4}{\pi}A\,|\,\sin(\pi\tau/T)|$$

$$d_3(x,y) = 4T^{-1}\int_0^{|\tau|} \mathrm{d}t = 4|\tau|/T \quad -T/2 \leqslant \tau \leqslant T/2$$

$$d_4(x,y) = 2A|\sin(\pi\tau/T)| \cdot \sup\{|\sin[\pi(2t - \tau)/T]|;\, t \in T\}$$
$$= 2A|\sin(\pi\tau/T)|$$

By choosing coefficient $K = 1/T$ in (3.3), (3.4), and (3.5), we obtain a uniform description of the various measurements of distance, since they all have the same dimension as signals $x(t)$ and $y(t)$. Thus, we can make valid comparisons among them. This is graphically done in figure 3.2.

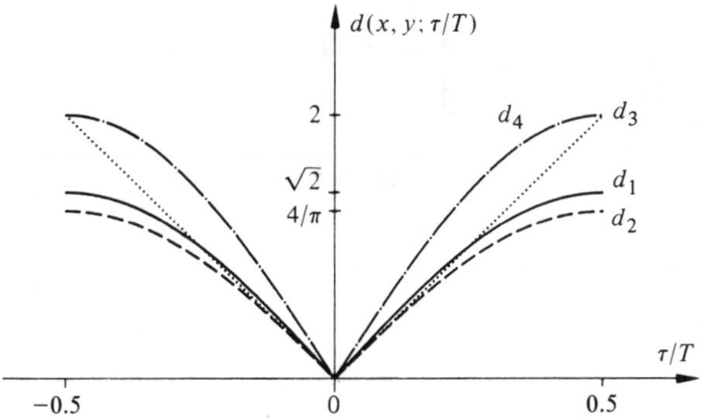

Fig. 3.2

3.1.6 Example 2

Let us consider the four pulse signals, with limited duration T, as represented in figure 3.3. The distances $d_1(x_i,x_j)$ and $d_2(x_i,x_j)$, calculated for $K = 1/T$, respectively give:

$d_1(x_1,x_2) = d_1(x_1,x_3) = d_1(x_1,x_4) = d_1(x_2,x_4) = d_1(x_3,x_4) = \sqrt{2}$
$d_1(x_2,x_3) = 2$
$d_2(x_1,x_2) = d_2(x_1,x_3) = d_2(x_1,x_4) = d_2(x_2,x_4) = d_2(x_3,x_4) = 1$
$d_2(x_2,x_3) = 2$

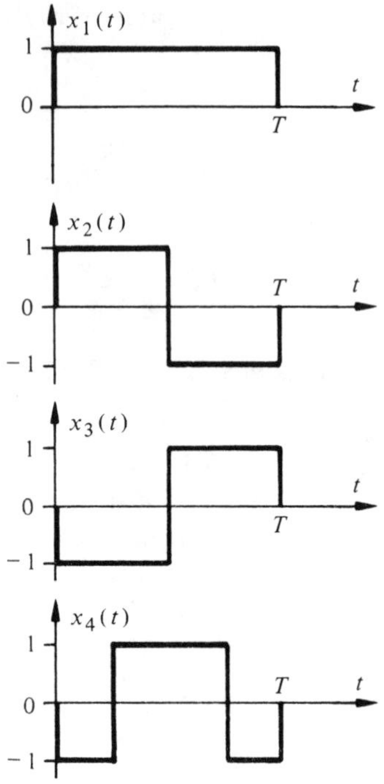

Fig. 3.3

Of course, we have for any i:

$$d_1(x_i,x_i) = d_2(x_i,x_i) = 0$$

3.1.7 Example 3

Let us determine the distances $d_1(x,y)$ *and* $d_2(x,y)$ between the two signals $x(t) = \exp(-at) \cdot \epsilon(t)$ and $y(t) = x(t - \tau)$, where $\epsilon(t)$ denotes the unit-step function (sub-sect. 1.3.3) and $a > 0$. The definition interval is infinite for this case.

We will use the measurements of distance (3.3) and (3.4) assuming K = 1. The measurements obtained are not homogeneous, and thus cannot be compared quantitatively.

For $\tau > 0$, we have

$$|x(t) - y(t)| = \begin{cases} \exp(-at) \cdot \epsilon(t) & -\infty \leqslant t \leqslant \tau \\ \exp(-at) \cdot [\exp(a\tau) - 1] & \tau \leqslant t \leqslant \infty \end{cases}$$

For $\tau < 0$, we have

$$|x(t) - y(t)| = \begin{cases} \exp(-at) \cdot \epsilon(t - \tau) \cdot \exp(a\tau) & -\infty < t < 0 \\ \exp(-at) \cdot [1 - \exp(a\tau)] & 0 \leqslant t < \infty \end{cases}$$

Then

$$d_1(x,y) = \left\{\frac{1}{a}[1 - \exp(-a|\tau|)]\right\}^{1/2}$$

and

$$d_2(x,y) = \frac{2}{a}[1 - \exp(-a|\tau|)]$$

3.1.8 L^2 space of finite energy signals

The set of all signals (real or complex functions of time defined on an interval $[t_1,t_2]$) that are square integrable (sub-sect. 2.3.3) form a functional space, denoted $L^2(t_1,t_2)$, the norm of which is

$$\|x\| = \left[\int_{t_1}^{t_2} |x(t)|^2 \mathrm{d}t\right]^{1/2} \tag{3.8}$$

The square of the norm corresponds to the energy of the signal $x(t)$.

The distance between two signals $x(t)$ and $y(t)$ of $L^2(t_1,t_2)$ is the Euclidian distance (3.3), for $K = 1$:

$$d(x,y) = \|x - y\| = \left[\int_{t_1}^{t_2} |x(t) - y(t)|^2 \mathrm{d}t\right]^{1/2} \tag{3.9}$$

In L^2, a signal $y(t)$ converges to $x(t)$ in the *mean square* sense when their distance $d(x,y)$ tends toward zero. The exponent 2 appearing in the symbolic notation of the considered functional space is not an indication of its dimension, but of the implied convergence criterion.

The origin of this space is the signal, which is null almost everywhere, i.e., except on a limited set of finite discontinuities (isolated points).

3.1.9 Inner product of signals

The *inner product* (also called *scalar* or *dot product*) of two vectors $\boldsymbol{x} = (x_1,x_2, \ldots, x_n)$ and $\boldsymbol{y} = (y_1,y_2, \ldots, y_n)$ with real or complex coordinates is defined by

$$x \cdot y = \sum_{i=1}^{n} x_i y_i^* \tag{3.10}$$

It is related to the norm by the identity

$$x \cdot x = \|x\|^2 \tag{3.11}$$

By analogy, we define the *inner product of two signals*—real or complex functions of time—$x(t)$ and $y(t)$ of $L^2(t_1,t_2)$ by

$$< x,y^* > = \int_{t_1}^{t_2} x(t)y^*(t)\mathrm{d}t \tag{3.12}$$

which is related to the norm (3.8) by the identity

$$< x,x^* > = \|x\|^2 \tag{3.13}$$

It can be demonstrated that the space $L^2(t_1,t_2)$, with its inner product inducing a norm, is complete. In mathematics, such a space is called a *Hilbert space*.

The inner product has *hermitian symmetry*

$$< x,y^* > = < y,x^* > ^* \tag{3.14}$$

3.1.10 Inner product notation

The functional notation $< x,y^*>$ used in this book, where the asterisk indicates the complex conjugation, is somewhat unusual, but it reminds the reader of the need for complex conjugation. In most reference books, the inner product is frequently referred to as $<x,y>$ or (x,y). The notation $<x,y>$ here will be reserved for the case of real signals.

3.1.11 Orthogonal functions

In Euclidian geometry, two vectors are orthogonal when their inner product is zero. By analogy, $x(t)$ and $y(t)$ are *orthogonal functions* on interval $[t_1,t_2]$ if their inner product:

$$< x,y^* > = \int_{t_1}^{t_2} x(t)y^*(t)\mathrm{d}t = 0 \tag{3.15}$$

The specification of interval $[t_1,t_2]$ is important. Two orthogonal functions on $[t_1,t_2]$ are not necessarily orthogonal on any other interval $[t_3,t_4]$.

3.1.12 Example

The four signals of figure 3.3 are orthogonal by pairs on interval $[0,T\,]$, except for the pair $x_2(t)$ and $x_3(t)$, the inner product of which is equal to $<x_2,x_3^*> = T$.

3.1.13 Relationship between inner product and Euclidian distance

The concept of inner product is, like that of distance, important in signal theory. Moreover, these concepts are closely related. Indeed, in $L^2(t_1,t_2)$, we have according to (3.9), (3.12), (3.13), and (3.14):

$$d(x,y) = \left| \int_{t_1}^{t_2} |x(t) - y(t)|^2 dt \right|^{\frac{1}{2}}$$

$$d^2(x,y) = \int_{t_1}^{t_2} [x(t) - y(t)][x^*(t) - y^*(t)]dt$$

$$= <x,x^*> + <y,y^*> - <x,y^*> - <x,y^*>* \qquad (3.16)$$

$$= \|x\|^2 + \|y\|^2 - 2\text{Re} <x,y^*>$$

For two orthogonal functions, we obtain from (3.15) the general formulation of the Pythagorean theorem:

$$d^2(x,y) = \|x\|^2 + \|y\|^2 \qquad \left[\begin{array}{l} <x,y^*> = 0 \\ <x,y^*>* = 0 \end{array} \right] \qquad (3.17)$$

The inner product is directly linked to the correlation concept (sub-sect. 4.2.10), which is also used to compare two signals (sect. 13.2).

3.1.14 Schwarz inequality

Let us consider the functions $x(t)$ and $ky(t)$ of $L^2(t_1,t_2)$, where k is a real or complex arbitrary constant. By definition, the distance $d(x,ky)$ is positive or zero. A development similar to (3.16) yields

$$d^2(x,ky) = <x,x^*> + |k|^2 <y,y^*> - k^* <x,y^*> - k<x,y^*>* \geq 0 \qquad (3.18)$$

By inserting the arbitrary value $k = <x,y^*> / <y,y^*>$ which must satisfy (3.18), we obtain

$$<x,x^*> - \frac{|<x,y^*>|^2}{<y,y^*>} \geq 0 \qquad (3.19)$$

from which we extract the *Schwarz inequality*:

$$|<x,y^*>|^2 \leq <x,x^*> \cdot <y,y^*> \qquad \textbf{(3.20)}$$

Taking into account (3.8), (3.12), and (3.13), this inequality can also be written as

$$\left| \int_{t_1}^{t_2} x(t)y^*(t)dt \right|^2 \leq \int_{t_1}^{t_2} |x(t)|^2 dt \cdot \int_{t_1}^{t_2} |y(t)|^2 dt \qquad (3.21)$$

Equality is only achieved when $d(x,ky) = 0$, i.e., when

$$x(t) = ky(t) \qquad (3.22)$$

The Schwarz inequality is often used to solve optimization problems (see chap. 4,7, or 13).

3.2 LEAST SQUARE APPROXIMATION

3.2.1 Optimal approximation of a signal in L^2

Let us consider a signal of a n-dimensional L^2 space and a set $\{\psi_k(t)\}$ of $m < n$ functions, linearly independent in L^2, forming a basis of a L^2 subspace E_m.

We can define an m-order approximation $\tilde{x}(t)$ in E_m of signal $x(t)$ by the linear combination

$$\tilde{x}(t) = \sum_{k=1}^{m} \alpha_k \psi_k(t) \tag{3.23}$$

The difference

$$e(t) = x(t) - \tilde{x}(t) \tag{3.24}$$

is an approximation error signal, the norm of which is, according to (3.9), equal to the distance $d(x,\tilde{x})$:

$$\|e\| = d(x,\tilde{x}) \tag{3.25}$$

The square of the norm $\|e\|$ is called the *mean square error.* The approximation $\tilde{x}(t)$ of $x(t)$ is optimal in the least square sense if the coefficients α_k are chosen so that distance $d(x,\tilde{x})$ is minimized.

If $x(t) \in E_m$, $d(x,\tilde{x}) = 0$ and $\tilde{x}(t) \equiv x(t)$.

3.2.2 Projection theorem

The distance $d(x,\tilde{x})$ between a function $x(t)$ and its approximation (3.23) is minimized when the approximation error $e(t) = x(t) - \tilde{x}(t)$ is orthogonal to the functions $\psi_k(t)$, i.e., when

$$<e,\psi_k^*> = 0 \tag{3.26}$$

The theorem is a generalization of the very well-known result of Euclidian geometry: the shortest distance between a point and a plane is the length of the line drawn perpendicularly from the point to the plane.

3.2.3 Demonstration

Assume $\tilde{x}(t) = \Sigma\alpha_k\psi_k(t)$ is an m-order approximation of $x(t)$ so that the error $e(t) = x(t) - \tilde{x}(t)$ complies with the orthogonality condition (3.26). Let us consider another arbitrary m-order approximation $\hat{x}(t) = \Sigma\beta_k\psi_k(t)$ of $x(t)$. We can write with (3.16)

$$
\begin{aligned}
d^2(x,\hat{x}) &= d^2(x - \tilde{x},\hat{x} - \tilde{x}) \\
&= \|x - \tilde{x}\|^2 + \|\hat{x} - \tilde{x}\|^2 - 2\mathrm{Re} < x - \tilde{x},\hat{x}^* - \tilde{x}^* >
\end{aligned}
\tag{3.27}
$$

Now $< x - \tilde{x},\hat{x}^* - \tilde{x}^* > = 0$, since $\hat{x}(t) - \tilde{x}(t) = \Sigma(\beta_k - \alpha_k)\psi_k(t)$ and $< x - \tilde{x},\psi_k^* > = 0$ for any k. Relation (3.27), therefore, becomes $d^2(x,\hat{x}) = \|x - \tilde{x}\|^2 + \|\hat{x} - \tilde{x}\|^2$, which is clearly minimal when $\hat{x}(t) = \tilde{x}(t)$.

3.2.4 Minimum mean square error

If the condition (3.26) is met, then

$$
< x,\tilde{x}^* > = < \tilde{x},\tilde{x}^* > = \|\tilde{x}\|^2 \tag{3.28}
$$

because $< x,\tilde{x}^* > = < \tilde{x} + e,\tilde{x}^* >$ and $< e,\tilde{x}^* > = 0$.

Then, with (3.16), (3.25), and (3.28), the minimum mean square error is given by

$$
\|e\|^2 = d^2(x,\tilde{x}) = \|x\|^2 - \|\tilde{x}\|^2 \tag{3.29}
$$

or, in other words, using (3.8), (3.16), and (3.23)

$$
\begin{aligned}
\|e\|^2 &= \int_{t_1}^{t_2} |x(t) - \tilde{x}(t)|^2 \mathrm{d}t \\
&= \int_{t_1}^{t_2} |x(t)|^2 \mathrm{d}t - \sum_{k=1}^{m} \sum_{l=1}^{m} \alpha_k \alpha_l^* < \psi_k,\psi_l^* >
\end{aligned}
\tag{3.30}
$$

3.2.5 Determination of the α_k coefficients

If the condition (3.26) is met, by analogy with (3.28), we have

$$
< x, \psi_l^* > = < \tilde{x}, \psi_l^* > \tag{3.31}
$$

The substitution of $\tilde{x}(t)$ according to the development of (3.23) leads to a set of m equations:

$$
\sum_{k=1}^{m} < \psi_k, \psi_l^* > \alpha_k = < x, \psi_l^* > \; ; l = 1, 2 \ldots, m \tag{3.32}
$$

the solution of which is the set of optimum coefficients $\{\alpha_k\}$.

In matrix notation, the system of equation (3.32) becomes simply

$$
\mathbf{\Lambda} \cdot \boldsymbol{\alpha} = \mathbf{\Gamma} \tag{3.33}
$$

where $\mathbf{\Lambda}$ is the $m \times m$ matrix of the inner products of the basis functions:

$$\lambda_{kl} = < \psi_k, \psi_l^* > = \int_{t_1}^{t_2} \psi_k(t)\psi_l^*(t) \, \mathrm{d}t \tag{3.34}$$

and $\mathbf{\Gamma}$ is the column vector of the inner products (projections) of the signal with the various basis functions:

$$\gamma_l = < x, \psi_l^* > = \int_{t1}^{t_2} x(t)\psi_l^*(t) \, \mathrm{d}t \tag{3.35}$$

The vector $\boldsymbol{a}$ of coefficients α_k complying with (3.26) is then given by the matrix relation:

$$\boldsymbol{\alpha} = \mathbf{\Lambda}^{-1} \cdot \mathbf{\Gamma} \tag{3.36}$$

We can see by (3.34) that the Λ matrix is diagonal if the basis functions are orthogonal.

3.2.6 Example

Let us consider the approximation of a rectangular pulse $x(t) = \mathrm{rect}\,(t - \frac{1}{2})$ by a linear combination of $m = 3$ exponentially decaying functions $\psi_k(t) = \exp[-kt]$, with $k = 1$ to 3, defined on interval $[0, \infty]$.

We get (the first result indicates that these functions are not orthogonal):

$$\lambda_{kl} = < \psi_k, \psi_l^* > = \int_0^{\infty} \exp\,[-\,(k + l)t] \, \mathrm{d}t = \frac{1}{k + l} \qquad k, l = 1,2,3$$

and

$$\gamma_l = < x, \psi_l^* > = \int_0^{\infty} x(t) \exp\,(-lt) \, \mathrm{d}t = \int_0^{1} \exp\,(-lt) \, \mathrm{d}t$$

$$= \frac{1}{l}\,[1 - \exp\,(-l)]$$

We deduce that the matrix can be written as:

$$\mathbf{\Lambda} = \begin{bmatrix} \frac{1}{2} & \frac{1}{3} & \frac{1}{4} \\ \frac{1}{3} & \frac{1}{4} & \frac{1}{5} \\ \frac{1}{4} & \frac{1}{5} & \frac{1}{6} \end{bmatrix}$$

and

$$\mathbf{\Lambda}^{-1} = \begin{bmatrix} 72 & -240 & 180 \\ -240 & 900 & -720 \\ 180 & -720 & 600 \end{bmatrix}$$

and

$$\mathbf{\Gamma} = \begin{bmatrix} 0.63212 \\ 0.43233 \\ 0.31674 \end{bmatrix}$$

Hence, the solution is

$$\begin{bmatrix} \alpha_1 \\ \alpha_2 \\ \alpha_3 \end{bmatrix} = \mathbf{\Lambda}^{-1} \cdot \mathbf{\Gamma} = \begin{bmatrix} -1.234 \\ 9.338 \\ -7.454 \end{bmatrix}$$

Thus, the best approximation of $x(t) = \text{rect}\,(t - \frac{1}{2})$ on interval $0 < t < \infty$ with a linear combination of functions $\psi_k(t) = \exp[-kt]$ is (fig. 3.4):

$$\tilde{x}\,(t) = -1.234 \cdot \exp(-t) + 9.338 \cdot \exp(-2t) - 7.454 \cdot \exp(-3t)$$

The mean square value of the approximation error, in this case, according to (3.30), is $\|e\|^2 = 1 - 0.896 = 0.104$

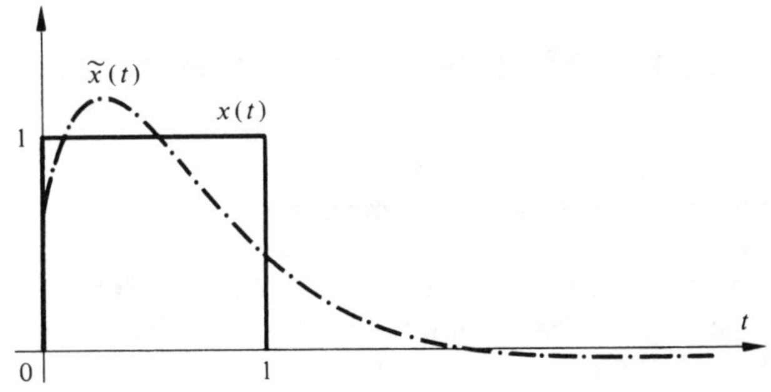

Fig. 3.4

3.2.7 Approximation quality

The mean square error $||e||^2$ is an absolute measure of our m-order approximation imperfection. Generally, the quality of this approximation is best expressed as a relative value, given by the ratio:

$$\xi = \frac{||x||^2}{||e||^2} = \frac{||x||^2}{||x||^2 - ||\tilde{x}||^2} = \frac{1}{1 - ||\tilde{x}||^2/||x||^2} \tag{3.37}$$

where

$$||x||^2 = \int_{t1}^{t_2} |x(t)|^2 \mathrm{d}t \tag{3.38}$$

is the signal energy, and

$$||\tilde{x}||^2 = \sum_k \sum_l \alpha_k \, \alpha_l^* < \psi_k, \psi_l^* > \tag{3.39}$$

is the approximation energy.

The ratio (3.37) can be considered as a kind of ***signal-to-noise*** (approximation) ***ratio***. Other examples of this concept will be shown in subsequent chapters. It is convenient to express this ratio in decibels by using relation (1.2):

$$\xi_{\mathrm{dB}} = 10 \log_{10} \xi \quad \mathrm{dB} \tag{3.40}$$

3.2.8 Example

The quality of the rectangular signal approximation realized in example 3.2.6 with a linear combination of three exponential functions defined on interval $[0, \infty]$ is characterized by values

$$\xi = 1/0.104 = 9.62 \text{ or } \xi_{\mathrm{dB}} = 10 \log 9.62 = 9.8 \text{ dB}$$

3.3 SIGNAL REPRESENTATION BY ORTHOGONAL SERIES EXPANSION

3.3.1 Conditions for coefficient independence

The system of equation (3.32) is considerably simplified if the functions $\psi_k(t)$ form an orthogonal basis on interval $[t_1, t_2]$, i.e., if their inner products (3.12) comply with the condition (3.15):

$$\lambda_{kl} = < \psi_k, \psi_l^* > = 0 \quad \forall \, k \neq l \tag{3.41}$$

Then, we get a ***representation in series of orthogonal functions***:

$$\tilde{x}(t) = \sum_{k=1}^{m} \alpha_k \psi_k(t) \tag{3.42}$$

the coefficients of which are ***independent*** and determined by the general equation (with $\lambda_{kk} = \lambda_k$):

$$\alpha_k = \frac{1}{\lambda_k} < x, \psi_k^* > = \frac{1}{\lambda_k} \int_{t1}^{t2} x(t)\psi_k^*(t)\,\mathrm{d}t \tag{3.43}$$

where

$$\lambda_k = < \psi_k, \psi_k^* > = \|\psi_k\|^2 = \int_{t1}^{t2} |\psi_k(t)|^2\mathrm{d}t \tag{3.44}$$

is a ***real scalar***.

3.3.2 Orthonormal functions

The functions $\psi_k(t)$ are said to be orthonormal when

$$\lambda_k = 1 \quad \forall\, k \tag{3.45}$$

The normalization can always be obtained, when $\lambda_k \neq 1$, by introducing the set of weighted functions:

$$\Upsilon_k(t) = \frac{\psi_k(t)}{\|\psi_k\|} = \frac{1}{\sqrt{\lambda_k}}\,\psi_k(t) \tag{3.46}$$

3.3.3 Approximation error

An m-order approximation of signal $x(t)$ by an expansion in series of orthogonal functions (3.42) generates a mean square error value, which, after inserting (3.41) into (3.30), is given by

$$\|e\|^2 = \int_{t1}^{t2} |x(t)|^2\mathrm{d}t - \sum_{k=1}^{m} |\alpha_k|^2 \cdot \lambda_k \tag{3.47}$$

Since $\|e\| > 0$, we obtain the inequality:

$$\sum_{k=1}^{m} |\alpha_k|^2 \cdot \lambda_k \leqslant \int_{t1}^{t2} |x(t)|^2\mathrm{d}t = \|x\|^2 \quad \forall m \tag{3.48}$$

3.3.4 Parseval identity

Since $\lambda_k = \|\psi_k\|^2$ is always positive, we deduce from relation (3.48) that it is sufficient to increase m—indefinitely, if necessary—to cancel the mean square error, i.e., the distance $d(x,\tilde{x})$. Thus, the approximation $\tilde{x}(t)$ converges toward $x(t)$ in the mean square sense. In the limit as $m \to \infty$, relation (3.48) yields the *Parseval identity*:

$$\int_{t_1}^{t_2} |x(t)|^2 \mathrm{d}t = \sum_{k=1}^{\infty} |\alpha_k|^2 \cdot \lambda_k \tag{3.49}$$

3.3.5 Definition: complete set of orthogonal functions

The set of orthogonal functions $\{\psi_k(t),\ k = 1, 2, \ldots, m\}$ is *complete* when it is possible to approximate any function $x(t) \in \mathrm{L}^2$ with a mean square error approaching zero when m tends toward infinity.

3.3.6 Energy interpretation

Each component $\alpha_k\ \psi_k(t)$ of the complete series expansion in orthogonal functions of $x(t) \in \mathrm{L}^2(t_1, t_2)$

$$x(t) = \sum_{k=1}^{\infty} \alpha_k \psi_k(t) \tag{3.50}$$

has an energy given by

$$\int_{t_1}^{t_2} |\alpha_k \psi_k|^2 \mathrm{d}t = |\alpha_k|^2 \cdot \lambda_k \tag{3.51}$$

The left-hand side of equation (3.49) represents the ***total energy*** of signal $x(t)$. So thanks to the Parseval identity, the development (3.50) ***analyzes*** the signal $x(t)$ in such a way that the signal total energy is equal to the sum of the component energies.

3.3.7 Representation of periodic signals

Any periodic signal, with period T, and finite average power, can be considered (sub-sect. 1.3.14) as the repetition $\text{rep}_T\,\{x(t,T)\}$ of a finite energy signal $x(t,T)$.

Any set of orthogonal functions on interval $[t_1, t_1 + T]$, which also has a T-periodic structure, is orthogonal on the whole interval $[-\infty, \infty]$ and fits the representation of a T-periodic signal of finite average power.

Relations (3.42), (3.43), and (3.44) can be directly applied by simply writing $t_2 = t_1 + T$.

The mean square approximation error is deduced from (3.47) again by writing $t_2 = t_1 + T$ and by dividing each term by T: it then measures the average power and not the energy (infinite in this case) of the error:

$$P_e = \frac{1}{T}\int_{t_1}^{t_1+T} |x(t)|^2 \mathrm{d}t - \sum_{k=1}^{m} |\alpha_k|^2 \cdot \lambda_k/T \tag{3.52}$$

The Parseval identity, obtained when the error power is zero (complete set) takes here the following form, which expresses in two equivalent ways the average power of the periodic signal:

$$P_x = \frac{1}{T}\int_{t_1}^{t_1+T} |x(t)|^2 \mathrm{d}t = \frac{1}{T}\sum_{k=1}^{\infty} |\alpha_k|^2 \cdot \lambda_k \tag{3.53}$$

3.3.8 Fundamental principle of signals analyzers and synthesizers

The fundamental relations used in signal analysis—or synthesis—based on the expansion in series of orthogonal functions are summarized in table 3.5

The basic structure of a signal analyzer can be directly deduced (fig. 3.6).

Conversely, we obtain the basic structure of a signal synthesizer (fig. 3.7).

Table 3.5

Complete expansion:

$$x(t) = \sum_{k=1}^{\infty} \alpha_k \, \psi_k(t) \quad \in \mathrm{L}^2(t_1, t_2)$$

with

$$\lambda_{kl} = \langle \psi_k, \psi_l^* \rangle = \int_{t_1}^{t_2} \psi_k(t)\psi_l^*(t)\mathrm{d}t = 0 \qquad \forall \; k \neq l$$

and

$$\alpha_k = \frac{1}{\lambda_k} \int_{t_1}^{t_2} x(t)\, \psi_k^*(t)\,\mathrm{d}t$$

where

$$\lambda_k = \langle \psi_k, \psi_k^* \rangle = \int_{t_1}^{t_2} |\psi_k(t)|^2 \mathrm{d}t$$

Approximation:

$$\tilde{x}(t) = \sum_{k=1}^{m} \alpha_k \, \psi_k(t)$$

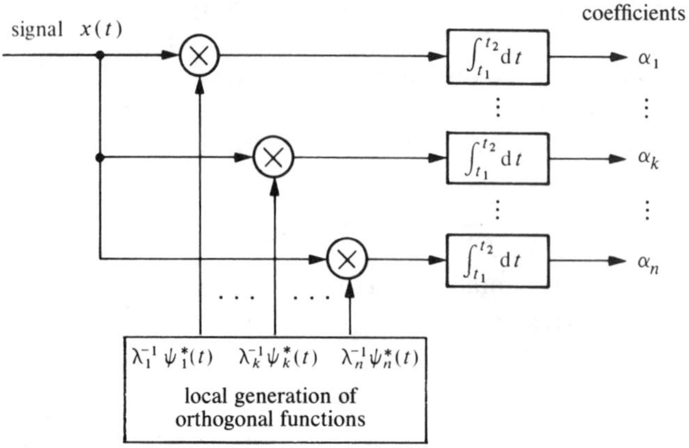

Fig. 3.6 Principle of signal analyzer

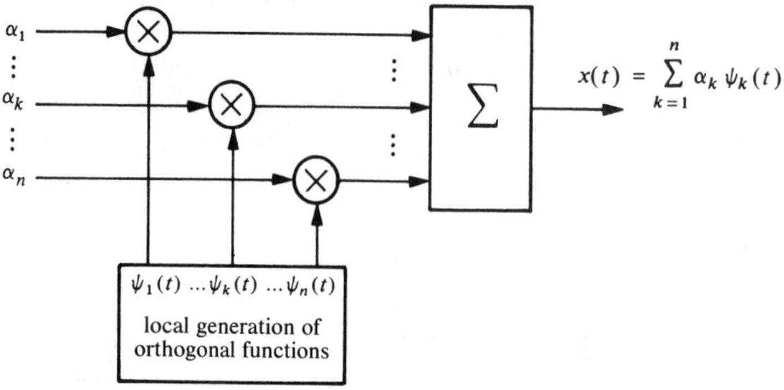

Fig. 3.7 Principle of signal synthesizers

3.3.9 Gram-Schmidt orthogonalization procedure

The following iterative procedure allows us to build an orthonomal basis $\{\psi_k(t),\ k = 1, 2 \ldots, n\}$ from a set of n linearly independent functions $\{v_k(t),\ k = 1,2, \ldots n\}$ of the given space:

$$\psi_k(t) = \frac{w_k(t)}{\|w_k(t)\|} \tag{3.54}$$

with

$$w_1(t) = v_1(t) \tag{3.55}$$

and the general term:

$$w_k(t) = v_k(t) - \sum_{i=1}^{k-1} < v_k, \psi_i^* > \psi_i(t) \tag{3.56}$$

Equation (3.56) expresses the function (vector) $w_k(t)$ as the difference between the function (vector) $v_k(t)$ and its projection on the subspace of dimension $k - 1$, generated by the predefined orthogonal function set $\{\psi_k(t), i = 1, 2, \ldots, k - 1\}$. Thus, $w_k(t)$ is perpendicular to all the functions of this set.

If the functions $v_k(t)$ are not linearly independent, the described orthogonalization procedure remains applicable, but yields $m < n$ non-zero orthogonal functions; then m represents the signal space dimension spanned by $\{v_k(t)\}$.

3.3.10 Example

Let us assume the following set of linearly independent functions of $L^2(0, \infty)$:

$$v_k(t) = \exp(-kt); \quad k = 1,2, \ldots$$

Applying the Gram-Schmidt orthogonalization procedure, we obtain for interval $0 \leqslant t < \infty$:

$$w_1(t) = v_1(t) = \exp(-t)$$

$$\psi_1(t) = \sqrt{2} \exp(-t)$$

$$w_2(t) = \exp(-2t) - (2/3) \exp(-t)$$

$$\psi_2(t) = 6 \exp(-2t) - 4 \exp(-t)$$

$$w_3(t) = \exp(-3t) - (6/5) \exp(-2t) + (3/10) \exp(-t)$$

$$\psi_3(t) = \sqrt{6}[10 \exp(-3t) - 12 \exp(-2t) + 3 \exp(-t)]$$

et cetera

We can easily check (problem 3.59) that functions $w_k(t)$ and $\psi_k(t)$ are respectively orthogonal and orthonormal on $[0, \infty]$. Functions $v_k(t)$ and $\psi_k(t)$ for $k = 1,2,3$, are represented in figure 3.8.

3.4 PRINCIPAL SETS OF ORTHOGONAL FUNCTIONS

3.4.1 Shifted rectangular pulses

A trivial example of expansion in series of orthogonal functions, on interval $-\infty < t < \infty$, is obtained by choosing rectangular pulses of width ΔT as basis

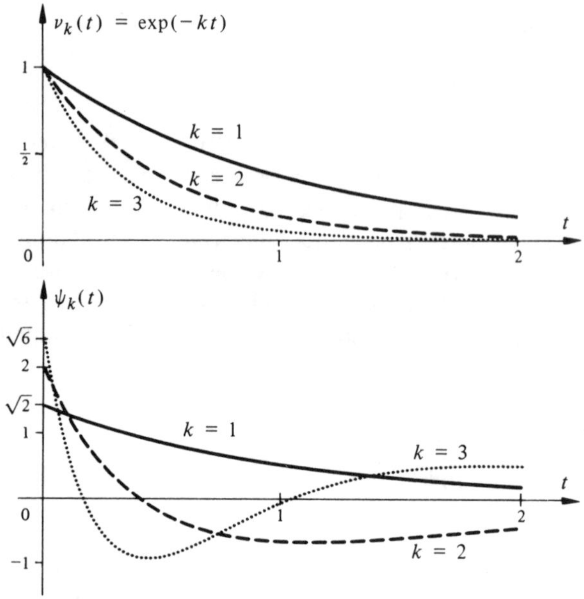

Fig. 3.8

functions $\psi_k(t)$, centered on $t = k \cdot \Delta\tau$ with $k = 0, \pm 1, \pm 2, \ldots, \pm \infty$:

$$\psi_k(t) = \text{rect}\left(\frac{t - k \cdot \Delta\tau}{\Delta\tau}\right) \tag{3.57}$$

which comply with the orthogonalization condition (3.41)

From (3.44), we obtain $\lambda_k = \Delta\tau$, and from (3.43) the development coefficients are

$$\begin{aligned} \alpha_k &= \frac{1}{\Delta\tau} \int_{-\infty}^{\infty} x(t)\, \text{rect}\left(\frac{t - k \cdot \Delta\tau}{\Delta\tau}\right) \mathrm{d}t \\ &= \frac{1}{\Delta\tau} \int_{(k-1/2)\Delta\tau}^{(k+1/2)\Delta\tau} x(t)\mathrm{d}t \end{aligned} \tag{3.58}$$

Thus, the α_k coefficients correspond to the ***average values of the signal*** measured on the intervals $\Delta\tau$ centered on $t = k \cdot \Delta\tau$:

$$\alpha_k \cong x(k \cdot \Delta\tau) \tag{3.59}$$

An approximative model for signal $x(t)$ is then obtained by combining equations (3.42) and (3.59):

$$\tilde{x}(t) = \sum_{k=-\infty}^{\infty} x(k\Delta\tau)\, \text{rect}\left(\frac{t - k \cdot \Delta\tau}{\Delta\tau}\right) \tag{3.60}$$

This approximation (fig. 3.9) corresponds in a way to a pulse amplitude modulation (sect. 11.4).

The set of orthogonal functions $\{\psi_k(t)\}$ defined by (3.57) is obviously not complete. However, the smaller the $\Delta\tau$ interval, the better is the approximation quality (3.60). In the limit as $\Delta\tau \to 0$, the rectangular pulse weighted by $1/\Delta\tau$ becomes a delta function (sub-sect. 1.3.9), and we obtain the true description (with $k \cdot \Delta\tau \to \tau$):

$$\begin{aligned} x(t) &= \lim_{\Delta\tau \to 0} \tilde{x}(t) \\ &= \lim_{\Delta\tau \to 0} \sum_{k=-\infty}^{\infty} x(k \cdot \Delta\tau) \frac{1}{\Delta\tau} \operatorname{rect}\left(\frac{t - k \cdot \Delta\tau}{\Delta\tau}\right) \Delta\tau \\ &= \int_{-\infty}^{\infty} x(\tau)\delta(t - \tau)\mathrm{d}\tau \end{aligned} \tag{3.61}$$

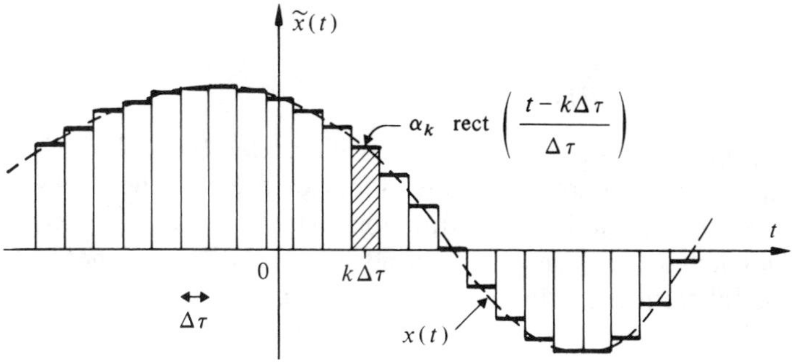

Fig. 3.9

This is the convolution integral (1.47) that is basic to signal theory and linear systems theory because it allows us to think of a signal $x(t)$ as an infinite weighted sum of delta functions.

3.4.2 Rademacher functions

A very simple, incomplete set of orthogonal functions of rectangular type is built with the *Rademacher functions* (fig. 3.10). These are binary functions, with a periodic structure, having only +1 or −1 values.

A convenient definition of Rademacher functions, with index i and period T, is given by

$$\operatorname{rad}(0, t/T) = 1$$

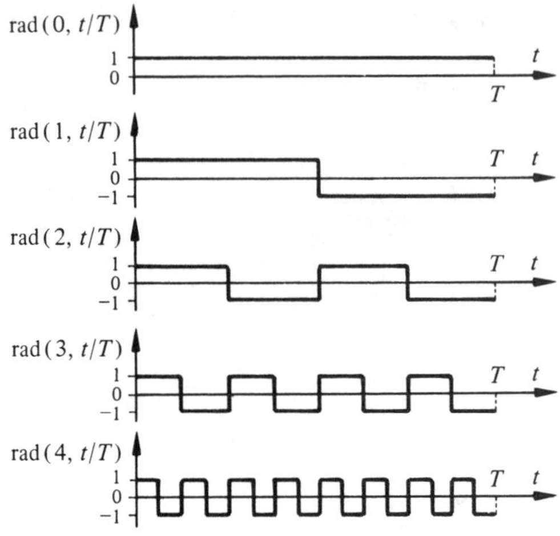

Fig. 3.10 Rademacher functions

and

$$\text{rad}(i,t/T) = \text{sgn}\{\sin(2^i \pi t/T)\};\ i = 1, 2, \ldots \tag{3.62}$$

The incomplete character of this set is illustrated by problem 3.5.12.

The Rademacher function is the model of the so-called *periodic square wave*. Except for the index zero function, which is a simple constant, all the other Rademacher functions correspond to signals easily generated by a cascade of digital frequency dividers, the input of which is the signal of highest index.

3.4.3 Walsh functions with sequential classification

A complete set of binary orthogonal functions including Rademacher functions is given by the *Walsh functions*. These functions take only +1 or −1 values and change their sign k times on the open interval $0 < t < T$. They are generally considered as T-periodic. They can be defined as follows [60]:

$$\psi_k(t) = \text{wal}(k,t/T) = \prod_{i=0}^{r-1} \text{sgn}\,\{\cos^{k_i}(2^i \pi t/T)\} \tag{3.63}$$

where k_i is the ith coefficient (equal to zero or one) of k expressed in base 2 as:

$$k = \sum_{i=0}^{r-1} k_i \cdot 2^i \quad (k_i = 0, 1) \tag{3.64}$$

and r is given by $2^{r-1} < k \leqslant 2^{r}$.

The Rademacher and Walsh functions are linked by the relation:

$$\mathrm{rad}(i,t/T) = \mathrm{wal}(2^i - 1,t/T) \tag{3.65}$$

The Walsh functions defined by (3.63) are classified in *sequential order* when they are ranked according to increasing values of index k. The *sequence* concept [61] is similar to the frequency concept of a sine function. It characterizes the mean number of sign-change pairs (cycles) by time unit. Thus, $\mathrm{wal}(1,t/T)$ and $\mathrm{wal}(2,t/T)$ functions have the same sequence $s = 1/T$, as $\sin(2\pi t/T)$ and $\cos(2\pi t/T)$ functions have the same frequence $f = 1/T$. As a general rule, the sequence of a Walsh function with index k is given by

$$s_k = \begin{cases} 0 & k = 0 \\ (k+1)/2 & k \text{ odd} \\ k/2 & k \text{ even} \end{cases} \tag{3.66}$$

Another notation emphasizes odd and even symmetries of Walsh functions by defining—again in analogy with sine and cosine—sal and cal functions as follows:

$$\begin{aligned} \mathrm{sal}(s_k,t/T) &= \mathrm{wal}(k,t/T) \quad k \text{ odd} \\ \mathrm{cal}(s_k,t/T) &= \mathrm{wal}(k,t/T) \quad k \text{ even} \end{aligned} \tag{3.67}$$

The first eight Walsh functions classified in sequential order are displayed in figure 3.11. We easily verify their orthogonal property on interval $[0,T]$, with $\lambda_k = T$. An example of the approximation of a continuous function by a weighted sum of Walsh functions is given in figure 3.12.

3.4.4 Usual generation procedure

The generation of signals representing Walsh functions can be achieved with square waves (Rademacher functions) by using the following property:

$$\mathrm{wal}(k \oplus l,t/T) = \mathrm{wal}(k,t/T) \cdot \mathrm{wal}(l,t/T) \tag{3.68}$$

where $\oplus$ indicates the modulo-2 addition of the k and l indices represented in base 2 (e.g., $3 \oplus 5 = 6$ as $3 \leftrightarrow 011$, $5 \leftrightarrow 101$ and $6 \leftrightarrow 110 = 011 \oplus 101$). Moreover, by representing the $+1$ and -1 levels by the logic states 0 and 1, respectively, the multiplication of Walsh functions becomes a modulo-2 addition, which is easily realized with an exclusive-OR logic operator.

The set of orthogonal (or orthonormal for $T = 1$) Walsh functions has been successfully applied [62] in signal and image processing as well as in pattern recognition. This is due to their ease of generation and logic manipulation, on one hand, and, on the other hand, to the ease with which the corresponding α_k coefficients can be calculated.

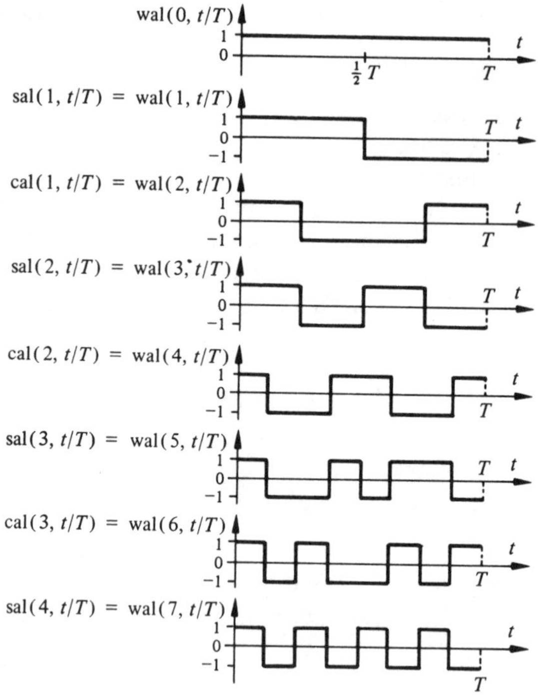

Fig. 3.11 Walsh functions

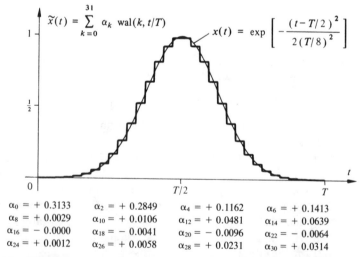

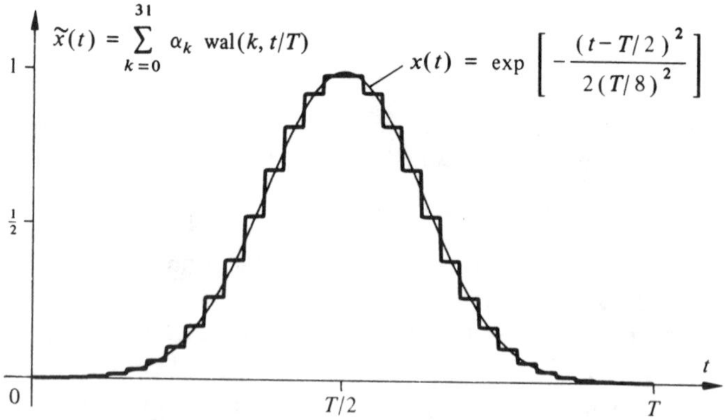

$\alpha_0 = +0.3133$	$\alpha_2 = +0.2849$	$\alpha_4 = +0.1162$	$\alpha_6 = +0.1413$
$\alpha_8 = +0.0029$	$\alpha_{10} = +0.0106$	$\alpha_{12} = +0.0481$	$\alpha_{14} = +0.0639$
$\alpha_{16} = -0.0000$	$\alpha_{18} = -0.0041$	$\alpha_{20} = -0.0096$	$\alpha_{22} = -0.0064$
$\alpha_{24} = +0.0012$	$\alpha_{26} = +0.0058$	$\alpha_{28} = +0.0231$	$\alpha_{30} = +0.0314$

Fig. 3.12 Example of approximation by Walsh functions ($\alpha_k = 0$ for k odd)

3.4.5 Walsh functions in natural order

A quick method to set up Walsh functions is obtained by considering their relationship with *Hadamard matrices* H_N. These are square matrices, of $N \times N$ order, the elements of which are equal to $+1$ or -1, and the lines (or the columns) of which are mutually orthogonal. They are of primary interest in digital signal and image processings (vol. XX).

Each line of a Hadamard matrix of $N = 2^n$ order contains k sign changes and corresponds to the discrete version (sampling of 2^n values regularly distributed between 0 and T) of the wal$(k, t/T)$ Walsh function.

The following recurrence formula allows quick set-up of the Hadamard matrix H_{2N} with $N = 2^n$:

$$H_{2N} = H_2 \otimes H_N = \begin{bmatrix} H_N & H_N \\ H_N & -H_N \end{bmatrix} \tag{3.69}$$

where

$$H_1 = [1] \tag{3.70}$$

and $\otimes$ indicates the *Kronecker product* (or tensor product).

Matrices H_4 and H_8 are represented on fig. 3.13 with some corresponding Walsh functions.

$$H_4 = \begin{bmatrix} 1 & 1 & 1 & 1 \\ 1 & -1 & 1 & -1 \\ 1 & 1 & -1 & -1 \\ 1 & -1 & -1 & 1 \end{bmatrix} \qquad H_8 = \begin{bmatrix} 1 & 1 & 1 & 1 & 1 & 1 & 1 & 1 \\ 1 & -1 & 1 & -1 & 1 & -1 & 1 & -1 \\ 1 & 1 & -1 & -1 & 1 & 1 & -1 & -1 \\ 1 & -1 & -1 & 1 & 1 & -1 & -1 & 1 \\ 1 & 1 & 1 & 1 & -1 & -1 & -1 & -1 \\ 1 & -1 & 1 & -1 & -1 & 1 & -1 & 1 \\ 1 & 1 & -1 & -1 & -1 & -1 & 1 & 1 \\ 1 & -1 & -1 & 1 & -1 & 1 & 1 & -1 \end{bmatrix} \begin{matrix} \\ \longleftrightarrow \mathrm{wal}(7, t/T) \\ \\ \\ \\ \\ \longleftrightarrow \mathrm{wal}(2, t/T) \\ \\ \end{matrix}$$

Fig. 3.13

The classification of Walsh functions according to the line number of the H_N matrix is called the *natural order*. We can link the natural classification to the sequential one by using the following procedure: translate the natural classification line index into its binary form; by reverse reading (i.e., starting the reading with the least significant bit), we get a new binary number which is translated into a decimal number using Gray code (sub-sect. 10.4.3); this decimal number corresponds to the k index of the sequential classification (e.g., line index $i = 6 \leftrightarrow 110$ in binary form; by reverse reading, we get $011 \leftrightarrow 2 = k$ using Gray code).

3.4.6 Sine functions

The set of harmonic functions:

$$\psi_k(t) = \sqrt{\frac{2}{T}}\sin(2\pi kt/T) \tag{3.71}$$

is orthonormal on interval $[0,T]$. However, it is incomplete; no function with even symmetry on interval $[0,T]$ can be exactly represented by a linear combination of the $\psi_k(t)$. The set is completed by adding the cosine harmonic functions; this yields the classical Fourier series expansion.

3.4.7 Fourier series

Among all the orthogonal expansion of a signal on an interval of duration T, the Fourier series is, theoretically speaking, the most important one.

Any signal $x(t) \in L^2(t_1,t_1 + T)$ can be expressed as a linear combination of the following complex exponential functions (here index k is replaced by n positive, zero, or negative):

$$\psi_n(t) = \exp(j2\pi n\, t/T) \tag{3.72}$$

which form a complete set of orthogonal functions on interval $(t_1,t_1 + T)$ with

$$\lambda_n = T \quad \forall n \tag{3.73}$$

By convention, the nth coefficient of the complex Fourier series of signal $x(t)$ will be denoted X_n (instead of α_k).

Thus, on interval $[t_1,t_1 + T]$, we get the following expression:

$$x(t) = \sum_{n=-\infty}^{\infty} X_n \exp(j2\pi nt/T) \tag{3.74}$$

where, according to (3.43) and (3.73), we have

$$X_n = \frac{1}{T}\int_{t_1}^{t_1+T} x(t)\exp(-j2\pi nt/T)\mathrm{d}t \tag{3.75}$$

These coefficients offer the advantage of being directly ranked according to the harmonic order of frequencies $f_n = n/T$ and lead to the classical concept of signal frequency (bilateral) spectrum. When $x(t)$ is a real function, X_n and X_{-n} are complex conjugated values.

The limit of the $X_n \cdot T$ product reached by indefinitely increasing T is the $X(f)$ Fourier transform of the signal (sub-sect. 4.1.6).

Let us point out that when the signal has a ***period T***, i.e., if $x(t) = x(t + mT)$, with m integer, the validity of (3.74) is extended over the whole real axis:

$-\infty < t < \infty$. Then, the Parseval identity (3.53) allows us to express the signal average power in two equivalent ways:

$$P_x = \frac{1}{T}\int_{t_1}^{t_1+T} |x(t)|^2 \mathrm{d}t = \sum_{n=-\infty}^{\infty} |X_n|^2 \qquad \textbf{(3.76)}$$

3.4.8 Expansion of a frequency function

By analogy with the development of a time function $x(t)$ on an interval $[t_1, t_1 + T]$, the Fourier series development of a frequency function $Z(f)$ on ***frequency interval*** $[f_1, f_1 + F]$ gives:

$$Z(f) = \sum_{k=-\infty}^{\infty} z_k \exp(\mathrm{j}2\pi kf/F) \qquad (3.77)$$

where the coefficients:

$$z_k = \frac{1}{F}\int_{f_1}^{f_1+F} Z(f)\exp(-\mathrm{j}2\pi kf/F)\mathrm{d}f \qquad (3.78)$$

are generally complex, except if $Z(f)$ is an even function.

3.4.9 Sampled representation of a lowpass band-limited signal

Let us consider a signal $x(t)$, the Fourier transform of which is zero for $|f| > f_2$. For a lowpass spectrum, and taking definition (2.15) into account, the bandwidth B and the cut-off frequency f_2 are equal:

$B = f_2 - f_1 = f_2$, since $f_1 = 0$.

The relation between the signal and its transform is:

$$x(t) = \int_{-B}^{B} X(f)\exp(\mathrm{j}2\pi ft)\mathrm{d}f \qquad (3.79)$$

The Fourier series expansion of function $Z(f) = \exp(\mathrm{j}2\pi ft)$ on frequency interval $[-B, B]$ is given by

$$Z(f) = \sum_{k=-\infty}^{\infty} z_k(t)\exp(\mathrm{j}2\pi kf/2B) \qquad (3.80)$$

where [according to (1.58)]:

$$z_k(t) = \frac{1}{2B}\int_{-B}^{B} \exp(\mathrm{j}2\pi[t - k/2B]f)\mathrm{d}f \tag{3.81}$$

$$= \frac{\sin[\pi(2Bt - k)]}{\pi(2Bt - k)} = \mathrm{sinc}(2Bt - k)$$

These functions are represented in figure 3.14.

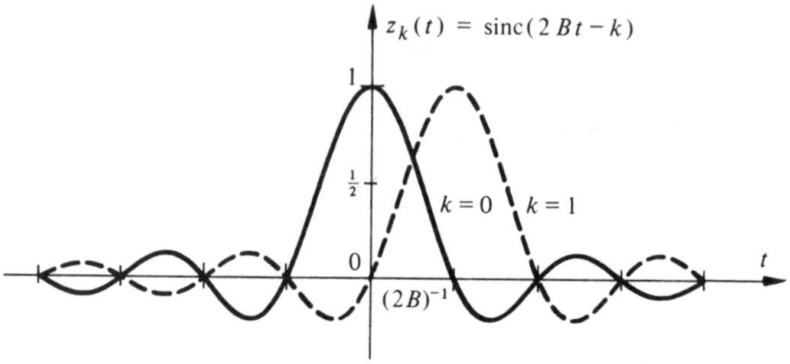

Fig. 3.14

Introducing (3.80) and (3.81) into (3.79), and inverting the order of summation and integration, we get

$$x(t) = \sum_{k=-\infty}^{\infty} \left[\int_{-B}^{B} X(f)\exp(\mathrm{j}2\pi kf/2B)\mathrm{d}f \right] z_k(t) \tag{3.82}$$

$$= \sum_{k=-\infty}^{\infty} x(k/2B)\ \mathrm{sinc}(2Bt - k)$$

Functions $z_k(t) = \mathrm{sinc}(2Bt - k)$ are orthogonal (see problem 4.7.19) and generate a complete functional space of band-limited signals. On the other hand, the set of these functions is not complete on L^2, since signals with a bandwidth greater than B are not included in the generated subspace.

The development coefficients (3.82) have the remarkable property of being exactly the ***signal sampled values*** taken periodically every $(2B)^{-1}$ seconds. This result will be interpreted in chapter 9.

3.4.10 Other orthogonal sets of functions

Other orthogonal functions are described in more specialized references [21,49,63].

Let us especially mention (the definition interval being indicated in square brackets):

- Legendre polynomials [−1,+1]
- Legendre functions [0,∞]
- Laguerre polynomials [0,∞]
- Laguerre functions [0,∞]
- Chebyshev polynomials [−1,+1]
- Chebyshev functions [0,∞]
- Hermite polynomials [−∞,∞]
- Hermite functions [−∞,∞]
- Haar functions [0,1]
- *slant* functions [0,1]

Haar functions and slant functions (staircase-shaped), like Walsh functions, are of rectangular structure. They have found specific applications in signal or image coding.

3.5 PROBLEMS

3.5.1 Find the distances $d_1(x_k,x_l)$ defined by (3.3) with $K = 1$, between signals $x_k(t) = A\sin(2\pi kt/T)\cdot\text{rect}[(t - T/2)/T]$ and $x_l(t) = A\cos(2lt/T)\cdot\text{rect}[(t - T/2)/T]$, where k and l are positive integers. Introduce the result into relation (3.16) and interpret it.

3.5.2 Compare distances $d_1(x_i,x_j)$ and $d_2(x_i,x_j)$, respectively defined by (3.3) and (3.4), with $K = 1/T$, for the three signals represented in figure 3.15.

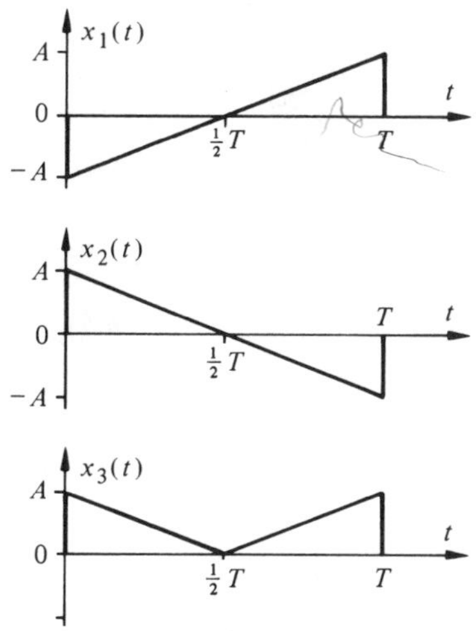

Fig. 3.15

3.5.3 Determine the inner product evolution of signals $x(t) = \cos(2\pi t/T - \theta_1)$ *and* $y(t) = \sin(2\pi t/T - \theta_2)$ in function of parameter values $\Delta\theta = \theta_2 - \theta_1$ and for the definition domain $t_1 \leqslant t \leq t_1 + T$. For which values of $\Delta\theta$ are these two functions orthogonal?

3.5.4 Define the best mean square approximation of $x(t) = \text{rect}[(t - 0.5)/0.5]$ by the sum $\tilde{x}(t) = \Sigma_{k=1}^{3} \alpha_k \exp(-kt)$, defined on interval $0 \leqslant t < \infty$. Calculate the mean square value of the resulting error.

3.5.5 Determine the best mean square approximation of $x(t) = \sin \pi t \cdot \text{rect}(t/2)$ that is achievable with function set $\{\psi_k(t) = \text{rect}(t - k/2; k = \pm 1, \pm 3, \pm 5, \ldots\}$ and calculate the mean square value of the residual error.

3.5.6 Demonstrate that the function set $\{\psi_k(t) = (2/T)^{1/2} \cdot \cos(2\pi kt/T) \cdot \text{rect}(t/T); k = 0, \pm 1, \pm 2, \ldots\}$ is an incomplete orthonormal set on interval $[-T/2, T/2]$.

3.5.7 Assume $x(t)$ and $y(t)$ are two signals of L^2 and $\{\psi_k(t); k = 1, 2, \ldots\}$ is a complete set of orthonormal functions. Demonstrate the Parseval general relation:

$$< x,y^* > = \sum_{k=1}^{\infty} < x,\psi_k^* > < \psi_k,y^* > \tag{3.83}$$

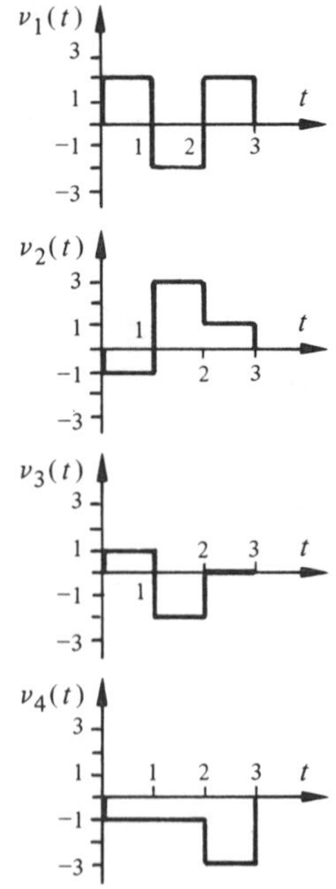

Fig. 3.16

3.5.8 Demonstrate that functions $w_k(t)$ obtained through relation (3.56) are orthogonal.

3.5.9 Verify that functions $w_k(t)$ and $\psi_k(t)$ of example 3.3.10 are respectively orthogonal and orthonormal.

3.5.10 Complete example 3.3.10 by defining the orthonormal functions $\psi_4(t)$ and $\psi_5(t)$.

3.5.11 Demonstrate that the four functions $v_k(t)$; $k = 1$ to 4 of figure 3.16 are not a set of linearly independent functions. Applying the Gram-Schmidt orthogonalization procedure, show that this set defines a bidimensional space. Find the orthonormal components of the $v_k(t)$ functions.

3.5.12 Expand signal $x(t) = \text{tri}[(t - 0{,}5)/0{,}5]$ in (a) Rademacher and (b) Walsh orthogonal series. Define in each case the minimum number m of coefficients needed to get an approximation quality (see relation (3.40)) better than 20 dB. Compare graphically $x(t)$ and $\tilde{x}(t)$.

3.5.13 Expand in Fourier series signal $x(t) = (At/T)\text{rect}[(t - T/2)/T]$ on interval $[0,T]$.

Chapter 4

Deterministic Signals

4.1 FOURIER TRANSFORM

4.1.1 Introduction

The main tool of signal theory (sub-sect. 1.1.7) is harmonic analysis. The Fourier transform provides us with the spectral representation of deterministic signals. It gives a frequency representation of signal amplitude, phase, and energy or power. As will be demonstrated, the integral Fourier transform may be considered as an extension of the concept of the orthogonal Fourier series expansion (sub-sect. 3.4.7).

Because the Fourier transform is the main topic of many mathematical books [22,23], we will only summarize its main definitions, notations and properties, the importance of which is essential to signal theory.

A table displaying the main transforms is given in the appendix (sect. 15.4).

4.1.2 Definition

Let $x(t)$ be a deterministic signal, its Fourier transform is a function (normally complex) of the real variable f and is given by

$$X(f) = \mathrm{F}\{x(t)\} = \langle x, \exp(-\mathrm{j}2\pi ft)\rangle$$

$$= \int_{-\infty}^{\infty} x(t)\exp(-\mathrm{j}2\pi ft)\mathrm{d}t \tag{4.1}$$

The inverse transform is given by

$$x(t) = \mathrm{F}^{-1}\{X(f)\} = \langle X, \exp(\mathrm{j}2\pi ft)\rangle \tag{4.2}$$

$$= \int_{-\infty}^{\infty} X(f)\exp(\mathrm{j}2\pi ft)\mathrm{d}f$$

4.1.3 Notations

In signal theory, the Fourier transform of a signal is generally expressed as a function of frequency f, rather than as a function of its angular frequency

$\omega = 2\pi f$, as is done in circuit theory, where a similarity with the Laplace transform is preferred.

The use of the $\langle f \rangle$ variable yields a remarkable symmetry between the direct and inverse transforms, and is dictated by practical considerations, since the experimental results are always expressed as functions of the measurable parameter f.

Shorthand notation for the direct and inverse transform operators are $F\{\ \}$ and $F^{-1}\{\ \}$, respectively. The reciprocal pair of transforms is often indicated with a bidirectional arrow:

$$x(t) \leftrightarrow X(f) \tag{4.3}$$

4.1.4 Time-frequency duality

The symmetry of the direct (4.1) and inverse (4.2) transforms establishes the existence of a duality between time domain (t variable) and frequency domain (f variable). This duality is of prime importance in most signal processing techniques. This will be further elaborated.

4.1.5 Existence of a Fourier transform

Thanks to the Plancherel theorem [58], all of the square integrable functions (i.e., all signals with finite energy) have a Fourier transform (mean square convergence) which is also a square integrable function.

Nonetheless, all of the functions having a Fourier transform are not necessarily square integrable (i.e., physical signals). Most theoretical signal models (signals with finite average power, delta function, *et cetera*) also have Fourier transforms.

4.1.6 Heuristic justification of the Fourier transform

The orthogonal series expansion (sect. 3.3) of a signal on interval $[t_1,t_2]$ is a valuable analysis tool. In particular, the Fourier series expansion (sub-sect. 3.4.7) gives a discrete frequency representation, whereby each coefficient X_n is located at a frequency $f_n = nf_1 = n/(t_2 - t_1)$.

This description and that of a periodic repetition, with a period $T = t_2 - t_1$, of that signal are identical.

In order to get a frequency representation of a nonperiodic deterministic signal on interval $-\infty < t < +\infty$, we start with a description on a finite interval, T, and then we look for its limit when T tends toward infinity.

Let $x(t)$ be a signal (fig. 4.1), including a segment $x(t,T)$, of duration T, centered on the origin:

$$x(t,T) = x(t) \cdot \mathrm{rect}(t/T) \tag{4.4}$$

The Fourier series of $x(t,T)$ is

$$x(t,T) = \sum_{n=-\infty}^{\infty} X_n \exp(\,\mathrm{j}2\pi n f_1 t) \tag{4.5}$$

where

$$X_n = \frac{1}{T} \int_{-T/2}^{T/2} x(t,T)\exp(-\mathrm{j}2\pi n f_1 t)\mathrm{d}t \tag{4.6}$$

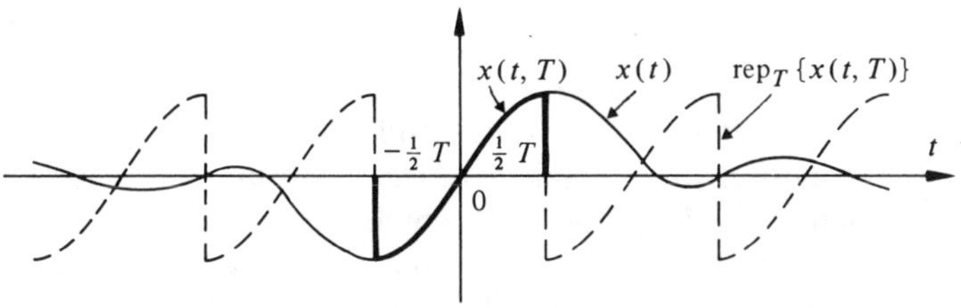

Fig. 4.1

Outside the interval $[-T/2, T/2]$, this expansion no longer corresponds to $x(t)$, but to the periodic repetition of $x(t,T)$. However, when T tends toward infinity, $x(t,T)$ and $x(t)$ become identical everywhere and the definition interval is the complete real axis.

$$\lim_{T\to\infty} x(t,T) = x(t) \tag{4.7}$$

The frequency representation of $x(t,T)$ is a ***discrete frequency spectrum.*** The interval between two consecutive frequencies is $f_1 = 1/T$. Thus, when T increases, the density of the discrete spectrum also increases. This ***spectral density*** can be expressed as

$$\frac{X_n}{f_1} = X_n \cdot T = \int_{-T/2}^{T/2} x(t,T)\exp(-\mathrm{j}2\pi n f_1 t)\mathrm{d}t \tag{4.8}$$

When $T \to \infty$, the frequency interval becomes infinitesimally small and the number of coefficients in a given frequency interval becomes infinite: $f_1 =$

$1/T \to \mathrm{d}f$, $nf_1 = n/T \to f$ (continuous variable) and $\Sigma \to \int$. This yields

$$x(t) = \lim_{T\to\infty} x(t,T) = \lim_{T\to\infty} \sum_{n=-\infty}^{\infty} X_n \exp(\,\mathrm{j}2\pi n f_1 t)$$

$$= \lim_{T\to\infty} \sum_{n=-\infty}^{\infty} \left\{ \frac{1}{T} \int_{-T/2}^{T/2} x(t,T)\exp(-\mathrm{j}2\pi n f_1 t)\mathrm{d}t \right\} \exp(\,\mathrm{j}2\pi n f_1 t)$$

$$= \int_{-\infty}^{\infty} \mathrm{d}f \underbrace{\left\{ \int_{-\infty}^{\infty} x(t)\exp(-\mathrm{j}2\pi f t)\mathrm{d}t \right\}}_{X(f)} \exp(\,\mathrm{j}2\pi f t) \tag{4.9}$$

This result is identical to relations (4.1) and (4.2). The transform $X(f)$ appears as the limit of the discrete spectral density, as introduced above.

4.1.7 Interpretation

The similarity between Fourier series and Fourier transform leads to the conclusion that function $X(f)$ "analyzes" $x(t)$ as an infinite collection of sine complex coefficients, the amplitudes of which are given by $|X(f)|\mathrm{d}f$. Hence, function $X(f)$ provides information about the ***frequency distribution*** of the amplitude, phase, and energy or power of signal $x(t)$.

4.1.8 Fourier transform of a real signal

The Fourier transform of a real signal is generally a complex function of the f variable. Assume $x_p(t)$ and $x_i(t)$, respectively, are the even and odd components of the signal. Let us denote their Fourier transforms as $X_p(f)$ and $X_i(f)$. We can now write (4.1) as

$$X(f) = \int_{-\infty}^{\infty} x(t)\exp(-\mathrm{j}2\pi f t)\mathrm{d}t$$

$$= \int_{-\infty}^{\infty} x(t)\cos(2\pi f t)\mathrm{d}t - \mathrm{j}\int_{-\infty}^{\infty} x(t)\sin(2\pi f t)\mathrm{d}t$$

$$= X_p(f) + X_i(f)$$

$$= \mathrm{Re}\{X(f)\} + \mathrm{j}\mathrm{Im}\{X(f)\}$$

$$= |X(f)|\exp[\mathrm{j}\vartheta_x(f)] \tag{4.10}$$

The modulus $|X(f)|$ is called the *amplitude spectrum* and the argument $\vartheta_x(f) = \arg X(f)$ is called the *phase spectrum*.

The amplitude spectrum is an even function, while the phase spectrum is an odd one.

The transform of a real even signal is a real even function, while the transform of a real odd signal is an imaginary odd function.

The real part of $X(f)$ is the transform of the even part of the signal, while the transform of the odd part of the signal is equal to the imaginary part of $X(f)$ multiplied by j.

4.1.9 Properties

The importance of the Fourier transform to signal theory is mainly due to its remarkable properties. These are summarized in the tables of section 15.3.

The main properties are hereafter given without demonstrations, which are left as exercises for the reader. Thus,

$$x^*(t) \leftrightarrow X^*(-f) \tag{4.11}$$

$$ax(t) + by(t) \leftrightarrow aX(f) + bY(f) \tag{4.12}$$

$$\mathrm{d}^n x/\mathrm{d}t^n \leftrightarrow (\mathrm{j}2\pi f)^n X(f) \tag{4.13}$$

$$x(t) * y(t) \leftrightarrow X(f) \cdot Y(f) \tag{4.14}$$

$$x(t) \cdot y(t) \leftrightarrow X(f) * Y(f) \tag{4.15}$$

$$x(t - t_0) \leftrightarrow X(f) \cdot \exp(-\mathrm{j}2\pi f t_0) \tag{4.16}$$

$$x(t) \cdot \exp(\mathrm{j}2\pi f_0 t) \leftrightarrow X(f - f_0) \tag{4.17}$$

$$x(at) \leftrightarrow |a|^{-1} X(f/a) \tag{4.18}$$

Property (4.12) indicates that the Fourier transform is a ***linear operation.*** Properties (4.14) and (4.15) show that ***to every product in the time domain there corresponds a convolution product in the frequency domain and vice versa.*** Relation (4.16) is often referred to as the *time-shift theorem*.

4.1.10 Complex convolution

Formula (4.15) establishes the correspondence between a product of signals and the convolution product of their Fourier transforms, which are generally complex functions of the real variable f. Separating the signals (sub-sect. 2.55) into their even and odd parts yields

$$\begin{aligned} z(t) &= x(t) \cdot y(t) = [x_p(t) + x_i(t)] \cdot [y_p(t) + y_i(t)] \\ &= x_p(t)y_p(t) + x_i(t)y_i(t) + x_p(t)y_i(t) + x_i(t)y_p(t) \end{aligned} \tag{4.19}$$

and

$$Z(f) = X(f) * Y(f)$$
$$= \underbrace{X_p(f) * Y_p(f) + X_i(f) * Y_i(f)}_{Z_p(f) = \mathrm{Re}\{Z(f)\}} + \underbrace{X_p(f) * Y_i(f) + X_i(f) * Y_p(f)}_{Z_i(f) = \mathrm{jIm}\{Z(f)\}} \quad (4.20)$$

This expression is highly simplified when signals are either even or odd.

4.2 FINITE ENERGY SIGNALS

4.2.1 Amplitude spectrum and phase spectrum

Signals with finite energy as defined in sub-section 2.3.3 comply with

$$W_x = \int_{-\infty}^{\infty} |x(t)|^2 \mathrm{d}t < \infty \quad (4.21)$$

where W_x represents the total normalized energy. The condition (4.21) implies that signals have a transient behavior.

Any signal of this kind has a Fourier transform.

$$X(f) = \mathrm{F}\{x(t)\} = |X(f)| \exp[\mathrm{j}\vartheta_x(f)] \quad (4.22)$$

where $|X(f)|$ is the amplitude spectrum and $\vartheta_x(f)$ is the phase spectrum of the signal $x(t)$. When $x(t)$ is measured in volts, $|X(f)|$ is given in V/Hz and $\vartheta_x(f)$ in radians (rad).

4.2.2 Example: exponentially decaying signal

Let $x(t) = \exp(-at)\,\epsilon(t)$; hence, its Fourier transform is

$$X(f) = \int_0^{\infty} \exp[-(a + \mathrm{j}\,2\pi f)t] \mathrm{d}t = \frac{1}{a + \mathrm{j}\,2\pi f}$$

$$|X(f)| = 1/\sqrt{a^2 + (2\pi f)^2}$$

$$\vartheta_x(f) = -\arctan(2\pi f/a)$$

This signal is shown in Fig. 4.2, and its amplitude and phase spectra are shown in Fig. 4.3.

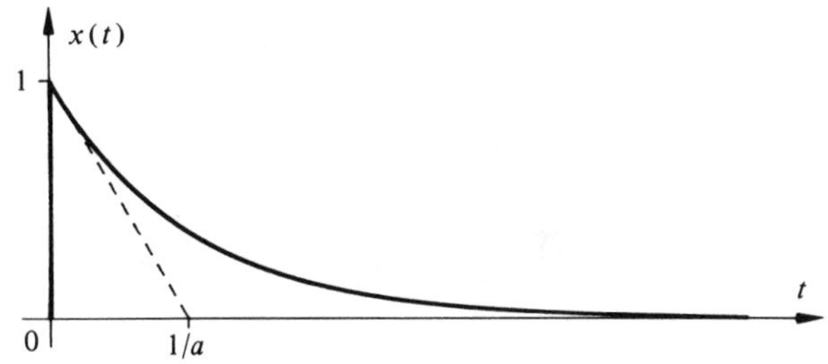

Fig. 4.2

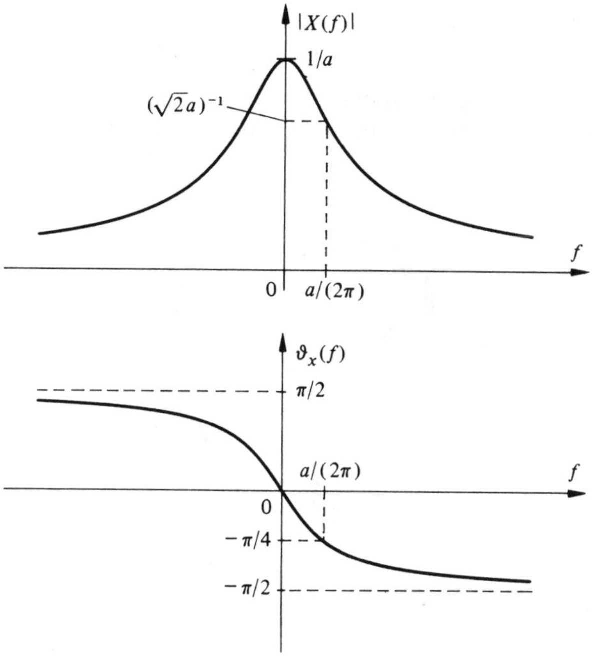

Fig. 4.3

4.2.3 Example

Assume (fig. 4.4) the product $z(t) = x(t) \cdot y(t)$, where $x(t)$ is as given in example 4.2.2 and $y(t) = \cos(2\pi f_0 t) = 1/2[\exp(\, j2\pi f_0 t) + \exp(-j2\pi f_0 t)]$.

From (4.17), we get $Z(f) = 1/2\,X(f - f_0) + 1/2\,X(f + f_0)$, where $X(f)$ is as given in example 4.2.2.

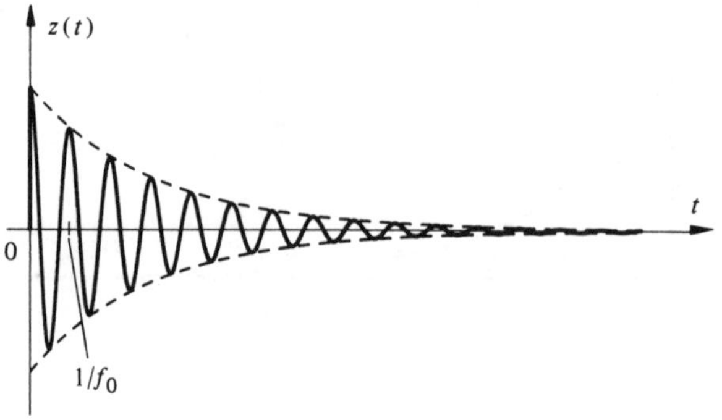

Fig. 4.4

The amplitude and phase spectra of $z(t)$ are shown in fig. 4.5 for $f \gg a$. This result can be directly obtained by way of (4.15) as shown in sub-section 4.4.5.

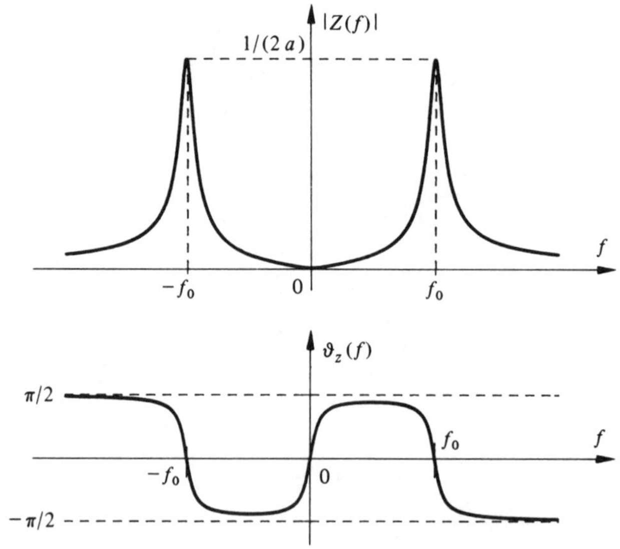

Fig. 4.5

4.2.4 Example: rectangular pulse

The following pair of transforms plays a major role in signal theory:

$$\text{rect}(t) \leftrightarrow \text{sinc}(f) \tag{4.23}$$

and, symmetrically,

$$\text{sinc}(t) \leftrightarrow \text{rect}(f) \tag{4.24}$$

Because of property (4.18), this result can be generalized to a rectangular pulse with amplitude A and width T (fig. 4.6):

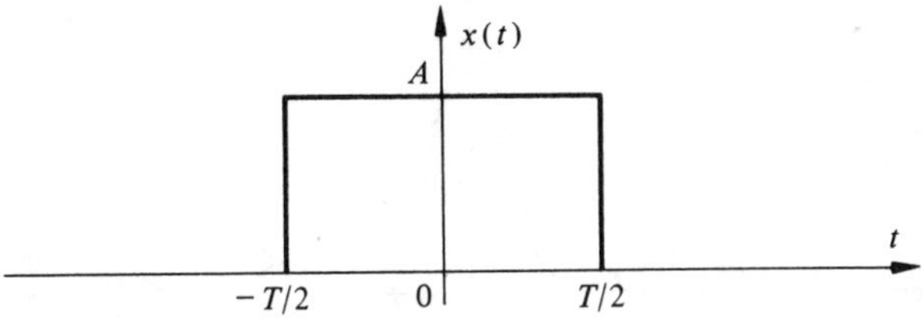

Fig. 4.6

$$x(t) = A\ \text{rect}(t/T) \tag{4.25}$$

the Fourier transform of which is (fig. 4.7):

$$X(f) = AT\ \text{sinc}(Tf) \tag{4.26}$$

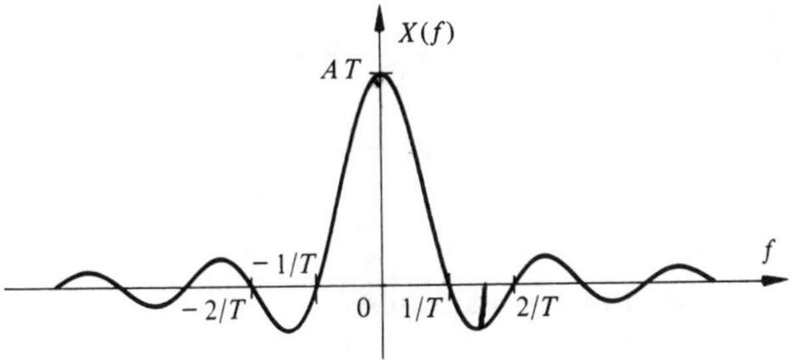

Fig. 4.7

Using the sinc function definition (sub-sect. 1.3.15), we check this result.

$$\begin{aligned} \text{F}\{A\ \text{rect}(t/T)\} &= A \int_{-T/2}^{T/2} \exp(-\text{j}2\pi f t)\text{d}t \\ &= \frac{A}{2\text{j}\pi f}[\exp(\text{j}\pi Tf) - \exp(-\text{j}\pi Tf)] \\ &= \frac{A}{\pi f}\sin(\pi Tf) \\ &= AT\ \text{sinc}(Tf) \end{aligned} \tag{4.27}$$

The signal amplitude and phase spectra (4.25) are shown in fig. 4.8. The phase spectrum, alternately taking the $-\pi$, 0, $+\pi$ values, is drawn so that we get an odd function (sub-section 4.1.8).

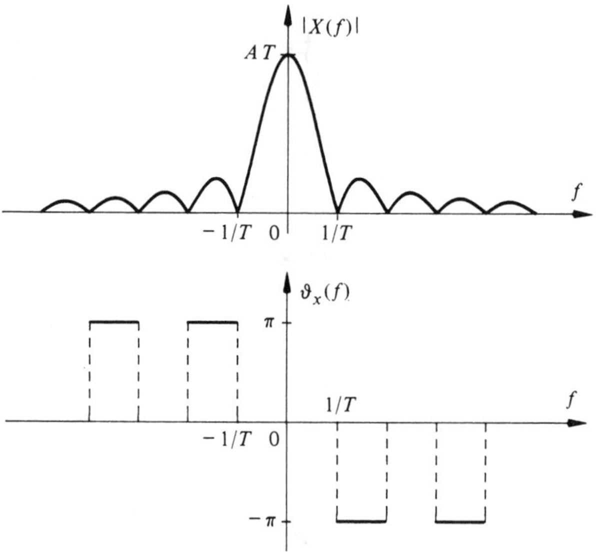

Fig. 4.8

4.2.5 Example: delayed rectangular pulse

Let us consider a rectangular pulse with amplitude A and width T, centered on some position $t_0 \neq 0$ (fig. 4.9):

$$x(t) = A \operatorname{rect}[(t - t_0)/T] \tag{4.28}$$

Its Fourier transform, as a result of (4.16) and (4.27), is

$$X(f) = AT \operatorname{sinc}(Tf) \exp(-j2\pi f t_0) \tag{4.29}$$

Hence, its amplitude and phase spectra are:

$$|X(f)| = AT\,|\operatorname{sinc}(Tf)| \tag{4.30}$$

$$\vartheta_x(f) = \begin{cases} -2\pi t_0 f & \text{for } \operatorname{sinc}(Tf) > 0 \\ -2\pi t_0 f \pm \pi & \text{for } \operatorname{sinc}(Tf) < 0 \end{cases} \tag{4.31}$$

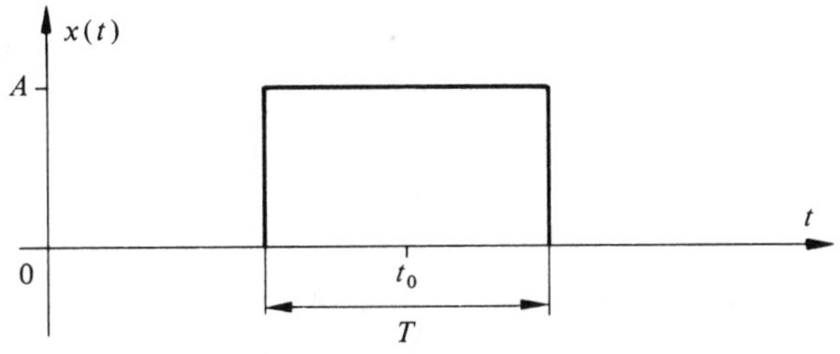

Fig. 4.9

The amplitude spectrum is invariant with respect to any shift. The only effect of a time shift is to add a term to the phase spectrum that varies linearly with frequency. Figure 4.10 shows examples of three different t_0 delays. We choose the phase jumps so that $\vartheta_x(f)$ is odd.

4.2.6 Example: triangular pulse

The triangular pulse is equal to the convolution of two rectangular pulses (1.33)

$$\text{tri}(t) = \text{rect}(t) * \text{rect}(t) \tag{4.32}$$

Thanks to (4.14), we have

$$\text{tri}(t) \leftrightarrow \text{sinc}^2(f) \tag{4.33}$$

or, in a more general way, as a result of (4.18), any even triangular pulse with height A and base $2T$ (fig. 4.11):

$$x(t) = A\ \text{tri}(t/T) \tag{4.34}$$

has a Fourier transform (fig. 4.12):

$$X(f) = AT\ \text{sinc}^2(Tf) \tag{4.35}$$

Because the phase spectrum is zero ($X(f)$ is real), $X(f)$ directly represents the amplitude spectrum of $x(t)$.

4.2.7 Example

Let us consider a trapezoidal shaped signal in order to display the efficiency of the Fourier transform properties.

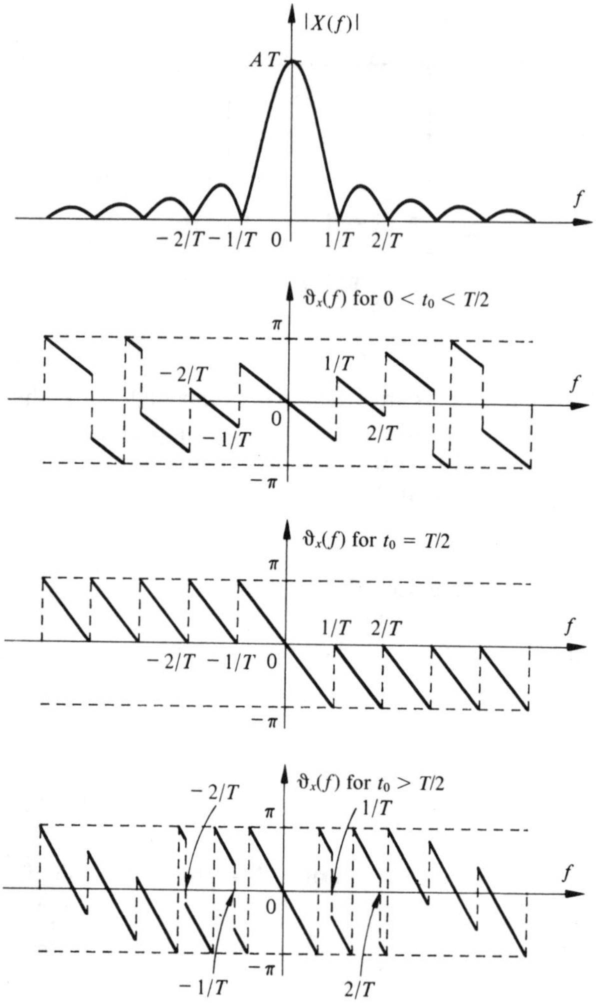

Fig. 4.10

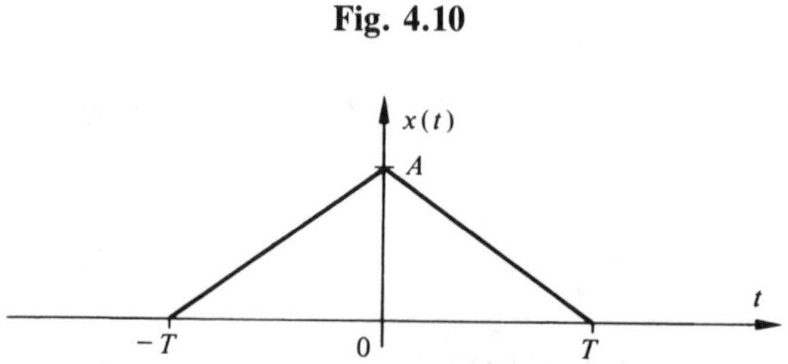

Fig. 4.11

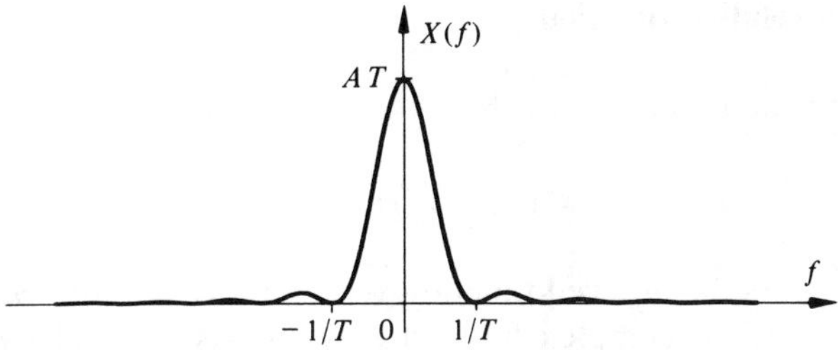

Fig. 4.12

Let $x(t)$ be such a signal; thus, its derivative is

$$\frac{dx}{dt} = \frac{A}{b-a}\left\{ \text{rect}\left[\frac{t+(b+a)/2}{b-a}\right] - \text{rect}\left[\frac{t-(b+a)/2}{b-a}\right]\right\}$$

as shown in fig. 4.13.

With (4.13), (4.16), and (4.29), we get

$$F[dx/dt] = j\,2\pi f \cdot X(f)$$
$$= 2A\,j\,\text{sinc}[(b-a)f] \cdot \sin[\pi(b+a)f]$$

and, thus,

$$X(f) = A(b+a)\,\text{sinc}[(b-a)f]\text{sinc}[(b+a)f]$$

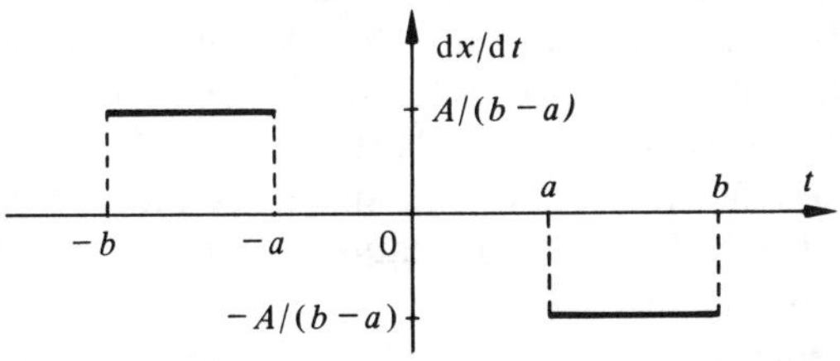

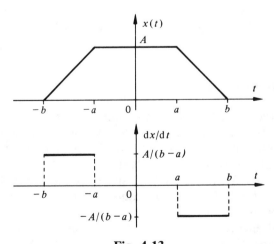

Fig. 4.13

4.2.8 Crosscorrelation function

The inner product (sub-sect. 3.1.9)

$$\mathring{\varphi}_{xy}(\tau) = < x^*, y_\tau > = \int_{-\infty}^{\infty} x^*(t)y(t + \tau)\mathrm{d}t \quad \textbf{(4.36)}$$

where y_τ represents the delayed function $y(t + \tau)$ is called the *crosscorrelation function* of the real or complex finite energy signals $x(t)$ and $y(t)$. According to (3.15), those signals are ***orthogonal***—or ***uncorrelated—for each value of*** τ ***that cancels the crosscorrelation function.***

According to the Hermitian property (3.14), we can easily show that

$$\mathring{\varphi}_{xy}(\tau) = \mathring{\varphi}^*_{yx}(-\tau) \quad (4.37)$$

4.2.9 Autocorrelation function

Replacing $x(t + \tau)$ for $y(t + \tau)$ in (4.36), we get the *autocorrelation function* of a finite energy signal $x(t)$:

$$\mathring{\varphi}_{x}(\tau) = < x^*, x_\tau > = \int_{-\infty}^{\infty} x^*(t)x(t + \tau)\mathrm{d}t \quad \textbf{(4.38)}$$

with the Hermitian symmetry: $\mathring{\varphi}_x(\tau) = \mathring{\varphi}^*_x(-\tau)$. Thus, we can deduce that ***the real part of a correlation function is an even function while the imaginary part is an odd function.***

For $\tau = 0$, ***the correlation function has the value of the signal energy***:

$$\mathring{\varphi}_x(0) = < x^*, x > = \|x\|^2 = W_x = \int_{-\infty}^{\infty} |x(t)|^2\mathrm{d}t \quad (4.39)$$

4.2.10 Interpretation

The inner product of two vectors is proportional to the projection of one onto the other. It is maximum, in absolute value, when the two vectors are parallel. It is zero when they are orthogonal. The inner product can be seen as a measure of the two vectors' similarity.

By analogy, the inner product of two signals is a measure of their similarity in shape and location. The crosscorrelation function represents the evolution of this similarity as τ changes.

In the signal space, the variation of τ is analogous to a rotation of the considered vector. Hence, there is a link between the Euclidian distance of two signals and their crosscorrelation, which can be established with (3.14),

(3.16), (4.36), and (4.39):

$$d^2(x,y_\tau) = \mathring{\varphi}_x(0) + \mathring{\varphi}_y(0) - 2\,\mathrm{Re}\{\mathring{\varphi}_{xy}(\tau)\} \tag{4.40}$$

Introducing $y(t + \tau) = x(t + \tau)$ yields

$$d^2(x,x_\tau) = 2[\mathring{\varphi}_x(0) - \mathrm{Re}\{\mathring{\varphi}_x(\tau)\}] \tag{4.41}$$

4.2.11 Special notation

In this book, we use $\mathring{\varphi}$ to denote the correlation function of two signals, at least one of which has finite energy, in order to distinguish it from the correlation function of two finite-power functions, simply noted as φ.

Some authors swap indices in definition (4.36).

4.2.12 Notation for real signals

The cross- and autocorrelation functions are mostly used in connection with real signals. Therefore, the following simplified notation will be adopted in this book when the real character of the signals is obvious.

$$\mathring{\varphi}_{xy}(\tau) = \langle x,y_\tau \rangle = \int_{-\infty}^{\infty} x(t)y(t + \tau)\mathrm{d}t \tag{4.42}$$

$$\mathring{\varphi}_x(\tau) = \langle x,x_\tau \rangle = \int_{-\infty}^{\infty} x(t)x(t + \tau)\mathrm{d}t \tag{4.43}$$

With a variable transformation, they can also be written as

$$\mathring{\varphi}_{xy}(\tau) = \int_{-\infty}^{\infty} x(t - \tau)y(t)\mathrm{d}t \tag{4.44}$$

$$\mathring{\varphi}_x(\tau) = \int_{-\infty}^{\infty} x(t - \tau)x(t)\mathrm{d}t \tag{4.45}$$

Cross- and autocorrelation functions of real signals are also real. When $x(t)$ and $y(t)$ are measured in volts, $\mathring{\varphi}_{xy}(\tau)$ and $\mathring{\varphi}_x(\tau)$ are in $\mathrm{V}^2 \cdot \mathrm{s}$.

4.2.13 Property

For real signals, (4.37) yields

$$\mathring{\varphi}_{yx}(\tau) = \mathring{\varphi}_{xy}(-\tau) \tag{4.46}$$

$$\mathring{\varphi}_x(\tau) = \mathring{\varphi}_x(-\tau) \tag{4.47}$$

The autocorrelation function of a real signal is a real even function.

4.2.14 Property

Inserting (4.36) and (4.38) into the Schwarz inequality (3.20) yields

$$|\mathring{\varphi}_{xy}(\tau)|^2 \leqslant \mathring{\varphi}_x(0)\mathring{\varphi}_y(0) \tag{4.48}$$

and for autocorrelation, this becomes

$$|\mathring{\varphi}_x(\tau)| \leqslant \mathring{\varphi}_x(0) \tag{4.49}$$

Thus, the absolute value of the autocorrelation function is bounded by the signal energy.

4.2.15 Relationship between convolution and correlation

Using the variable transformation $t' = -t$ in the crosscorrelation function of two signals (4.36) allows us to write it as the convolutional product of $x^*(-t)$ and $y(t)$:

$$\begin{aligned} \mathring{\varphi}_{xy}(\tau) &= \int_{-\infty}^{\infty} x^*(t)y(t+\tau)\mathrm{d}t \\ &= \int_{-\infty}^{\infty} x^*(-t')y(\tau - t')\mathrm{d}t' \\ &= x^*(-\tau)*y(\tau) \end{aligned} \tag{4.50}$$

Similarly, for autocorrelation, we have

$$\mathring{\varphi}_x(\tau) = x^*(-\tau) * x(\tau) \tag{4.51}$$

When signals are real, we obtain $\mathring{\varphi}_{xy}(\tau) = x(-\tau) * y(\tau)$ and $\mathring{\varphi}_x(\tau) = x(-\tau) * x(\tau)$.

4.2.16 Illustration

Although the mathematical expressions for convolution and correlation seem almost the same, they yield quite different results when the signal shapes are asymmetrical. Equation (4.50) shows that convolution and correlation are identical only when one of the two signals is even. Fig. 4.14 exhibits graphically the correlation and convolution computational procedure for real signals $x(t) = \exp(-at)\,\epsilon(t)$ and $y(t) = \exp(-2at)\,\epsilon(t)$. We leave the demonstration of the following results as exercises:

$$x(\tau) * y(\tau) = (1/a)[\exp(-a\tau) - \exp(-2a\tau)]\epsilon(\tau)$$

$$\mathring{\varphi}_{xy}(\tau) = (3a)^{-1}\exp(a\tau)\,\epsilon(-\tau) + (3a)^{-1}\exp(-2a\tau)\epsilon(\tau)$$

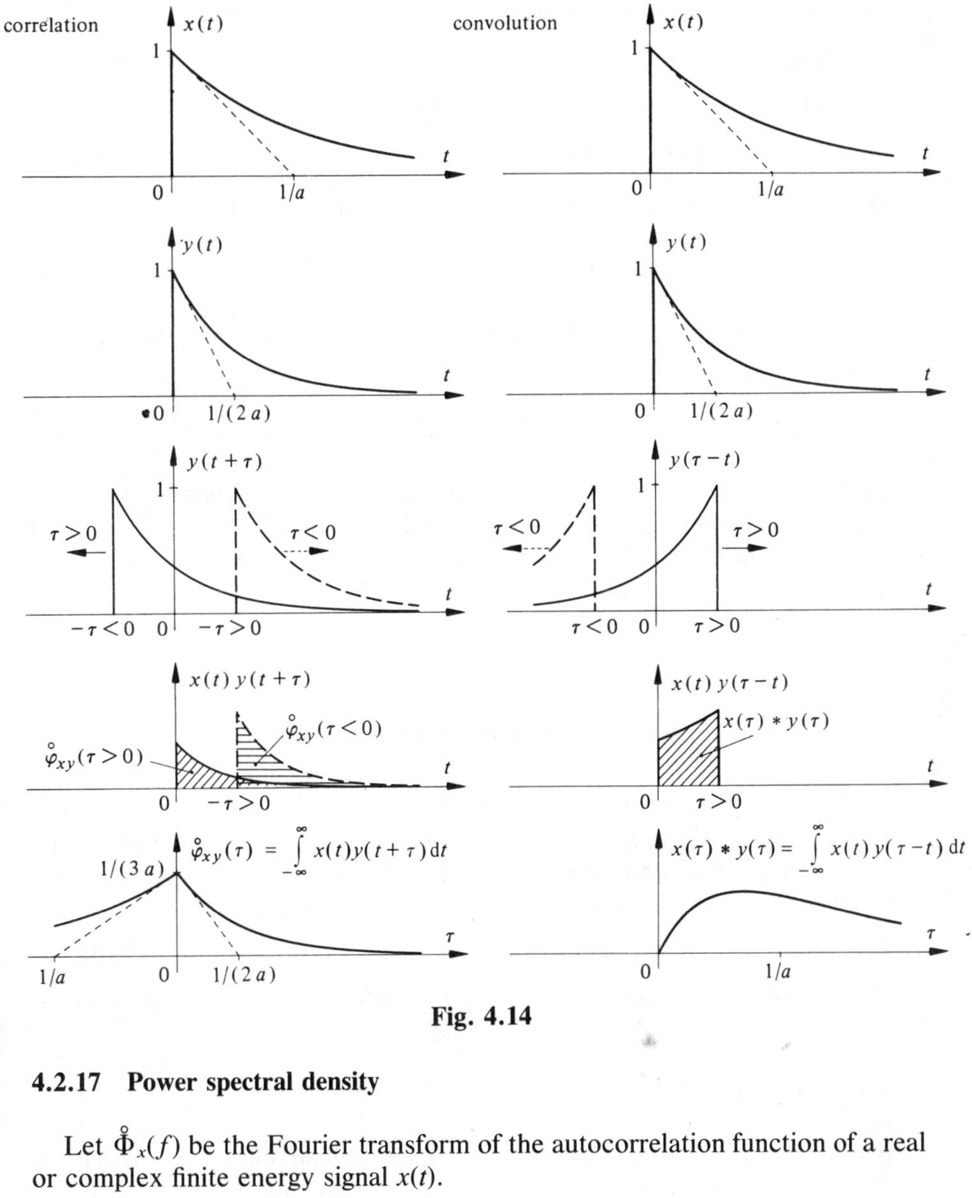

Fig. 4.14

4.2.17 Power spectral density

Let $\mathring{\Phi}_x(f)$ be the Fourier transform of the autocorrelation function of a real or complex finite energy signal $x(t)$.

$$\mathring{\Phi}_x(f) = \mathrm{F}\,\{\mathring{\varphi}_x(\tau)\} = \int_{-\infty}^{\infty} \mathring{\varphi}_x(\tau)\exp(-\mathrm{j}2\pi f\tau)\mathrm{d}\tau \qquad \textbf{(4.52)}$$

With (4.51), we have

$$\mathring{\Phi}_x(f) = \mathrm{F}\,\{x^*(-\tau) * x(\tau)\} \qquad (4.53)$$

but, as a result of (4.11) and (4.18):

$$\mathrm{F}\,\{x^*(-\tau)\} = X^*(f) \tag{4.54}$$

The following result is drawn from property (4.14):

$$\mathring{\Phi}_x(f) = X^*(f)X(f) = |X(f)|^2 \tag{4.55}$$

Inserting $\tau = 0$ in the inverse Fourier transform of $\mathring{\Phi}_x(f)$:

$$\mathring{\varphi}_x(\tau) = \int_{-\infty}^{\infty} \mathring{\Phi}_x(f)\exp(\,\mathrm{j}2\pi f\tau)\mathrm{d}f \tag{4.56}$$

and, taking into account (4.39), we get a special expression of the Parseval identity (sub-sect. 3.3.4)

$$W_x = \mathring{\varphi}_x(0) = \int_{-\infty}^{\infty} |x(t)|^2\mathrm{d}t = \int_{-\infty}^{\infty} \mathring{\Phi}_x(f)\mathrm{d}f \tag{4.57}$$

The total energy W_x of a signal can be computed by integrating either its time distribution $|x(t)|^2$, or its frequency distribution $\mathring{\Phi}_x(f)$. This is why $\mathring{\Phi}_x(f)$ is called the *energy spectral density*—or simply *energy spectrum*—of signal $x(t)$.

4.2.18 Properties of the energy spectral density

From (4.55), we deduce that the energy spectral density is

- independent of the signal phase spectrum, hence, due to the time-shift theorem (4.16), insensitive to any signal delay.
- a ***non-negative*** real function:

$$\mathring{\Phi}_x(f) \geqslant 0 \tag{4.58}$$

The part of the energy of signal $x(t)$ distributed between f_1 and f_2 is given by

$$\Delta W_x = \int_{-f_2}^{-f_1} \mathring{\Phi}_x(f)\mathrm{d}f + \int_{f_1}^{f_2} \mathring{\Phi}_x(f)\mathrm{d}f = 2\int_{f_1}^{f_2} \mathring{\Phi}_x(f)\mathrm{d}f \tag{4.59}$$

When signal $x(t)$ is real, by (4.47), the autocorrelation function is a real even function. Its Fourier transform is also a real even function:

$$\mathring{\Phi}_x(f) = \mathring{\Phi}_x(-f) \tag{4.60}$$

If a signal is measured in volts, its energy spectral density is in $\mathrm{V}^2 \cdot \mathrm{s}^2 = \mathrm{V}^2 \cdot \mathrm{s/Hz}$.

4.2.19 Example

Let $x(t) = \exp(-at) \cdot \epsilon(t)$, its autocorrelation function, from (4.43) and (4.47), is

$$\mathring{\varphi}_x(\tau) = \frac{1}{2a} \exp(-a|\tau|) \tag{4.61}$$

and its energy spectral density is

$$\mathring{\Phi}_x(f) = \mathrm{F}\{\mathring{\varphi}_x(\tau)\} = \frac{1}{a^2 + (2\pi f)^2} \tag{4.62}$$

and it is equal to the square of the amplitude spectrum of $x(t)$ given in example 4.2.2.

All three functions are shown in fig. 4.15.

4.2.20 Example

The autocorrelation function and the energy spectral density of signal $x(t) = A\,\mathrm{rect}(t/T)$ are respectively (fig. 4.16):

$$\mathring{\varphi}_x(\tau) = A^2T\,\mathrm{tri}(\tau/T) \tag{4.63}$$

$$\mathring{\Phi}_x(f) = (AT)^2\,\mathrm{sinc}^2(Tf) \tag{4.64}$$

Relation (4.63) is analogous to (1.33), as a result of (4.51), and can be easily obtained graphically by following the procedure described in fig. 4.14. We can easily check that $\mathring{\varphi}_x(0)$ corresponds to the signal energy.

4.2.21 Cross energy spectral density

The Fourier transform of the crosscorrelation function of two finite energy signals is called the *cross energy spectral density,* sometimes also referred to as *mutual spectral density.*

According to (4.14), (4.50), and (4.54), for real or complex signals we have

$$\mathring{\Phi}_{xy}(f) = \mathrm{F}\{\mathring{\varphi}_{xy}(\tau)\} = X^*(f)Y(f) \tag{4.65}$$

The cross energy spectral density is generally an asymmetrical complex function.

From (4.37) and (4.54), we get

$$\mathring{\Phi}_{xy}(f) = \mathring{\Phi}^*_{yx}(f) \tag{4.66}$$

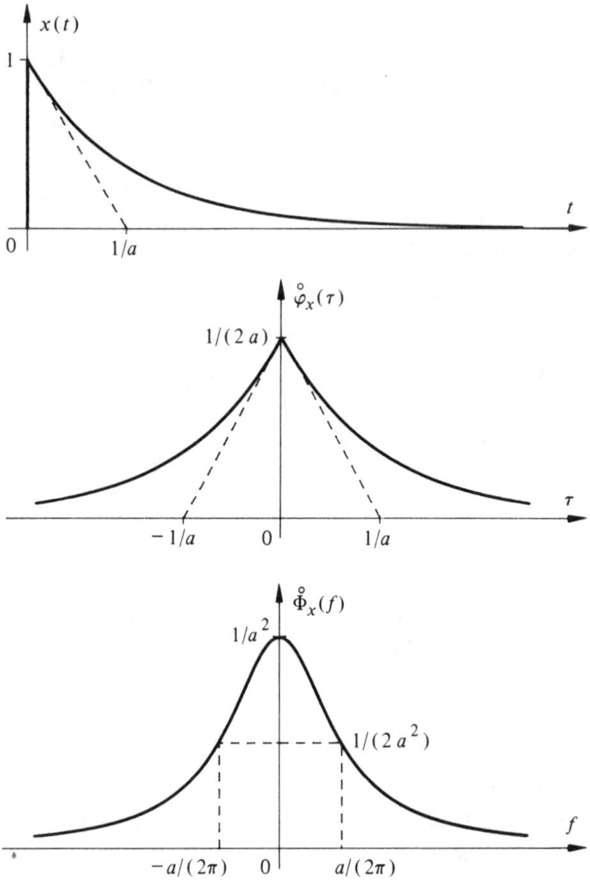

Fig. 4.15

4.2.22 Product theorem

Inserting $\tau = 0$ in the crosscorrelation function (4.36), the general Parseval identity (3.83) takes the following form, which is often called the *product theorem*:

$$\mathring{\varphi}_{xy}(0) = \mathrm{F}^{-1}\{\mathring{\Phi}_{xy}(f)\}|_{\tau=0}$$

$$= \int_{-\infty}^{\infty} x^*(t)y(t)\mathrm{d}t = \int_{-\infty}^{\infty} X^*(f)Y(f)\mathrm{d}f \qquad \textbf{(4.67)}$$

This expresses the ***inner product conservation*** between time domain and frequency domain.

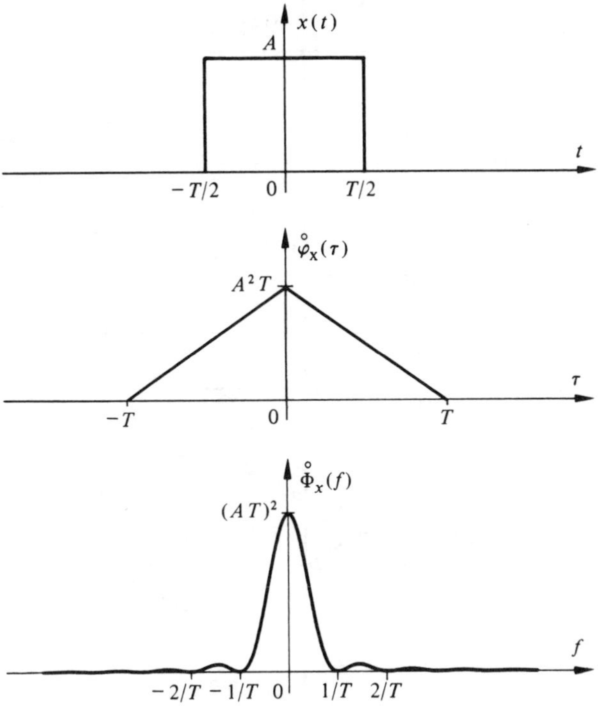

Fig. 4.16

4.2.23 Example

The real signals

$x(t) = A\ \mathrm{rect}(t/T)$

$y(t) = B\ \mathrm{rect}[(t + T/4)/(T/2)] - B\ \mathrm{rect}[(t - T/4)/(T/2)]$

and their crosscorrelation functions $\overset{\circ}{\varphi}_{xy}(\tau)$ and $\overset{\circ}{\varphi}_{yx}(\tau)$ are shown in fig. 4.17. The crosscorrelation functions can be obtained graphically as depicted in fig. 4.14. The cross energy spectral density is

$$\overset{\circ}{\Phi}_{xy}(f) = \mathrm{j}ABT^2\ \mathrm{sinc}(Tf)\ \mathrm{sinc}(Tf/2)\ \sin(\pi Tf/2)$$
$$= \mathrm{j}AB(T^2/2)[\mathrm{sinc}^2(Tf/2)\ \sin(\pi fT)]$$

The reader will check this result as an exercise either by way of (4.65) or directly computing the Fourier transform of $\overset{\circ}{\varphi}_{xy}(\tau)$. He or she will find that the signals are orthogonal, since (4.67) is zero.

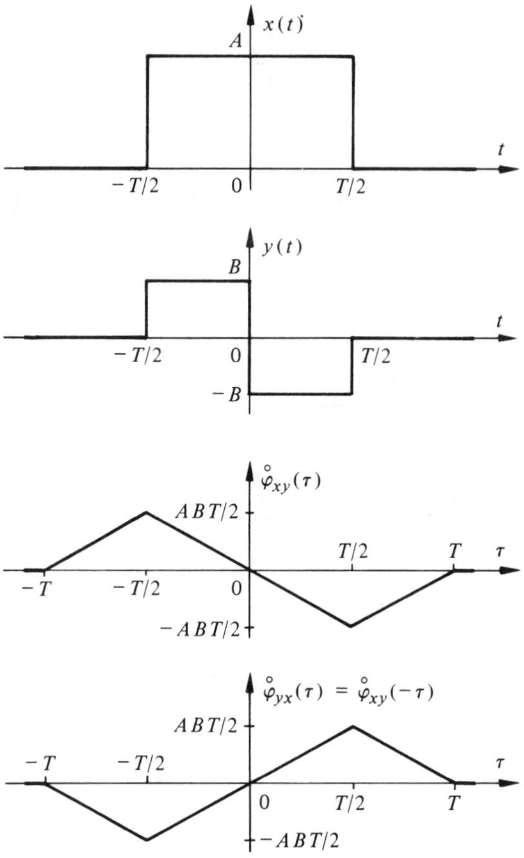

Fig. 4.17

4.2.24 Sufficient condition of orthogonality of two signals

As a result of (4.67), two signals with non-overlapping spectra are orthogonal. However, this condition is not necessary as the previous example demonstrates.

4.2.25 Correlation function derivative

Equations (4.50) and (4.65) demonstrate that

$$\mathring{\varphi}_{xy}(\tau) = x^*(-\tau) * y(\tau) \leftrightarrow \mathring{\Phi}_{xy}(f) = X^*(f)Y(f) \tag{4.68}$$

Taking advantage of the Fourier transform property (4.13), we easily obtain the crosscorrelation function derivative:

$$\mathring{\varphi}'_{xy}(\tau) = \mathrm{d}\mathring{\varphi}_{xy}/\mathrm{d}t \leftrightarrow \mathrm{j}2\pi f X^*(f)Y(f) \tag{4.69}$$

By adjunction of the ($\mathrm{j}2\pi f$) factor to either $X^*(f)$ or $Y(f)$, we obtain the inverse transform:

$$\mathring{\varphi}'_{xy}(\tau) = -x^{*\prime}(-\tau) * y(\tau) = x^*(-\tau) * y'(\tau) \tag{4.70}$$

or

$$\mathring{\varphi}'_{xy}(\tau) = -\mathring{\varphi}_{x'y}(\tau) = \mathring{\varphi}_{xy'}(\tau) \tag{4.71}$$

More generally, the $(j+k)$th derivative of the crosscorrelation function can be deduced from the $x(t)$ jth derivative and the $y(t)$ kth derivative, as follows:

$$\mathring{\varphi}^{(j+k)}_{xy}(\tau) = (-1)^j\mathring{\varphi}_{x(j)y(k)}(\tau) \tag{4.72}$$

4.2.26 Example

Let $x(t)$ and $y(t)$ be two real signals as shown with their crosscorrelation function in fig. 4.17.
The $y(t)$ derivative is (sub-sect. 1.3.8):

$$y'(t) = B\,\delta(t + T/2) - 2B\,\delta(t) + B\,\delta(t - T/2)$$

The derivative of the crosscorrelation function $\mathring{\varphi}_{xy}(\tau)$ can be directly deduced from (4.70) using the delta function convolution property (1.48):

$$\begin{aligned}\mathring{\varphi}'_{xy}(\tau) &= \mathring{\varphi}_{xy'}(\tau) = x(-\tau) * y'(\tau)\\ &= AB\,\mathrm{rect}[(\tau + T/2)/T] - 2AB\,\mathrm{rect}(\tau/T) + AB\,\mathrm{rect}[(\tau - T/2)/\mathrm{T}]\\ &= AB\,\{\mathrm{rect}[(\tau + 3T/4)/(T/2)] - \mathrm{rect}(\tau/T) + \mathrm{rect}[(\tau - 3T/4)/(T/2)]\}\end{aligned}$$

This function is illustrated in fig. 4.18. It can be checked to verify that its integral is as shown in fig. 4.17.

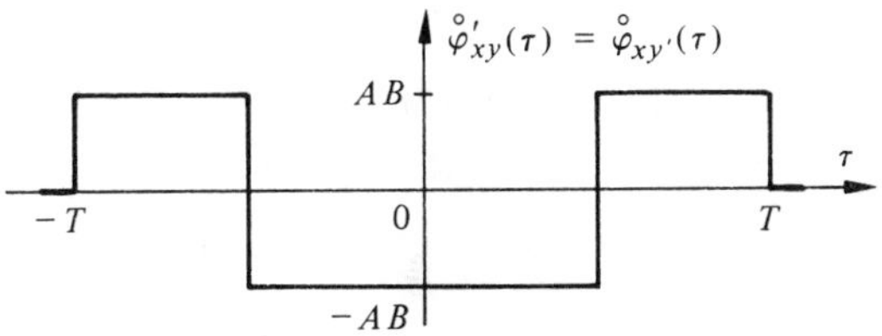

Fig. 4.18

4.3 FINITE POWER SIGNALS

4.3.1 Models for physically unrealizable signals

Signals with non-zero finite average power have been defined in sub-section 2.3.4 as complying with

$$0 < P_x = \lim_{T\to\infty} \frac{1}{T} \int_{-T/2}^{T/2} |x(t)|^2 \mathrm{d}t < \infty \tag{4.73}$$

where P_x is the total normalized power.

In chapter 2, we saw that these signals have infinite energy and thus are physically unrealizable. However, they provide a convenient abstraction for the generation of models of important categories of signals with quasipermanent behavior, such as the constant (e.g., dc in electrical engineering), the unit-step, the signum functions, and all the usual periodic or quasiperiodic signals. This abstraction is also valuable in modeling random signals (chap. 5).

4.3.2 Limiting cases of Fourier transform

Finite power signals do not comply with the usual Fourier transform convergence criteria. This analysis method can only be used, strictly speaking, if the application field of the Fourier transform is extended to encompass the so-called distributions [22].

For signal analysis, the main result of the distribution theory is the correspondence:

$$\delta(t) \leftrightarrow 1 \tag{4.74}$$

deduced from the delta function fundamental property (1.35).

According to the symmetry of the direct and inverse transforms, we also have

$$1 \leftrightarrow \delta(f) \tag{4.75}$$

Direct application of (4.16) and (4.17) yields

$$\delta(t - t_0) \leftrightarrow \exp(-\mathrm{j}2\pi f t_0) \tag{4.76}$$

and

$$\exp(\mathrm{j}2\pi f_0 t) \leftrightarrow \delta(f - f_0) \tag{4.77}$$

4.3.3 Application: constant signal and average value

The Fourier transform of a constant value signal $x(t) = C$ is a delta function with a weight C (fig. 4.19):

$$x(t) = C \leftrightarrow X(f) = C\delta(f) \tag{4.78}$$

By analogy, this result can be extended to a signal average value:

$$\bar{x} = \lim_{T\to\infty} \frac{1}{T} \int_{-T/2}^{T/2} x(t)\mathrm{d}t \tag{4.79}$$

$$\bar{x} \leftrightarrow \bar{x}\delta(f) \tag{4.80}$$

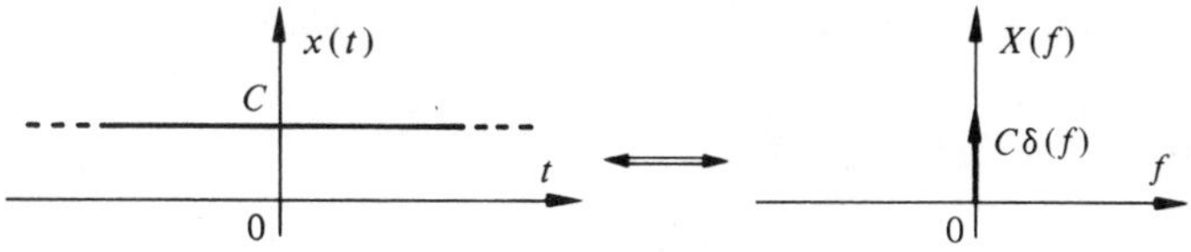

Fig. 4.19

4.3.4 Fourier transform of signals with non-zero time average

Any finite power signal can be viewed as the sum of its average value and a term with zero average value:

$$x(t) = \bar{x} + x_0(t) \tag{4.81}$$

with

$$\bar{x}_0(t) = \overline{x(t) - \bar{x}} = 0 \tag{4.82}$$

Differentiating $x(t)$, we get

$$x'(t) = \mathrm{d}x/\mathrm{d}t = x_0'(t) \tag{4.83}$$

for any constant $\bar{x}$.

In other words, we can write

$$x(t) = \int_{-\infty}^{t} x_0'(t)\mathrm{d}t + \bar{x} \tag{4.84}$$

where $\bar{x}$ is the integration constant.

According to (4.13), we have

$$\mathrm{F}\,\{x_0'(t)\} = \mathrm{j}2\pi f X_0(f) \qquad \forall\bar{x} \tag{4.85}$$

hence,

$$\mathrm{F}\{x_0(t) = \int_{-\infty}^{t} x_0'(t)\mathrm{d}t\} = \frac{1}{\mathrm{j}2\pi f}\,\mathrm{F}\{x_0'(t)\} \tag{4.86}$$

Taking into account (4.80), (4.84), and (4.86), the Fourier transform of the finite power signal $x(t)$ can be expressed as

$$X(f) = \frac{1}{\mathrm{j}2\pi f}\,\mathrm{F}\{x_0'(t)\} + \bar{x}\delta(f) \tag{4.87}$$

4.3.5 Example: unit-step signal

Let $x(t) = \epsilon(t) : \bar{x} = 1/2$, $x_0(t) = 1/2\,\mathrm{sgn}(t)$ and $x_0'(t) = \delta(t)$. Following (4.87), we obtain the transform pair:

$$\epsilon(t) \leftrightarrow \frac{1}{\mathrm{j}2\pi f} + \frac{1}{2}\delta(f) \tag{4.88}$$

4.3.6 Example: signum signal

Let $x(t) = \mathrm{sgn}(t)$: $\bar{x} = 0$ and $x_0'(t) = 2\,\delta(t)$; hence,

$$\mathrm{sgn}(t) \leftrightarrow \frac{1}{\mathrm{j}\pi f} \tag{4.89}$$

4.3.7 Correlation of finite power signals

The inner product (4.36), which defines the crosscorrelation of finite energy signals, cannot be applied to finite power signals, the integral of which is not defined. However, we can substitute for it with the time average:

$$\begin{aligned}\varphi_{xy}(\tau) &= \langle x^*, y_\tau \rangle \\ &= \lim_{T\to\infty} \frac{1}{T}\int_{-T/2}^{T/2} x^*(t)y(t+\tau)\mathrm{d}t\end{aligned} \tag{4.90}$$

which defines the *crosscorrelation function of finite power signals*. It also measures the similarity of the shape and position of the signals.

By analogy, the *autocorrelation function of finite power signals* is defined by the average:

$$\begin{aligned}\varphi_{x}(\tau) &= \langle x^*, x_\tau \rangle \\ &= \lim_{T\to\infty} \frac{1}{T}\int_{-T/2}^{T/2} x^*(t)x(t+\tau)\mathrm{d}t\end{aligned} \tag{4.91}$$

When the signals are real, so are the cross- and autocorrelation functions. If $x(t)$ and $y(t)$ are measured in volts, $\varphi_{xy}(\tau)$ and $\varphi_x(\tau)$ are measured in V^2.

4.3.8 Properties

The cross- and autocorrelation functions $\varphi_x(\tau)$ and $\varphi_{xy}(\tau)$ have the same properties as $\mathring{\varphi}_{xy}(\tau)$ and $\mathring{\varphi}_x(\tau)$:

$$\varphi_{yx}(\tau) = \varphi^*_{xy}(-\tau) \tag{4.92}$$

$$\varphi_x(\tau) = \varphi^*_x(-\tau) \tag{4.93}$$

$$|\varphi_{xy}(\tau)|^2 \leqslant \varphi_x(0)\varphi_y(0) \tag{4.94}$$

$$|\varphi_x(\tau)| \leqslant \varphi_x(0) = P_x \tag{4.95}$$

$$\varphi'_{xy}(\tau) = -\varphi_{x'y}(\tau) = \varphi_{xy'}(\tau) \tag{4.96}$$

The value at origin of the autocorrelation function is the signal normalized power. When signals are real: $\varphi_{xy}(\tau) = \varphi_{yx}(-\tau)$ and $\varphi_x(\tau) = \varphi_x(-\tau)$. ***The autocorrelation function of a real signal is always an even real function.***

Relations, which are similar to (4.50) and (4.51), tie the correlation functions of finite power signals to their convolution product, which is itself defined as a time average (notation $\overline{*}$):

$$x(\tau) \overline{*} y(\tau) = \lim_{T\to\infty} \frac{1}{T}\int_{-T/2}^{T/2} x(t)y(\tau - t)\mathrm{d}t \tag{4.97}$$

This yields the equivalences:

$$\varphi_{xy}(\tau) = x^*(-\tau) \overline{*} y(\tau) \tag{4.98}$$

$$\varphi_x(\tau) = x^*(-\tau) \overline{*} x(\tau) \tag{4.99}$$

For every value of τ where the crosscorrelation function is zero, signals $x(t)$ and $y(t + \tau)$ are ***orthogonal.***

4.3.9 Power spectral density

The Fourier transform of the autocorrelation function (4.91) is called the *power spectral density* of the finite mean power signal $x(t)$:

$$\Phi_x(f) = \mathrm{F}\{\varphi_x(\tau)\} = \int_{-\infty}^{\infty} \varphi_x(\tau)\exp(-\mathrm{j}2\pi f\tau)\mathrm{d}\tau \tag{4.100}$$

The inverse transform is

$$\varphi_x(\tau) = \int_{-\infty}^{\infty} \Phi_x(f)\exp(\mathrm{j}2\pi f\tau)\mathrm{d}f \tag{4.101}$$

For $\tau = 0$, we get the special form of the Parseval identity:

$$P_x = \varphi_x(0) = \lim_{T\to\infty} \frac{1}{T} \int_{-T/2}^{T/2} |x(t)|^2 \mathrm{d}t = \int_{-\infty}^{\infty} \Phi_x(f)\mathrm{d}f \qquad \textbf{(4.102)}$$

Hence, function $\Phi_x(f)$ is the frequency distribution of the total signal power. Thus, it is a non-negative function:

$$\Phi_x(f) \geqslant 0 \qquad (4.103)$$

If $x(t)$ is in volts, $\varphi_x(\tau)$ is in V^2 and $\Phi_x(f)$ is in V^2/Hz.

In the case of a ***real*** signal $x(t)$, $\varphi_x(\tau)$ is an even real function and the power spectral density is also an ***even real function***:

$$\Phi_x(f) = \Phi_x(-f).$$

4.3.10 Important observation

Contrary to the energy spectral density $\mathring{\Phi}_x(f)$, the power spectral density $\Phi_x(f)$ ***is not equal to the modulus square of the signal Fourier transform.***

However, another relation can be established. Let $x(t)$ be a finite power signal and $x(t,T) = x(t) \cdot \mathrm{rect}(t/T)$ a finite energy signal, the Fourier transform of which is $X(f,T)$:

$$x(t) = \lim_{T\to\infty} x(t,T) \qquad (4.104)$$

From (4.55), the energy spectral density of $x(t,T)$ can be written as

$$\mathring{\Phi}_x(f,T) = |X(f,T)|^2 \qquad (4.105)$$

When $T \to \infty$, the signal energy becomes infinite while its power remains finite, and hence (4.57) yields the last equality:

$$P_x = \lim_{T\to\infty} \frac{1}{T} \int_{-T/2}^{T/2} |x(t)|^2 \mathrm{d}t = \lim_{T\to\infty} \frac{1}{T} \int_{-\infty}^{\infty} |x(t,T)|^2 \mathrm{d}t$$
$$= \lim_{T\to\infty} \frac{1}{T} \int_{-\infty}^{\infty} |X(f,T)|^2 \mathrm{d}f \qquad (4.106)$$

Comparing (4.102) and (4.106), we see that the power spectral density can also be defined by the limit (if it exists)

$$\Phi_x(f) = \lim_{T\to\infty} \frac{1}{T} |X(f,T)|^2 \qquad (4.107)$$

4.3.11 Examples

Autocorrelation functions of sgn(t) and unit-step $\epsilon(t)$ signals are respectively equal to 1 and ½. From (4.100) and (4.78), their power spectral densities are respectively equal to $\delta(f)$ and ½$\delta(f)$. These results cannot be inferred from (4.88) or (4.89).

For the unit-step signal $x(t) = \epsilon(t)$, using (4.29), we get

$$x(t,T) = \text{rect}[(t - T/4)/(T/2)] \leftrightarrow X(f,T) = (T/2)\,\text{sinc}(fT/2) \cdot \exp(-j\pi f T/2)$$

thus

$$\Phi_x(f) = \lim_{T\to\infty} \frac{1}{T}\,|X(f,T)|^2 = \frac{1}{2} \lim_{T\to\infty} \frac{T}{2}\,\text{sinc}^2(fT/2) = \frac{1}{2}\,\delta(f)$$

However, if $x(t) = \text{sgn}(t)$, we have

$$x(t,T) = \text{rect}\left(\frac{t + T/4}{T/2}\right) - \text{rect}\left(\frac{t - T/4}{T/2}\right) \leftrightarrow X(f) = jT\;\text{sinc}(fT/2) \cdot \sin(\pi f T/2)$$

but its limit (4.107) is undefined.

4.3.12 Cross power spectral density

The Fourier transform of the crosscorrelation function $\varphi_{xy}(\tau)$ is called the *cross power spectral density:*

$$\Phi_{xy}(f) = \text{F}\,\{\varphi_{xy}(\tau)\} \qquad \textbf{(4.108)}$$

It is normally a complex, asymmetrical function.

4.4 PERIODIC SIGNALS

4.4.1 Fourier transform of a periodic signal

The Fourier transform of a periodic signal can be directly deduced from the Fourier series (3.74), using (4.77). Indeed, if $x(t) = x(t + mT)$, where m is an integer:

$$x(t) = \sum_{n=-\infty}^{\infty} X_n \exp(\,j2\pi nt/T) \qquad (4.109)$$

hence, with $f_n = n/T$

$$X(f) = \sum_{n=-\infty}^{\infty} X_n \delta\left(f - \frac{n}{T}\right) = \sum_{n=-\infty}^{\infty} X_n \delta(f - f_n) \qquad \textbf{(4.110)}$$

Thus, a periodic signal has a Fourier transform which is a linear combination of specific weights placed at the discrete frequencies $f_n = nf_1 = n/T$. This is the mathematical model for a ***line spectrum*** (fig. 4.20).

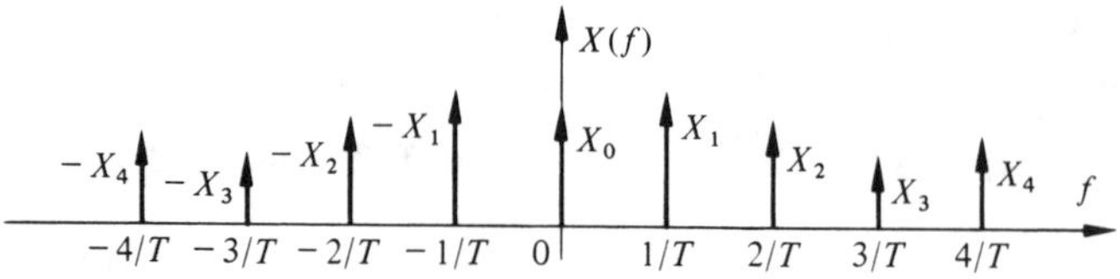

Fig. 4.20

4.4.2 Amplitude spectrum and phase spectrum of a periodic signal

The X_n coefficient is the complex weight of the line at frequency f_n. The set of their modules $\{|X_n|\}$ provides information on the signal component amplitudes and the set of their arguments $\{\arg X_n\}$ on their phases.

To maintain consistent notation, we will denote the amplitude and phase spectra of a periodic signal by the symbols $|X(f)|$ and $\vartheta_x(f)$, respectively, with

$$|X(f)| \triangleq \sum_{n=-\infty}^{\infty} |X_n| \, \delta\left(f - \frac{n}{T}\right) \tag{4.111}$$

$$\vartheta_x(f) \triangleq \arg X_n \Big|_{f=n/T} \tag{4.112}$$

For any real signal, coefficients X_n and X_{-n} are complex conjugates:

$$|X_n| = |X_{-n}| \tag{4.113}$$

$$\arg X_n = -\arg X_{-n} \tag{4.114}$$

4.4.3 Example: sinusoidal signal

Let

$$x(t) = A\cos(2\pi f_0 t - \alpha)$$
$$= \tfrac{1}{2}A\exp[-j(2\pi f_0 t - \alpha)] + \tfrac{1}{2}A\exp[j(2\pi f_0 t - \alpha)]$$

From (4.77), we have

$$X(f) = \tfrac{1}{2}A\exp(j\alpha)\delta(f + f_0) + \tfrac{1}{2}A\exp(-j\alpha)\delta(f - f_0) \tag{4.115}$$

hence,

$$X_{-1} = \tfrac{1}{2}A \exp(\,\mathrm{j}\alpha); \quad X_1 = \tfrac{1}{2}A \exp(-\mathrm{j}\alpha)$$

$$|X_{-1}| = |X_1| = \tfrac{1}{2}A$$

$$\arg X_1 = -\arg X_{-1} = -\alpha \tag{4.116}$$

Inserting $\alpha = 0$ and $\alpha = \pi/2$, we get the following transform pairs (fig. 4.21):

$$x_1(t) = A\cos(2\pi f_0 t) \leftrightarrow \tfrac{1}{2}A[\delta(f + f_0) + \delta(f - f_0)] \tag{4.117}$$

$$x_2(t) = A\sin(2\pi f_0 t) \leftrightarrow \tfrac{1}{2}A[\delta(f + f_0) - \delta(f - f_0)] \tag{4.118}$$

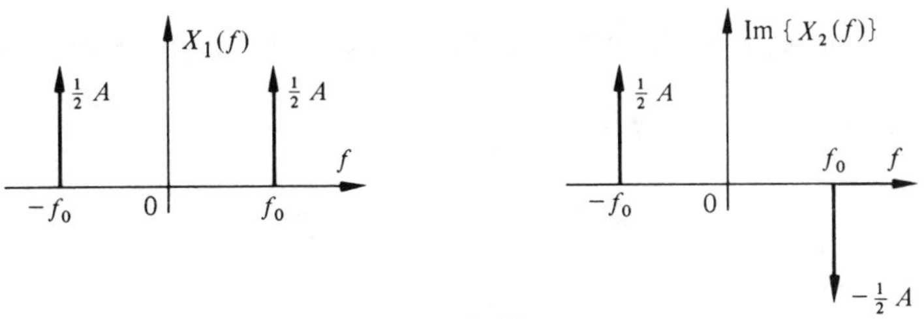

Fig. 4.21

4.4.4 Product of a periodic function and a finite energy signal

The signal $z(t) = x(t) \cdot y(t)$, where $x(t)$ is a finite energy signal and $y(t) = y(t + mT)$ is a periodic function, also has finite energy.

According to (4.15), (4.110), and (1.48), the Fourier transform of the product is

$$\begin{aligned} Z(f) &= X(f) * Y(f) \\ &= X(f) * \sum_{n=-\infty}^{\infty} Y_n \delta(f - n/T) \\ &= \sum_{n=-\infty}^{\infty} Y_n X(f - n/T) \end{aligned} \tag{4.119}$$

The result is a linear combination of $x(t)$ spectra, shifted by $1/T$ steps along the frequency axis.

4.4.5 Example

Let us return to example 4.2.3 where we have the product $z(t) = x(t) \cdot y(t)$, where $y(t) = \cos(2\pi f_0 t)$. From (4.117) and (4.119), we immediately obtain

$$Z(f) = \tfrac{1}{2} X(f + f_0) + \tfrac{1}{2} X(f - f_0)$$

4.4.6 Product of a periodic function and a finite power signal

The result is a finite power signal, the Fourier transform of which is obtained by convolving the respective transforms.

4.4.7 Example

As an exercise, we will let the reader check the following correspondences:

$$\epsilon(t)\cos(2\pi f_0 t) \leftrightarrow \frac{1}{4}[\delta(f + f_0) + \delta(f - f_0)] + \frac{1}{j4\pi(f + f_0)} + \frac{1}{j4\pi(f - f_0)}$$

$$\mathrm{sgn}(t)\sin(2\pi f_0 t) \leftrightarrow \frac{1}{2\pi(f + f_0)} - \frac{1}{2\pi(f - f_0)}$$

$$\cos(2\pi f_1 t)\cos(2\pi f_2 t) \leftrightarrow \tfrac{1}{4}[\delta(f - f_1 - f_2) + \delta(f - f_1 + f_2) + \delta(f + f_1 - f_2) + \delta(f + f_1 + f_2)]$$

4.4.8 Periodic sequence of delta functions

The periodic distribution:

$$\delta_T(t) = \sum_{k=-\infty}^{\infty} \delta(t - kT) \tag{4.120}$$

does not represent a finite power signal, but is often used in modeling periodic signals as well as sampling processes.

This distribution can be developed in Fourier series:

$$\delta_T(t) = \sum_{n=-\infty}^{\infty} \Delta_n \exp(j2\pi nt/T) \tag{4.121}$$

with, from (3.75) and (1.35):

$$\Delta_n = \frac{1}{T}\int_{-T/2}^{T/2} \delta_T(t)\exp(-j2\pi nt/T)\mathrm{d}t = \frac{1}{T} \tag{4.122}$$

Its Fourier transform is then deduced from (4.121), (4.122), and (4.77):

$$\mathrm{F}\,\{\delta_T(t)\} = \frac{1}{T}\sum_{n=-\infty}^{\infty} \delta(f - n/T) = \frac{1}{T}\,\delta_{1/T}(f) \tag{4.123}$$

which is also a periodic sequence of delta functions with weight $1/T$ and period $1/T$ along the frequency axis. The sequence $\delta_T(t)$ and its Fourier transform are shown on fig. 4.22.

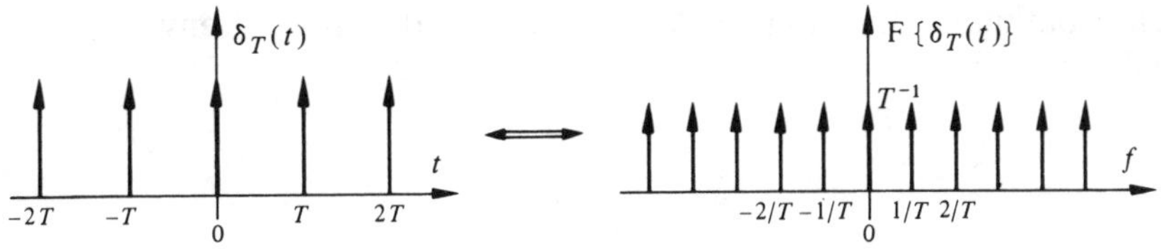

Fig. 4.22

4.4.9 Spectral envelope

A periodic signal can be seen as the cyclical repetition of an elementary signal $x(t,T)$ representing its fundamental period. As a result of the delta-function convolution property, we have

$$x(t) = x(t + mT) = \mathrm{rep}_T\{x(t,T)\} = x(t,T) * \delta_T(t) \tag{4.124}$$

Consequently, from (4.14) and (4.123):

$$X(f) = \frac{X(f,T)}{T}\cdot\delta_{1/T}(f) = \sum_{n=-\infty}^{\infty}\frac{1}{T}X\left(\frac{n}{T},T\right)\delta\left(f-\frac{n}{T}\right) \tag{4.125}$$

The continuous function $(1/T)X(f,T)$ is the *spectral envelope* of the weights of $\delta_{1/T}(f)$. This envelope depends only (except for the $1/T$ multiplier) on the shape of the signal's fundamental period.

Comparing (4.110) and (4.125), we get

$$X_n = (1/T)X(n/T,T) \tag{4.126}$$

The Fourier series coefficients of a periodic signal $x(t)$ can be found by dividing the fundamental period transform by T and evaluating the result for each discrete frequency $f_n = n/T$.

The set of values $X(n/T,T) = T\cdot X_n$ forms a discrete representation of the Fourier transform of the signal $x(t,T)$, similar to that of sub-section 9.3.11.

4.4.10 Example: periodic sequence of rectangular pulses

The following periodic signal plays an important role in theory and practice:

$$\begin{aligned} x(t) &= \text{rep}_T \{A \operatorname{rect}(t/\Delta)\} \\ &= A \operatorname{rect}(t/\Delta) * \delta_T(t) \end{aligned} \tag{4.127}$$

According to (4.26) and (4.125), its Fourier transform is

$$X(f) = A \frac{\Delta}{T} \operatorname{sinc}(\Delta f) \cdot \delta_{1/T}(f) = \sum_{n=-\infty}^{\infty} A \frac{\Delta}{T} \operatorname{sinc}(n\Delta/T)\delta(f - n/T) \tag{4.128}$$

The continuous function $(A\ \Delta/T)\operatorname{sinc}(\Delta f)$ is the spectral envelope of this signal (fig. 4.23).

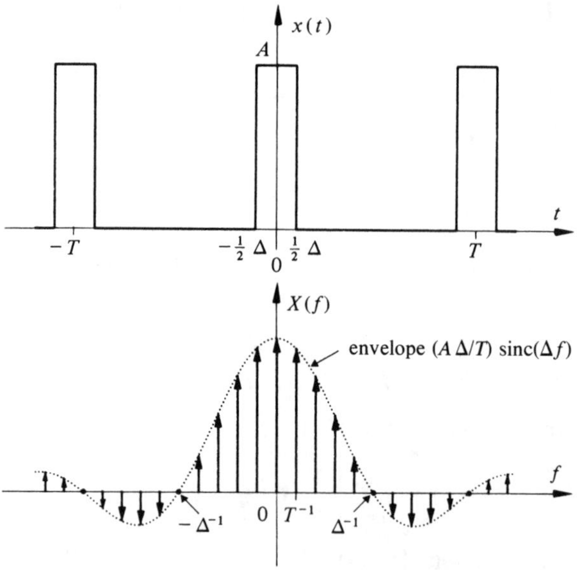

Fig. 4.23

4.4.11 Example

Consider the saw-toothed signal $x(t)$, illustrated by fig. 4.24, and its fundamental period $x(t,T)$. The derivative:

$$x'(t,T) = \frac{A}{T} \operatorname{rect}(t/T) - A\delta(t) + \frac{A}{2} [\delta(t + T/2) - \delta(t - T/2)]$$

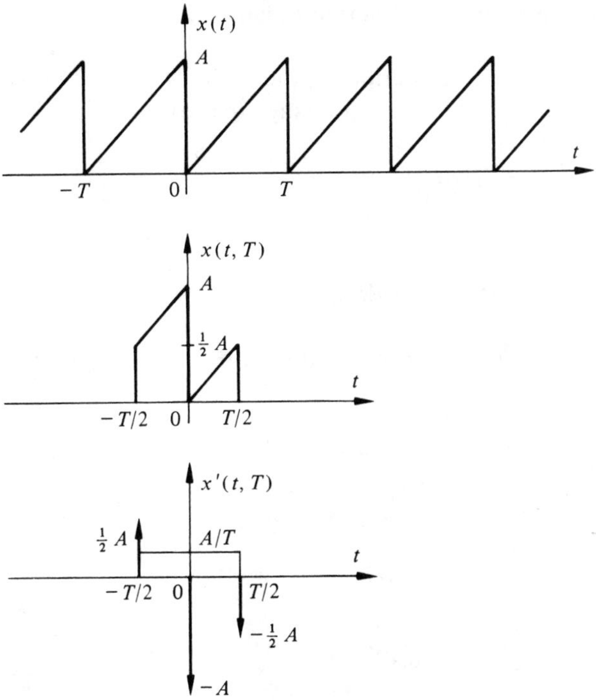

Fig. 4.24

has a Fourier transform:

$$F\{x'(t)\} = A[\text{sinc}(Tf) - 1 + \text{j}\sin(\pi Tf)]$$

From (4.13) and (4.125), we get

$$\frac{1}{T} X(f,T) = \frac{A}{\text{j}2\pi f T} |\text{sinc}(Tf) - 1 + \text{j}\sin(\pi Tf)|$$

hence,

$$X(f) = \sum_n X_n \delta(f - n/T)$$

with

$$X_n = \frac{1}{T} X(n/T, T) = \begin{cases} A/2 & \text{for } n = 0 \\ \text{j} \dfrac{A}{2\pi n} = \dfrac{A}{2\pi|n|} \exp\left[\text{j} \dfrac{\pi}{2} \cdot \text{sgn}(n)\right] & \text{for } n \neq 0 \end{cases}$$

4.4.12 Correlation function of periodic signals

For periodic signals with period T, the limits (4.90) and (4.91) are identical to the average values evaluated on a single period:

$$\varphi_{xy}(\tau) = \frac{1}{T}\int_{-T/2}^{T/2} x^*(t)y(t+\tau)\mathrm{d}t \qquad \textbf{(4.129)}$$

and

$$\varphi_{x}(\tau) = \frac{1}{T}\int_{-T/2}^{T/2} x^*(t)x(t+\tau)\mathrm{d}t \qquad \textbf{(4.130)}$$

By the Fourier series of $x(t+\tau)$ and $y(t+\tau)$, we can write these correlation functions as

$$\varphi_{xy}(\tau) = \sum_{n=-\infty}^{\infty} X_n^* Y_n \exp(\mathrm{j}2\pi n\tau/T) \qquad (4.131)$$

and

$$\varphi_{x}(\tau) = \sum_{n=-\infty}^{\infty} |X_n|^2 \exp(\mathrm{j}2\pi n\tau/T)$$

$$= X_0^2 + 2\sum_{n=1}^{\infty} |X_n|^2 \cos(2\pi n\tau/T) \qquad (4.132)$$

The demonstration is left as an exercise. For $\tau = 0$, we find again the Parseval identity (3.76).

Relations (4.131) and (4.132) show that ***cross- and autocorrelation functions of periodic signals, with period T, are also periodic functions having the same period.***

Inserting the fundamental period $x(t,T)$ into (4.129), and taking into account (4.50) and (4.124), we can write:

$$\varphi_{xy}(\tau) = \frac{1}{T}\int_{-\infty}^{\infty} x^*(t,T)y(t+\tau)\mathrm{d}t = \frac{1}{T}\, x^*(-\tau,T) * y(\tau)$$

$$= \frac{1}{T}\,[x^*(-\tau,T) * y(\tau,T)] * \delta_T(\tau) \qquad (4.133)$$

The convolution in square brackets is the crosscorrelation function of the fundamental periods of the two signals, defined by

$$\mathring{\varphi}_{xy}(\tau,T) = x^*(-\tau,T) * y(\tau,T) \qquad (4.134)$$

The crosscorrelation function of periodic signals can also be expressed as

$$\varphi_{xy}(\tau) = T^{-1}\,\mathring{\varphi}_{xy}(\tau,T) * \delta_T(\tau)$$

$$= \mathrm{rep}_T\,\{T^{-1}\,\mathring{\varphi}_{xy}(\tau,T)\} \qquad (4.135)$$

and for the autocorrelation, it is simply

$$\varphi_x(\tau) = T^{-1}\,\mathring{\varphi}_x(\tau,T) * \delta_T(\tau)$$
$$= \text{rep}_T\,\{T^{-1}\,\mathring{\varphi}_x(\tau,T)\} \tag{4.136}$$

where $\mathring{\varphi}_x(\tau,T)$ is the autocorrelation of $x(t,T)$.

Contrary to $x(t,T)$ or $y(t,T)$, $\mathring{\varphi}_{xy}(\tau,T)$ and $\mathring{\varphi}_x(\tau,T)$ are non-zero outside the fundamental period of $\varphi_{xy}(\tau)$ or $\varphi_x(\tau)$. Generally, their non-zero range is $[-T,T]$.

4.4.13 Example

Let a periodic sequence of rectangular pulses be

$$x(t) = \text{rep}_T\,\{A\,\text{rect}(t/\Delta)\}$$

From (4.63)

$$\mathring{\varphi}_x(\tau,T) = A^2\,\Delta\,\text{tri}(\tau/\Delta)$$

and

$$\varphi_x(\tau) = A^2\,\Delta\,T^{-1}\,\text{rep}_T\,\{\text{tri}(\tau/\Delta)\}$$

This autocorrelation function is shown with $x(t)$ in fig. 4.25, on one hand, for $\Delta_1 < T/2$ and, on the other hand, for $T/2 < \Delta_2 < T$.

4.4.14 Autocorrelation function of a periodic sequence of delta-functions

According to (4.51), the autocorrelation function of a delta-function is similar to its convolution (1.47), i.e., it is a delta function:

$$\mathring{\varphi}_\delta(\tau) = \int_{-\infty}^{\infty} \delta(t)\delta(t+\tau)\mathrm{d}t = \delta(\tau) \tag{4.137}$$

In the case of a periodic sequence of delta-functions, from (4.130), we get

$$\varphi_{\delta_T}(\tau) = \frac{1}{T}\int_{-T/2}^{T/2} \delta(t)\delta_T(t+\tau)\mathrm{d}t = \frac{1}{T}\,\delta_T(\tau) \tag{4.138}$$

This result is consistent with (4.136).

4.4.15 Power spectral density of a periodic signal

As was previously shown, the power spectral density is, by definition, the Fourier transform of the signal autocorrelation function.

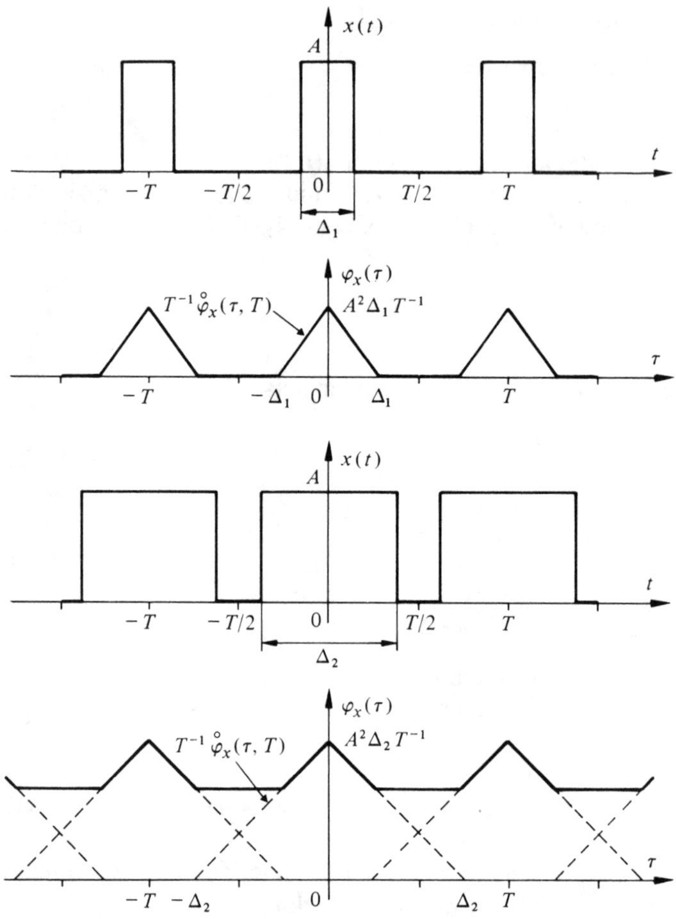

Fig. 4.25

According to (4.132) and (4.136), we obtain equivalent equations:

$$\Phi_x(f) = \mathrm{F}\,\{\varphi_x(\tau)\}$$

$$= \sum_{n=-\infty}^{\infty} |X_n|^2 \delta\left(f - \frac{n}{T}\right)$$

$$= \frac{1}{T^2}\,\mathring{\Phi}_x(f,T)\cdot\delta_{1/T}(f) \qquad \textbf{(4.139)}$$

where

$$\mathring{\Phi}_x(f,T) = \mathrm{F}\{\mathring{\varphi}_x(\tau,T)\} = |X(f,T)|^2 \qquad (4.140)$$

The continuous function $(1/T^2)\, \mathring{\Phi}_x(f,T)$ is the power spectral envelope of the lines $\delta_{1/T}(f)$ of the signal spectrum, and

$$|X_n|^2 = (1/T)^2 \mathring{\Phi}_x(n/T,T) = (1/T)^2 |X(n/T,T)|^2 \tag{4.141}$$

which is consistent with (4.126).

The signal total power is obtained by the Parseval identity:

$$P_x = \varphi_x(0) = \frac{1}{T}\int_{-T/2}^{T/2} |x(t)|^2 \mathrm{d}t = \int_{-\infty}^{\infty} \Phi_x(f)\mathrm{d}f = \sum_{n=-\infty}^{\infty} |X_n|^2 \tag{4.142}$$

4.4.16 Example: sinuosidal signal

Let $x(t) = A\cos(2\pi f_0 t - \alpha)$, following (4.116), (4.132), (4.139), and (4.142), we derive (fig. 4.26):

$$\varphi_x(\tau) = (A^2/2)\cos(2\pi f_0 \tau) \tag{4.143}$$

$$\Phi_x(f) = (A^2/4)[\delta(f + f_0) + \delta(f - f_0)] \tag{4.144}$$

$$P_x = \varphi_x(0) = \int_{-\infty}^{\infty} \Phi_x(f)\mathrm{d}f = A^2/2 \tag{4.145}$$

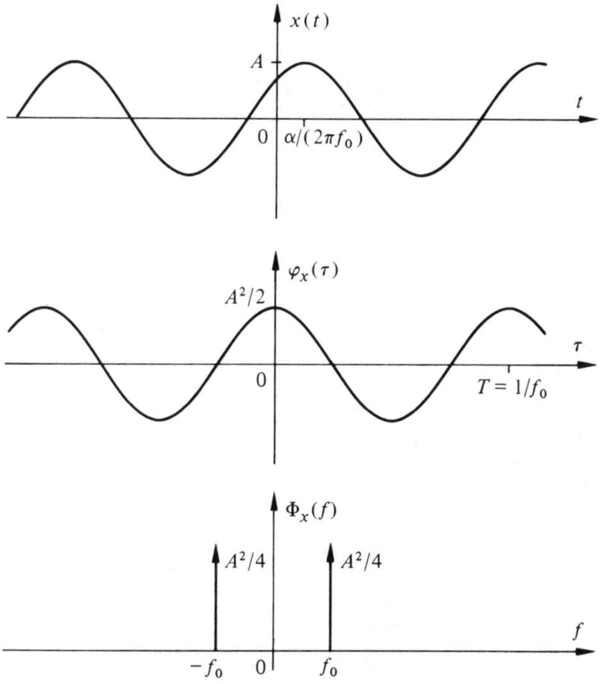

Fig. 4.26

4.4.17 Example

The power spectral density of a periodic sequence of rectangular pulses $x(t) = \text{rep}_T\{A\ \text{rect}(t/\Delta)\}$, is, from sub-sections 4.4.10 or 4.4.13, given by

$$\Phi_x(f) = (A\Delta/T)^2\ \text{sinc}^2(\Delta f) \cdot \delta_{1/T}(f)$$

$$= \sum_{n=-\infty}^{\infty} (A\Delta/T)^2\ \text{sinc}^2(\Delta n/T)\ \delta\left(f - \frac{n}{T}\right)$$

4.4.18 Cross power spectral density of periodic signals

By analogy with (4.139) and (4.140), we obtain for two periodic signals having the same period T:

$$\Phi_{xy}(f) = F\{\varphi_{xy}(\tau)\}$$

$$= \sum_{n=-\infty}^{\infty} X_n^* Y_n \delta\left(f - \frac{n}{T}\right) \tag{4.146}$$

$$= \frac{1}{T^2}\ \mathring{\Phi}_{xy}(f,T) \cdot \delta_{1/T}(f)$$

where

$$\mathring{\Phi}_{xy}(f,T) = \text{F}\{\mathring{\varphi}_{xy}(\tau,T)\} = X^*(f,T) \cdot Y(f,T) \tag{4.147}$$

and

$$X_n^* = T^{-1}X^*(n/T,T) \tag{4.148}$$

$$Y_n = T^{-1}Y(n/T,T) \tag{4.149}$$

4.4.19 Application

Let us determine the crosscorrelation function and the cross power spectral density of signals $x(t) = B\ \sin(2\pi t/T)$ and $y(t) = \text{rep}_T\{A\ \text{rect}(t/\Delta)\}$.

From (4.118), (4.128), (4.131), and (4.146):

$$\varphi_{xy}(\tau) = \frac{AB\Delta}{T}\ \text{sinc}(\Delta/T) \cdot \sin(2\pi\tau/T)$$

and

$$\Phi_{xy}(f) = \frac{\text{j}AB\Delta}{2T}\ \text{sinc}(\Delta/T)\left[\delta\left(f + \frac{1}{T}\right) - \delta\left(f - \frac{1}{T}\right)\right]$$

We can easily check that the crosscorrelation function of any T-periodic signal with a sinewave of the same period is always a sine function.

4.4.20 Quasiperiodic signals

A quasiperiodic signal is a linear combination of sine waves with incommensurable frequencies:

$$x(t) = \sum_k A_k \sin(2\pi f_k t + \alpha_k) \tag{4.150}$$

where $f_{k+1} = \lambda f_k$ with λ irrational.

Such a signal has properties both similar to and different from those of periodic signals.

The autocorrelation function of a quasiperiodic signal is

$$\varphi_x(\tau) = \sum_k (A_k^2/2)\cos(2\pi f_k \tau) \tag{4.151}$$

Indeed, applying definition (4.91), the only one valid here, we get

$$\begin{aligned}\varphi_x(\tau) &= \lim_{T\to\infty} \frac{1}{T}\int_{-T/2}^{T/2} \sum_k \sum_\ell A_k A_\ell \sin(2\pi f_k t + \alpha_k)\sin[2\pi f_\ell(t+\tau) + \alpha_\ell]\mathrm{d}t \\ &= \sum_k \sum_\ell A_k A_\ell I_{k\ell}\end{aligned} \tag{4.152}$$

$$\begin{aligned}I_{k\ell} &= \lim_{T\to\infty} \frac{1}{T}\int_{-T/2}^{T/2} \sin(2\pi f_k t + \alpha_k)\sin[2\pi f_\ell(t+\tau) + \alpha_\ell]\mathrm{d}t \\ &= \lim_{T\to\infty} \frac{1}{2T}\int_{-T/2}^{T/2} \cos[2\pi(f_k - f_\ell)t - 2\pi f_\ell \tau + \alpha_k - \alpha_\ell]\mathrm{d}t \\ &\quad - \lim_{T\to\infty} \frac{1}{2T}\int_{-T/2}^{T/2} \cos[2\pi(f_k + f_\ell)t + 2\pi f_\ell \tau + \alpha_k + \alpha_\ell]\mathrm{d}t \\ &= \begin{cases} 0 & \text{for } k \neq \ell \\ \frac{1}{2}\cos(2\pi f_k \tau) & \text{for } k = \ell \end{cases}\end{aligned} \tag{4.153}$$

From (4.151) and (4.144), we derive the power spectral density:

$$\Phi_x(f) = \sum_k (A_k^2/4)\left[\delta(f + f_k) + \delta(f - f_k)\right] \tag{4.154}$$

and the total power:

$$P_x = \varphi_x(0) = \int_{-\infty}^{\infty} \Phi_x(f)\mathrm{d}f = \sum_k A_k^2/2 \tag{4.155}$$

4.5 BILATERAL AND UNILATERAL SPECTRAL REPRESENTATIONS

4.5.1 Bilateral representations

As already mentioned in chapter 2, the complex representation of the cosine function:

$$\cos(2\pi ft) = \frac{1}{2}\exp(\,\mathrm{j}2\pi ft) + \frac{1}{2}\exp(-\mathrm{j}2\pi ft) \tag{4.156}$$

displays two terms with amplitude ½, the frequency of which is associated with a positive sign for one and with a negative sign for the other. This representation splits the spectral components of a signal symmetrically with respect to the origin on the f axis. This is why we usually speak of positive and negative frequencies.

All of the spectral representations that we have seen up to this point are bilateral: X_n,$X(f)$, $\mathring{\Phi}_x(f)$, $\mathring{\Phi}_{xy}(f)$, $\Phi_x(f)$, $\Phi_{xy}(f)$. Therefore, to compute the energy or, respectively, the power of a real signal in bandwidth $[f_1,f_2]$, we must write

$$W_x(f_1,f_2) = \int_{-f_2}^{-f_1} \mathring{\Phi}_x(f)\mathrm{d}f + \int_{f_1}^{f_2} \mathring{\Phi}_x(f)\mathrm{d}f = 2\int_{f_1}^{f_2} \mathring{\Phi}_x(f)\mathrm{d}f \tag{4.157}$$

$$P_x(f_1,f_2) = 2\int_{f_1}^{f_2} \Phi_x(f)\mathrm{d}f \tag{4.158}$$

The bilateral spectral representation is thus somewhat abstract. It is a natural consequence of the time-frequency duality introduced by the Fourier transform. Its advantage is to facilitate the determination of some properties of signal processing such as amplitude modulation or sampling.

4.5.2 Unilateral representations

It is sometimes more useful to have a unilateral spectral representation, displaying only the positive part of the frequency axis. Although tricky to do, this is possible, thanks to the complex conjugated symmetry of spectra for ***real*** signals:

$$X_n = X^*_{-n};\ X(f) = X^*(-f);\ \Phi_{xy}(f) = \Phi^*_{xy}(-f) \tag{4.159}$$

The modules are real even functions:

$$|X_n| = |X_{-n}|;\ |X(-f)| = |X(f)|;\ |\Phi_{xy}(f)| = |\Phi_{xy}(-f)| \tag{4.160}$$

and the arguments are odd functions.

Moreover, the energy or power spectral densities of ***real*** signals are real even functions:

$$\mathring{\Phi}_x(f) = \mathring{\Phi}_x(-f);\ \Phi_x(f) = \Phi_x(-f) \tag{4.161}$$

A unilateral spectral representation is obtained by multiplying the bilateral one by the function $1 + \text{sgn}(f) = 2\epsilon(f)$, in order to double the $f > 0$ contribution, to maintain its value for $f = 0$, and to cancel the negative frequency part:

$$\Phi_x^+(f) = 2\epsilon(f)\Phi_x(f) = \begin{cases} 0 & f < 0 \\ \Phi_x(0) & f = 0 \\ 2\Phi_x(f) & f > 0 \end{cases} \tag{4.162}$$

This method, since it is only valid for real signals, should be used with caution. It can yield an incorrect evaluation of the spectral contribution at $f = 0$. Moreover, the very useful concept of spectral envelope (sub-sect. 4.4.9) can no longer be used.

Another approach is introduced in chapter 7, which associates the real signal with a complex signal (called analytic signal), the Fourier transform of which is zero for negative frequencies.

4.5.3 Unilateral Fourier series

The complex Fourier series expansion (4.109) is equal to the sum of the real terms:

$$\begin{aligned} x(t) &= \sum_{n=-\infty}^{\infty} X_n \exp(\mathrm{j}2\pi f_n t) \\ &= A_0 + \sum_{n=1}^{\infty} A_n \cos(2\pi f_n t + \vartheta_n) \end{aligned} \tag{4.163}$$

with $f_n = n/T$ and the correspondences:

$$A_0 = X_0;\ A_n = 2|X_n|;\ \vartheta_n = \arg X_n \tag{4.164}$$

4.5.4 Example

The bilateral $\Phi_x(f)$ and unilateral $\Phi_x^+(f)$ power spectral densities of a periodic series of rectangular pulses are shown in fig. 4.27.

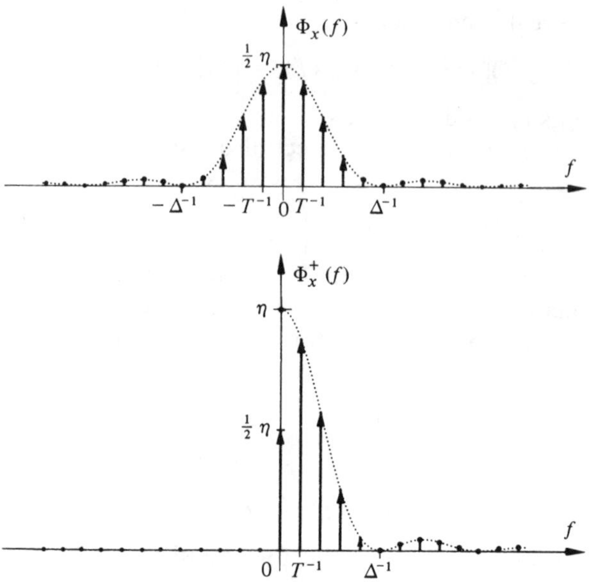

Fig. 4.27

4.6 PROBLEMS

4.6.1 Demonstrate the Fourier transform of a real even [odd] signal is a real even [imaginary odd] function.

4.6.2 Show that the Fourier transform of a complex signal is a real even [imaginary odd] function when the signal's real part is even [odd] and its imaginary part is odd [even].

4.6.3 Find the results (4.23) and (4.33) by using (4.13).

4.6.4 Show that $\text{sinc}(t) * \text{sinc}(t) = \text{sinc}(t)$.

4.6.5 Show that

$$\int_{-\infty}^{\infty} \text{sinc}^3(f)\text{d}f = 3/4$$

$$\int_{-\infty}^{\infty} \text{sinc}^4(f)\text{d}f = 2/3$$

4.6.6 Evaluate the integral

$$\nu(t) = \int_{-\infty}^{\infty} \frac{\sin 3\tau}{\tau} \cdot \frac{\sin(t - \tau)}{t - \tau} \text{d}\tau$$

4.6.7 Determine the Fourier transforms of the $x(t)$ and $y(t)$ signals illustrated in fig. 4.28.

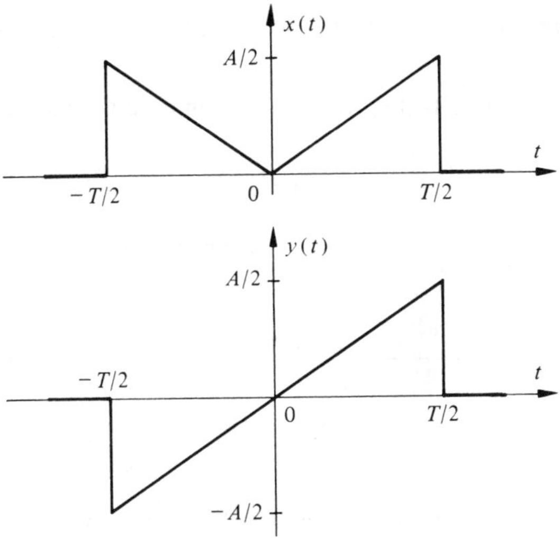

Fig. 4.28

4.6.8 Demonstrate that $\mathrm{d}X/\mathrm{d}f \leftrightarrow (-\mathrm{j}2\pi t)x(t)$.

4.6.9 Use the previous result and property (4.13) to demonstrate that $\mathrm{ig}(t) = \exp(-\pi t^2) \leftrightarrow \mathrm{ig}(f)$.

4.6.10 Show that the convolution product and the crosscorrelation function of $x(t) = \mathrm{ig}(t/T_1)$ and $y(t) = \mathrm{ig}(t/T_2)$ are also Gaussian functions of the kind $K \cdot \mathrm{ig}(t/T)$, where $T = [T_1^2 + T_2^2]^{1/2}$.

4.6.11 For the signal depicted in fig. 4.29, determine:

- the analytical expression of its amplitude spectrum
- the graph of its autocorrelation function
- the analytical expression of its energy spectral density

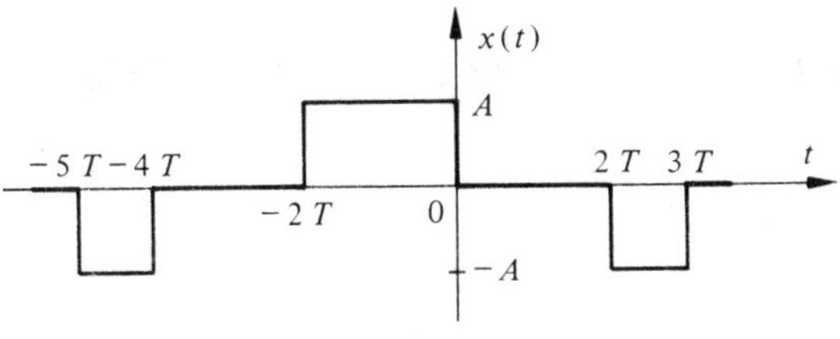

Fig. 4.29

4.6.12 Evaluate the Fourier transform of $x(t) = A \sin[2\pi(t - T/2)/T] \cdot \text{tri}[(t - T/2)/T]$ and plot its amplitude spectrum, phase spectrum, and energy spectral density.

4.6.13 Determine the crosscorrelation function of $x(t) = A\,\text{rect}(t/T)$ and $y(t) = B\,\text{tri}(2t/T)$.

4.6.14 Find the Fourier transform of the signal depicted in fig. 4.30.

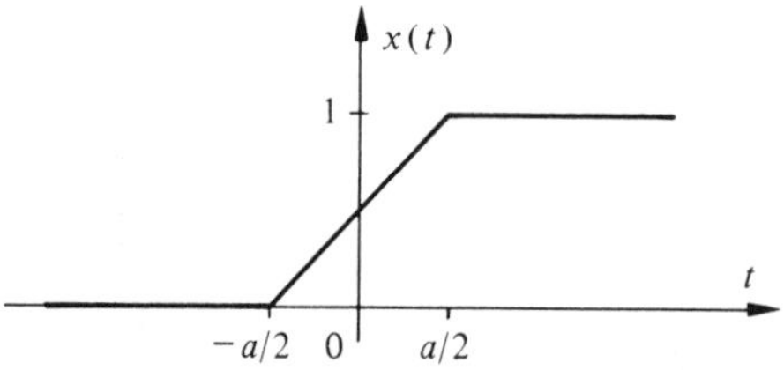

Fig. 4.30

4.6.15 Evaluate and plot the autocorrelation function and power spectral density of $x(t) = A\,\text{rep}_T\{\text{rect}[(t + T/4)/\Delta] - \text{rect}[(t - T/4)/\Delta]\}$ for $\Delta < T/4$ and $T/4 < \Delta < T/2$.

4.6.16 Evaluate and plot the power spectral density of $x(t) = m(t)p(t)$, where $m(t) = \text{rep}_T\{\text{rect}(t/\Delta)]$ and $p(t) = A\cos(2\pi f_0 t)$ for $\Delta/T = 0{,}2$ and $f_0 = k/T >> 1/\Delta$ with k integer.

4.6.17 Show that

$$\sum_{n=-\infty}^{\infty} (\Delta/T)\,\text{sinc}^2(n\Delta/T) = 1$$

4.6.18 Let $x(t) = A\sin(2\pi f_0 t)$ and $y(t) = (A/2)\cos(2f_0 t + \pi/4)$. Evaluate:

- their crosscorrelation function $x(t)$ and $y(t)$;
- the autocorrelation function of $z(t) = x(t) + y(t)$;
- the power spectral density of $z(t)$;
- the total power of $z(t)$.

4.6.19 Demonstrate that functions $z_k(t) = \text{sinc}(t - k)$ form an orthogonal set when $k = 0, \pm 1, \pm 2$, *et cetera*.

Chapter 5

Random Signals

5.1 STATISTICAL MODEL: STOCHASTIC PROCESS

5.1.1 Introduction

By definition (sub-sect. 2.2.1), a signal is said to be random if it depends on probabilistic laws. Such signals have unpredictable instantaneous values and cannot be described by analytical time models. However, they can be characterized by their statistical and spectral properties. The basic concepts of probability theory, used in describing statistically random signals, are summarized in chap. 14.

Random signals form a very important group because only signals with random behavior can transmit information (fundamental axiom of information theory). Also, their importance stems from the large variety of cases where we try to avoid random disturbing effects or try to identify and measure a phenomenon embedded as a weak signal in a strong noise environment (signal detection).

An observed random signal must be seen as a ***particular experimental realization*** of a ***set*** (referred to as the ***ensemble***) of similar signals that can all be produced by the same phenomenon or stochastic process.

5.1.2 Stochastic process

Mathematically [24,64,149], a *stochastic* (or *random*) *process* can be defined as an ensemble of real or complex functions of two variables, noted $\{\mathbf{x}(t,\zeta)\}$ or simply, $\mathbf{x}(t)$. In signal theory, the variable t is usually set for time.

Variable ζ expresses the random nature of the process: ζ is an element of the sample space (set of all possible outcomes of a statistical experiment) and depends on probabilistic laws (fig. 5.1).

The statistical description of a stochastic process is generally achieved by means of marginal and joint probabilities (chap. 14) of the random variables representing the process at prescribed times. Depending on whether those variables are discrete or continuous, the process is said to be either *discrete* or *continuous*.

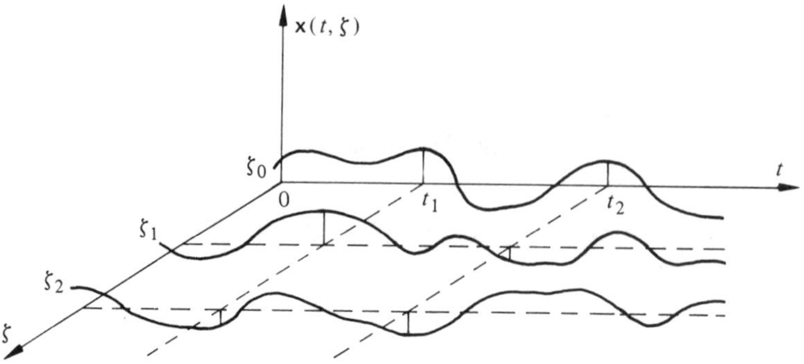

Fig. 5.1

When time is discrete (e.g., sampling), we deal with *stochastic time series.* When the stochastic event comes by discontinuously at random times, the process is said to be a *point process.*

Three stochastic process models are very important:

- Gaussian processes: model for continuous processes such as thermal noise;
- Poisson processes: model of point processes such as shot noise;
- Markov processes: model for many information-carrying signals, as an example.

5.1.3 Random signal

For each ζ_i, the process $\mathbf{x}(t,\zeta)$ provides one unique ensemble member from all possible realizations. We call this the *random signal,* which is written as $x_i(t)$, or more simply $x(t)$.

Such a signal will be considered, by convention, as a finite average power signal.

Observing that signal allows us to define, by analysis or measurement, some interesting time averages, such as the average value (dc value), the mean square value (power), *et cetera*. A time average can be written in the form:

$$\overline{f[x(t)]} = \lim_{T\to\infty} \frac{1}{T} \int_{-T/2}^{T/2} f[x(t)]\mathrm{d}t \tag{5.1}$$

5.1.4 Random variable

For each time t_i, process $\mathbf{x}(t,\zeta)$ becomes only a random variable $\mathbf{x}(t_i)$, or more simply, $\mathbf{x}_i$, the statistical behavior of which is described by its cumulative distribution $F(x;t_i)$, or its probability density $p(x;t_i)$. The main moments (statistical averages = mathematical expectations, i.e., theoretically expected values) of a variable can be drawn from those probabilistic laws.

An ***ensemble*** average, computed along the ζ-axis (i.e., expectation evaluated on the complete set of possible realizations) is given by relation:

$$\mathrm{E}[f(\mathbf{x})] = \int_{-\infty}^{\infty} f(x)p(x)\mathrm{d}x \tag{5.2}$$

5.1.5 Notations

Throughout this book, stochastic processes or random variables will be designated by boldface type (e.g., **x,y,z**); while the states taken by a variable are noted as usually (i.e., *x,y,z*).

Some authors note a stochastic process or a random variable with an uppercase character.

5.1.6 Example

A turbulent flow in a fluid is a perfect example of a continuous stochastic process. If we run simultaneously, under similar experimental conditions, many turbulent flow experiments, we will get a set of different results, each one representing a unique realization of the process. Measuring, during each experiment, let us say, the instantaneous local speed provides us with a set (ensemble) of random signals.

If we record all the values of those signals at time t_i, we get a sampling of all the possible states of the considered random variable (i.e., speed at time t_i).

If the number of simultaneous observations is large enough (ideally infinite), some process characteristics at time t_i can be deduced (histogram, experimental mean value, and variance, *et cetera*).

Unfortunately, the simultaneous observation of such a set is unrealistic. Therefore, we must generally deal with a single signal corresponding to a particular realization of the considered process. The analysis of such a typical signal provides us with information that, under certain assumptions (sub-sect. 5.1.14) can be considered as characteristic of the general process.

A classical example of continuous stochastic process is the noise generated in electric circuits by the random motion of electrons in conductors (sect. 6.2). Associated with other stochastic phenomena, this process accounts, for instance, for the noise perceived in audio equipment.

5.1.7 Example

An example of a stochastic point process is given by the emission of a radioactive material, as measured by a Geiger-Muller counter. The number of particles emitted by time unit is a discrete random variable.

Many electronic phenomena (thermo-electric emission or photo-electricity and charges flowing through a semiconductor junction) are stochastic point processes. These are the sources of shot noise, studied in section 6.3. Calls arriving at a switchboard or exchange and failures disabling an installation are other point process examples.

5.1.8 Random vector

If we consider k instants, $t_1, t_2, \ldots, t_k$, we can define k random variables $\mathbf{x}_1, \mathbf{x}_2, \ldots, \mathbf{x}_k$, and thus a *random vector* with k components $\boldsymbol{x} = (\mathbf{x}_1, \mathbf{x}_2, \ldots, \mathbf{x}_k)$, characterized by a joint probability density with k dimensions $p(x_1, x_2, \ldots, x_k)$. This joint probability density function defines a *kth order statistics* of the process.

In practice, the first-order and second-order statistics are generally adequate to characterize accurately enough the process behavior and its time dependence.

5.1.9 First-order statistics

Let $\mathbf{x}_i$ be a real random variable at time t_i. Its *cumulative distribution* expresses the probability that $\mathbf{x}_i$ is not greater than a given x value. It is generally a function of t_i:

$$F(x;t_i) = \text{Prob}(\mathbf{x}_i \leqslant x) \tag{5.3}$$

The *probability density* is simply the x-derivative of the cumulative distribution:

$$p(x;t_i) = \mathrm{d}F(x;t_i)/\mathrm{d}x \tag{5.4}$$

The *statistical mean value* of a random variable $\mathbf{x}_i$ is the first-degree moment:

$$\mu_x(t_i) = \mathrm{E}[\mathbf{x}_i] = \int_{-\infty}^{\infty} x_i p(x;t_i)\mathrm{d}x_i \tag{5.5}$$

The moments of higher degree are similarly defined as

$$m_{xn}(t_i) = \mathrm{E}[\mathbf{x}_i^n] = \int_{-\infty}^{\infty} x_i^n p(x,t_i)\mathrm{d}x_i \tag{5.6}$$

The central second-degree moment or *variance* is the second-degree moment of the difference $\mathbf{x}_i - \mu_x(t_i)$ between the random variable and its mean value:

$$\sigma_x^2(t_i) = \mathrm{E}\{[\mathbf{x}_i - \mu_x(t_i)]^2\} = \mathrm{E}[\mathbf{x}_i^2] - \mu_x^2(t_i)$$

$$= \int_{-\infty}^{\infty} [x_i - \mu_x(t_i)]^2 p(x,t_i)\mathrm{d}x_i \tag{5.7}$$

It is a measure of the dispersion of this difference; its square root is the *standard deviation* of the variable.

5.1.10 Second-order statistics

Consider the two real random variables $\mathbf{x}_1 = \mathbf{x}(t_1)$ and $\mathbf{x}_2 = \mathbf{x}(t_2)$. The *joint cumulative distribution* represents the probability that $\mathbf{x}_1$ and $\mathbf{x}_2$, respectively, are not greater than x_1 and x_2:

$$F(x_1,x_2;t_1,t_2) = \mathrm{Prob}(\mathbf{x}_1 \leqslant x_1, \mathbf{x}_2 \leqslant x_2) \tag{5.8}$$

The derivative of (5.8) with respect to x_1 and x_2 is the *joint probability density* of variables $\mathbf{x}_1$ and $\mathbf{x}_2$:

$$p(x_1,x_2;t_1,t_2) = \frac{\partial^2 F(x_1,x_2;t_1,t_2)}{\partial x_1\, \partial x_2} \tag{5.9}$$

The moments of $\mathbf{x}_1$ and $\mathbf{x}_2$ are, by definition, the statistical ensemble averages of functions $\mathbf{x}_1^n \cdot \mathbf{x}_2^m$, where n and m are positive (or zero) integers.

The moment corresponding to $m = n = 1$ is called the *statistical autocorrelation function* of process $\mathbf{x}(t)$ and is denoted by R:

$$R_x(t_1,t_2) = \mathrm{E}[\mathbf{x}(t_1)\mathbf{x}(t_2)] = \int_{-\infty}^{\infty}\int_{-\infty}^{\infty} x_1 x_2 p(x_1,x_2;t_1,t_2)\mathrm{d}x_1\mathrm{d}x_2 \tag{5.10}$$

If instead of considering the random variables $\mathbf{x}_1$ and $\mathbf{x}_2$, we had considered the differences between those variables and their mean values, respectively, we would have gotten the *autocovariance function,* denoted by C:

$$C_x(t_1,t_2) = E\,\{[\mathbf{x}(t_1) - \mu_x(t_1)][\mathbf{x}(t_2) - \mu_x(t_2)]\}$$

$$= R_x(t_1,t_2) - \mu_x(t_1) \cdot \mu_x(t_2) \tag{5.11}$$

When $t_2 = t_1$, (5.11) is identical to (5.7).

5.1.11 Stationarity

A stochastic process is said to be *strictly stationary* if all its statistical properties are time-invariant.

It is said to be *second-order stationary* if only its first-order and second-order statistics are time-invariant. In this case, we have

$$p(x;t_i) = p(x) \quad \forall t_i \tag{5.12}$$

hence, $\mu_x(t_i) = \mu_x$ and $\sigma_x^2(t_i) = \sigma_x^2$ are time-invariant characteristics, and thus they are constants for a given process.

Furthermore, the joint probability density depends only on the time difference:

$$\tau = t_2 - t_1 \tag{5.13}$$

thus,

$$p(x_1,x_2;t_1,t_2) = p(x_1,x_2;\tau) \tag{5.14}$$

and the autocorrelation and autocovariance are only functions of the time difference τ:

$$R_x(t_1,t_2) = R_x(\tau) \tag{5.15}$$

$$C_x(t_1,t_2) = C_x(\tau) = R_x(\tau) - \mu_x^2 \tag{5.16}$$

A stochastic process is said to be *wide-sense stationary* when only its mean value (5.5) and its autocorrelation function are time invariant. Obviously, a second-order stationary process is also wide-sense stationary. The reverse is not necessarily true.

5.1.12 Observations

A physical stochastic process is never totally stationary. It always has a transient phase when it is created, before it stabilizes. It is also influenced from time to time by the evolution of the system to which it is associated.

Hence, the stationary process concept is only a simplified model. However, it can be widely used and applied in practice when measurements of reasonable duration, during which the observed phenomenon has a permanent behavior, are sufficient.

5.1.13 Ergodism

A stochastic process is said to be *ergodic* if it exhibits identical statistical averages (5.2) and time averages (5.1) of the same degree and order.

As for stationarity, various levels of ergodism can be defined.

5.1.14 Consequences

Let

$$\overline{x^n} = \lim_{T\to\infty} \frac{1}{T} \int_{-T/2}^{T/2} x^n(t)\mathrm{d}t \tag{5.17}$$

be the nth degree time average, and

$$\mathrm{E}[\mathbf{x}^n] = \int_{-\infty}^{\infty} x^n p(x)\mathrm{d}x \tag{5.18}$$

the nth degree ensemble average—or moment—of a stationary stochastic process. The ergodic assumption leads to the identities shown in table 5.2.

The standard deviation σ_x for random signals is the equivalent of the rms value introduced for sine waves or other deterministic signals.

On some instruments, which are able to measure this magnitude, it is called the *true rms value*.

5.1.15 Observations

Therefore, when we can make the ***ergodic assumption,*** we can thus evaluate the statistical properties of process $\mathbf{x}(t,\zeta)$ through a time analysis of the sole $x(t)$ signal (unique observation of the set of all possible functions). Obviously, this procedure is much more practical and easy to implement.

Stationarity does not necessarily mean ergodism. Assume, for example, a process $\mathbf{x}(t,\zeta)$, each realization of which has the form $x(t) = x_0(t) + \bar{x}$, where $x_0(t)$ represents a stationary and ergodic random signal with a zero mean value. Let $\bar{x}$ be a random variable [thus, generally different for each $x(t)$], which also has a zero mean value. In this case, $\mu_x = 0$ and the process $\mathbf{x}(t,\zeta)$ is obviously stationary, but not ergodic, since $\mu_x \neq \bar{x}$.

While ergodism does not strictly imply stationarity [69], the definition used in sub-sections 5.1.13 implies that the considered ***ergodic processes form a subset of stationary processes.***

The ergodic assumption is often very difficult to verify. It is generally accepted under theoretical considerations. In practice, we admit that most usual random signals stem from ergodic processes. For a deeper analysis of ergodic conditions, the reader can refer to [24, 50, 57, 65].

5.1.16 Example: sine wave with random phase

Consider a process producing a sine wave with constant amplitude A, and constant angular frequency ω, but with an initial phase, $\boldsymbol{\alpha}$, that has an uniform

Table 5.2

Name	Statistical expression	Time expression	Interpretation
Mean value	$\mu_x = \mathrm{E}[\mathbf{x}] = \int_{-\infty}^{\infty} x p(x)\mathrm{d}x$	$\equiv \bar{x} = \lim_{T\to\infty} \frac{1}{T} \int_{-T/2}^{T/2} x(t)\mathrm{d}t$	dc value
Squared mean value	μ_x^2	$\equiv \bar{x}^2$	dc power
Mean square value	$\mathrm{E}[\mathbf{x}^2] = \int_{-\infty}^{\infty} x^2 p(x)\mathrm{d}x$	$\equiv P_x = \overline{x^2} = \lim_{T\to\infty} \frac{1}{T} \int_{-T/2}^{T/2} x^2(t)\mathrm{d}t$	total power
Variance	$\sigma_x^2 = \mathrm{E}[(\mathbf{x} - \mu_x)^2] = \int_{-\infty}^{\infty} (x - \mu_x)^2 p(x)\mathrm{d}x$	$\equiv \overline{(x - \bar{x})^2} = \lim_{T\to\infty} \frac{1}{T} \int_{-T/2}^{T/2} [x(t) - \bar{x}]^2\mathrm{d}t$	ac power
Standard deviation	σ_x	$\equiv \sqrt{\overline{(x - \bar{x})^2}} = \sqrt{P_x - \bar{x}^2}$	ac-rms value
Autocorrelation	$R_x(\tau) = \mathrm{E}[\mathbf{x}(t)\mathbf{x}(t + \tau)]$	$\equiv \varphi_x(\tau) = x(-\tau) \bar{*} x(\tau)$	
	$= \int_{-\infty}^{\infty} \int_{-\infty}^{\infty} x_1 x_2 p(x_1, x_2; \tau)\mathrm{d}x_1\mathrm{d}x_2$	$= \lim_{T\to\infty} \frac{1}{T} \int_{-T/2}^{T/2} x(t)x(t + \tau)\mathrm{d}t$	
Autocovariance	$C_x(\tau) = R_x(\tau) - \mu_x^2$	$\equiv \varphi_x(\tau) - \bar{x}^2$	
Special relations	$R_x(0) = \mathrm{E}[\mathbf{x}^2] = \sigma_x^2 + \mu_x^2$	$\equiv P_x = \varphi_x(0)$	
	$C_x(0) = \sigma_x^2$	$\equiv \varphi_x(0) - \bar{x}^2$	

random distribution on interval $[0, 2\pi]$. Such a random signal can be mathematically expressed as

$$x(t) = A \sin(\omega t + \alpha) \tag{5.19}$$

The random variable $\mathbf{x}_0$, observed at instant t_0, depends only on variable $\boldsymbol{\alpha}' = t + \boldsymbol{\alpha}$, which also has a uniform distribution:

$$\mathbf{x}_0 = \mathbf{x}(t_0) = A \sin \boldsymbol{\alpha}' \tag{5.20}$$

Conversely,

$$\boldsymbol{\alpha}' = \text{Arcsin}(\mathbf{x}_0/A) \tag{5.21}$$

The cumulative distribution $F(x)$ of the stochastic variable $\mathbf{x}_0$ is easily determined from fig. 5.3.

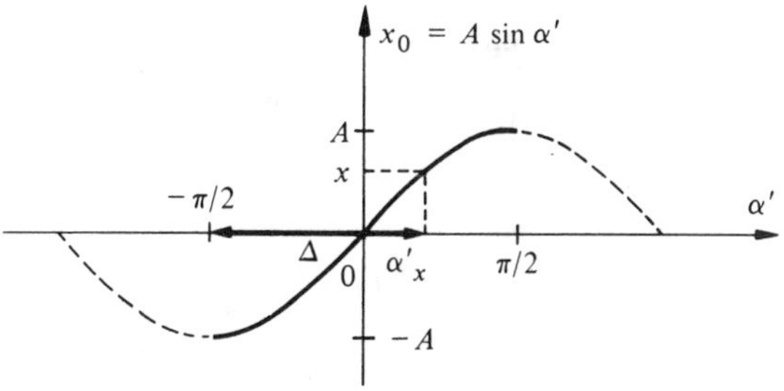

Fig. 5.3

On the half-period $[-\pi/2, \pi/2]$, $\mathbf{x}_0$ is inferior to a given value x on interval $\Delta = \alpha'_x + \pi/2$. Because the distribution of $\boldsymbol{\alpha}'$ is uniform, the probability of $\mathbf{x}_0$ being inferior or equal to x is simply the ratio of interval Δ to the half period:

$$F(x) = \text{Prob}(\mathbf{x}_0 \leqslant x) = \frac{\Delta}{\pi} = \frac{1}{\pi}\left[\text{Arcsin}\left(\frac{x}{A}\right) + \frac{\pi}{2}\right]$$

$$= \frac{1}{\pi}\text{Arcsin}\left(\frac{x}{A}\right) + \frac{1}{2} \tag{5.22}$$

Differentiating (5.22), we get the probability density:

$$p(x) = \frac{\mathrm{d}F(x)}{\mathrm{d}x} = \frac{1}{\pi\sqrt{A^2 - \mathrm{x}}}, \quad |x| \leqslant A \tag{5.23}$$

Those statistical laws are independent of the considered instant t_0; hence, the process is at least first-order stationary.

The statistical mean value:

$$\mu_x = \int_{-\infty}^{\infty} xp(x)\mathrm{d}x = 0 \tag{5.24}$$

since $p(x)$ is an even function.

The time average value is given by

$$\bar{x} = \lim_{T\to\infty} \frac{1}{T} \int_{-T/2}^{T/2} A \sin(\omega t + \alpha)\mathrm{d}t = 0 \tag{5.25}$$

hence, $\mu_x = \bar{x}$.

The statistical mean square value (which here is identical to the variance since $\mu_x = 0$) is

$$\mathrm{E}[\mathbf{x}^2] = \int_{-\infty}^{\infty} x^2 p(x)\mathrm{d}x = \frac{1}{\pi} \int_{-A}^{A} \frac{x^2}{\sqrt{A^2 - x}}\mathrm{d}x = \frac{A^2}{2} \tag{5.26}$$

and is also equal to the time mean square value (or total power) of the signal:

$$P_x = \overline{x^2} = \lim_{T\to\infty} \frac{1}{T} \int_{-T/2}^{T/2} x^2(t)\mathrm{d}t = \lim_{T\to\infty} \frac{1}{T} \int_{-T/2}^{T/2} A^2 \sin^2(\omega t + \alpha)\mathrm{d}t = \frac{A^2}{2} \tag{5.27}$$

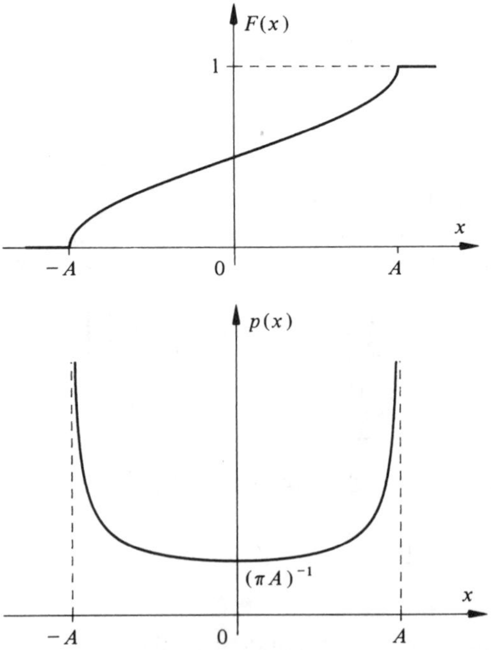

Fig. 5.4

Since $\mu_x = 0$, the statistical autocorrelation function here is identical to the autocovariance function.

$$\begin{aligned} R_x(t_1,t_2) &= \mathrm{E}[\mathbf{x}(t_1)\mathbf{x}(t_2)] \\ &= A^2\,\mathrm{E}[\sin(\omega t_1 + \boldsymbol{\alpha})\sin(\omega t_2 + \boldsymbol{\alpha})] \\ &= \tfrac{1}{2}A^2\,\mathrm{E}[\cos\{\omega(t_2 - t_1)\} - \cos\{\omega(t_1 + t_2) + 2\boldsymbol{\alpha}\}] \end{aligned} \tag{5.28}$$

The first term in brackets is not stochastic, but a constant for the given time lap $\tau = t_2 - t_1$; the second one is a function similar to $x(t)$ and, therefore, its mean value is zero. The statistical autocorrelation function of the sine wave with random phase is, finally,

$$R_x(t_1,t_2) = R_x(\tau) = \frac{A^2}{2}\cos(\omega\tau) \tag{5.29}$$

and depends only on the difference $\tau = t_2 - t_1$. The process is thus wide-sense stationary.

The time autocorrelation function is easily obtained and is identical to (5.29):

$$\begin{aligned} \varphi_x(\tau) &= \lim_{T\to\infty}\frac{1}{T}\int_{-T/2}^{T/2}\sin(\omega t + \alpha)\sin[\omega(t + \tau) + \alpha]\,\mathrm{d}t \\ &= \frac{A^2}{2}\cos(\omega\tau) \end{aligned} \tag{5.30}$$

Hence, this process is ergodic. We also check that the value of the autocorrelation function at $\tau = 0$ is equal to results (5.26) and (5.27).

Eventually, we observe that the autocorrelation function of a sine wave with random initial phase is identical to that of a deterministic sine wave (4.143).

5.1.17 Example

Consider a process producing a random binary signal $y(t)$. Its two amplitude levels, $y_1 = -2$ volts and $y_2 = 5$ volts, respectively occur with probabilities Prob $(y_1) = 1/3$ and Prob $(y_2) = 2/3$. Let us assume the process is stationary and ergodic.

Its cumulative distribution is

$$F(y) = \sum_{i=1}^{2}\mathrm{Prob}(y_i)\cdot\epsilon(y - y_i) \tag{5.31}$$

and its probability density (fig. 5.5) is

$$p\,(y) = \frac{\mathrm{d}F\,(y)}{\mathrm{d}y} = \sum_i \text{Prob}\,(y_i) \cdot \delta\,(y - y_i) \tag{5.32}$$

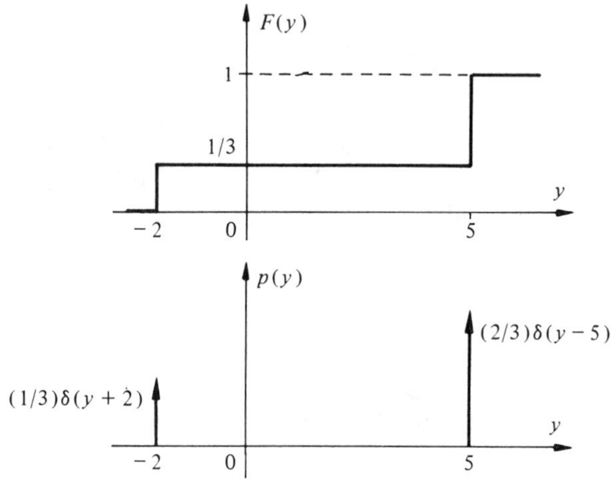

Fig. 5.5

where $\epsilon\,(y)$ and $\delta\,(y)$ are the unit-step and delta functions.

The dc component and the total power of such a signal cannot be mathematically evaluated by computing their time average values.

However, since we assumed the process $\mathbf{y}(t, \zeta)$ is ergodic, the dc component is equal to the statistical mean value and the total power to the statistical mean square value:

$$\bar{y} \equiv \mu_y = \int_{-\infty}^{\infty} y \cdot p\,(y)\,\mathrm{d}y = \sum_i y_i\,\text{Prob}\,(y_i)$$

$$= -2 \cdot 1/3 + 5 \cdot 2/3 = 8/3 = 2.67\text{ V}$$

$$P_y \equiv \mathrm{E}\,[\mathbf{y}^2] = \int_{-\infty}^{\infty} y^2 p\,(y)\,\mathrm{d}y = \sum_i y_i^2\,\text{Prob}\,(y_i)$$

$$= 4 \cdot 1/3 + 25 \cdot 2/3 = 54/3 = 18\text{ V}^2$$

5.1.18 Independent stochastic processes

Two stationary stochastic processes $\mathbf{x}(t)$ and $\mathbf{y}(t)$ are *independent* when their joint probability density is equal to the product of their marginal probability densities:

$$p_{xy}(x,y) = p_x(x)\, p_y(y) \tag{5.33}$$

This relation is consistent with (14.36).

5.1.19 Observation

Statistically speaking, independence means that we cannot learn anything about one process by observing the other one.

In signal processing, we often deal with combinations of independent processes; for example, noise or interference contaminating a signal during its transmission or amplification. This is also the case in modulation systems, where the carrying information-signal is mixed with an auxiliary signal.

When there is independence, condition (5.33) simplifies the mathematical relationships and allows us to reach very general results.

5.1.20 Transformation of random vectors

When the signal shape is modified by a processing system, its statistical characteristics change. Here, we shall see the effects of such a transformation on random variables or vectors.

Let a random vector (sub-sect. 5.1.8) of k components (random variables) $\mathbf{x}_1, \mathbf{x}_2, \ldots, \mathbf{x}_k$ be transformed as follows:

$$\begin{aligned} \mathbf{y}_1 &= f_1(\mathbf{x}_1, \mathbf{x}_2, \ldots, \mathbf{x}_k) \\ \mathbf{y}_2 &= f_2(\mathbf{x}_1, \mathbf{x}_2, \ldots, \mathbf{x}_k) \\ &\vdots \\ \mathbf{y}_k &= f_k(\mathbf{x}_1, \mathbf{x}_2, \ldots, \mathbf{x}_k) \end{aligned} \tag{5.34}$$

The joint probability density of the random variables $\mathbf{y}_i$ can be deduced from the joint probability density of $\mathbf{x}_i$ due to the ***probabilistic equivalence condition*** which introduces the Jacobian J of the transformation [58]:

$$p_y(y_1,y_2, \ldots, y_k) = |J| \cdot p_x(x_1,x_2, \ldots, x_k) \Big|_{x_i = g_i(y_1,y_2, \ldots, y_k)} \tag{5.35}$$

$$J = \frac{\partial(x_1,x_2, \ldots, x_k)}{\partial(y_1,y_2, \ldots, y_k)} = \begin{vmatrix} \partial x_1/\partial y_1 & \ldots & \partial x_1/\partial y_k \\ \vdots & & \vdots \\ \partial x_k/\partial y_1 & \ldots & \partial x_k/\partial y_k \end{vmatrix} \tag{5.36}$$

The absolute value of the Jacobian is necessary to comply with the condition: $p_y(y_1, y_2, \ldots, y_k) \geqslant 0$.

5.1.21 Example: instantaneous linear transformation

Consider the simple instantaneous (i.e., only the current value is taken into account) linear transformation of a single variable:

$$\mathbf{y} = a\mathbf{x} + b \tag{5.37}$$

Here, the Jacobian is simply

$$J = \mathrm{d}x/\mathrm{d}y = 1/a \tag{5.38}$$

thus,

$$p_y\,(y) = \frac{1}{|a|}\, p_x\left(x = \frac{y - b}{a}\right) \tag{5.39}$$

Hence, the probability density of **y** is homothetical to the one of **x**, but with a b-shift and a different scale-factor, a.

5.1.22 Example: quadratic transformation

Squaring is an operation used in all energy or power computations. We also find it in spectral analysis, signal detection *et cetera*.

Let the nonlinear transformation be

$$\mathbf{y} = a\mathbf{x}^2 \qquad a > 0 \tag{5.40}$$

This is not a one-to-one relationship, since two values of **x** may yield the same value of **y**. The probabilistic equivalence condition allows us to write

$$p_y\,(y) = \left|\frac{\mathrm{d}x_-}{\mathrm{d}y}\right| \cdot p_x\left(x_- = -\sqrt{y/a}\right) + \left|\frac{\mathrm{d}x_+}{\mathrm{d}y}\right| \cdot p_x\left(x_+ = \sqrt{y/a}\right) \tag{5.41}$$

where

$$x_+ = \sqrt{y/a} \text{ and } \mathrm{d}x_+/\mathrm{d}y = \frac{1}{2\sqrt{ay}}\,;\; x_- = -\sqrt{y/a} \text{ and } \mathrm{d}x_-/\mathrm{d}y = \frac{-1}{2\sqrt{ay}} \tag{5.42}$$

$$p_y\,(y) = \frac{1}{2\sqrt{ay}}\left[\, p_x\left(x_- = -\sqrt{y/a}\right) + p_x\left(x_+ = \sqrt{y/a}\right)\right];\; y \geqslant 0 \tag{5.43}$$

Let us examine a particular case: when $p_x\,(x)$ is an even function, $p_x\,(x) = p_x\,(-x)$ and

$$p_y\,(y) = \frac{1}{\sqrt{ay}}\, p_x\,(\sqrt{y/a}) \tag{5.44}$$

For example (fig. 5.6), for the quadratic transformation defined by (5.40), if $p_x(x) = 1/A$ for $|\mathbf{x}| \leqslant A/2$ (centered uniform distribution): $p_y(y) = 1/(A\sqrt{ay})$ with $0 \leqslant \mathbf{y} \leqslant aA^2/4$.

As a general rule, nonlinear transformations modify the shape of the probability density.

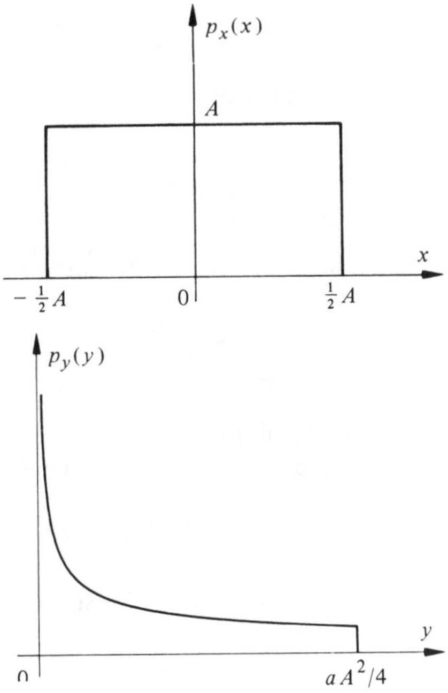

Fig. 5.6

5.1.23 Example: orthogonal to polar coordinate transformation

This transformation is found in many applications, such as determination of the statistics of signals in complex form: random phasor, analytic signal (chap. 7).

The transformation is defined by relations:

$$\mathbf{r} = \sqrt{\mathbf{x}^2 + \mathbf{y}^2}$$

$$\boldsymbol{\phi} = \arctan \mathbf{y}/\mathbf{x} \tag{5.45}$$

with

$$J = \begin{vmatrix} \partial x/\partial r & \partial x/\partial \phi \\ \partial y/\partial r & \partial y/\partial \phi \end{vmatrix} = \begin{vmatrix} \cos\phi & -r\sin\phi \\ \sin\phi & r\cos\phi \end{vmatrix} = r \tag{5.46}$$

thus, from (5.35), we have

$$p_{r\phi}(r,\phi) = r \cdot p_{xy}(x = r\cos\phi, y = r\sin\phi) \tag{5.47}$$

Let us consider, for example, two independent variables **x** and **y** with Gaussian marginal probability densities and zero mean values (chap. 14):

$$p_x(x) = \frac{1}{\sqrt{2\pi}\,\sigma_x}\exp\left[-\frac{x^2}{2\sigma_x^2}\right];\ p_y(y) = \frac{1}{\sqrt{2\pi}\,\sigma_y}\exp\left[-\frac{y^2}{2\sigma_y^2}\right] \tag{5.48}$$

As a result of (5.33), the joint probability density is given, in this case, by the product of marginal densities:

$$p_{xy}(x,y) = p_x(x)p_y(y) = \frac{1}{2\pi\sigma_x\sigma_y}\exp\left[\frac{1}{2}\left(\frac{x^2}{\sigma_x^2} + \frac{y^2}{\sigma_y^2}\right)\right] \tag{5.49}$$

The joint probability density of the polar coordinates is then given, with $\mathbf{r} \geqslant 0$ and $0 \leqslant \boldsymbol{\phi} < 2\pi$, by

$$p_{r\phi}(r,\phi) = \frac{r}{2\pi\sigma_x\sigma_y}\exp\left[-\frac{1}{2}\left(\frac{r^2\cos^2\phi}{\sigma_x^2} + \frac{r^2\sin^2\phi}{\sigma_y^2}\right)\right] \tag{5.50}$$

This is a generalized Rayleigh distribution.

In the case where $\sigma_x = \sigma_y = \sigma$, we obtain a Rayleigh distribution:

$$p_{r\phi}(r,\phi) = \frac{r}{2\pi\sigma^2}\exp\left[-\frac{r^2}{2\sigma^2}\right]\ ;\ r \geqslant 0,\ 0 \leqslant \phi < 2\pi \tag{5.51}$$

We deduce from this the marginal probability densities:

$$p_r(r) = \int_0^{2\pi} p_{r\phi}(r,\phi)\mathrm{d}\phi = \frac{r}{\sigma^2}\exp\left[-\frac{r^2}{2\sigma^2}\right]\ ;\ r \geqslant 0 \tag{5.52}$$

$$p_\phi(\phi) = \int_0^{\infty} p_{r\phi}(r,\phi)\mathrm{d}r = \frac{1}{2\pi}\ ;\ 0 \leqslant \phi < 2\pi \tag{5.53}$$

hence,

$$p_{r\phi}(r,\phi) = p_r(r) \cdot p_\phi(\phi) \tag{5.54}$$

which demonstrates that the random variables **r** and $\boldsymbol{\phi}$ are also statistically independent.

The Rayleigh distribution (5.52) is illustrated in fig. 14.15.

5.2 AUTOCORRELATION AND AUTOCOVARIANCE FUNCTIONS

5.2.1 Autocorrelation function

The ***statistical autocorrelation function*** of a real stationary stochastic process $\mathbf{x}(t)$ has been defined in sub-sections 5.1.10 and 5.1.11 as the mathematical expectation of the product of $\mathbf{x}(t)$ by $\mathbf{x}(t + \tau)$:

$$R_x(\tau) = \mathrm{E}[\mathbf{x}(t)\mathbf{x}(t + \tau)] = \int_{-\infty}^{\infty}\int_{-\infty}^{\infty} x_1 x_2 p(x_1,x_2;\tau)\mathrm{d}x_1\mathrm{d}x_2 \tag{5.55}$$

where x_1 and x_2 represent the values that the process can take at instants t and $t + \tau$, respectively. The adaptation of this definition for a process represented by a complex function is given in sub-section 5.2.5.

When the process is discrete, the joint probability density can be written as

$$p(x_1,x_2;\tau) = \sum_i \sum_j \delta(x_1 - x_{1i},x_2 - x_{2j})\mathrm{Prob}(x_{1i},x_{2j};\tau) \tag{5.56}$$

and the double integration becomes a double summation over indices i and j representing the various discrete states:

$$R_x(\tau) = \sum_i \sum_j x_{1i}x_{2j}\mathrm{Prob}(x_{1i},x_{2j};\tau) \tag{5.57}$$

The ***time autocorrelation function*** of a real stationary stochastic signal $x(t)$ is given by the time average value of the product of $x(t)$ and $x(t + \tau)$:

$$\varphi_x(\tau) = \overline{x(t)x(t + \tau)} = \lim_{T\to\infty} \frac{1}{T}\int_{-T/2}^{T/2} x(t)x(t + \tau)\mathrm{d}t \tag{5.58}$$

By analogy with (4.99), this relation can be expressed as a convolution product (finite average power signals: notation $\overline{*}$):

$$\varphi_x(\tau) = x(-\tau)\overline{*}x(\tau) \tag{5.59}$$

Under the ergodic assumption (sub-sect. 5.1.13), we have the identity:

$$R_x(\tau) \equiv \varphi_x(\tau) \tag{5.60}$$

The two notations can then be used interchangeably.

5.2.2 Remarks

Relations (5.55) or (5.57) provide us with a means to compute the autocorrelation function of a stationary process, the second-order statistics of which are analytically known.

Relation (5.58) is identical to (4.91) for finite average power deterministic signals. However, except in some particular cases, a random signal is not analytically known, and thus this relation cannot be used. Nonetheless, it shows us how we can experimentally (and ideally because of the limit) obtain the autocorrelation function.

The feasible experiment is to measure, during a finite period T, the function:

$$\tilde{\varphi}_x(\tau) = \frac{1}{T}\int_0^T x(t)x(t+\tau)\mathrm{d}t \tag{5.61}$$

with

$$\varphi_x(\tau) = \lim_{T\to\infty} \tilde{\varphi}_x(\tau) \tag{5.62}$$

5.2.3 Autocovariance function

The ***autocovariance function***, defined by (5.11) and (5.16) is equal to the autocorrelation function of the centered process, $\mathbf{x}(t) - \mu_x$:

$$C_x(\tau) = \mathrm{E}\,\{[\mathbf{x}(t) - \mu_x][\mathbf{x}(t+\tau) - \mu_x]\} = R_x(\tau) - \mu_x^2 \tag{5.63}$$

For any zero mean value process, autocorrelation and autocovariance functions are identical.

The *normalized autocovariance function* (or *correlation coefficient*) is the ratio:

$$\rho_x(\tau) = C_x(\tau)/\sigma_x^2 \tag{5.64}$$

5.2.4 Properties

We can easily check that the autocorrelation and autocovariance functions of real processes are even:

$$R_x(\tau) = R_x(-\tau) \tag{5.65}$$

$$C_x(\tau) = C_x(-\tau) \tag{5.66}$$

For $\tau = 0$, ***the autocovariance function is equal to the variance*** σ_x^2. Hence,

$$C_x(0) = \sigma_x^2 \tag{5.67}$$

$$R_x(0) = \sigma_x^2 + \mu_x^2 \tag{5.68}$$

$$\rho_x(0) = 1 \tag{5.69}$$

The value at origin of the autocovariance function of an ergodic and stationary process corresponds to the ac power, and the value at origin of the autocorrelation function to the total power (see table 5.2).

Furthermore,

$$|R_x(\tau)| \leqslant R_x(0) \tag{5.70}$$

for the mean square value of the following sum or difference is necessarily either positive or zero:

$$\mathrm{E}\{[\mathbf{x}(t) \pm \mathbf{x}(t+\tau)]^2\} = 2[R_x(0) \pm R_x(\tau)] \geqslant 0 \tag{5.71}$$

hence,

$$-R_x(0) \leqslant R_x(\tau) \leqslant R_x(0) \tag{5.72}$$

This result is similar to those established for deterministic signals in sub-sections 4.2.14 and 4.3.8.

Using an identical derivation for centered processes, we get

$$|C_x(\tau)| \leqslant C_x(0) \tag{5.73}$$

and, from (5.64) and (5.67), we have

$$|\rho_x(\tau)| \leqslant 1 \tag{5.74}$$

Because the square of the mean value is always positive, inequalities (5.72) can be replaced by more strict conditions:

$$\mu_x^2 - \sigma_x^2 \leqslant R_x(\tau) \leqslant \mu_x^2 + \sigma_x^2 \tag{5.75}$$

Relation (5.55) shows that the autocorrelation function—as well as the covariance function—depends on the joint probability density of $\mathbf{x}(t)$ and $\mathbf{x}(t + \tau)$. For a stochastic process ***without periodic components***, the interdependence of $\mathbf{x}(t)$ and $\mathbf{x}(t+\tau)$ decreases when $|\tau|$ increases indefinitely. Ultimately, those variables become statistically independent. Introducing the condition (5.33), we get

$$\lim_{|\tau|\to\infty} R_x(\tau) = \mu_x^2 \tag{5.76}$$

$$\lim_{|\tau|\to\infty} C_x(\tau) = 0 \tag{5.77}$$

A typical example of an autocorrelation function of such a process is illustrated in fig. 5.7.

As an obvious counter example, consider a sine wave with random phase (sub-sect. 5.1.16), where the autocorrelation function (5.29) is periodic.

5.2.5 Notation for the complex representation of a process

If the process $\mathbf{x}(t)$ is represented as a complex function, the statistical and time autocorrelation functions, respectively, are defined by

$$R_x(\tau) = \mathrm{E}[\mathbf{x}^*(t)\mathbf{x}(t+\tau)] \tag{5.78}$$

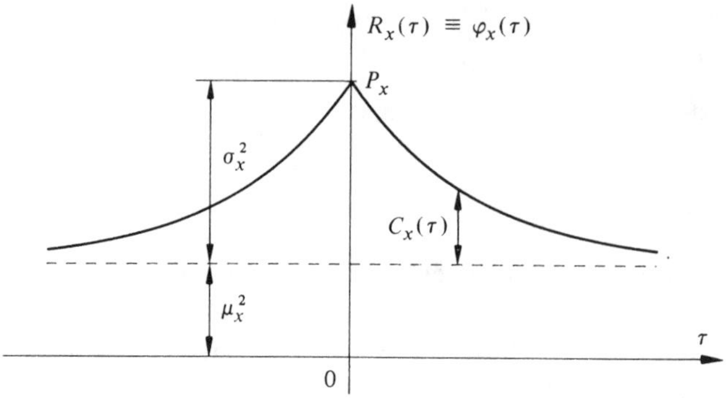

Fig. 5.7

$$\varphi_x(\tau) = \lim_{T\to\infty} \frac{1}{T} \int_{-T/2}^{T/2} x^*(t)x(t+\tau)\mathrm{d}t \tag{5.79}$$

and relations (5.65) and (5.66) are replaced (thanks to the Hermitian symmetry) by

$$R_x(-\tau) = R_x^*(\tau);\ C_x(-\tau) = C_x^*(\tau) \tag{5.80}$$

5.2.6 Interpretation

The statistical autocorrelation and autocovariance functions of a stochastic process can be interpreted in a similar way to those of a deterministic signal (sub-sect. 4.2.10).

Relations (5.55), (5.63), and (5.78) are special expressions of the inner product of random variables.

The statistical mean square value of the difference $\mathbf{x}(t) - \mathbf{x}(t+\tau)$ corresponds to the square of the distance between those two variables. Assuming $\mu_x = 0$ for simplicity, with shorthand notation $\mathbf{x}_1 = \mathbf{x}(t)$ and $\mathbf{x}_2 = \mathbf{x}(t+\tau)$, we have

$$\begin{aligned} \mathrm{E}\{[\mathbf{x}(t) - \mathbf{x}(t+\tau)]^2\} = d^2(\mathbf{x},\mathbf{x}_\tau) &= \iint (x_1 - x_2)^2 p(x_1,x_2;\tau)\,\mathrm{d}x_1\,\mathrm{d}x_2 \\ &= 2[\sigma_x^2 - C_x(\tau)] \\ &= 2\sigma_x^2[1 - \rho_x(\tau)] \end{aligned} \tag{5.81}$$

This distance vanishes for $\rho_x = 1$. Variables $\mathbf{x}(t)$ and $\mathbf{x}(t+\tau)$ are ***uncorrelated*** for the values of τ yielding $C_x(\tau) = 0$, and ***orthogonal*** for those yielding $R_x(\tau) = 0$.

5.2.7 Example

Let us consider a stochastic process $\mathbf{x}(t)$ where the values of x change after each interval of T seconds, while staying constant during that interval.

Two possible realizations $\mathbf{x}_i(t)$ and $\mathbf{x}_j(t)$ of this process are illustrated in fig. 5.8.

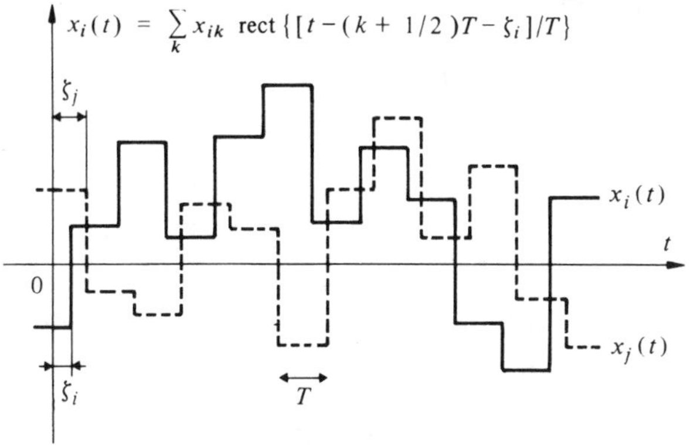

Fig. 5.8

The position ζ of the first change after the origin is not the same for every realization of the process, but is uniformly distributed over interval T: $p(\zeta) = 1/T$ for $0 \leqslant \zeta < T$. The probability density $p(x)$ is, therefore, time-invariant and the process $\mathbf{x}(t)$ is stationary.

Let us assume that values x_k, taken on the various intervals, are ***independent*** of each other and the probability density distribution $p(x)$ is arbitrary.

For $|\tau| > T$, the variables $\mathbf{x}_1 = \mathbf{x}(t)$ and $\mathbf{x}_2 = \mathbf{x}(t+\tau)$ are independent and, consequently, from (5.55) and (5.33):

$$R_x(|\tau| > T) = \mathrm{E}[\mathbf{x}(t)\mathbf{x}(t+\tau)] = \mathrm{E}[\mathbf{x}_1]\,\mathrm{E}[\mathbf{x}_2] = \mu_x^2 \tag{5.82}$$

We deduce from (5.63)

$$C_x(|\tau| > T) = 0 \tag{5.83}$$

For $|\tau| \leqslant T$, we must consider two mutually exclusive cases: the states given by variables $\mathbf{x}_1 = \mathbf{x}(t)$ and $\mathbf{x}_2 = \mathbf{x}(t+\tau)$ are either different or identical. So,

$$R_x(|\tau| \leqslant T) = \begin{cases} \mathrm{E}[\mathbf{x}^2] = \sigma_x^2 + \mu_x^2 & \text{if } \mathbf{x}_1 = \mathbf{x}_2 \\ \mathrm{E}[\mathbf{x}_1]\,\mathrm{E}[\mathbf{x}_2] = \mu_x^2 & \text{if } \mathbf{x}_1 \neq \mathbf{x}_2 \end{cases} \tag{5.84}$$

Or, in another form:

$$R_x(|\tau| \leqslant T) = (\sigma_x^2 + \mu_x^2)\,\mathrm{Prob}(x_1 = x_2) + \mu_x^2\,\mathrm{Prob}(x_1 \neq x_2) \tag{5.85}$$

The two probabilities can be evaluated by taking into account the uniform distribution of the transition instants on interval T. Let the relative random position of those transitions be ζ.

$$p(\zeta) = 1/T \quad 0 \leqslant \zeta < T \tag{5.86}$$

Event $x_1 \neq x_2$ appears when a transition occurs during interval $|\tau|$, i.e., when $\zeta < |\tau|$. We then get for $|\tau| \leqslant T$:

$$\text{Prob}\,(x_1 \neq x_2) = \text{Prob}\,(\zeta < |\tau|) = F_\zeta(|\tau|) = \int_{-\infty}^{|\tau|} p(\zeta)\mathrm{d}\zeta = \frac{|\tau|}{T} \tag{5.87}$$

On the other hand

$$\text{Prob}\,(x_1 = x_2) + \text{Prob}\,(x_1 \neq x_2) = 1 \tag{5.88}$$

thus,

$$\text{Prob}\,(x_1 = x_2) = 1 - |\tau|/T \tag{5.89}$$

Relation (5.85) becomes

$$\begin{aligned} R_x(|\tau| \leqslant T) &= (1 - |\tau|/T)(\sigma_x^2 + \mu_x^2) + (|\tau|/T)\mu_x^2 \\ &= (1 - |\tau|/T)\sigma_x^2 + \mu_x^2 \end{aligned} \tag{5.90}$$

and

$$C_x(|\tau| \leqslant T) = (1 - |\tau|/T)\sigma_x^2 \tag{5.91}$$

These various results can be combined for any τ using the notation of (1.34):

$$R_x(\tau) = \sigma_x^2 \operatorname{tri}(\tau/T) + \mu_x^2 \tag{5.92}$$

$$C_x(\tau) = \sigma_x^2 \operatorname{tri}(\tau/T) \tag{5.93}$$

and, from (5.64):

$$\rho_x(\tau) = \operatorname{tri}(\tau/T) \tag{5.94}$$

The result here is independent of the density $p(x)$ of the process. The autocorrelation and autocovariance functions of this process are illustrated by fig. 5.9.

Other examples of the theoretical evaluation of $R_x(\tau)$ are given in sections 5.3 and 5.8, or are suggested as problems for the reader.

5.2.8 Correlation and covariance matrices

In the case of a random series (or vector), where the components are random variables associated with the same process, the set of autocorrelation or co-

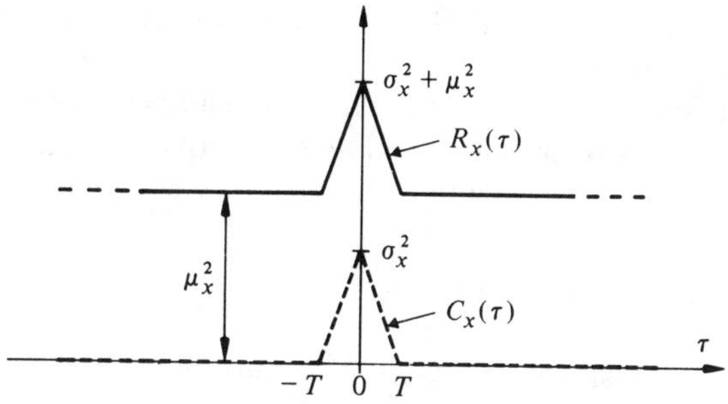

Fig. 5.9

variance (normalized or not) values can be displayed in an array, which is called the *correlation or covariance matrix*. A vector with k components $(\mathbf{x}_1, \ldots, \mathbf{x}_k)$ generates a $k \times k$ matrix with k^2 elements:

$$R_x(t_i,t_j) = \mathrm{E}[\mathbf{x}(t_i)\,\mathbf{x}\,(t_j)^{\mathrm{T}}] \tag{5.95}$$

or

$$C_x(t_i,t_j) = \mathrm{E}\,\{[\mathbf{x}(t_i) - \mu_x(t_i)][\mathbf{x}(t_j) - \mu_x(t_j)]^{\mathrm{T}}\} \tag{5.96}$$

When the vector components come from the periodic sampling (chap. 9) of a stationary process, the correlation or covariance values depend only on the difference $\tau_{ij} = (i - j)\Delta\tau$, where $\Delta\tau$ is the sampling interval and $\mu_x(t_i) = \mu_x \forall i$:

$$R_x = \begin{bmatrix} R_x(0) & R_x(\Delta\tau) & \cdots & R_x([k-1]\Delta\tau) \\ R_x(\Delta\tau) & \ddots & \ddots & \vdots \\ \vdots & \ddots & \ddots & R_x(\Delta\tau) \\ R_x([k-1]\Delta\tau & \cdots & R_x(\Delta\tau) & R_x(0) \end{bmatrix} \tag{5.97}$$

Such a matrix is symmetrical and each diagonal is composed of identical elements (Toepliz matrix).

Taking into account (5.64), the covariance matrix can be written as

$$C_x = \sigma_x^2 \cdot \begin{vmatrix} 1 & \rho_1 & \cdots & \rho_{k-1} \\ \rho_1 & \ddots & \ddots & \vdots \\ \vdots & \ddots & \ddots & \rho_1 \\ \rho_{k-1} & \cdots & \rho_1 & 1 \end{vmatrix} \tag{5.98}$$

where $\rho_{|i-j|} = \rho(|i - j|\Delta\tau)$.

Remember that the matrices (5.97) and (5.98) are identical when the process has a zero mean value ($\mu_x = 0$).

An interesting special case occurs when the samples are uncorrelated: the correlation coefficients ρ_i are cancelled and the covariance matrix becomes a simple diagonal matrix of elements σ_x^2.

5.2.9 Karhunen-Loeve expansion

The principle of a signal representation on some interval T by a linear combination of orthogonal functions (presented in chap. 3) can be extended to a stochastic process $\mathbf{x}(t)$.

The expansion (3.50):

$$\mathbf{x}(t) = \sum_{k=1}^{\infty} \boldsymbol{\alpha}_k \, \psi_k(t) \quad , \qquad t \in T \tag{5.99}$$

where the $\psi_k(t)$ functions are chosen orthonormal for simplicity sake, in this case has random coefficients (3.43):

$$\boldsymbol{\alpha}_k = \int_T \mathbf{x}(t)\psi_k^*(t)\mathrm{d}t \tag{5.100}$$

since the result depends on the realization $x(t)$ of the considered process on interval T.

The statistical properties of the $\boldsymbol{\alpha}_k$ coefficients depend on the choice of the orthogonal functions $\psi_k(t)$. For example, we could use a Fourier series (3.74), but it can be demonstrated [24] that those coefficients are correlated.

A better (i.e., faster converging) representation can be obtained when coefficients are uncorrelated (linearly independent). As shown in sub-section 5.2.6, being uncorrelated implies the covariance cancellation. Consider, for simplification, a zero mean value process, for which covariance and correlation, on the one hand, and orthogonality and no correlation, on the other hand, are identical. The problem now consists of defining the $\psi_k(t)$ functions, so that the orthogonality of the $\boldsymbol{\alpha}_k$ coefficients is ensured:

$$\mathrm{E}[\boldsymbol{\alpha}_k\boldsymbol{\alpha}_l^*] = \begin{cases} \sigma_k^2 & \text{if} \quad k = l \\ 0 & \text{if} \quad k \neq l \end{cases} \tag{5.101}$$

If we insert (5.100) into (5.101), exchange the integration and mathematical expectation operations, and consider that the $\psi_k(t)$ functions are orthonomal, then the orthogonality of the $\boldsymbol{\alpha}_k$ coefficients, as defined by (5.101), is achieved for a stationary process only when

$$\int_T R_x(t - \tau)\psi_k(\tau)\,\mathrm{d}\tau = \sigma_k^2\psi_k(t) \tag{5.102}$$

Such a representation with $\psi_k(t)$ functions complying with the (5.102) condition is called a Karhunen-Loeve expansion. Its interest is mainly theoretical (optimal convergence). The resolution of the integral equation (5.102) is sophisticated [25,26]. It can be achieved digitally [vol. XX].

5.3 POWER SPECTRAL DENSITY

5.3.1 Periodogram and power spectal density of a random process

Consider a member $x_i(t)$ of a set of random signals, which are different realizations of the same stationary stochastic process ($i = 1,2, \ldots$). The Fourier transform of $x_i(t)$ does not generally exist, since the convergence condition:

$$\int_{-\infty}^{\infty} |x_i(t)| \mathrm{d}t < \infty \tag{5.103}$$

is not satisfied.

However, if we consider (fig. 5.10) an aperiodic function $x_i(t,T)$, equal to $x_i(t)$ on interval $-T/2<t<T/2$ and equal to zero for any other value of t, we can write as in sub-section 4.1.6:

$$x_i(t) = \lim_{T\to\infty} x_i(t,T) \tag{5.104}$$

with

$$x_i(t,T) = x_i(t) \cdot \mathrm{rect}(t/T) \tag{5.105}$$

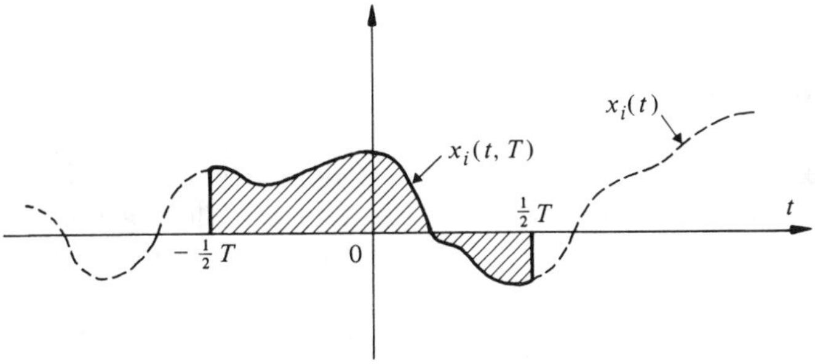

Fig. 5.10

The function $x_i(t,T)$ is generally Fourier transformable.

$$X_i(f,T) = \int_{-\infty}^{\infty} x_i(t,T)\exp(-j2\pi ft)dt$$
$$= \int_{-T/2}^{T/2} x_i(t)\exp(-j2\pi ft)dt \quad (5.106)$$

The signal average power contained on interval T is given by

$$P_{xi}(T) = \frac{1}{T}\int_{-\infty}^{\infty} |X_i(f,T)|^2 df = \int_{-\infty}^{\infty} \Phi_{xi}(f,T)df \quad (5.107)$$

The power spectral density of $x_i\ (t,\ T)$:

$$\Phi_{xi}\ (f,\ T) = T^{-1}\ |\ X_i\ (f,\ T)\ |^2 \quad \mathbf{(5.108)}$$

is sometimes referred to as *periodogram*. This periodogram is a ***random variable,*** since it is different for each member of the signal set $\{x_i(t)\}$ of the process.

In order to characterize the spectral properties of the ensemble (complete set), measured on interval T, we shall introduce the concept of spectral density expectation:

$$\Phi_x(f,T) = E[\Phi_{xi}(f,T)] \quad (5.109)$$

So, for each signal $x_i(t,T)$, $i = 1, 2, \ldots$, we have a periodogram $\Phi_{xi}(f,T)$. Hence, there corresponds to the set of $x_i(t,T)$, for ***each frequency value f,*** a set of $\Phi_{xi}(f,T)$ periodogram values ***randomly distributed around the theoretical mean value*** $\Phi_x(f,T)$.

The *power spectral density* (sometimes simply called *power spectrum*) of a stochastic process $\mathbf{x}(t)$ can be finally defined as the limit for $T \to \infty$ the statistical mean $\Phi_x(f,T)$:

$$\Phi_x(f) = \lim_{T\to\infty} \{\Phi_x(f,T) = E[\Phi_{xi}(f,T)]\} = \lim_{T\to\infty} E\left[\frac{|X_i(f,T)|^2}{T}\right] \quad \mathbf{(5.110)}$$

5.3.2 Observations

If, for example, $x(t)$ is a voltage, the periodogram and power spectral density (5.110) unit is V^2/Hz. In fact, as a result of introducing the mathematical expectation concept, definition (5.110) is similar to (4.107) for deterministic signals.

Obviously, we cannot experimentally measure $\Phi_x(f)$. Thus, it is just a theoretical concept. Only the periodogram $\Phi_{xi}(f,T)$ can be experimentally measured. Because of its random nature, the $\Phi_{xi}(f,T)$ measures, obtained from different realizations of a stochastic process, yield slightly different results. If the observed signal is stationary and ergodic, it is possible to get a good approximation of the spectral density $\Phi_x(f,T)$ by averaging various periodograms measured on different segments T of a single signal.

Some aspects of the experimental spectral analysis of a stochastic process are studied in chapter 12. The more specific problems related to the digital approach are presented in [vol. XX].

5.3.3 Wiener-Kinchin theorem

The power spectral density of a wide-sense stationary stochastic process is the Fourier transform of its statistical autocorrelation function:

$$\Phi_x(f) = \mathrm{F}\{R_x(\tau)\} = \int_{-\infty}^{\infty} R_x(\tau)\exp(-\mathrm{j}2\pi f\tau)\mathrm{d}\tau \qquad \mathbf{(5.111)}$$

5.3.4 Demonstration

The periodogram (5.108) of a real random signal, taking into account (5.105), can be written as

$$\Phi_{xi}(f,T) = \frac{1}{T} X_i^*(f,T)X_i(f,T) \qquad (5.112)$$

$$= \frac{1}{T}\int_{-\infty}^{\infty}\int_{-\infty}^{\infty} x_i(t)x_i(t')\mathrm{rect}(t/T)\mathrm{rect}(t'/T)\exp[-\mathrm{j}2\pi f(t' - t)]\mathrm{d}t\mathrm{d}t'$$

Because the process is stationary, only the difference $\tau = t' - t$ is significant. Substituting t' by $t + \tau$ and $\mathrm{d}t'$ by $\mathrm{d}\tau$, and exchanging integration and mathematical expectation operations, we get (5.109) as

$$\mathrm{E}[\Phi_{xi}(f,T)] = \frac{1}{T}\int_{-\infty}^{\infty}\int_{-\infty}^{\infty} \mathrm{E}[x_i(t)x_i(t + \tau)]\mathrm{rect}\left(\frac{t}{T}\right)\mathrm{rect}\left(\frac{t+\tau}{T}\right) \exp(-\mathrm{j}2\pi f\tau)\mathrm{d}t\mathrm{d}\tau \qquad (5.113)$$

However, by definition

$$\mathrm{E}[x_i(t)x_i(t + \tau)] = \mathrm{E}[\mathbf{x}(t)\mathbf{x}(t + \tau)] = R_x(\tau) \qquad (5.114)$$

Moreover, the autocorrelation function of a rectangular pulse (see sub-sect. 4.2.20) is

$$\frac{1}{T}\int_{-\infty}^{\infty} \mathrm{rect}\left(\frac{t}{T}\right)\mathrm{rect}\left(\frac{t+\tau}{T}\right)\mathrm{d}t = \mathrm{tri}\left(\frac{\tau}{T}\right) \qquad (5.115)$$

Hence,

$$\Phi_x(f,T) = \mathrm{E}[\Phi_{xi}(f,T)] = \int_{-\infty}^{\infty} R_x(\tau)\mathrm{tri}(\tau/T)\exp(-\mathrm{j}2\pi f\tau)\mathrm{d}\tau \tag{5.116}$$

and, due to (5.110), we have

$$\Phi_x(f) = \lim_{T\to\infty} \mathrm{E}[\Phi_{xi}(f,T)] = \int_{-\infty}^{\infty} R_x(\tau)\exp(-\mathrm{j}2\pi f\tau)\mathrm{d}\tau \tag{5.117}$$

5.3.5 Observations

Thanks to the Wiener-Kinchin theorem, signal theory has an uniform definition of the (energy or power) spectral density of a signal, be it deterministic or random or any combination of the two: it is its autocorrelation function Fourier transform. This explains the importance of the autocorrelation concept. Table 5.11 summarizes the fundamental relations of spectral analysis.

When, for a random signal, $\Phi_x(f)$ is known, its autocorrelation function can be deduced by means of the inverse Fourier transform:

$$R_x(\tau) = \int_{-\infty}^{\infty} \Phi_x(f)\exp(\mathrm{j}2\pi f\tau)\mathrm{d}f \tag{5.118}$$

A result analogous to (4.102) for finite average power deterministic signals is obtained for $\tau = 0$. With (5.68) and the ergodic condition (5.60), the average power of a random signal is equivalently given by

$$P_x = \lim_{T\to\infty} \frac{1}{T}\int_{-T/2}^{T/2} |x(t)|^2\mathrm{d}t = R_x(0) = \sigma_x^2 + \mu_x^2 = \int_{-\infty}^{\infty} \Phi_x(f)\mathrm{d}f \tag{5.119}$$

Function $\Phi_x(f)$ is the frequency distribution of the random signal power $P_x = \sigma_x^2 + \mu_x^2$. Hence, this is a real non-negative function:

$$\Phi_x(f) \geqslant 0 \tag{5.120}$$

Because of the Hermitian symmetry (5.80) of the autocorrelation function, this property is valid, whether the signal is real or complex.

When the signal is ***real,*** $\mathrm{R}_x(\tau) = R_x(-\tau)$ and $\Phi_x(f)$ is also an even function:

$$\Phi_x(f) = \Phi_x(-f) \tag{5.121}$$

We can, thus, also introduce a unilateral spectral representation (see sub-sect. 4.5.2) by writing:

$$\Phi_x^+(f) = 2\epsilon(f)\Phi_x(f) \tag{5.122}$$

Table 5.11

Finite energy signal

$$x(t) \xleftrightarrow{\text{TF}} X(f)$$

$$\downarrow \qquad \downarrow$$

$$\mathring{\varphi}_x(\tau) = \int_{-\infty}^{\infty} x(t)x(t+\tau)\mathrm{d}t \xleftrightarrow{\text{TF}} \mathring{\Phi}_x(f) = |X(f)|^2$$

Finite average power signal

- deterministic signal

$$x(t) \xleftrightarrow{\text{TF?}} X(f)$$

$$\downarrow$$

$$\varphi_x(\tau) = \lim_{T\to\infty} \frac{1}{T}\int_{-T/2}^{T/2} x(t)x(t+\tau)\mathrm{d}t \xleftrightarrow{\text{TF}} \Phi_x(f) \overset{?}{=} \lim_{T\to\infty} \frac{1}{T}|X(f,T)|^2$$

$$X(f,T) = \mathrm{F}\{x(t,T) = x(t)\mathrm{rect}(t/T)\}$$

- particular case: T-periodic signal

$$x(t) = x(t+mT) \xleftrightarrow{\text{TF}} X(f) = \sum_n X_n\delta(f - n/T)$$

$$\downarrow$$

$$\varphi_x(\tau) = \frac{1}{T}\int_{-T/2}^{T/2} x(t)x(t+\tau)\mathrm{d}t \qquad X_n = \frac{1}{T}\int_{-T/2}^{T/2} x(t)\exp(-\mathrm{j}2\pi nt/T)\mathrm{d}t$$

$$= \sum_n |X_n|^2\exp(\mathrm{j}2\pi n\tau/T) \xleftrightarrow{\text{TF}} \Phi_x(f) = \sum_n |X_n|^2\delta(f - n/T)$$

- random signal (stationary and ergodic)

$$x(t)$$

$$\downarrow$$

$$\varphi_x(\tau) = \lim_{T\to\infty} \frac{1}{T}\int_{-T/2}^{T/2} x(t)x(t+\tau)\mathrm{d}t$$

$$\equiv R_x(\tau) = \mathrm{E}[\mathbf{x}(t)\mathbf{x}(t+\tau)] \xleftrightarrow{\text{TF}} \Phi_x(f) = \lim_{T\to\infty} \mathrm{E}\left[\frac{1}{T}|X(f,T)|^2\right]$$

$$X(f,T) = \mathrm{F}\{x(t,T) = x(t)\mathrm{rect}(t/T)\}$$

From (5.63), we can write the autocorrelation function as the sum of the autocovariance functions and a constant μ_x^2:

$$R_x(\tau) = C_x(\tau) + \mu_x^2 \qquad (5.123)$$

From (5.11), we have then

$$\Phi_x(f) = \mathrm{F}\{C_x(\tau)\} + \mu_x^2\delta(f) \qquad (5.124)$$

Because the autocovariance function is the autocorrelation function of a centered process $\mathbf{x}_0(t) = \mathbf{x}(t) - \mu_x$, i.e., with a zero mean value, its Fourier transform has no delta function at the origin (except for some special cases as the example in sub-sect. 5.1.15, which is not ergodic [50]). The term $F\{C_x(\tau)\}$, that can be noted $\Phi_{x0}(f)$, corresponds to the power spectral density of the process random fluctuations ***relative*** to the mean value μ_x. The presence in the power spectrum of a delta function $\mu_x^2\delta(f)$ at the origin reveals the existence of a non-zero mean value (i.e., dc value).

It can be demonstrated [25] that, if the statistical mean value of $\Phi_x(f,T)$, when $T \rightarrow \infty$, ultimately tends toward the power spectral density, as evidenced by (5.110), the variance of the periodogram does not necessarily tend toward zero. Therefore, during a random (even ergodic) signal observation, the measured periodogram is not necessarily a good spectral estimation, even if we increase the observation duration T (sub-sect. 12.1.8). It is thus more advisable to run many (if possible, independent) measurements and to average the obtained periodogram. This method is used in digital spectral analysis systems (sub-sect. 12.2.3). Another approach, used in both analog and digital analyzers, is to smooth the measured periodogram with an appropriate filter (sub-sect. 12.1.11). Those various methods must, however, be applied with some care, if we want to avoid biased measurements.

The effect of a limited observation duration T can be directly deduced from the Fourier transform of the $R(\tau) \cdot \mathrm{tri}(\tau/T)$ product. The statistical mean value $\Phi_x(f,T)$ of the periodogram thus corresponds to the ***convolution*** (analogous to filtering) of the theoretical spectral density $\Phi_x(f)$ and a $T\mathrm{sinc}^2(Tf)$ function. Thus, the spectral resolution directly depends on the observation duration (sub-sect. 12.1.5).

5.3.6 Example

The sine wave with a random phase $x(t) = A \sin(2\pi f_0 t + \alpha)$ of sub-section 5.1.16 has an autocorrelation function (5.29):

$$R_x(\tau) = \tfrac{1}{2}A^2 \cos(2\pi f_0 \tau) \tag{5.125}$$

From (5.111) and (4.117), its power spectral density is

$$\Phi_x(f) = (A^2/4)[\delta(f + f_0) + \delta(f - f_0)] \tag{5.126}$$

and

$$P_x = R_x(0) = \int_{-\infty}^{\infty} \Phi_x(f)\mathrm{d}f = \frac{A^2}{2} \tag{5.127}$$

The spectral density $\Phi_x(f,T)$ of a limited observation is equal to

$\Phi_x(f) * T\ \mathrm{sinc}^2(Tf) = \tfrac{1}{4}A^2T\{\mathrm{sinc}^2[T(f + f_0)] + \mathrm{sinc}^2[T(f - f_0)]\}$.

5.3.7 Example: clocked NRZ binary signal

This kind of signal (fig. 5.12) plays an important role, especially in telecommunications, since it translates into its simplest form (NRZ mode = non-return to zero mode) binary coded information, as provided by a digital system. Binary symbols 0 and 1 are generated at a $1/T$ bit/s rate and respectively correspond to the x_0 and x_1 signal levels.

If the time origin is unknown, i.e., if the ζ variable is uniformly distributed over the clock period T, this signal is stationary.

Assuming that successive symbols are statistically independent of each other, the signal $x(t)$ (fig. 5.12) is a special case of the type studied in subsection 5.2.7. Its probability density is, hence, from (5.32):

$$p(x) = \text{Prob}(x_0)\delta(x - x_0) + \text{Prob}(x_1)\delta(x - x_1) \tag{5.128}$$

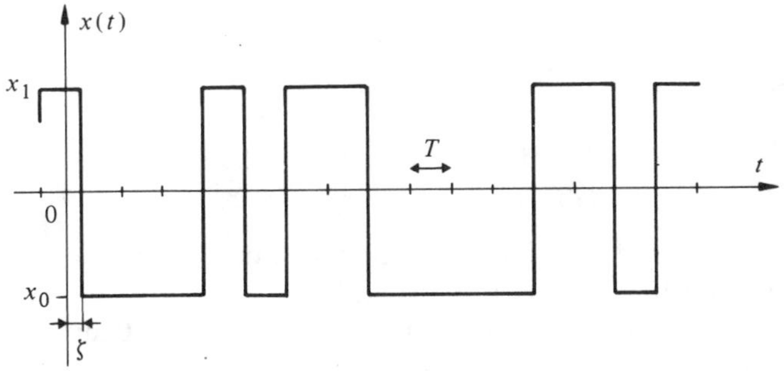

Fig. 5.12

with $\text{Prob}(x_0) + \text{Prob}(x_1) = 1$.

We derive:

$$\mu_x = \text{E}[\mathbf{x}] = x_0\text{Prob}(x_0) + x_1\text{Prob}(x_1) \tag{5.129}$$

$$P_x = \text{E}[\mathbf{x}^2] = \sigma_x^2 + \mu_x^2 = x_0^2\text{Prob}(x_0) + x_1^2\text{Prob}(x_1) \tag{5.130}$$

The autocorrelation function, given by (5.92) and illustrated in fig. 5.9, has a triangular shape and depends only on the clock period T, on the squared mean value μ_x^2, and on the variance σ_x^2:

$$R_x(\tau) = \sigma_x^2\text{tri}(\tau/T) + \mu_x^2 \tag{5.131}$$

The power spectral density of such a signal, represented in fig. 5.13, from (5.111) and (4.35), is

$$\Phi_x(f) = \sigma_x^2 T\,\text{sinc}^2(fT) + \mu_x^2\delta(f) \tag{5.132}$$

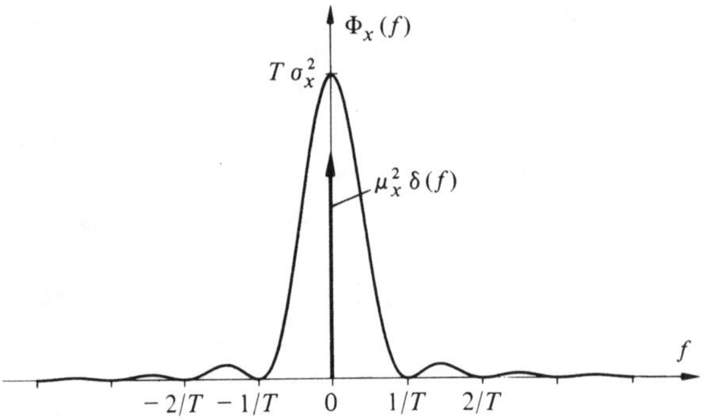

Fig. 5.13

An interesting special case occurs when binary symbols 0 and 1 are equally likely, i.e., $\text{Prob}(x_0) = \text{Prob}(x_1) = 1/2$; thus,

$$\mu_x = \tfrac{1}{2}(x_0 + x_1) \tag{5.133}$$

$$\mu_x^2 = \tfrac{1}{4}(x_0 + x_1)^2 \tag{5.134}$$

$$\sigma_x^2 = \tfrac{1}{4}(x_0 - x_1)^2 \tag{5.135}$$

$$P_x = \tfrac{1}{2}(x_0^2 + x_1^2) \tag{5.136}$$

Moreover, if the signal is antipolar, i.e., if $x_1 = -x_0 = A$: $\mu_x = 0$; the dc value vanishes and $\sigma_x^2 = A^2 = P_x$.

If, on the contrary, the signal is unipolar: $x_0 = 0$ and $x_1 = A$, and we get a dc value $\mu_x = A/2$, a variance $\sigma_x^2 = A^2/4$, and a total power $P_x = A^2/2$.

Examination of fig. 5.13 shows that 91 percent of the clocked NRZ binary signal power is contained in the $0 < |f| < 1/T$ frequency range.

5.3.8 Example: clocked diphase binary signal

The spectrum of a signal carrying binary information can be modified by the representation mode of symbols 0 and 1. The so-called diphase mode (or Manchester code) is such a modification. Symbols 0 and 1 are here represented by the sign of the transition which is always present at the center of the corresponding clock period: negative, for example, for symbol 1 and positive for symbol 0. To simplify, consider an antipolar signal $y_1 = -y_0 = A$ (fig. 5.14).

Assume symbols 0 and 1 are again independent and equally likely. As was the case previously, the signal comes from a stationary process if variable ζ is uniformly distributed over clock period T.

The autocorrelation function of this signal can be evaluated by relation (5.57), where indices 1 and 2 represent the variables separated by interval τ,

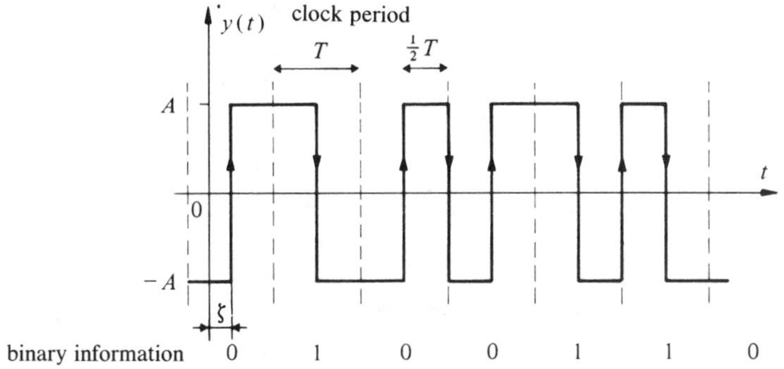

Fig. 5.14

and indices i and j are the two possible states

$$R_y(\tau) = \sum_{i=0}^{1} \sum_{j=0}^{1} y_{1i}y_{2j} \text{ Prob } (y_{1i},y_{2j};\tau) \tag{5.137}$$

with

$$y_{1i}y_{2j} = \begin{cases} A^2 & \text{if } i = j \\ -A^2 & \text{if } i \neq j \end{cases} \tag{5.138}$$

Or, otherwise stated,

$$\begin{aligned} R_y(\tau) &= A^2[\text{Prob}(y_{1i} = y_{2j};\tau) - \text{Prob}(y_{1i} \neq y_{2j};\tau)] \\ &= A^2[1 - 2 \text{ Prob}(y_{1i} \neq y_{2j};\tau)] \end{aligned} \tag{5.139}$$

For $|\tau| > T$, the two probabilities are obviously equal because of the independence and equal likelihood of symbols 0 and 1. Hence,

$$R_x(|\tau| > T) = 0 \tag{5.140}$$

For $|\tau| \leqslant T$, the probability that variables $\mathbf{y}_1$ and $\mathbf{y}_2$ have different values is equal to the probability of having a single transition on interval τ.

Consider the two time domains $|\tau| \leqslant T/2$ and $T/2 \leqslant |\tau| \leqslant T$. For $|\tau| \leqslant T/2$, the probability of having a single transition is zero if $\mathbf{y}_1$ is already in the second half of a symbol clock period and if the following binary symbol is different. The first case appears with a probability of ½ because of the uniform distribution of ζ over a clock period. The probability of a symbol change is also ½ because we assumed equally likely symbols. The joint probability of those two independent events is therefore ¼. The complementary situation, therefore, has a probability of ¾. In this case, the probability of having a transition is proportional to $|\tau|$ and varies from zero to one for $|\tau|$ varying from zero to $T/2$. Thus,

$$\text{Prob}(y_{1i} \neq y_{2j}; |\tau| \leqslant T/2) = \tfrac{3}{4}|\tau|/(\tfrac{1}{2}T) = (3/2)|\tau|/T \quad (5.141)$$

and, by insertion into (5.139)

$$R_y(|\tau| \leqslant T/2) = A^2(1 - 3|\tau|/T) \quad (5.142)$$

For $T/2 \leqslant |\tau| \leqslant T$, we can have zero, one, or even two transitions. The probability of having a single transition is equal to one minus the probabilities of having zero or two. The first case can only occur if $\mathbf{y}_1$ is in the second half of a symbol clock period and if the following symbol is different. Its probability is ¼ for $|\tau| = T/2$ and linearly decreases with $|\tau|$ down to zero when $|\tau| = T$.

$$\text{Prob}(\text{zero transition in } T/2 \leqslant |\tau| \leqslant T)$$

$$= \tfrac{1}{4}(T - |\tau|)/\tfrac{1}{2}T = \tfrac{1}{2}(1 - |\tau|/T) \quad (5.143)$$

Two transitions in $T/2 \leqslant |\tau| \leq T$ can only occur if the same symbol is repeated. The corresponding probability is zero for $|\tau| = T/2$ and linearly increases with $|\tau|$ to reach ½ for $|\tau| = T$.

$$\text{Prob}(\text{two transitions in } T/2 \leqslant |\tau| \leqslant T) = |\tau|/T - \tfrac{1}{2} \quad (5.144)$$

Thus,

$$\text{Prob}(y_{1i} \neq y_{2j}; T/2 \leqslant |\tau| \leqslant T) = 1 - \tfrac{1}{2}|\tau|/T \quad (5.145)$$

and, from (5.139):

$$R_y(T/2 \leqslant |\tau| \leqslant T) = A^2[|\tau|/T - 1] \quad (5.146)$$

Results (5.140), (5.142), and (5.146) are represented in fig. 5.15. They can be combined, due to (1.34), into a general relation:

$$R_y(\tau) = A^2[2\text{tri}(2\tau/T) - \text{tri}(\tau/T)] \quad (5.147)$$

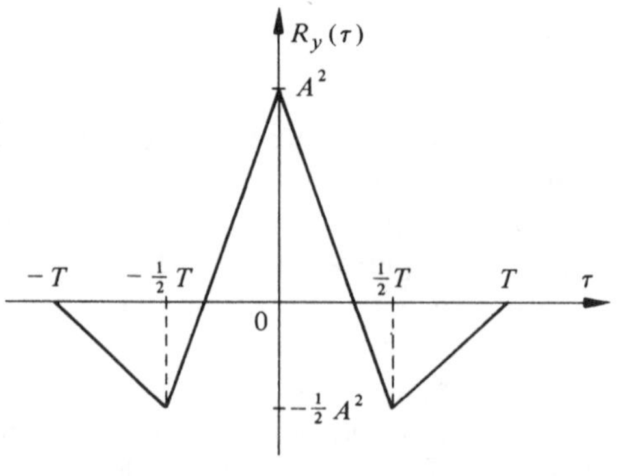

Fig. 5.15

The power spectral density is easily deduced from (5.147) according to (4.35):

$$\Phi_y(f) = A^2T[\mathrm{sinc}^2(fT/2) - \mathrm{sinc}^2(fT)] \tag{5.148}$$

In fig. 5.16, we compare this spectral density with $\Phi_x(f) = A^2T\,\mathrm{sinc}^2(fT)$ obtained in (5.132) for the antipolar NRZ mode.

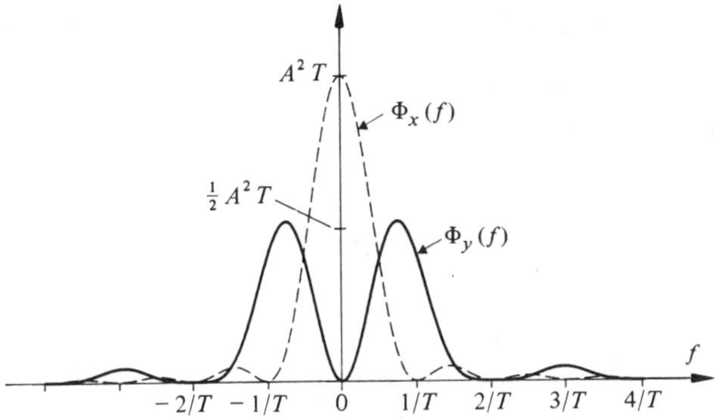

Fig. 5.16

5.3.9 General relation for data transmission signals

A general expression for the spectrum of a signal $x(t)$ carrying binary coded information can be established [66], assuming a statistical independence of the symbols. If $s_0(t) = \mathrm{F}^{-1}[S_0(f)]$ and $s_1(t) = \mathrm{F}^{-1}[S_1(f)]$ are finite energy signals, of duration T equal to the clock period, corresponding to logic symbols 0 and 1, respectively, and if p_0 and p_1 are their probabilities of occurrence:

$$\Phi_x(f) = \frac{1}{T}(p_0|S_0(f)|^2 + p_1|S_1(f)|^2) - \frac{1}{T}|p_0S_0(f) + p_1S_1(f)|^2 + \frac{1}{T^2}|p_0S_0(f) + p_1S_1(f)|^2 \cdot \sum_{n=-\infty}^{\infty} \delta(f - n/T) \tag{5.149}$$

In the special case of equally likely symbols and antipolar associated signals $[s_1(t) = -s_0(t) = s(t)]$, relation (5.149) becomes

$$\Phi_x(f) = T^{-1}|S(f)|^2 \tag{5.150}$$

5.3.10 White noise

A stochastic process $\mathbf{x}(t)$, the power spectral density of which is constant for any value of f, is called *white noise*, by analogy with the white light made up of radiations of all wavelengths:

$$\Phi_x(f) = \tfrac{1}{2}\eta \text{ for } |f| < \infty \tag{5.151}$$

Such a model is purely theoretical because such a process cannot exist. Its average power is, indeed, infinite. However, this concept is quite useful in many cases where the exact spectrum can, as a first approximation, be replaced by a constant spectrum (chap. 6).

According to (4.74), the white noise autocorrelation function (fig. 5.17) is a delta function of weight $\tfrac{1}{2}\eta$:

$$R_x(\tau) = \mathrm{F}^{-1}\{\tfrac{1}{2}\eta\} = \tfrac{1}{2}\eta\delta(\tau) \tag{5.152}$$

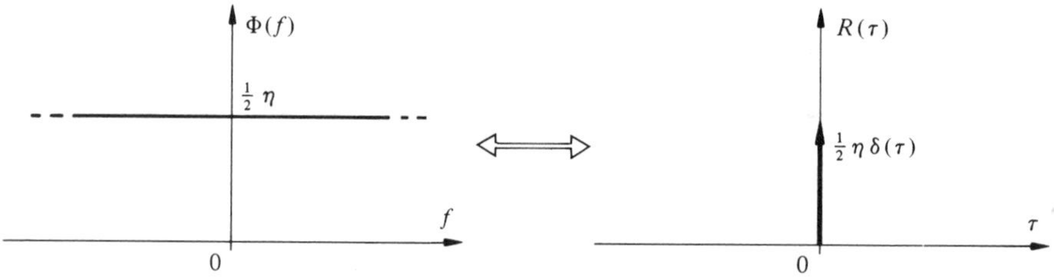

Fig. 5.17

This result shows that, for white noise, stochastic variables $\mathbf{x}(t_1)$ and $\mathbf{x}(t_2)$ are totally uncorrelated, regardless of $\tau = t_2 - t_1 \neq 0$. If the process is Gaussian (sect. 5.7), being ***uncorrelated*** implies a statistical independence.

We also call *band-limited white noise* a stochastic process having a spectrum which is constant over a finite frequency band and zero elsewhere:

$$\Phi(f) = \begin{cases} \tfrac{1}{2}\eta & \text{for } f_1 < |f| < f_2 \\ 0 & \text{otherwise} \end{cases} \tag{5.153}$$

We can consider such a process as the output of an ideal $B = f_2 - f_1$ bandpass filter excited by an ideal white noise with a spectral density equal to $\eta/2$.

The mean value of this process is zero and its total power is finite:

$$P = \sigma^2 = R(0) = \int_{-\infty}^{\infty} \Phi(f)\mathrm{d}f = \eta B \tag{5.154}$$

This spectrum is of the ***lowpass*** kind if $f_1 = 0$ and $f_2 = B$:

$$\Phi_1(f) = \tfrac{1}{2}\eta \ \mathrm{rect}[f/(2B)] \tag{5.155}$$

hence, from (4.26):

$$R_1(\tau) = \mathrm{F}^{-1}\{\Phi_1(f)\} = \eta B \ \mathrm{sinc}(2B\tau) \tag{5.156}$$

Here, variables $\mathbf{x}(t_1)$ and $\mathbf{x}(t_2)$ are uncorrelated if they are separated by an interval τ equal to $K/(2B)$ with $K = 1, 2, \ldots$

This spectrum is of the ***bandpass*** kind if $f_1 = f_0 - B/2$ and $f_2 = f_0 + B/2$:

$$\Phi_2(f) = \tfrac{1}{2}\eta \ \mathrm{rect}[(f + f_0)/B] + \tfrac{1}{2}\eta \ \mathrm{rect}[(f - f_0)/B] \tag{5.157}$$

hence,

$$R_2(\tau) = \eta B \ \mathrm{sinc}\ (B\tau) \cos\ (2\pi f_0 \tau) \tag{5.158}$$

The spectral densities and autocorrelation functions of those two band-limited white noises are represented in fig. 5.18.

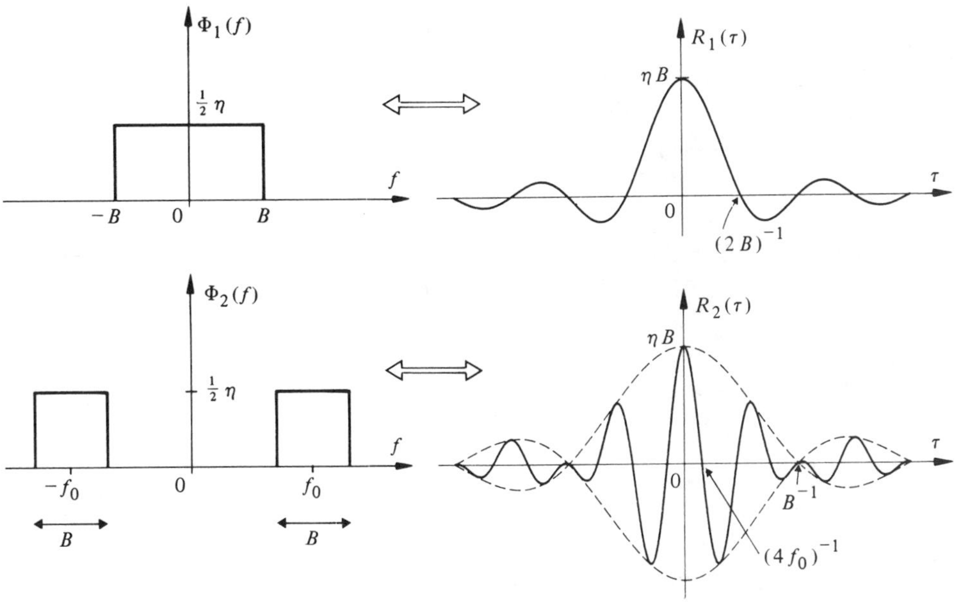

Fig. 5.18

5.4 CROSSCORRELATION FUNCTIONS AND CROSS-POWER SPECTRAL DENSITIES

5.4.1 Crosscorrelation functions

Consider two stationary stochastic processes $\mathbf{x}(t)$ and $\mathbf{y}(t)$. Let $\mathbf{x}_1 = \mathbf{x}(t)$, $\mathbf{x}_2 = \mathbf{x}(t + \tau)$, $\mathbf{y}_1 = \mathbf{y}(t)$, and $\mathbf{y}_2 = \mathbf{y}(t + \tau)$.

The joint distribution of the two stochastic variables $\mathbf{x}_1$ and $\mathbf{y}_2$ depends only on the time difference τ: $p(x_1,y_2;t_1,t_2) = p(x_1,y_2;\tau)$.

The *statistical crosscorrelation functions* of $\mathbf{x}(t)$ and $\mathbf{y}(t)$ are defined as follows when the processes are described by ***complex*** functions:

$$R_{xy}(\tau) = \mathrm{E}[\mathbf{x}^*(t)\mathbf{y}(t + \tau)] = \mathrm{E}[\mathbf{x}_1^*\mathbf{y}_2] \tag{5.159}$$

$$R_{yx}(\tau) = \mathrm{E}[\mathbf{y}^*(t)\mathbf{x}(t + \tau)] = \mathrm{E}[\mathbf{y}_1^*\mathbf{x}_2] \tag{5.160}$$

They are also complex functions.

Usually, the processes are ***real,*** and hence these functions are also real and equal to

$$R_{xy}(\tau) = \mathrm{E}[\mathbf{x}_1\mathbf{y}_2] = \int_{-\infty}^{\infty}\int_{-\infty}^{\infty} x_1y_2p(x_1,y_2;\tau)\mathrm{d}x_1\mathrm{d}y_2 \tag{\textbf{5.161}}$$

$$R_{yx}(\tau) = \mathrm{E}[\mathbf{y}_1\mathbf{x}_2] = \int_{-\infty}^{\infty}\int_{-\infty}^{\infty} y_1x_2p(y_1,x_2;\tau)\mathrm{d}y_1\mathrm{d}x_2 \tag{\textbf{5.162}}$$

For ergodic processes, definitions (5.161) and (5.162) are identical to time-crosscorrelation functions:

$$R_{xy}(\tau) \equiv \varphi_{xy}(\tau) = \lim_{T\to\infty} \frac{1}{T}\int_{-T/2}^{T/2} x(t)y(t + \tau)\mathrm{d}t = x(-\tau) \mathbin{\overline{*}} y(\tau) \tag{5.163}$$

and

$$R_{yx}(\tau) \equiv \varphi_{yx}(\tau) = \lim_{T\to\infty} \frac{1}{T}\int_{-T/2}^{T/2} y(t)x(t + \tau)\mathrm{d}t = y(-\tau) \mathbin{\overline{*}} x(\tau) \tag{5.164}$$

with the relation:

$$R_{xy}(\tau) = R_{yx}(-\tau) \tag{5.165}$$

In the case of complex processes, the Hermitian symmetry gives

$$R_{xy}(\tau) = R_{yx}^*(-\tau)$$

5.4.2 Cross-power spectral density

The Fourier transforms of crosscorrelation functions are generally complex functions [$R_{xy}(\tau)$ and $R_{yx}(\tau)$ are not even functions of τ]:

$$\Phi_{xy}(f) = \int_{-\infty}^{\infty} R_{xy}(\tau)\, e^{-j2\pi f\tau} d\tau \tag{5.166}$$

$$\Phi_{yx}(f) = \int_{-\infty}^{\infty} R_{yx}(\tau)\, e^{-j2\pi f\tau} d\tau \tag{5.167}$$

with $\Phi_{xy}(f) = \Phi_{yx}^*(f)$.

These transforms are called *cross-power spectral densities,* or simply *cross-power spectra.*

By analogy with (5.110), the cross-power spectral density corresponds to the limit:

$$\Phi_{xy}(f) = \lim_{T\to\infty} E\left[\frac{|X^*(f,T)Y(f,T)|}{T}\right] \tag{5.168}$$

5.4.3 Crosscovariance functions

The *crosscovariance functions* are the crosscorrelation functions of centered processes $\mathbf{x}(t) - \mu_x$ and $\mathbf{y}(t) - \mu_y$:

$$C_{xy}(\tau) = E[(\mathbf{x}_1 - \mu_x)(\mathbf{y}_2 - \mu_y)] = R_{xy}(\tau) - \mu_x\mu_y \tag{5.169}$$

$$C_{yx}(\tau) = E[(\mathbf{y}_1 - \mu_x)(\mathbf{x}_2 - \mu_y)] = R_{yx}(\tau) - \mu_x\mu_y \tag{5.170}$$

The *crosscorrelation coefficient* is defined as

$$\rho_{xy}(\tau) = C_{xy}(\tau)/(\sigma_x\sigma_y) \tag{5.171}$$

and it measures the degree of linear dependence between variables $\mathbf{x}_1$ and $\mathbf{y}_2$.

5.4.4 Properties

By analogy with sub-section 5.2.4, we show that

$$|R_{xy}(\tau)| \leqslant \sqrt{R_x(0)R_y(0)} \tag{5.172}$$

$$|C_{xy}(\tau)| \leqslant \sqrt{C_x(0)C_y(0)} = \sigma_x\sigma_y \tag{5.173}$$

and

$$|\rho_{xy}(\tau)| \leqslant 1 \tag{5.174}$$

For values of $\tau = t_2 - t_1$ where $C_{xy}(\tau) = 0$, random variables $\mathbf{x}(t_1)$ and $\mathbf{y}(t_2)$ are ***uncorrelated***. They are ***orthogonal*** for τ values that imply $R_{xy}(\tau) = 0$. We have orthogonality and uncorrelation simultaneously if at least one of the processes has a zero mean value.

If $\mathbf{x}(t)$ and $\mathbf{y}(t)$ are statistically ***independent,*** we have for any τ:

$$C_{xy}(\tau) = C_{yx}(\tau) = \rho_{xy}(\tau) = 0 \tag{5.175}$$

and

$$R_{xy}(\tau) = R_{yx}(\tau) = \mu_x \mu_y \tag{5.176}$$

On the contrary, ***crosscovariance cancellation does not necessarily imply independence*** [except if the processes are Gaussian (sect. 5.7)].

For centered processes, we can show [65] that

$$|\Phi_{xy}(f)|^2 \leqslant \Phi_x(f)\Phi_y(f) \tag{5.177}$$

5.4.5 Coherence function

The *coherence function* is the ratio:

$$\Gamma_{xy}(f) = \frac{|\Phi_{xy}(f)|^2}{\Phi_x(f)\Phi_y(f)} \tag{5.178}$$

From (5.177), we have

$$0 \leqslant \Gamma_{xy}(f) \leqslant 1 \tag{5.179}$$

The coherence function plays a role in the frequency domain analogous to the correlation coefficient in the time domain: the $\mathbf{x}(t)$ and $\mathbf{y}(t)$ processes are noncoherent, i.e., uncorrelated for frequency f_* if $\Gamma_{xy}(f_*) = 0$. If $\Gamma_{xy}(f) = 1$ for any f, the processes are totally coherent.

The coherence function can be used [65, 67] to check the linearity of a system binding $\mathbf{y}(t)$ to $\mathbf{x}(t)$. It is a measure of the output signal power portion, at frequency f, that is due to the input signal. For perfect linearity, $\Gamma_{xy}(f) = 1$ when no disturbing noise appears (see problem 8.5.8).

A coherence function that is not 1, indicates either the existence of additional noise, a nonlinear relationship between $\mathbf{x}(t)$ and $\mathbf{y}(t)$, or that $\mathbf{y}(t)$ does not depend on the sole $\mathbf{x}(t)$ excitation.

5.4.6 Crosscorrelation and crosscovariance matrices

In the case of random vectors, as in sub-section 5.2.8, we can gather the crosscorrelation and crosscovariance values in a matrix.

5.5 SUM OF RANDOM SIGNALS

5.5.1 Probability density of the sum of two random variables

Let three random variables **x, y, z** be related by:

$$\mathbf{z} = \mathbf{x} + \mathbf{y} \tag{5.180}$$

Assume we know the joint probability density of variables **x** and **y**: $p_{xy}(x,y)$. The **z**-marginal probability density can be obtained by finding the joint density of (**x**,**z**) by the method depicted in sub-section 5.1.20 and by integration with respect to **x**:

$$p_z(z) = \int_{-\infty}^{\infty} p_{xy}(x, z - x)\mathrm{d}x \tag{5.181}$$

When variables **x** and **y** are ***statistically independent***: $p_{xy}(x,y) = p_x(x)p_y(y)$, and the probability density of the sum is equal to the ***convolution*** of their respective densities, then we have

$$p_z(z) = \int_{-\infty}^{\infty} p_x(x)p_y(z - x)\mathrm{d}x = p_x(z) * p_y(z) \tag{5.182}$$

The characteristic function (14.64) of a variable is the inverse Fourier transform of its probability density:

$$\Pi_x(\upsilon) = \mathrm{F}^{-1}\{p_x(x)\} \tag{5.183}$$

From property (4.15), we have for independent variables:

$$\Pi_z(\upsilon) = \Pi_x(\upsilon) \cdot \Pi_y(\upsilon) \tag{5.184}$$

These relations (5.182) and (5.184) can be generalized for n independent random variables. If

$$\mathbf{z} = \sum_{k=1}^{n} \mathbf{x}_k \tag{5.185}$$

$$p_z(z) = p_1(z) * p_2(z) * \ldots * p_n(z) \tag{5.186}$$

and

$$\Pi_z(\upsilon) = \Pi_1(\upsilon) \cdot \Pi_2(\upsilon) \ldots \Pi_n(\upsilon) \tag{5.187}$$

5.5.2 Example

Let $x(t)$ and $y(t)$ be two independent stochastic signals with uniform distributions between 0 and 1: $p_x(x) = \mathrm{rect}(x - \tfrac{1}{2})$ and $p_y(y) = \mathrm{rect}(y - \tfrac{1}{2})$. The probability density of their sum (fig. 5.19) has a triangular shape: $p_z(z = x + y) = \text{tri}(z - 1)$.

This result can be obtained by evaluation of the convolution (5.182), or the indirect transform $p_z(z) = \text{F}[\Pi_z(\upsilon)]$, with $\Pi_z(\upsilon)$ given by (5.184). Applying here property (4.17), $\Pi_x(\upsilon) = \Pi_y(\upsilon) = \text{sinc}(\upsilon)\exp(\text{j}\pi)$.

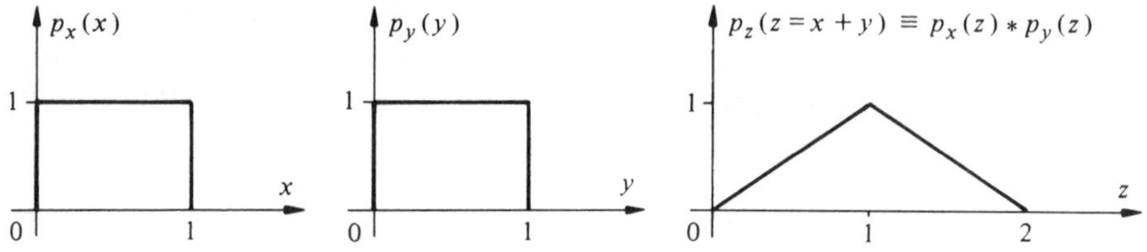

Fig. 5.19

5.5.3 Central limit theorem

This theorem, which is not demonstrated here [68], is a consequence of the multiple convolution (5.186). It can be stated as follows: ***The statistical distribution of the sum of*** n ***independent random variables, having the same probability distribution, asymptotically tends toward a Gaussian distribution when*** $n \to \infty$, ***whatever the distribution of the individual terms.***

Figure 5.20 illustrates this phenomenon.

This theorem is valid even if the individual distributions are different as long as the variance of each of the individual terms is very small with respect to the variance of the sum.

5.5.4 Importance of the central limit theorem

The central limit theorem plays a fundamental role in stochastic processes theory [69] and in signal theory [70]. This is due to the fact that many stochastic phenomena (for example, some noise sources presented in chap. 6) are the result of the sum of many individual contributions of unknown distribution. It is then possible to assimilate the resulting distribution to a normal distribution if the contributions can be considered as independent. Moreover, Gaussian processes have special properties allowing the resolution of otherwise unsolvable problems.

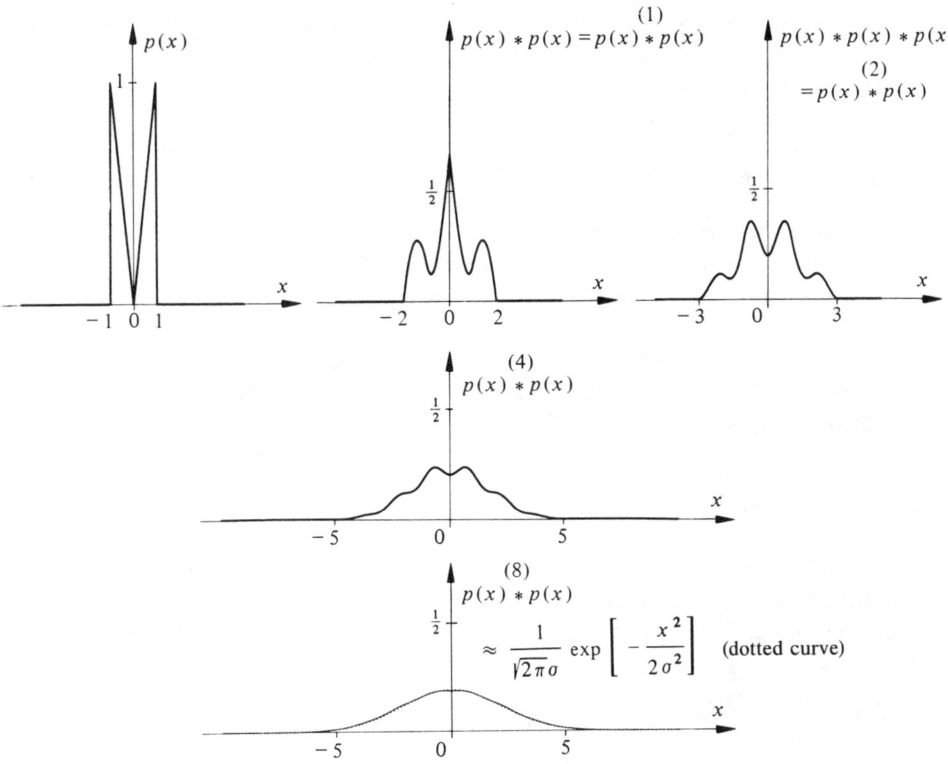

Fig. 5.20

5.5.5 Autocorrelation function and spectral density of the sum of random signals

Consider the linear combination:

$$z(t) = ax(t) + by(t) \tag{5.188}$$

The autocorrelation function and the spectral density of $z(t)$, according to (5.55), (5.111), (5.161), and (5.166), are

$$\begin{aligned} R_z(\tau) &= \mathrm{E}[\mathbf{z}(t)\mathbf{z}(t+\tau)] \\ &= a^2R_x(\tau) + b^2R_y(\tau) + ab[R_{xy}(\tau) + R_{yx}(\tau)] \end{aligned} \tag{5.189}$$

$$\Phi_z(f) = a^2\Phi_x(f) + b^2\Phi_y(f) + ab[\Phi_{xy}(f) + \Phi_{yx}(f)] \tag{5.190}$$

If $x(t)$ and $y(t)$ are statistically independent, according to (5.176): $R_{xy}(\tau) = R_{yx}(\tau) = \mu_x\mu_y$ and $\Phi_{xy}(f) = \Phi_{yx}(f) = \mu_x\mu_y\delta(f)$.

If, moreover, at least one of the signal has zero mean value:

$$R_z(\tau) = a^2R_x(\tau) + b^2R_y(\tau) \tag{5.191}$$

$$\Phi_z(f) = a^2\Phi_x(f) + b^2\Phi_y(f) \tag{5.192}$$

So, the autocorrelation function [the spectral density] of the sum, or difference, of two independent random variables, ***where at least one of them has a zero mean value,*** is the sum of their autocorrelation functions [spectral densities].

This result is also valid for any pair of orthogonal signals. A special consequence is that the variance of the sum of independent random, or simply orthogonal, signals is the sum of the variances of each term (sub-sect. 14.3.10).

5.5.6 Example 1

Consider a signal $z(t)$ sum of a sine wave $x(t)$ with random phase (sub-sect. 5.1.16) and a step signal $y(t)$, like the one presented in sub-sect. 5.2.7. Assuming the independence of those two signals, the autocorrelation function of $z(t)$ is (fig. 5.21):

$$R_z(\tau) = \tfrac{1}{2}A^2\cos(2\pi f_0\tau) + \sigma_y^2\text{tri}(\tau/T) + \mu_y^2$$

and the power spectral density (fig. 5.22):

$$\Phi_z(f) = \tfrac{1}{4}A^2[\delta(f + f_0) + \delta(f - f_0)] + \mu_y^2\delta(f) + \sigma_y^2T\,\text{sinc}^2(Tf)$$

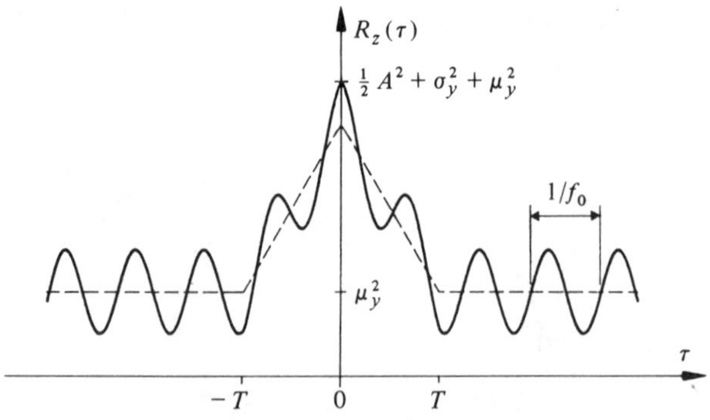

Fig. 5.21

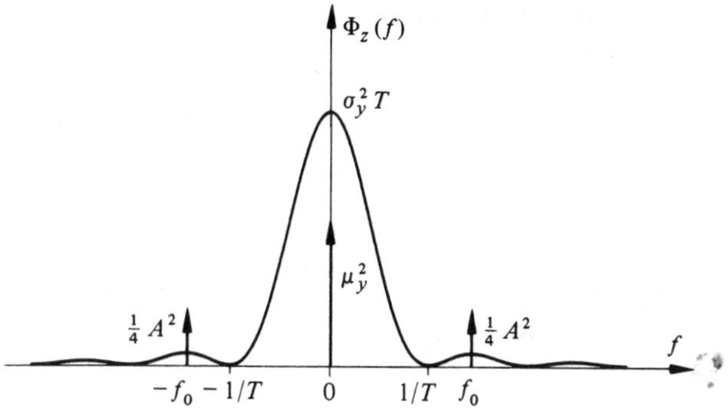

Fig. 5.22

This result suggests that correlation, as well as spectral analysis, can be used to detect hidden periodicities (sect. 13.2).

The total power of signal $z(t)$ is given by

$$P_z = R_z(0) = \sigma_z^2 + \mu_z^2 = \tfrac{1}{2} A^2 + \sigma_y^2 + \mu_y^2$$

with

$$\sigma_z^2 = A^2/2 + \sigma_y^2$$

and

$$\mu_z = \mu_y$$

5.5.7 Example 2

An analytical illustration of the central limit theorem can be obtained in the following case: consider a random variable $\mathbf{z} = \Sigma \mathbf{x}_k$, where the $\mathbf{x}_k$ are independent variables, uniformly distributed with a probability density $p_x(x) = \text{rect}(x)$. According to (14.95), the variance is $\sigma_x^2 = 1/12$. The corresponding characteristic function is $\Pi_x(\upsilon) = \text{sinc}\,(\upsilon)$ and from (5.187), $\Pi_z(\upsilon) = \text{sinc}^n(\upsilon)$ if the summation has n terms.

Taking the natural logarithm of $\Pi_z(\upsilon)$ and inserting the series (1.59) of sinc(υ), we get: $\ln \Pi_z(\upsilon) = n \ln[1 - (\pi\upsilon)^2/3! + (\pi\upsilon)^4/5! - \ldots]$.

For $n \gg 1$ and $\upsilon \ll 1$: $\ln \Pi_2(\upsilon) \cong n \ln[1 - (\pi\upsilon)^2/3!] = -n(\pi\upsilon)^2/6$.

So, $\Pi_z(\upsilon) = \exp[-n\pi^2\upsilon^2/6] = \exp[-nu^2/24]$, where $u = 2\pi\upsilon$. According to (14.96) and (14.97), the probability density of variable z is approximately $p_z(z) = (2\pi)^{-1/2}\, \sigma_z^{-1} \exp[-\tfrac{1}{2} z^2/\sigma_z^2]$, where, according to (14.62), $\sigma_z^2 = n\sigma_x^2$.

5.6 PRODUCT OF RANDOM SIGNALS

5.6.1 Probability density of the product of two random variables

Let

$$\mathbf{z} = \mathbf{xy} \tag{5.193}$$

Once again applying the method of sub-section 5.1.20 to derive the probability density of a pair of variables $(\mathbf{x},\mathbf{z})$ from $p_{xy}(x,y)$ and integration with respect to x, we get

$$p_z(z) = \int_{-\infty}^{\infty} \frac{1}{|x|} p_{xy}\left(x, \frac{z}{x}\right) \mathrm{d}x \tag{5.194}$$

If $\mathbf{x}$ and $\mathbf{y}$ are independent variables, (5.194) becomes

$$p_z(z) = \int_{-\infty}^{\infty} \frac{1}{|x|} p_x(x) p_y\left(\frac{z}{x}\right) \mathrm{d}x \tag{5.195}$$

To use (5.195), some care has to be taken on the evaluation of the integration limits.

5.6.2 Example

Let $\mathbf{x}$ and $\mathbf{y}$ be two independent random variables, uniformly distributed on interval [0, 1]. The product $\mathbf{z} = \mathbf{xy}$ is also distributed on [0, 1] and its probability density can be evaluated from (5.194). To keep variable $\mathbf{y} = \mathbf{z}/\mathbf{x}$ less than or equal to one, for a given value of $\mathbf{z}$ less than one, $\mathbf{x}$ must be greater than or equal to $\mathbf{z}$ (fig. 5.23).

The probability density of $\mathbf{z}$ becomes (fig. 5.24)

$$p_z(z) = \int_z^1 \frac{1}{x} \mathrm{d}x = -\ln z$$

Its mean value is

$$\mu_z = \mathrm{E}[\mathbf{z}] = \int_0^1 - z \ln z \, \mathrm{d}z = 1/4$$

5.6.3 Autocorrelation function and spectral density of a product of independent random signals

Let the product of real signals be

$$z(t) = x(t)y(t) \tag{5.196}$$

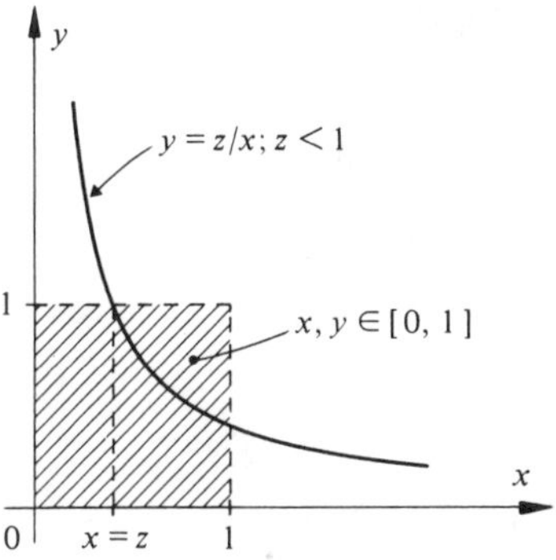

Fig. 5.23

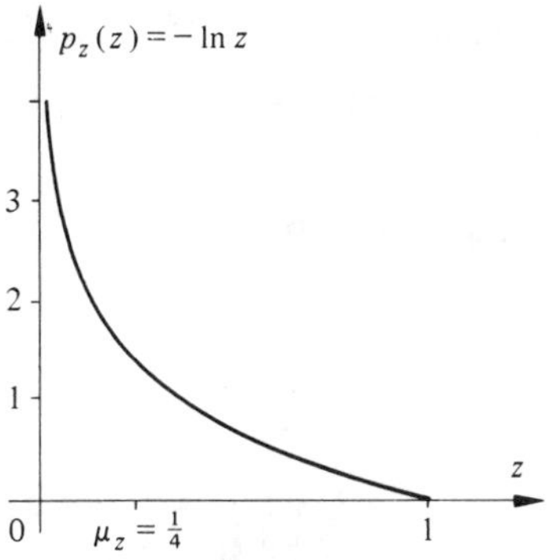

Fig. 5.24

According to definition (5.55), the statistical autocorrelation function is [with $x_1 = x(t)$, $x_2 = x(t + \tau)$, $y_1 = y(t)$, $y_2 = y(t + \tau)$]:

$$\begin{aligned} R_z(\tau) &= \mathrm{E}[\mathbf{z}(t)\mathbf{z}(t+\tau)] \\ &= \mathrm{E}[\mathbf{x}(t)\mathbf{y}(t)\mathbf{x}(t+\tau)\mathbf{y}(t+\tau)] \\ &= \iiiint x_1 x_2 y_1 y_2 \, p_{xy}(x_1,x_2,y_1,y_2)\mathrm{d}x_1\mathrm{d}x_2\mathrm{d}y_1\mathrm{d}y_2 \end{aligned} \quad (5.197)$$

If the two signals $x(t)$ and $y(t)$ are ***independent,*** according to (5.33): $p_{xy}(x_1,x_2,y_1,y_2) = p_x(x_1,x_2)p_y(y_1,y_2)$ and, consequently,

$$\begin{aligned} R_z &= \mathrm{E}[\mathbf{x}(t)\mathbf{x}(t+\tau)]\mathrm{E}[\mathbf{y}(t)\mathbf{y}(t+\tau)] \\ &= R_x(\tau) \cdot R_y(\tau) \end{aligned} \quad (5.198)$$

The power spectral density is, according to (4.15):

$$\Phi_z(f) = \Phi_x(f) * \Phi_y(f) \quad (5.199)$$

Hence, ***the autocorrelation of a product of independent signals is simply the product of the respective autocorrelation functions and the resulting power spectral density is obtained by convolving the respective spectral densities.***

5.6.4 Example

Consider a product $z(t) = x(t)y(t)$ of a sine wave $x(t)$ with random phase (sub-sect. 5.1.16), frequency f_0 and amplitude A and of an independent NRZ binary signal $y(t)$ (sub-sect. 5.3.7), with antipolar and equally likely levels $\pm V$ of duration T.

According to (5.125), (5.126), (5.131), (5.132), (5.198), and (5.199):

$R_z(\tau) = \frac{1}{2} A^2 V^2 \cos(2\pi f_0 \tau) \cdot \text{tri}(\tau/T)$
$\Phi_z(f) = \frac{1}{4} A^2 V^2 T \{\text{sinc}^2[T(f + f_0)] + \text{sinc}^2[T(f - f_0)]\}$

The signal $z(t)$, its autocorrelation function, and its power spectral density are represented in fig. 5.25.

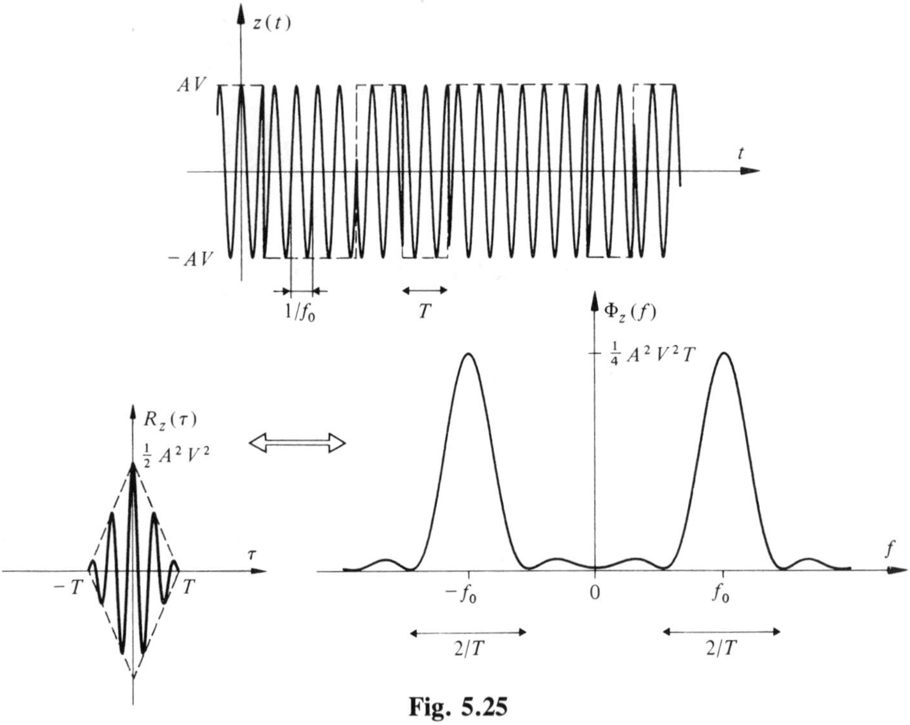

Fig. 5.25

5.7 GAUSSIAN PROCESSES

5.7.1 Definition

A stochastic process is said to be *Gaussian* if for any time sequence $\{t_i\}$, the corresponding random vector $\boldsymbol{x} = (\mathbf{x}_1, \mathbf{x}_2, \ldots, \mathbf{x}_n)$ with $\mathbf{x}_i = \mathbf{x}(t_i)$ has a joint multidimensional Gaussian probability density. Using matrix notations, this probability density is as follows (sub-sect. 14.4.5):

$$\mathrm{p}(\boldsymbol{x}) = \frac{1}{(2\pi)^{n/2}|\mathbf{C}_x|^{1/2}} \exp\left[-\frac{1}{2}(\boldsymbol{x} - \boldsymbol{\mu}_x)\mathbf{C}_x^{-1}(\boldsymbol{x} - \boldsymbol{\mu}_x)^{\mathrm{T}}\right] \tag{5.200}$$

where $(\boldsymbol{x} - \boldsymbol{\mu}_x)$ is an n-dimensional row vector, $(\boldsymbol{x} - \boldsymbol{\mu}_x)^{\mathrm{T}}$ is its transpose, and $\mathbf{C}_x$ is the covariance matrix (5.98).

The associated characteristic function is

$$\Pi_x(\boldsymbol{u}) = \exp[\mathrm{j}\boldsymbol{\mu}_x^{\mathrm{T}}\boldsymbol{u} - \tfrac{1}{2}\boldsymbol{u}\mathbf{C}_x\boldsymbol{u}^{\mathrm{T}}] \tag{5.201}$$

Each variable $\mathbf{x}_i$ has a marginal probability density (fig. 14.14):

$$p_{xi}(x) = \frac{1}{\sqrt{2\pi}\sigma_{xi}} \exp\left[-\frac{(x - \mu_{xi})^2}{2\sigma_{xi}^2}\right] \tag{5.202}$$

and a characteristic function:

$$\Pi_{xi}(u = 2\pi\upsilon) = \mathrm{F}^{-1}\{p_{xi}(x)\} = \exp[\mathrm{j}\mu_{xi}u - \tfrac{1}{2}\sigma_{xi}^2u^2] \tag{5.203}$$

If the variables $\mathbf{x}_i$ are mutually uncorrelated, $\mathbf{C}_x$ and $\mathbf{C}_x^{-1}$ are simple diagonal matrices of elements σ_{xi}^2 and $1/\sigma_{xi}^2$, respectively. The joint distribution (5.200) then complies with the independence condition (5.33):

$$p(\boldsymbol{x}) = p(x_1)\,p(x_2)\cdots p(x_n)$$

5.7.2 Bidimensional case

Let $(\mathbf{x},\mathbf{y})$ be a pair of Gaussian random variables. Their covariance matrix, from (5.171), is

$$C_{xy} = \begin{bmatrix} \sigma_x^2 & \rho\sigma_x\sigma_y \\ \rho\sigma_x\sigma_y & \sigma_y^2 \end{bmatrix} \tag{5.204}$$

thus,

$$|C_{xy}| = \sigma_x^2\sigma_y^2(1 - \rho^2) \tag{5.205}$$

$$\begin{aligned} C_{xy}^{-1} &= \frac{1}{|C_{xy}|}\begin{bmatrix} \sigma_y^2 & -\rho\sigma_x\sigma_y \\ -\rho\sigma_x\sigma_y & \sigma_x^2 \end{bmatrix} \\ &= \frac{1}{1 - \rho^2}\begin{bmatrix} 1/\sigma_x^2 & -\rho/(\sigma_x\sigma_y) \\ -\rho/(\sigma_x\sigma_y) & 1/\sigma_y^2 \end{bmatrix} \end{aligned} \tag{5.206}$$

The joint probability density (fig. 5.26) of these two variables is obtained by inserting (5.205) and (5.206) into (5.200):

$$p(x,y) = \frac{1}{2\pi\sigma_x\sigma_y\sqrt{1 - \rho^2}} \exp\left[\frac{-1}{2(1 - \rho^2)}\left\{\frac{(x - \mu_x)^2}{\sigma_x^2} - \frac{2\rho(x - \mu_x)(y - \mu_y)}{\sigma_x\sigma_y} + \frac{(y - \mu_y)^2}{\sigma_y^2}\right\}\right] \tag{5.207}$$

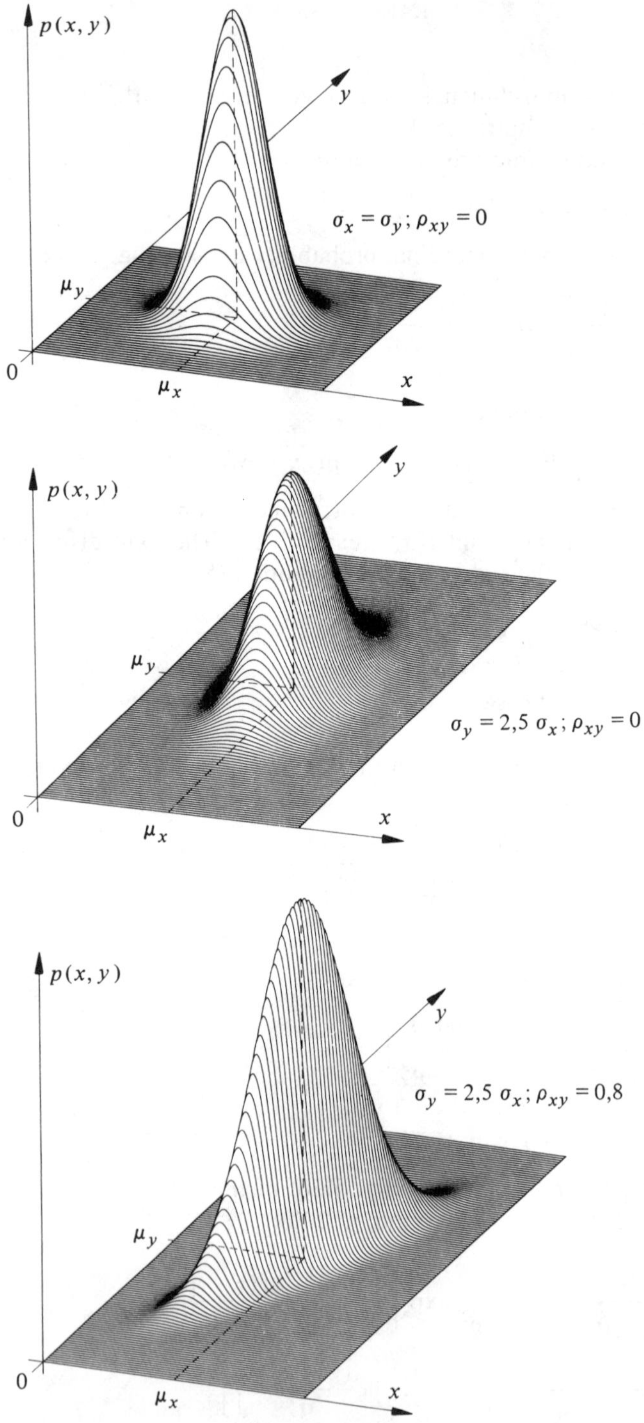

$p(x, y)$
y
$\sigma_x = \sigma_y ; \rho_{xy} = 0$
μ_y
0
μ_x
x
$p(x, y)$
y
μ_y
$\sigma_y = 2{,}5\ \sigma_x ; \rho_{xy} = 0$
0
μ_x
x
$p(x, y)$
y
$\sigma_y = 2{,}5\ \sigma_x ; \rho_{xy} = 0{,}8$
μ_y
0
μ_x
x

Fig. 5.26

5.7.3 Properties

The importance of Gaussian processes stems, on one hand, from the central limit theorem, making it the asymptotic model for many natural phenomena, and, on the other hand, from the following properties.

According to definition (5.200), a Gaussian process is totally defined by its first-degree (mean value) and second-degree (covariance) moments. So, when we know the autocorrelation function, or the spectrum, of a Gaussian signal, we automatically know its statistical distribution.

Also, as a result of this property, when the process is wide-sense stationary, it is also strictly stationary (sub-sect. 5.1.11).

The uncorrelation of Gaussian variables yields their statistical independence. Indeed, if we write $\rho = 0$ in (5.207), we satisfy condition (5.33): $p(x,y) = p_x(x) \cdot p_y(y)$. If the statistical independence always implies the uncorrelation (sub-sect. 5.4.4), the reverse is generally not true, but, for Gaussian processes, the reverse is always true.

Any linear transformation (e.g., filtering) ***of a Gaussian process produces a Gaussian process*** [69]. This property can easily be checked (problem 5.11.34), since any linear combination of Gaussian variables is also Gaussian, and since the convolution integral defining any linear system is a weighted sum.

5.7.4 Example

Consider a random Gaussian signal $x(t)$, with the spectral characteristics of a band-limited white noise (sub-sect. 5.3.10) plus a positive dc value (fig. 5.27). Its power spectral density (fig. 5.28) is

$$\Phi_x(f) = \tfrac{1}{2}\eta \, \mathrm{rect}[f/(2B)] + A\delta(f)$$

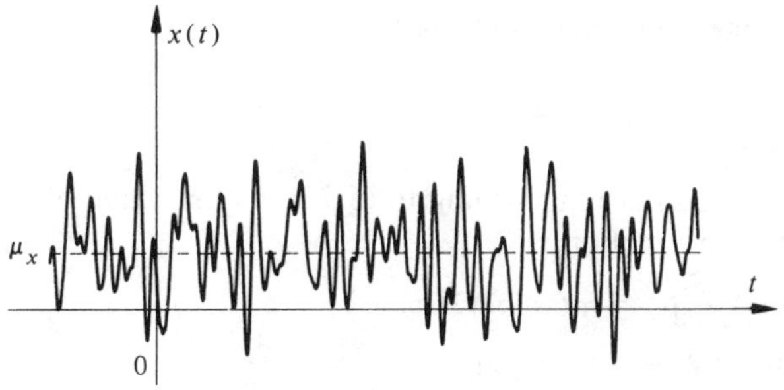

Fig. 5.27

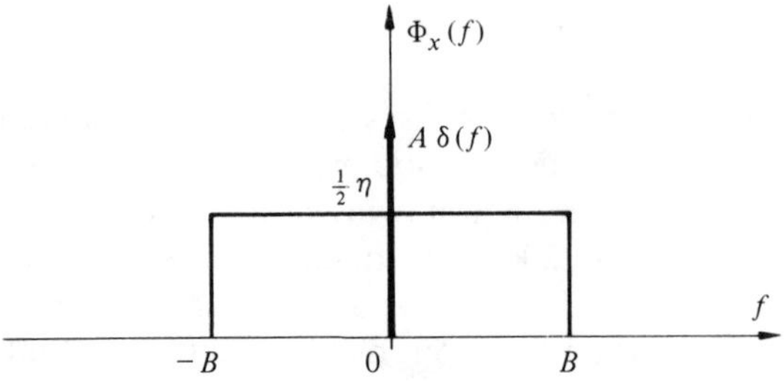

Fig. 5.28

By inverse Fourier transform, the autocorrelation function (fig. 5.29) is

$$R_x(\tau) = \eta B \operatorname{sinc}(2B\tau) + A$$

and, from (5.119), the signal total power is equal to $P_x = R_x(0) = \eta B + A$.

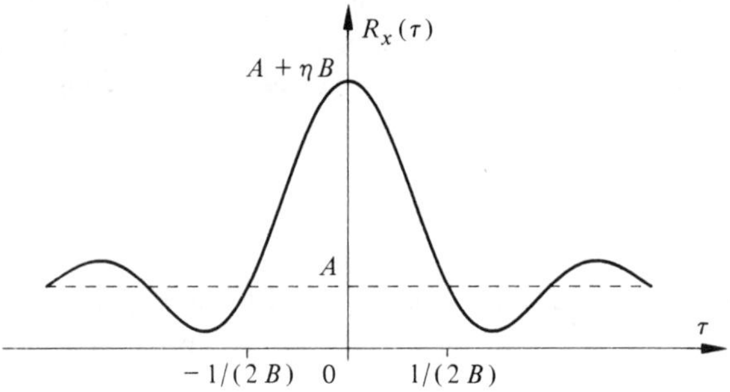

Fig. 5.29

The square of the mean value μ_x is equal to the weight of the discrete spectral component at frequency $f = 0$:

$$\mu_x^2 = \lim_{\epsilon \to 0} \int_{-\epsilon}^{\epsilon} \Phi_x(f)\mathrm{d}f = A$$

This can be defined here, according to (5.76), as the value of $R_x(\tau \to \infty)$.

The covariance function is

$$C_x(\tau) = R_x(\tau) - \mu_x^2 = \eta B \operatorname{sinc}(2B\tau)$$

or, in normalized form (fig. 5.30):

$$\rho_x(\tau) = C_x(\tau)/C_x(0) = \operatorname{sinc}(2B\tau)$$

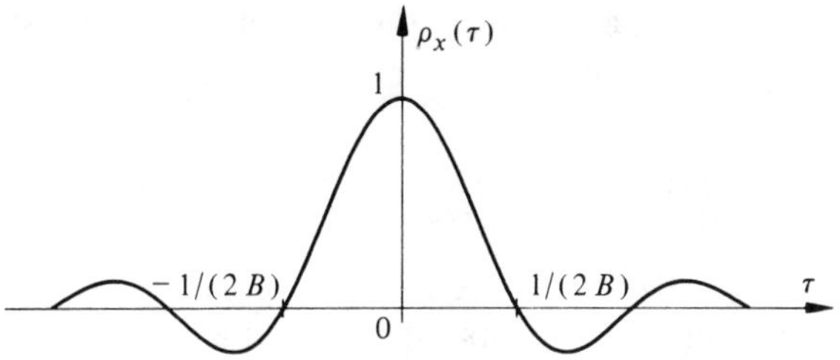

Fig. 5.30

The function is zero for $\tau = k/(2B)$ with $k = \pm 1, \pm 2, \pm 3, \ldots$ Thus, the samples periodically taken every $\Delta t = 1/(2B)$ are random variables, not only uncorrelated ($\rho_x = 0$), but also ***independent*** because we are dealing with a Gaussian process.

The process variance is

$$\sigma_x^2 = C_x(0) = \eta B$$

and the signal amplitude probability density, from (5.202), becomes

$$p(x) = \frac{1}{\sqrt{2\pi\eta B}} \exp\left[-\frac{(x - \sqrt{A})^2}{2\eta B}\right]$$

Figure 5.31 represents this distribution for $A = 4\ \text{V}^2$, $\eta = 9 \cdot 10^{-4}\ \text{V}^2/\text{Hz}$ and $B = 10^4$ Hz; thus, $\mu_x = 2$ V and $\sigma_x = 3$ V.

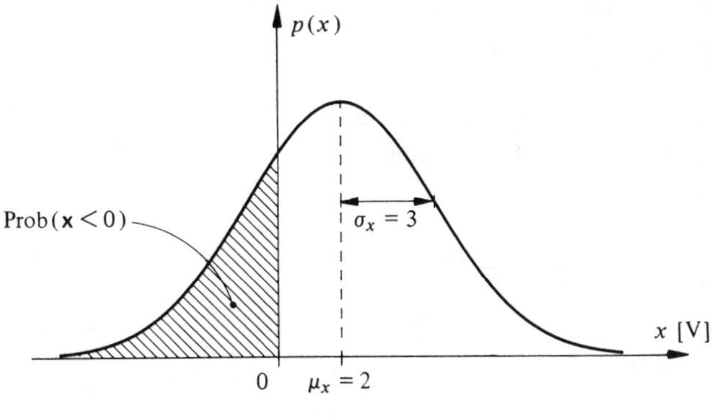

Fig. 5.31

If we want to know, for instance, the probability of observing negative amplitudes of $x(t)$, we will calculate

$$\text{Prob}\ (\mathbf{x} < 0) = \int_{-\infty}^{0} p(\mathbf{x})\, \mathrm{d}x = \frac{1}{\sqrt{2\pi}} \int_{-\infty}^{-2/3} \exp(-z^2/2)\, \mathrm{d}z = 0.252$$

by inserting the standardized variable $\mathbf{z} = (\mathbf{x} - \mu_x)/\sigma_x = (\mathbf{x} - 2)/3$, in order to use a normalized numerical table, like the one reproduced in section 15.8.

5.7.5 Representation of a Gaussian random signal

An approximate representation [14] of a Gaussian random signal can be obtained by considering it as the sum of an infinite number of sine waves with random phases α_n, independent and uniformly distributed on interval $[0, 2\pi]$:

$$x(t) \cong \sum_{n=0}^{\infty} a_n \cos(2\pi f_n t + \alpha_n) \tag{5.208}$$

with $f_n = n \cdot \Delta f$ and $a_n = 2\sqrt{\Phi_x(f_n)\Delta f}$. Signal $x(t)$ is Gaussian due to the central limit theorem (sub-sect. 5.5.3) if the rms value of each term is very small compared to the rms value of the set.

The representation, as an orthogonal function series, of a lowpass band-limited white noise, stems from (3.82) by writing

$$x(t) = \sum_{k=-\infty}^{\infty} x_k \operatorname{sinc}(2Bt - \mathrm{k}) \tag{5.209}$$

where x_k are the values taken by Gaussian random variables $\mathbf{x}_k$, statistically independent, with zero mean value and variance σ_x^2. The $x(t)$ signal obviously belongs to a Gaussian process because it stems from the weighted sum of Gaussian processes. Moreover, the power spectral density and the autocorrelation function correspond to (5.155) and (5.156):

$$R_x(\tau) = \sigma_x^2 \operatorname{sinc}(2B\tau) \leftrightarrow \Phi_x(f) = (2B)^{-1}\sigma_x^2 \operatorname{rect}[f/2B)] \tag{5.210}$$

because $R_x(\tau) = \mathrm{E}[\mathbf{x}(t)\mathbf{x}(t + \tau)] = \Sigma\, \Sigma\, \mathrm{E}[\mathbf{x}_k\mathbf{x}_l] \operatorname{sinc}(2Bt - k) \operatorname{sinc}[2\mathrm{B}(\mathrm{t} + \tau) - l] = \sigma_x^2 \operatorname{sinc}(2\mathrm{B}\tau)$ due to the orthogonality (sub-sect. 3.5.9) of functions $z_k(t) = \operatorname{sinc}(2Bt - k)$.

5.8 POISSON PROCESSES

5.8.1 Definition

The *Poisson process* is the simplest point process (counting). This model, however, permits the study of many phenomena resulting from time distributed random events: phone calls, accidents, equipment failures, creation of electron-hole pairs in a semiconductor junction, *et cetera*. Such phenomena can be represented by a random sequence of independent events that can occur at any time with the same chance.

The formal derivation of the Poisson distribution is established in subsection 14.4.3.

If the average number of events per time unit is λ, the probability (14.80) of seeing a single event occuring during the infinitesimal time interval $\mathrm{d}\tau$ is $\lambda \mathrm{d}\tau$. Since the probability of having more than one event in that interval is negligible, the probability (14.81) of having no event during $\mathrm{d}\tau$ is $(1 - \lambda \mathrm{d}\tau)$ Finally, the probability of counting exactly N events during a given interval τ, from (14.88), is

$$\text{Prob}(N,\tau) = \mu^N \exp(-\mu)/N! \tag{5.211}$$

where parameter $\mu = \lambda\tau$ is the average number of events occurring during interval τ.

If $\mathbf{z}$ is the random continuous variable representing the time interval separating two successive events, we obtain for this variable an exponential distribution (14.90):

$$p(z) = \lambda \exp(-\lambda z); \quad z \geqslant 0 \tag{5.212}$$

5.8.2 Representation

A point process $x(t)$ with a Poisson distribution can be represented (fig. 5.32) as a sequence of delta functions of weights α_i positioned at random instants t_i:

$$x(t) = \Sigma\alpha_i\delta(t - t_i) \tag{5.213}$$

Considering $x(t)$ as the input of a linear filter of impulse response $g(t)$, the output $y(t)$ of this filter represents a ***filtered Poisson process*** (shot-noise study, for instance):

$$y(t) = x(t) * g(t) = \Sigma\alpha_i g(t - t_i) \tag{5.214}$$

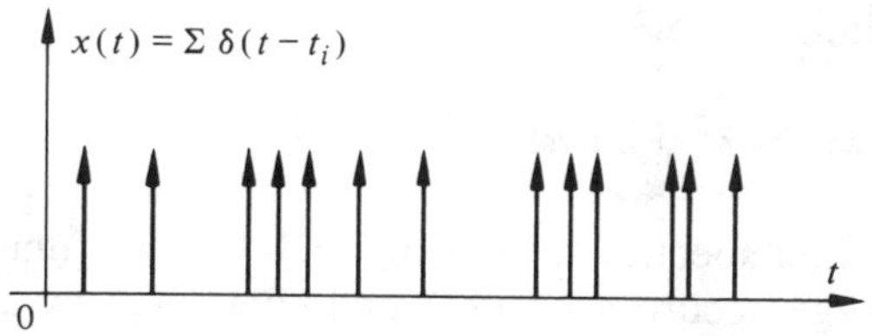

Fig. 5.32

5.8.3 Autocorrelation function and spectral density of a random sequence of delta-functions

Consider the random sequence (5.213) with $\alpha_i = 1$. To define the autocorrelation function and the spectral density of such a sequence, we will consider the delta-function as the limit of a rectangular pulse (1.44):

$$\delta(t) = \lim_{\epsilon \to 0} \frac{1}{\epsilon} \operatorname{rect}(t/\epsilon) \tag{5.215}$$

For a small enough interval ϵ, the probability of finding a single pulse is, according to (14.80):

$$\operatorname{Prob}(1, \epsilon) = \lambda\epsilon \tag{5.216}$$

where λ is the average number of pulses per unit time.

The desired autocorrelation function is given, from (5.57), by

$$R_x(\tau) = \lim_{\epsilon \to 0} \tilde{R}_x(\tau) \tag{5.217}$$

with

$$\tilde{R}_x(\tau) = \sum_i \sum_j x_{1i} x_{2j} \operatorname{Prob}(x_{1i}, x_{2j}; \tau) \tag{5.218}$$

where $x = 1/\epsilon$ or 0. The product $x_{1i}\, x_{2j} = x(t)\, x(t + \tau)$ is zero only if pulses are simultaneously present at times t and $t + \tau$.

For $|\tau| > \epsilon$: pulses are statistically independent. So,

$$\tilde{R}_x(|\tau| > \epsilon) = (1/\epsilon^2) \operatorname{Prob}(x_1 = 1/\epsilon) \operatorname{Prob}(x_2 = 1/\epsilon) = \lambda^2 \tag{5.219}$$

For $|\tau| \leqslant \epsilon$, the $x_{1i}\, x_{2j}$ product is non-zero if the rectangular pulse present at time t lasts at least until time $t + \tau$. According to (14.6), the corresponding joint probability can be expressed as a product: $\operatorname{Prob}(x_{1i}, x_{2j};\tau) = \operatorname{Prob}(x_{1i})$ $\operatorname{Prob}(x_{2j}|x_{1i};\tau)$, where $\operatorname{Prob}(x_{1i}) = \lambda\epsilon$ by (5.216) and $\operatorname{Prob}(x_{2j}|x_{1i};\tau) = \operatorname{tri}(\tau/\epsilon)$ by analogy with (5.87). Finally, we have

$$\tilde{R}_x(|\tau| \leqslant \epsilon) = (\lambda/\epsilon) \operatorname{tri}(\tau/\epsilon) \tag{5.220}$$

The autocorrelation function of the random sequence of delta-functions is obtained as the limit (fig. 5.33):

$$R_x(\tau) = \lim_{\epsilon \to 0} \tilde{R}_x(\tau) = \lambda^2 + \lambda\delta(\tau) \tag{5.221}$$

The corresponding spectral density (fig. 5.34) is then uniform with a value equal to the average occurrence rate λ and completed by a delta-function at

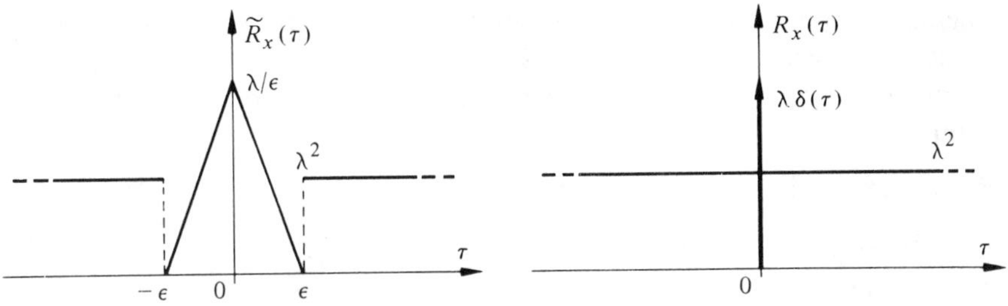

Fig. 5.33

origin with a weight of λ^2 representing the square of the dc component (mean value) of the signal.

$$\Phi_x(f) = \lambda^2\delta(f) + \lambda \tag{5.222}$$

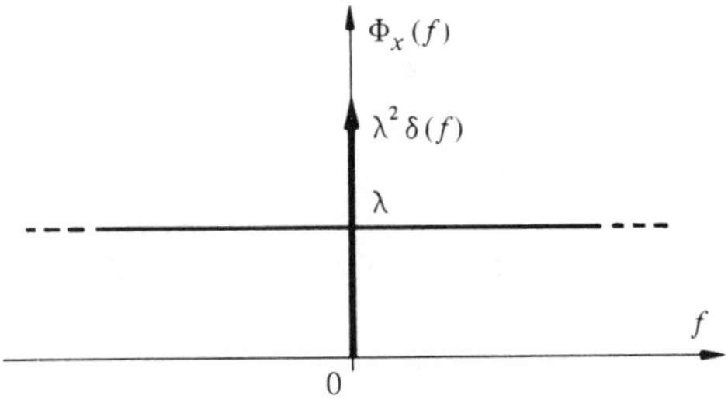

Fig. 5.34

5.8.4 Binary signal with random transitions

This signal (fig. 5.35) is also called a random telegraphic signal. It can, at any instant t and with an equal probability, take values $x(t) = 0$ or $x(t) = A$. The transitions from one state to the other are independent, and, according to our assumptions, their number N in interval τ is Poisson distributed (5.211). The mean value is obviously: $\mu_x = 0 \cdot \text{Prob}(x = 0) + A \cdot \text{Prob}(x = \text{A}) = A/2$.

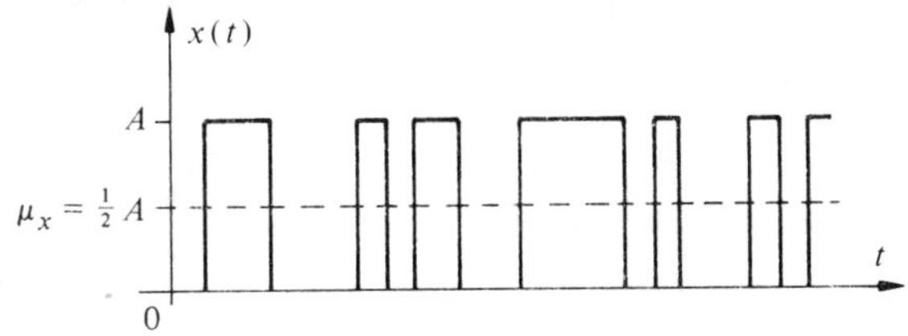

Fig. 5.35

The autocorrelation function is easily obtained, from (5.57), by taking into account the distribution of N transitions on $|\tau|$. With $x_1 = x(t)$ and $x_2 = x(t + \tau)$, the $x_1 x_2$ product is non-zero and equal to A^2 only if N is even or zero on $|\tau|$:

$$\begin{aligned} R_x(\tau) &= A^2 \operatorname{Prob}(x_1 = x_2 = A; \tau) \\ &= A^2 \operatorname{Prob}(x_1 = A) \operatorname{Prob}(N \text{ even or zero on } |\tau|) \\ &= \frac{A^2}{2} \exp(-\lambda|\tau|) \sum_{\substack{N=0 \\ (N \text{ even})}}^{\infty} \frac{(\lambda|\tau|)^N}{N!} \end{aligned} \tag{5.223}$$

However, the sum can be evaluated by

$$\begin{aligned} \sum_{\substack{N=0 \\ (N \text{ even})}}^{\infty} \frac{(\lambda|\tau|)^N}{N!} &= \frac{1}{2}\left[\sum_{N=0}^{\infty} \frac{(\lambda|\tau|)^N}{N!} + \sum_{N=0}^{\infty} \frac{(-\lambda|\tau|)^N}{N!}\right] \\ &= \frac{1}{2}[\exp(\lambda|\tau|) + \exp(-\lambda|\tau|)] \end{aligned} \tag{5.224}$$

then, finally (fig. 5.36),

$$R_x(\tau) = \tfrac{1}{4}A^2[1 + \exp(-2\lambda|\tau|)] \tag{5.225}$$

and, by Fourier transform (cf. example 4.2.19), the power spectral density is

$$\Phi_x(f) = \tfrac{1}{4}A^2[\delta(f) + \lambda/(\lambda^2 + \pi^2 f^2)] \tag{5.226}$$

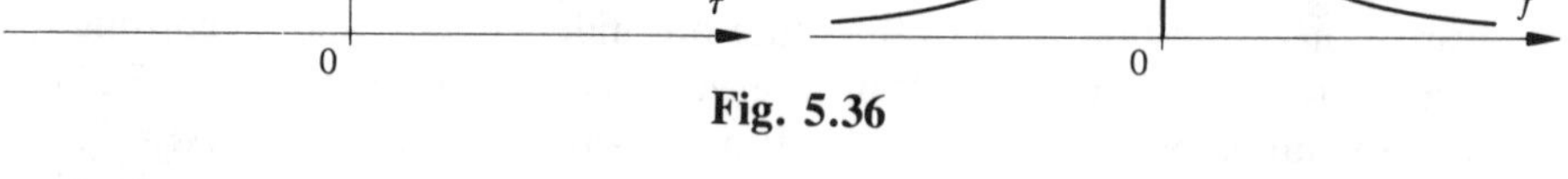

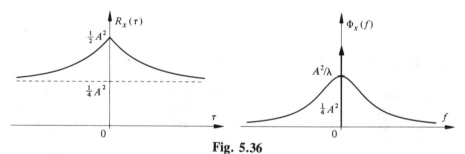

Fig. 5.36

The signal total power is $P_x = R_x(0) = A^2/2$ and the variance $\sigma_x^2 = C_x(0) = R_x(0) - \mu_x^2 = A^2/4$.

5.9 MARKOV PROCESSES

5.9.1 Definitions

The study of the mathematical properties of Markov processes is beyond the scope of this book. Here we shall only give a short definition in order to introduce the reader to this very important concept. For deeper study, the reader can refer himself or herself, for example, to [68, 71, 72].

In certain instances, the future evolution of a stochastic process does not depend on its past, but solely on its current state. The processes having such a property are called *Markov* (or Markoff) *processes*.

In other words, let a process take arbitrary values $x_1, x_2, \ldots, x_m, x_{m+1}$, at instants $t_1 < t_2 < t_m < t_{m+1}$. This process is a strictly Markov process if the conditional probability density at time t_{m+1}:

$$p(x_{m+1}|x_m, x_{m-1}, \ldots, x_1) = p(x_{m+1}|x_m) \quad (5.227)$$

By extension, we talk of Markov processes of nth order when the next state depends on the n previous states.

When the process is discrete, we call it a *Markov chain*.

5.9.2 Illustration

The Markovian model can be applied to many natural phenomena: particles in motion, genetic evolution, wearing process, *et cetera*. In engineering, the evolution of a dynamic system (automatic control, for example), of a signal, or of the sequence of information (written language) can often be assimilated to a Markov process.

The Markovian character is generally combined with other statistical characteristics: Gauss-Markov process, for instance.

Some data transmission signals are Markovian [73].

5.10 PSEUDORANDOM SIGNALS

5.10.1 Introduction

Pseudorandom signals or variables are, in fact, truly deterministic, but their behavior appears random and has clearly defined statistical properties. They are used in particular to simulate (on computer or in laboratories) random phenomena or to generate signals with very tight autocorrelation functions, to encode and synchronize information in telecommunications [75], or to detect radar echoes [27].

5.10.2 Maximum length binary sequences

The most current method for generation of pseudorandom signals is based on the so-called theory of maximum length binary sequences. Such sequences are easily electronically produced by a shift register, which is built with n

serial flip-flops and a feedback circuit sending, as an input to the first flip-flop, the modulo-2 sum (exclusive-OR gate) of some other flip-flop states. This circuit is clocked at rate $f_H = 1/T_H$. An elementary example of such a generator ($n = 3$) is given, with the chronological sequence of its states, in fig. 5.37.

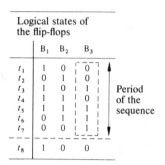

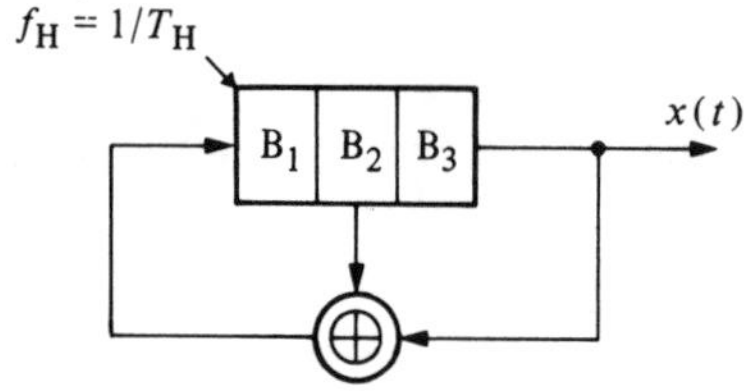

	Logical states of the flip-flops B_1	B_2	B_3	
t_1	1	0	0	Period of the sequence
t_2	0	1	0	
t_3	1	0	1	
t_4	1	1	0	
t_5	1	1	1	
t_6	0	1	1	
t_7	0	0	1	
t_8	1	0	0	

Fig. 5.37

The register is set to provide an initial condition having at least a single one, then it will successively take all the possible states, except the all-zero one. The theory defining the kind of feedback to use for a register of length n is beyond the scope of this book and can be found in [76]. Table 5.38 gives some examples in which the feedback is the exclusive-OR of the sole output n and output m or $n - m$.

Table 5.38

n	$2^n - 1$	m	$n - m$
3	7	1	2
5	31	2	3
7	127	1 or 3	6 or 4
10	1 023	3	7
15	32 767	1, 4 or 7	14, 11, or 8
20	1 048 575	3	17
22	4 194 303	1	21
25	33 554 431	3 or 7	22 or 18
28	268 435 455	3, 9 or 13	25, 19 or 15
29	536 870 911	2	27
31	2 147 483 647	3, 6, 7 or 13	28, 25, 24 or 18
33	8 589 934 591	13	20
39	$5{,}5 \cdot 10^{11}$	4, 8 or 14	35, 31 or 25

5.10.3 Properties

The binary sequences with maximum length have the following main properties:

- the generated sequence is ***periodic*** with period

$$T_s = (2^n - 1)T_H \tag{5.228}$$

- during a period T_s, always 2^{n-1} ones and $(2^{n-1} - 1)$ zeros are counted. So, for large enough n, these symbols can be considered practically independent and equally likely;
- by comparing bit to bit a period of the sequence with any of its circular permutations, we get a Hamming distance (sub-sect. 3.14) equal to 2^{n-1}. In other words, the number of coinciding symbols is equal to the number of noncoinciding symbols minus one.

5.10.4 Autocorrelation function and spectral density of a pseudorandom binary signal

Consider a binary signal $x(t)$ with two levels $+A$ and $-A$ representing the symbols 1 and 0, outputs of a maximum length sequence generator. Its autocorrelation function (fig. 5.39) is easily found, taking into account the properties mentioned in the previous sub-section.

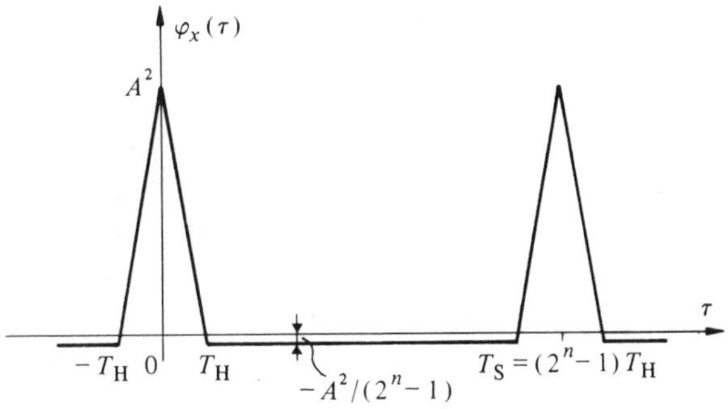

Fig. 5.39

The sequence being T_s-periodic, its autocorrelation function, consequently, is also T_s-periodic (sub-sect. 4.4.12). The comparison of the sequence with any of its circular permutations yields for $kT_s + T_H < |\tau| < (k + 1)T_s - T_H$, $\varphi_x(\tau) = -A^2/(2^n - 1)$ and for $kT_s - T_H < |\tau| < kT_s + T_H$,

$\varphi_x(\tau) = A^2\{[2^n/(2^n - 1)]\,\text{tri}(\tau/T_H) - 1/(2^n - 1)\}$. These results can be expressed, using chapter 1 shorthand notations, in a more general form:

$$\varphi_x(\tau) = A^2[2^n/(2^n - 1)]\,\text{rep}_{T_S}[\text{tri}(\tau/T_H)] - A^2/(2^n - 1) \tag{5.229}$$

The corresponding power spectral density (fig. 5.40) is a line spectrum given by the equation:

$$\Phi_x(f) = A^2[2^n/(2^n - 1)^2]\,\text{sinc}^2(T_H f) \cdot \delta_{1/T_S}(f) - [A^2/(2^n - 1)]\delta(f) \tag{5.230}$$

The line at the origin has a weight $A^2/(2^n - 1)^2$, corresponding to the power of the dc value $\bar{x} = \pm A/(2^n - 1)$, the sign depending on which level is assigned the symbol 1. This dc value vanishes almost completely for $n \gg 1$.

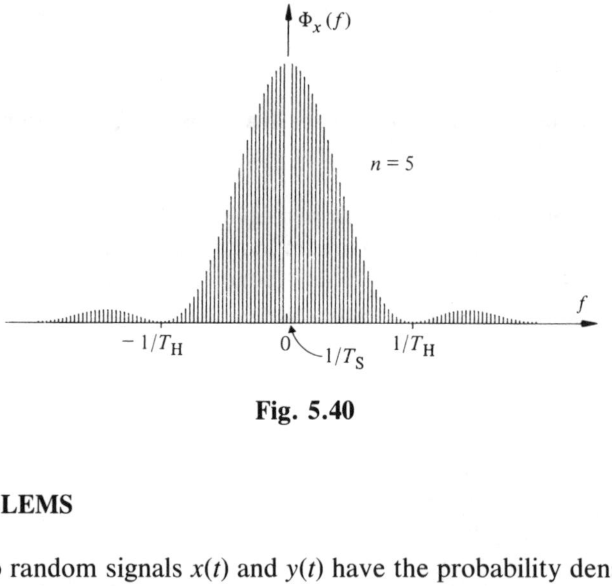

Fig. 5.40

5.11 PROBLEMS

5.11.1 Two random signals $x(t)$ and $y(t)$ have the probability densities $p(x) = a \cdot \text{rect}[(x - 2)/2]$ and $p(y) = b(2 - y)\text{rect}[(y - 1)/2]$, respectively. Evaluate their statistical mean values and their variances.

5.11.2 A random signal $z(t)$ has the probability density shown in fig. 5.41. Calculate the probability of having $|z(t)| < 1.5$.

5.11.3 Demonstrate relation (5.11).

5.11.4 Find with which probability the instantaneous value of a sine wave with random phase is superior to the half of its peak value (amplitude).

5.11.5 Evaluate the mean value and total power of a three-state random signal that takes the three values $x_1 = -2$ V, $x_2 = 0.5$ V, and $x_3 = 3$ V with probabilities $\text{Prob}(x_1) = 1/4$, $\text{Prob}(x_2) = 5/8$ and $\text{Prob}(x_3) = 1/8$, respectively.

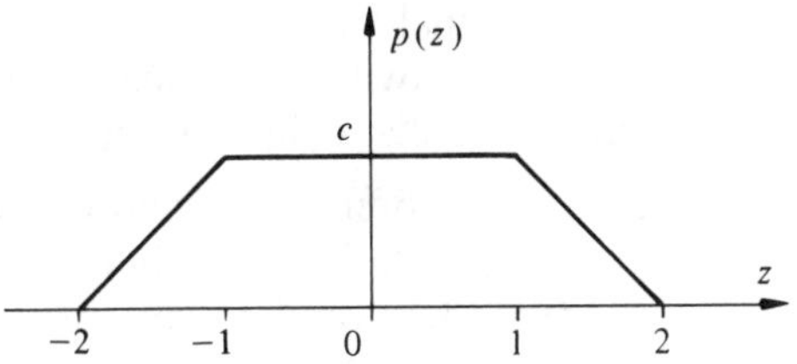

Fig. 5.41

5.11.6 Check whether random variables **x** and **y** having the joint probability density hereafter described are independent or not:

$$p_{xy}(x,y) = \begin{cases} 1 & 0 \leqslant x \leqslant \sqrt{2};\ 0 \leqslant y \leqslant x \\ 0 & \text{otherwise} \end{cases}$$

5.11.7 Knowing that **x** and **y** variables have the joint probability density $p_{xy}(x,y) = 1/(ab)$ with $|y| \leqslant b$, demonstrate that they are statistically independent.

5.11.8 An amplifier transforms an input signal $x(t)$ into an output signal $y(t) = Ax(t)$. Find the mean value and the variance of the output signal with those of the input signal and the amplifier gain A.

5.11.9 Derive the expression of the probability density of the variable $\mathbf{y} = a/\mathbf{x}$ if $p(x)$ is known.

5.11.10 Show that if a transmission attenuation factor (expressed here in nepers) $\alpha = \ln(P_2/P_1)$ randomly fluctuates around a mean value μ with a Gaussian distribution of variance σ^2, the power ratio $\beta = P_2/P_1$ has a log-normal probability density:

$$p(\beta) = (\beta\sigma\sqrt{2\pi})^{-1} \exp[-½(\ln \beta - \mu)^2/\sigma^2];\ \beta > 0$$

5.11.11 What is the probability density of a rectifier output, defined by $y = x \cdot \epsilon(x)$, in function of the input probability density.

5.11.12 Let $x(t)$ be a random signal, measured in volts, uniformly distributed on interval $[0, A]$. Find and plot the probability density and the cumulative distribution of its normalized instantaneous power. Deduce the probability of the instantaneous power being greater than $A^2/2$.

5.11.13 Show that if $x(t)$ is a zero mean value signal with a Gaussian distribution, the distribution of $y(t) = ax^2(t)$ is an χ^2 distribution with one degree of freedom (sub-sect. 14.4.7).

5.11.14 Signal $r(t)$ has the distribution (5.52): find the equation of its cumulative distribution, and deduce the probability of having $r > 3\sigma$.

5.11.15 Show that the random signal $x(t) = y \cos \omega t + z \sin \omega t$, where **y** and **z** are independent random variables with zero mean value and the same cumulative distribution, is wide-sense stationary, but not strictly stationary.

5.11.16 Let $x(t)$ be a stationary random signal and ω a constant. Show that $y(t) = x(t) \cos \omega t$ is not wide-sense stationary, while $z(t) = x(t) \cos(\omega t + \alpha)$ is stationary when α is an independent random variable uniformly distributed on interval $[0, 2\pi]$.

5.11.17 We want to predict the evolution of a stationary and ergodic signal $x(t)$ on the basis of its past values $x(t - T)$ by the linear approximation $\tilde{x}(t) = ax(t - T)$. Find the optimal value which minimizes the approximation error power.

5.11.18 Rework example 5.2.7, assuming the process has a probability density $p(x) = a \exp(-ax)$ for $x > 0$ and $a > 0$.

5.11.19 Consider a three-state clocked signal which takes the independent values $+1$, 0, or -1 during each successive T-interval with probabilities ¼, ½, and ¼, respectively. Find the power and the power spectral density of this signal.

5.11.20 Demonstrate property (5.70) using the Wiener-Kinchin theorem (subsect. 5.3.3).

5.11.21 Assume a NRZ binary signal which takes during each successive T-interval the independent values $x_0 = A$ and $x_1 = -A$ with probabilities $\text{Prob}(x_0) = 2/3$ and $\text{Prob}(x_1) = 1/3$, respectively. Find its power spectral density.

5.11.22 An ergodic stationary random signal $y(t)$ is generated by modulating the amplitudes of a periodic sequence of rectangular pulses (random PAM signal) with a stationary random signal $x(t)$, the mean value of which is μ_x and the variance is σ_x^2. Assuming that the pulse amplitudes are statistically independent and that the pulsewidth Δ and period T are bound by relation $\Delta < T/2$, find the power spectral density of signal $y(t)$ and plot the result. Compare it with example 5.2.7 and result (5.132).

5.11.23 Derive results (5.132) and (5.148) from the general equation (5.149).

5.11.24 Find the power spectral density of a binary data transmission signal, with data rate $1/T$, where each logic symbol 0 corresponds to a zero level and each symbol 1 is represented by a rectangular pulse of amplitude A and width equal to $T/2$ followed by a $T/2$ interval at level zero (RZ mode). Assume symbols to be independent and equally likely.

5.11.25 Show that for white noise, equation (5.102) is satisfied by any set of orthogonal functions.

5.11.26 Demonstrate relation (5.181) using the method of sub-sect. 5.1.20.

5.11.27 Random signal $z(t)$ is the sum of two independent signals $x(t)$ and $y(t)$. The signal $x(t)$ is an antipolar binary signal taking value A with probability of 2/3 and value $-A$ with probability 1/3. The signal $y(t)$ is Gaussian, having a power spectral density $\Phi_y(f) = \frac{1}{2}\eta\,\text{tri}(f/B)$. Find the probability density, the mean value and the variance of z, and the correlation coefficient of x and y.

5.11.28 Signal $z(t)$ is the sum of two independent signals $x(t)$ and $y(t)$. Amplitudes of $x(t)$ are uniformly distributed with a mean value $\mu_x = 0$ V and a variance $\sigma_x^2 = 1/3$ V^2. Amplitudes of $y(t)$ are also uniformly distributed with $\mu_y = 2.5$ V and $\sigma_y^2 = 3/4$ V^2. Find the probability of $z(t)$ being greater than 3 V.

5.11.29 Demonstrate relation (5.194) using the method described in sub-sect. 5.1.20.

5.11.30 Find the probability density of the product of two independent stochastic variables uniformly distributed on interval $[a, a + 1]$ with $a > 0$.

5.11.31 Let $z(t) = x(t)y(t)$, where $x(t)$ is the process described in sub-sect. 5.2.7 and $y(t)$ is a periodic series of delta functions independent of $x(t)$ but with same period T. Find the autocorrelation function and the spectral density of $z(t)$.

5.11.32 Let $x(t)$ be a Gaussian random signal with a negative mean value and autocorrelation function $R_x(\tau) = A + B\,\text{sinc}^2(\tau/T)$. Find the probability of having the signal's instantaneous value between 2 V and 3 V and the probability of it being greater than 3 V if $A = 4$ V^2 and $B = 9$ V^2.

5.11.33 For which values of τ, are samples $x(t_0)$ and $x(t_0 + \tau)$ of the previous signal two independent random variables.

5.11.34 Demonstrate that any linear combination of Gaussian variables is also a Gaussian variable.

5.11.35 Express the conditional probability density $p(x|y)$ associated with Gaussian random variables with zero mean value.

5.11.36 We can demonstrate [69] that the product $z(t) = x(t)y(t)$ of two Gaussian random signals with zero mean value has an autocorrelation function:

$$R_z(\tau) = R_{xy}^2(0) + R_x(\tau)R_y(\tau) + R_{xy}(\tau)R_{yx}(\tau) \tag{5.231}$$

Check that this relation yields (5.198) when $x(t)$ and $y(t)$ are independent and deduce the power spectral density of $z(t) = x^2(t)$.

5.11.37 Find the autocorrelation function and the power spectral density of an antipolar ($\pm A$) binary signal with random transitions (sub-sect. 5.8.4).

Chapter 6

Background Noise

6.1 NOISE SOURCES

6.1.1 Noise and interference

In the broadest sense, any unwanted signal which more or less limits the integrity and intelligibility of a desired signal can be called noise: the term is used by analogy with the acoustical phenomenon bearing the same name.

Nonetheless, it is advisable to make a distinction between the noise generated by purely random—and hence unpredictable—disturbances and the interference caused by unwanted reception of other useful signals (such as those produced by the coupling of close transmission lines) or poor suppression of parasitic periodical signal components (e.g., hum in electroacoustical systems caused by insufficient filtering of power supplies or by a poor grounding).

Noise sources can be classified in two large groups:

- noise sources located outside a given processing system and acting on it by susceptibility
- noise sources internal to that system, generating a noise independent of external conditions

Although it is always possible to improve a processing system design to reduce interferences down to an acceptable level, and while it is generally possible—but hard—to eliminate the effects of external noise sources thanks to adequate technical means (shielding, EMC, *et cetera*), it is absolutely impossible to eliminate the contribution of internal noise sources.

Internal noise sources ultimately limit the system performances. Their study is, therefore, of prime importance, both in telecommunication and instrumentation.

Usually, a system will run properly only when the useful signal power level is higher, by many orders of magnitude, than the noise power (signal-to-noise ratio of a few tens of decibels). However, some elaborate processing methods (see chapt. 13) allow us to work with very small signal-to-noise ratios, thanks to the optimal use of any *a priori* information available on the useful signal.

Moreover, although noise is by and large considered as a prejudicial phenomenon, it may carry information on its origins (e.g., radio astronomy, machine-tool vibration analysis, *et cetera*). It is also sometimes necessary to generate noise intentionally (sect. 6.7) in order to test experimentally the sensitivity of a system or estimate its state by statistical methods.

6.1.2 External noise sources

External noise sources can be divided into two classes [77]:

- man-made noise
- natural noise

Man-made noise is mainly caused by disturbances generated by industrial environment (switches, relays, thyristors, electrical motors, arc welding, high voltage lines, *et cetera*), or even non-industrial environment (domestic appliances, engine ignition, fluorescent lights, *et cetera*).

The intensity of these disturbances is highly dependent on location; it is maximum in urban areas, but it usually decreases as wavelength increases and is practically negligible for $\lambda < 3$ m (i.e., $f > 100$ MHz). The effect of some of these disturbances can be drastically reduced by appropriate interference suppression methods.

Natural noise is associated with atmospheric (e.g., electrical storms) or cosmic (solar bursts, EM galactic radiations) phenomena.

The atmospheric noise sources are important mainly in subtropical regions. The average intensity decreases as wavelength increases and is practically negligible for $\lambda < 10$ m ($f > 30$ MHz). For $\lambda > 200$ m, storms generate effects that can be perceived over great distances. Cosmic noise may affect systems with directional antennas.

In addition to natural noise, we have the fading of radio-communication signals caused by fluctuating propagation conditions. This disturbance has a multiplying effect.

The influence of external noise sources on processing systems depends on various specific factors such as location, electromagnetic environment, system architecture, *et cetera*. The performance evaluation can thus be made only by heuristic formulas based on *in situ* measurements.

Sometimes, pulse-like disturbances can be considered as filtered (sub-sect. 5.8.2 and problem 8.5.9) or compound Poisson processes: pulses generated in bursts according to a given statistical law and bursts occurring according to a Poisson distribution.

Because of the large variety of causes, it is impossible to set a general model.

6.1.3 Internal noise sources

Internal noise sources can also be divided into two classes:

- pulse-like disturbances generated by current switching (logic circuits, electronic comparators, and switches)
- *background noise* generated in cables, wires, and electronic components because of the statistical nature of electrical conduction mechanisms.

Sound circuit design and careful manufacturing will generally reduce—and sometimes eliminate—the effects of the pulse-like disturbances.

Unfortunately, background noise cannot be removed. It has many origins [78]. As a general statement, we can say that it stems from the random movement of charged particles in thermal equilibrium (Brownian motion) or under the influence of applied fields. Under stable conditions, it can be seen as an ergodic stationary process. Its three main components in electronic devices are

- thermal noise
- shot noise
- flicker ($1/f$) noise

6.2 THERMAL NOISE

6.2.1 Generating process

Above absolute zero, the paths of free electrons in a conducting material are subject to random oscillations (thermal agitation). These erratic movements cause (even without any electric field applied) a random fluctuation of the instantaneous value of the voltage. When a current is generated by an applied electrical field, it corresponds to a slow drift of these electrons.

This *thermal noise* (or *Johnson noise*) exists in any active or passive component having some electrical resistance R.

6.2.2 Spectral density and autocorrelation function

Based on statistical thermodynamics, it can be shown that the bilateral spectral distribution of the maximum available power generated by this phenomenon is given by

$$\Phi_{th}(f) = \frac{1}{2} \frac{h|f|}{\exp(h|f|/kT) - 1} \quad \text{W/Hz} \tag{6.1}$$

where

- $h = 6.62 \times 10^{-34}$ J · s = *Planck constant*
- $k = 1.38 \times 10^{-23}$ J · K^{-1} = *Boltzmann constant*

At ambient temperature ($T_a = 290$ K $\triangleq \theta_a = 17$°C), the product kT has the value

$$kT = \cong 4 \cdot 10^{-21} \text{ J} \tag{6.2}$$

which yields, by integrating (6.1) over the entire $|f|$ domain, a maximum available power of some 40 nW. At ultra-high frequencies ($h|f| > kT$), the thermal noise spectral density (6.1) vanishes. It is superseded by another noise term generated by quantum effects [79] having a spectral density equal to $\frac{1}{2}h|f|$. On the other hand, for usual frequencies ($h|f| \ll$ kT on $|f| < 100$ GHz), relation (6.1) becomes (fig. 6.1):

$$\Phi_{th}(f) \cong \tfrac{1}{2}kT \qquad \text{W/Hz} \tag{6.3}$$

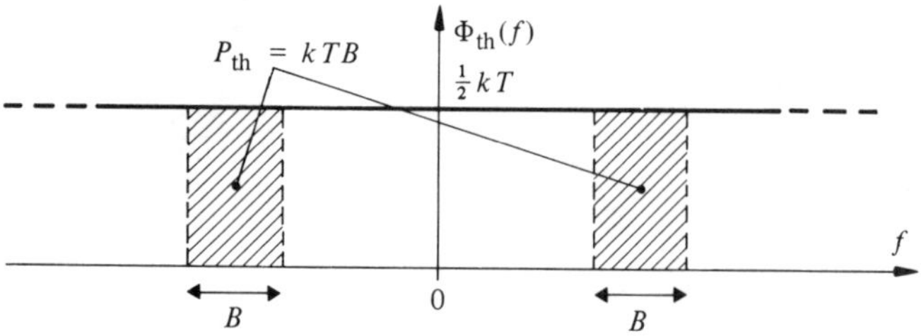

Fig. 6.1

In this frequency range, thermal noise, hence, is a ***white noise*** (sub-sect. 5.3.10). The total thermal noise power available in a frequency bandwidth B, is simply

$$P_{th} = kTB \qquad \text{W} \tag{6.4}$$

The autocorrelation function of the thermal noise, in first approximation, is a delta function:

$$R_{th}(\tau) \cong \tfrac{1}{2}kT\delta(\tau) \tag{6.5}$$

6.2.3 Voltage and current fluctuations in a resistor *R*

The charged-particle thermal motion in a device with a resistance $R = 1/G$ causes a random voltage fluctuation $u(t)$ with zero mean value and variance σ_u^2 (alias normalized power: see table 5.2), the value of which depends on the considered bandwidth B. A noise current $i(t)$, with zero mean value and variance σ_i^2, can be associated with it. Standard deviations (alias rms values) are related to one another by Ohm's law:

$$\sigma_i = \sigma_u/R = G\sigma_u \qquad \text{A} \qquad (6.6)$$

and their product corresponds to the total available power (6.4):

$$\sigma_u\sigma_i = kTB \qquad \text{W} \qquad (6.7)$$

Combining (6.6) and (6.7), the total available power can be expressed (in normalized form) as

$$\sigma_u^2 = kTRB \qquad \text{V}^2 \qquad (6.8)$$

$$\sigma_i^2 = kTGB \qquad \text{A}^2 \qquad (6.9)$$

The associated power spectral densities are

$$\Phi_u(f) = \tfrac{1}{2}kTR \qquad \text{V}^2/\text{Hz} \qquad (6.10)$$

$$\Phi_i(f) = \tfrac{1}{2}kTG \qquad \text{A}^2/\text{Hz} \qquad (6.11)$$

6.2.4 Equivalent model of a noisy resistor

To analyze the thermal noise effects on an electric circuit, it is convenient to have an equivalent model of a noisy resistor R. Thanks to the Thevenin-Norton theorems, the noisy resistor can be depicted either as a random voltage or random current generator with a R internal and noiseless resistance (or a $G = 1/R$ internal conductance). These are schematically represented in fig. 6.2.

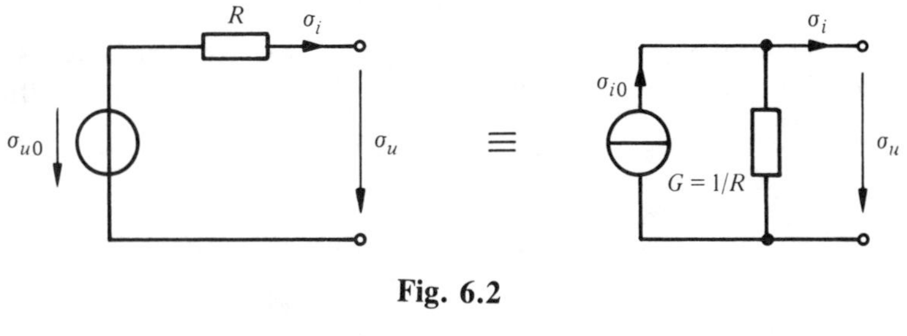

Fig. 6.2

The rms values σ_{u0} of the open-circuit voltage and σ_{i0} of the short-circuit current are found by taking into account that the maximum power is available when the generator load is a resistor whose value is equal to the internal resistance (power matching):

$$P_{\text{th}} = \sigma_u^2/R = R\sigma_i^2 = \sigma_{u0}^2/4R = R\sigma_{i0}^2/4 \qquad \text{W} \qquad (6.12)$$

hence,

$$\sigma_{u0}^2 = 4kTRB \qquad \mathrm{V}^2 \qquad \mathbf{(6.13)}$$

$$\sigma_{i0}^2 = 4kTGB \qquad \mathrm{A}^2 \qquad \mathbf{(6.14)}$$

Then, the normalized power spectral densities are

$$\Phi_{u0}(f) = 2kTR \qquad \mathrm{V}^2/\mathrm{Hz} \qquad (6.15)$$

$$\Phi_{i0}(f) = 2kTG \qquad \mathrm{A}^2/\mathrm{Hz} \qquad (6.16)$$

6.2.5 Examples

The rms value of the open-circuit voltage caused by thermal noise in a resistor R at ambient temperature is approximately

- $R = 10\ \mathrm{k\Omega}$ and $B = 20\ \mathrm{kHz}$: $\sigma_{u0} = 1.8\ \mu\mathrm{V}$
- $R = 1\ \mathrm{M\Omega}$ and $B = 20\ \mathrm{MHz}$: $\sigma_{u0} = 566\ \mu\mathrm{V}$

The rms values of the short circuit current are $\sigma_{i0} = 180$ pA and $\sigma_{i0} = 566$ pA, respectively.

6.2.6 Thermal noise statistics

Thermal noise is generated by the independent random motion of many electrons. The global effect stems from the sum of those elementary contributions. The statistical distribution of the resulting noise (voltage or current) is thus Gaussian (fig. 6.3), thanks to the central limit theorem (sub-sect. 5.5.3):

$$p(u) = \frac{1}{\sqrt{2\pi}\sigma_u} \exp\left[-\frac{u^2}{2\sigma_u^2}\right] \qquad (6.17)$$

Using a Gaussian probability table (sect. 15.8), we can check that the amplitude of such a signal is on interval $[-3\ \sigma_u, 3\ \sigma_u]$ 99.7 percent of the time.

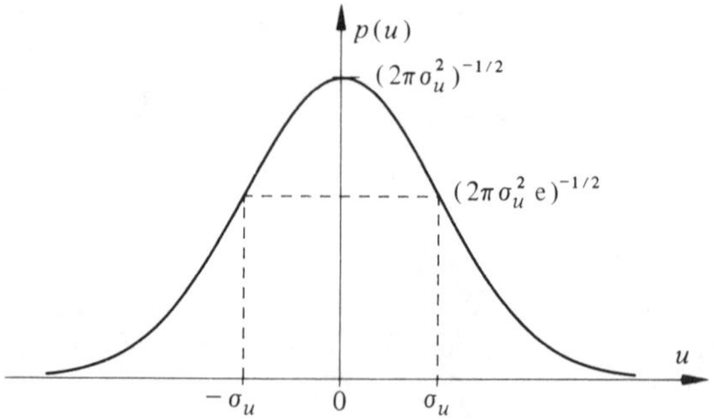

Fig. 6.3

The envelope (sub-sect. 7.3.8) of such a noise voltage has a Rayleigh distribution (5.52).

6.2.7 Serial or parallel network of noisy resistors

The thermal noise fluctuations in two different resistors R_1 and R_2 are statistically independent. Consequently (sub-sect. 5.5.5), the variance of their sum is the sum of their individual variances.

For two serial resistors:

$$\sigma_{u0}^2 = \sigma_{u01}^2 + \sigma_{u02}^2 \tag{6.18}$$

For two parallel resistors:

$$\sigma_{i0}^2 = \sigma_{i01}^2 + \sigma_{i02}^2 \tag{6.19}$$

If both resistors are at the same temperature, relations (6.18) and (6.19) are equivalent to (6.13) and (6.14) when substituting $R = R_1 + R_2$ for the series and $G = G_1 + G_2$ for the parallel connection.

6.2.8 Thermal noise in an impedance

Any impedance can be written as

$$\underline{Z}(f) = R(f) + \mathrm{j}X(f) \tag{6.20}$$

The thermal noise effect applies only to the resistive term $R(f)$. Thus, the power spectral density (6.15) becomes

$$\Phi_{u0}(f) = 2kTR(f) \qquad \mathrm{V^2/Hz} \tag{6.21}$$

and varies with respect to frequency as the real part of the impedance.

The variance of the open-circuit voltage of the thermal noise is found by integrating (6.21) on the considered bandwidth $B = f_2 - f_1$. Since $\Phi_{u0}(f)$ is even, we can write:

$$\sigma_{u0}^2 = 2\int_{f_1}^{f_2} \Phi_{u0}(f)\,\mathrm{d}f = 4kT\int_{f_1}^{f_2} R(f)\,\mathrm{d}f \tag{6.22}$$

6.2.9 Example: parallel *RC* network

The impedance of a parallel *RC* network is

$$Z(f) = \frac{R}{1 + \mathrm{j}2\pi fRC} = \frac{R}{1 + (f/f_c)^2} - \mathrm{j}\,\frac{Rf/f_c}{1 + (f/f_c)^2} \tag{6.23}$$

where $f_c = 1/(2\pi RC)$ is the cut-off frequency (−3 dB attenuation).

The open-circuit voltage variance evaluated over an infinite frequency range is

$$\sigma_{u0}^2 = 4kTR \int_0^\infty \frac{1}{1 + (f/f_c)^2} \, df = \frac{kT}{C} \tag{6.24}$$

The variance does not depend on the value of the resistor. This is not surprising, since, in such a RC network, the cut-off frequency is inversely proportional to R.

The power spectral density of the open-circuit voltage here is (fig. 6.4):

$$\Phi_{u0}(f) = 2kTR[1 + (f/f_c)^2]^{-1} \tag{6.25}$$

The corresponding autocorrelation function (fig. 6.5) is found by inverse Fourier transform:

$$R_{u0}(\tau) = \frac{kT}{C} \exp(-|\tau|/RC) \tag{6.26}$$

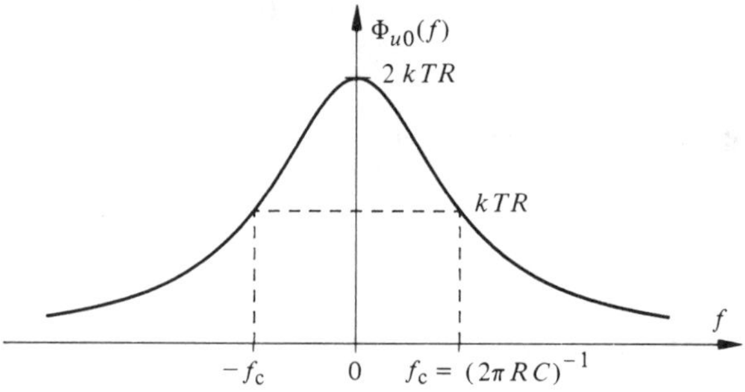

Fig. 6.4

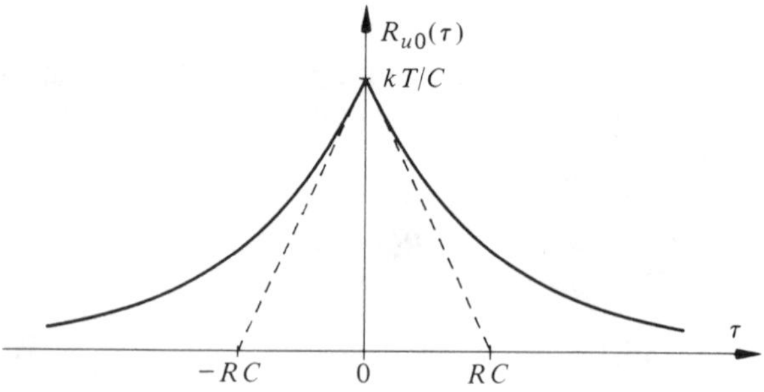

Fig. 6.5

6.2.10 Application example

The input impedance of an oscilloscope amplifier is the result of the parallel connection of a 1 MΩ resistor with a 50 pF parasitic capacitor. Assuming that the amplifier has a bandwidth $B \gg f_c = (2\pi RC)^{-1} \cong 3.2$ kHz and an internal noise which is negligible compared to the impedance thermal noise, the rms value of the input noise open-circuit voltage is roughly $\sigma_{u0} = 9\mu V$.

6.2.11 Thermal noise reduction

The total thermal noise power (6.4) depends only on bandwidth B and temperature T. Once the bandwidth is set at its absolute minimum, the only way to reduce the thermal noise further is to cool the circuit. This approach is sometimes used in special-purpose communication receivers and amplifiers.

6.2.12 Concept of noise equivalent bandwidth

The normalized total noise power in a RC network can be compared to the one generated by a single resistor R measured in a certain bandwidth, the so-called equivalent noise bandwidth, B_{eq}. Equating (6.13) and (6.24) yields

$$B_{eq} = (4RC)^{-1} = (\pi/2) f_c \qquad (6.27)$$

This concept can be extended to any linear operator (sub-sect. 8.2.23).

6.2.13 Equivalent noise temperature

Thermal noise is often used as reference, although the observed noise is the combination of effects of different origins, some of them being temperature-independent and different from white noise.

Thus, an *equivalent noise temperature* T_{en} can be associated with any noise source that has an effective power P_n. This fictional temperature is defined as that of a purely thermal noise source having the same power. From (6.4):

$$T_{en} = \frac{P_n}{kB} \quad \text{K} \qquad (6.28)$$

6.3 SHOT NOISE

6.3.1 Definition

We call *shot noise* the statistical fluctuations of the population of charged particles (electrons or holes) participating in the creation of a current crossing a potential barrier. All semiconductor *pn* junctions exhibit such a potential barrier. It can also be found in thermo-electric (vacuum tubes) and photo-electric emissions.

Unlike thermal noise, which exists independently of the presence of an average current in a resistor, shot noise results from the existence of an average current to which it is added (fig. 6.6).

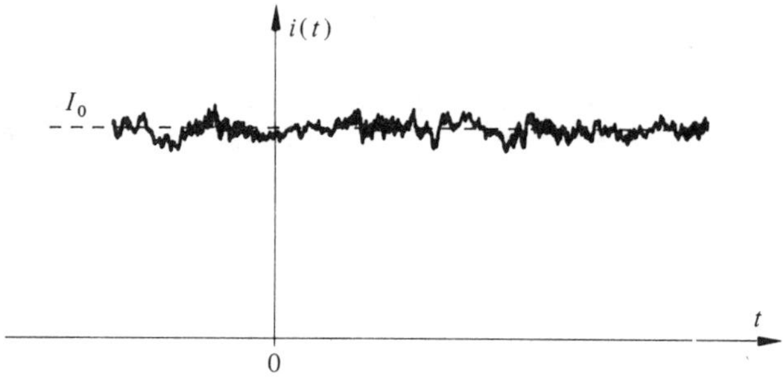

Fig. 6.6

6.3.2 Spectral density and autocorrelation function

To the extent that the charged particles can be considered independent, the number of barrier crossings per unit time follows a Poisson distribution (5.2.11). We can, as a first approximation (assuming negligible transit time of the carriers through the potential barrier), model the carrier flow as a random sequence of current pulses represented by delta functions (sub-sect. 5.8.2) weighted by the electron charge e. The equation for the current becomes, from (5.213):

$$i(t) = \sum_k e\,\delta(t - t_k) \qquad \text{A} \tag{6.29}$$

where t_k are the random instants when a carrier crosses the potential barrier and $e = 0.16 \times 10^{-18}$ C.

The current autocorrelation function is deduced from (5.221) by taking into account the charge e carried by each pulse:

$$R_i(\tau) = (e\lambda)^2 + e^2\lambda\delta(\tau) = I_0^2 + eI_0\delta(\tau) \qquad \text{A}^2 \tag{6.30}$$

where λ is the average number of charges per time unit and $I_0 = \lambda e$ is the current mean value.

Thus, the normalized power spectral density is

$$\Phi_i(f) = \Phi_{i0}(f) + \Phi_{ig}(f) = I_0^2\delta(f) + eI_0 \qquad \text{A}^2/\text{Hz} \tag{6.31}$$

The first spectral term is due to the dc component and the second results from the current fluctuations generated by shot noise (fig. 6.7)

$$\Phi_{ig}(f) = eI_0 \qquad \text{A}^2/\text{Hz} \tag{6.32}$$

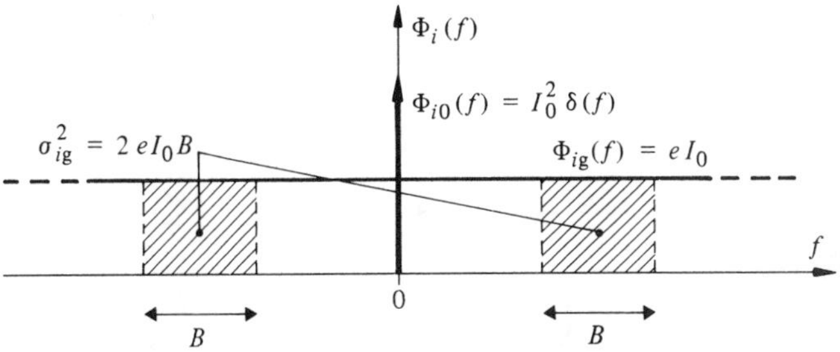

Fig. 6.7

Under our assumption, the shot noise is a ***white noise***. We can, however, take into account a more realistic shape for the current pulses and use the representation of a filtered Poisson process (5.214) in which the function $g(t)$ has a rectangular shape (mobility of carriers in solids) or a saw-toothed shape (uniform acceleration in a vacuum). The real fluctuation spectrum (problem 8.5.11) then decreases with frequencies larger than the inverse of the transit time.

The variance of shot noise current in a given bandwidth B is simply, integrating (6.32)

$$\sigma_{ig}^2 = 2eI_0B \qquad \mathrm{A}^2 \qquad \textbf{(6.33)}$$

In vacuum tubes, a space charge in the neighborhood of the cathode tends to smooth the current fluctuations. This can be taken into account through multiplying (6.33) by a noise reduction factor.

6.3.3 Equivalent network

In order to analyze shot noise in a circuit, we must introduce in parallel with the considered element or *pn* junction, a random current-dependent source, with a variance given by (6.33) (fig. 6.8).

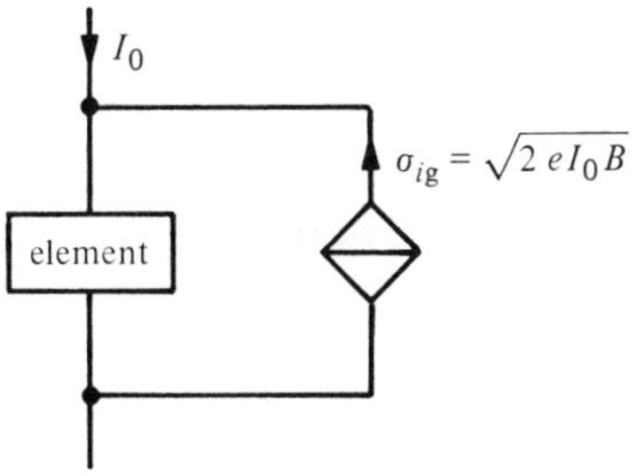

Fig. 6.8

If this current, with rms value σ_{ig}, flows into a resistor R (fig. 6.9), it produces a random voltage with a rms value:

$$\sigma_{ug} = R \cdot \sigma_{ig} \qquad (6.34)$$

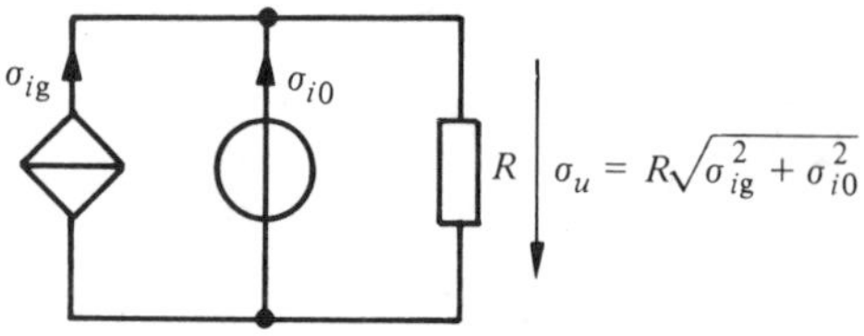

Fig. 6.9

which is added to the independent thermal noise voltage, the rms value of which is $\sigma_{u0} = \sqrt{4k\,TRB}$, and to the mean value $U_0 = R\,I_0$.

If we are only interested in the random fluctuations (fig. 6.9), we get a resulting rms value of

$$\sigma_u = \sqrt{\sigma_{ug}^2 + \sigma_{u0}^2} = R\sqrt{\sigma_{ig}^2 + \sigma_{i0}^2} \qquad (6.35)$$

An application example is described in sub-section 6.6.7 for the noise equivalent network of a transistor in linear mode.

6.3.4 Shot noise statistics

At microscopic scale, where the electrical current can be seen as a flow of individual electrons, the distribution of the population of charges crossing the potential barrier obeys Poisson's law. At normal macroscopic scales, where the mean electrical current is the result of the flow of an enormous number of charges ($\lambda = I_0/e = 6.25\mathrm{I}_0 \times 10^{18}$ electrons/s), the asymptotic limit of Poisson's law is a Gaussian law. The probability density of the total current $i(t)$ is then given by (fig. 6.10):

$$p(i) = \frac{1}{\sqrt{2\pi}\sigma_{ig}} \exp\left[-\frac{(i - I_0)^2}{2\sigma_{ig}^2}\right] \qquad (6.36)$$

6.3.5 Example: shot noise in junction diode

The voltage-current characteristic of a pn junction is approximately

$$I = I_s\left[\exp\left(\frac{U}{nU_T}\right) - 1\right] \qquad \mathrm{A} \qquad (6.37)$$

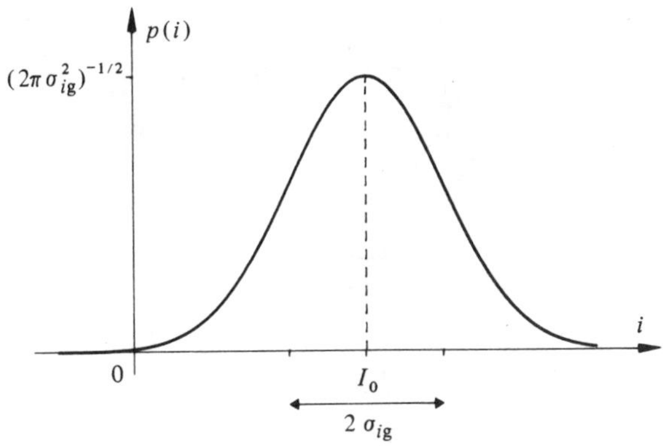

Fig. 6.10

where I_s is the reverse saturation current, n is a coefficient included on interval [1, 2], and $U_T = kT/e$ is the thermodynamical voltage of about 25 mV at ambient temperature.

In fact, the shot noise is produced by two independent contributors—direct and reverse currents [80]:

$$\sigma_{ig}^2 = \left[2eI_s \exp\left(\frac{U}{nU_T}\right) + 2eI_s\right] B$$

$$= 2e(I + 2I_s)B \qquad \text{A}^2 \tag{6.38}$$

Under normal direct bias, $I \gg I_s$; hence, (6.38) becomes

$$\sigma_{ig}^2 = 2eIB \qquad \text{A}^2 \tag{6.39}$$

For a current $I = 1$ mA and a bandwidth B respectively equal to 20 kHz and 20 MHz, the shot noise current variance is $\sigma_{ig}^2 = 6.4 \times 10^{-18}$ A^2 and $\sigma_{ig}^2 = 6.4 \times 10^{-15}$ A^2.

The equivalent network (fig. 6.11) of shot noise in a diode or *pn* junction in linear mode is a current source with rms value σ_{ig} in parallel with the differential conductance (dynamic characteristic; i.e., without thermal noise):

$$g_d = \frac{dI}{dU} = \frac{I_s}{nU_T} \exp\left(\frac{U}{nU_T}\right) = \frac{e(I + I_s)}{nkT} \tag{6.40}$$

When current $i_g(t)$ flows through the differential conductance g_d, it produces a noise voltage whose rms value is

$$\sigma_{ug} = \sigma_{ig}/g_d \tag{6.41}$$

For $I = 1$ mA, $n = 1$, $T = 293$ K and $B = 20$ MHz: $\sigma_{ug} = 2.0$ μV.

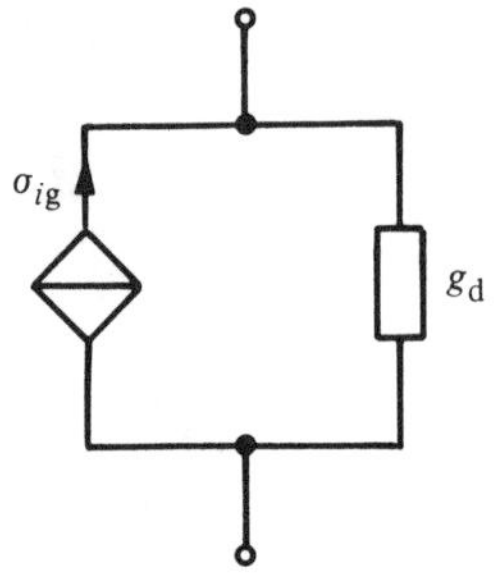

Fig. 6.11

Combining (6.38) and (6.40), we get

$$\sigma_{ig}^2 = n2kTg_dB\,\frac{I + 2I_s}{I + I_s} \approx n2kTg_dB \tag{6.42}$$

For $I = 0$ and $n = 1$, (6.42) is equal to the thermal noise in a normal conductance g_d. In direct bias ($I \gg I_s$), this variance is half that of the thermal noise in such a conductance. When a strong direct bias is applied, the diode equivalent network must be completed with a serial noisy resistor.

6.4 FLICKER (1/*f*) NOISE

6.4.1 Random conduction modulation in some electronic devices

At frequencies higher than a few kilohertz or tens of kilohertz, the background noise of electric and electronic devices is mainly white, and depends almost completely on thermal and shot noises. At lower frequencies, the power spectral density increases as frequency decreases. This phenomenon has not yet received a clear and complete explanation, and seems to be due to conduction fluctuations caused by surface defects or inhomogeneity of the conducting material:

- transient modifications of surface state of cathodes in vacuum tubes
- change in the surface recombination rate of electron-hole pairs in semiconductors because of inherent defects
- changes in the contact resistance between grains in carbon resistors and charcoal microphones, *et cetera*

This type of additional noise is highly dependent on technology and surface passivation methods.

6.4.2 Spectral density

There is no model that yields a precise theoretical expression for the spectral density of this additional noise. It has been experimentally observed that this density approximately varies as the inverse of frequency (fig. 6.12):

$$\Phi(f) = k \frac{f_a}{|f|^{\alpha}} \tag{6.43}$$

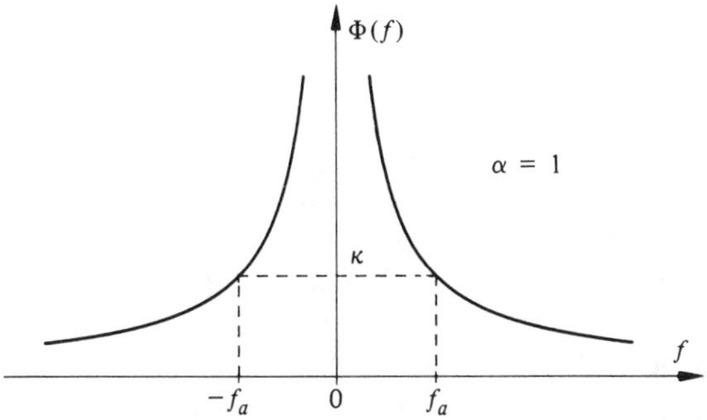

Fig. 6.12

where $0 < \alpha < 2$ and is generally close to 1; this is why *flicker noise* is also called *low-frequency* or *1/f noise*.

Relation (6.43) has been experimentally checked down to frequencies as low as 10^{-6} Hz. If we assume this to be a stationary process, an upper bound must be attained as frequency tends toward zero, otherwise the noise power would be infinite. However, this is overcome by assuming a nonstationary process in which the influence of the past on the present is great [81]. The low-frequency noise imposes severe limitations on the direct amplification of dc or very low frequency signals. In some critical applications (physics, bioengineering, *et cetera*), an indirect amplification method (lock-in amplifiers) based on synchronous modulation-demodulation (sub-sect. 13.2.10) is used.

6.4.3 Other processes with $1/f$ spectral behavior

Many other random processes, unrelated to conduction problems, produce fluctuations with spectrum similar to (6.43):

- frequency of quartz oscillators
- seasonal temperature variations
- biological parameters

- economic data
- musical phenomena
- *et cetera*

This suggests that the nonstationary model [81] of the low frequency noise is likely to be applicable to many natural phenomena.

6.5 OTHER NOISE SOURCES

6.5.1 Partition noise

This kind of noise occurs when the current is subdivided into two or more subcurrents. For instance, it can be observed in a bipolar transistor, where carriers injected from the emitter go either into the base (recombination) or into the collector, or in vacuum tubes with more than one positive electrode. In a transistor, the carrier on its way to the collector may or may not be recombined in the base; similarly, in a pentode, an electron may be equally captured by any positive electrode.

This arbitrary current repartition generates a random fluctuation superimposed on the shot noise.

6.5.2 Avalanche noise

An important type of noise is generated by *pn* junctions working at their reverse breakdown voltage (Zener diodes). It is associated with the avalanche phenomenon that the crystal atom ionizations produce when they collide with charge carriers, accelerated by the applied electric field.

It can, in fact, be considered as an amplified shot noise.

6.5.3 Induced noise

At high frequencies, the noise current flowing through some electronic devices (triodes, FETs) induces a noise current in the gate by capacitive coupling.

6.5.4 Secondary emission noise

In devices based on current amplification by secondary emission (photomultipliers), the number of secondary electrons emitted by every primary electron varies randomly. This yields an additional noise.

6.5.5 Additional information

Our presentation of background noise sources was very short and incomplete. For more information, the reader should refer himself or herself to [78, 80]. Noise measurement methods are described in [82]. The basic rules for designing low-noise circuits are described in [83, 84].

6.6 NOISE FACTOR IN LINEAR SYSTEMS

6.6.1 Signal-to-noise ratio

If $\Phi_s(f)$ is the power spectral density of a useful signal, its power over a bandwidth B (excluding any dc value) is

$$P_s = \sigma_s^2 = \int_B \Phi_s(f)\mathrm{d}f \tag{6.44}$$

Let $\Phi_n(f)$ be the noise power spectral density arising from the addition of all independent disturbances, then the total background noise power

$$P_n = \sigma_n^2 = \sum_{k=1}^{N} \sigma_k^2 \tag{6.45}$$

in the bandwidth B, corresponds to

$$P_n = \int_B \Phi_n(f)\,\mathrm{d}f \tag{6.46}$$

A measure of the signal contamination by the background noise is the *signal-to-noise ratio* defined in sub-section 1.1.5:

$$\xi = P_s/P_n \tag{6.47}$$

which is generally expressed in decibels in the form

$$\xi_{\mathrm{dB}} = 10\log(P_s/P_n) \qquad \mathrm{dB} \tag{6.48}$$

6.6.2 Average noise factor

The *noise factor F*—also called *noise figure*—at conventional temperature T_0 = 290 K defines the quality of a linear system (fig. 6.13) with respect of its own noise: it is the ratio of its input and output signal-to-noise ratios ξ_1 and ξ_2:

$$F = \xi_1/\xi_2 \tag{6.49}$$

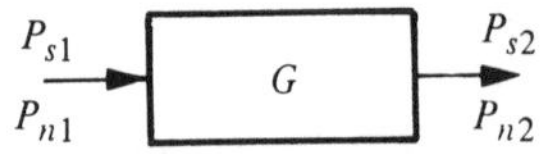

Fig. 6.13

This average factor measures the signal-to-noise ratio deterioration. If P_{ni} is the system's internal noise power and $G = P_{s2}/P_{s1}$ is its power gain, then

$$P_{n2} = GP_{n1} + P_{ni} \tag{6.50}$$

and

$$F = \frac{P_{n2}}{GP_{n1}} = 1 + \frac{P_{ni}}{GP_{n1}} \geqslant 1 \tag{6.51}$$

or, expressed in decibels:

$$F_{dB} = 10 \log F \geqslant 0 \text{ dB} \tag{6.52}$$

The term

$$\frac{P_{ni}}{GP_{n1}} = F - 1 \tag{6.53}$$

is called the *excess noise factor.* The input noise power used as reference is conventionally noted, according to (6.4), $P_{n1} = kT_0B_{eq}$, where B_{eq} is the considered noise equivalent bandwidth.

The ratio $P_{ni}/G = P_{n1}(F - 1)$ is the ***equivalent input noise.*** For an ideal linear system (noiseless): $P_{ni} = 0$ and $F = 1$.

6.6.3 Cascaded linear systems

Consider (fig. 6.14) two systems in cascade with power gains G_1 and G_2, and noise factors F_1 and F_2 defined for the same input noise power $P_{n1} = kT_0B_{eq}$. From (6.51), we have

$$P_{n2} = F_1G_1P_{n1} \tag{6.54}$$

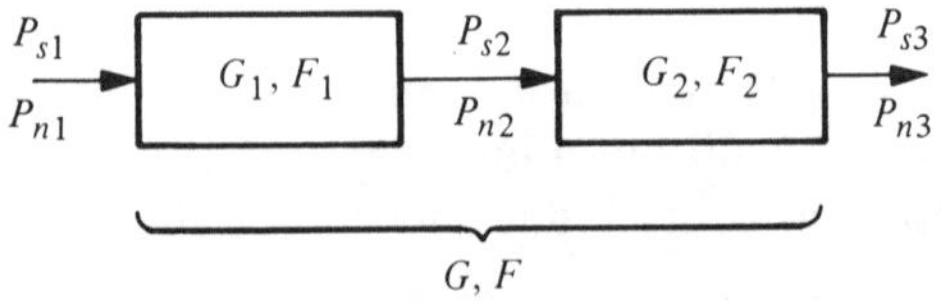

Fig. 6.14

and from (6.50) and (6.53):

$$P_{n3} = G_2P_{n2} + P_{ni2} = F_1G_1G_2P_{n1} + (F_2 - 1)G_2P_{n1} \tag{6.55}$$

However, we also have for the total system:

$$P_{n3} = FGP_{n1} = FG_1G_2P_{n1} \tag{6.56}$$

Combining (6.55) and (6.56), we get

$$F = F_1 + \frac{F_2 - 1}{G_1} \tag{6.57}$$

which can be easily generalized for any number m of cascaded systems:

$$F = F_1 + \frac{F_2 - 1}{G_1} + \frac{F_3 - 1}{G_1G_2} + \cdots + \frac{F_m - 1}{\prod_{i=1}^{m-1} G_i} \tag{6.58}$$

If $G \gg F_2 - 1$, $F \cong F_1$; hence, ***the total system noise is mainly determined by its first stage.*** We must, therefore, optimize its design to reduce the internal noise.

6.6.4 Spectral noise factor

Taking into account the real power spectral densities, we can define a noise factor, as a function of frequency, called the *spectral noise factor*:

$$F(f) = \frac{\Phi_{s1}(f)/\Phi_{n1}(f)}{\Phi_{s2}(f)/\Phi_{n2}(f)} \tag{6.59}$$

The graph of this spectral noise factor (fig. 6.15) can be split into three parts: a negative slope segment for low frequencies, where the flicker noise dominates below frequency f_a; a positive slope segment for high frequencies, resulting from the progressive gain $G(f)$ decrease, varying approximately as $1/f^2$ above a cut-off frequency f_b; and a constant part between f_a and f_b, depending only on thermal noise and shot noise.

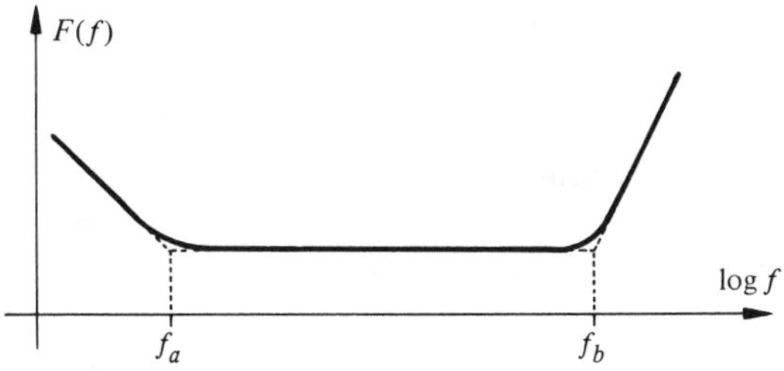

Fig. 6.15

6.6.5 Noise equivalent network of an amplifier stage

The main noise sources in an elementary amplifier stage built around a three-pole active component (bipolar transistor, JFET, MOST, vacuum tube) are the shot noises of the control current I_β and of the controlled current I_α, and the thermal noises of the serial resistor r_β of the control electrode and of the load resistor R_L (fig. 6.16). Representing, as a first approximation, the component dynamic behavior in linear mode by its transconductance $g_m = dI_\alpha/dU_\beta$, the shot noise current source associated with current $I_{\alpha 0}$ can be replaced by an input equivalent voltage source with variance:

$$\sigma_{u0\alpha}^2 = \frac{2eI_{\alpha 0}B}{g_m^2} \tag{6.60}$$

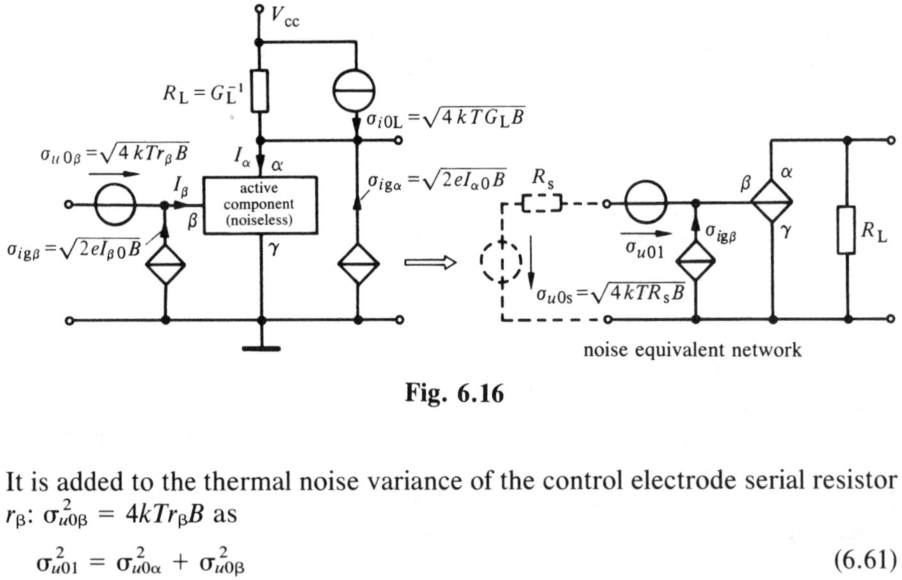

Fig. 6.16

It is added to the thermal noise variance of the control electrode serial resistor r_β: $\sigma_{u0\beta}^2 = 4kTr_\beta B$ as

$$\sigma_{u01}^2 = \sigma_{u0\alpha}^2 + \sigma_{u0\beta}^2 \tag{6.61}$$

Since g_m is generally proportional to $I_{\alpha 0}$, variance (6.60) is, in fact, inversely proportional to g_m.

If the dc bias voltage across R_L is greater than 50 mV (see problem 6.8.4), the thermal noise contribution is very small when compared to the shot noise associated with I_α.

The equivalent input noise sources are chiefly defined by the control current and the transconductance.

This simple model permits us to analyze, as a first approximation, the noise performances of the amplifier stage in bandwidth $f_b - f_a$. The thermal noise of an input biasing resistor network can be combined with the thermal noise of the signal source internal resistor. The contribution of a feedback resistor R_γ, between electrode γ and ground, adds the following term to (6.61): $4kTB(R_\gamma g_m^2)^{-1}$.

6.6.6 Amplifier stage noise factor

Let us connect a signal source with internal resistance R_s to the amplifier stage of fig. 6.16. The open circuit thermal noise voltage in R_s has a variance of

$$\sigma_{u0s}^2 = 4kTR_sB \tag{6.62}$$

The amplifier input equivalent noise voltage variance is

$$\sigma_{u0n}^2 = \sigma_{u01}^2 + \sigma_{ig\beta}^2 \cdot R_s^2 \tag{6.63}$$

The ratio of (6.63) to (6.62) corresponds to the excess noise factor (6.53); thus,

$$F = 1 + \frac{\sigma_{u01}^2 + \sigma_{ig\beta}^2 R_s^2}{4kTR_sB} = 1 + \frac{r_\beta}{R_s} + \frac{eI_{\beta 0}R_s}{2kT} + \frac{eI_{\alpha 0}}{2kTR_s g_m^2} \tag{6.64}$$

where σ_{u01}^2 is given by (6.61) and $\sigma_{ig\beta}^2 = 2eI_{\beta 0}B$.

Thus, the noise factor F depends on the source resistance. It is minimum for an optimal value R_{so} found under condition $dF/dR_s = 0$:

$$R_{so} = \sigma_{u01} / \sigma_{ig\beta} = g_m^{-1} \sqrt{\frac{I_{\alpha 0}}{I_{\beta 0}} \left[1 + \frac{2\,kTr_\beta\, g_m^2}{e\, I_{\alpha 0}}\right]} \tag{6.65}$$

This optimal resistance does not directly depend on the amplifier input resistance R_e. It does not generally correspond to the one required for maximum power transfer ($R_s = R_e$).

Then, the minimum noise factor is

$$F_{min} = 1 + \frac{\sigma_{u01} \cdot \sigma_{ig\beta}}{2kTB} \tag{6.66}$$

6.6.7 Example: amplifier stage with bipolar transistor

In an amplifier stage with a transistor in common-emitter configuration, ratio $I_{\alpha 0}/I_{\beta 0}$ is the current gain $\beta = I_{C0}/I_{\beta 0}$. The transconductance at ambient temperature is

$$g_m = I_{C0}/U_T = eI_{C0}/(kT) = 40 \cdot I_{C0} \qquad \text{AV}^{-1} \tag{6.67}$$

and r_β, usually denoted $r_{bb'}$, is the intrinsic base resistance, which is sometimes introduced in the π-hybrid equivalent network of the transistor (fig. 6.17):

$$\sigma^2_{ig\beta} = 2eBI_{C0}/\beta \tag{6.68}$$

and

$$\sigma^2_{u01} = 4kTB(r_{bb'} + g_m^{-1}/2) \tag{6.69}$$

The optimal source resistance, from (6.65), is

$$R_{so} = g_m^{-1}\sqrt{\beta(1 + 2r_{bb'} \cdot g_m)} \tag{6.70}$$

and the minimum noise factor, from (6.66), is

$$F_{min} = 1 + \sqrt{\beta^{-1}(1 + 2r_{bb'} \cdot g_m)} \tag{6.71}$$

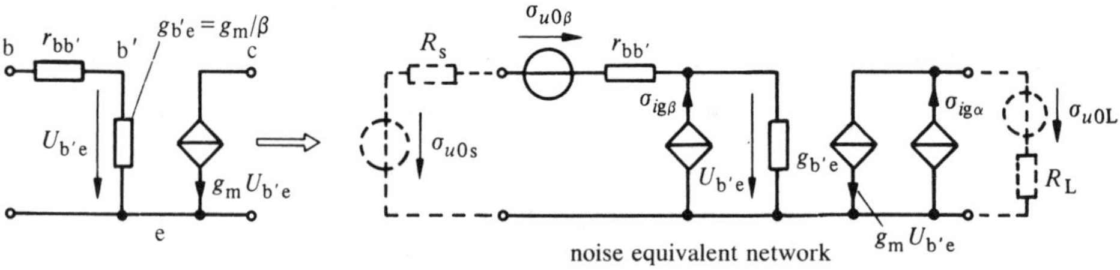

Fig. 6.17

If $r_{bb'} = 100\ \Omega$, $\beta = 100$, and $I_{C0} = 1$ mA: $g_m = 40 \times 10^{-3}$ AV^{-1} and the optimal source resistance $R_{so} = 750\ \Omega$. In this case, the minimum noise factor $F_{min} = 1.3$, or 1.14 dB. The evolution of F as a function of R_s is represented in fig. 6.18.

Alternatively, we could look for the current I_{C0} minimizing the noise factor for a given source resistance.

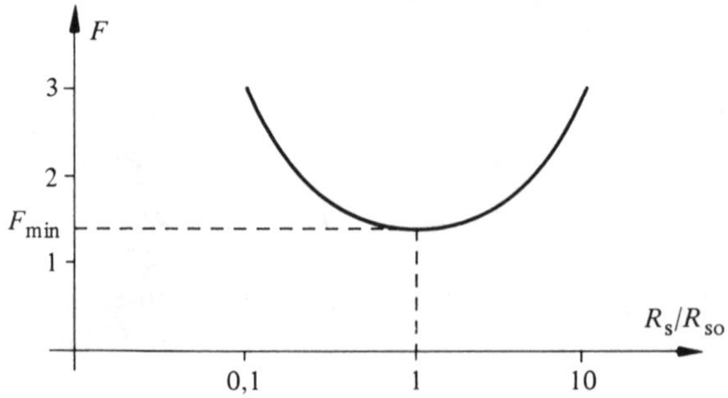

Fig. 6.18

6.6.8 Usual noise specification for transistors or amplifiers

We are generally provided with only a global noise figure. Under the assumption that the noise spectral density (in its unilateral form) is constant, the equivalent input noise voltage variance is given by

$$\sigma_{u0n}^2 = \int_B \Phi_{u0n}^+(f)\,\mathrm{d}f = \Phi_{u0n}^+ \cdot B \qquad \mathrm{V}^2 \tag{6.72}$$

and the rms value is

$$\sigma_{u0n} = \sqrt{\Phi_{u0n}^+} \cdot \sqrt{B} \tag{6.73}$$

The following specification is usually given in order to be independent of the bandwidth:

$$\sqrt{\Phi_{u0n}^+} \qquad \mathrm{V}/\sqrt{\mathrm{Hz}} \tag{6.74}$$

For example, $\sqrt{\Phi_{u0n}^+} = 32\ \mathrm{nV}/\sqrt{\mathrm{Hz}}$ corresponds to $\Phi_{u0n}^+ = 1024 \times 10^{-18}$ V^2/Hz and with a bandwidth $B = 10^6$ Hz, we get $\sigma_{u0n}^2 = 1024 \times 10^{-12}\ \mathrm{V}^2$ and $\sigma_{u0n} = 32\ \mu\mathrm{V}$.

6.7 NOISE GENERATORS

6.7.1 Use of a random excitation

Noise is generally seen as an undesirable effect. However, there are situations where it is convenient to have a source generating noise of controlled spectrum and power [85]; for example, when we want to test the noise sensitivity of an instrument or telecommunication equipment. Other examples are random phenomenon simulation, speech synthesis, dynamic identification of linear systems, *et cetera*.

6.7.2 Generator principles

There are two kinds of noise generators: those simulating noise with a pseudorandom signal (sect. 5.10) and those amplifying the background noise provided by an adequate component (Zener diode, gas tube, *et cetera*).

The interested reader can find more information in the literature [86].

6.8 PROBLEMS

6.8.1 Find the total thermal noise power available at ambient temperature $T_a = 290$ K for bandwidth $B_1 = 3.1$ kHz (telephone line) and $B_2 = 5$ MHz (TV broadcasting).

6.8.2 Find the rms value of the thermal noise voltage across the bipole in fig. 6.19 when temperature is 290 K and the bandwidth is 100 kHz.

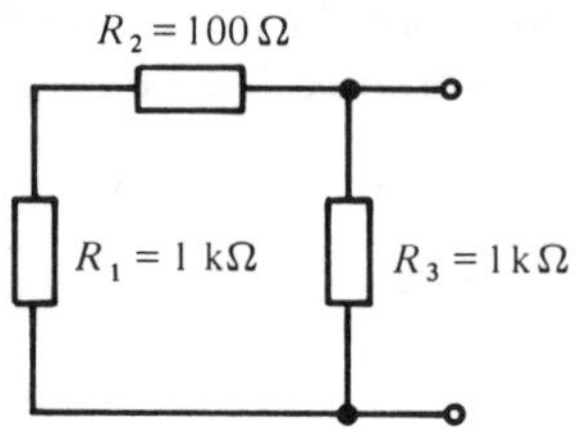

Fig. 6.19

6.8.3 Find the rms value of the total noise voltage across the circuit in fig. 6.20, working in linear mode, when $U_0 = 12$ V, temperature is 290 K, the bandwidth is 1 MHz, and the diode characteristic equation is $I = I_s [\exp(U/U_T) - 1]$ with $I_s = 43$ pA.

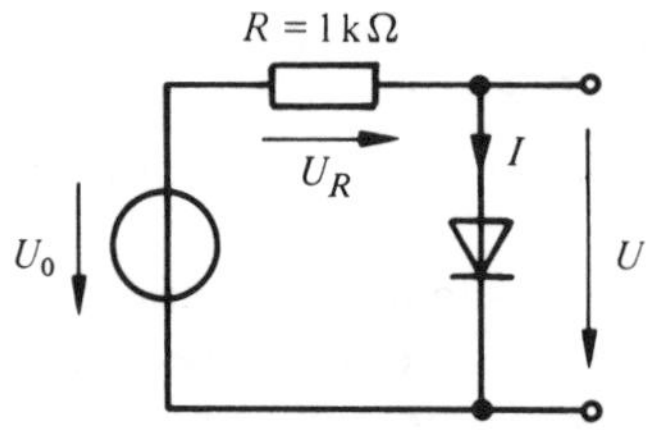

Fig. 6.20

6.8.4 Find the value of U_R for which the thermal noise and shot noise contributions are equal in the circuit of fig. 6.20.

Chapter 7

Analytic Signal and Complex Envelope

7.1 HILBERT TRANSFORM

7.1.1 Generalization of the phasor concept

In electrical engineering, the complex instantaneous value of a cosine wave $u(t) = \hat{U}\cos(\omega_0 t + \alpha)$ is denoted

$$\underline{u}(t) = \hat{U}\exp[j(\omega_0 t + \alpha)] = \hat{U}\exp(j\alpha)\exp(j\omega_0 t)$$
$$= \hat{U}\cos(\omega_0 t + \alpha) + j\hat{U}\sin(\omega_0 t + \alpha) \tag{7.1}$$

where $\hat{U} = |\underline{u}|$ is the modulus, $(\omega_0 t + \alpha) = \arg \underline{u}$ is the instantaneous phase, and α is the initial phase of $\underline{u}$. The real part of $\underline{u}(t)$ corresponds to $u(t)$.

According to (4.77), the Fourier transform of $\underline{u}(t)$ becomes (fig. 7.1), with $f_0 = \omega_0/(2\pi)$:

$$F\{\underline{u}(t)\} = \hat{U}\exp(j\alpha)\delta(f - f_0)$$
$$= \underline{\hat{U}}\delta(f - f_0) \tag{7.2}$$

where $\underline{\hat{U}} = \hat{U}\exp(j\alpha)$ is the *phasor* carrying the amplitude and initial phase information.

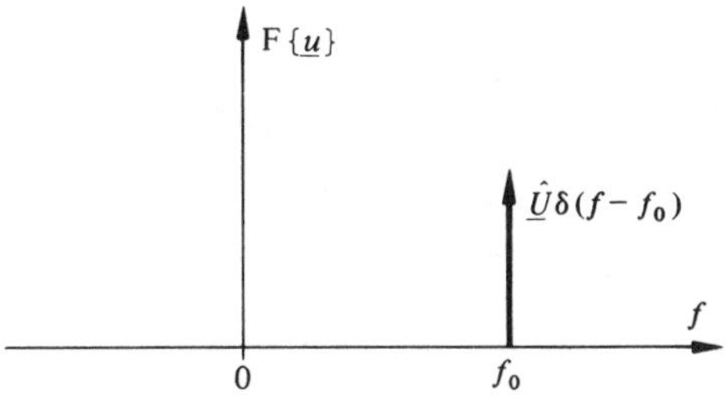

Fig. 7.1

The Fourier transform of the complex function $\underline{u}(t)$***, hence, is zero for negative frequencies.***

The Fourier transform of the real function $u(t)$, from (4.115), is

$$F\{u(t) = \hat{U}\cos(\omega_0 t + \alpha)\} = \tfrac{1}{2}\hat{U}\exp(j\alpha f/f_0) \cdot [\delta(f + f_0) + \delta(f - f_0)] \tag{7.3}$$

Consequently, (7.2) corresponds to the unilateral form of (7.3). Thus, from (4.162), we have

$$\mathrm{F}\{\underline{u}(t)\} = 2\epsilon(f)\mathrm{F}\{u(t)\}, \quad f \geqslant 0 \tag{7.4}$$

By analogy, this complex representation can be extended to a deterministic, or random, signal $x(t)$ by considering $x(t)$ to be the real part of a complex signal [10, 17]. This representation can only be applied to zero mean value signals, the spectrum of which contains no line at the origin.

7.1.2 Analytic signal

We call *analytic signal* (fig. 7.2) a complex time-function the Fourier transform of which is the unilateral form of the Fourier transform of its real part. Let us designate the real part by $x(t)$, the imaginary part by $\check{x}(t)$ and the analytic signal by $\underline{\underline{x}}(t)$; then,

$$\underline{\underline{x}}(t) = x(t) + \mathrm{j}\check{x}(t) \tag{7.5}$$

with

$$\underline{\underline{X}}(f) = \mathrm{F}\{\underline{\underline{x}}(t)\} = X^{+}(f) = 2\epsilon(f)X(f) \tag{7.6}$$

The analytic signal is sometimes referred to as the signal *pre-envelope* in the English literature.

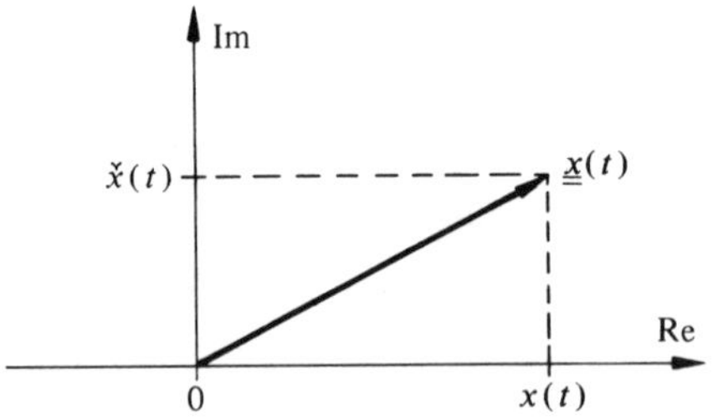

Fig. 7.2

7.1.3 Hilbert transform

Designating the Fourier transform of the imaginary part $\check{x}(t)$ by $\check{X}(f)$, according to (7.5), (7.6), and (1.20), we have

$$\underline{\underline{X}}(f) = X(f) + \mathrm{j}\check{X}(f) = 2\epsilon(f)X(f) = [1 + \mathrm{sgn}(f)]X(f) \tag{7.7}$$

Hence, $\text{sgn}(f) = j\check{X}(f)/X(f)$, thus,

$$\check{X}(f) = -j\,\text{sgn}(f)X(f) = j\,\text{sgn}(-f)X(f) \tag{7.8}$$

The imaginary part is obtained by inverse Fourier transform thanks to (4.14), (4.89), and the quasisymmetry (sect. 15.3) of the direct and inverse transforms (4.1) and (4.2), implying that if $X(f) = F\{x(t)\}:X(t) = F^{-1}\{x(-f)\}$:

$$\check{x}(t) = \frac{1}{\pi t} * x(t) = \frac{1}{\pi}\int_{-\infty}^{\infty} \frac{x(\tau)}{t-\tau}\,d\tau \tag{7.9}$$

Relation (7.9) is the *Hilbert transform* of $x(t)$ (sometimes denoted $\check{x}(t) = \text{H}\{x(t)\}$). It is also encountered in circuit theory, where it establishes a link between the real and imaginary parts of a causal system transfer function.

The inverse transform $x(t) = \text{H}^{-1}\{\check{x}(t)\}$ is obtained from (7.8) and by taking into account that

$$X(f) = j\,\text{sgn}(f)\check{X}(f) = -j\,\text{sgn}(-f)\check{X}(f) \tag{7.10}$$

thus,

$$x(t) = -\frac{1}{\pi t} * \check{x}(t) = -\frac{1}{\pi}\int_{-\infty}^{\infty} \frac{\check{x}(\tau)}{t-\tau}\,d\tau \tag{7.11}$$

Transforms (7.9) and (7.11) are defined as Cauchy principal values.

7.1.4 Example

According to (4.117), $x(t) = A\cos(2\pi f_0 t) \leftrightarrow \frac{1}{2}A[\delta(f+f_0) + \delta(f-f_0)]$. According to (7.8): $\check{X}(f) = (j/2)A[\delta(f+f_0) - \delta(f-f_0)]$; thus, from (4.118), $\check{x}(t) = A\sin(2\pi f_0 t)$. The corresponding analytic signal is $\underline{\underline{x}}(t) = A\exp(j2\pi f_0 t)$, which justifies the analogy of sub-section 7.1.1.

7.1.5 Example

Let $x(t) = A\,\text{rect}(t/T)$; from (7.9):

$$\check{x}(t) = \frac{A}{\pi}\int_{-\infty}^{\infty} \text{rect}(\tau/T)\cdot[t-\tau]^{-1}\,d\tau = \frac{A}{\pi}\int_{-T/2}^{T/2}(t-\tau)^{-1}\,d\tau = \frac{A}{\pi}\ln\left|\frac{t+T/2}{t-T/2}\right|$$

These two functions are shown in fig. 7.3.

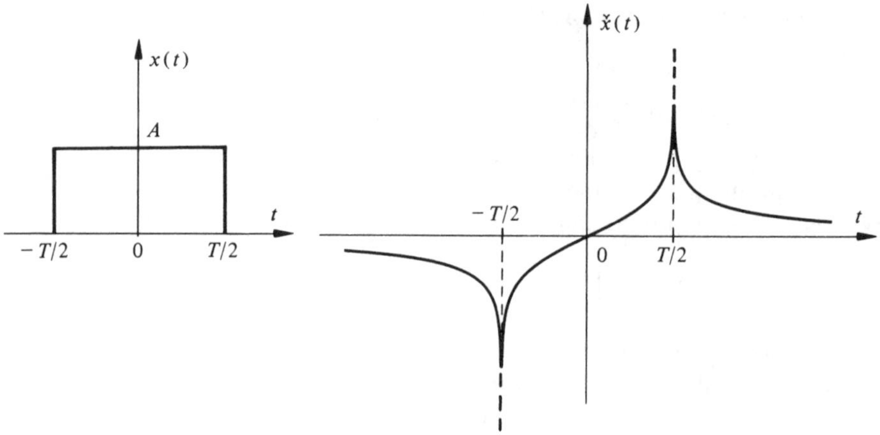

Fig. 7.3

7.2 MAIN PROPERTIES

7.2.1 Linearity

The Hilbert transform $\check{x}(t)$ of signal $x(t)$ is equal to the convolution product of $x(t)$ with function $g(t) = (\pi t)^{-1}$. Hence, it is a linear operation (sub-sect. 8.2.18) corresponding to the filtering of $x(t)$ by a linear system (fig. 7.4) with impulse response $g(t)$.

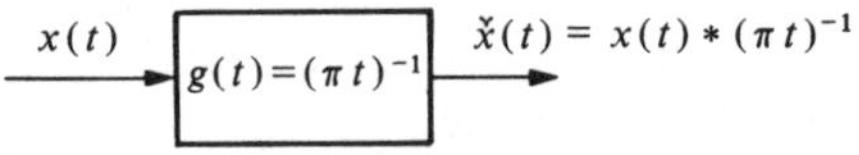

Fig. 7.4

7.2.2 Ideal phase-shifter

The Fourier transforms of a signal and its Hilbert transform are related by (7.8). Therefore, their amplitude spectra are identical:

$$|\check{X}(f)| = |X(f)| \tag{7.12}$$

while their phase spectra differ from one another only by a constant:

$$\arg \check{X}(f) = \arg X(f) - \frac{\pi}{2}\,\mathrm{sgn}(f) \tag{7.13}$$

The system in fig. 7.4 behaves like an ideal $\pm\pi/2$ phase-shifter. The signal $x(t)$ and its Hilbert transform $\check{x}(t)$ are said to be ***in quadrature.***

7.2.3 Orthogonality

The inner product $<x,\check{x}>$ is zero. Indeed, inserting (7.8) into the product theorem (4.67), we get

$$<x,\check{x}> = \varphi_{x\check{x}}(0) = \int_{-\infty}^{\infty} x(t)\check{x}(t)\mathrm{d}t = \int_{-\infty}^{\infty} X^*(f)\check{X}(f)\mathrm{d}f$$

$$= \mathrm{j}\int_{-\infty}^{\infty} |X(f)|^2 \mathrm{sgn}(-f)\mathrm{d}f = 0 \tag{7.14}$$

The signal $x(t)$ and its Hilbert transform $x(t)$ are orthogonal functions on interval $-\infty < t < +\infty$. This property can be checked with fig. 7.3.

7.2.4 Convolution product

Let $z(t) = x(t) * y(t)$, then

$$\check{z}(t) = x(t) * \check{y}(t) = \check{x}(t) * y(t) \tag{7.15}$$

Indeed, from (4.14), $Z(f) = X(f)Y(f)$ and via (7.8):

$$\check{Z}(f) = \mathrm{j}\ Z(f)\ \mathrm{sgn}(-f) = \mathrm{j}\ X(f)Y(f)\ \mathrm{sgn}(-f)$$

$$= X(f)\check{Y}(f) = \check{X}(f)Y(f) \tag{7.16}$$

We also have

$$x(t) * y(t) = -\check{x}(t) * \check{y}(t) \tag{7.17}$$

7.2.5 Product of two signals with non-overlapping bandwidth when one is of the lowpass kind

Let $z(t) = a(t)y(t)$ with $A(f) = 0$ for $|f| > B$ and $Y(f) = 0$ for $|f| < B$, then

$$\check{z}(t) = a(t)\check{y}(t) \tag{7.18}$$

Indeed, we can easily check that, in this case, we have

$$\check{Z}(f) = \mathrm{j}\ \mathrm{sgn}(-f)Z(f) = \mathrm{j}\ \mathrm{sgn}(-f)[A(f) * Y(f)]$$

$$= A(f) * \mathrm{j}\ \mathrm{sgn}(-f)Y(f) \tag{7.19}$$

7.2.6 Crosscorrelation

Let $\mathring{\varphi}_{xy}(\tau) = x(-\tau) * y(\tau)$ or $\varphi_{xy}(\tau) = x(-\tau) \overline{*} y(\tau)$ be the crosscorrelation function of two finite energy or finite mean power signals. With a method similar to that of sub-section 7.2.4, we can see that its Hilbert transform is equal to

$$\check{\mathring{\varphi}}_{xy}(\tau) = \mathring{\varphi}_{x\check{y}}(\tau) = -\mathring{\varphi}_{\check{x}y}(\tau);\ \check{\varphi}_{xy}(\tau) = \varphi_{x\check{y}}(\tau) = -\varphi_{\check{x}y}(\tau) \tag{7.20}$$

Especially, if $y(t) = x(t)$, we get

$$\check{\mathring{\varphi}}_{x}(\tau) = \mathring{\varphi}_{x\check{x}}(\tau) = -\mathring{\varphi}_{\check{x}x}(\tau);\ \check{\varphi}_{x}(\tau) = \varphi_{x\check{x}}(\tau) = -\varphi_{\check{x}x}(\tau) \tag{7.21}$$

7.2.7 Causal signal

According to (2.25) and (4.89), the Fourier transforms of the even and odd parts of a causal real signal, except for a j factor, are Hilbert transforms of each other.

7.2.8 Functional representation of the negative and positive frequency parts of the spectrum

Let $x(t)$ be a zero mean value signal

$$x(t) = \int_{-\infty}^{\infty} X(f) \exp(\mathrm{j}2\pi f t)\mathrm{d}f \tag{7.22}$$

with

$$x_-(t) = \int_{-\infty}^{0} X(f) \exp(\mathrm{j}2\pi f t)\mathrm{d}f \tag{7.23}$$

$$x_+(t) = \int_{0}^{\infty} X(f) \exp(\mathrm{j}2\pi f t)\mathrm{d}f \tag{7.24}$$

These two components of signal $x(t)$ are complex conjugates $x_+(t) = x_-^*(t)$. They are directly related to the analytic signal as a result of (7.6):

$$x_+(t) = \tfrac{1}{2}\underline{\underline{x}}(t) = \tfrac{1}{2}[x(t) + \mathrm{j}\,\check{x}(t)] \tag{7.25}$$

$$x_-(t) = \tfrac{1}{2}\underline{\underline{x}}^*(t) = \tfrac{1}{2}[x(t) - \mathrm{j}\,\check{x}(t)] \tag{7.26}$$

Hence,

$$x(t) = x_-(t) + x_+(t) = \tfrac{1}{2}[\underline{\underline{x}}(t) + \underline{\underline{x}}^*(t)] \tag{7.27}$$

$$\check{x}(t) = \mathrm{j}[x_-(t) - x_+(t)] = \frac{\mathrm{j}}{2}[\underline{\underline{x}}^*(t) - \underline{\underline{x}}(t)] \tag{7.28}$$

7.2.9 Autocorrelation and spectrum of finite energy signals

The energy spectral densities and autocorrelation functions of a finite energy signal $x(t)$ and its Hilbert transform are identical. Indeed, from (4.55) and (7.12), we have

$$\mathring{\Phi}_x(f) = |X(f)|^2 = |\check{X}(f)|^2 = \mathring{\Phi}_{\check{x}}(f) \tag{7.29}$$

and by inverse Fourier transform:

$$\mathring{\varphi}_x(\tau) = \mathring{\varphi}_{\check{x}}(\tau) \tag{7.30}$$

Similar results are established for deterministic and random finite mean power signals.

7.2.10 Random signals

Although definition 7.1.2 cannot be strictly applied to random signals (the Fourier transform of which is not defined), the analytic signal and Hilbert transform concepts can be extended straightforwardly to this kind of signal. The analytic signal (7.5) of a stationary, ergodic, zero mean value, random signal $x(t)$ is found by adding to it an imaginary part defined by convolution (7.9). We then have, as with (7.29) and (7.30), an identity of autocorrelation functions and of power spectral densities:

$$R_{\check{x}}(\tau) = R_x(\tau) \tag{7.31}$$

thus,

$$\Phi_{\check{x}}(f) = \Phi_x(f) \tag{7.32}$$

Moreover, by analogy with (7.21):

$$\check{R}_x(\tau) = R_{x\check{x}}(\tau) = -R_{\check{x}x}(\tau) \tag{7.33}$$

with the particular value at origin

$$R_{x\check{x}}(0) = R_{\check{x}x}(0) = 0 \tag{7.34}$$

which imply that signal $x(t)$ and its Hilbert transform $\hat{x}(t)$ are uncorrelated (and, hence, statistically independent in the case of a Gaussian distribution [see sub-sect. 5.7.3], since the Hilbert transform is linear, thus conserving the Gaussian property).

From (7.8) and (7.33), we draw the equivalence:

$$\text{F}\{\check{R}_x(\tau)\} = \text{j}\ \text{sgn}(-f)\Phi_x(f) = \Phi_{x\check{x}}(f) = -\Phi_{\check{x}x}(f) \tag{7.35}$$

7.2.11 Demonstrations

The results of the previous sub-section are easily found because, for an ergodic stationary signal, due to (5.59), we have

$$R_{\check{x}}(\tau) = \check{x}(-\tau) \overline{*} \check{x}(\tau) = x(-\tau) * (-\pi\tau)^{-1} \overline{*} x(\tau) * (\pi\tau)^{-1}$$
$$= R_x(\tau) * [(-\pi\tau)^{-1} * (\pi\tau)^{-1}] = R_x(\tau) * \mathrm{F}^{-1}\{\mathrm{j}\,\mathrm{sgn}(f) \cdot \mathrm{j}\,\mathrm{sgn}(-f)\}$$
$$= R_x(\tau) * \mathrm{F}^{-1}\{1\} = R_x(\tau) * \delta(\tau) = R_x(\tau) \quad (7.36)$$

and, taking (7.17) into account,

$$\check{R}_x(\tau) = R_x(\tau) * (\pi\tau)^{-1} = x(-\tau) \overline{*} x(\tau) * (\pi\tau)^{-1}$$
$$= x(-\tau) \overline{*} \check{x}(\tau) = -\check{x}(-\tau) \overline{*} x(\tau)$$
$$= R_{x\check{x}}(\tau) = -R_{\check{x}x}(\tau) \quad (7.37)$$

Therefore,

$$R_{x\check{x}}(0) = \int_{-\infty}^{\infty} \mathrm{F}\{R_x(\tau) * (\pi\tau)^{-1}\}\mathrm{d}f = \mathrm{j}\int_{-\infty}^{\infty} \Phi_x(f)\mathrm{sgn}(-f)\mathrm{d}f = 0 \quad (7.38)$$

since $\Phi_x(f)$ is an even function and $\mathrm{sgn}(-f)$ is an odd one.

7.2.12 Double Hilbert transform

As a general result, we have

$$\check{\check{x}}(t) = -x(t) \quad (7.39)$$

which can be drawn from (7.15) and (7.17).

7.2.13 Autocorrelation function and spectral density of the analytic signal

The autocorrelation function of the analytic signal $\underline{\underline{x}}(t)$ associated with a real random signal $x(t)$ is given by

$$R_{\underline{\underline{x}}}(\tau) = 2[R_x(\tau) + \mathrm{j}\check{R}_x(\tau)] \quad \mathbf{(7.40)}$$

where $\check{R}_x(\tau) = \mathrm{H}\{R_x(\tau)\}$. Expression (7.40) also has a format like an analytic signal.

The spectral density of the analytic signal $\underline{\underline{x}}(t)$ is also bound to the random signal $x(t)$ spectrum by

$$\Phi_{\underline{\underline{x}}}(f) = 4\epsilon(f)\Phi_x(f) = 2\Phi_x^{+}(f) \quad \mathbf{(7.41)}$$

where $\epsilon(f)$ is the unit-step function defined by (1.20). This relation is illustrated by fig. 7.5.

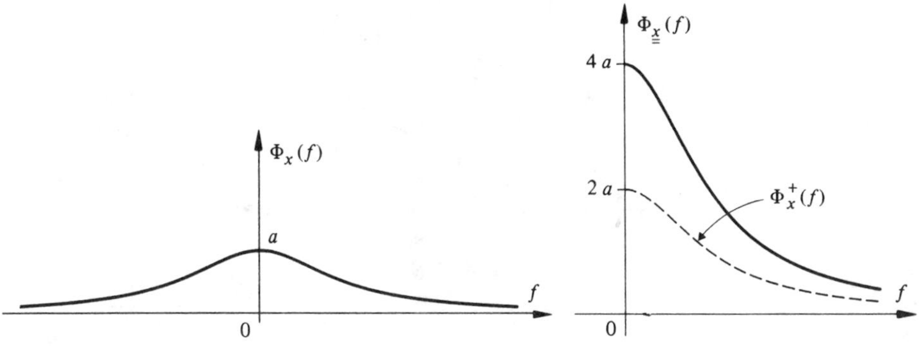

Fig. 7.5

7.2.14 Demonstration

The analytic signal being a complex function, its autocorrelation function, due to (5.78), (7.31), and (7.33), is

$$\begin{aligned} R_{\underline{x}}(\tau) &= \mathrm{E}[\underline{x}^*(t)\underline{x}(t+\tau)] \\ &= R_x(\tau) + R_{\check{x}}(\tau) + \mathrm{j}\, R_{x\check{x}}(\tau) - \mathrm{j}\, R_{\check{x}x}(\tau) \\ &= 2[R_x(\tau) + \mathrm{j}\, \check{R}_x(\tau)] \end{aligned} \tag{7.42}$$

By Fourier transform and (7.35):

$$\begin{aligned} \Phi_{\underline{x}}(f) &= 2[1 + \mathrm{sgn}(f)]\Phi_x(f) \\ &= 4\epsilon(f)\Phi_x(f) = 2\Phi_x^+(f) \end{aligned} \tag{7.43}$$

7.3 REAL ENVELOPE AND PHASE OF A SIGNAL

7.3.1 Polar form of the analytic signal

The Cartesian form (7.5) of the analytic signal is equivalent to the polar form (fig. 7.6):

$$\underline{x}(t) = r_x(t) \exp[\mathrm{j}\, \phi_x(t)] \tag{7.44}$$

the modulus of which is

$$r_x(t) = |\underline{x}(t)| = \sqrt{x^2(t) + \check{x}^2(t)} \tag{7.45}$$

and the argument of which is

$$\phi_x(t) = \arg \underline{x}(t) = \arctan[\check{x}(t)/x(t)] \tag{7.46}$$

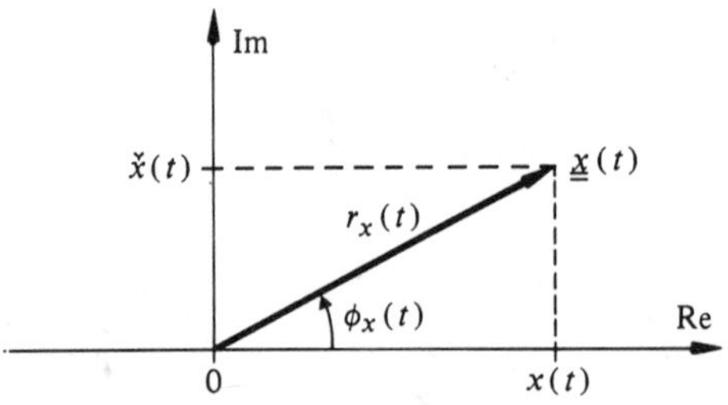

Fig. 7.6

7.3.2 Signal envelope and instantaneous phase models

A model of the envelope of signal $x(t)$ is, by definition, the modulus $r_x(t)$ of the analytic signal $\underline{\underline{x}}(t)$. Other definitions are also suggested [87].

The model of the instantaneous phase of signal $x(t)$ is, by definition, the argument $\phi_x(t)$ of the analytic signal $\underline{\underline{x}}(t)$.

7.3.3 Instantaneous frequency

The instantaneous angular frequency is the time-derivative of the instantaneous phase:

$$\omega_i(t) = \mathrm{d}\phi_x(t)/\mathrm{d}t \tag{7.47}$$

and, consequently, the instantaneous frequency is

$$f_i(t) = \frac{\omega_i(t)}{2\pi} = \frac{1}{2\pi}\frac{\mathrm{d}\phi_x(t)}{\mathrm{d}t} = \frac{x(t)\dot{\breve{x}}(t) - \dot{x}(t)\breve{x}(t)}{2\pi r_x^2(t)} \tag{7.48}$$

where $\dot{x}(t) = dx/dt$.

This function approximately corresponds to the output of a frequency discriminating device.

7.3.4 Example: sine wave

For a cosine signal $x(t) = A\cos(\omega_0 t + \alpha)$, according to sub-section 7.1.4, we have $\breve{x}(t) = A\sin(\omega_0 t + \alpha)$ and $\underline{\underline{x}}(t) = A\exp[\mathrm{j}(\omega_0 t + \alpha)]$; thus, the envelope is $r_x(t) = |A|$ and the instantaneous phase is $\phi_x(t) = \omega_0 t + \alpha$. The instantaneous angular frequency is $\omega_i(t) = d\phi_x(t)/\mathrm{d}t = \omega_0$ and the instantaneous frequency $f_i(t) = \omega_i(t)/(2\pi) = f_0$.

The definitions 7.3.2 and 7.3.3 are thus consistent with the classical concepts of phase, angular frequency, and frequency. Moreover, the proposed model signal envelope is consistent with the intuitive idea we can have of it.

By (7.41), the spectral density of the analytic signal associated with a sinewave is $\Phi_{\underline{x}}(f) = \delta(f - f_0)$ and its autocorrelation function is $\varphi_{\underline{x}}(\tau) = \exp(\,j2\pi f_0\tau)$.

7.3.5 Example: modulated sine wave

Consider a sine wave, whose amplitude $a(t)$ is a low-pass spectrum time function of zero value for $|f| > B$ with $B < f_0 = \omega_0/(2\pi)$: $x(t) = a(t)\cos(\omega_0 t + \alpha)$. Its Hilbert transform, from (7.18) and previous example, is $\check{x}(t) = a(t)\sin(\omega_0 t + \alpha)$. Hence, the corresponding analytic signal is $\underline{x}(t) = a(t)\exp[j(\omega_0 t + \alpha)]$; thus, the envelope is $r_x(t) = |a(t)|$ and the instantaneous phase is $\phi_x(t) = \omega_0 t + \alpha$.

This signal and its envelope are represented in figure 7.7.

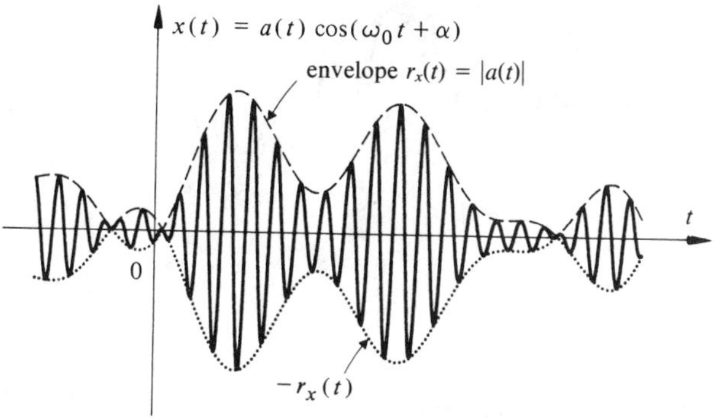

Fig. 7.7

7.3.6 Example: transient signal

Figure 7.8 represents a transient signal $y(t)$ generated by the start of an electrical motor and its envelope, calculated according to (7.45) by a computer.

7.3.7 Example: random signal

Figure 7.9 represents a random signal (in reality, it is a computer-simulated pseudorandom signal) and its computed envelope.

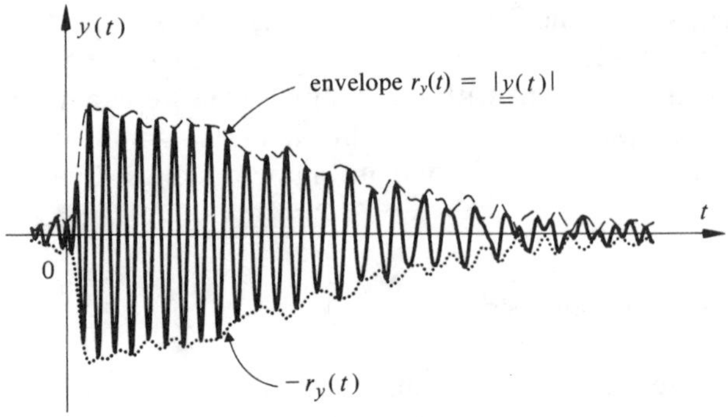

Fig. 7.8

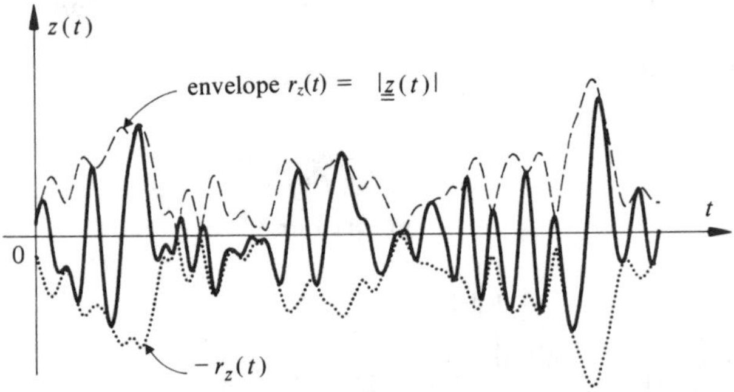

Fig. 7.9

7.3.8 Statistics of the envelope and instantaneous phase of a Gaussian noise

If $n(t)$ is a Gaussian noise with zero mean value and σ_n^2 variance its Hilbert transform $\check{n}(t) = n(t) * (\pi t)^{-1}$ is also a Gaussian noise with same variance, thanks to (7.31), uncorrelated, and thus ***independent***, because of (7.34).

Thus, the envelope $r_n(t) = [n^2(t) + \check{n}^2(t)]^{1/2}$ is also random and its statistical distribution is easily found by orthogonal-to-polar coordinate transformation (sub-sect. 5.1.23).

If $n(t)$ and $\check{n}(t)$ are Gaussian, the probability distribution of the envelope is a Rayleigh distribution (5.52):

$$p_r(r_n) = \frac{r_n}{\sigma_n^2} \exp\left[-\frac{r_n^2}{2\sigma_n^2}\right]; \quad r_n \geqslant 0 \tag{7.49}$$

The instantaneous phase is a random variable with a uniform distribution, according to (5.53), and independent of the envelope, according to (5.54):

$$p_\phi(\phi) = \frac{1}{2\pi}\,; \quad 0 \leq \phi < 2\pi \tag{7.50}$$

7.3.9 Distribution of the envelope and instantaneous phase of a signal disturbed by additive Gaussian noise

Let a noisy signal be

$$x(t) = s(t) + n(t) \tag{7.51}$$

where $s(t)$ is the useful signal and $n(t)$ is a Gaussian noise with a zero mean value and a variance σ_n^2.

The noisy signal envelope is given by

$$r_x(t) = \sqrt{x^2(t) + \check{x}^2(t)} = \sqrt{[s(t) + n(t)]^2 + [\check{s}(t) + \check{n}(t)]^2} \tag{7.52}$$

With

$$x = s + n = r_x \cos \phi; \quad \check{x} = \check{s} + \check{n} = r_x \sin \phi \tag{7.53}$$

and, using the orthogonal-to-polar coordinate transformation (sub-sect. 5.1.23), we get, by inserting n and $\check{n}$ from (7.53) into (5.49) and evaluating (5.47):

$$\begin{aligned} p_{r\phi}(r_x,\phi_x) &= \frac{r_x}{2\pi\sigma_n^2} \exp\left[-\frac{(s - r_x\cos\phi_x)^2 + (\check{s} - r_x \sin\phi_x)^2}{2\sigma_n^2}\right] \\ &= \frac{r_x}{2\pi\sigma_n^2} \exp\left[-\frac{r_x^2 + r_s^2}{2\sigma_n^2}\right] \exp\left[\frac{r_x r_s \cos(\phi_x - \phi_s)}{\sigma_n^2}\right] \end{aligned} \tag{7.54}$$

where

$$r_s(t) = \sqrt{s^2(t) + \check{s}^2(t)} \tag{7.55}$$

is the envelope of the useful signal alone, with $s = r_s \cos \phi_s$ and $\check{s} = r_s \sin \phi_s$.

Inserting the modified Bessel function of the first kind and zero order [88]:

$$I_0(\alpha) = \frac{1}{2\pi}\int_0^{2\pi} \exp(\alpha \cos\phi)\mathrm{d}\phi \tag{7.56}$$

and by integrating (7.54) with respect to $\phi = \phi_x - \phi_s$, we get the probability density of the envelope $r_x(t)$, called the *Rice-Nakagami distribution* (fig. 7.10):

$$p_r(r_x) = \frac{r_x}{\sigma_n^2} \exp\left[-\frac{r_x^2 + r_s^2}{2\sigma_n^2}\right] \cdot I_0\left(\frac{r_x r_s}{\sigma_n^2}\right)\,; \quad r_x \geq 0 \tag{7.57}$$

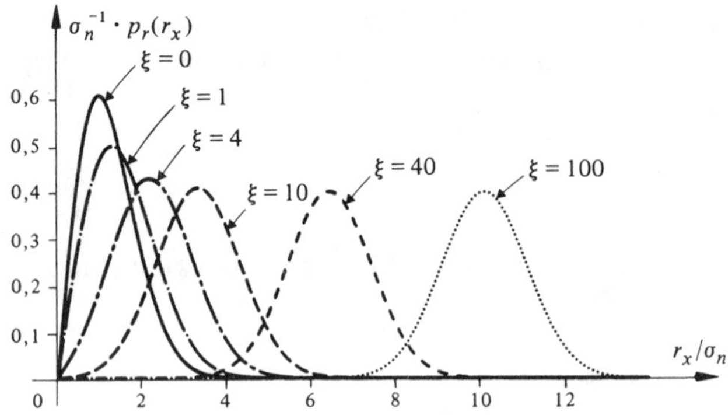

Fig. 7.10 Rice-Nakagami distribution for various values of the signal-to-noise ratio $\xi = r_s^2/\sigma_n^2$.

In the absence of a useful signal: $s(t) = 0$, $r_s(t) = 0$, and $r_x(t) = r_n(t)$. The value $I_0(0) = 1$, and (7.57) is identical to (7.49).

In the case of a simple cosine signal $s(t) = A\cos(\omega_0 t + \alpha)$, $r_s(t) = |A|$.

The Rice-Nakagami distribution is used to evaluate the performances of telecommunication or radar systems using the envelope detection method. For a high signal-to-noise ratio $\xi = r_s^2/\sigma_n^2$, it is close to a Gaussian law.

The instantaneous phase distribution is obtained by integrating (7.54) with respect to r_x[24]:

$$p_\phi(\phi) = (2\pi)^{-1}\exp[-r_s^2/(2\sigma_n^2)] \cdot \{1 + \gamma\sqrt{\pi}\exp(\gamma^2)[1 + \operatorname{erf}\gamma]\} \tag{7.58}$$

where $\gamma = r_s \cos\phi/(\sqrt{2}\,\sigma_n)$ and erf γ is the error function defined by (14.108).

7.3.10 Random phasor

Some random phenomena can be represented by a vectorial sum of random components. Each component is then described by a vector of the kind (7.44):

$$\mathbf{x}_k = \mathbf{r}_k \exp(j\boldsymbol{\phi}_k) \tag{7.59}$$

where the modulus $\mathbf{r}_k$ and instantaneous phase $\boldsymbol{\phi}_k$ are random variables. Such a vector is called *random phasor*.

The study of a phenomenon corresponding to such a model can be seen as the study of the behavior of the resulting phasor:

$$\mathbf{r}\exp(j\boldsymbol{\phi}) = \sum_{k=0}^{n} \mathbf{r}_k \exp(j\boldsymbol{\phi}_k) \tag{7.60}$$

having a random modulus $\mathbf{r}$ and a random phase $\boldsymbol{\phi}$ (fig. 7.11).

We sometimes call such a phenomenon a *random walk*.

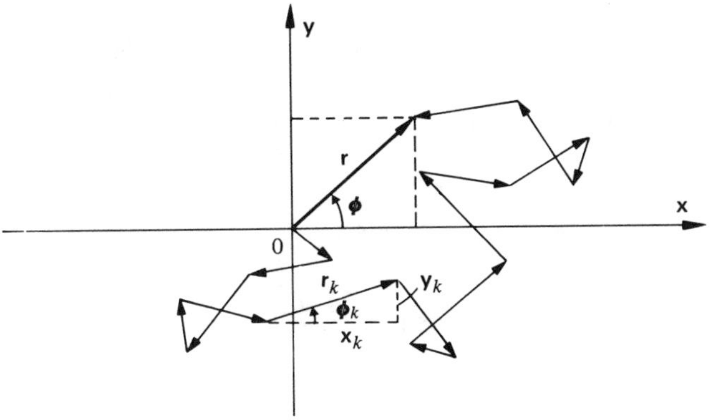

Fig. 7.11

7.3.11 Applications

This model allows us to study the behavior of the envelope of a signal of given amplitude and phase with additive noise by a sum of random phasors model. It can also be used to characterize the fluctuations disturbing a radio-electric signal transmission when many signals with same frequency interfere with one another. These interferences can be produced by a series of reflections on the ionized atmospheric layers (fig. 7.12), or by atmospherics inducing local variations of the refractive index, generating the so-called fading phenomenon.

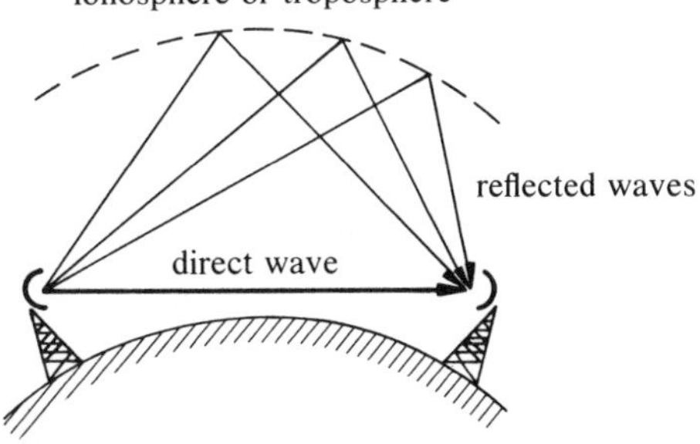

Fig. 7.12

An approximating model of this kind of disturbance is achieved by expressing the received signal as the sum of the respective contributions of the direct wave and the many reflected waves:

$$\mathbf{r}\exp(\,j\phi) = \underbrace{\mathbf{r}_0\exp(\,j\phi_0)}_{\text{direct wave}} + \sum_{k=1}^{n}\underbrace{\mathbf{r}_k\exp(\,j\phi_k)}_{\text{reflected waves}} \tag{7.61}$$

To simplify, the direct wave phase ϕ_0 can be taken as reference phase (fig. 7.13).

The general method of resolution again calls on the transformation of the orthogonal coordinates **x** and **y** into the polar ones, **r** and $\boldsymbol{\phi}$ (sub-sect. 5.1.23), with

$$\mathbf{x} = \mathbf{r} \cos \boldsymbol{\phi} = \sum_{k=0}^{n} \mathbf{x}_k = \sum_{k=0}^{n} \mathbf{r}_k \cos \boldsymbol{\phi}_k \tag{7.62}$$

$$\mathbf{y} = \mathbf{r} \sin \boldsymbol{\phi} = \sum_{k=0}^{n} \mathbf{y}_k = \sum_{k=0}^{n} \mathbf{r}_k \sin \boldsymbol{\phi}_k \tag{7.63}$$

Knowing the statistical properties of the $\mathbf{r}_k$ and $\boldsymbol{\phi}_k$, we can derive the joint probability density of **x** and **y**, $p(x,y)$. Taking into account (5.47), we can then find the joint probability density of **r** and $\boldsymbol{\phi}$, $p(r,\phi)$.

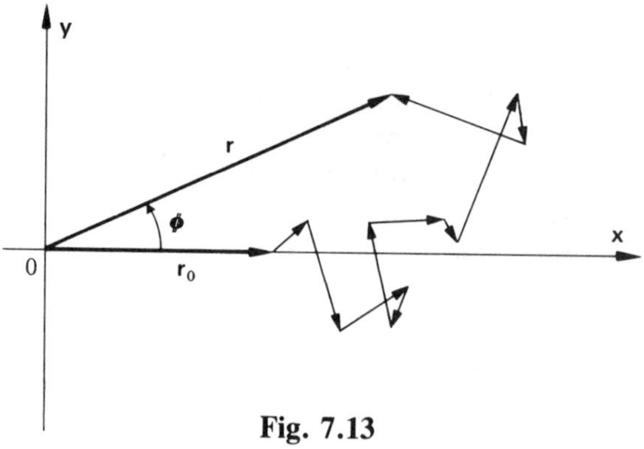

Fig. 7.13

Then, the marginal probability densities are

$$p_r(r) = \int_0^{2\pi} p(r,\phi)\, \mathrm{d}\phi \tag{7.64}$$

$$p_\phi(\phi) = \int_0^{\infty} p(r,\phi)\, \mathrm{d}r \tag{7.65}$$

7.3.12 Sum of independent random phasors: Rayleigh phasor

Consider the sum (with $\mathbf{r}_0 = 0$, reflected waves alone):

$$\mathbf{r} \exp(\mathrm{j}\boldsymbol{\phi}) = \sum_{k=1}^{n} \mathbf{r}_k \exp(\mathrm{j}\boldsymbol{\phi}_k) \tag{7.66}$$

in which all the phasors $\mathbf{x}_k = \mathbf{r}_k \exp(\,j\boldsymbol{\phi}_k)$ are statistically independent. Moreover, assume that the variables $\mathbf{r}_k$ have the same distribution and the probability density of phases $\boldsymbol{\phi}_k$ is uniform:

$$p_\phi(\phi) = \frac{1}{2\pi}, \quad 0 \leqslant \phi < 2\pi \tag{7.67}$$

and, finally, assume also variables $\mathbf{r}_k$ and $\boldsymbol{\phi}_k$ are independent.

If n in (7.66) is large enough, the random variables:

$$\mathbf{x} = \sum_{k=1}^{n} \mathbf{r}_k \cos \boldsymbol{\phi}_k \tag{7.68}$$

$$\mathbf{y} = \sum_{k=1}^{n} \mathbf{r}_k \sin \boldsymbol{\phi}_k \tag{7.69}$$

are nearly Gaussian, as a result of the central limit theorem (sub-sect. 5.53). Moreover, these two variables are uncorrelated:

$$\mathrm{E}[\mathbf{xy}] = \sum_{k=1}^{n} \sum_{l=1}^{n} \mathrm{E}[\mathbf{r}_k\mathbf{r}_l]\mathrm{E}[\cos \boldsymbol{\phi}_k \sin \boldsymbol{\phi}_l] = 0 \tag{7.70}$$

because $\mathrm{E}[\cos \boldsymbol{\phi}_k \sin \boldsymbol{\phi}_l] = 0$ for all k and l. Hence, the Gaussian variables $\mathbf{x}$ and $\mathbf{y}$ are statistically independent. Their mean values are zero:

$$\mathrm{E}[\mathbf{x}] = \sum_{k=1}^{n} \mathrm{E}[\mathbf{r}_k]\mathrm{E}[\cos \boldsymbol{\phi}_k] = 0 \tag{7.71}$$

$$\mathrm{E}[\mathbf{y}] = \sum_{k=1}^{n} \mathrm{E}[\mathbf{r}_k]\mathrm{E}[\sin \boldsymbol{\phi}_k] = 0 \tag{7.72}$$

because $\mathrm{E}[\cos \boldsymbol{\phi}_k] = \mathrm{E}[\sin \boldsymbol{\phi}_k] = 0$, and their variance is

$$\sigma^2 = \mathrm{E}[\mathbf{x}^2] = \mathrm{E}[\mathbf{y}^2] = \frac{n}{2}\,\mathrm{E}[\mathbf{r}_k^2] \tag{7.73}$$

since

$$\mathrm{E}[\mathbf{x}^2] = \sum_{k=1}^{n} \mathrm{E}[\mathbf{r}_k^2]\mathrm{E}[\cos^2 \boldsymbol{\phi}_k] \tag{7.74}$$

and

$$\mathrm{E}[\cos^2 \boldsymbol{\phi}_k] = \frac{1}{2\pi} \int_0^{2\pi} \cos^2 \phi_k \, d\phi_k = \frac{1}{2} \tag{7.75}$$

The transformation into polar coordinates shows that the envelope $\mathbf{r}$ and the phase $\boldsymbol{\phi}$ respectively have, as in sub-sect. 7.3.8, a Rayleigh distribution (7.49) and a uniform distribution (7.50). The result $\mathbf{r} \exp(\,j\boldsymbol{\phi})$ is called the *Rayleigh phasor.*

7.3.13 Sum of a constant and a Rayleigh phasor

If, in (7.61), we take r_0 constant and $\phi_0 = 0$ (i.e., ϕ_0 is the reference phase), we get

$$\mathbf{r} \exp(\mathrm{j}\boldsymbol{\phi}) = r_0 + \sum_{k=1}^{n} \mathbf{r}_k \exp(\mathrm{j}\boldsymbol{\phi}_k) \tag{7.76}$$

This corresponds to a transmission with a direct-wave component. The joint probability density of random variables $\mathbf{x}$ and $\mathbf{y}$ becomes

$$p_{xy}(x,y) = \frac{1}{2\pi\sigma^2} \exp\left[-\frac{(x - r_0)^2 + y^2}{2\sigma^2}\right] \tag{7.77}$$

In polar coordinates, we get

$$p_{r\phi}(r,\phi) = \frac{1}{2\pi\sigma^2} \exp\left[-\frac{(r\cos\phi - r_0)^2 + r^2 \sin^2\phi}{2\sigma^2}\right] \tag{7.78}$$

$$= \frac{r}{2\pi\sigma^2} \exp\left[-\frac{r^2 + r_0^2}{2\sigma^2}\right] \exp\left[\frac{r_0 r \cos\phi}{\sigma^2}\right]$$

with $\mathbf{r} > 0$ and $0 \leqslant \boldsymbol{\phi} < 2\pi$.

By integration on ϕ and inserting the modified Bessel function integral (7.56), we find that the envelope has a Rice-Nakagami distribution (7.57):

$$p_r(r) = \frac{r}{\sigma^2} \exp\left[-\frac{r^2 + r_0^2}{2\sigma^2}\right] I_0\left(\frac{r r_0}{\sigma^2}\right) \tag{7.79}$$

The distribution of the phase $\boldsymbol{\phi}$ is identical to (7.58).

7.4 COMPLEX ENVELOPE AND REPRESENTATION OF BANDPASS SIGNALS

7.4.1 Definition of the complex envelope

Let $\underline{\underline{x}}(t) = x(t) + \mathrm{j}\check{x}(t)$ be the analytic signal and $\omega_0 = 2\pi f_0$ an arbitrary angular frequency.

The expression:

$$\underline{r}(t) = \underline{\underline{x}}(t) \exp(-\mathrm{j}\omega_0 t) \tag{7.80}$$

is called *complex envelope* of the real signal $x(t)$. The real and imaginary parts are designated by $a(t)$ and $b(t)$, its modulus and argument are the real envelope $r(t)$ and an instantaneous phase $\alpha(t)$, which are related to the analytic signal phase $\phi(t)$ and dependent on the choice of ω_0 (fig. 7.14):

$$\underline{r}(t) = a(t) + \mathrm{j}b(t) = r(t)\exp[\mathrm{j}\alpha(t)] \quad \mathbf{(7.81)}$$

with

$$\alpha(t) = \phi(t) - \omega_0 t = \arctan[b(t)/a(t)] \quad (7.82)$$

and

$$r(t) = \sqrt{a^2(t) + b^2(t)} \quad (7.83)$$

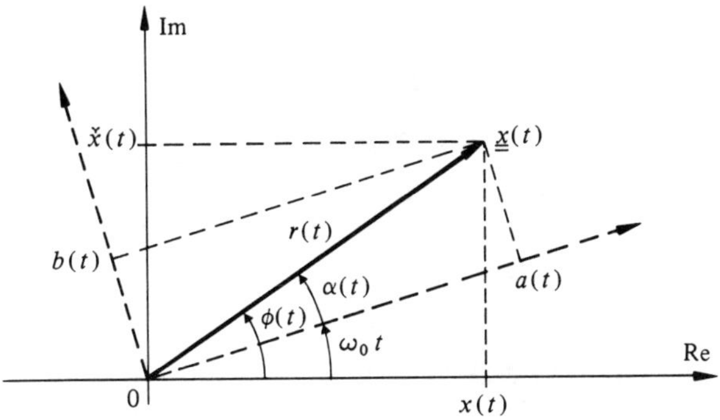

Fig. 7.14

By insertion into (7.80) and identification, we get

$$a(t) = r(t)\cos[\alpha(t)] = x(t)\cos(\omega_0 t) + \check{x}(t)\sin(\omega_0 t) \quad (7.84)$$

$$b(t) = r(t)\sin[\alpha(t)] = \check{x}(t)\cos(\omega_0 t) - x(t)\sin(\omega_0 t) \quad (7.85)$$

or, conversely,

$$\begin{aligned} x(t) &= \mathrm{Re}\{\underline{r}(t)\exp(\mathrm{j}\omega_0 t)\} \\ &= r(t)\cos[\omega_0 t + \alpha(t)] \\ &= a(t)\cos(\omega_0 t) - b(t)\sin(\omega_0 t) \end{aligned} \quad \mathbf{(7.86)}$$

and

$$\begin{aligned} \check{x}(t) &= r(t)\sin[\omega_0 t + \alpha(t)] \\ &= b(t)\cos(\omega_0 t) + a(t)\sin(\omega_0 t) \end{aligned} \quad (7.87)$$

The $a(t)$ and $b(t)$ functions are respectively called the *in-phase* and *quadrature* components of $x(t)$. If $x(t)$ is random and wide-sense stationary, $a(t)$ and $b(t)$ are also stationary random signals.

The complex envelope, as defined in (7.80), is related to the analytic signal. This representation is not unique, but it is considered as optimal [89].

7.4.2 Theorem

Let $x(t)$ be a signal with a bandpass spectrum $\Phi_x(f)$ (fig. 7.15), i.e., zero except on the frequency interval $f_1 < |f| < f_2$, with $0 < f_1 < f_2 < \infty$. Then $x(t)$ can be expressed, according to (7.86), as

$$\Phi_x(f) = \tfrac{1}{4}[\Phi_{\underset{\sim}{r}}(-f - f_0) + \Phi_{\underset{\sim}{r}}(f - f_0)] \tag{7.88}$$

where

$$\Phi_{\underset{\sim}{r}}(f) = \Phi_{\underline{\underline{x}}}(f + f_0) = 2\Phi_x^+(f + f_0) = 2[\Phi_a(f) + \mathrm{j}\Phi_{ab}(f)] \tag{7.89}$$

is the complex envelope spectral density. It is a real non-negative, but not necessarily even, function.

The spectral densities of $a(t)$ and $b(t)$ are identical:

$$\begin{aligned}\Phi_a(f) = \Phi_b(f) &= \tfrac{1}{4}[\Phi_{\underset{\sim}{r}}(f) + \Phi_{\underset{\sim}{r}}(-f)]\\ &= \tfrac{1}{2}[\Phi_x^+(f + f_0) + \Phi_x^+(f_0 - f)]\end{aligned} \tag{7.90}$$

They have a bandwidth $2B$ centered on $f = 0$, with

$$B = \max(|f_1 - f_0|, |f_2 - f_0|) \tag{7.91}$$

If $f_1 < f_0 < f_2$: $a(t)$ and $b(t)$ are lowpass signals. If $f_0 \leqslant f_1$, the complex envelope is itself an analytic signal: $\underset{\sim}{r}(t) = \underline{\underline{a}}(t)$ and $b(t) = \check{a}(t)$ for $\Phi_{\underset{\sim}{r}}(f) = 2\Phi_a^+(f) = \Phi_{\underline{\underline{a}}}(f)$.

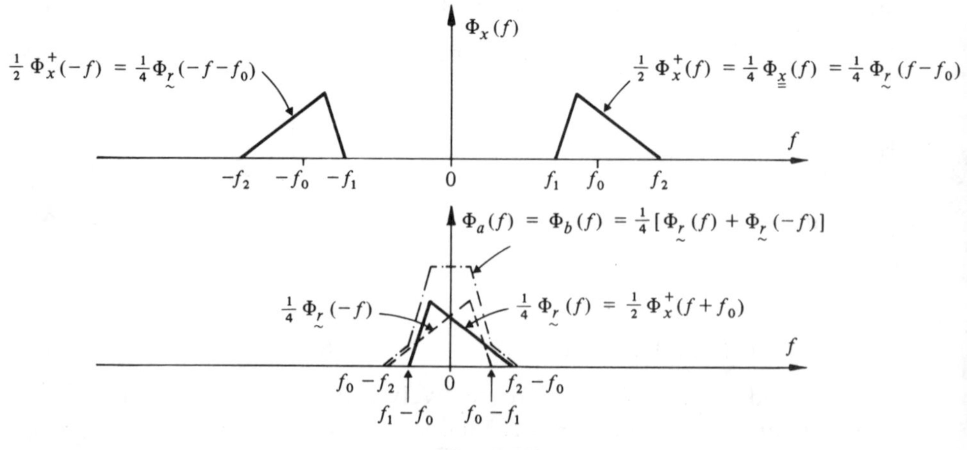

Fig. 7.15

7.4.3 Demonstration

Thanks to (5.78), (5.198), (7.80), and (7.81), and by analogy with (7.31) and (7.33), we write

$$R_{\underset{\sim}{r}}(\tau) = \mathrm{E}[\underset{\sim}{\mathbf{r}}^*(t)\underset{\sim}{\mathbf{r}}(t+\tau)] = 2[\mathrm{R}_a(\tau) + \mathrm{j}R_{ab}(\tau)]$$

$$= R_{\underline{\underline{x}}}(\tau)\exp(-\mathrm{j}2\pi f_0\tau) \quad (7.92)$$

and by (4.17) and (7.41)

$$\Phi_{\underset{\sim}{r}}(f) = 2[\Phi_a(f) + \mathrm{j}\Phi_{ab}(f)] = \Phi_{\underline{\underline{x}}}(f+f_0) = 2\Phi_x^+(f+f_0) \quad (7.93)$$

Thus,

$$\Phi_x(f) = \tfrac{1}{2}[\Phi_x^+(-f) + \Phi_x^+(f)]$$

$$= \tfrac{1}{4}[\Phi_{\underset{\sim}{r}}(-f-f_0) + \Phi_{\underset{\sim}{r}}(f-f_0)] \quad (7.94)$$

The complex envelope spectral density is real and non-negative because of the Hermitian symmetry of the autocorrelation function of a complex signal. It is an even function of f if components $a(t)$ and $b(t)$ are uncorrelated.

The spectral density of $a(t)$ or $b(t)$ is found, from (7.84) or (7.85), by Fourier transform of the corresponding autocorrelation function. For example,

$$R_a(\tau) = \mathrm{E}[a(t)a(t+\tau)]$$

$$= \tfrac{1}{2}[R_x(\tau) + R_{\check{x}}(\tau)]\cos(\omega_0\tau) + \tfrac{1}{2}[R_{x\check{x}}(\tau) - R_{\check{x}x}(\tau)]\sin(\omega_0\tau) \quad (7.95)$$

From (7.31) and (7.33), $R_x(\tau) = R_{\check{x}}(\tau)$, $R_{x\check{x}}(\tau) = -R_{\check{x}x}(\tau)$, and from (7.35), $\Phi_{\check{x}x}(f) = -\Phi_{x\check{x}}(f) = \mathrm{j}\Phi_x(f)\,\mathrm{sgn}(f)$, we get

$$\Phi_a(f) = \Phi_x(f) * \tfrac{1}{2}[\delta(f+f_0) + \delta(f-f_0)]$$

$$+ \Phi_x(f)\,\mathrm{sgn}(f) * \tfrac{1}{2}[\delta(f+f_0) - \delta(f-f_0)]$$

$$= \tfrac{1}{2}\Phi_x(f+f_0) + \tfrac{1}{2}\Phi_x(f-f_0) + \tfrac{1}{2}\Phi_x(f+f_0)\,\mathrm{sgn}(f+f_0)$$

$$- \tfrac{1}{2}\Phi_x(f-f_0)\,\mathrm{sgn}(f-f_0)$$

$$= \tfrac{1}{2}[\Phi_x^+(f+f_0) + \Phi_x^+(f_0-f)] = \tfrac{1}{4}[\Phi_{\underset{\sim}{r}}(f) + \Phi_{\underset{\sim}{r}}(-f)] \quad (7.96)$$

and the occupied bandwidth is $2B$ with $B = \max(|f_1 - f_0|, |f_2 - f_0|)$, as shown in fig. 7.16, where $f_1 < f_0 < f_2$.

7.4.4 Observation

A bandpass signal is thus uniquely defined by its in-phase and quadrature components $a(t)$ and $b(t)$, and hence by its complex envelope $\underset{\sim}{r}(t) = a(t) + \mathrm{j}b(t) = r(t)\exp[\mathrm{j}\alpha(t)]$, which is ***analogous to the phasor*** $\underline{U} = \hat{U}\exp(\mathrm{j}\alpha)$ ***associated with a sinewave.***

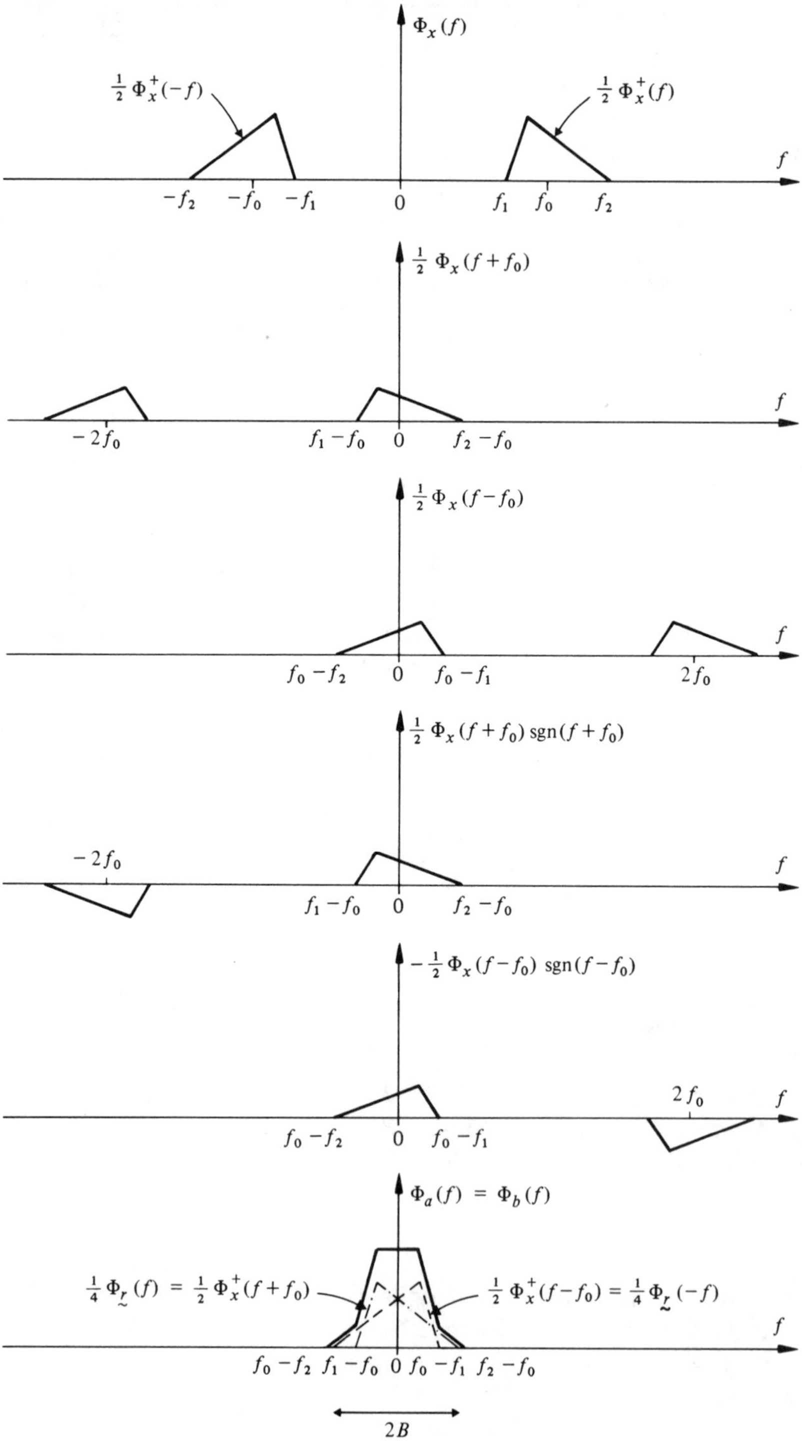

$\Phi_x(f)$
$\frac{1}{2}\Phi_x^+(-f)$
$\frac{1}{2}\Phi_x^+(f)$
f
$-f_2$ $-f_0$ $-f_1$ 0 f_1 f_0 f_2
$\frac{1}{2}\Phi_x(f+f_0)$
$-2f_0$ f_1-f_0 0 f_2-f_0
$\frac{1}{2}\Phi_x(f-f_0)$
f_0-f_2 0 f_0-f_1 $2f_0$
$\frac{1}{2}\Phi_x(f+f_0)\,\mathrm{sgn}(f+f_0)$
$-2f_0$ f_1-f_0 0 f_2-f_0
$-\frac{1}{2}\Phi_x(f-f_0)\,\mathrm{sgn}(f-f_0)$
$2f_0$ f_0-f_2 0 f_0-f_1
$\Phi_a(f) = \Phi_b(f)$
$\frac{1}{4}\Phi_{\underset{\sim}{z}}(f) = \frac{1}{2}\Phi_x^+(f+f_0)$
$\frac{1}{2}\Phi_x^+(f-f_0) = \frac{1}{4}\Phi_{\underset{\sim}{z}}(-f)$
f_0-f_2 f_1-f_0 0 f_0-f_1 f_2-f_0
$2B$

Fig. 7.16

If $f_1 < f_0 < f_2$, these components are of the lowpass kind, with a maximum frequency B. Each of them can be represented as a series of sinc functions as in (3.82). This property is used to establish the sampling theorem for bandpass signals (see sub-sect. 9.3.8).

From (7.90), we derive that the powers (variances) of the $a(t)$ and $b(t)$ components are identical and equal to the power of the signal $x(t)$.

7.4.5 Particular case

If the spectral density $\Phi_x(f)$ is locally symmetrical with respect to $\pm f_0$, i.e., if

$$\Phi_x^+(f + f_0) = \Phi_x^+(f_0 - f) \tag{7.97}$$

then (7.96) is simplified in the form

$$\Phi_a(f) = \Phi_x^+(f + f_0) = \tfrac{1}{2}\Phi_{\underset{\sim}{r}}(f) \tag{7.98}$$

hence,

$$\begin{aligned}\Phi_x(f) &= \tfrac{1}{2}\Phi_x^+(f) + \tfrac{1}{2}\Phi_x^+(-f)\\ &= \tfrac{1}{2}\Phi_a(\mathrm{f} - \mathrm{f}_0) + \tfrac{1}{2}\Phi_a(f + f_0)\\ &= \Phi_a(f) * \tfrac{1}{2}[\delta(f - f_0) + \delta(f + f_0)]\end{aligned} \tag{7.99}$$

By inverse transform, the autocorrelation function becomes (fig. 7.17):

$$R_x(\tau) = R_a(\tau) \cdot \cos(\omega_0\tau) \tag{7.100}$$

The $a(t)$ and $b(t)$ components here are ***uncorrelated***: $R_{ab}(\tau) = 0$ and $\Phi_{ab}(f) = 0$.

Thus, a signal with a locally symmetrical bandpass spectrum is equivalent to a sine wave of frequency f_0, which is multiplied by an independent lowpass component, $a(t)$. This is the modulated sine wave of example 7.3.5, the real envelope of which is $|a(t)|$.

7.4.6 Narrowband signals and the Gaussian case

If $f_2 - f_1 \ll f_0$ with $f_1 < f_0 < f_2$, the in-phase and quadrature components $a(t)$ and $b(t)$ have slow variations. This is also valid for the real envelope $r(t)$ and phase $\alpha(t)$.

The instantaneous frequency (7.48) becomes

$$f_i(t) = f_0 + \frac{1}{2\pi}\frac{\mathrm{d}\alpha(t)}{\mathrm{d}t} = f_0 + \frac{a(t)\dot{b}(t) - \dot{a}(t)b(t)}{2\pi[a^2(t) + b^2(t)]} \tag{7.101}$$

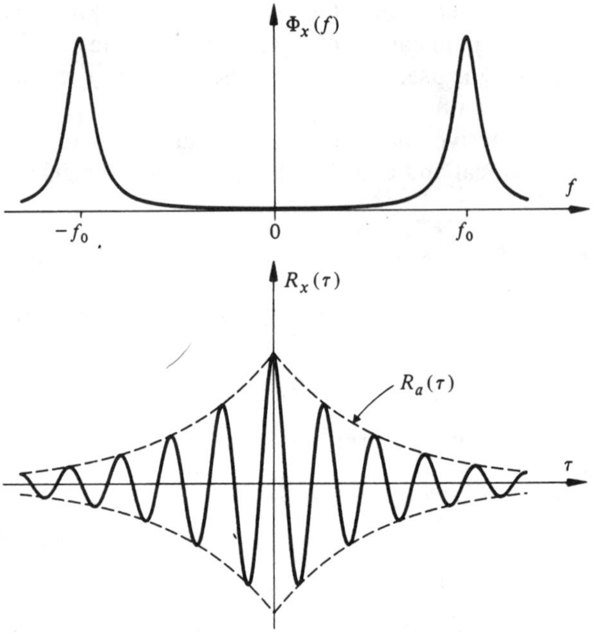

Fig. 7.17

When dealing with a Gaussian random signal (or noise) with a narrowband of variance σ^2, components $a(t)$ and $b(t)$ are also independent Gaussian signals with variance σ^2 and a lowpass spectrum. Phase $\alpha(t)$ is uniformly distributed on interval $[0, 2\pi]$, and the real envelope $r(t)$ has a Rayleigh distribution (7.49).

If the signal has a spectrum $\Phi_x(f) = \frac{1}{2}\eta\{\text{rect}[(f + f_*\}/B] + \text{rect}[(f - f_*)/B]\}$ we have, from (7.90), with $f_0 = f_*$, $\Phi_a(f) = \Phi_b(f) = \eta\ \text{rect}(f/B)$ with $\sigma_x^2 = \sigma_a^2 = \sigma_b^2 = \eta B$.

7.4.7 Application to the vectorial representation of a noisy signal

Consider a cosine signal $s(t) = A\cos(\omega_0 t + \alpha)$ with frequency f_0 and random phase α, disturbed by an additive noise $n(t)$ with a narrowband spectrum centered on f_0. As per the previously developed model, we have (fig. 7.18):

$$n(t) = a_n(t)\cos(\omega_0 t + \alpha) - b_n(t)\sin(\omega_0 t + \alpha) \tag{7.102}$$

and

$$\begin{aligned} x(t) &= s(t) + n(t) \\ &= [A + a_n(t)]\cos(\omega_0 t + \alpha) - b_n(t)\sin(\omega_0 t + \alpha) \end{aligned} \tag{7.103}$$

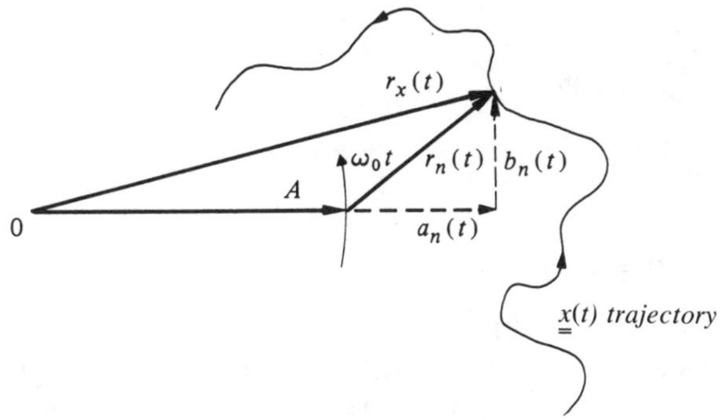

Fig. 7.18

Vector $\underline{\underline{x}}(t)$ describes a random trajectory in the complex plane. When $n(t)$ has a narrowband spectrum, the relative speed of the extremity of the $\underline{\underline{x}}(t)$ vector is small compared to the angular velocity ω_0.

This representation is highly useful when evaluating the influence of additive noise on modulated signals. The in-phase component is primarily used in the study of amplitude demodulation and the quadrature component in angular (phase or frequency) demodulation.

More generally, the (7.86) representation of bandpass signals is an efficient tool for the theoretical study of sine wave modulation techniques (chap. 11).

7.4.8 Application: radar ambiguity function

Consider a radar pulse emission described by

$$s(t) = r_s(t) \cos[\omega_0 t + \alpha(t)] \tag{7.104}$$

the envelope $r_s(t)$ and the phase $\alpha(t)$ of which are real functions, slowing varying *vis-a-vis* $\omega_0 t$.

The perceived echo signal is delayed with respect to the emitted signal by a time t_0, which is proportional to the distance between the antenna and the target. Its frequency is shifted, by Doppler effect, at a frequency f_d (called Doppler frequency), which is proportional to the target radial velocity. Its amplitude is decreased by an attenuation factor k.

Using the analytic signal notations, we can write the emitted and received signals, respectively, as

$$\underline{\underline{s}}(t) = \underset{\sim}{r}(t) \exp[j\omega_0 t] \tag{7.105}$$

and

$$\underline{\underline{s}}_r(t) = k\underline{\underline{s}}(t - t_0) \exp[j2\pi f_d(t - t_0)]$$
$$= k\underset{\sim}{r}(t - t_0) \exp[j2\pi(f_0 + f_d)(t - t_0)] \quad (7.106)$$

where

$$\underset{\sim}{r}(t) = a(t) + jb(t) = r_s(t) \exp[j\alpha(t)] \quad (7.107)$$

is the complex envelope (7.80).

If many targets are present in the radar beam, a resolution problem arises, since many echoes must be identified. Moreover, the radar signal is generally formed by a periodic sequence of pulses, rather than a single pulse, bringing additional ambiguity. Indeed, if two targets are separated by a distance corresponding to a multiple of the time interval separating two pulses, then the two targets will be hard to discriminate. Another such ambiguity exists for frequencies that are multiples of frequence f_0, encountered when evaluating the radial velocities measured by Doppler shift.

Assume that the emitted signal is represented by the analytic signal $\underline{\underline{s}}(t)$ and the returned received signal, generated by the sum of the echoes $\underline{\underline{s}}_{r1}(t)$ and $\underline{\underline{s}}_{r2}(t)$ of two targets is as follows (one of the echoes is taken as time reference, the attenuations are assumed to be equal, and, for simplicity's sake, only the Doppler-shift difference ν is taken into account):

$$\underline{\underline{s}}_r(t) = k\underline{\underline{s}}(t) \exp(-j2\pi\nu t) + k\underline{\underline{s}}(t + \tau) \quad (7.108)$$

To obtain good resolution, the signal $\underline{\underline{s}}(t)$ must be selected (i.e. its complex envelope $\underset{\sim}{r}(t)$) so that the contributions of the two echoes are as different as possible for a wide range of τ and ν. Hence, we try to maximize the Euclidian distance (3.3):

$$d(\underline{\underline{s}}_{r1},\underline{\underline{s}}_{r2}) = k^2 \int_{-\infty}^{\infty} |\underline{\underline{s}}(t) \exp(-j2\pi\nu t) - \underline{\underline{s}}(t + \tau)|^2 \, dt \quad (7.109)$$
$$= 2k^2 \left\{ \int_{-\infty}^{\infty} |\underline{\underline{s}}(t)|^2 \, dt - \text{Re} \left[\int_{-\infty}^{\infty} \underline{\underline{s}}^*(t)\underline{\underline{s}}(t + \tau) \exp(\,j2\pi\nu t) \, dt \right] \right\}$$

The first integral in (7.109) is the envelope energy and the second one is an inner product: a kind of bidimensional autocorrelation function, the modulus of which must be minimized for $\tau \neq 0$ and $\nu \neq 0$ in order to obtain good resolution.

With $\underline{\underline{s}}(t) = \underset{\sim}{r}(t) \exp(\,j\omega_0 t)$, this second integral becomes

$$I(\tau,\nu) = \exp(\,j\omega_0\tau) \cdot \chi(\tau,\nu) \quad (7.110)$$

where

$$\chi(\tau,\nu) = \int_{-\infty}^{\infty} \underset{\sim}{r}^*(t)\underset{\sim}{r}(t + \tau) \exp(\,j2\pi\nu t) \, dt \quad (7.111)$$

According to [11, 27, 67, 90, 91, 92], the function $\chi(\tau,\nu)$ and its complex conjugate or its squared modulus $|\chi(\tau,\nu)|^2$ are called the *ambiguity function* of the signal.

The meaning of this function is the following: two targets, the echoes of which differ by a delay τ and a Doppler shift ν, cannot be distinguished from one another if $|\chi(\tau,\nu)|^2$ equals, or is nearly equal to, $|\chi(0,0)|^2$.

According to the product theorem (4.67), function $\chi(\tau,\nu)$ can also be written

$$\chi(\tau,\nu) = \int_{-\infty}^{\infty} \underset{\sim}{R}(f)\underset{\sim}{R}^*(f+\nu)\exp(\,\mathrm{j}2\pi f\tau)\,\mathrm{d}f \tag{7.112}$$

where $\underset{\sim}{R}(f) = F\{\underset{\sim}{r}(t)\}$.

Through (7.111), we get the projection:

$$\chi(\tau,0) = \int_{-\infty}^{\infty} \underset{\sim}{r}^*(t)\underset{\sim}{r}(t+\tau)\mathrm{d}t = \mathring{\varphi}_{\underset{\sim}{r}}(\tau) \tag{7.113}$$

which is the autocorrelation function (4.38) of the complex envelope. For good range resolution, this function should be as close as possible to a delta function.

Similarly, through (7.112), we get

$$\chi(0,\nu) = \int_{-\infty}^{\infty} \underset{\sim}{R}(f)\underset{\sim}{R}^*(f+\nu)\mathrm{d}f = \mathring{\varphi}^*_{\underset{\sim}{R}}(\nu) \tag{7.114}$$

which is the ***frequency autocorrelation function*** of the envelope spectrum. Again, for good radial velocity resolution, this function should be as close as possible to a delta function.

The value at origin:

$$\chi(0,0) = \int_{-\infty}^{\infty} |\underset{\sim}{r}(t)|^2\,\mathrm{d}t = \int_{-\infty}^{\infty} |\underset{\sim}{R}(f)|^2\,\mathrm{d}f = W \tag{7.115}$$

is the envelope energy equal to twice the real signal energy.

The $|\chi(\tau,\nu)|^2$ function describes a surface above the (τ,ν) plane, with maximum value $|\chi(0,0)|^2$. An indication of the time-frequency combined resolution is given by the Δ area of the base of a cylinder of height $W^2 = |\chi(0,0)|^2$ and of volume equal to the volume beneath the $|\chi(\tau,\nu)|^2$ surface. The Δ area is called the *effective area of ambiguity*:

$$\Delta = W^{-2}\cdot\iint |\chi(\tau,\nu)|^2\,\mathrm{d}\tau\,\mathrm{d}\nu \tag{7.116}$$

It can be shown (problem 7.6.10) that this area is not dependent on the signal selection and is ***always equal to one***. This result sets a theoretical limit for time-frequency joint resolution.

We also define a *time* (or *range*) *resolution constant*:

$$T_\mathrm{d} = W^{-2}\cdot\int_{-\infty}^{\infty} |\chi(\tau,0)|^2\,\mathrm{d}\tau = W^{-2}\int_{-\infty}^{\infty} |\varphi_{\underset{\sim}{r}}(\tau)|^2\,\mathrm{d}\tau \tag{7.117}$$

and a *frequency* (or *velocity*) *resolution constant*:

$$F_v = W^{-2} \cdot \int_{-\infty}^{\infty} |\chi(0,\nu)|^2 \, d\nu = W^{-2} \int_{-\infty}^{\infty} |\mathring{\varphi}_{\underline{R}}(\nu)|^2 \, d\nu \tag{7.118}$$

The locations of two targets with similar echo intensities will be hardly differentiated if the echoes are received with a time delay of less than T_d. Similarly, the radial velocities of those targets will be hard to discriminate if the difference between their respective Doppler shifts is less than F_v. The inverses of T_d and F_v are respectively referred to as the *effective bandwidth* and the *effective duration* of the signal.

7.4.9 Some ambiguity functions

To a sinusoidal pulse with a frequency f_0 and a rectangular envelope of duration T:

$$s_1(t) = \text{rect}\left(\frac{t - T/2}{T}\right) \cdot \cos(2\pi f_0 t) \tag{7.119}$$

there corresponds an ambiguity function (fig. 7.19):

$$|\chi_1(\tau,\nu)|^2 = T^2 \, \text{sinc}^2[\nu(T - |\tau|)] \, \text{tri}^2(\tau/T) \tag{7.120}$$

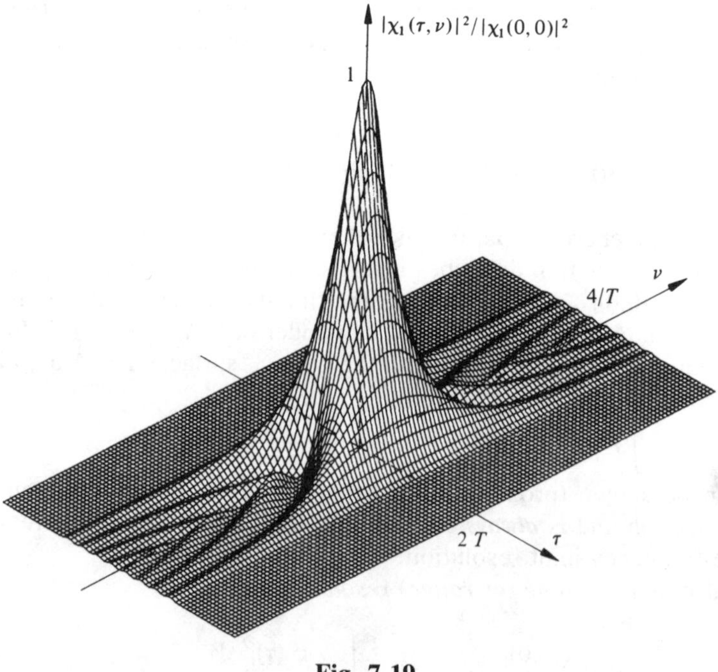

Fig. 7.19

In order to improve the time resolution, a "chirp" signal (linear frequency modulated pulse) with a rectangular envelope is often used in radars:

$$s_2(t) = \text{rect}\left(\frac{t - T/2}{T}\right) \cdot \cos[2\pi(f_0 + \beta t)t] \tag{7.121}$$

to which corresponds the complex envelope:

$$\underset{\sim}{r}(t) = \text{rect}\left[\frac{t - T/2}{T}\right] \exp(\,j2\pi\beta t^2) \tag{7.122}$$

and the ambiguity function:

$$|\chi_2(\tau,\nu)|^2 = T^2 \,\text{sinc}^2[(2\beta\tau + \nu)(T - |\tau|)] \cdot \text{tri}^2(\tau/T) \tag{7.123}$$

represented in fig. 7.20 for $\beta = 2/T^2$Hz/s.

When a periodic sequence of pulses is emitted, the ambiguity function becomes too locally periodic along both the τ and ν axes.

7.4.10 Other applications of the ambiguity function

The ambiguity function was introduced in radar theory [11]. However, it also finds application in other fields [67] where the time and frequency shifts

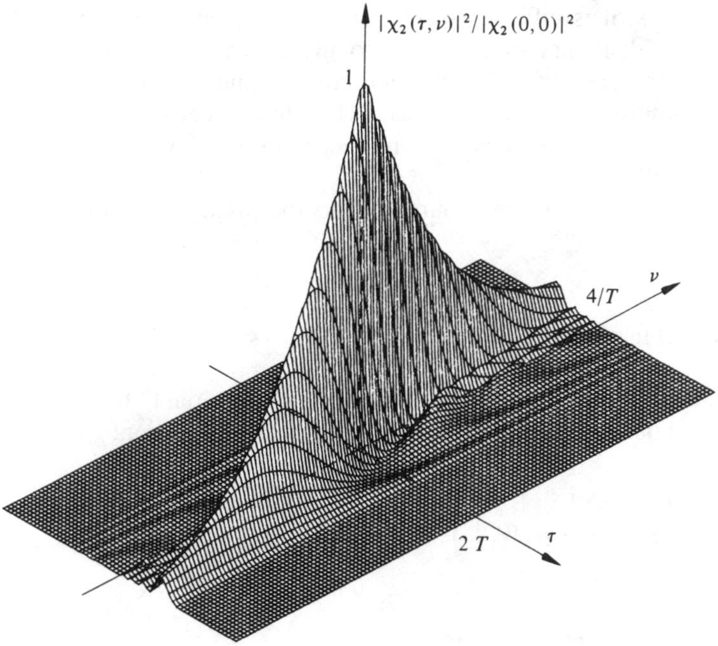

Fig. 7.20

of two signals are to be evaluated; for example, sonar signals (submarine acoustics, bats). The ambiguity function is also linked to the complex signal representation (especially with strong phase or frequency modulation) in a time-frequency plane leading to the *instantaneous spectrum* concept [93].

By analogy with (7.111), we can also define an cross-ambiguity function $\chi_{xy}(\tau,\nu)$ for two signals $x(t)$ and $y(t)$. It has been used in identifying time-varying linear systems.

7.5 OCCUPIED BANDWIDTH AND SIGNAL DURATION

7.5.1 Time dispersion and spectral dispersion

For physical signals, i.e., with finite energy, the time distribution $x^2(t)$ and frequency distribution $\overset{\circ}{\Phi}_x(f) = |X(f)|^2$ of this energy necessarily tend towards zero when $|t|$ and $|f|$ tend towards infinity. The dispersion breadth of this energy along the time or frequency axis brings useful information.

Thanks to the Fourier transform property (4.18), to any signal time compression, there corresponds an expansion of its spectrum and *vice versa*. This suggests the existence of a relationship between the time and frequency dispersions of these energy distributions [10].

Nonetheless, it is difficult to measure these dispersions because there are many ways of defining them, all being totally arbitrary. The choice of definition depends on the ease of use in a given context and on the signal shape. This problem is addressed in detail in [23, 91]. It has a special importance in radar theory [90] and in data transmission where intersymbol interference must be minimized [49].

In this book, we will limit ourselves to the presentation of rather general definitions.

7.5.2 Signal localization

The time dispersion of signal energy is time-invariant. It is, however, advisable to choose in advance, as origin, an average location:

$$t_0 = \frac{\int_{-\infty}^{\infty} t \cdot x^2(t)\mathrm{d}t}{\int_{-\infty}^{\infty} x^2(t)\mathrm{d}t} = W_x^{-1} \int_{-\infty}^{\infty} t \cdot x^2(t)\mathrm{d}t \tag{7.124}$$

where W_x is the signal total energy.

The average location is the ***centroid*** of the energy time distribution. It is similar to the mean value (mathematical expectation) of a random variable **t** with probability density $p(t) = W_x^{-1}x^2(t)$.

7.5.3 Example 1

Let a rectangular signal on interval $[t_1,t_2]$ be

$$x(t) = A \operatorname{rect}\left(\frac{t - \tau}{t_2 - t_1}\right)$$

From (7.124), we have

$$t_0 = \frac{(t_2^2 - t_1^2)/2}{(t_2 - t_1)} = \frac{t_1 + t_2}{2} = \tau$$

7.5.4 Example 2

Consider a delayed exponentially decaying signal

$$x(t) = \epsilon(t - \tau) \exp[-a(t - \tau)]$$

$$t_0 = 1/(2a) + \tau$$

7.5.5 Useful duration

In statistics (chap. 14), the amplitude dispersion of a random variable is characterized by its standard deviation σ, the square root of its variance. By analogy, we can define a variance σ_t^2 of the energy time distribution of a signal, once the average location defined by (7.124) has been chosen, for the sake of simplicity, as origin:

$$\sigma_t^2 = W_x^{-1} \int_{-\infty}^{\infty} t^2 x^2(t) \mathrm{d}t \tag{7.125}$$

The signal useful duration D_u is proportional to the standard deviation σ_t (fig. 7.21):

$$D_u = 2\alpha\sigma_t \tag{7.126}$$

The choice of α is arbitrary. We can take, for instance, $\alpha = 1$ [57].

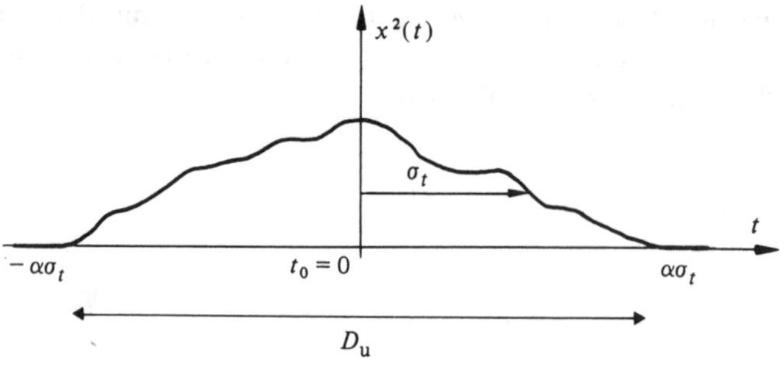

Fig. 7.21

7.5.6 Useful bandwidth

Similarly, the dispersion of the energy spectral density can be characterized by the standard deviation σ_f.

For a lowpass spectrum, the variance is given by

$$\sigma_f^2 = \frac{\int_{-\infty}^{\infty} f^2 \, \mathring{\Phi}_x(f) \, \mathrm{d}f}{\int_{-\infty}^{\infty} \mathring{\Phi}_x(f) \, \mathrm{d}f} = W_x^{-1} \int_{-\infty}^{\infty} f^2 \, \mathring{\Phi}_x(f) \, \mathrm{d}f \tag{7.127}$$

Again, the signal useful bandwidth B_u is proportional to the standard deviation σ_f (fig. 7.22):

$$B_u = k\alpha\sigma_f \tag{7.128}$$

with $k = 1$ for a lowpass spectrum.

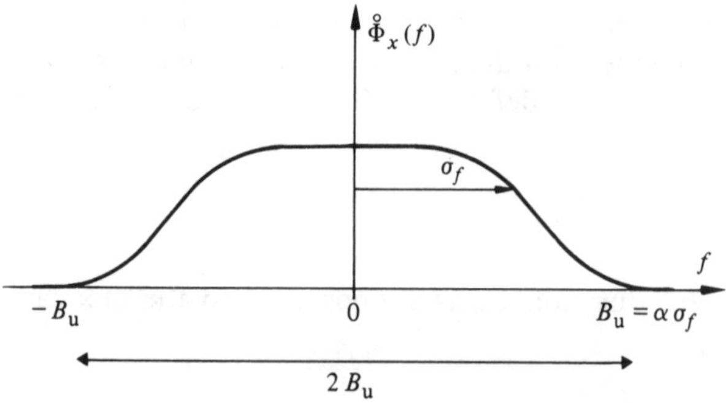

Fig. 7.22

7.5.7 Useful duration and bandwidth of bandpass signals

For a bandpass spectrum, the bilateral energy spectral density is similar to (7.88). It has two symmetrical terms centered on frequencies $\pm f_0$, corresponding to the centroids of the energy spectral density of each term (fig. 7.23):

$$\mathring{\Phi}_x(f) = \tfrac{1}{4}[\mathring{\Phi}_{\underset{\sim}{r}}(-f - f_0) + \mathring{\Phi}_{\underset{\sim}{r}}(f - f_0)] \qquad (7.129)$$

We can also use (7.127) and (7.128), provided that we replace $\mathring{\Phi}_x(f)$ by $\tfrac{1}{2}\Phi_r(f)$ and set $k = 2$.

Similarly, the average location (7.124) and the time variance (7.125) can be redefined by replacing $x^2(t)$ by $\tfrac{1}{2}|\underset{\sim}{r}_x^2(t)|^2$. The coefficient takes into account that the energy of the complex envelope $\underset{\sim}{r}_x(t)$ is twice that of the signal $x(t)$.

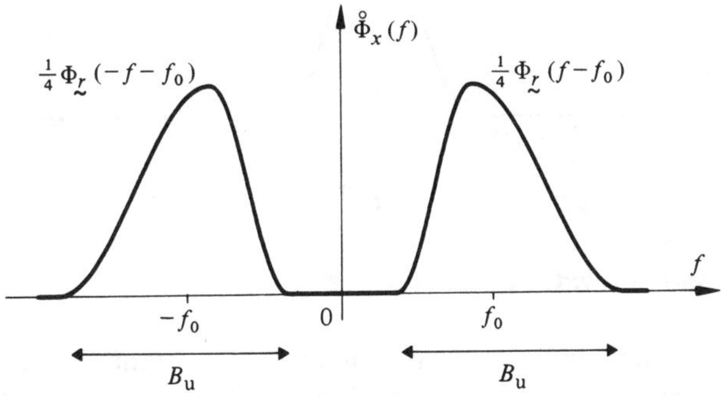

Fig. 7.23

7.5.8 Example

Let $x(t) = A\ \mathrm{tri}(t/T)$. By (7.125), the time standard deviation is $\sigma_t = T/\sqrt{10}$. By (4.35) and (4.55), the energy spectral density of this signal is $\mathring{\Phi}_x(f) = (AT)\ \mathrm{sinc}^4(Tf)$. We can calculate the frequency standard deviation, from (7.127), taking into account problem 2.6.1 and the result [94]:

$$\int_0^\infty \frac{\sin^4\alpha}{\alpha^2}\, d\alpha = \pi/4$$

Eventually, we get: $\sigma_f = \sqrt{3}/(2\pi T)$.

The time and frequency distributions of this signal are given in fig. 7.24, with their respective standard deviations.

It can be observed that product $\sigma_t \cdot \sigma_f$ is independent of the signal total duration $2T$ and is equal to $1/(3.65 \times \pi)$:

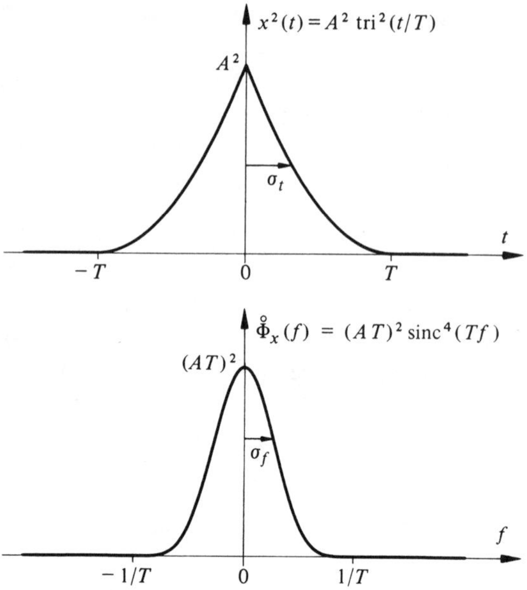

Fig. 7.24

7.5.9 Duration-bandwidth product

A general relationship linking time and frequency standard deviations—and, consequently, the useful duration and bandwidth—of a signal $s(t)$ can be established by the Schwarz inequality (3.21) with $x(t) = \mathrm{d}s/\mathrm{d}t$ and $y(t) = t \cdot s(t)$:

$$\left| \int_{-\infty}^{\infty} ts^*(t) \frac{\mathrm{d}s}{\mathrm{d}t} \mathrm{d}t \right|^2 \leqslant \int_{-\infty}^{\infty} t^2 |s(t)|^2 \mathrm{d}t \cdot \int_{-\infty}^{\infty} \left| \frac{\mathrm{d}s}{\mathrm{d}t} \right|^2 \mathrm{d}t \tag{7.130}$$

Integrating by parts, we get

$$\int_{-\infty}^{\infty} ts^*(t) \frac{\mathrm{d}s}{\mathrm{d}t} \mathrm{d}t = [\tfrac{1}{2}t|s(t)|^2]_{-\infty}^{\infty} - \tfrac{1}{2} \int_{-\infty}^{\infty} |s(t)|^2 \mathrm{d}t \tag{7.131}$$

Limiting our analysis to the sole signals having an energy time distribution $|s(t)|^2$ decreasing faster than $1/t$ when $|t| \to \infty$, we get

$$\left| \int_{-\infty}^{\infty} ts^*(t) \frac{\mathrm{d}s}{\mathrm{d}t} \mathrm{d}t \right|^2 = \frac{1}{4} W_s^2 \tag{7.132}$$

where W_s is the total energy of signal $s(t)$.

On the other hand, from (4.13), (4.55), (4.57), and (7.127), we have

$$\int_{-\infty}^{\infty} \left| \frac{\mathrm{d}s}{\mathrm{d}t} \right|^2 \mathrm{d}t = 4\pi^2 \int_{-\infty}^{\infty} f^2 \overset{\circ}{\Phi}_s(f) \mathrm{d}f = 4\pi^2 W_s \sigma_f^2 \tag{7.133}$$

Finally, inserting (7.125), (7.132), and (7.133) into inequality (7.130) leads to

$$\sigma_t \cdot \sigma_f \geqslant 1/(4\pi) \tag{7.134}$$

This expression is called the ***uncertainty relation,*** by analogy with the Heisenberg's uncertainty principle of quantum mechanics (wave-particle duality).

For instance, choosing $\alpha = \sqrt{2\pi}$, we get from (7.126), (7.128), and (7.134), for a lowpass signal:

$$D_{\mathrm{u}} \cdot B_{\mathrm{u}} \geqslant 1 \tag{7.135}$$

The duration-bandwidth product has a lower bound.

7.5.10 Observation

Relation (7.133) shows that a signal with fast fluctuations must have a large bandwidth. Conversely, it can be shown that large fluctuations in an amplitude or phase spectrum imply a signal with long duration.

7.5.11 Minimal duration-bandwidth product of a signal

The Schwarz inequality (3.21) becomes an equality when (3.22) is valid, which yields

$$\frac{\mathrm{d}s}{\mathrm{d}t} = \lambda t s(t) \tag{7.136}$$

thus,

$$\frac{\mathrm{d}s}{s} = \lambda t \, \mathrm{d}t \tag{7.137}$$

Integrating, we get

$$s(t) = C \exp(\lambda t^2/2) \tag{7.138}$$

where C is a constant and $\lambda < 0$ to comply with the finite energy condition.

Therefore, ***the signal that has a minimal duration-bandwidth product is the Gaussian pulse.***

This signal has a remarkable property: the Fourier transform of a Gaussian pulse is also a Gaussian pulse (problem 4.6.9). With sub-section 1.3.16 notations, we have

$$\mathrm{ig}(t') = \exp(-\pi t'^2) \leftrightarrow \mathrm{ig}(f') = \exp(-\pi f'^2) \tag{7.139}$$

and

$$\mathrm{ig}(t/T) \leftrightarrow T \cdot \mathrm{ig}(Tf) \tag{7.140}$$

From (7.125) and (7.127), we get the equivalences $\sigma_t = T(2\sqrt{\pi})^{-1}$ and $\sigma_f = (2\sqrt{\pi}T)^{-1}$, the product of which corresponds effectively to the limit (7.134). With $\alpha = \sqrt{2\pi}$ in (7.126) and (7.128), $D_u = B_u^{-1} = T\sqrt{2}$.

7.5.12 Other definitions: efficient duration and bandwidth

The standard deviations defined by (7.125) and (7.127) as measures of the dispersion of energy time and frequency distributions are not universally applicable. Some distributions have no finite variance. As an exercise, we can check that this is the case for signals as simple as $x(t) = A\text{rect}(t/T)$ and $y(t) = \epsilon(t)\exp(-at)$. The first has a time standard deviation equal to $T/(2\sqrt{3})$ and an infinite frequency standard deviation. For the second, the time standard deviation is $1/(2a)$ and the frequency standard deviation is undefined.

This can be solved by defining, with arguments stemming for instance from information theory [91], an *efficient bandwidth*:

$$B_e = \frac{1}{2}\frac{\left[\int_{-\infty}^{\infty}\mathring{\Phi}(f)\mathrm{d}f\right]^2}{\int_{-\infty}^{\infty}\mathring{\Phi}^2(f)\mathrm{d}f} = \frac{1}{2}\frac{\mathring{\varphi}^2(0)}{\int_{-\infty}^{\infty}\mathring{\varphi}^2(\tau)\mathrm{d}\tau} \tag{7.141}$$

and an *efficient duration*:

$$D_e = \frac{\left[\int_{-\infty}^{\infty}\mathring{\varphi}(\tau)\mathrm{d}\tau\right]^2}{\int_{-\infty}^{\infty}\mathring{\varphi}^2(\tau)\mathrm{d}\tau} = \frac{\mathring{\Phi}^2(0)}{\int_{-\infty}^{\infty}\mathring{\Phi}^2(f)\mathrm{d}f} \tag{7.142}$$

The efficient bandwidth (7.141) is half the effective bandwidth defined in subsection 7.4.8 as the inverse of the time resolution constant (7.117) in radars.

The advantage of definition (7.141) is that it supplies an efficent bandwidth $B_e = B$ to a rectangular spectral density $\mathring{\Phi}(f) = \text{rect}[f/(2B)]$.

The evaluation of B_e and D_e for signals $x(t)$ and $y(t)$ is left as exercise.

Definitions (7.141) and (7.142) are also valid for zero mean value random signals, replacing $\mathring{\Phi}(f)$ by $\Phi(f)$ and $\mathring{\varphi}(\tau)$ by $R(\tau) = C(\tau)$.

7.5.13 Correlation duration and approximate bandwidth

For random signals, the concept of useful or efficient duration is not very pertinent. A more relevant measure is the *correlation duration*, D_τ, as defined below:

$$D_\tau = C^{-1}(0)\int_{-\infty}^{\infty}|C(\tau)|\mathrm{d}\tau = \int_{-\infty}^{\infty}|\rho(\tau)|\mathrm{d}\tau \tag{7.143}$$

where $\rho(\tau) = C(\tau)/C(0)$ is the normalized autocovariance function.

The inverse B_τ of the correlation duration is also an approximative measure of the signal spectral dispersion, or *approximative bandwidth*.

For all signals with ***non-negative*** autocovariance function, relation (7.143) becomes

$$D_\tau = B_\tau^{-1} = \Phi(0)/C(0) \tag{7.144}$$

Relations (7.143) and (7.144) are also valid for finite energy deterministic signals, replacing the autocovariance function $C(\tau)$ by the autocorrelation function $\mathring{\varphi}(\tau)$.

7.5.14 Examples

For the NRZ binary signal of sub-section 5.3.7, $\rho(\tau) = \text{tri}(\tau/T)$, and from (7.143): $D_\tau = B_\tau^{-1} = T$. For the diphase binary signal of sub-section 5.3.8, $\rho(\tau) = 2\,\text{tri}(2\tau/T) - \text{tri}(\tau/T)$ leading to $D_\tau = B_\tau^{-1} = 2T/3$.

For a rectangular signal $x(t) = A\,\text{rect}(t/T)$, we easily obtain from (4.63), (4.64), and (7.143): $D_\tau = T$ and $B_\tau = 1/T$.

For an exponentially decaying signal $y(t) = \epsilon(t)\exp(-at)$, we obtain from (4.61), and (4.62): $D_\tau = 2/a$ and $B_\tau = a/2$.

For the sake of comparison, these correlation durations and approximation bandwidths can be drawn on figs. 4.15, 4.16, 5.9, 5.13, 5.15, and 5.16.

7.6 PROBLEMS

7.6.1 Find the Hilbert transform of signal $x(t) = A\sin(\omega t + \alpha)$.

7.6.2 Find the Hilbert transform and the analytic signal of $x(t) = 2B\,\text{sinc}(2Bt)$.

7.6.3 If P_x is the power of signal $x(t)$, calculate the power of analytic signal $\underline{\underline{x}}(t)$.

7.6.4 Find the crosscorrelation function $\varphi_{xy}(\tau)$ and its Hilbert transform $\check{\varphi}_{xy}(\tau)$ for $x(t) = A\cos(\omega_0 t + \alpha)$ and $y(t) = A\sin(\omega_0 t + \alpha)$.

7.6.5 Determine with which probability the envelope of a Gaussian random signal with zero mean value and variance σ_n^2 remains less than threshold $V_0 = 2\sigma_n$.

7.6.6 A signal $x(t)$ has the spectral density shown in fig. 7.25. Find for which value of f_0 that $x(t)$ can be expressed by equation (7.86) with
1) $\Phi_a(f) = \frac{1}{2}\eta\ \mathrm{rect}[f/(f_2 - f_1)]$, and
2) $\Phi_a(f) = \frac{1}{2}\eta\ \mathrm{tri}[f/(f_2 - f_1)]$.

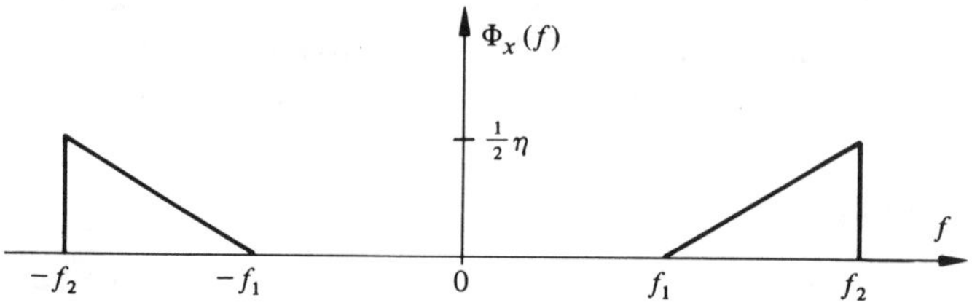

Fig. 7.25

7.6.7 Show that the Fourier transforms of the even and odd parts of a causal real signal are Hilbert transforms of one another, except for a factor j.

7.6.8 Let $x(t)$ and $y(t)$ be two real bandpass signals with complex envelopes $\underset{\sim}{r}_x(t)$ and $\underset{\sim}{r}_y(t)$. Show that the inner product $<x,y> = \frac{1}{2}\mathrm{Re}<\underset{\sim}{r}_x,\underset{\sim}{r}_y^*>$.

7.6.9 Verify (7.120) and (7.123).

7.6.10 Show that

$$\int_{-\infty}^{\infty}\int_{-\infty}^{\infty} |\chi(\tau,\nu)|^2 \mathrm{d}\tau \mathrm{d}\nu = \chi^2(0,0) \tag{7.145}$$

knowing that

$$\int_{-\infty}^{\infty} \exp[\mathrm{j}2\pi\nu(t_1 - t_2)]\mathrm{d}\nu = \delta(t_1 - t_2) \tag{7.146}$$

7.6.11 Find the correlation duration D_τ and the approximative bandwidth B_τ of signal $x(t) = A\ \mathrm{tri}(t/T)$.

Chapter 8

Functional Operators

8.1 SIGNAL PROCESSING SYSTEM MODELING

8.1.1 Signal processing system

A *signal processing system* is a device that performs some basic operations such as amplification, filtering, nonlinear transformation, modulation, detection, parameter estimation, *et cetera*, on an internally or externally generated signal. The result is either another signal or measurements displayed in analog or digital form. This kind of device is often called a *signal processor.*

Signal processors can be divided in three groups (fig. 8.1):

- *signal generators*: only an output signal;
- *signal transformers*: an input signal and an output signal;
- *signal analyzers*: only an input signal, the output being a display of the analysis results.

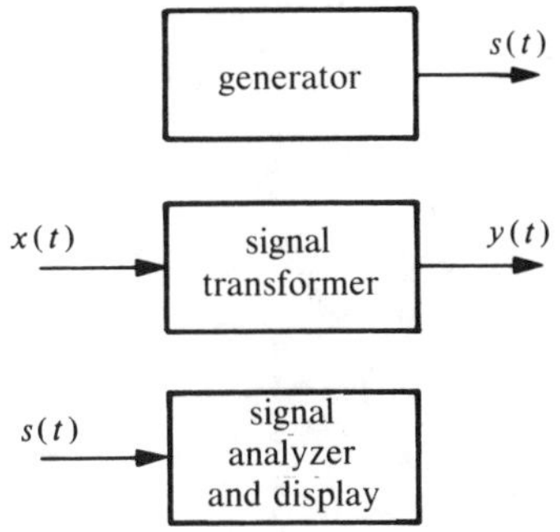

Fig. 8.1

The input or output signal is generally an analog signal (fig. 8.2). The processor can be either analog, digital, or hybrid: the translation of the analog signal into a digital signal is carried out by an analog-to-digital (A/D) converter and the reverse translation is carried out by a digital-to-analog (D/A) converter. The required operating conditions for this kind of conversion are detailed in chapters 9 and 10. The electronic design of such converters is described in [95].

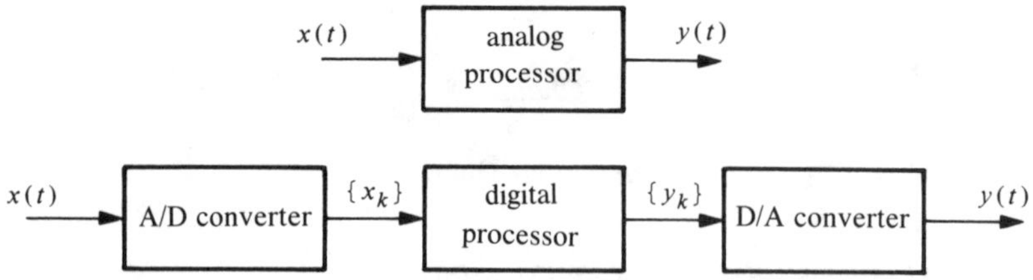

Fig. 8.2

When the results are provided at the input signal rate, the system is said to be *real-time processing*. This term is mainly used to characterize digital processors with an execution time compatible with the sampling rate of the A/D converter.

8.1.2 Functional block diagram description

Owing to its often complex nature, the processing system is best described in such a way that each basic operation is explicitly displayed. This is the principle of block diagram representation (sub-sect. 1.2.4).

The advantage of such a representation is that it generally corresponds to the hardware or software internal structure of the processor and it enlightens its modularity.

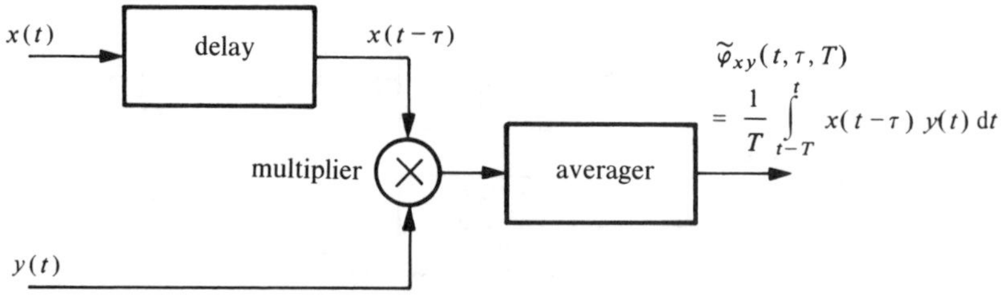

Fig. 8.3

Figure 8.3 gives an example of block diagram, which corresponds to the principle of a crosscorrelator. Figure 8.4 gives another example, which is the principle of a power spectral density measurement. In the first example, the system computes the local average value, over time T, of the product of signal $y(t)$ and the delayed signal $x(t - \tau)$. The basic operations are delay, multiplication, and averaging. Repeating the measurement for various τ values yields

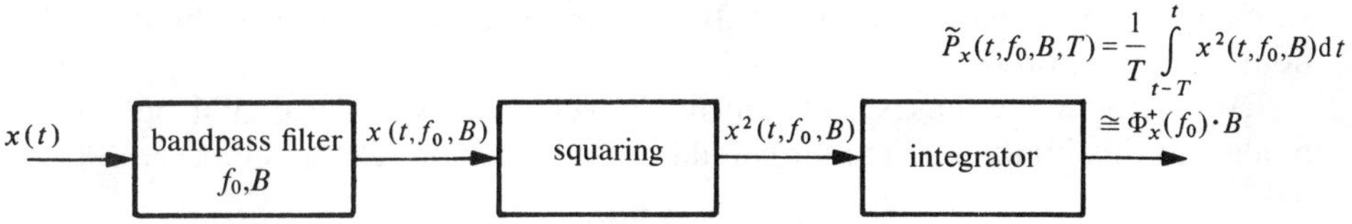

Fig. 8.4

an estimation of the crosscorrelation function. In the second example, input signal $x(t)$ is first applied to a narrow bandpass filter, with a bandwidth B centered on f_0. The average power of its output $x(t,f_0,B)$, corresponding to the part of $x(t)$ contained in the frequency interval: $[f - B/2, f + B/2]$, is evaluated by first squaring it and then time-averaging it over time T. The evaluated power roughly corresponds to the product of the unilateral power spectral density for $f = f_0$ by the filter bandwidth, B.

8.1.3 Functional operator

Most functional blocks feature an output which is dependent on an input or a combination of inputs.

The theoretical model of a *block* is called a *functional operator*: the correspondence (mapping) rule between two sets of functions. The transform of entity x into entity y by operator S is symbolically denoted by (fig. 8.5):

$$y = \mathrm{S}\{x\} \tag{8.1}$$

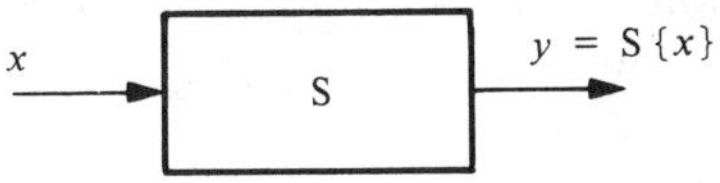

Fig. 8.5

When we deal with analog signals, we write

$$y(t) = \mathrm{S}\{x(t)\} \tag{8.2}$$

where t is a continuous variable, set for time.

When signals are digital (or, more generally, with a discrete time variable), the input and output signal values are denoted by $\{x_k\}$ and $\{y_k\}$, or $x(k)$ and $y(k)$, where k is an integer:

$$y(k) = \mathrm{S}\{x(k)\} \tag{8.3}$$

In many cases, equation (8.3) is only a discrete representation of the continuous time relation (8.2).

Except for a few cases, we will not develop the specific digital operator models in this book. The reader should consult volume XX or references [44–48].

8.1.4 Classification and definitions

There is a wide variety of operators, but they can be divided into three main groups:

- *invariant linear operators* featuring properties of additivity (superposition principle), homogeneity, and time stationarity;
- *parametric operators,* which are dependent on time through an auxiliary control entity or signal;
- *nonlinear operators* forming a broad group with no universal representation.

The models of the delay circuit, bandpass filter, and averager of figures 8.3 and 8.4 are invariant linear operators. The multiplier in figure 8.3 is a parametric operator and the squaring operator in figure 8.4 is nonlinear.

8.1.5 Implementation of functional blocks

The practical implementation of a functional block requires the choice of an adequate technical solution in order to build an acceptable approximation of the corresponding operator.

There are three main approaches:

- the hardware analog approach, for example, in the low and medium frequency domain, drawing on the various options provided by electronic devices based on operational amplifiers or other functional integrated circuits (multipliers, nonlinear circuits, *et cetera*);
- the hardware digital approach, based on the combination of elementary or complex logic circuits such as gates, flip-flops, shift registers, memories, arithmetic and logic units (ALU), *et cetera*;
- the computer programming approach, where the function is realized by appropriate software, the storage and execution of which are achieved by specialized circuits (memories, microprocessors, digital signal processors, *et cetera*) for real-time processing, or by a general-purpose computer if the processing or simulation do not require real time.

The description of these technical approaches is beyond the scope of this book. Volumes VIII and XIV and other books [96–98] can be referred to for further information.

8.2 INVARIANT LINEAR OPERATORS

8.2.1 Basic properties

Let S be an invariant linear operator. Linearity and homogeneity imply that if

$$x(t) = \sum_i a_i\, x_i(t) \tag{8.4}$$

then

$$\mathrm{S}\{x(t)\} = \sum_i a_i\, \mathrm{S}\{x_i(t)\} \tag{8.5}$$

If $y(t) = \mathrm{S}\{x(t)\}$, then

$$y(t - \tau) = \mathrm{S}\{x(t - \tau)\} \tag{8.6}$$

means that S is time-invariant, and is thus independent of the time origin.

8.2.2 Convolution operator and orthogonal transform operator

There are two kinds of invariant linear operators:

- *orthogonal transform operators,* for which the inputs and outputs are functions of different variables, such as time t and frequency f, in a Fourier transform;
- *convolution operators,* for which the inputs and outputs are function of the same independent variable, usually time.

8.2.3 Example: Fourier transform operator

A Fourier transform operator is characterized by an operator F that gives a function $X(f)$, defined by (4.1), for each signal $x(t)$, stated as

$$X(f) = \int_{-\infty}^{\infty} x(t)\exp(-\mathrm{j}2\pi ft)\mathrm{d}t \tag{8.7}$$

In the case of sampled or digital signals, operator F is defined by (sub-sect. 9.3.11 and vol. XX):

$$X(n) = \sum_{k=k_0}^{k_0+N-1} x(k) W_N^{-nk} \tag{8.8}$$

where N is the number of samples of the considered signal, W_N is the Nth principle root of unity:

$$W_N = \exp(\,\mathrm{j}2\pi/N) \tag{8.9}$$

and k and n are the time and frequency discrete indices, respectively.

The inverse transform operator is given by F^{-1} (fig. 8.6), which is defined in the continuous case by

$$x(t) = \int_{-\infty}^{\infty} X(f)\exp(\,\mathrm{j}2\pi ft)\mathrm{d}f \tag{8.10}$$

and in the discrete case by

$$x(k) = N^{-1} \sum_{n=-N/2}^{N/2-1} X(n) W_N^{nk} \tag{8.11}$$

Relations (8.8) and (8.11) are, in fact, the vector product of an $N \times N$ transform matrix.

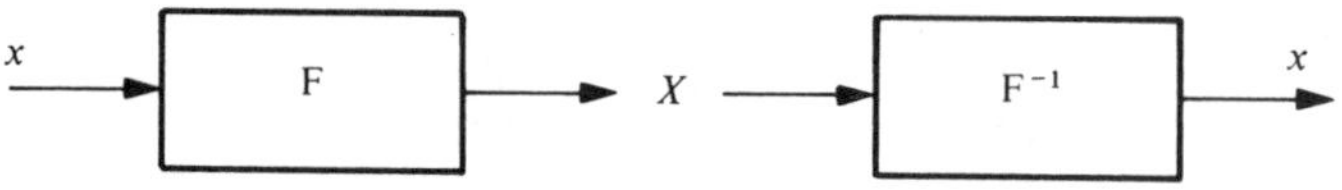

Fig. 8.6

8.2.4 Direct description of convolution operators

A convolution operator (fig. 8.7) is totally defined by its impulse response, noted $g(t)$ or $h(t)$ in the continuous case: it is the system output corresponding to a delta-function input (sub-sect. 1.3.12). In the discrete case, it is written $g(k)$ or $h(k)$, and corresponds to the system response to a unit sample defined by $d(k) = 1$ for $k = 0$ and $d(k) = 0$ for $k \neq 0$.

The response to any kind of input is given by the convolution product $y = x * g$, which is defined in the continuous case by

$$y(t) = x(t) * g(t) = \int_{-\infty}^{\infty} x(\tau)g(t - \tau)\mathrm{d}\tau \tag{8.12}$$

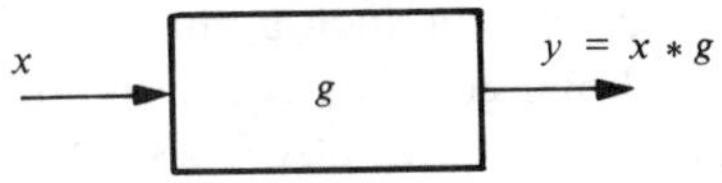

Fig. 8.7

and in the discrete case by

$$y(k) = x(k) * g(k) = \sum_{l=-\infty}^{\infty} x(l)g(k - l) \tag{8.13}$$

The convolution product is a weighted sum of (continuous or sampled) values of the input signal: the weighting function being the impulse response. A graphic interpretation is given by figure 1.5.

8.2.5 Indirect description of convolution operators

If input and output signals are deterministic and therefore have a Fourier transform, a product of the transforms can be associated with the convolution product (8.12) or (8.13), because of property (4.14):

$$Y(f) = X(f) \cdot G(f) \tag{8.14}$$

in the continuous case, and

$$Y(n) = X(n) \cdot G(n) \tag{8.15}$$

in the discrete case.

The function $G = \mathrm{F}\{g\} = \mathrm{Y/X} = |G| \exp(\mathrm{j}\vartheta_g)$ is the *harmonic transfer function* (or *frequency response function*) of the convolution operator. Its modulus $|G|$ is the *amplitude response*, and its argument $\vartheta_g = \arg G$ is the *phase response* of the operator.

8.2.6 Transfer function of a lumped linear system

A complete description of the behavior of an invariant linear system modeled with lumped parameter elements is given by the differential equation:

$$\sum_{i=0}^{n} a_i \frac{\mathrm{d}^{n-i}}{\mathrm{d}t^{n-i}} y(t) = \sum_{i=0}^{m} b_i \frac{\mathrm{d}^{m-i}}{\mathrm{d}t^{m-i}} x(t) \tag{8.16}$$

where coefficients a_i and b_i are constant.

In the case of electrical circuits, this equation is obtained by applying Kirchhoff's laws.

We know that the solution $y(t)$ of such a differential equation is the sum of the general solution of the homogeneous equation and a particular solution of equation (8.16). This solution $y(t)$ can also be seen as the sum of a transient term and a permanent term. The transient term is the portion of $y(t)$ that vanishes (or indefinitely grows) as time passes by. The permanent term is either a constant or a combination of sine waves.

If, instead of solving (8.16), we take the Fourier transform of each term, using property (4.13):

$$\frac{d^k x(t)}{dt^k} \leftrightarrow (j2\pi f)^k X(f) \tag{8.17}$$

we get

$$\left[\sum_{i=0}^{n} a_i (j2\pi f)^i\right] Y(f) = \left[\sum_{i=0}^{m} b_i (j2\pi f)^i\right] X(f) \tag{8.18}$$

As a result of the Fourier transform, the differential equation becomes an algebraic equation, expressing the behavior of the linear circuit in the frequency domain. The transfer function can then be expressed as a function of the system constants a_i and b_i as

$$G(f) = \frac{Y(f)}{X(f)} = \frac{\sum_{i=0}^{m} b_i (j2\pi f)^i}{\sum_{i=0}^{n} a_i (j2\pi f)^i} = G_0 \frac{\prod_{i=1}^{m} (j2\pi f + z_i)}{\prod_{i=1}^{n} (j2\pi f + p_i)} \tag{8.19}$$

where $G_0 = b_m/a_n$, and z_i and p_i are the roots of the numerator and denominator, respectively, i.e., the zeros and poles of the transfer function.

8.2.7 *z*-transform

In the study of discrete-time systems and signals based on a periodic sampling, it is more convenient to use the indirect z-transform description than that of the Fourier transform. The z-transform is defined by

$$G(z) = \sum_{k=-\infty}^{\infty} g(k) z^{-k} \tag{8.20}$$

where z is a complex variable.

The z-transform is related to the bilateral Laplace transform of a periodically sampled function. It is a generalization of the Fourier transform, these two transforms being identical on the unit circle $z = \exp(j2\pi f)$. For more detailed information, the reader should refer himself or herself to volume XX.

8.2.8 Relationships between correlation functions

From (4.50), (4.98), or (5.59), the time correlation function, whatever the signal, can be written as a convolution product:

$$\varphi_{xy}(\tau) = x(-\tau) * y(\tau) \tag{8.21}$$

Combining this result with (8.12) and making use of the associative and commutative properties of the convolution, we get the following relations, if $x(t)$ and $y(t)$ are input and output signals, respectively, of a linear operator with impulse response $g(t)$:

$$\varphi_y(\tau) = \varphi_x(\tau) * \mathring{\varphi}_g(\tau) \tag{8.22}$$

where $\mathring{\varphi}_g(\tau)$ is the autocorrelation function of the impulse response, and

$$\varphi_{xy}(\tau) = \varphi_x(\tau) * g(\tau) \tag{8.23}$$

8.2.9 Spectral relationships

Relation (8.14) is only valid for deterministic signals. However, an expression that is also valid for random signals is obtained via the Fourier transform of (8.22):

$$\Phi_y(f) = \Phi_x(f) \cdot |G(f)|^2 \tag{8.24}$$

for $\mathrm{F}\{\mathring{\varphi}_g(\tau)\} = \mathrm{F}\{g(-\tau) * g(\tau)\} = G^*(f)G(f) = |G(f)|^2$. From the Fourier transform of (8.23), we get

$$\Phi_{xy}(f) = \Phi_x(f) \cdot G(f) \tag{8.25}$$

8.2.10 Application in linear system identification

As a result of the delta-function property (1.47), if $x(t)$ is a white noise (subsect. 5.3.10) with autocorrelation function:

$$\varphi_x(\tau) \equiv R_x(\tau) = \tfrac{1}{2}\eta\delta(\tau) \tag{8.26}$$

then the impulse response of a linear system can be found by evaluating the crosscorrelation function (8.23):

$$\varphi_{xy}(\tau) = \tfrac{1}{2}\eta g(\tau) \tag{8.27}$$

The evaluation, under the same conditions, of the cross-spectral density (8.25) allows us to get the transfer function:

$$\Phi_{xy}(f) = \tfrac{1}{2}\eta G(f) \tag{8.28}$$

8.2.11 Application: deconvolution

When a signal is perceived by a linear sensor, with a transfer function $G_1(f)$, the original information can be affected by the sensor itself. The input and output signals of the sensor being related by the convolution (8.12), the operation to compensate for the sensor effect is called *deconvolution*.

Theoretically, a perfect correction would be achieved by cascading (fig. 8.8) the sensor with an inverse operator defined by

$$G_2(f) = G_1^{-1}(f) = \frac{\Phi_x(f)}{\Phi_{xy}(f)} \tag{8.29}$$

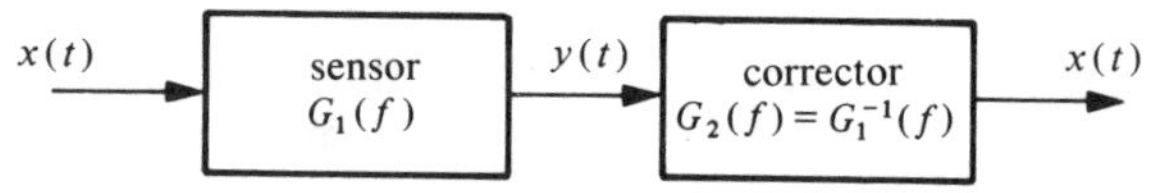

Fig. 8.8

The practical implementation of this principle is tricky, since the correction system is not necessarily stable and strongly amplifies all the additional disturbances in the frequency ranges around the zeros of the function $G_1(f)$.

8.2.12 Indirect simulation of convolution or correlation

An indirect way to implement a convolution or crosscorrelation operator is shown in figures 8.9 and 8.10. It is based on the use of Fourier transforms and properties (4.14) and (4.18). This solution is often employed in digital signal analyzers.

The first case requires the preliminary storage of the desired G response.

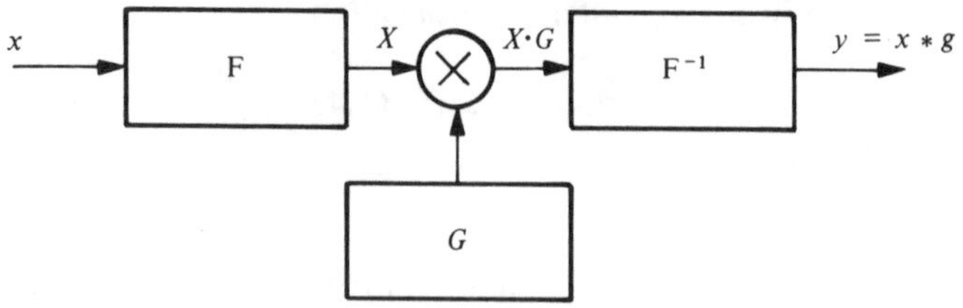

Fig. 8.9

8.2.13 Convolution operators in cascade

It is generally admitted that the input of an operator does not affect the output of the previous operator. This assumption is true for digital systems,

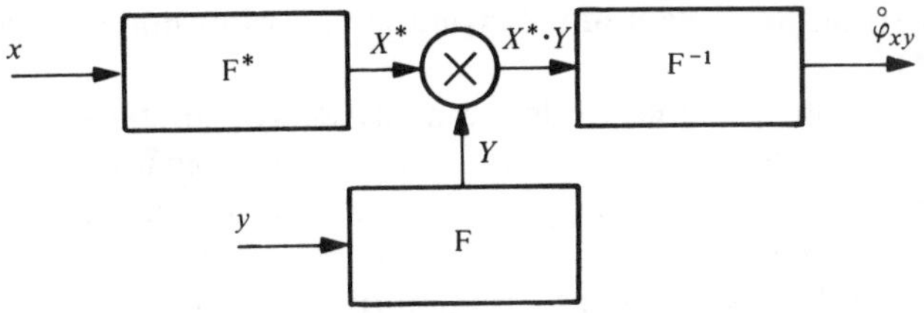

Fig. 8.10

but may not always be true for analog systems (load effect). When it is true, the total impulse response of a cascade of convolution operators (fig. 8.11) is equal to the multiple convolution of the partial impulse responses:

$$y(t) = x(t) * g(t) \tag{8.30}$$

with

$$g(t) = g_1(t) * g_2(t) * \ldots * g_n(t) \tag{8.31}$$

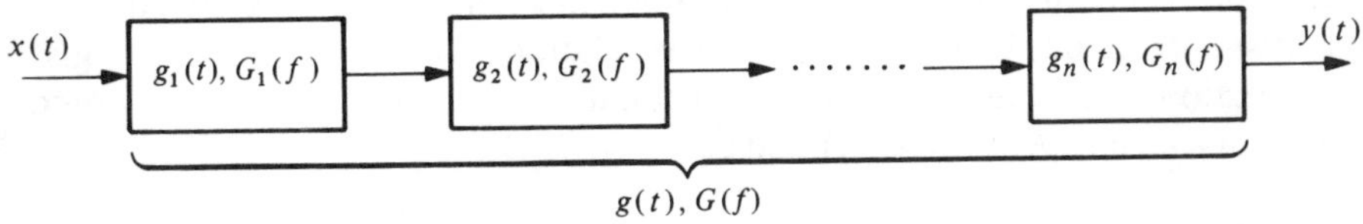

Fig. 8.11

Thus, the total transfer function is simply the product of the partial transfer functions:

$$Y(f) = X(f) \cdot G(f) \tag{8.32}$$

with

$$G(f) = G_1(f)G_2(f) \ldots G_n(f) = \prod_{i=1}^{n} G_i(f) \tag{8.33}$$

This assumption of independence is not valid in some practical circumstances, such as passive networks in cascade. It can be easily shown that two *RC* networks in cascade, similar to the filter of example 8.2.24, do not yield a global transfer function equal to the square of the transfer function of a single RC network.

8.2.14 Statistical description of a convolution-operator output signal

Usually, it is not possible to define analytically the statistical distribution of a convolution-operator output signal, each output signal value being a linear combination of the input signal (passed, when the operator is causal) values, as shown by the convolution equation (8.12) or (8.13). These values are usually not independent. To estimate the probability of finding the output signal amplitude in a given range, we need to know the multidimensional joint probability distribution of all the input signal values! This distribution is usually unknown, except when the process is Gaussian (sect. 5.7) or when those values are independent. When they are independent, the output statistical distribution tends to be Gaussian because of the central limit theorem (sub-sect. 5.5.3). When the input process is Gaussian, it can be shown (problem 5.11.34) that any linear combination of Gaussian variables is also a Gaussian variable. Thus, ***a convolution operator output signal is Gaussian when its input signal is also Gaussian.***

Experimentally, it can be observed that the output statistical distribution of a bandpass filter that is not too narrow is close to Gaussian for many non-Gaussian input signals.

Although we cannot always define the analytical representation of the output signal statistical distribution, its main moments can be evaluated: mean value, mean square value, variance, and autocorrelation. The autocorrelation is given by equation (8.22), rewritten hereafter taking into account that for an ergodic stationary signal: $R_x(\tau) \equiv \varphi_x(\tau)$ and $R_y(\tau) \equiv \varphi_y(\tau)$:

$$R_y(\tau) = R_x(\tau) * \mathring{\varphi}_g(\tau) = \int_{-\infty}^{\infty} R_x(\tau')\mathring{\varphi}_g(\tau - \tau')\mathrm{d}\tau' \tag{8.34}$$

We know that the correlation function value at origin is the mean square value (i.e., total power), equal to the sum of the variance and the square of the mean value:

$$\begin{aligned} P_y &= R_y(0) = \mathrm{E}[\mathbf{y}^2] = \sigma_y^2 + \mu_y^2 \\ &= \int_{-\infty}^{\infty} R_x(\tau')\mathring{\varphi}_g(\tau')\mathrm{d}\tau' = \int_{-\infty}^{\infty} \Phi_x(f) \cdot |G(f)|^2 \mathrm{d}f \end{aligned} \tag{8.35}$$

The second integral is obtained from the product theorem (4.67).

The mean value μ_y is given by

$$\mu_y = \mathrm{E}[\mathbf{y}] = \mathrm{E}[\mathbf{x}(t) * g(t)] = \mathrm{E}[\mathbf{x}] \cdot \int_{-\infty}^{\infty} g(t)\mathrm{d}t \tag{8.36}$$

hence,

$$\mu_y = \mu_x \int_{-\infty}^{\infty} g(t)\mathrm{d}t = \mu_x G(0) \tag{8.37}$$

This result shows that the output mean value of a linear system is equal to the product of the input mean value and the system static gain $G(0)$.

Combining (8.35) and (8.37), the output signal variance may be expressed in the following ways:

$$\begin{aligned}\sigma_y^2 &= \int_{-\infty}^{\infty} [\Phi_x(f) - \mu_x^2\delta(f)] \cdot |G(f)|^2 \mathrm{d}f \\ &= \int_{-\infty}^{\infty} \Phi_x(f)|G(f)|^2 \mathrm{d}f - \mu_x^2 G^2(0) \\ &= \int_{-\infty}^{\infty} R_x(\tau)\mathring{\varphi}_g(\tau)\mathrm{d}\tau - \mu_x^2 \left[\int_{-\infty}^{\infty} g(\tau)\mathrm{d}\tau\right]^2 \\ &= \int_{-\infty}^{\infty} C_x(\tau)\mathring{\varphi}_g(\tau)\mathrm{d}\tau \end{aligned} \tag{8.38}$$

where $C_x(\tau) = R_x(\tau) - \mu_x^2$ is the input signal autocovariance function.

8.2.15 Special cases

Usually, for a lowpass filter: $G(0) = 1$ and $\mu_y = \mu_x$. For a highpass or bandpass filter: $G(0) = 0$ and $\mu_y = 0$ whatever μ_x.

If the input signal is a white noise with a spectral density $\eta/2$ (infinite input variance): $R_x(\tau) = C_x(\tau) = \frac{1}{2}\eta\delta(\tau)$ and $\sigma_y^2 = \frac{1}{2}\eta\mathring{\varphi}_g(0) < \infty$. Other special cases will be presented in the following examples.

8.2.16 Constant multiplier operator

Consider the relation graphically symbolized in fig. 8.12:

$$y(t) = Kx(t) \tag{8.39}$$

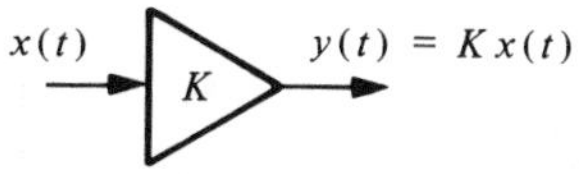

Fig. 8.12

where K is a constant. This is the model of the following linear systems:

- $K > 1$: ideal amplifier (with no bandwidth limitation)
- $K = 1$: closed switch, all-pass circuit (identity operator)
- $0 < K < 1$: attenuator

- $K = 0$: open switch
- $K = -1$: inverter
- $K < 0$: ideal inverting amplifier or attenuator

The properties of the constant multiplier operator are summarized in table 8.13.

Table 8.13

$$g(t) = K\delta(t)$$

$$G(f) = K \text{ with } \begin{cases} |G(f)| = |K| \\ \vartheta_g(f) \begin{cases} 0 & \text{for } K > 0 \\ -\pi \,\mathrm{sgn}(f) & \text{for } K < 0 \end{cases} \end{cases}$$

$$\mathring{\varphi}_g(\tau) = K^2\delta(\tau)$$

It can be shown that $\Phi_y(f) = K^2\Phi_x(f)$, $\varphi_y(\tau) = K^2\varphi_x(\tau)$, $\mu_y = K\mu_x$, $\sigma_y^2 = K^2\sigma_x^2$, and $P_y = K^2P_x$. From (5.39), the output signal distribution is directly drawn from that of the input: $p_y(y) = |K|^{-1}p_x(y/K)$.

8.2.17 Delay operator

The relation symbolized in fig. 8.14:

$$y(t) = x(t - t_0) \tag{8.40}$$

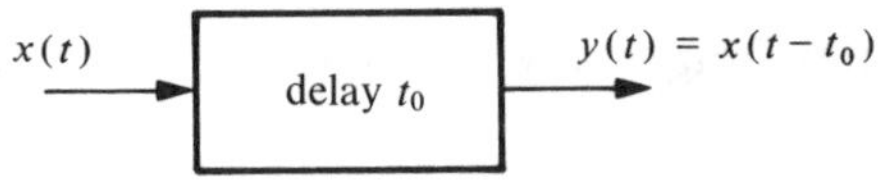

Fig. 8.14

is the model of any system propagating a signal without distortion, but featuring a delay t_0 (ideal transmission line, delay line, shift-register circuit, cyclical memory, magnetic tape device with spaced read and write heads, *et cetera*).

Its properties are summarized in table 8.15

For stationary signals, all the input statistical characteristics are obviously equal to those of the output. In the frequency domain, the delay operator acts (time-shift theorem) like a ***perfect linear phase shifter.*** Hence, any system featuring $|G(f)| = K$, but without a linear phase response, produces a distortion (sub-sect. 8.2.25).

Table 8.15

$g(t) = \delta(t - t_0)$	
$G(f) = \exp(-\mathrm{j}2\pi f t_0)$ with	$\begin{cases} \lvert G(f) \rvert = 1 \\ \vartheta_g(f) = -2\pi f t_0 \end{cases}$
$\mathring{\varphi}_g(\tau) = \delta(\tau)$	

In digital systems (or, more appropriately, in discrete-time systems) we often take into account operators with a sampling interval delay, called a *unit delay operator* (fig. 8.16). Using the z-transform introduced in sub-sect. 8.2.7, it is characterized by the transfer function $G(z) = z^{-1}$.

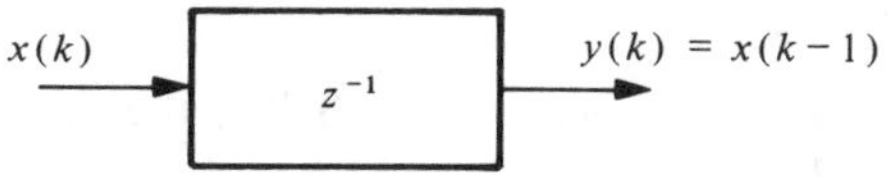

Fig. 8.16

8.2.18 Hilbert operator

The Hilbert transform of a signal was introduced in chapter 7, with respect to the analytic signal concept. A *Hilbert operator* H (fig. 8.17) implements this transform:

$$y(t) = \check{x}(t) = \frac{1}{\pi} \int_{-\infty}^{\infty} \frac{x(\tau)}{t - \tau} \, d\tau \tag{8.41}$$

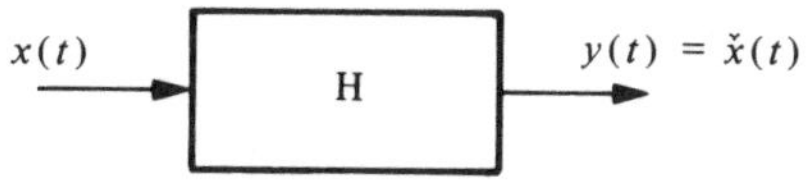

Fig. 8.17

It is the model of a ***perfect*** $\pm 90°$ ***phase shifter,*** as shown by its properties drawn from sub-sect. 7.13 relations and summarized in table 8.18.

The input spectral density, autocorrelation function, and, thus, total power, variance, and mean value are equal to those of the output. The practical implementation of the transform (8.41) can only be approximate, since the impulse response $g(t)$ is not causal. An almost constant and nearly 90° phase shift can only be achieved on a limited bandwidth.

Table 8.18

$g(t)$	$= (\pi t)^{-1}$
$G(f)$	$= -\mathrm{j}\,\mathrm{sgn}(f)$ with $\begin{cases} \lvert G(f)\rvert = 1 \\ \vartheta_g(f) = -\dfrac{\pi}{2}\,\mathrm{sgn}(f) \end{cases}$
$\mathring{\varphi}_g(\tau)$	$= \delta(\tau)$

8.2.19 Time average operator

The estimation of a time average is a very common operation in signal processing. The running average (1.11) is the output of a linear operator:

$$y(t) = \bar{x}(t,T) = \frac{1}{T}\int_{t-T}^{t} x(\tau)\mathrm{d}\tau \tag{8.42}$$

which is called *perfect time averager* (fig. 8.19), the characteristics of which are summarized in table 8.20.

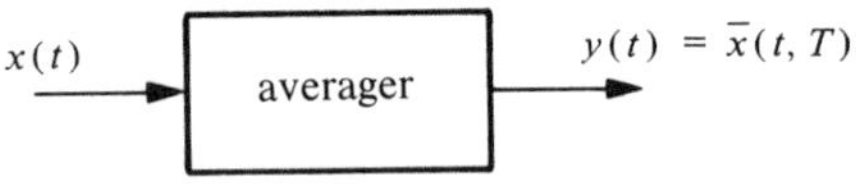

Fig. 8.19

Table 8.20

$g(t)$	$= T^{-1}\,\mathrm{rect}\,[(t - T/2)/T]$
$G(f)$	$= \mathrm{sinc}\,(Tf)\exp(-\mathrm{j}\pi fT)$
$\lvert G(f)\rvert$	$= \lvert\mathrm{sinc}\,(Tf)\rvert$
$\mathring{\varphi}_g(\tau)$	$= T^{-1}\,\mathrm{tri}\,(\tau/T)$

The impulse response of the averager is rectangular (fig. 8.21). It is the model of an integrator circuit, the approximation of which is a single lowpass filter (sub-sect. 8.2.24). An almost perfect implementation (with, however, a sampled output and a periodic reset) can be achieved with an operational amplifier. For a discrete-time system, equation (8.42) is replaced by a simple sum.

The main statistical parameters of the output signal can be deduced from table 8.20 and from (8.37) and (8.38). The mean value is

$$\mu_y = \mu_x \tag{8.43}$$

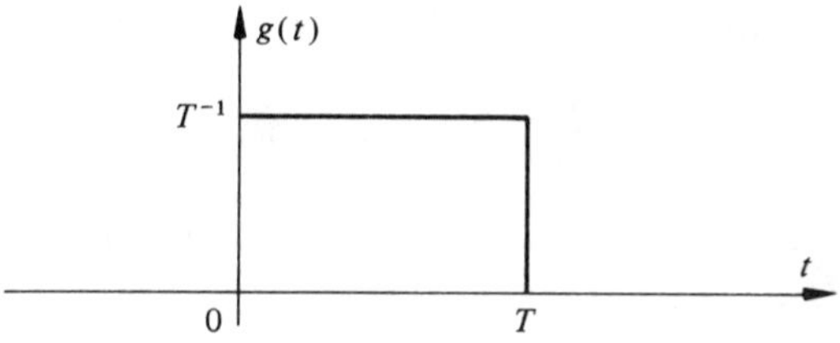

Fig. 8.21

The output variance is equal to the power of the ***mean value estimation error***, and is therefore used as a quality criterion (sub-sect. 13.1.24):

$$\sigma_y^2 = \frac{1}{T}\int_{-\infty}^{\infty} C_x(\tau)\ \mathrm{tri}(\tau/T)\mathrm{d}\tau$$

$$= \int_{-\infty}^{\infty} [\Phi_x(f) - \mu_x^2\delta(f)]\ \mathrm{sinc}^2(Tf)\mathrm{d}f$$

$$= \int_{-\infty}^{\infty} \Phi_x(f)\ \mathrm{sinc}^2(Tf)\mathrm{d}f - \mu_x^2 \qquad \textbf{(8.44)}$$

It obviously decreases as the integration duration T increases.

8.2.20 Example: estimation of a mean value embedded in white noise

Consider a signal $x(t)$, composed of a constant value to be estimated and either an additive white noise, with a spectral density $\Phi_n(f) = \eta/2$, or a lowpass band-limited white noise (sub-sect. 5.3.10).

The autocovariance function $C_x(\tau)$ is equal to the noise autocorrelation function. For white noise, we have

$$C_{x1}(\tau) = \tfrac{1}{2}\eta\delta(\tau) \qquad (8.45)$$

Alternatively, the spectral density of $x(t)$ minus the continuous component contribution is equal to the noise spectral density $\Phi_n(f)$. For a band-limited white noise, with a spectral density (5.155), we get

$$\Phi_{x2}(f) - \mu_x^2\delta(f) = \tfrac{1}{2}\eta\ \mathrm{rect}[f/(2B)] \qquad (8.46)$$

In the first case, the estimation variance is easily obtained by inserting (8.45) into (8.44):

$$\sigma_{y1}^2 = \tfrac{1}{2}\eta/T \qquad (8.47)$$

Although the input signal variance is theoretically infinite, the estimation variance is finite and inversely proportional to the measurement time T.

Inserting (8.46) into (8.44), we get for the second case:

$$\sigma_{y2}^2 = \frac{\eta}{2} \int_{-B}^{B} \text{sinc}^2(Tf)\text{d}f = \frac{\eta}{2T} \int_{-BT}^{BT} \text{sinc}^2(\alpha)\text{d}\alpha \tag{8.48}$$

For $BT = 1$, $\sigma_{y2}^2 \cong 0.45\ \eta/T$ and progressively tends toward the limit (8.47) when BT increases.

8.2.21 Ideal filter operator

An *ideal filter* is an operator transferring ***without distortion*** all the input signal components contained in a spectral bandwidth B, defined by (2.15). It destroys the other components.

For a lowpass filter (fig. 8.22), taking into account an arbitrary delay t_0 as in (8.40), we get the characteristics summarized in table 8.23.

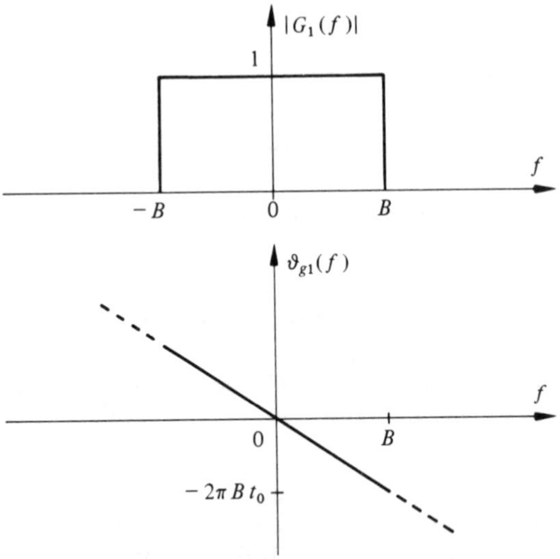

Fig. 8.22

Table 8.23

$G_1(f)$	$= \text{rect}[f/(2B)] \cdot \exp(-\text{j}2\pi f t_0)$ with	$\begin{cases} \lvert G_1(f)\rvert = \text{rect}[f/(2B)] \\ \vartheta_{g1}(f) = -2\pi f t_0 \end{cases}$
$g_1(t)$	$= 2B\,\text{sinc}[2B(t - t_0)]$	
$\mathring{\varphi}_{g1}(\tau)$	$= 2B\,\text{sinc}(2B\tau)$	

Such an ideal filter is a non-causal operator, since $g_1(t) \neq 0$ for $t < 0$ if the delay t_0 is not infinite (fig. 8.24).

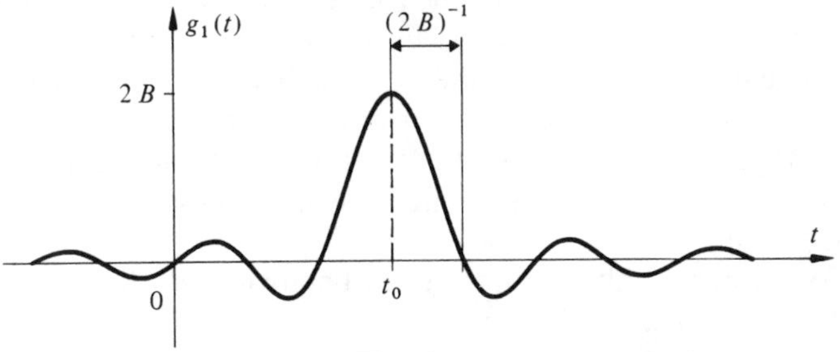

Fig. 8.24

We can deduce the ideal bandpass filter characteristics (fig. 8.25) from those of a $B/2$ lowpass filter, according to the delta-function shift property:

$$G_2(f) = \text{rect}(f/B)\exp(-j2\pi f t_0) * [\delta(f+f_0) + \delta(f-f_0)] \tag{8.49}$$

Results are summarized in table 8.26.

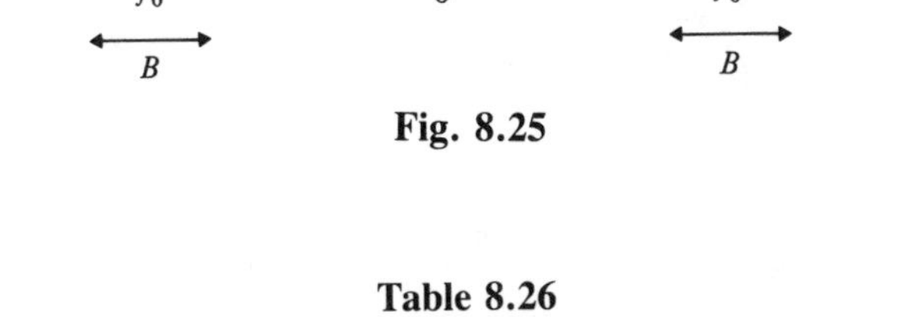

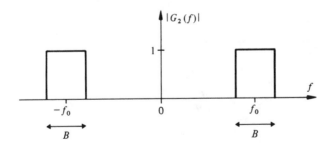

Fig. 8.25

Table 8.26

$\lvert G_2(f)\rvert$	$= \text{rect}[(f+f_0)/B] + \text{rect}[(f-f_0)/B]$
$g_2(t)$	$= 2B\,\text{sinc}[B(t-t_0)]\cdot\cos(2\pi f_0 t)$
$\mathring{\varphi}_{g_2}(\tau)$	$= 2B\,\text{sinc}(B\tau)\cos(2\pi f_0\tau)$

8.2.22 Ideal filter rise time

The unit-step response $\gamma(t)$ is the integral of the impulse response (sub-sect. 1.3.12). For a lowpass filter, it allows us to characterize directly the reaction time (inertia) of the filter, arbitrarily measured by a *rise time* t_m. For a bandpass filter, the response to $\epsilon(t)\cos(2\pi f_0 t)$, where f_0 is the center of the bandwidth, yields the same information. It is, however, equivalent and simpler to consider, in this case, the integral $\gamma_{rg2}(t)$ of the envelope $r_{g2}(t)$ of the impulse response.

This envelope is found through (7.45). Taking into account (7.18) and example 7.1.4, it is given for an ideal bandpass filter of bandwidth B by $r_{g2}(t) = 2B\,\text{sinc}[B(t - t_0)]$.

For an ideal lowpass filter with the same bandwidth B, the impulse response (table 8.23) is $g_1(t) = 2B\,\text{sinc}[2B(t - t_0)]$.

With result (1.65), through integration, we get

$$\gamma_1(t) = \tfrac{1}{2} + \pi^{-1}\text{Si}[2\pi B(t - t_0)]; \quad \tfrac{1}{2}\gamma_{rg2}(t) = \tfrac{1}{2} + \pi^{-1}\text{Si}[\pi B(t - t_0)] \tag{8.50}$$

These functions are plotted in fig. 8.27. The oscillation displayed by these responses to a discontinuity is known as the Gibbs phenomenon.

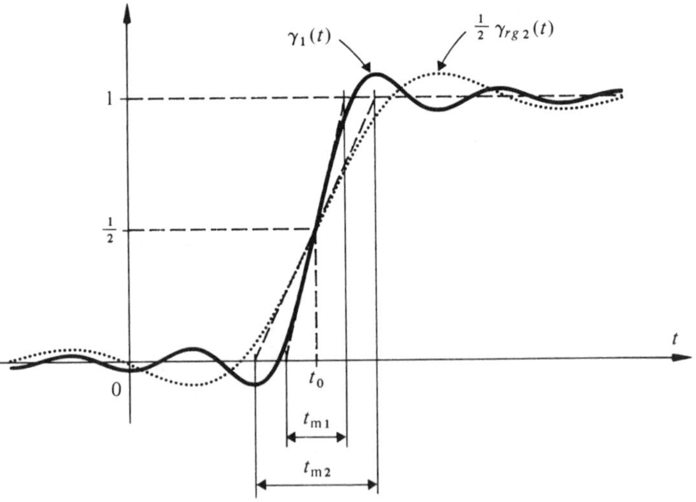

Fig. 8.27

By arbitrarily defining respective rise times t_{m1} and t_{m2} as the interval between the intersections of the derivatives of $\gamma_1(t)$ and $\tfrac{1}{2}\gamma_{rg2}(t)$ at $t = t_0$ with ordinates 0 and 1, we get

$$\begin{aligned} t_{m1}^{-1} &= \mathrm{d}\gamma_1/\mathrm{d}t|_{t=t_0} = g_1(t_0) = 2B \\ t_{m2}^{-1} &= \tfrac{1}{2}\mathrm{d}\gamma_{rg2}/\mathrm{d}t|_{t=t_0} = \tfrac{1}{2}r_{g2}(t_0) = B \end{aligned} \tag{8.51}$$

The rise time thusly defined is the inverse of the bandwidth for an ideal bandpass filter and the inverse of twice the bandwidth for an ideal lowpass filter.

This result provides us with a useful order of magnitude. It is approximately applicable to real filters if B is taken as the equivalent bandwidth defined in sub-sect. 8.2.23.

Experimentally, the rise time is usually defined as the time needed by the unit-step response to go from 10% to 90% of its final value.

8.2.23 Bandwidth of a real filter

A real filter is a causal operator with frequency-selective properties. The modulus of its transfer function cannot have discontinuities. Hence, the bandwidth definition is somewhat arbitrary.

In electronics, for the sake of easy measurement, we often use the *bandwidth at −3 dB* (fig. 8.28), denoted here $B_{-3\mathrm{dB}}$ and defined as the range of positive frequencies that comply with

$$\frac{|G(f)|^2}{G_{\max}^2} \geqslant \frac{1}{2} \tag{8.52}$$

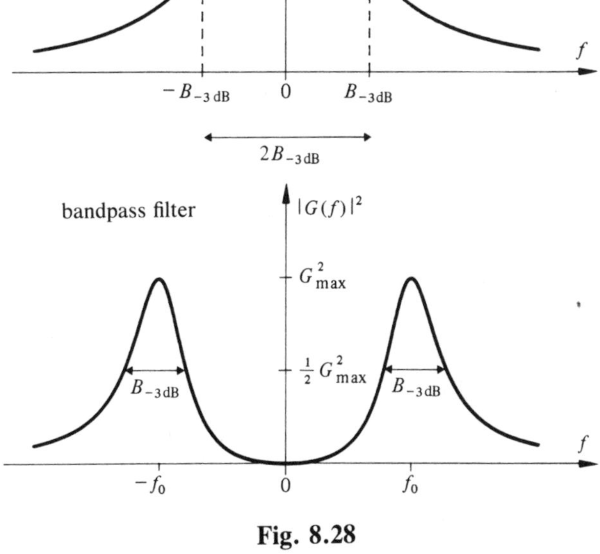

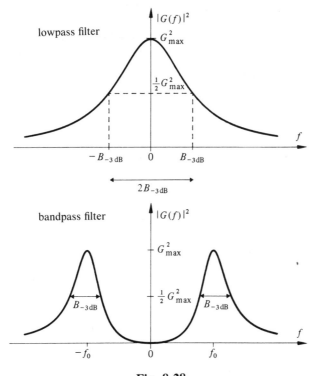

Fig. 8.28

The concept of *noise equivalent bandwidth* is introduced in signal processing because of the importance of stochastic processes:

$$B_{eq} = \frac{1}{G_{max}^2} \int_0^\infty |G(f)|^2 df = \frac{\mathring{\varphi}_g(0)}{2G_{max}^2} \tag{8.53}$$

The noise equivalent bandwidth is the bandwidth of an ideal filter, the output signal of which has the same power as that of the real filter when both inputs have the same white noise.

8.2.24 Illustration

A lowpass *RC* filter (fig. 8.29), with an infinite resistive load, has a transfer function:

$$G(f) = \frac{1}{1 + j2\pi fRC} = \frac{1}{1 + jf/f_c} \tag{8.54}$$

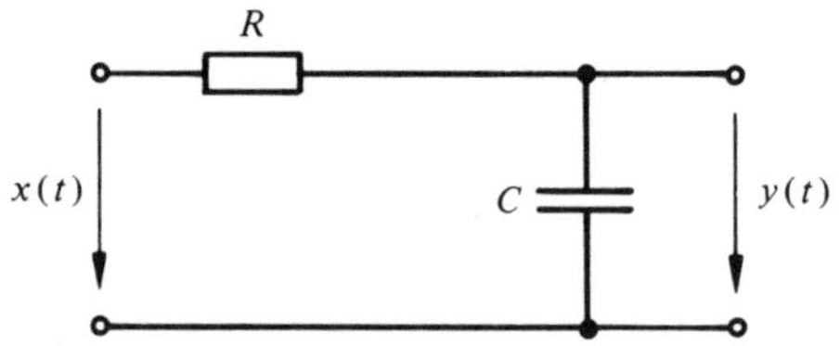

Fig. 8.29

where $f_c = (2\pi RC)^{-1}$ is the cut-off frequency.

We easily get $G(f) = [1 + (f/f_c)^2]^{-1/2}$, $\vartheta_g(f) = -\arctan(f/f_c)$, and

$$|G(f)|^2 = \frac{1}{1 + (f/f_c)^2} \tag{8.55}$$

thus,

$$g(t) = \frac{1}{RC} \exp[-t/(RC)] \cdot \epsilon(t) \tag{8.56}$$

$$\mathring{\varphi}_g(\tau) = \frac{1}{2RC} \exp[-|\tau|/(RC)] \tag{8.57}$$

$$B_{-3dB} = f_c \tag{8.58}$$

$$B_{eq} = \frac{1}{4RC} = \frac{\pi}{2} B_{-3dB} \tag{8.59}$$

According to definition (8.51), the rise time is

$$t_{m1} = (2B_{eq})^{-1} = 2RC \tag{8.60}$$

Hence, it is equal to twice the network time constant.

Such a filter is an imperfect integrator often used to implement an approximate averager (see problem 8.5.6 and sub-sect. 13.1.24).

For a perfect averager (sub-sect. 8.2.19), the equivalent bandwidth (8.53) is given, with $|G(f)|^2 = \text{sinc}^2(Tf)$ and (1.63), by $B_{eq} = (2T)^{-1}$. Thus, the rise time $t_{m1} = T$, which corresponds rather well with reality.

8.2.25 Linear distortion

Every linear system with a non-constant amplitude response, or with a nonlinear phase response inflicts distortions on the input signal, called *amplitude distortion* or *phase distortion,* respectively.

Phase distortion stems from the different delays that affect each signal frequency component. This distortion has no audible effect in electro-acoustic systems or in telephony because our ears are insensitive to this phenomenon. By contrast, it must be reduced in many other instances: data or TV signal transmission, radar signal receiving, *et cetera*.

8.2.26 Phase delay and group delay

Consider a linear system input signal with a narrowband spectrum centered around frequency f_0. Using the complex notation of section 7.4, this signal can be represented by the analytic signal:

$$\underline{\underline{x}}(t) = \underline{r}(t) \exp(\mathrm{j}2\pi f_0 t) \tag{8.61}$$

where $\underline{r}(t)$ is the complex envelope (sub-sect. 7.4.1).

If the system phase response is not linear, its behavior in the frequency range around f_0 can be approximately described by the first two members of a Taylor-series expansion:

$$\vartheta(f) = \vartheta(f_0) + (f - f_0)\left.\frac{\mathrm{d}\vartheta}{\mathrm{d}f}\right|_{f=f_0} = -2\pi f_0 t_\phi - 2\pi(f - f_0)t_g \tag{8.62}$$

where t_ϕ and t_g are two constants, with time dimension, defined by

$$t_\phi = -\frac{\vartheta(f_0)}{2\pi f_0} \tag{8.63}$$

$$t_g = -\frac{1}{2\pi}\left.\frac{\mathrm{d}\vartheta}{\mathrm{d}f}\right|_{f=f_0} \tag{8.64}$$

They are respectively referred to as the *phase delay* and the *group delay*.

Assuming, for the sake of simplicity, that the amplitude response is constant and equal to one on the considered bandwidth, the system transfer function becomes

$$G(f) = \exp\{-j2\pi[f_0(t_\phi - t_g) + ft_g]\} \tag{8.65}$$

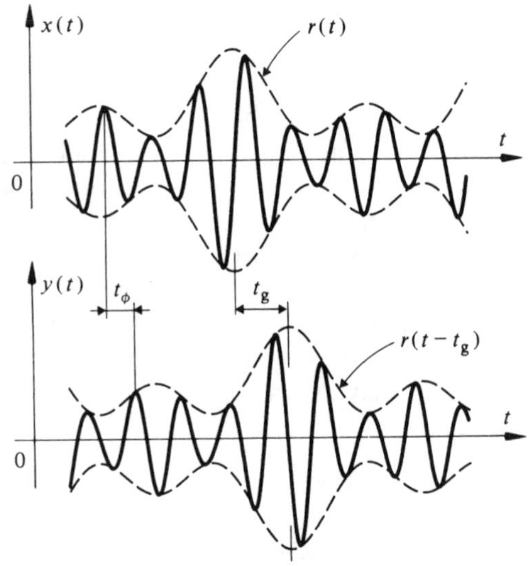

Fig. 8.30

The Fourier transform of the analytic output signal is then given by (8.14):

$$\underline{\underline{Y}}(f) = \underline{\underline{X}}(f)G(f)$$

$$= \underline{\underline{X}}(f) \exp(-j2\pi ft_g) \exp[-j2\pi f_0(t_\phi - t_g)] \tag{8.66}$$

By inverse transform and taking into account (8.61), we easily get

$$\underline{y}(t) = \underline{x}(t - t_g) \exp[-j2\pi f_0(t_\phi - t_g)]$$

$$= \underset{\sim}{r}(t - t_g) \exp[j2\pi f_0(t - t_\phi)] \tag{8.67}$$

Thus, the signal envelope is delayed by t_g, while the sinusoidal auxiliary component with frequency f_0 is delayed by t_ϕ. These two delays are equal only when the phase response varies linearly with frequency.

8.2.27 Matched filter or correlation operator

A *matched filter* (sect. 13.4) is a particular case of a linear filter designed to optimize the signal-to-noise ratio when attempting to detect a signal $s(t)$ of known shape and duration T, embedded in background noise. In the presence of white noise, its impulse response is the time-reversal, shifted by T, of the signal to be detected:

$$g(t) = ks(T - t) \tag{8.68}$$

where k is an arbitrary constant.

The response of a matched filter to the sole input $x(t) = s(t)$ is equal to the shifted autocorrelation function of the signal:

$$y(t) = ks(T - t) * s(t) = k\mathring{\varphi}_s(t - T) \tag{8.69}$$

The matched filter behaves as a correlator with respect to the signal to be detected. We can show that in presence of a white noise the signal-to-noise ratio is optimal for $t = T$, and only depends on the signal energy and the noise spectral density. This result demonstrates the role of correlation in detection and identification processes.

The characteristics of this operator are summarized in table 8.31.

Table 8.31

$g(t)$	$= ks(T-t)$
$G(f)$	$= kS^*(f)\exp(-j2\pi fT)$
$\lvert G(f)\rvert^2$	$= k^2\mathring{\Phi}_s(f)$
$\mathring{\varphi}_g(\tau)$	$= k^2\mathring{\varphi}_s(\tau)$

8.2.28 Differentiating operator

The linear transform

$$y(t) = \frac{\mathrm{d}}{\mathrm{d}t}x(t) \tag{8.70}$$

defines an operator, called a *differentiator,* the characteristics of which are summarized in table 8.32. Such an operator is not physically achievable, since its amplitude response increases linearly and infinitely with frequency. A high-pass filter, like the differentiating RC network of fig. 8.33, is an approximate implementation.

Table 8.32

$g(t)$	$= \dfrac{\mathrm{d}}{\mathrm{d}t}\,\delta(t) = \delta'(t)$
$G(f)$	$= \mathrm{j}2\pi f$
$\lvert G(f)\rvert^2$	$= (2\pi f)^2$

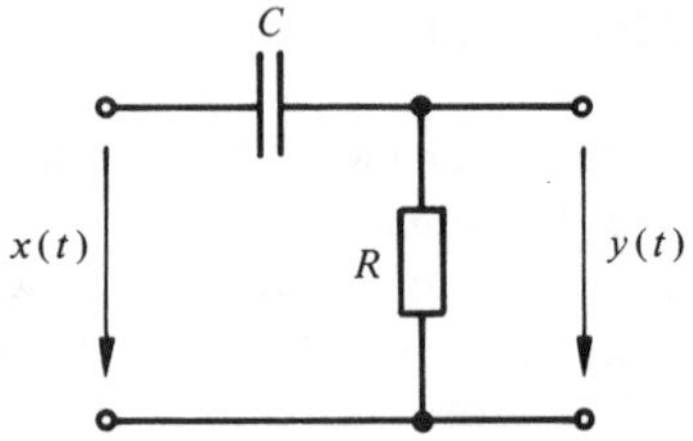

Fig. 8.33

From table 8.32, we can deduce that the spectral density of the derivative (when it exists) of a random process is

$$\Phi_{x'}(f) = (2\pi f)^2 \Phi_x(f) \tag{8.71}$$

Thus, as a result of the Wiener-Khinchin theorem (sub-sect. 5.3.3), we have the following relation for the autocorrelations:

$$R_{x'}(\tau) = -\frac{\mathrm{d}^2 R_x(\tau)}{\mathrm{d}\tau^2} \tag{8.72}$$

According to the mean square convergence, the existence of a derivative is only possible if the second derivative of the process autocorrelation function exists.

8.3 PARAMETRIC OPERATORS

8.3.1 Definition

What we call a *parametric operator* is any nonstationary operator that is dependent on an auxiliary signal or control. The output signal:

$$y(t) = \mathrm{S}\{x(t), u(t)\} \tag{8.73}$$

is thus a function of two signals $x(t)$ and $u(t)$, one being considered as an input signal and the other as a control signal (fig. 8.34).

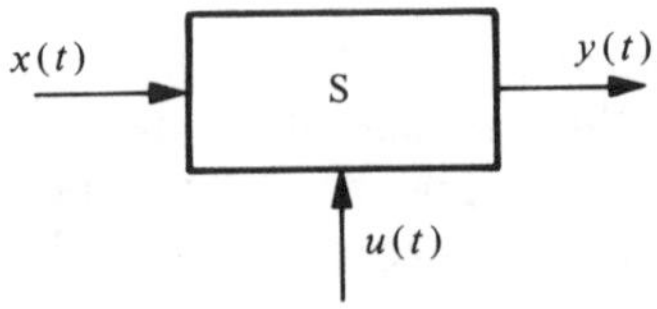

Fig. 8.34

8.3.2 General description of linear parametric operators

An operator is linear, but nonstationary, if condition (8.5) is satisfied, but condition (8.6) is not satisfied.

The output signal $y(t)$ of such an operator is expressed as a function of the input signal $x(t)$ by the relation [49, 53]:

$$y(t) = \int_{-\infty}^{\infty} h(t,\tau)x(\tau)\mathrm{d}\tau \tag{8.74}$$

where $h(t,\tau)$ is the impulse response, depending on the auxiliary signal $u(t)$. It is the time response to a delta-function applied to the input at time τ. The operator is causal if $h(t,\tau) = 0$ for $\tau > t$.

A special case of this general class is the invariant linear operator for which $h(t,\tau) = g(t - \tau)$.

An indirect description of the nonstationary linear operator is the response function:

$$H(\nu,t) = \int_{-\infty}^{\infty} h(t,\tau) \exp(-\mathrm{j}2\pi\nu\tau)\mathrm{d}\tau \tag{8.75}$$

which also depends on time. Usually, an expression depending on two frequencies f and ν obtained by a second Fourier transform is preferable, such that

$$H(f,\nu) = \iint_{-\infty}^{\infty} h(t,\tau) \exp[\mathrm{j}2\pi(\nu\tau - ft)]\mathrm{d}\tau\mathrm{d}t \tag{8.76}$$

We can check (problem 8.5.12) that the Fourier transforms of the input and output signals $x(t)$ and $y(t)$ are related by the integral equation:

$$Y(f) = \int_{-\infty}^{\infty} H(f,\nu)X(\nu)\mathrm{d}\nu \tag{8.77}$$

8.3.3 Separable operator

An operator is said to be *separable* when the impulse response can be written as

$$h(t,\tau) = u(t)g(t - \tau) \tag{8.78}$$

or, in an equivalent manner,

$$h(t,\tau) = u(t)g(t - \tau) \tag{8.79}$$

The transfer function (8.76), for an impulse response like (8.78), becomes

$$H(f,\nu) = U(f-\nu)G(\nu) \tag{8.80}$$

and, for (8.79),

$$H(f,\nu) = U(f-\nu)G(f) \tag{8.81}$$

8.3.4 Multiplier operator

Inserting $g(t-\tau) = \delta(t-\tau)$ into (8.78), we get

$$h(t,\tau) = u(t)\delta(t-\tau) \tag{8.82}$$

and

$$H(f,\nu) = U(f-\nu) \tag{8.83}$$

hence, replacing in (8.74) and (8.77), we write

$$y(t) = u(t) \cdot x(t) \tag{8.84}$$

and

$$Y(f) = U(f) * X(f) \tag{8.85}$$

The *multiplier* (fig. 8.35) is thus a linear parametric operator (identical to the stationary operator of sub-sect. 8.2.16 if $u(t)$ is constant).

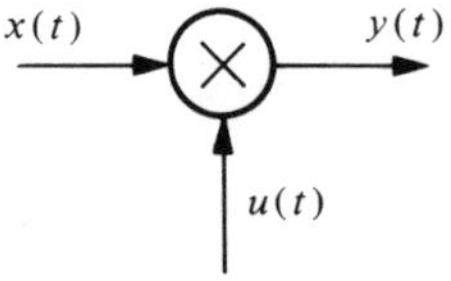

Fig. 8.35

The relations (8.84) and (8.85) reproduce property (4.15). In the case of ***independent random signals*** $x(t)$ and $u(t)$, the statistical and spectral properties of the output signal $y(t)$ can be deduced from (5.195), (5.198), and (5.199), which become

$$p_y(y) = \int_{-\infty}^{\infty} \frac{1}{|x|} p_x(x) p_u\left(\frac{y}{x}\right) \mathrm{d}x \tag{8.86}$$

$$R_y(\tau) = R_x(\tau) \cdot R_u(\tau) \tag{8.87}$$

$$\Phi_y(f) = \Phi_x(f) * \Phi_u(f) \tag{8.88}$$

8.3.5 Particular case: ideal sampling operator

Inserting $u(t) = \delta_T(t)$ into (8.82), and taking into account (1.54), the output signal becomes

$$y(t) = x(t)\delta_T(t) = \sum_{k=-\infty}^{\infty} x(kT)\delta(t - kT) \tag{8.89}$$

It is the model of the idealized sampled signal described in chapter 9.

8.3.6 Particular case: periodic switch

The nonstationary operator symbolized (fig. 8.36) by a switch periodically closed at a rate $f = 1/T$ for a duration Δ is a multiplier operator, for which (see notation 1.56)

$$u(t) = \text{rep}_T\{\text{rect}(t/\Delta)\} \tag{8.90}$$

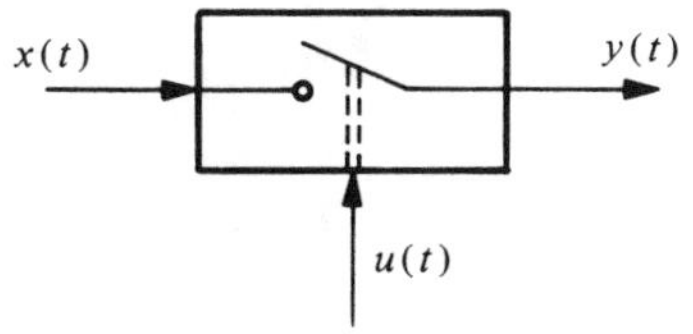

Fig. 8.36

8.3.7 Particular case: uniform weighting operator

The operator symbolized by figure 8.36 allows us to model an observation $y(t) = x(t,T)$ of a finite duration T of a signal $x(t)$ by giving to the control function the shape of a weighting rectangular time window:

$$u(t) = \text{rect}(t/T) \tag{8.91}$$

The finite observation duration yields a limited spectral resolution (problem 8.5.14) and the presence of oscillations (Gibbs phenomenon) around the $x(t)$ spectral discontinuities.

Other weighting functions (time windows) are sometimes used (sub-sect. 12.1.6 and vol. XX) to reduce this problem.

8.3.8 Particular case: periodic inverter

The operator symbolized by a periodic inverter (multiplying by ± 1) corresponds to multiplication by an auxiliary function (in this case with a 50% duty cycle):

$$u(t) = \text{sgn}\{\cos(2\pi t/T)\}$$
$$= \text{rep}_T\{\text{rect}(2t/T) - \text{rect}[(2t - T)/T]\} \tag{8.92}$$

8.3.9 Example: prefiltered multiplier operator

The impulse response (8.78) defines an operator composed of a cascade of an invariant linear operator followed by a multiplier (fig. 8.37). Indeed, inserting $h(t,\tau) = u(t)g(t - \tau)$ into (8.74), we obtain for the output signal:

$$y(t) = u(t) \cdot \int_{-\infty}^{\infty} g(t - \tau)x(\tau)\mathrm{d}\tau$$
$$= u(t) \cdot [x(t) * g(t)] \tag{8.93}$$

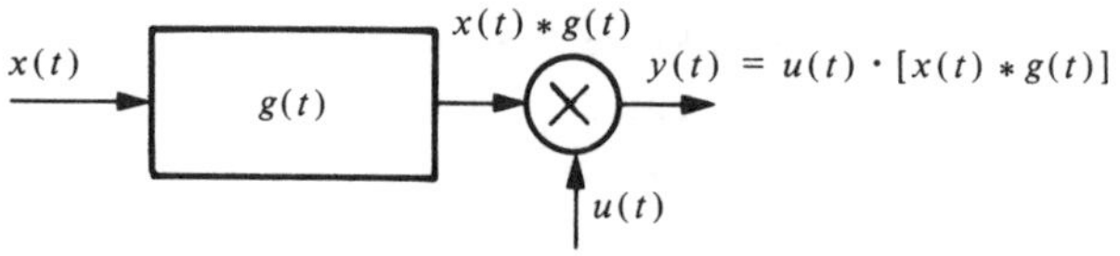

Fig. 8.37

The spectral properties of signal $y(t)$ are easily deduced thusly

$$Y(f) = U(f) * [X(f)G(f)] \tag{8.94}$$

$$\Phi_y(f) = \Phi_u(f) * [\Phi_x(f)|G(f)|^2] \tag{8.95}$$

8.3.10 Example: postfiltered multiplier operator

From the impulse response (8.79), the output signal becomes

$$y(t) = \int_{-\infty}^{\infty} g(t - \tau)u(\tau)x(\tau)\mathrm{d}\tau = [x(t)u(t)] * g(t) \tag{8.96}$$

This is the response of an operator (fig. 8.38) combining a multiplier followed by an invariant linear filter.

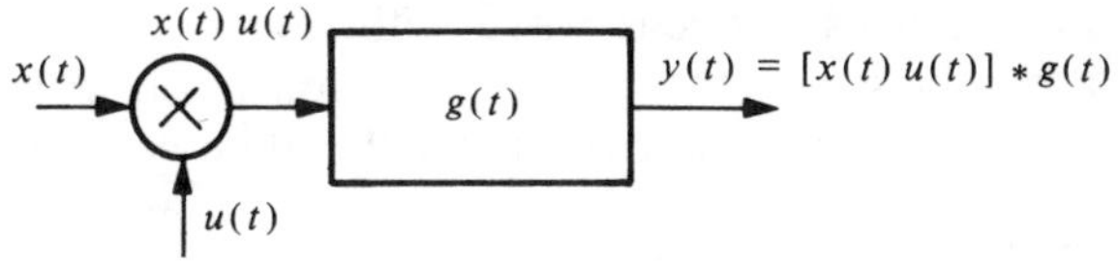

Fig. 8.38

The spectral properties of the output signal here are

$$Y(f) = [X(f) * U(f)]G(f) \tag{8.97}$$

$$\Phi_y(f) = [\Phi_x(f) * \Phi_u(f)] \cdot |G(f)|^2 \tag{8.98}$$

Such an operator is used to model the real sampling, the reconstruction of a continuous signal from a sampled signal, the amplitude modulation, and the frequency translation.

8.3.11 Particular case: real sampling operator

Inserting $u(t) = \delta_T(t)$ into (8.96), we get

$$y(t) = [x(t)\delta_T(t)] * g(t) = \sum_{k=-\infty}^{\infty} x(kT)g(t - kT) \tag{8.99}$$

Each sample here is the multiplying factor of a delayed pulse with a $g(t)$ shape. For a delta-function, $g(t) = \delta(t)$, and we again find the ideal sampling (8.89).

8.3.12 Modulation operator

A *modulator* (fig. 8.39) is typically a parametric linear or nonlinear device, producing an output signal with one or more parameters varying with the input signal.

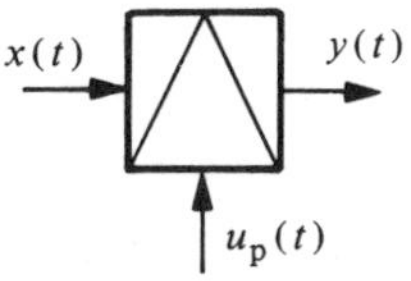

Fig. 8.39

The auxiliary signal is called the carrier $u_p(t)$ in telecommunications. The input and output signals are called the modulating (or primary) signal and the modulated (or secondary) signal, respectively.

The carrier $u_p(t)$ is usually a sine wave with frequency f_p, and the output signal $y(t)$ has a bandpass spectrum. Using the concept of complex envelope $\underset{\sim}{r}(t)$, introduced in section 7.4, the output signal can be written

$$y(t) = \text{Re}\{\underset{\sim}{r}(t)\exp(\ j2\pi f_p t)\} \tag{8.100}$$

where $\underset{\sim}{r}(t) = f\{x(t)\}$ is a function of the input signal characteristic of each kind of modulation (chap. 11).

Because the complex envelope depends only on the input signal, we can simplify the notations by defining [99] a modulation operator (fig. 8.40) according to the relation $\underset{\sim}{r}(t) = S_m\{x(t)\}$. Conversely, we can define a demodulation operator (fig. 8.41) by relation $x(t) = S_d\{\underset{\sim}{r}(t)\}$. This one can be either parametric or nonlinear.

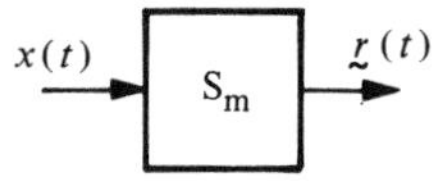

Fig. 8.40

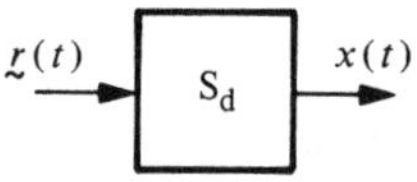

Fig. 8.41

8.3.13 Sum operator

An *adder* (fig. 8.42) or a *subtractor* (fig. 8.43) is an operator that adds or subtracts $u(t)$ to or from the input signal $x(t)$:

$$y(t) = \text{S}\{x(t),\ u(t)\} = x(t) \pm u(t) \tag{8.101}$$

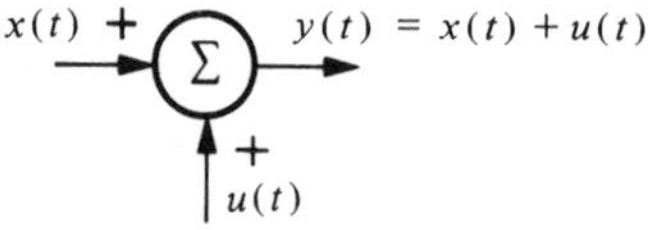

Fig. 8.42

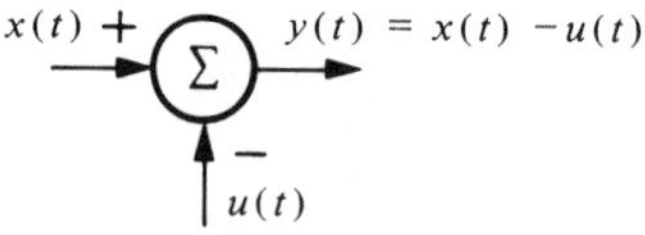

Fig. 8.43

For the input signal, this operator is not a homogeneous linear one because it does not comply with (8.5), and it is not stationary if $u(t)$ is not a constant. It is thus logical to classify it as a parametric operator, although, in some applications, it may be better to describe it as linear vector operator.

According to (4.12), for deterministic signals, we get

$$Y(f) = X(f) \pm U(f) \tag{8.102}$$

As a general rule, which is also valid for random signals, from (5.189) and (5.190), we have

$$R_y(\tau) = R_x(\tau) + R_u(\tau) \pm R_{xu}(\tau) \pm R_{ux}(\tau) \tag{8.103}$$

and

$$\Phi_y(f) = \Phi_x(f) + \Phi_u(f) \pm \Phi_{xu}(f) \pm \Phi_{ux}(f) \tag{8.104}$$

In these two expressions, the last two terms disappear when signals $x(t)$ and $u(t)$ are independent, and when at least one has a zero mean value (subsect. 5.5.5).

8.4 INVARIANT NONLINEAR OPERATORS

8.4.1 Definitions and classification

The modeling of nonlinear systems comes up against enormous difficulties. There exists no general theory, as in the linear case, which allows us to define easily the relations binding a system's inputs and outputs. It is a difficult field that, despite the long time it has been studied, still needs a huge research effort. Some partial approaches have been tried [100–102]. The topic is so complex that it cannot be treated in depth here.

A way to relate the input $x(t)$ to the output $y(t)$ of a nonlinear operator—assumed here to be invariant—is the *Volterra functional series representation*:

$$y(t) = \sum_{n=1}^{\infty} \int_{-\infty}^{\infty} \ldots \int_{-\infty}^{\infty} h_n(\tau_1, \ldots, \tau_n) x(t - \tau_1) \ldots x(t - \tau_n) d\tau_1 \ldots d\tau_n \tag{8.105}$$

It is an extension of the integral representation of linear operators (8.12), this case corresponding to the first series term. It implies that the integrals exist and that the series converges in the mean square sense. The general functional:

$$\int_{-\infty}^{\infty} \ldots \int_{-\infty}^{\infty} h_n(\tau_1, \ldots, \tau_n) x(t - \tau_1) \ldots x(t - \tau_n) d\tau_1 \ldots d\tau_n \tag{8.106}$$

defines an *n*th-*order homogeneous nonlinear operator* characterized (fig. 8.44) by kernel $h_n(t_1, \ldots, t_n)$.

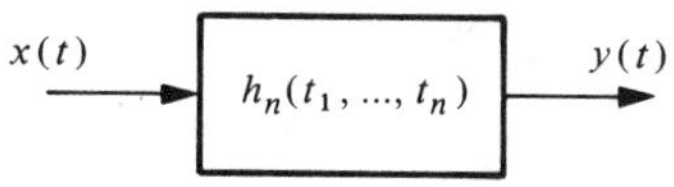

Fig. 8.44

It can be shown [102] that the response of such an operator to one permanent sine-wave input with frequency f_0 is composed of sine waves with frequencies nf_0, $(n-2)f_0, \ldots, f_0$ (n odd), or 0 (n even). If the order n is odd, the response contains only odd harmonics; and, if n is even, it contains only even harmonics and a continuous component. An example of this result is exhibited in subsections 8.4.4 and 8.4.5.

Some nonlinear circuits (e.g., with a nonlinear reactance) are able to generate sub-harmonics of the input frequency. They cannot, therefore, be represented by this model.

Relation (8.105) suggests a representation (fig. 8.45) by parallel homogeneous operators with added outputs.

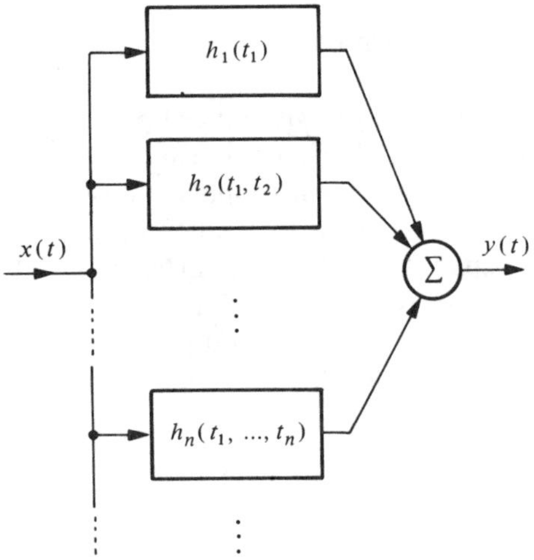

Fig. 8.45

Relation (8.105) describes a special nonlinear operator when the output signal can be expressed as a power series:

$$y(t) = \sum_{n=1}^{\infty} \alpha_n x^n(t) \tag{8.107}$$

The kernel becomes simply

$$h_n(t_1, \ldots, t_n) = \alpha_n \delta(t_1) \ldots \delta(t_n) \tag{8.108}$$

Relation (8.107) describes a wide variety of *amnesic* (or *without memory, instantaneous*) *nonlinear operators*: the output signal $y(t)$ at time t depends only on the input at that very same time, and not on any past value. In electrical engineering, a nonlinear circuit is, in principle, of this kind if it does not contain any reactive element.

It is often possible [103] to model a nonlinear system by a cascade of two or three operators (fig. 8.46): for example, a linear operator followed by an amnesic nonlinear operator and a second linear operator. This is called a *separable system*.

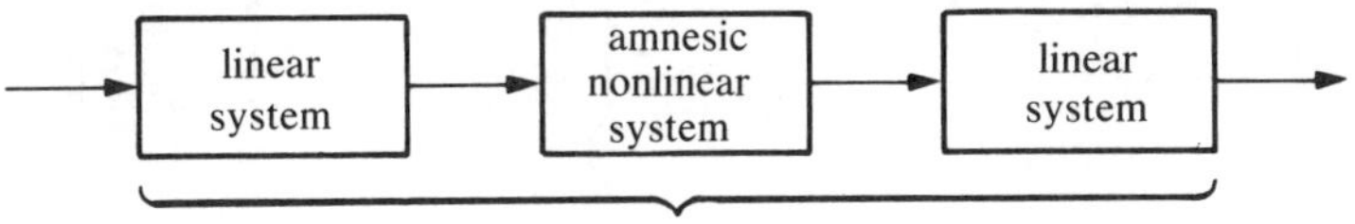

Fig. 8.46

We can thus consider the following classification:

- amnesic nonlinear operators
- separable nonlinear operators
- nonseparable nonlinear operators

The last group will not be treated. The second is hence a cascade of operators in which the inertia (memory) is taken into account by the linear operator, while the nonlinearity is represented by the amnesic operator. Hereafter we will only study some of the properties of these operators.

8.4.2 Amnesic nonlinear operators

An amnesic nonlinear operator is characterized by an instantaneous relationship between input and output signals (fig. 8.47), which is the equation of the nonlinearity:

$$y = g(x) \tag{8.109}$$

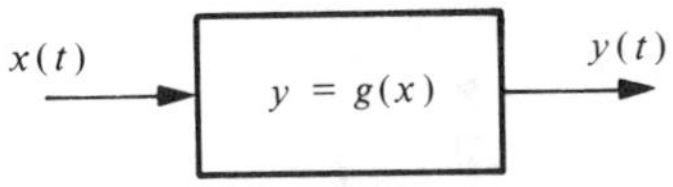

Fig. 8.47

This is the model of many electronic devices implemented in analog designs with diodes, operational amplifiers, comparators, multipliers, *et cetera* and implemented in digital designs by look-up tables stored in memory or by an

Table 8.48

Functions	Characteristics	Graphic symbols
Bipolar rectifier (absolute value operator)	$y\ (t) = \lvert x(t)\rvert$	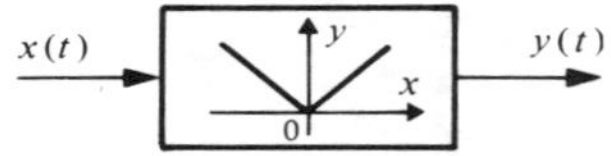
Unipolar rectifier	$y\ (t) = \epsilon(x) \cdot \lvert x(t)\rvert$ $= \tfrac{1}{2}[x(t) + \lvert x(t)\rvert]$	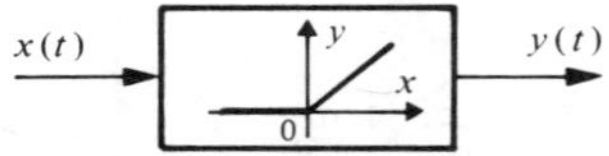
Sign detector	$y(t) = A \cdot \operatorname{sgn}\{x(t)\}$	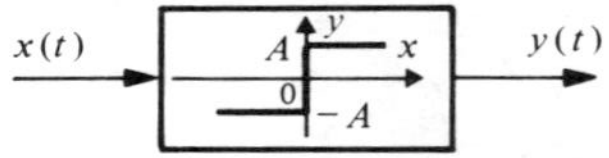
Comparator	$y(t) = \operatorname{sgn}\{x(t) - a\}$	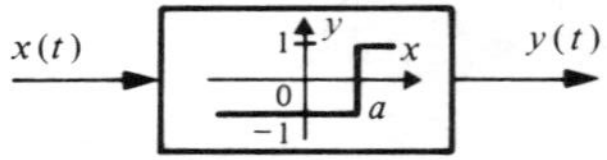
Clipper	$y\ (t) = \begin{cases} a & \text{if } x < a \\ x(t) & \text{if } a < x < b \\ b & \text{if } x > b \end{cases}$	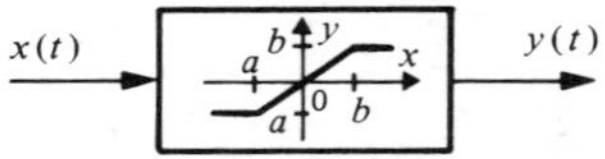
Squaring operator	$y\ (t) = x^2(t)$	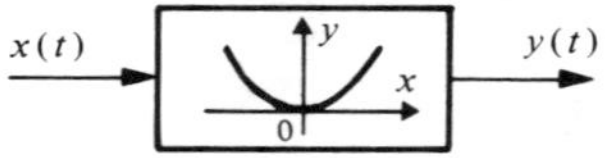
Square-root operator	$y\ (t) = \sqrt{x(t)};\ x \geqslant 0$	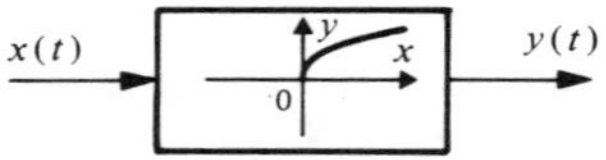
Logarithmic operator	$y\ (t) = \log_a\{x(t)\};\ x > 0$	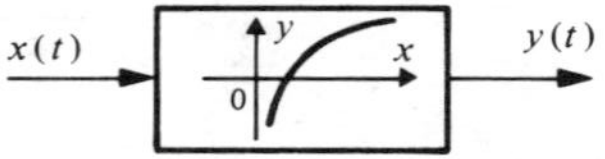
Trigonometric operator	for example: $y\ (t) = \sin\{x(t)\}$	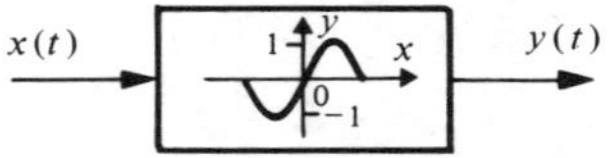

adequate algorithm. The most common ones are summarized in table 8.48. If the output is delayed (reaction time, computation time), we take it into account by cascading an ideal amnesic operator (without delay) and a delay operator (sub-sect. 8.2.17).

8.4.3 Nonlinearity expanded in Taylor series

The Taylor-series expansion (8.107) does not converge around discontinuities. It can thus only be used when dealing with nonlinear characteristics $g(x)$ with progressive variation. In this case, coefficients α_n are related to the nonlinearity $g(x)$ by

$$\alpha_n = \frac{1}{n!} \left. \frac{d^n g(x)}{dx^n} \right|_{x=0} \tag{8.110}$$

The Fourier transform of the output signal of such an operator excited by a deterministic signal simply becomes, from (4.15):

$$Y(f) = \sum_{n=1}^{\infty} \alpha_n \mathop{\boxed{*}}_{i=1}^{n} X(f) \tag{8.111}$$

where

$$\mathop{\boxed{*}}_{i=1}^{n} X(f) = \underbrace{X(f) * X(f) * \ldots * X(f)}_{n \text{ times}} \tag{8.112}$$

denotes a $(n - 1)$th order multiconvolution.

Hence, the complex spectrum of the output signal contains a first member proportional to the input signal spectrum $X(f)$, then a member proportional to the autoconvolution of $X(f)$, *et cetera*. Because of the central limit theorem (sub-sect. 5.5.3), the spectral contributions (8.112) of high order are close to a Gaussian distribution when $X(f)$ is real (or imaginary).

A typical property of nonlinear systems is to create spectral components that are totally absent from the input spectrum.

This property is sometimes undesired and interpreted as a distortion source (sub-sect. 8.4.6) of the input signal. On the other hand, it can be put to work in some applications (modulation, frequency translation and multiplication, *et cetera*).

8.4.4 Example: squaring operator

If

$$y(t) = x^2(t) \tag{8.113}$$

the (8.107) expansion is reduced to a single quadratic term. The Fourier transform of the output signal is here a simple convolution product:

$$Y(f) = X(f) * X(f) \tag{8.114}$$

When the input signal is a simple sine wave $x(t) = A\cos(2\pi f_0 t)$: $X(f) = \frac{1}{2}A[\delta(f + f_0) + \delta(f - f_0)]$ and, from (1.50) and ((8.114), $Y(f) = \frac{1}{2}A^2\delta(f) + \frac{1}{4}A^2[\delta(f + 2f_0) + \delta(f - 2f_0)]$. This result (fig. 8.49), which can be easily checked with trigonometric relations, corresponds to an output signal $y(t) = \frac{1}{2}A^2 + \frac{1}{4}A^2\cos(4\pi f_0 t)$. The continuous component $\frac{1}{2}A^2$ measures the input signal power.

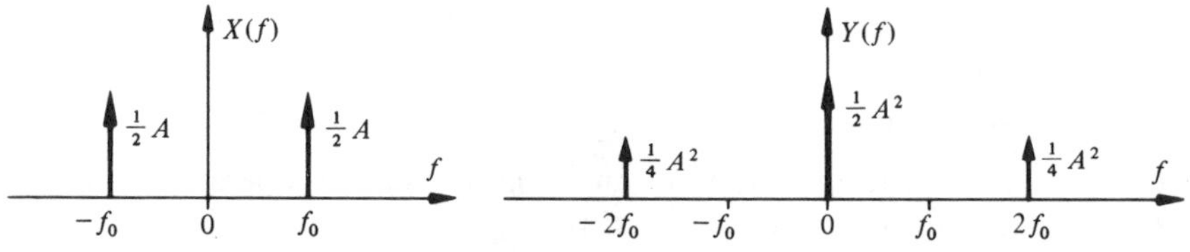

Fig. 8.49

If the input is the sum of two sine waves: $x(t) = A_1\cos(2\pi f_1 t) + A_2\cos(2\pi f_2 t)$, we get $X(f) = \frac{1}{2}A_1[\delta(f + f_1) + \delta(f - f_1)] + \frac{1}{2}A_2[\delta(f + f_2) + \delta(f - f_2)]$.

By convolution (fig. 8.50), the output signal transform here becomes: $Y(f) = \frac{1}{2}(A_1^2 + A_2^2)\delta(f) + \frac{1}{4}A_1^2[\delta(f + 2f_1) + \delta(f - 2f_1)] + \frac{1}{4}A_2^2[\delta(f + 2f_2) + \delta(f - 2f_2)] + \frac{1}{2}A_1A_2[\delta(f + f_1 + f_2) + \delta(f - f_1 - f_2) + \delta(f - f_1 + f_2) + \delta(f + f_1 - f_2)]$. The first term again represents the average power of the input signal. The last term is the result of the so-called *intermodulation products*: it displays components with frequencies $(f_2 - f_1)$ and $(f_2 + f_1)$.

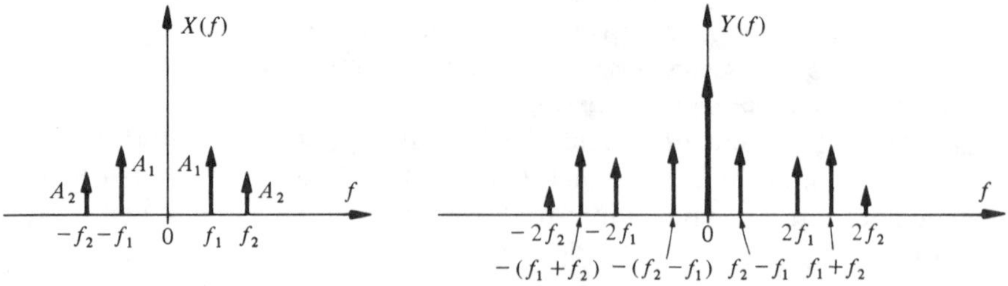

Fig. 8.50

In a more general way, if

$$X(f) = X_1(f) + X_2(f) \tag{8.115}$$

then

$$Y(f) = X_1(f) * X_1(f) + X_2(f) * X_2(f) + 2X_1(f) * X_2(f) \tag{8.116}$$

The third term characterizes the intermodulation phenomenon.

Consider (fig. 8.51) a finite energy input signal $x(t) = 2aB$ sinc $(2Bt)$ with $X(f) = a \cdot \text{rect}[f/(2B)]$. From (1.33), we get here $Y(f) = 2a^2B \cdot \text{tri}[f/(2B)]$. The value at origin represents the input signal energy.

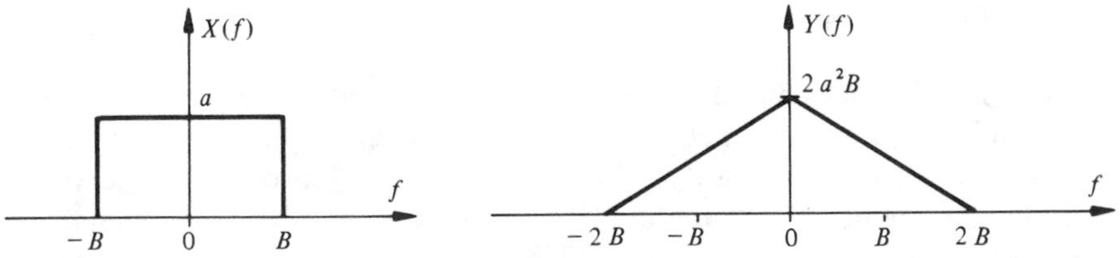

Fig. 8.51

Finally (fig. 8.52), consider a bandpass input signal with $X(f) = a\{\text{rect}[(f + f_0)/B] + \text{rect}[(f - f_0)/B]\}$. The output signal transform becomes: $Y(f) = a^2B\{2\text{tri}(f/B) + \text{tri}[(f + 2f_0)/B] + \text{tri}[(f - 2f_0)/B]\}$.

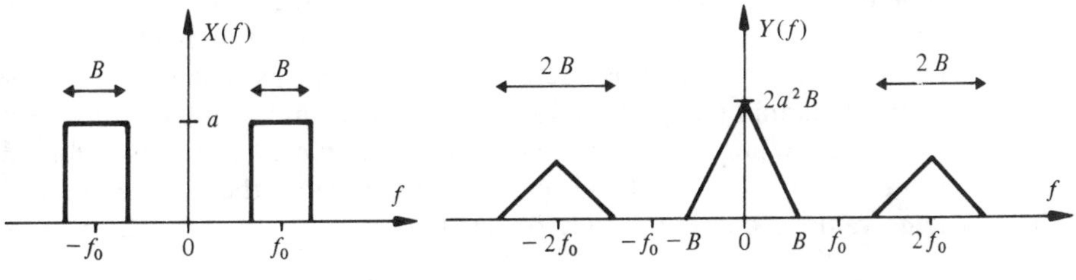

Fig. 8.52

8.4.5 Example: cubic operator

For a cubic operator, we have

$$y(t) = x^3(t) \tag{8.117}$$

$$Y(f) = \mathop{\boxed{*}}_{i=1}^{n} X(f) = X(f) * X(f) * X(f) \tag{8.118}$$

If the input is a simple sine wave $x(t) = A \cos(2\pi f_0 t)$, the output Fourier transform becomes $Y(f) = (3A^3/8)[\delta(f + f_0) + \delta(f - f_0)] + (A^3/8)[\delta(f + 3f_0) + \delta(f - 3f_0)]$ corresponding to $y(t) = (3A^3/4)\cos(2\pi f_0 t) + (A^3/4)\cos(6\pi f_0 t)$. In the case of a sum of two sine waves at the input, the intermodulation products appear at frequencies $2f_1 + f_2$, $2f_1 - f_2$, $2f_2 + f_1$ and $2f_2 - f_1$.

The determination of $Y(f)$ for other input signals is left as an exercise for the reader.

8.4.6 Nonlinear distortion

Many signal processing devices used in instrumentation systems or in telecommunications are assumed to be linear. However, in practice, they always display some slight nonlinearity. This is due to many reasons: nonlinearity of electronic components' characteristics, saturation or clamping phenomena caused by bounded linear ranges (e.g., by power supplies in operational amplifiers or other electronic circuits), *et cetera*. The input and output signals of these devices are not exactly isomorphic: additional spectral terms resulting from the nonlinear x^2, x^3, *et cetera* contributions have been added in the output signal spectrum. This is the so-called *nonlinear distortion*.

We can distinguish between

- harmonic distortion
- intermodulation distortion

Harmonic distortion is due to the generation, when the input is a pure sine wave of frequency f_0, of components with harmonic frequencies kf_0. We can then define a harmonic distortion factor, indexed by k, as being the ratio of the kth harmonic component amplitude to that of the fundamental component (usually expressed in percent). A *global harmonic distortion factor* is obtained as the ratio of the rms value of the output signal without the fundamental to the total rms value. According to (3.76), we have

$$d = \left[\frac{\sum_{k=2}^{\infty} |Y_k|^2}{\sum_{k=1}^{\infty} |Y_k|^2}\right]^{1/2} = \left[1 - \frac{|Y_1|^2}{\sum_{k=1}^{\infty} |Y_k|^2}\right]^{1/2} \leqslant 1 \qquad (8.119)$$

Intermodulation distortion is tied to the existence of intermodulation products. Intermodulation products appear when the input is a sum of terms. If they are sines with harmonic frequencies f_i, f_j, *et cetera* (quasiperiodic input signal), the output signal will contain, besides the harmonic components, intermodulation terms with frequency $(mf_i \pm nf_j)$.

8.4.7 Output signal statistical distribution

When the input signal of an amnesic nonlinear system with characteristic $y = g(x)$ is random with a known probability density, the output signal distribution can be found by applying the transformation rule introduced in subsection 5.1.20, where J denotes the Jacobian value:

$$p_y(y) = |J| p_x[x = g^{-1}(y)] \tag{8.120}$$

If the nonlinearity obeys a $g(x)$ law and if the probability distribution of the input signal is $p(x)$, all the first-order moments (sub-sect. 14.3.3) of the output signal are given by

$$m_{yn} = \mathrm{E}[\mathbf{y}^n] = \int_{-\infty}^{\infty} g^n(x) p_x(x) \mathrm{d}x \tag{8.121}$$

and, as a general rule, the mean value of any function $f(y)$ of the output signal is expressed by the mathematical expectation:

$$\mathrm{E}[f(\mathbf{y})] = \int_{-\infty}^{\infty} f(y) p_y(y) \mathrm{d}y = \int_{-\infty}^{\infty} f[g(x)] p_x(x) \mathrm{d}x \tag{8.122}$$

8.4.8 Output signal autocorrelation function

A key problem in the analysis of the effects of a nonlinear operator in a processing system is to determine the output signal spectral density when the input is random. This spectral density, as a result of the Wiener-Khinchin theorem (sub-sect. 5.3.3), is the Fourier transform of the autocorrelation function $R_y(\tau)$ of signal $y(t)$. Hence, the problem is now to evaluate this second-order moment. By definition with $\mathbf{x}_1 = \mathbf{x}(t)$, $\mathbf{x}_2 = \mathbf{x}(t + \tau)$, $\mathbf{y}_1 = g(\mathbf{x}_1)$ and $\mathbf{y}_2 = g(\mathbf{x}_2)$:

$$\begin{aligned} R_y(\tau) &= \mathrm{E}[\mathbf{y}(t)\mathbf{y}(t + \tau)] = \mathrm{E}[\mathbf{y}_1\mathbf{y}_2] \\ &= \mathrm{E}[g(\mathbf{x}_1)g(\mathbf{x}_2)] \\ &= \iint_{-\infty}^{\infty} g(x_1)g(x_2)p(x_1,x_2;\ \tau)\mathrm{d}x_1\mathrm{d}x_2 \end{aligned} \tag{8.123}$$

To solve this integral, we need to know the nonlinear relationship and the second-order statistics [joint probability density of $\mathbf{x}(t)$ and $\mathbf{x}(t + \tau)$] of the input signal. Unfortunately, this law is seldomly known, except in special cases such as Gaussian processes (sect. 5.7).

Finding $R_y(\tau)$ by solving (8.123) is called the *direct method*. Sometimes, especially when the nonlinearity exhibits discontinuities, it is better to use an *indirect method* for which the second-order statistics are given by the characteristic function (14.66).

Assume that the nonlinearity $y = g(x)$ has a Fourier transform $G(\upsilon)$. Then inserting $g(x) = \mathrm{F}^{-1}\{G(\upsilon)\}$ into (8.123) with $u = 2\pi\upsilon$, we get

$$R_y(\tau) = \frac{1}{4\pi^2}\,\mathrm{E}\left\{\int_{-\infty}^{\infty}\int_{-\infty}^{\infty} G(u)G(v)\exp[\mathrm{j}(u\mathbf{x}_1 + v\mathbf{x}_2)]\mathrm{d}u\mathrm{d}v\right\}$$

$$= \frac{1}{4\pi^2}\int_{-\infty}^{\infty}\int_{-\infty}^{\infty} G(u)G(v)\Pi_x(u,v)\mathrm{d}u\mathrm{d}v \tag{8.124}$$

where

$$\Pi_x(u,v) = \mathrm{E}\{\exp[\mathrm{j}(u\mathbf{x}_1 + v\mathbf{x}_2)]\}$$

$$= \int_{-\infty}^{\infty}\int_{-\infty}^{\infty} p(x_1,x_2;\tau)\exp[\mathrm{j}(ux_1 + vx_2)]\mathrm{d}x_1\mathrm{d}x_2 \tag{8.125}$$

is precisely the second-order characteristic function of the input process.

8.4.9 Gaussian input signal with zero mean value

It is theoretically possible to solve (8.123) or (8.124) when the input signal is Gaussian. The joint probability density can then be written, from (14.98) and with $\mu_x = 0$:

$$p(x_1,x_2,\tau) = \frac{1}{2\pi[R_x^2(0) - R_x^2(\tau)]^{1/2}} \times \exp\left\{-\frac{R_x(0)[x_1^2 - x_2^2] + 2R_x(\tau)x_1x_2}{2[R_x^2(0) - R_x^2(\tau)]}\right\} \tag{8.126}$$

where $R_x(0) \equiv C_x(0) = \sigma_x^2$ and $R_x(\tau) \equiv C_x(\tau) = \sigma_x^2\rho_x(\tau)$.

The corresponding second-order characteristic function, according to (14.104), is

$$\Pi_x(u,v) = \exp\{-\tfrac{1}{2}R_x(0)[u^2 + v^2] - R_x(\tau)uv\} \tag{8.127}$$

8.4.10 Example: unipolar rectifier

This operator has (tab. 8.48) a characteristic

$$g(x) = x \cdot \epsilon(x) \tag{8.128}$$

Inserting it into (8.123), we get

$$R_y(\tau) = \int_0^{\infty}\int_0^{\infty} x_1x_2p(x_1,x_2;\tau)\mathrm{d}x_1\mathrm{d}x_2 \tag{8.129}$$

The rectified signal $y(t)$ then has, for a Gaussian input with zero mean value, the autocorrelation function (problem 8.5.24):

$$R_y(\tau) = \frac{1}{2\pi}\left\{[R_x^2(0) - R_x^2(\tau)]^{1/2} + R_x(\tau)\arccos\frac{-R_x(\tau)}{R_x(0)}\right\} \tag{8.130}$$

8.4.11 Example: x^n nonlinearity

When

$$y = g(x) = x^n \tag{8.131}$$

it is easier to take advantage of relation (14.70), binding the second-order moment $E[\mathbf{x}_1^n\mathbf{x}_2^n]$ with the value at origin of the (n,n) degree partial derivative of the characteristic function:

$$R_y(\tau) = E[\mathbf{x}_1^n\mathbf{x}_2^n] = j^{-2n}\left.\frac{\partial^{2n}\Pi_x(u,v)}{\partial u^n \partial v^n}\right|_{u=v=0} \tag{8.132}$$

If the input signal $x(t)$ is Gaussian with a zero mean value, the characteristic function is given by (8.127). The general solution of (8.132) can be expressed as [27, 104]:

$$R_y(\tau) = \sum_k^n \frac{\{n!/[(n-k)/2]!\}^2}{2^{n-k}\cdot k!} R_x^{n-k}(0)R_x^k(\tau) \tag{8.133}$$

where $k = 0, 2, 4, \ldots$, when n is even, and $k = 1, 3, 5, \ldots$, when n is odd (and $0! = 1$).

Table 8.53 summarizes results for some values of n.

Table 8.53

n	$R_y(\tau)$ for $y = x^n$
1	$R_x(\tau)$
2	$2R_x^2(\tau) + R_x^2(0)$
3	$6R_x^3(\tau) + 9R_x^2(0)R_x(\tau)$
4	$24R_x^4(\tau) + 72R_x^2(0)R_x^2(\tau) + 9R_x^4(0)$
5	$120R_x^5(\tau) + 600R_x^2(0)R_x^3(\tau) + 225R_x^4(0)R_x(\tau)$

As an example, for a cubic transformation, we have

$$R_y(\tau) = 6R_x^3(\tau) + 9R_x^2(0)R_x(\tau) \tag{8.134}$$

the Fourier transform of which is the spectral density:

$$\Phi_y(f) = 6 \overset{3}{\underset{i=1}{\ast}} \Phi_x(f) + 9R_x^2(0)\Phi_x(f) \tag{8.135}$$

This result can be compared with (8.118), valid for deterministic inputs only.

8.4.12 Price theorem

The following theorem, based on the indirect approach, is especially suitable when the nonlinear $g(x)$ law, or any of its derivatives, is piecewise linear. If the input signal $x(t)$ is Gaussian with a zero mean value, the autocorrelation function $R_y(\tau)$ of the output signal can be defined by integrating

$$\frac{\partial^k R_y(\tau)}{\partial R_x^k(\tau)} = \mathrm{E}[g^{(k)}(\mathbf{x}_1)g^{(k)}(\mathbf{x}_2)] \tag{8.136}$$

where $g^{(k)}(x)$ is the kth derivative of $g(x)$ with respect to x.

In a similar way, the crosscorrelation function $R_{y1y2}(\tau)$ of the output signals $y_1(t)$ and $y_2(t)$ of two nonlinear systems $g_1(x_1)$ and $g_2(x_2)$ with zero mean value input Gaussian signals $x_1(t)$ and $x_2(t)$, respectively, is found by integrating [with $\mathbf{x}_{11} = \mathbf{x}_1(t)$ and $\mathbf{x}_{22} = \mathbf{x}_2(t + \tau)$]:

$$\frac{\partial^k R_{y1y2}(\tau)}{\partial R_{x1x2}^k(\tau)} = \mathrm{E}[g_1^{(k)}(\mathbf{x}_{11})g_2^{(k)}(\mathbf{x}_{22})] \tag{8.137}$$

8.4.13 Demonstration

Inserting (8.127) into (8.124) and deriving k times with respect to $R_x(\tau)$, we get

$$\frac{\partial^k R_y(\tau)}{\partial R_x^k(\tau)} = \frac{1}{4\pi^2}\iint G(u)G(v)(-1)^k u^k v^k \Pi_x(u,v)\mathrm{d}u\mathrm{d}v \tag{8.138}$$

Substituting $\Pi_x(u,v)$ with the mathematical expectation (8.125), writing $(-1)^k = \mathrm{j}^{2k}$, separating the u and v variables, and taking property (4.13) into account here with $u = 2\pi\upsilon$ and $v = 2\pi\nu$, we get

$$\frac{\partial^k R_y(\tau)}{\partial R_x^k(\tau)} = \frac{1}{4\pi^2}\mathrm{E}\left[\int G(u)\mathrm{j}^k u^k \exp(\mathrm{j}u\mathbf{x}_1)\mathrm{d}u \cdot \int G(v)\mathrm{j}^k v^k \exp(\mathrm{j}v\mathbf{x}_2)\mathrm{d}v\right] \tag{8.139}$$

$$= \mathrm{E}[g^{(k)}(\mathbf{x}_1) \cdot g^{(k)}(\mathbf{x}_2)]$$

Relation (8.137) is similarly demonstrated.

8.4.14 Other general relation valid for Gaussian inputs

Using the series expansion (8.127), we can write the output signal autocorrelation function by a similar approach [69] as

$$R_y(\tau) = \sum_{k=0}^{\infty} \frac{1}{k!} \mathrm{E}[g^{(k)}(\mathbf{x}_1)]\mathrm{E}[g^{(k)}(\mathbf{x}_2)]R_x^k(\tau) \tag{8.140}$$

In contrast to the Price theorem, this relation is not handy for the study of nonlinearities with derivatives containing discontinuities. However, it is quite suitable when the nonlinearity is expandable in power series.

8.4.15 Example: sign detector

Such an operator is listed in table 8.48. Its characteristic is

$$y = g(x) = A\ \mathrm{sgn}(x) \tag{8.141}$$

If $x(t)$ is Gaussian, with $\mu_x = 0$, we can apply the Price theorem here with

$$g^{(1)}(x) = 2A\delta(x) \tag{8.142}$$

thus, taking (8.126) and (1.35) into account:

$$\begin{aligned} \frac{\partial R_y(\tau)}{\partial R_x(\tau)} &= 4A^2\ \mathrm{E}[\delta(\mathbf{x}_1)\delta(\mathbf{x}_2)] \\ &= 4A^2 \iint \delta(x_1)\delta(x_2)p_x(x_1,x_2,\tau)\mathrm{d}x_1\mathrm{d}x_2 \\ &= 4A^2 p_x(0,0) \\ &= \frac{2A^2}{\pi[R_x^2(0) - R_x^2(\tau)]^{1/2}} \end{aligned} \tag{8.143}$$

By integrating, we have

$$R_y(\tau) = \frac{2A^2}{\pi} \arcsin \frac{R_x(\tau)}{R_x(0)} + K \tag{8.144}$$

The integration constant is equal to the particular value of $R_y(\tau)$ ***found when imposing the condition*** $R_x(\tau) = 0$. Since $x(t)$ is a zero mean value Gaussian signal (sub-sect. 5.7.3), $R_x(\tau) = 0$ implies the independence of $\mathbf{x}_1 = \mathbf{x}(t)$ and $\mathbf{x}_2 = \mathbf{x}(t + \tau)$. Thus, we have

$$K = R_y(\tau)|_{R_x(\tau)=0} = \mathrm{E}[g(\mathbf{x}_1)g(\mathbf{x}_2)]_{R_x(\tau)=0} = \mathrm{E}[g(\mathbf{x}_1)]\mathrm{E}[g(\mathbf{x}_2)] = 0 \tag{8.145}$$

because

$$\mathrm{E}[g(\mathbf{x})] = \int_{-\infty}^{\infty} \mathrm{sgn}(x)p(x)\mathrm{d}x = 0 \tag{8.146}$$

since sgn(x) is an odd function and $p(x)$ is here an even distribution. Finally,

$$R_y(\tau) = \frac{2A^2}{\pi} \arcsin \frac{R_x(\tau)}{R_x(0)} \tag{8.147}$$

The corresponding spectral density can be determined by series expansion of the arcsine function.

8.4.16 Example: squaring operator

The operator $y = x^2$ has already been considered in sub-section 8.4.4 for a deterministic input. The autocorrelation function of the output signal, obtained for a zero mean value Gaussian input, can be drawn from table 8.53:

$$R_y(\tau) = 2R_x^2(\tau) + R_x^2(0) \tag{8.148}$$

It can also be found (problem 8.5.26) by applying the Price theorem or relation (8.140).

The corresponding spectral density is simply

$$\Phi_y(f) = 2\Phi_x(f) * \Phi_x(f) + R_x^2(0)\delta(f) \tag{8.149}$$

From (8.148), we get the following parameters:

$$P_y = R_y(0) = 3R_x^2(0) = 3\sigma_x^4 \tag{8.150}$$

$$\mu_y = R_x(0) = \sigma_x^2 \tag{8.151}$$

$$\sigma_y^2 = 2R_x^2(0) = 2\sigma_x^4 \tag{8.152}$$

Figure 8.54 illustrates the case of a bandpass Gaussian input signal.

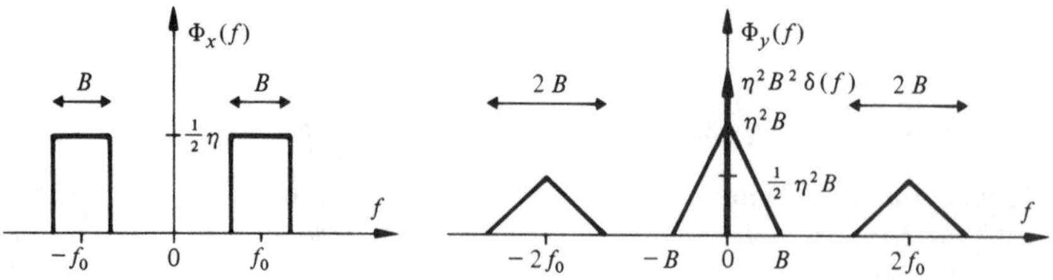

Fig. 8.54

8.4.17 Squaring operator with an input signal embedded in Gaussian noise

Let $x(t) = s(t) + n(t)$, where $s(t)$ and $n(t)$ are independent and have zero mean values; $s(t)$ is any signal, while $n(t)$ is a Gaussian noise. The squaring

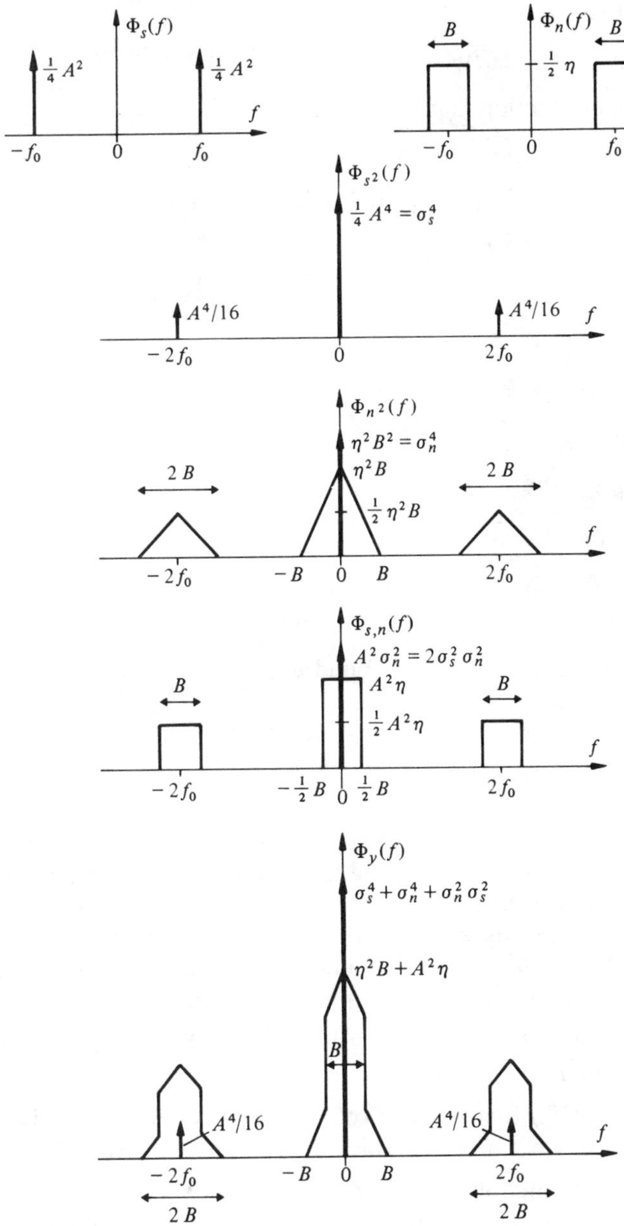
Φs(f)
¼ A²
¼ A²
f
−f0
0
f0
Φn(f)
B
B
½ η
f
−f0
0
f0
Φs²(f)
¼ A⁴ = σs⁴
A⁴/16
A⁴/16
f
−2f0
0
2f0
Φn²(f)
η²B² = σn⁴
η²B
2B
2B
½ η²B
f
−2f0
−B
0
B
2f0
Φs,n(f)
A²σn² = 2σs²σn²
A²η
B
B
½ A²η
f
−2f0
−½ B
0
½ B
2f0
Φy(f)
σs⁴ + σn⁴ + σn²σs²
η²B + A²η
B
A⁴/16
A⁴/16
f
−2f0
−B
0
B
2f0
2B
2B

Fig. 8.55

operator output is

$$y(t) = x^2(t) = s^2(t) + 2s(t)n(t) + n^2(t) \tag{8.153}$$

Its autocorrelation function is

$$\begin{aligned} R_y(\tau) &= \mathrm{E}[\mathbf{y}(t)\mathbf{y}(t+\tau)] = \mathrm{E}[\mathbf{y}_1\mathbf{y}_2] \\ &= \mathrm{E}[\mathbf{s}_1^2\mathbf{s}_2^2] + \mathrm{E}[\mathbf{n}_1^2\mathbf{n}_2^2] + \mathrm{E}[\mathbf{s}_1^2\mathbf{n}_2^2 + \mathbf{n}_1^2\mathbf{s}_2^2 + 4\mathbf{s}_1\mathbf{s}_2\mathbf{n}_1\mathbf{n}_2] \\ &= R_{s^2}(\tau) + R_{n^2}(\tau) + R_{s,n}(\tau) \end{aligned} \tag{8.154}$$

with

$$R_{s^2}(\tau) = \mathrm{E}[\mathbf{s}^2(t)\mathbf{s}^2(t+\tau)] \tag{8.155}$$

$$R_{n^2}(\tau) = 2R_n^2(\tau) + R_n^2(0) \tag{8.156}$$

thanks to (8.148), and

$$R_{s,n}(\tau) = 2R_s(0)R_n(0) + 4R_s(\tau)R_n(\tau) \tag{8.157}$$

Finally, the spectral density is

$$\begin{aligned} \Phi_y(f) = \Phi_{s^2}(f) &+ [R_n^2(0) + 2R_s(0)R_n(0)]\delta(f) \\ &+ 2\Phi_n(f) * \Phi_n(f) + 4\Phi_s(f) * \Phi_n(f) \end{aligned} \tag{8.158}$$

Figure 8.55 summarizes the results found when $s(t) = A\cos(2\pi f_0 t)$ and $n(t)$ is a bandlimited white Gaussian noise.

8.5 PROBLEMS

8.5.1 Let a linear system be characterized by the impulse response $g(t) = \omega_c \exp(-\omega_c t) \cdot \epsilon(t)$ with $\omega_c = 2\pi f_c$. Find the relationship between the input and output autocorrelation functions $\varphi_x(\tau)$ and $\varphi_y(\tau)$, and the noise equivalent bandwidth. If input $x(t)$ is a noise with a constant power spectral density $\Phi_x(f) = \eta/2$, find the autocorrelation function $R_y(\tau)$, the power spectral density $\Phi_y(f)$ and the total power P_y of the output signal $y(t)$.

8.5.2 Demonstrate that the fig. 8.56 system, where the sign Σ means addition, is also a linear system.

8.5.3 Find the transfer function $G(f)$ of the linear system depicted in fig. 8.57 and its special form when $t_0 = 0$. Which kind of filter is it then?

8.5.4 The transfer function of an audio signal amplifier is $G(f) = G_0(\mathrm{j}f/f_1)(1 + \mathrm{j}f/f_1)^{-1} \cdot (1 + \mathrm{j}f/f_2)^{-1}$, where $f_1 = 30\text{Hz}$ and $f_2 = 20$ kHz and $G_0 = 1000$. Find the impulse response $g(t)$. Calculate the output noise power P_y when input $x(t)$ is a white noise with spectral density $\Phi_x(f) = 10^{-7}\text{V}^2/\text{Hz}$. Also find the -3 dB bandwidth and the noise equivalent bandwidth of this amplifier.

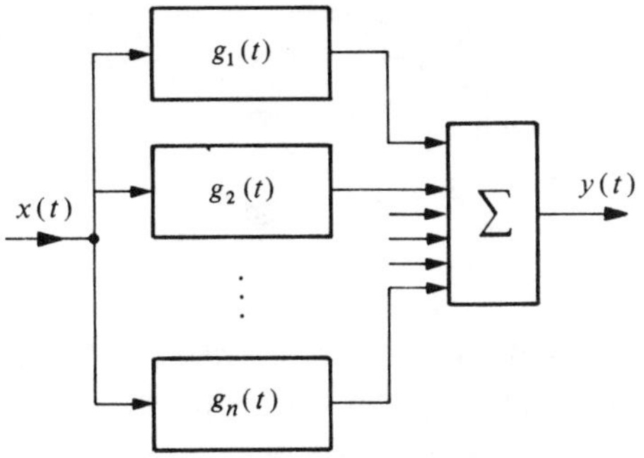

Fig. 8.56

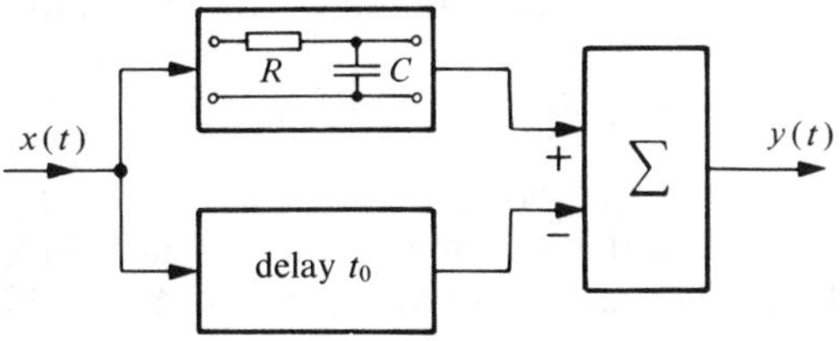

Fig. 8.57

8.5.5 Let $x(t) = A + n(t)$ be the input of a perfect time averager, defined by (8.42), where A is a constant to estimate and $n(t)$ is an additive noise with spectral density $\Phi_n(f) = \eta[1 + (f/B)^2]^{-1}$. Knowing that $\int x\exp(ax)dx = a^{-2}(ax - 1)\exp(ax)$, derive the expression in function of integration duration T of the output signal-to-noise ratio and calculate the value of T for which the output signal-to-noise ratio is 20 dB greater than that of the input when $A = 0.1$V, $\eta = 10^{-4}V^2/Hz$ and $B = 100/\pi$ Hz.

8.5.6 Rework problem 8.5.5 with an averager which is not perfect but can be approximately implemented by a lowpass RC filter, defined by (8.54), and substituting the integration duration with the filter time constant RC.

8.5.7 Find the phase and group delays of a lowpass first-order filter, defined by (8.54), and compare the results found for $f_0 = f_c$.

8.5.8 Show that the coherence function (5.178) between input $x(t)$ and output $y(t)$ of a linear system (without additional internal noise) is equal to one.

8.5.9 A model of an impulse noise with any kind of spectrum can be built with the block diagram of fig. 8.58, where $u_p(t) = \Sigma\delta(t - t_i)$ is a random sequence of delta-functions (Poisson process). The generated noise, $n(t)$, is a sum of individual pulses given by $\alpha_i g(t - t_i)$, where t_i is an arbitrary random

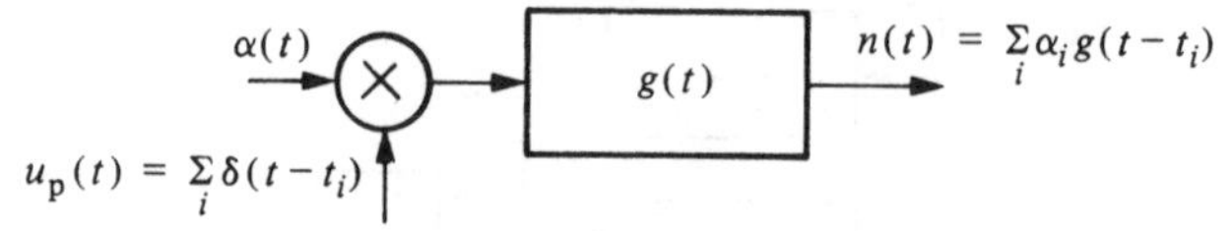

Fig. 8.58

instant and $\alpha_i = \alpha(t = t_i)$ is the sampled value of an independent function $\alpha(t)$. Establish the general equation of the noise $n(t)$ power spectral density if the average number of pulses per second is denoted by λ.

8.5.10 Find the conditions under which the result of problem 8.5.9 is identical to (6.31).

8.5.11 From the result of problem 8.5.9, find the spectrum of a Poisson sequence of triangular pulses with duration Δ and unit amplitude for $\alpha(t) = e$ (electron charge) and $\lambda e = I_0$.

8.5.12 Demonstrate relation (8.77).

8.5.13 From (8.90), derive the power spectral density of the output signal of a periodic switch, assuming that the input signal $x(t)$ and the control signal $u(t)$ are independent. Interpret graphically the result for $\Delta = T/2$ and for $x(t)$ being an ideal lowpass white noise with bandwidth $B = (2T)^{-1}$.

8.5.14 Find the energy spectral density of the output signal $y(t)$ of a weighting operator, defined by (8.91), when the input signal $x(t)$ is T_x-periodic. Plot the result for $T = T_x$ and $T = 10T_x$.

8.5.15 The quasiperiodic signal $x(t) = A_1 \cos(2\pi f_1 t) + A_2 \cos(2\pi f_2 t)$ is applied to the input of a weighting operator, the control signal of which is $u_i(t)$. Analytically find and graphically compare the amplitude spectra for $f_1 = 100$ Hz and $f_2 = 150$ Hz, $A_1 = 10$ V, $A_2 = 1$ V. For the three cases: $u_1(t) = \text{rect}(t/T)$, $u_2(t) = \text{tri}(t/T)$, $u_3(t) = \text{ig}(t/T)$ with $T = 50$ ms.

8.5.16 A rectified cosine signal can be considered to be the product of a cosine $A\cos(2\pi t/T)$ and its sign function. Find its Fourier transform.

8.5.17 What are the autocorrelation function, power spectral density, and total power of the output signal of the system represented in fig. 8.59, if $x(t)$ is a random signal with an autocorrelation function $R_x(\tau) = \sigma_x^2\delta(\tau) + \mu_x^2$ and if $u(t) = A\cos(2\pi f_0 t + \alpha)$ is independent of $x(t)$? Plot the result for $f_0 = 2/RC$.

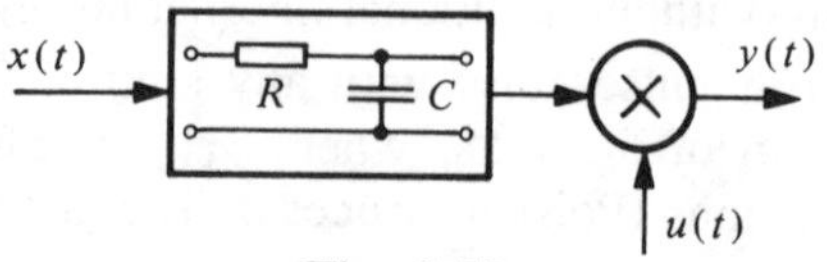

Fig. 8.59

8.5.18 Find the output signal $y(t)$ of the system shown in fig. 8.60 (Hartley modulator) and its autocorrelation function. Show that this signal has a single-sided spectrum.

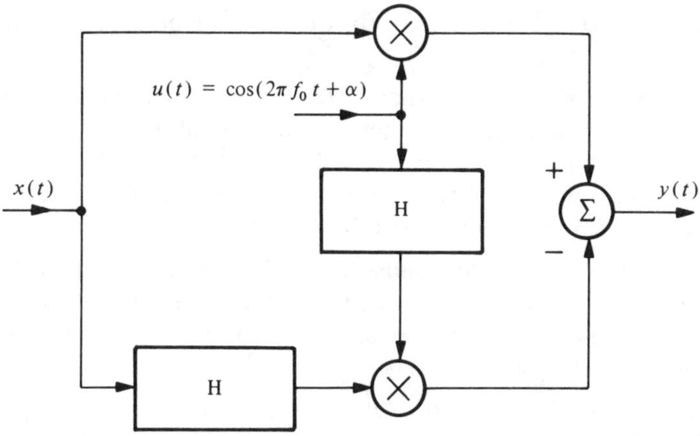

Fig. 8.60

8.5.19 Let $x(t) = A \text{ ig } (t/T)$ be the input signal of a squaring operator. Find and plot the output signal $y(t)$, its Fourier transform, and its energy spectral density.

8.5.20 Find and plot the Fourier transform of the output signal of a cubic operator, if (a) $x(t) = A_1 \cos(2\pi f_1 t) + A_2 \cos(2\pi f_2 t)$; (b) $x(t) = 2AB \text{ sinc } (2Bt)$.

8.5.21 The output current of an electronic device controlled by an input voltage u is given by equation $i = a + bu + cu^2$. Find the analytical expression of the distortion factor generated by the second-order harmonic when the input is a sine wave of amplitude $\hat{U}$. Estimate the value of this factor for $\hat{U} = 1$ V, if $i_{\min} = 0.45a$, $i_{\max} = 1.7a$, and if the output current without any input signal is 10 mA.

8.5.22 The real input-output characteristic of an amplifier can, around the origin, be written as $y(x) = AV \arctan(x/V)$, where $A = 100$ is the amplifier gain and $V = 1$ volt is the scale factor. Evaluate the global harmonic distortion factor of this amplifier for sine wave input signals with amplitudes: (a) 0.1 V and (b) 0.2 V.

8.5.23 Determine, as functions of parameter b, the mean value, the mean square value, and the variance of the output signal of the nonlinear operators defined by $y_1 = x^2$ and $y_2 = 1 + x^2 + x^3$, the input signal of which has an uniform distribution $p(x) = a \text{ rect } (x/b)$.

8.5.24 Check (8.130) with the Price theorem.

8.5.25 Find the output autocorrelation function $R_y(\tau)$ of a bipolar rectifier $y = |x|$, when the input is a Gaussian signal with zero mean value and known autocorrelation function $R_x(\tau)$.

8.5.26 Apply: (a) relation (8.133); (b) the Price theorem; (c) relation (8.140) to verify relation (8.148).

8.5.27 Let $y(t) = x(t) + x^3(t)$, where $x(t)$ is a zero mean value Gaussian signal with a known autocorrelation function $R_x(\tau)$. Find $R_y(\tau)$: (a) by the Price theorem (8.136), knowing that $E[\mathbf{x}^{2n}] = 1 \cdot 3 \ldots (2n - 1)R_x^2(0)$; (b) by taking the output $y(t)$ as the sum of two signals $y_1(t) = x(t)$ and $y_2(t) = x^3(t)$ and using relation (8.137); (c) by applying relation (8.140).

8.5.28 Two zero mean value Gaussian signals $x_1(t)$ and $x_2(t)$, with variances σ_{x1}^2 and σ_{x2}^2 and crosscorrelation function $R_{x1x2}(\tau)$, are applied to the inputs of two sign detectors. Find the crosscorrelation function of the output signals $y_1(t) = \text{sgn}[x_1(t)]$ and $y_2(t) = \text{sgn}[x_2(t)]$.

8.5.29 Show that the crosscorrelation function of signals $y_1(t) = x_1(t)$ and $y_2(t) = A\ \text{sgn}\ [x_2(t)]$, where $x_1(t)$ and $x_2(t)$ are zero mean value Gaussian signals, is proportional to $R_{x1x2}(\tau)$.

Chapter 9

Signal Sampling

9.1 INTRODUCTION

9.1.1 Sampling reasons

Primary information-carrying signals are almost always analog (continuous in both time and amplitude). On the other hand, computers, or any digital electronic system, only manipulate data, e.g., series of numbers. Apparently, this yields an incompatibility. If we want to process a signal digitally, we must therefore represent it as a sequence of regularly (fig. 9.1) or irregularly received values. Such an operation is called *sampling*. A sampling is said to be *regular* or *periodic* when values are picked up at a regular pace. The interval between two successive samples—the *sampling interval*—is then constant. *Irregular sampling* is seldomly used.

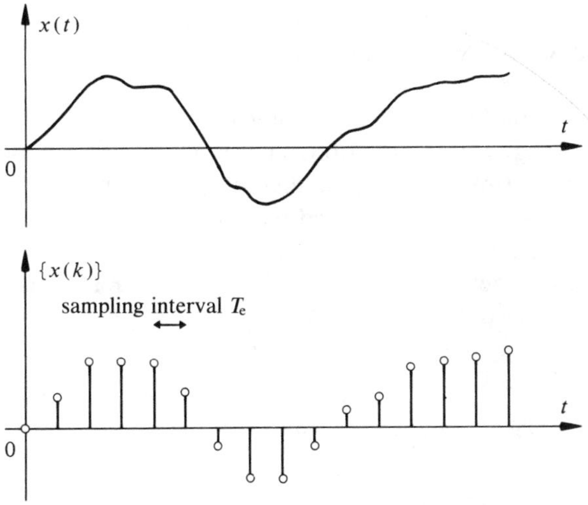

Fig. 9.1

To represent the samples digitally, we need to quantize and encode them. The nature and consequences of those operations will be studied in chapter 10. The total operation—sampling, quantizing, and encoding—realizes an analog-to-digital conversion.

There are also devices, such as charge-coupled devices (CCD) and switched-capacitor filters, working directly on nonquantized uncoded sampled signals.

9.1.2 Reversible and irreversible transformation

When a signal is to be sampled, the first problem to solve is knowing whether the realized transformation has to be reversible or not. Or, said in another way, shall we be able to reconstruct, if needed, the original analog signal with the samples, or not?

We do not need to reconstruct the original signal when only partial information on the signal nature has to be extracted in the process. A classical example is the computation of the signal histogram (sub-sect. 14.2.6) or other first-order statistics (mean value, power, *et cetera*.)

The reversibility of such a transformation is not straightforward. We shall hereafter show that only theoretical conditions, almost unachievable in practice, allow an accurate reconstruction of the analog signal from its samples. Sampling, thus, always introduces an unavoidable distortion that must be kept within acceptable boundaries.

9.1.3 Sampling and modulation

It is current practice in telecommunications textbooks to present sampling in a chapter along with modulation. The basis for this practice is grounded in historical reasons (invention of the digital signal representation method called pulse-coded modulation (PCM)) and pertinent practice. Indeed, the ideal sampling operator can be seen as an amplitude modulation operator working with a carrier that is a periodic sequence of delta-functions. All pulse modulation techniques use sampling (sect. 11.4).

9.1.4 Theoretical sampling and practice

The series of samples shown in fig. 9.1 cannot be physically achieved (zero duration = zero energy). It is, thus, only an abstract concept. Sampling a value implies a measurement operation requiring some time. In practice, sampled values $x(k) = x(t_k)$ are represented by a parameter (amplitude, area) of a sequence $x_e(t) = \Sigma g(t - t_k)$ of $g(t)$-shaped pulses with non-zero energy. An example is given in fig. 9.2 with rectangular pulses.

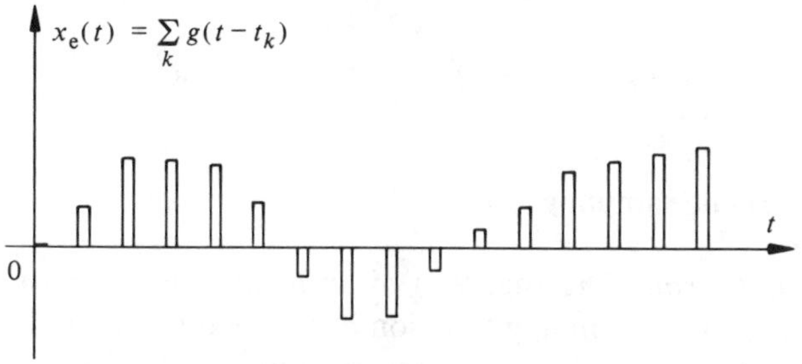

Fig. 9.2

9.2 SAMPLED SIGNAL MODELS

9.2.1 General sampling operator

The general model (fig. 9.3) of a sampling operator is a separable parametric operator (sub-sect. 8.3.10) composed of a multiplier followed by a pulse-shaping linear circuit with a $g(t)$ impulse response.

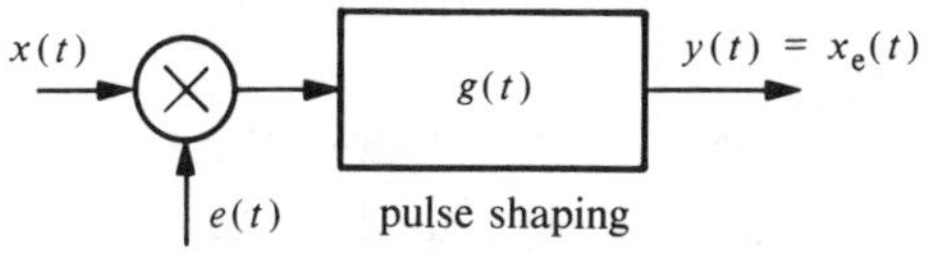

Fig. 9.3

The analog signal $x(t)$ is multiplied by a sampling function $e(t)$ and then filtered. The sample signal, $x_e(t)$, is related to $x(t)$ and $e(t)$ by equation (8.96), which becomes here

$$x_e(t) = [x(t) \cdot e(t)] * g(t) \tag{9.1}$$

For deterministic signals, the general form of the Fourier transform sampled signal, according to (8.97), becomes

$$X_e(f) = [X(f) * E(f)] \cdot G(f) \tag{9.2}$$

If $x(t)$ is a random signal statistically independent of the deterministic signal $e(t)$, from (8.22) and (8.87), we get

$$R_{xe}(\tau) = [R_x(\tau) \cdot \varphi_e(\tau)] * \mathring{\varphi}_g(\tau) \tag{9.3}$$

and

$$\Phi_{xe}(f) = [\Phi_x(f) * \Phi_e(f)] \cdot |G(f)|^2 \tag{9.4}$$

9.2.2 Real periodic sampling

A *real periodic sampling* (fig. 9.4) of an analog signal $x(t)$ is obtained by multiplying $x(t)$ with a sampling function $e(t)$, which is a periodic (period $T_e = 1/f_e$) *sequence of unit amplitude rectangular pulses with duration D*:

$$e(t) = \text{rep}_{T_e}\{\text{rect}(t/D)\} \tag{9.5}$$

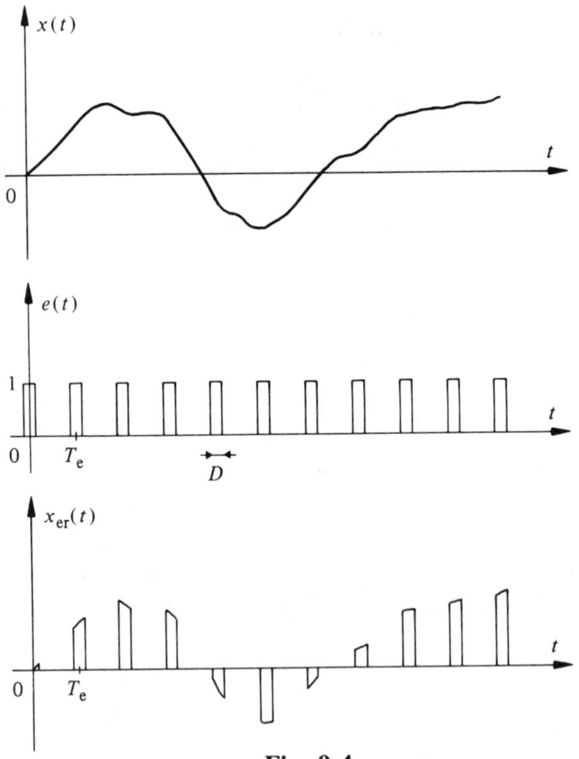

Fig. 9.4

In this case, the block diagram of fig. 9.3 is simplified and contains only a multiplier representing a periodic switch (sub-sect. 8.3.6). The linear filter is replaced by a short circuit with $g(t) = \delta(t)$. The real sampled signal is denoted $x_{er}(t)$.

By (4.128) and example 4.4.17, relations (9.2) and (9.4) in this case become

$$X_{er}(f) = X(f) * \sum_{n=-\infty}^{\infty} f_e D \operatorname{sinc}(nDf_e)\delta(f - nf_e)$$

$$= \sum_{n=-\infty}^{\infty} f_e D \operatorname{sinc}(nDf_e)X(f - nf_e) \tag{9.6}$$

and

$$\Phi_{xer}(f) = \Phi_x(f) * \sum_{n=-\infty}^{\infty} f_e^2 D^2 \operatorname{sinc}^2(nDf_e)\delta_x(f - nf_e)$$

$$= \sum_{n=-\infty}^{\infty} f_e^2 D^2 \operatorname{sinc}^2(nDf_e)\Phi_x(f - nf_e) \tag{9.7}$$

These results emphasize the importance of the sampling rate $f_e = 1/T_e$. The real sampled signal spectrum is hence obtained by a weighted sum of terms corresponding to a $T_e(=1/f_e)$ periodic repetition of the analog signal spectrum. The weighting factor depends on the pulse density $f_e \cdot D = D/T_e$, comprised of a value between 0 and 1, and varies as a function of sinc(n) for the Fourier transform and of sinc^2(n) for the power spectral density. This result is shown in figure 9.5.

A generalization of the real sampling is the multiplication of signal $x(t)$ by a periodic sequence of pulses with an arbitrary shape $g(t)$.

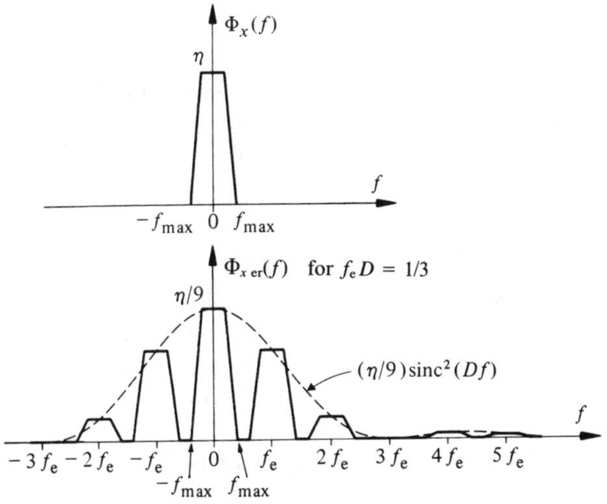

Fig. 9.5

9.2.3 Ideal periodic sampling

As we said in sub-section 9.1.4, the sequence of values $\{x_k\}$, where $x_k = x(t_k)$ cannot be a physically achievable signal. Indeed, we could try to get such a signal out of a real sampling by reducing the sampling pulse duration D to zero, but then we would end up with a zero power signal!

A theoretical way to show the spectral properties of such an ideal sequence is to multiply each sampling pulse by a $1/D$ factor. When D tends toward zero, each rectangular pulse is transformed into a delta-function.

We can, thus, theoretically identify the ideal sequence of samples taken at the $f_e(=1/T_e)$ rate with a signal $x_{ei}(t)$, resulting from the multiplication of the analog signal $x(t)$ with an ideal sampling function:

$$e_i(t) = \delta_{Te}(t) = \sum_{k=-\infty}^{\infty} \delta(t - kT_e) \tag{9.8}$$

The general model, shown in fig. 9.3, here becomes an *ideal sampling operator* (sub-sect. 8.3.5 and fig. 9.6), the output signal of which is

$$x_{ei}(t) = x(t)\delta_{Te}(t) = \sum_{k=-\infty}^{\infty} x(kT_e)\delta(t - kT_e) \tag{9.9}$$

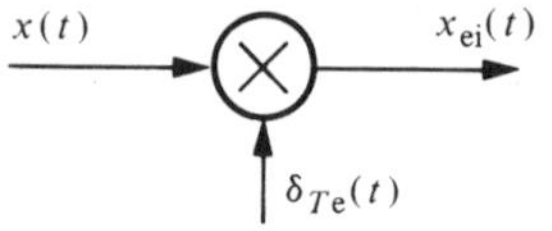

Fig. 9.6

The spectral properties of $x_{ei}(t)$ are found as previously. The Fourier transform and the autocorrelation function of a periodic sequence of delta-functions $\delta_{Te}(t)$ are respectively given by (4.123) and (4.138):

$$\mathrm{F}\{\delta_{Te}(t)\} = T_e^{-1}\delta_{1/Te}(f) = f_e\delta_{fe}(f) \tag{9.10}$$

$$\varphi_{\delta Te}(\tau) = T_e^{-1}\delta_{Te}(\tau) = f_e\delta_{Te}(\tau) \tag{9.11}$$

These two functions are also periodic sequences of delta-functions with a f_e weight and a f_e periodicity in the frequency domain, and a T_e periodicity in the time domain. By (1.57), (9.2) and (9.4), the Fourier transform and the spectral density of the ideal sampled signal become

$$X_{ei}(f) = X(f) * f_e\delta_{fe}(f) = \sum_{n=-\infty}^{\infty} f_eX(f - nf_e) = f_e \operatorname{rep}_{fe}\{X(f)\} \tag{9.12}$$

$$\Phi_{xei}(f) = \Phi_x(f) * f_e^2\delta_{fe}(f) = \sum_{n=-\infty}^{\infty} f_e^2\Phi_x(f - nf_e) = f_e^2 \operatorname{rep}_{fe}\{\Phi_x(f)\} \tag{9.13}$$

Hence, these two functions correspond to the periodic repetition (the period being equal to the sampling rate) of the Fourier transform or the spectral density of the analog signal, respectively (fig. 9.7).

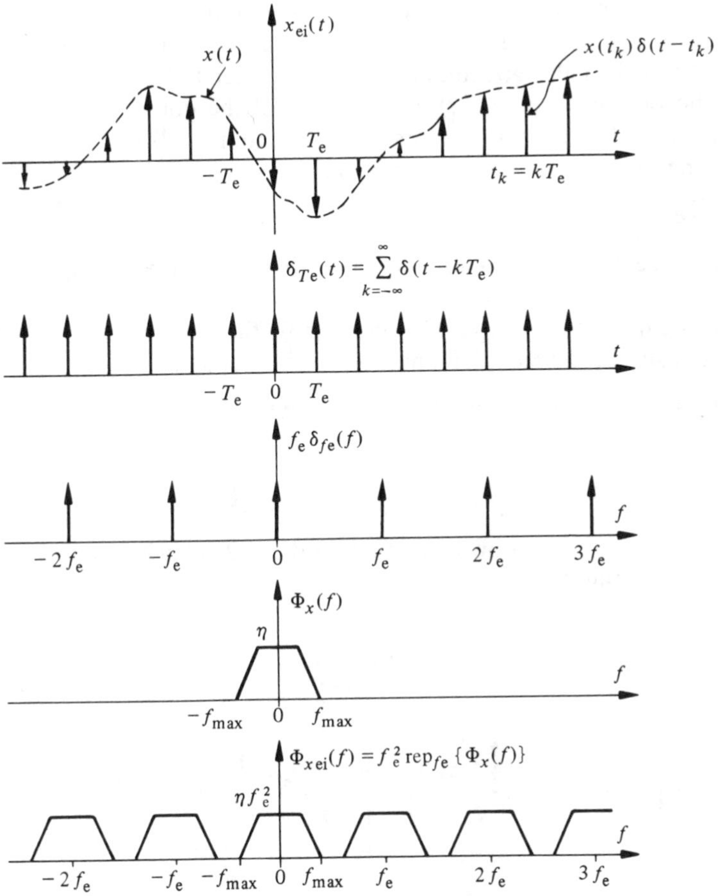

Fig. 9.7

We can easily see that the autocorrelation function (9.3) of the ideal sampled signal is simply an ideal sampled version of the analog signal autocorrelation function:

$$R_{xei}(\tau) = R_x(\tau) \cdot \varphi_{\delta Te}(\tau) = \sum_{k=-\infty}^{\infty} R_x(kT_e)\delta(t - kT_e) \tag{9.14}$$

9.2.4 Periodic sampling and holding

To *sample and hold* means that the instantaneously sampled value is stored temporarily as an analog value. This sampling procedure plays a very important role in practice.

The sampled and held signal model is easily obtained from fig. 9.3, assuming again that the sampling function $e(t)$ is a periodic sequence of delta-functions. The input signal of the $g(t)$-impulse response filter is thus the ideal sampled signal. The output signal $x_{em}(t)$ is the result of the convolution of the input signal and $g(t)$. The function representing the hold for D duration is obtained by choosing the filter impulse response as

$$g(t) = \text{rect}[(t - D/2)/D] \tag{9.15}$$

which is, except for a D factor, equal to that of a time averager (sub-sect. 8.2.19).

From (4.29), (9.2), (9.4), (9.12), and (9.13), the Fourier transform and the spectral density of the sampled and held signal become

$$X_{em}(f) = Df_e \,\text{sinc}(Df)\,\text{rep}_{fe}\{X(f)\}\exp(-j\pi fD) \tag{9.16}$$

and

$$\Phi_{xem}(f) = D^2 f_e^2 \,\text{sinc}^2(Df)\,\text{rep}_{fe}\{\Phi_x(f)\} \tag{9.17}$$

The $\exp(-j\pi fD)$ term shows the effect of the $D/2$ delay generated by holding the signal value (fig. 9.8).

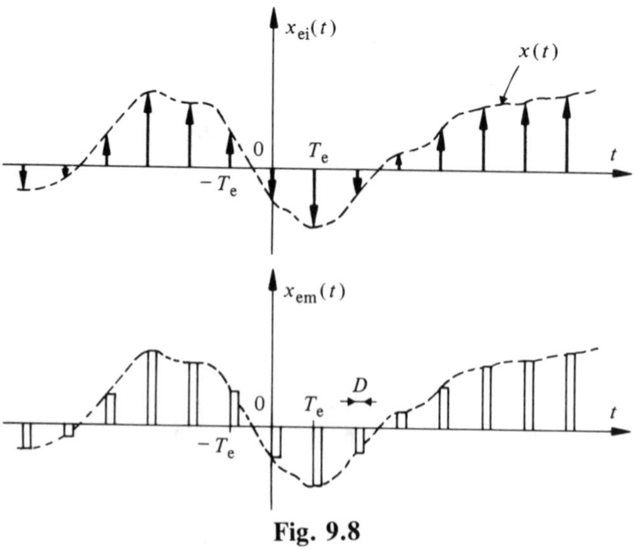

Fig. 9.8

Consequently, in case of sample and hold, the spectral density corresponds to that of the ideal case, ***continuously weighted*** by a sinc^2 function, the zeros of which depend on the hold duration D (fig. 9.9). It corresponds to a linear amplitude distortion (sub-sect. 8.2.25). Note that real sampling (fig. 9.5) yields a discrete weighting.

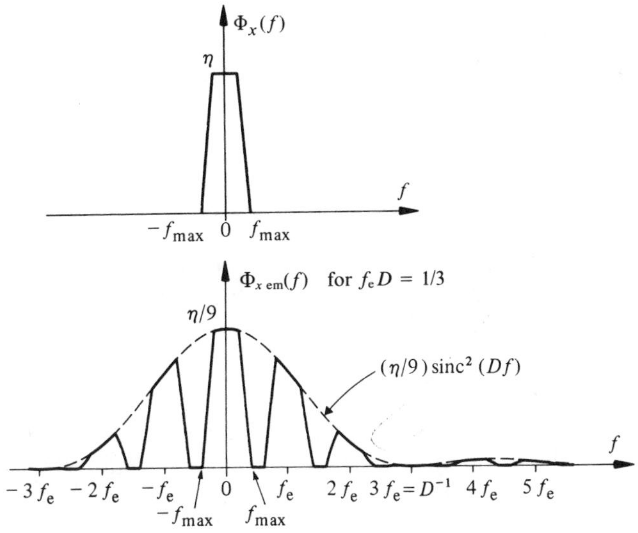

Fig. 9.9

9.2.5 Periodic averaged sampling

Averaged sampling is somewhere between real sampling and sampling and holding. The amplitude of each rectangular pulse of the sampled signal $x_{e\mu}(t)$ corresponds to the analog signal average value, measured on a time interval D. The model (fig. 9.10) of such a sampling is obtained by placing an averager, the impulse response of which is $g_1(t) = D^{-1}\text{rect}[(t - D/2)/D]$, in cascade with the sample and hold.

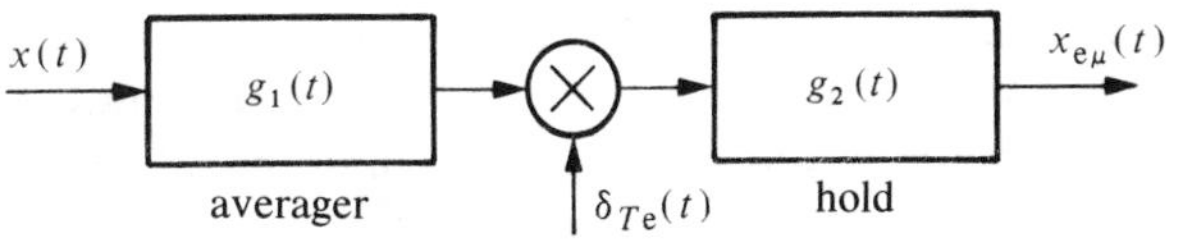

Fig. 9.10

The sampled averaged signal is

$$x_{e\mu}(t) = [x(t) * g_1(t)]\delta_{Te}(t) * g_2(t) \tag{9.18}$$

and, taking (9.15) into account, we get the spectral density:

$$\Phi_{xe\mu}(f) = D^2 f_e^2 \operatorname{sinc}^2(Df)\operatorname{rep}_{fe}\{\operatorname{sinc}^2(Df) \cdot \Phi_x(f)\} \tag{9.19}$$

9.2.6 Fourier transform and z-transform of an ideal sampled signal

According to (9.9), the ideal sampled signal can be written

$$x_{ei}(t) = \sum_{k=-\infty}^{\infty} x(kT_e)\delta(t - kT_e) \tag{9.20}$$

Taking the Fourier transform of (9.20), we get, from (4.76), the following expression, equivalent to (9.12):

$$X_{ei}(f) = \sum_{k=-\infty}^{\infty} x(kT_e)\exp(-j2\pi fkT_e) \tag{9.21}$$

It is common practice in digital signal processing textbooks to normalize the time axis with the sampling period. Inserting $T_e = 1$ into (9.21), we get the conventional Fourier transform definition of a digital signal:

$$X_{ei}(f) = \sum_{k=-\infty}^{\infty} x(k)\exp(-j2\pi fk) \tag{9.22}$$

Normalizing the result of (9.12) in the same way, we get the equivalent expression:

$$X_{ei}(f) = \operatorname{rep}_1\{X(f)\} \tag{9.23}$$

which best emphasizes its periodicity equal to one.

We already mention (sub-sect. 8.2.7) the advantage of the z-transform in the study of sampled signals and systems [105]. It can be written as a negative power series, thusly

$$X_{ei}(z) = \sum_{k=-\infty}^{\infty} x(k)\, z^{-k} \tag{9.24}$$

where z is a complex variable. This transform (vol. XX) is identical to the Fourier transform on the unit circle $z = \exp(\,j2\pi f)$. It allows us to represent a sampled signal by the poles and zeros of $X_{ei}(z)$ and to state easily the stability conditions of discrete (usually digital) processing systems.

9.3 SAMPLING THEOREMS AND CONSEQUENCES

9.3.1 Spectral aliasing

Each sampled signal spectrum is a function of the f_e periodic repetition (f_e being the sample rate) of the analog signal spectrum. Depending on the frequency range of the analog signal, the repeated spectra will either be completely disjoint (figs. 9.5, 9.7, 9.9), or they will partially overlap (fig. 9.11). This last phenomenon, called *aliasing,* prohibits reversibility of the transform. The sampling, even ideal, of a signal with a spectrum which is not strictly band-limited will generate a spectral aliasing effect, and, hence, its non-reversibility. To reduce the risk of spectral aliasing, it is common practice to employ appropriate filtering of the signal before sampling (sub-sect. 9.3.6).

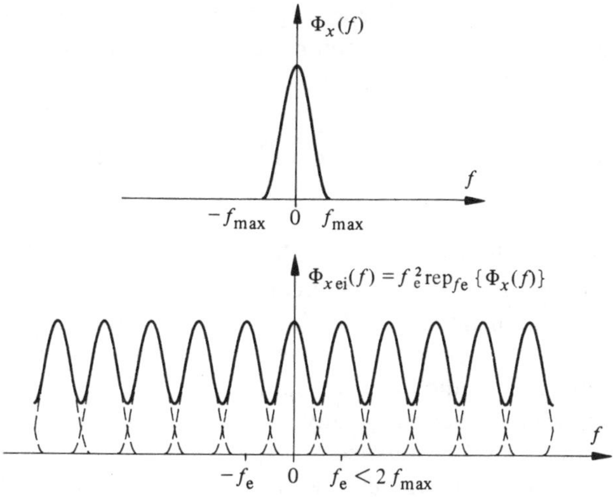

Fig. 9.11

9.3.2 Nonexistence of band-limited finite energy signals

Any finite energy signal can be represented as the product of a non-zero finite average power signal and a weighting function, such as (8.91). From (8.85) and (8.88), its spectrum is the result of a convolution product of two frequency functions, one of them being frequency-unlimited. Even if the non-zero finite average power signal is band-limited (extreme example: sine wave), the finite energy signal that stems from the product necessarily has a frequency-unlimited spectrum. ***Thus, no physically achievable signal can be strictly band-limited.*** This result is a corollary of the Paley-Wiener theorem [22].

The sampling of a physically achievable signal always yields some aliasing, excluding any perfect reversibility.

The finite energy condition, however, forces the spectrum to tend towards zero when $|f|$ tends toward infinity. Consequently, there exists a frequency beyond which the spectrum has almost totally vanished. Hence, it is possible to choose a sampling rate which leads to negligible aliasing and ensures acceptable reversibility.

9.3.3 Sampling theorem for band-limited lowpass spectrum signals

This theorem, first demonstrated by Shannon [5], can be stated as follows:

- any analog signal, with a lowpass spectrum of maximum frequency $f_{\max}$, is totally represented by the complete sequence of its instantaneous values $x(t_k)$ sampled at regular time intervals T, provided that T_e is less than or equal to $1/(2f_{\max})$.

In other words, the reversibility condition is satisfied if

$$f_e = 1/T_e \geqslant 2f_{\max} \tag{9.25}$$

The total bandwidth of such a signal is $B_m = f_{\max}$. Thus, (9.25) can also be expressed as $f_e \geqslant 2B_m$. A mathematical demonstration of this theorem, applicable to a deterministic signal $x(t)$ with a Fourier transform $X(f) = 0$ for $|f| > f_{\max} = B_m$, is given in sub-section 3.4.9. The case of signals with a bandpass spectrum is studied in sub-section 9.3.8.

A generalization, applicable to any kind of signal, can be established because, if condition (9.25) is satisfied, no spectrum aliasing can occur. Thus, it is theoretically possible to reconstruct a signal having a $\Phi_x(f)$ spectrum by filtering the sampled signal with the ideal lowpass filter defined in fig. 8.22 (real sampling and ideal sampling). In the case of sampling and holding or averaged sampling, the same recovery is possible if the filter compensates also for the unavoidable amplitude distortion encountered.

For ideal sampling followed by an ideal lowpass filter with an amplitude frequency response:

$$|G(f)| = T_e \operatorname{rect}(f/f_e) \tag{9.26}$$

we get the block-diagram shown in fig. 9.12.

As a result of (9.13) and (8.24), the output signal $y(t)$ has a spectral density:

$$\Phi_y(f) = \operatorname{rep}_{fe}\{\Phi_x(f)\} \cdot \operatorname{rect}(f/f_e) \tag{9.27}$$

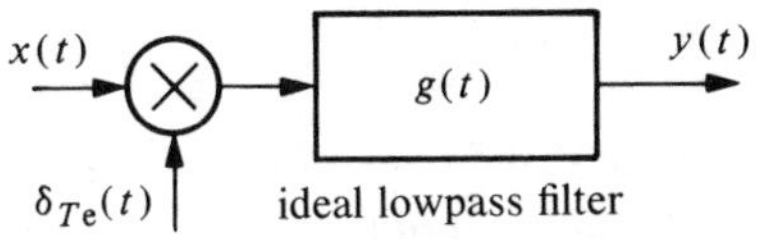

Fig. 9.12

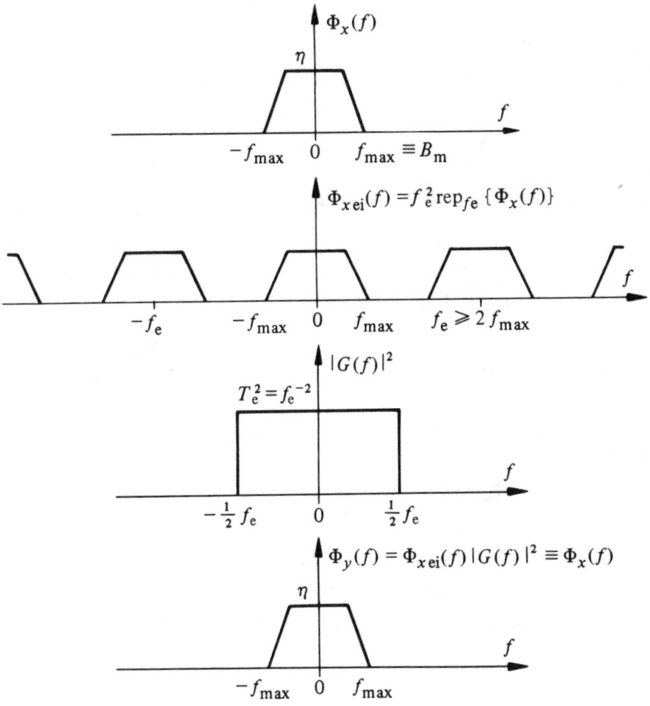

Fig. 9.13

which is equal to $\Phi_x(f)$ if this spectrum is bandlimited to $f = f_{\max} = B_m$ and if $f_e > 2f_{\max}$ (fig. 9.13).

The corresponding ideal filter impulse response is (tab. 8.23)

$$g(t) = \mathrm{sinc}[(t - t_0)/T_e] \tag{9.28}$$

and the output signal, assuming $t_0 = 0$ for simplicity, becomes

$$y(t) \equiv x(t) = [x(t)\delta_{Te}(t)] * g(t) = \sum_{k=-\infty}^{\infty} x(kT_e)\,\mathrm{sinc}[(t/T_e) - k] \tag{9.29}$$

This relation defines the ideal interpolation presented in the next section.

9.3.4 Remark

When sampling a sine wave of frequency f_0, the sampling rate cannot be exactly $2f_0$ because of the spectrum aliasing occuring in this theoretical case.

9.3.5 Practical considerations

The practical importance of the sampling theorem, expressed by relation (9.25), is enormous. It places a fundamental constraint on any real-time digital processing, transmitting, or recording system. To be able to recover the analog signal (sect. 9.4), the effective sampling rate usually must be much higher than the theoretical limit $f_e = 2f_{max}$.

For instance, a telephone signal with a maximum frequency of about 3400 Hz must be sampled at a rate higher than 6800 Hz (international standard: $f_e = 8000$ Hz). With a speech rate of a few words per second, a small sentence already needs tens of thousands of samples!

A European TV video signal has a spectrum extending to about 5 MHz. Hence, the sampling rate must be higher than 10 MHz (in practice, in the range of 18 MHz). A 20 ms frame alone needs many hundreds of thousands of samples! To store a single frame in a computer memory requires a huge memory space.

In contrast, it is useless to sample some signals stemming from physical phenomena of large inertia (e.g., temperature) at sampling rates higher than a few samples per second or even per minute.

Therefore, a good understanding of the spectral features of the analog signal which must be processed is key to the efficient selection of the appropriate sampling rate.

9.3.6 Anti-aliasing filter

In many cases, the spectrum of the signal to be sampled is not perfectly known. Often, it has a wideband component due to the background noise generated in the sensor, the amplifiers, *et cetera*. Thus, it is compulsory to filter the analog signal before it is sampled (fig. 9.14) in order to eliminate all risk of aliasing without requiring an excessive sampling rate.

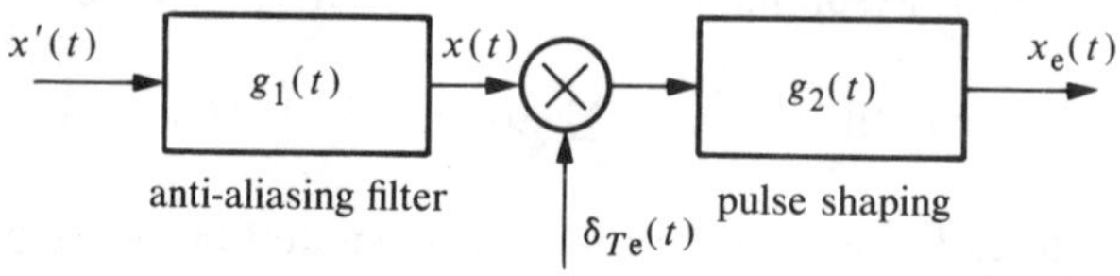

Fig. 9.14

The perfect *anti-aliasing filter* is an ideal lowpass filter with a bandwidth $B = f_e/2$. However, such a filter, as we know, cannot be implemented, since it is non-causal (sub-sect. 8.2.21). Any real anti-aliasing filter has a non-zero transition band (fig. 9.15). The maximum bandwidth B_m is, therefore, much larger than the effective bandwidth B. Thus, the minimal usable sampling rate must be equal to

$$f_{e,\min} = 2B_m > 2B \tag{9.30}$$

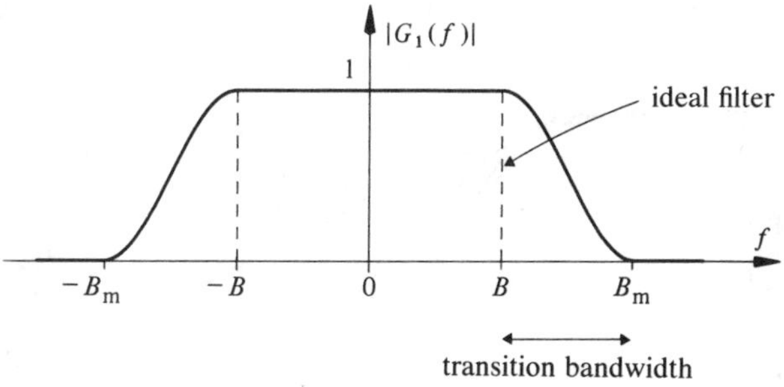

Fig. 9.15

The transition bandwidth depends on the kind of filter used and on how B_m is defined.

To design this filter, the nth degree *Butterworth approximation* is sometimes used (vol. XIX):

$$|G_1(f)|^2 = [1 + (f/f_c)^{2n}]^{-1} \tag{9.31}$$

It has an optimally flat response (fig. 9.16) in the passband with -3 dB attenuation for $f = f_c$ and an asymptotic attenuation slope of $-6n$ dB/octave ($-20n$ dB/decade) for $f > f_c$. The case $n = 1$ corresponds to the lowpass *RC* filter of sub-section 8.2.24.

If the spectrum of the input signal $x'(t)$ is constant, it can be shown [106], assuming $B = f_c$, that the ratio ξ_{rx} of the power P_r of the aliasing-induced error to the total power P_x of the filtered signal is approximately, for $(2f_c/f_e)^{2n} \ll 1$, given by

$$\xi_{rx} = \frac{P_r}{P_x} \cong 2^{2n} \frac{n}{\pi(2n-1)} \sin\left(\frac{\pi}{2n}\right)\left(\frac{f_c}{f_e}\right)^{2n-1} \tag{9.32}$$

Expressing this ratio in decibels ($\xi_{rxdB} = 10 \log_{10}\xi_{rx}$), we get the diagram of fig. 9.17:

$$\log_{10}\left(\frac{f_e}{f_c}\right) \cong \frac{1}{2n-1}\left[\log_{10}\frac{n2^{2n}\sin(\frac{1}{2}\pi/n)}{(2n-1)\pi} - \frac{\xi_{rxdB}}{10}\right] \tag{9.33}$$

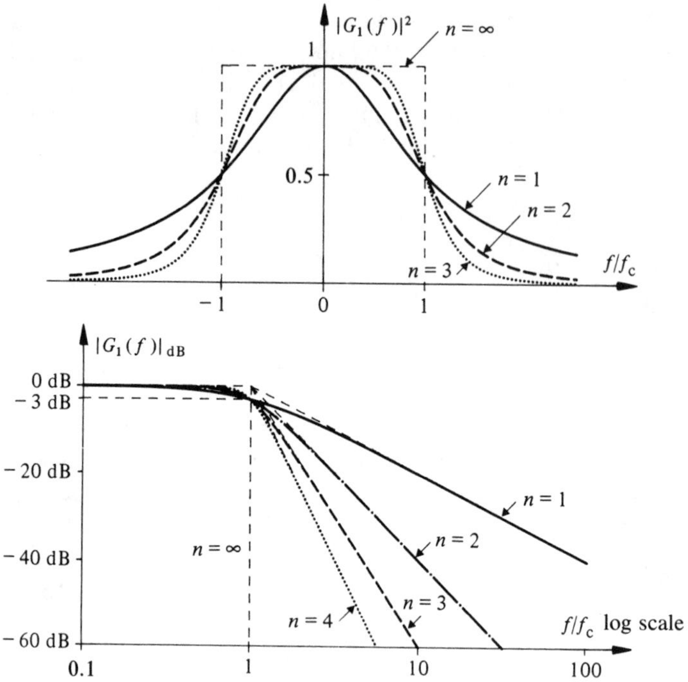

Fig. 9.16

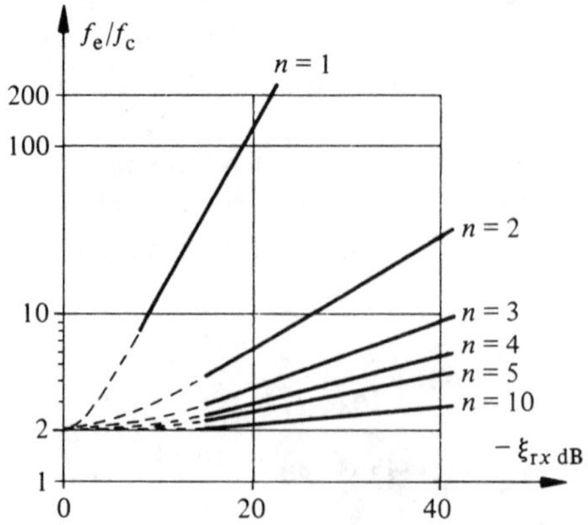

Fig. 9.17

For a simple *RC* filter ($n = 1$), a 1% power ratio ($\xi_{rxdB} = -20$ dB) requires a sampling rate $f_e \cong 127 f_c$. For $n = 2$ and $n = 4$, the same power ratio requires a sampling rate of $6 f_c$ and $3 f_c$, respectively. Hence, a trade-off exists between the sampling rate reduction (down to the theoretical limit) and the complexity of the anti-aliasing filter.

Other ideal filter approximations are possible (*Chebyshev filters, elliptic filters, et cetera*). Their transfer functions have a faster transition between passband and stopband, at the expense of a residual ripple in the passband, and, for the elliptic filters, also in the stopband.

9.3.7 Distortion due to spectrum aliasing

Figures 9.18 and 9.19 show two kinds of distortion that result in cases where the sampling theorem is not satisfied. In fig. 9.18, a sine wave with a frequency f_0 is sampled at a rate $f_e = 1/T_e = 1.25 f_0$. In fact, there is an infinite number of sine waves, with frequencies $nf_e \pm f_0$, passing through those samples. Without more information, we will systematically reconstruct the sine wave with the lowest frequency $f_0' = f_e - f_0$. This is what any recovering lowpass filter will automatically do.

Fig. 9.19 shows how the recovered signal $x'(t)$, even after ideal lowpass filtering, is affected when the initial analog signal $x(t)$ has a spectrum which is

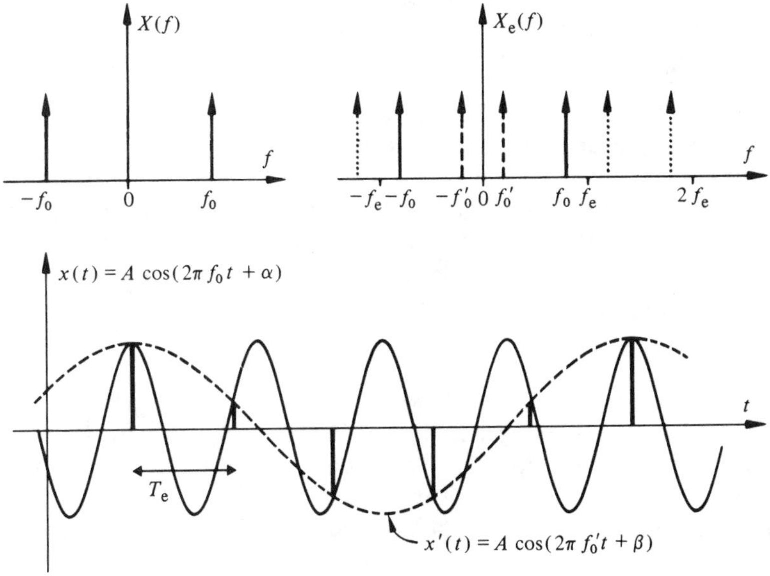

Fig. 9.18

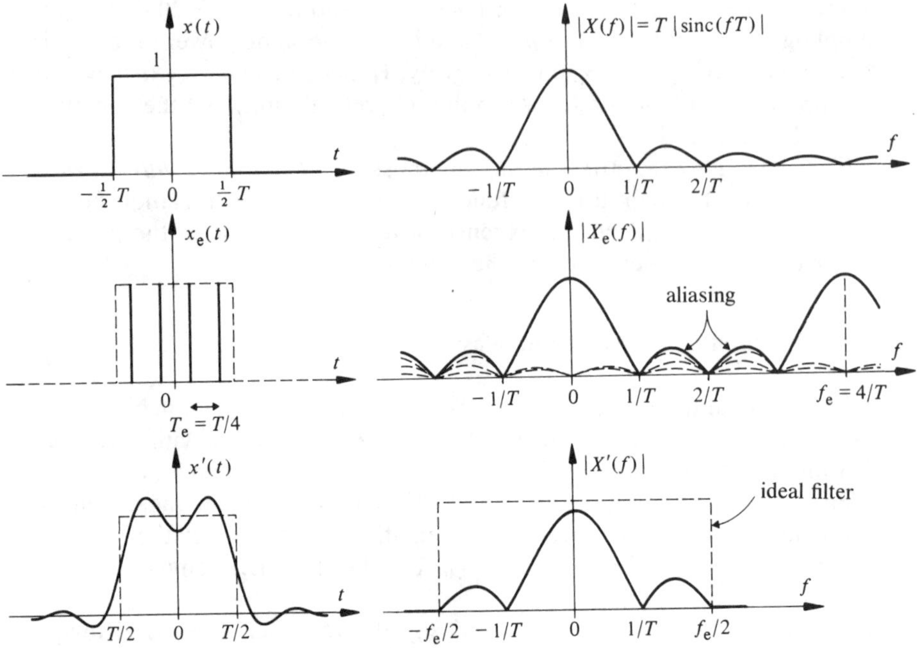

Fig. 9.19

not band-limited, and when the chosen sampling rate allows the existence of significant aliasing.

9.3.8 Sampling theorem for bandpass signals

When the analog signal has a bandpass spectrum with cut-off frequencies f_1 and f_2 ($f_2 > f_1$), a straightforward application of the lowpass sampling theorem (9.25) leads to sampling rate $f_e \geqslant 2f_2$.

In fact, the anti-aliasing condition is satisfied [35] for a minimum sampling rate:

$$f_e = 2f_2/m \tag{9.34}$$

where m is the largest integer less than or equal to f_2/B with $B = f_2 - f_1$. All the higher sampling rates are not necessarily usable. For $f_2 \gg B$, the minimum sampling rate tends toward $2B$.

This limit can, in fact, be reached theoretically regardless of the bandwidth B if we use the representation described in section 7.4:

$$\begin{aligned} x(t) &= a(t)\cos(2\pi f_0 t) - b(t)\sin(2\pi f_0 t) \\ &= \text{Re}\{\underline{r}(t)\exp(\,\text{j}2\pi f_0 t)\} \end{aligned} \tag{9.35}$$

Choosing here an arbitrary frequency $f_0 = \frac{1}{2}(f_1 + f_2)$, the in-phase $a(t)$ and quadrature $b(t)$ components have a lowpass spectrum, with $f_{max} = B/2$. Therefore, they can both be represented by samples taken at rate $f_e = 2f_{max} = B$, according to (9.25). We can thus state the following theorem:

- let $x(t)$ be a signal with a bandpass spectrum, the bandwidth of which is $B = f_2 - f_1$, and let $f_0 = \frac{1}{2}(f_2 + f_1)$ be the center frequency of this band; then $x(t)$ can be represented by the series:

$$x(t) = \sum_{k=-\infty}^{\infty} [a(kT_e)\cos(2\pi f_0 t) - b(kT_e)\sin(2\pi f_0 t)] \cdot \text{sinc}[(t/T_e) - k] \quad (9.36)$$

where

$$T_e = 1/f_e = 1/B \quad (9.37)$$

The difference between this theorem and that for the lowpass (sub-sect. 9.3.3) is that here, at each sampling instant, we pick up a sample ***on each of the two components*** $a(t)$ and $b(t)$. Each sample pair defines a sample value (real and imaginary parts, or modulus and argument) of the complex envelope $\underline{r}(t) = a(t) + \mathrm{j}b(t) = r(t)\exp \mathrm{j}\alpha(t)$. If the sampling rate is $f_e = B$, the number of real samples per second is still equal to $2B$.

The $a(t)$ and $b(t)$ components can be obtained by a system like the one of fig. 9.20.

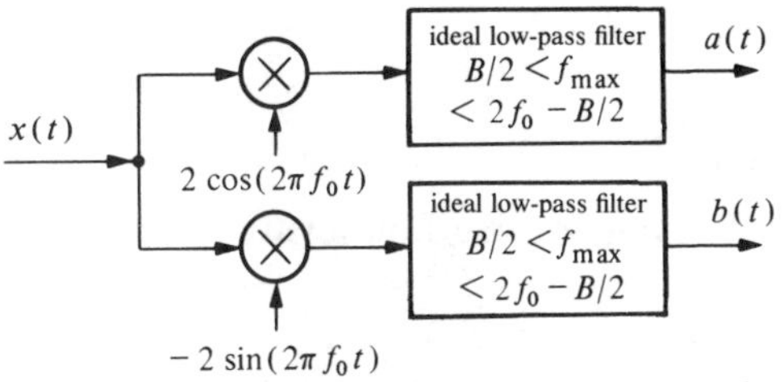

Fig. 9.20

9.3.9 Dimension of a band-limited signal of finite duration

Although such a signal cannot exist, it is a convenient abstract concept. The sampling theorems state that, if the signal duration is T and its spectrum bandwidth is B, the signal is completely defined by

$$N = 2BT \quad \textbf{(9.38)}$$

samples. This number N corresponds to the minimal quantity of information needed to recover the signal and can be seen as a measure of the signal's dimension.

In chapter 3, we showed that any signal of duration T can be expanded in a Fourier series with complex coefficients X_k associated with the discrete frequencies k/T. If the bandwidth is limited to B, we have $2BT$ information elements.

As an example, a one-minute signal, with a 10 kHz bandwidth, cannot be correctly represented by less than 1,200,000 information elements (if we disregard a possible dependence between samples).

9.3.10 Reciprocal theorem

Taking the time-frequency duality into account, we can easily see that the complex spectrum $X(f)$ of a signal $x(t,T)$ of finite duration T can only be defined by the series of its values sampled on the frequency axis at $\Delta f = 1/T$ intervals:

$$X(f) = \sum_{n=-\infty}^{\infty} X(n/T) \operatorname{sinc}(Tf - n) \tag{9.39}$$

By analogy with (9.12), the inverse Fourier transform of the ideal sampling function:

$$X(f) \cdot \delta_{\Delta f}(f) = \sum_{n=-\infty}^{\infty} X(n/T)\delta(f - n/T) \tag{9.40}$$

is a ***periodic signal***:

$$x_p(t) = T \cdot \operatorname{rep}_T\{x(t,T)\} \tag{9.41}$$

According to (4.77), this inverse Fourier transform can also be written as

$$x_p(t) = \sum_{n=-\infty}^{\infty} X(n/T) \exp(\,j2\pi nt/T) \tag{9.42}$$

which is the reciprocal of (9.21) and can be identified with the Fourier series expansion (3.74) by dividing the two members by T. The main period $x(t,T)$ can also be written

$$x(t,T) = \frac{1}{T} \sum_{n=-\infty}^{\infty} X(n/T) \exp(\,j2\pi nt/T) \tag{9.43}$$

9.3.11 Discrete Fourier transform

Combining the result of sub-sections 9.3.3 and 9.3.10, we see that a time sequence of $N = T/T_e$ values $x(kT_e)$ sampled on a duration T at a rate $f_e = 1/T_e$ corresponds to a frequency sequence of N complex values $X(n/T)$ sampled on a bandwidth f_e at intervals $\Delta f = 1/T = f_e/N$. From (9.43) and (9.21), we get the pair of transforms:

$$x(kT_e,T) = \frac{1}{NT_e} \sum_{n=n_0}^{n_0+N-1} X(n/T,f_e) \exp(\,j2\pi kn/N)$$

$$(k = k_0, \ldots, k_0 + N - 1) \quad (9.44)$$

$$X(n/T,f_e) = \sum_{k=k_0}^{k_0+N-1} x(kT_e,T) \exp(-j2\pi kn/N)$$

$$(n = n_0, \ldots, n_0 + N - 1) \quad (9.45)$$

This notation emphasizes the fact that (9.44) and (9.45) represent ***one period of a periodic sequence*** (fig. 9.21). If $x(t)$ is real, the complex frequency samples $X(n/T) = |X(n/T)| \exp j\vartheta_x(n/T)$ symmetrical with respect to $mf_e(m = 0, \pm 1, \pm 2, \ldots)$ are complex conjugates. Hence, to N information time elements, there always correspond N information frequency elements. Usually, we take

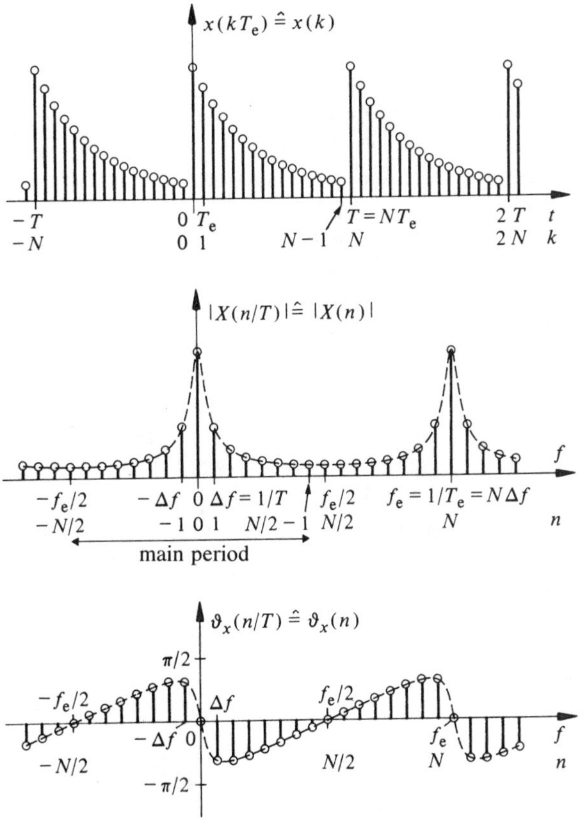

Fig. 9.21

$n_0 = -N/2$ for a convenient display of the ***main period*** of the complex spectrum. Similarly, we usually take k_0 as time origin.

Normalizing the time axis with the sampling period ($T_e = 1$), and using the shorthand notation:

$$x(k) \triangleq x(kT_e, T) \tag{9.46}$$

$$X(n) \triangleq X(n/T, f_e) \tag{9.47}$$

$$W_N \triangleq \exp(\,j2\pi/N) \tag{9.48}$$

we get the usual definition of the *discrete Fourier transform* (DFT) already mentioned in sub-section 8.2.3:

$$x(k) = N^{-1} \sum_{n=-N/2}^{N/2-1} X(n) W_N^{nk} \leftrightarrow X(n) = \sum_{k=k_0}^{k_0+N-1} x(k) W_N^{-nk} \tag{9.49}$$

with $k = k_0, \ldots, k_0 + N - 1$ and $n = -N/2, \ldots, N/2 - 1$.

Efficient algorithms, known as *fast Fourier transforms* (FFT) are described in specialized books on digital signal processing ([107], vol. XX). The efficiency comes from the suppression of the computational redundancies.

9.3.12 Periodic signal undersampling

Let T be the period of a periodic signal $x(t)$ and T_e the sampling interval. If $T_e > T$, only one sample, at best, will be taken in each period. It is, however, possible to recover $x(t)$ with a new time scale, if

$$T_e = mT + \epsilon \tag{9.50}$$

where m is a positive integer and ϵ is a sampling pseudoperiod (fig. 9.22). Indeed, a complete period of the signal can be reconstructed from the values sampled at a rate of $f_e = 1/T_e$ when $N = T/\epsilon$ samples are available.

This method is mainly used for the acquisition and display of very high frequency periodic phenomena (sampling oscilloscope) and in detection devices for periodic signals embedded in background noise, called boxcar integrators. It is somewhat similar to the stroboscopic effect used to monitor rotating movement.

9.4 RECONSTRUCTION BY INTERPOLATION OR EXTRAPOLATION

9.4.1 Interpolation and extrapolation

The reconstruction (fig. 9.23) of an analog signal $y(t)$ from a sequence of samples $\{y(kT_e)\}$ implies the recovery of all the intermediate values between

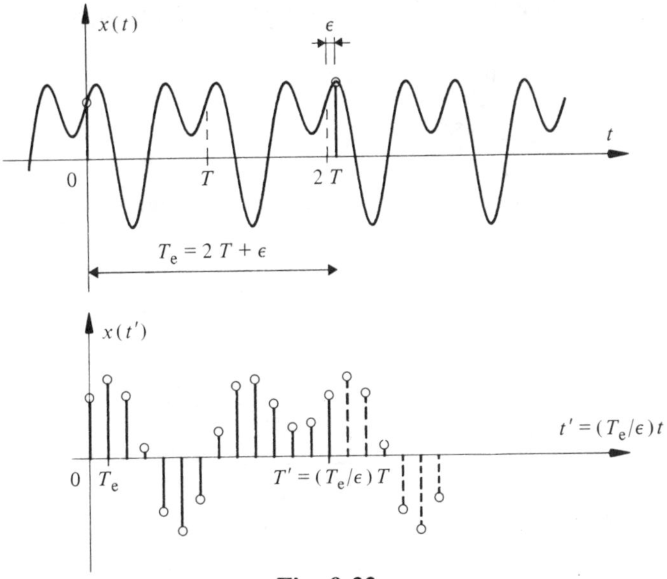

Fig. 9.22

two successive samples. This recovery, denoted $\tilde{y}(t)$, is generally an approximation and thus introduces a distortion $d(t) = y(t) - \tilde{y}(t)$. If the reconstruction system defines the signal values in the sampling interval following the sample of abscissa kT_e from this sample and the m previous samples, it is said to be an mth-*order extrapolator.*

If the recovered values are defined by $m + 1$ samples, with some of them of abscissa greater than kT_e, the reconstruction system is said to be a mth-*order interpolator.*

An interpolator does not seem to be a causal system, since it must know the future in order to compute the present. Nonetheless, we can make it causal by accepting some delay t_0 for an adequate recovery.

A delay is, in principle, objectionable only for real-time processing. Telecommunication systems, however, can tolerate rather long delays, but this is not the case in systems with feedback (e.g., automatic control) in which the induced delay can generate instability.

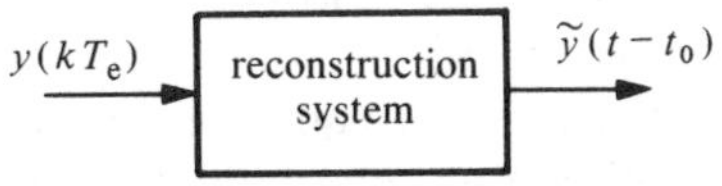

Fig. 9.23

9.4.2 Reconstruction with a linear operator: smoothing filter

According to (9.9), the ideal sampled signal is

$$y_{ei}(t) = y(t) \cdot \delta_{Te}(t) = \sum_{k=-\infty}^{\infty} y(kT_e)\delta(t - kT_e) \tag{9.51}$$

If this signal is the input of a linear operator (fig. 9.24) with an impulse response $g(t)$, the output is

$$\tilde{y}(t - t_0) = \sum_{k=-\infty}^{\infty} y(kT_e)g(t - kT_e) \tag{9.52}$$

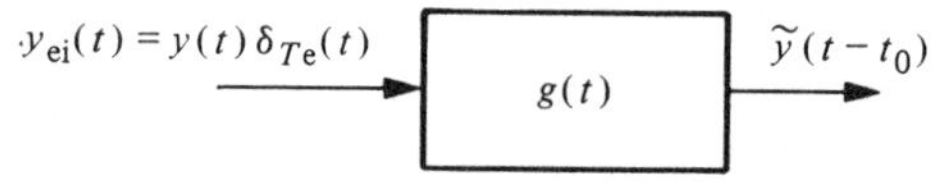

Fig. 9.24

The linear operator used for such a reconstruction is called a *smoothing* filter. If the signal has a lowpass spectrum, the filter is a lowpass filter, and if the signal has a bandpass spectrum, the filter is a bandpass filter. We will study later the lowpass filter case.

From (9.13) and (9.24), the spectral density of the recovered signal is given by (fig. 9.25)

$$\Phi_{\tilde{y}}(f) = f_e^2 \operatorname{rep}_{fe}\{\Phi_y(f)\} \cdot |G(f)|^2 \tag{9.53}$$

and the reconstruction mean square error is

$$\overline{d^2} = \int_{-\infty}^{\infty} [\Phi_{\tilde{y}}(f) - \Phi_y(f)]\mathrm{d}f \tag{9.54}$$

The quality of the reconstruction depends, as in sub-section 9.3.6, on the trade-off between the sampling rate and the smoothing filter complexity. A rate considerably greater than the theoretical limit $f_e = 2f_{max}$ is generally required to maintain both the aliasing distortion and the reconstruction distortion at an acceptable level.

9.4.3 Reconstruction by polynomial approximation

Only one polynomial of degree m can pass through $m + 1$ samples. This approach can be used to obtain an approximation of the analog signal to be recovered.

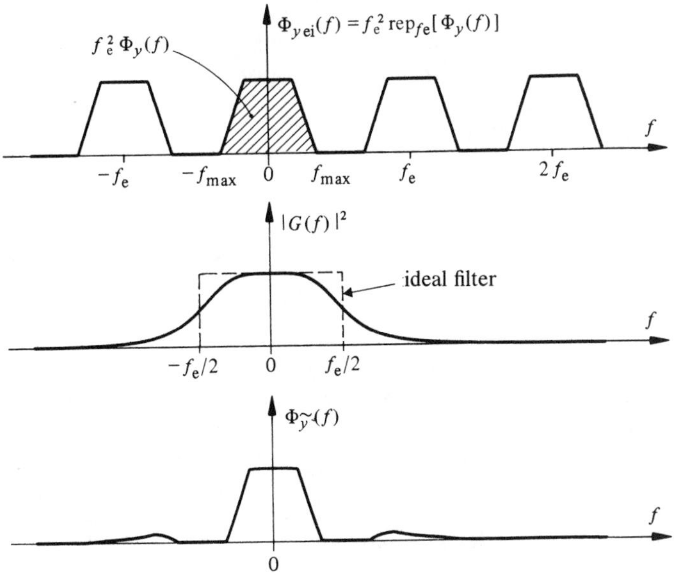

Fig. 9.25

If $m + 1$ samples $y[(n - m)T_e], \ldots, y(nT_e)$ are known, the approximation of $y(t - t_0)$ for $nT_e \leqslant t \leqslant (n + 1)T_e$ and $0 \leqslant r \leqslant m$ can be computed by using the Lagrange polynomial method (vol. XIX and [105, 108]):

$$\tilde{y}(t - rT_e - t_0') = \sum_{i=0}^{m} y[(i + n - m)T_e]q_{m,i}[t - (n + r - m)T_e] \tag{9.55}$$

with

$$q_{m,i}(t) = \prod_{\substack{l=0 \\ l \neq i}}^{m} \frac{t - lT_e}{(i - l)T_e} \quad \text{for } i = 0, \ldots, m \tag{9.56}$$

and

$$q_{0,0}(t) = 1 \tag{9.57}$$

For $0 < r \leqslant m$, we have an interpolation. The $r = 0$ case corresponds to an extrapolation that can be written as the Newton-Gregory relation [109]:

$$\tilde{y}(t - t_0') = y(nT_e) + \{y(nT_e) - y[(n - 1)T_e]\} \frac{t - nT_e}{T_e} + \{y(nT_e) - 2y[(n - 1)T_e] + y[(n - 2)T_e]\} \frac{(t - nT_e)t}{2T_e^2} + \ldots \tag{9.58}$$

The total delay $t_0 = rT_e + t_0'$ depends on the choice of r and m.

Equating (9.52) with (9.55), the polynomial reconstruction appears as a smoothing by a linear filter with impulse response:

$$g(t) = \sum_{l=0}^{m} g_{m,m-1}(t - lT_e) \tag{9.59}$$

where

$$g_{m,i}(t) = q_{m,i}[t - (r - m)T_e] \cdot \text{rect}[(t - T_e/2)/T_e] \tag{9.60}$$

Since $g_{m,i}(t)$ is zero outside interval $0 < t < T_e$, $g(t)$ is zero outside interval $0 < t < (m + 1)T_e$.

An mth-order polynomial reconstruction with a high value of m is only possible if appropriate computational means are available. In practice, a zero-order extrapolation, or a first-order extrapolation or interpolation, is often used, combined with additional smoothing performed by a conventional filter.

9.4.4 Example 1: zero-order extrapolator

The simplest reconstruction method (fig. 9.26) is to hold, during the whole interval $nT_e \leqslant t < (n + 1)T_e$, the value of sample $y(nT_e)$. This is a zero-order extrapolation corresponding to $r = 0$ and $m = 0$, and it is similar to the sampling and holding principle (sub-sect. 9.2.4).

The impulse response of the corresponding filter is (fig. 9.27):

$$g_0(t) = \text{rect}[(t - T_e/2)/T_e] \tag{9.61}$$

Its transfer function is (fig. 9.28)

$$G_0(f) = T_e \,\text{sinc}(T_e f) \exp(-j\pi T_e f) \tag{9.62}$$

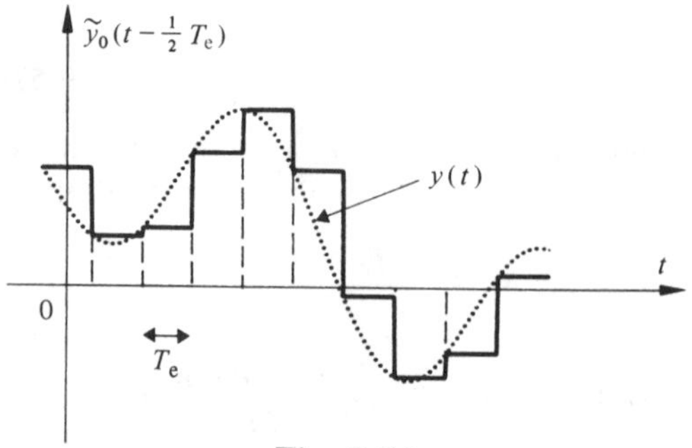

Fig. 9.26

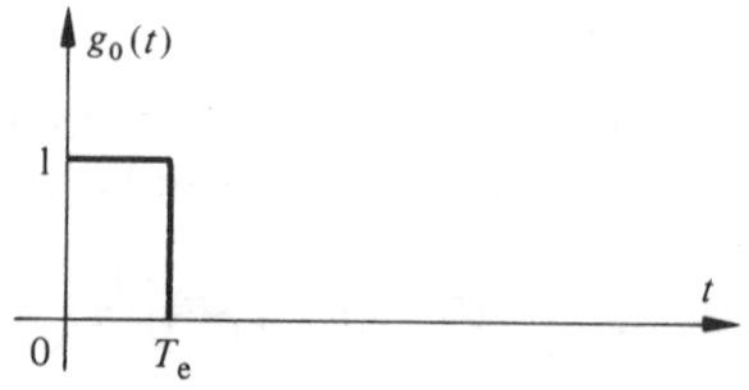

Fig. 9.27

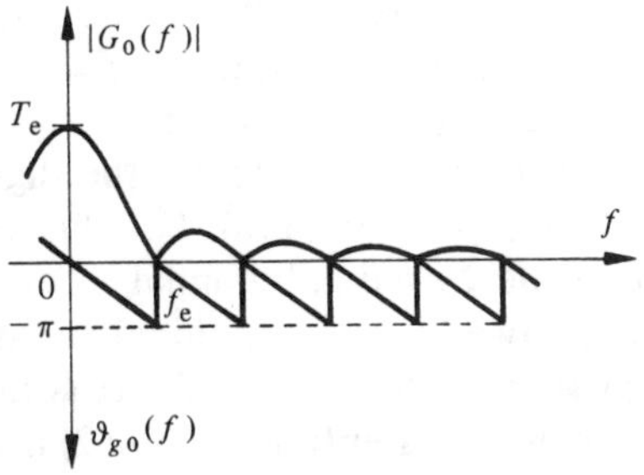

Fig. 9.28

which introduces a linear phase shift $-\pi T_e f$, yielding a delay $t_0 = T_e/2$. Thus, we get the approximation:

$$\tilde{y}_0(t - T_e/2) = \sum_{k=-\infty}^{\infty} y(kT_e) \operatorname{rect}\left(\frac{t - T_e/2 - kT_e}{T_e}\right) \tag{9.63}$$

with, from (9.53),

$$\Phi_{\tilde{y}0}(f) = \operatorname{rep}_{fe}\{\Phi_y(f)\} \cdot \operatorname{sinc}^2(T_e f) \tag{9.64}$$

An example of the resulting spectral density is given in figure 9.29.

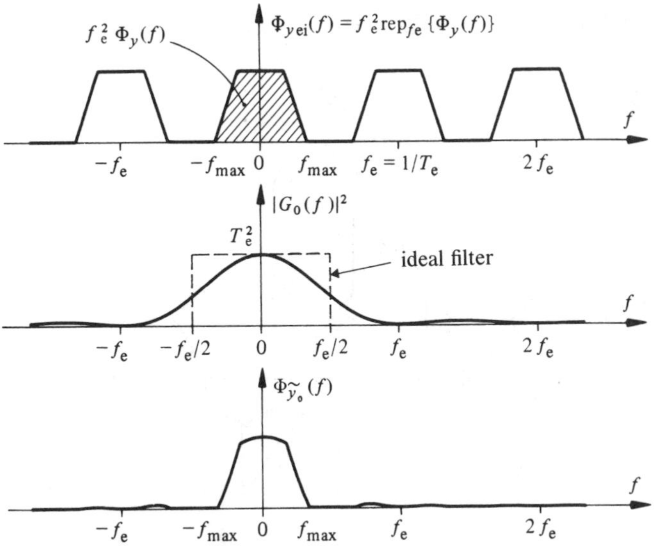

Fig. 9.29

We see that in order to have $\Phi_{\tilde{y}}(f) \cong \Phi_y(f)$, the sampling rate f_e must be ***much greater than*** twice the maximum frequency f_{max}.

The hold function required by (9.61) is automatically performed by a digital-to-analog converter (sub-sect. 10.1.3), whereby the digital values are held at the input during a whole sampling interval T_e. This is why the zero-order extrapolation is the most widely used polynomial reconstruction method. The hold circuit is generally followed by an additional smoothing filter that attenuates the undesired high frequencies (fig. 9.30). It is also possible to reduce the attenuation induced by the transfer function (9.62) in the desired bandwidth by requiring the second filter to have an equalizing transfer function with a modulus roughly proportional to $|G_0(f)|^{-1}$.

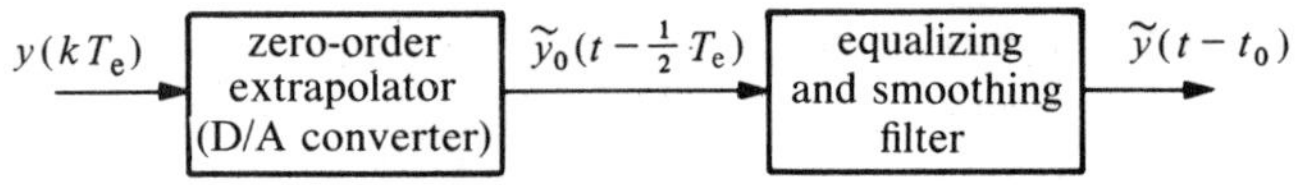

Fig. 9.30

9.4.5 Example 2: first-order extrapolator

Assuming $r = 0$ and $m = 1$ in approximation (9.55), we obtain a first-order extrapolation (fig. 9.31):

$$\tilde{y}_{10}(t - t_0') = y(nT_e) + \{y(nT_e) - y[(n-1)T_e]\}(t - nT_e)/T_e \tag{9.65}$$

for $nT_e \leqslant t < (n+1)T_e$.

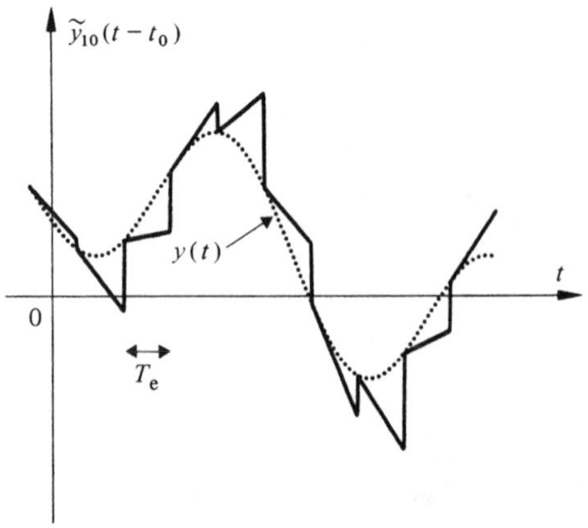

Fig. 9.31

The corresponding impulse response (fig. 9.32) is drawn from (9.60):

$$g_{10}(t) = \begin{cases} (t + T_e)/T_e & \text{for} \quad 0 \leqslant t < T_e \\ -(t - T_e)/T_e & \text{for} \quad T_e \leqslant t < 2T_e \\ 0 & \text{elsewhere} \end{cases} \tag{9.66}$$

and leads to the transfer function (fig. 9.33):

$$G_{10}(f) = T_e\sqrt{1 + (2\pi f T_e)^2}\,\mathrm{sinc}^2(T_e f)\exp\{-\mathrm{j}[2\pi f T_e - \arctan(2\pi f T_e)]\} \tag{9.67}$$

which introduces nonlinear phase response.

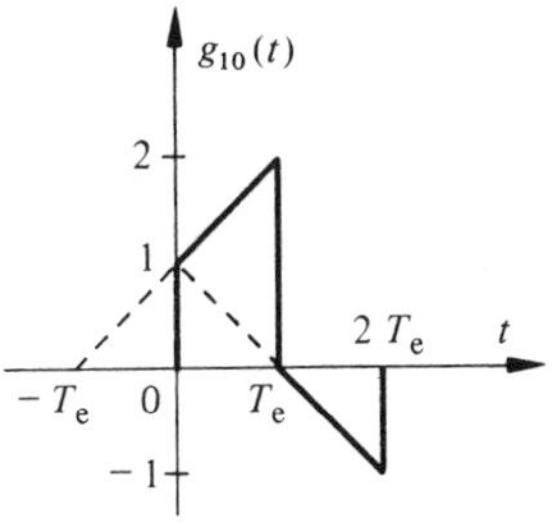

Fig. 9.32

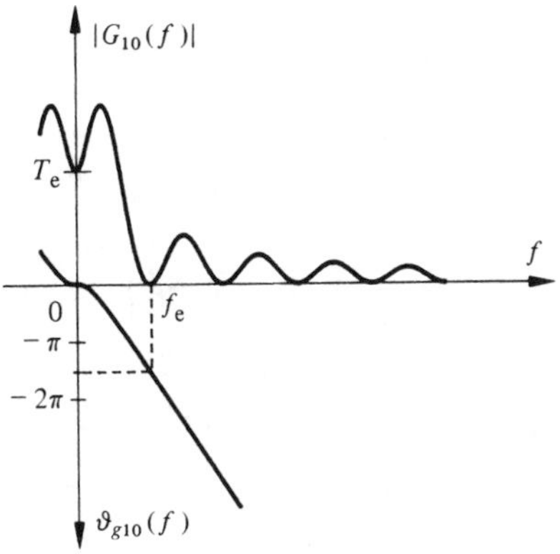

Fig. 9.33

This kind of extrapolator behaves approximately like a highpass filter in the considered bandwidth. Hence, it is not advisable to use it because of the distortion it induces, especially in the presence of noise.

This is sometimes replaced by a modified version halfway between the zero-order and the first-order extrapolator:

$$\tilde{y}(t - t_0') = y(nT_e) + a\{y(nT_e) - y[(n-1)T_e]\}\,(t - nT_e)/T_e \tag{9.68}$$

with $0 \leqslant a \leqslant 1$.

9.4.6 Example 3: first-order interpolator

When a longer delay t_0, equal to a sampling interval, is acceptable, a rather good approximation is obtained by a broken line fitted onto the samples. This is the principle of a ***linear interpolator*** (fig. 9.34) with $r = 1$ and $m = 1$; thus,

$$\tilde{y}_{11}(t - T_e) = y[(n-1)T_e] + \{y(nT_e) - y[(n-1)T_e]\}(t - nT_e)/T_e \tag{9.69}$$

for $nT_e \leqslant t \leqslant (n+1)T_e$.

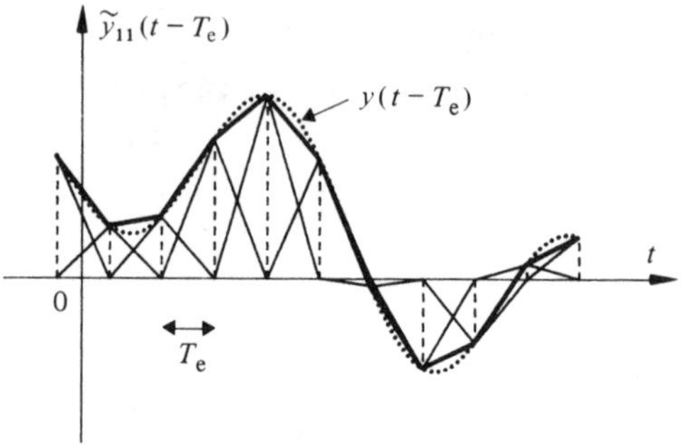

Fig. 9.34

The impulse response of the linear interpolator is (fig. 9.35)

$$g_{11}(t) = \text{tri}[(t - T_e)/T_e] \tag{9.70}$$

and its transfer function is (fig. 9.36)

$$G_{11}(f) = T_e \,\text{sinc}^2(T_e f)\exp(-j2\pi T_e f) \tag{9.71}$$

The spectral density of the reconstructed signal is

$$\Phi_{\tilde{y}}(f) = \text{rep}_{fe}\{\Phi_y(f)\} \cdot \text{sinc}^4(T_e f) \tag{9.72}$$

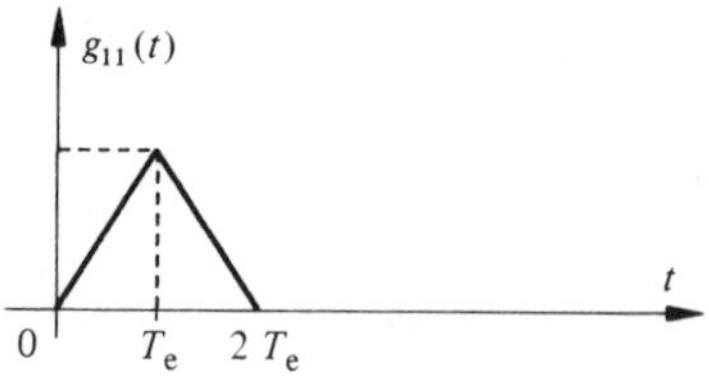

Fig. 9.35

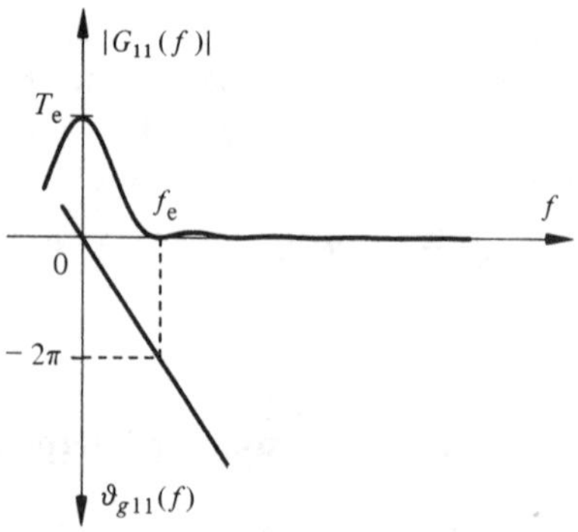

Fig. 9.36

The components with frequencies higher than f_e are more efficiently attenuated than in the case of a zero- or first-order extrapolator.

Once again, an additional smoothing and equalizing filter can be placed in cascade with the interpolator.

9.4.7 Ideal interpolator

Many other interpolation functions can be thought of that correspond to the impulse responses of various kinds of lowpass filters. An example is the Butterworth filter described in sub-section 9.3.6. As a rule of thumb, we can say that the higher the filter order, the better the approximation. However, it is advisable not to neglect the induced delay and the error generated by the unavoidable phase distortion.

As long as the sampling theorem is satisfied, the ideal interpolator is, of course, an ideal filter, the transfer function of which is (fig. 9.37)

$$G_i(f) = T_e \operatorname{rect}(f/f_e) \exp(-j2\pi f t_0) \tag{9.73}$$

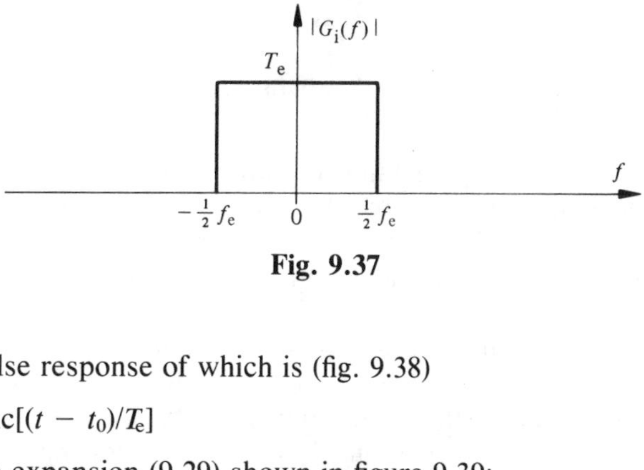

Fig. 9.37

and the impulse response of which is (fig. 9.38)

$$g_i(t) = \operatorname{sinc}[(t - t_0)/T_e] \tag{9.74}$$

leading to the expansion (9.29) shown in figure 9.39:

$$\tilde{y}_i(t - t_0) = \sum_{k=-\infty}^{\infty} y(kT_e) \operatorname{sinc}\left(\frac{t - kT_e - t_0)}{T_e}\right) \tag{9.75}$$

where the delay t_0 should be infinite if the filter is to be causal!

Although unrealistic, such an interpolation can be approximately implemented (usually in digital form) if it is limited to a truncated interpolation function (fig. 9.40):

$$\tilde{g}_i(t) = \operatorname{sinc}\left[\frac{t - mT_e/2}{T_e}\right] \cdot \operatorname{rect}\left[\frac{t - mT_e/2}{mT_e}\right] \tag{9.76}$$

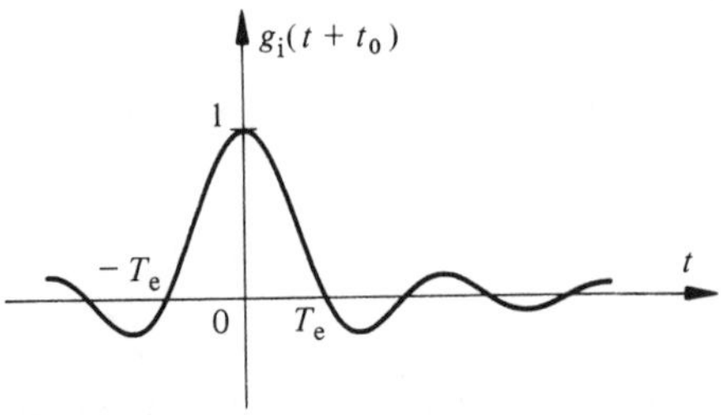

Fig. 9.38

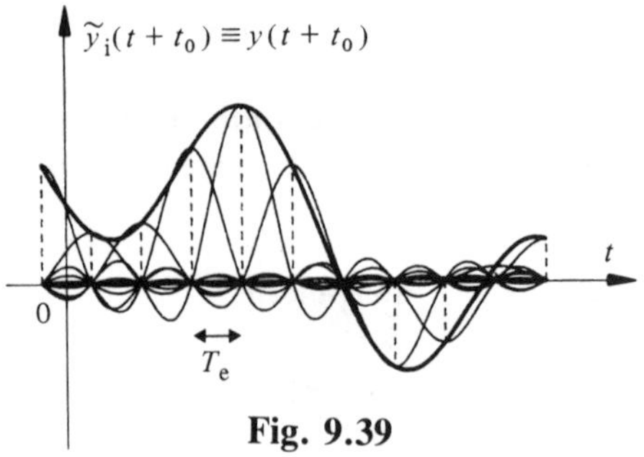

Fig. 9.39

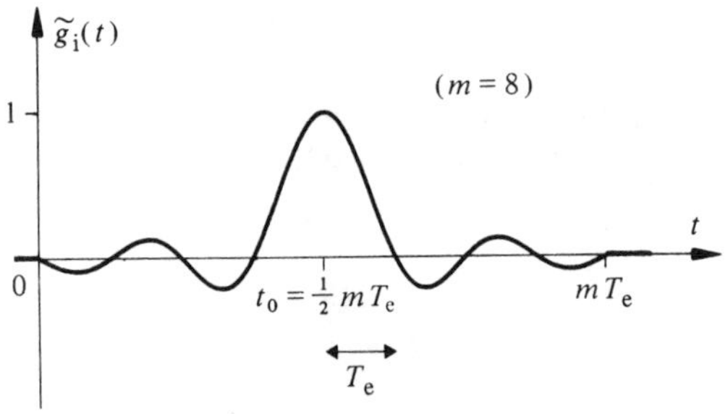

Fig. 9.40

Instead of a rectangular weighting window, it may be better to choose a window without discontinuities, the derivative of which is also without discontinuities, in order to limit the residual ripples of the transfer function modulus outside the considered bandwidth, however, at the expense of a larger transition band.

The analog signal reconstruction can then be done in two phases (fig. 9.41): (1) a quasi-ideal digital interpolator feeds at a rate $f_e' = Lf_e (L \gg 1)$ a zero-order extrapolator (digital-to-analog converter) to which (2) an additional smoothing filter can be connected in cascade. The oversampling provided by the interpolator replaces the repetition period f_e of the ideal sampled signal spectrum by Lf_e without modifying the limiting frequency f_{max}. The distortion

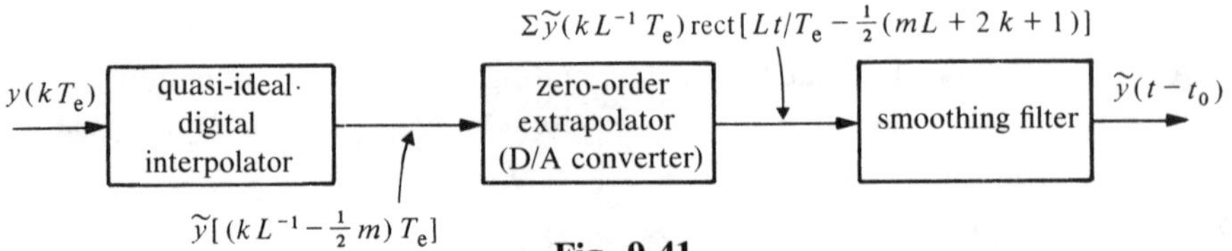

Fig. 9.41

introduced by the zero-order extrapolator and the smoothing filter becomes negligible for $L \gg 1$.

9.4.8 Representation of a Gaussian white noise

We have seen (sub-sect. 5.7.3) that the samples of a Gaussian process are independent if the interval T_e between them is such that the autocovariance function is zero. For a band-limited Gaussian white noise with a maximum frequency $f_{max} = B$, this condition is satisfied for

$$T_e = 1/(2B) \tag{9.77}$$

As in (9.75), we can build the following model of band-limited Gaussian white noise:

$$x(t) = \sum_{k=-\infty}^{\infty} x_k \operatorname{sinc}[2B(t - \tfrac{1}{2}k/B)] \tag{9.78}$$

where x_k are samples of Gaussian variables, statistically independent, with zero mean value and variance σ_x^2.

Signal $x(t)$ is obviously a Gaussian process, since it is the weighted sum of Gaussian processes (see problem 5.11.34).

Moreover, the correlation function is given by

$$\begin{aligned} R_x(\tau) &= \mathrm{E}[\mathbf{x}(t)\mathbf{x}(t + \tau)] \qquad (9.79) \\ &= \sum_{k=-\infty}^{\infty} \sum_{l=-\infty}^{\infty} \mathrm{E}[\mathbf{x}_k\mathbf{x}_l]\operatorname{sinc}[2B(t - \tfrac{1}{2}k/B)]\operatorname{sinc}[2B(t + \tau - \tfrac{1}{2}l/B)] \\ &= \sigma_x^2 \operatorname{sinc}(2B\tau) \end{aligned}$$

thus,

$$\Phi_x(f) = \frac{\sigma_x^2}{2B} \operatorname{rect}(\tfrac{1}{2}f/B) \tag{9.80}$$

with $\sigma_x^2/B = \eta$.

This result justifies the assumption of a band-limited stationary Gaussian white noise with a maximum frequency $f_{max} = B$.

9.4.9 Sample data system

A complete sample data system requires the cascade of basic operators shown in fig. 9.42.

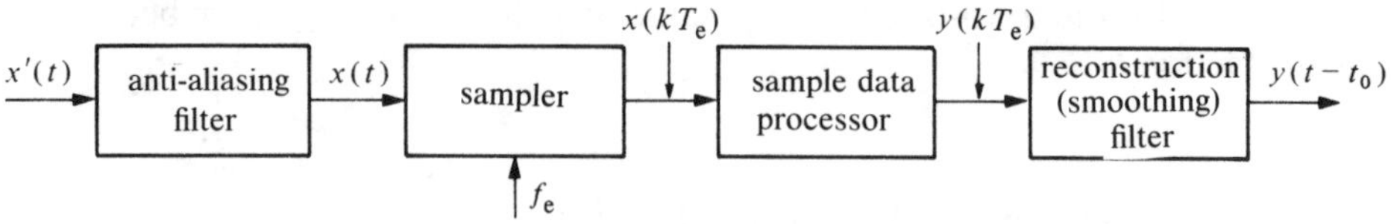

Fig. 9.42

9.5 PROBLEMS

9.5.1 A signal $x(t) = \exp(-a|t|)$ with $a = 2$ Hz flows through the system shown in figure 9.43

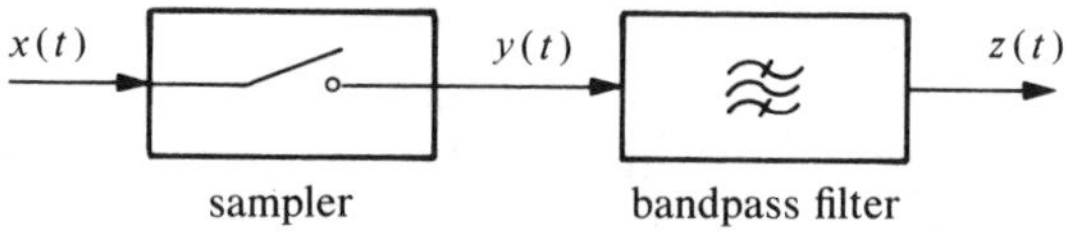

Fig. 9.43

Sampling is implemented by a switch operated at a $f_e = 20$ Hz rate with an on-time of $T_f = 20$ ms. The ideal bandpass filter has unity gain and a zero phase response in bandwidth $B = f_{max} - f_{min} = 10$ Hz centered on $f_e = 20$ Hz. The attenuation outside this bandwidth is total. Plot $y(t)$ and its amplitude spectrum $|Y(f)|$; plot $z(t)$ and give its approximate mathematical expression.

9.5.2 A signal $x(t)$, having the power spectral density $\Phi_x(f)$ shown in fig. 9.44, is sampled at a rate f_e. The (ideal) sampled signal has a spectral density $\Phi_{xei}(f)$, also shown in fig. 9.44. Find the sampling rate used and tell whether this choice is advisable to allow for signal reconstruction with an ideal filter (***justify*** your answer).

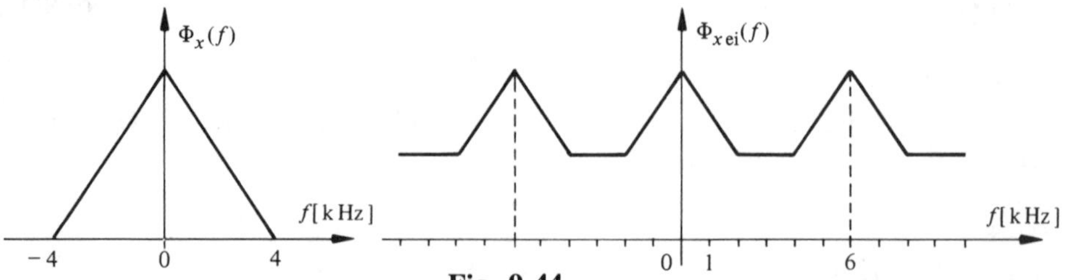

Fig. 9.44

9.5.3 Can we accurately reconstruct signal $x(t) = B\,\text{sinc}^2(t/\Delta t)$ if it is sampled at a rate $f_e = 1/\Delta t$?

9.5.4 Consider the processing system shown in fig. 9.45 with $X(f) = D\{\text{rect}[\tfrac{1}{2}(f + f_0)/B] + \text{rect}[\tfrac{1}{2}(f - f_0)/B]\}$, $G_1(f) = \text{rect}[f/(4B)]$ and $f_0 = 3B$.

- Find the minimum sampling rate f_e of signal $y(t)$ allowing its reconstruction without distortion by an ideal lowpass filter $G_2(f)$ with an appropriate bandwidth.

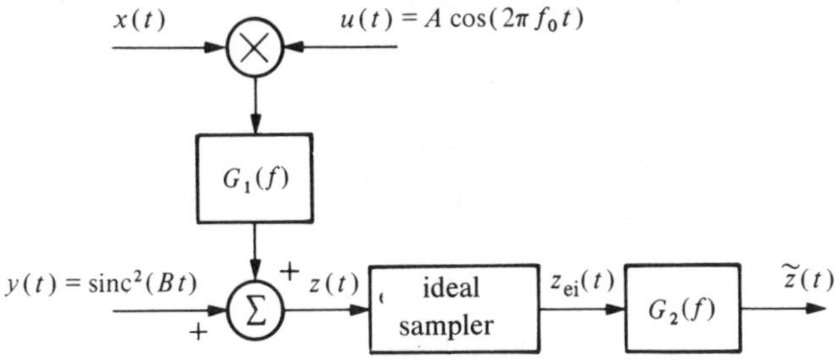

Fig. 9.45

- What sampling rate would you choose if filter $G_2(f)$ has the transfer function shown in fig. 9.46, assuming you want to recover $z(t)$ without distortion?
- Plot the amplitude spectrum at the ideal sampler output for the two cases given above.

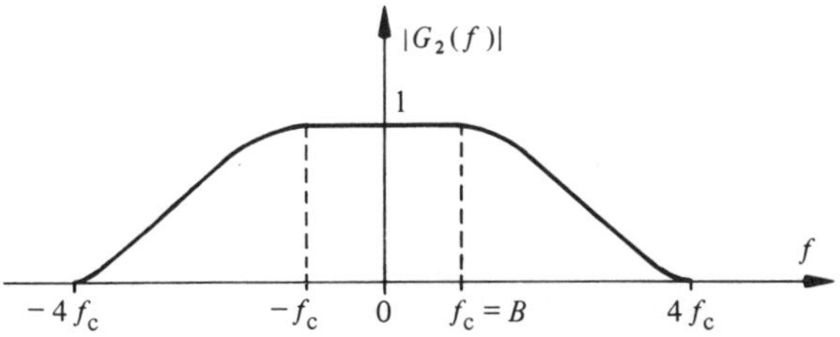

Fig. 9.46

9.5.5 Find the result (9.32) in which the aliasing error power P_r is defined as the power of signal components with frequencies higher than $f_e/2$, knowing that $\int_0^\infty (1 + x^m)^{-1} dx = \pi[m \sin(\pi/m)]^{-1}$.

9.5.6 A signal $x(t)$ with a uniform spectrum $\Phi_x(f) = \eta/2$ is filtered by a Butterworth filter and then sampled at a rate f_e. If f_c is the filter −3 dB cut-off frequency, find what the f_e/f_c ratio should be so that the aliasing error power-to-signal total power ratio $\xi_{r\,x\,dB}$ is −30 dB for $n = 1, 2, 3, 4$.

9.5.7 Demonstrate the result shown in fig. 9.20.

9.5.8 Find (neglecting insignificant terms) the power spectral density of the signal $y(t)$ from the schematic of fig. 9.47, knowing that

$$x(t) = \sum_i A \cos(2\pi f_i t)$$

for $i = 1, 2, 3$, and that $g(t) = \text{tri}[(t - \Delta)/\Delta]$ with $f_e = 1/T_e = 3$ kHz, $f_1 = 0.3$ kHz, $f_2 = 0.5$ kHz, $f_3 = 1.5$ kHz, $\Delta = 1$ ms, and $A = 10$ V.

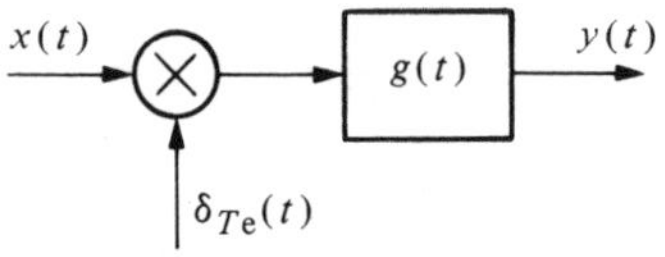

Fig. 9.47

9.5.9 A Gaussian random signal $n(t)$ has an autocorrelation function $R_n(\tau) = 4 + 16 \operatorname{sinc}^2(\tau/T)$ V^2, with $T = 1$ ms. Find the average value, the variance, and the power spectral density of this signal. What is the minimum sampling rate at which this signal can be ideally sampled and perfectly recovered? At which sampling rate do we get statistically independent samples?

Chapter 10

Signal Digitization

10.1 ANALOG-TO-DIGITAL AND DIGITAL-TO-ANALOG CONVERSION

10.1.1 Principle and definitions

Digital information processing systems deal with numbers. Any signal processing system based on a computer or a special-purpose digital processor thus requires a preliminary operation of *analog-to-digital conversion* (A/D). When the processed information must be restored in analog form, we proceed to the reverse *digital-to-analog conversion* (D/A) operation. The basic block diagram of a digital processing system of analog signals is represented in fig. 10.1.

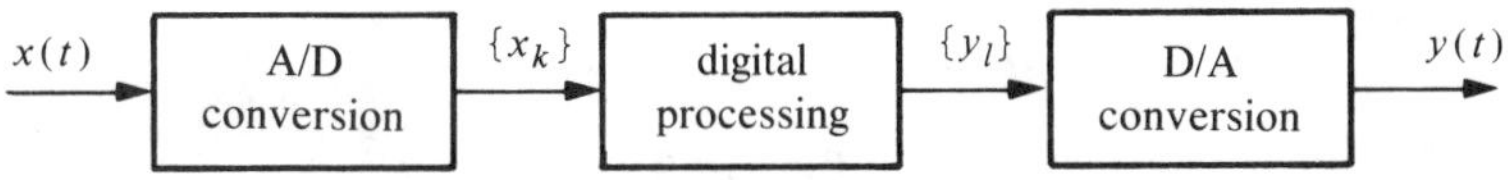

Fig. 10.1

Analog-to-digital conversion (fig. 10.2) transforms an analog input signal $x(t)$ into a sequence of numbers $\{x_k\}$, usually binary encoded. Each number corresponds to the amplitude $x(t_k)$ of a signal sample taken at a given time t_k. Generally, this sampling is made at regular time intervals T_e (chap. 9).

Since the determination of the digital value corresponding to the amplitude of a sample takes some time, it is often necessary to store the analog value between two successive samples.

Each sample can take an infinite number of values because of the analog nature of the signal. However, the accuracy with which these amplitudes must and can be known is necessarily limited by all kinds of practical considerations. We are allowed to replace the exact value of the sample by the nearest approximate value from a finite set of discrete values: this is called *quantization*. Each discrete value is denoted by a binary number in appropriate *coding*. This number is bound by two limiting values setting the *conversion range*. Each number x_k then represents a set of analog values belonging to an interval Δ_k called the *quantization step*. When the conversion range is split up into equal quantization steps, we speak of *uniform quantization* (sub-sect. 10.3.34).

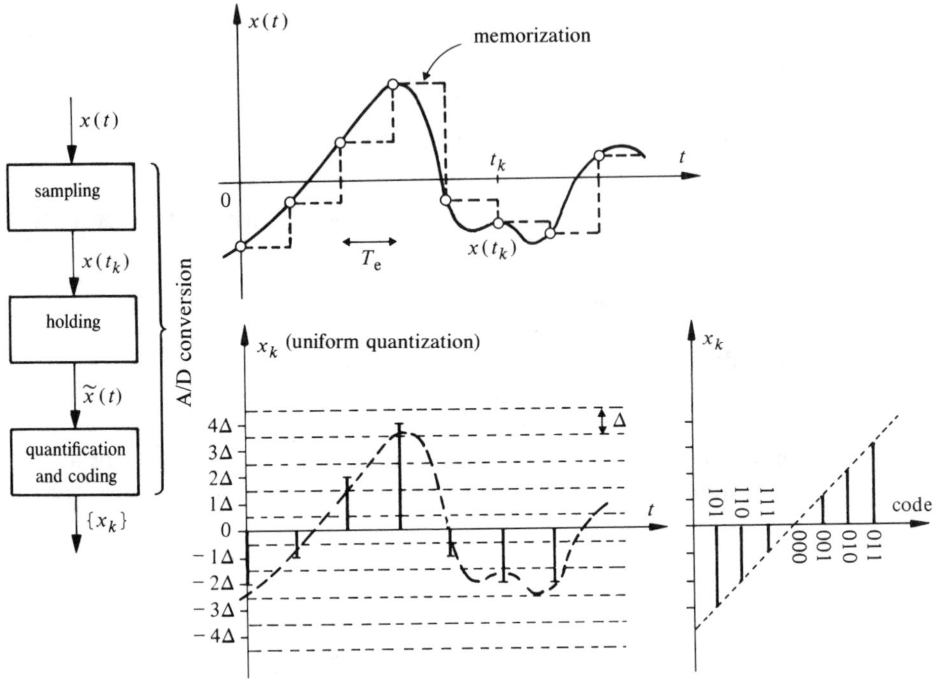

Fig. 10.2

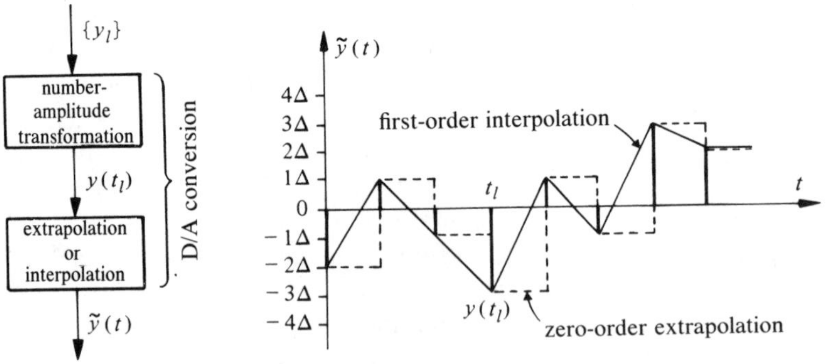

Fig. 10.3

The digital-to-analog conversion (fig. 10.3) is somewhat more straightforward. The sequence of numbers $\{y_l\}$ is transformed into a sequence of discrete amplitude samples $y(t_l)$.

The final reconstruction of the output analog signal $\tilde{y}(t)$ is then achieved by an extrapolation or interpolation operation between the samples (as indicated in sub-sect. 9.4.4). As mentioned, a simple zero-order extrapolation is often used. It is easily implemented by the digital memorization of each number at the D/A converter input. Analog post-filtering (smoothing) is generally added.

The standard graphic symbols of A/D and D/A converters are shown on fig. 10.4.

Fig. 10.4

10.1.2 Principal A/D conversion methods

There are *direct* and *indirect conversion* methods. A detailed study of these is beyond the scope of this book. The reader should refer himself or herself, for more details, to the specialized literature [95, 110, 111].

With indirect methods, the value of the input signal voltage is initially converted into a frequency by a *voltage-controlled oscillator* (or VCO), or into a proportional duration obtained through integration. In the first case, the corresponding digital value is defined by counting the number of periods of the oscillator signal during a prescribed time interval. In the second case, this digital value is given by counting the number of periods from an auxiliary clock between the beginning and the end of the integration.

Indirect methods lend themselves to the implementation of high accuracy A/D converters. They are, on the other hand, necessarily slow because of their serial nature and not suitable for usual digital signal processing applications. These methods are used in precision instruments (e.g., digital voltmeters, *et cetera*).

The direct A/D converter (the simplest in principle, but not in its technological implementation) is the *parallel (flash) converter* (fig. 10.5).

The input voltage is simultaneously compared to $2^n - 1$ values of the form:

$$\frac{k}{2^n} U_0 \text{ with } k = 1, 2, \ldots, 2^n - 1 \tag{10.1}$$

deduced from a reference voltage U_0 by a voltage-divider (2^n resistor-ladder). The state of the $2^n - 1$ comparator outputs (binary variables) is finally translated into a binary word of n bits (or binary digits, conventionally noted 0 and 1) by a logic decoder.

The advantage of flash converters is their speed, since the output binary word is determined almost instantaneously. The sampling then can be made after the conversion by periodically picking up digital values. However, the

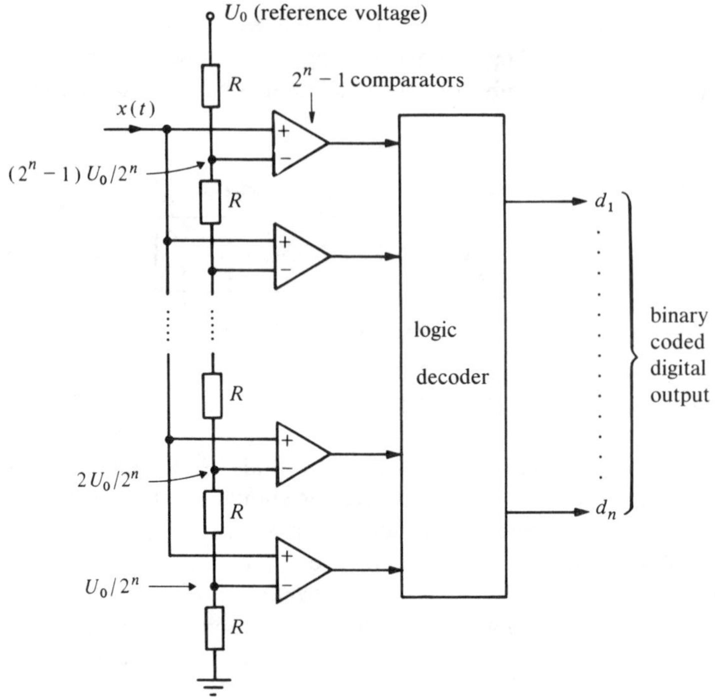

Fig. 10.5

number of comparators to be implemented limits the resolution to about 256 discrete levels ($n = 8$ bits). This type of converter allows the analog-to-digital conversion of high frequency signals (video, radar), which do not need too high an accuracy.

For reasons of complexity—hence, cost—and accuracy, *successive approximation converters* (fig. 10.6) are preferred when the required sampling rate is not too high. The input voltage here is successively compared to a convergent combination of the $2^n - 1$ weighted reference values $kU_0/2^n$. This is a closed-loop system (feedback) built around a digital-to-analog converter (sub-sect. 10.1.3).

Under internal clock control, the device initially determines the first bit d_1 by deciding whether the $x(t)$ analog input is higher ($d_1 = 1$) or lower ($d_1 = 0$) than half the U_0 reference voltage. During the following clock period, we determine whether $x(t) - d_1 \cdot U_0/2$ is higher ($d_2 = 1$) or lower ($d_2 = 0$) than $U_0/4$. So goes the procedure n times in an n-bit converter. The final binary word is then available at the output only after n internal clock periods. This type of converter allows higher accuracies (the maximum achievable value of

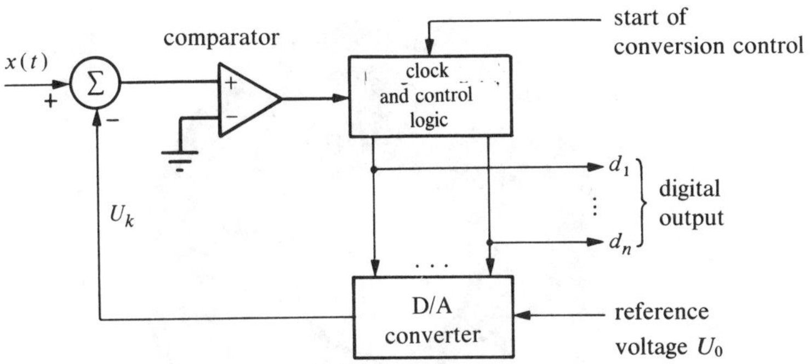

Fig. 10.6

n is on the order of 16) than with flash converters. On the other hand, they are at least n times slower; a new conversion being allowed only after the previous one is fully completed.

Hybrid structures can be designed for a good compromise between speed and accuracy.

Let us mention that besides the purely electronic converters previously described, electromechanical devices such as the angular encoding disk (shaft encoder) shown on fig. 10.7 are used in some applications.

The reading, in digital form, of the angular disk position is made with photoelectric detectors, one per track, located along a line passing through the disk rotation center. Each track is split into 2^i alternate transparent and opaque segments, with $i = 1$ (outer track in our example giving the d_1 bit) until n (inner track giving the d_n bit). The disk of fig. 10.7 has ten tracks: its angular resolution is then $360/2^{10} = 0.35$ degree.

If most converters divide the conversion range of the variable to be digitized into $q = 2^n$ equal intervals (uniform quantization), there are applications where it is more suitable to use nonuniform quantization [112]. The most classic example is conversion with logarithmic companding, which is used in the digital transmission of telephone signals [pulse coded modulation (vol. XVIII)].

10.1.3 D/A conversion

A digital-to-analog converter is a device producing a quantized output y that can take $q = 2^n$ different values. It is said to be uniform when these values are regularly distributed over a range of zero to $2^n \cdot \Delta$ according to

$$y = \Delta(d_1 2^{n-1} + \ldots + d_n 2^0) \tag{10.2}$$

where Δ is the quantization step. This output is generally either current or

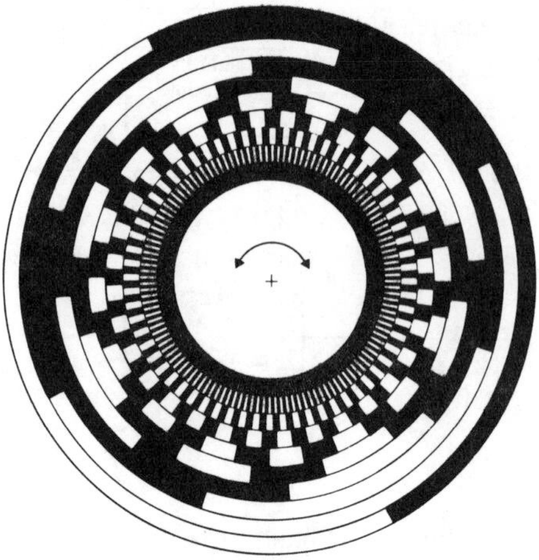

Fig. 10.7

voltage. The usual implementation mode is based on the principle illustrated in fig. 10.8.

Switches driven by the d_k binary variables ($d_k = 0$: open switch ; $d_k = 1$: closed switch) control the $I_0/2^k$ weighted currents, flowing from current sources dependent on a reference I_0, to a summation node (Kirchhoff). The resulting current corresponds to (10.2). A current-voltage conversion is then generally implemented, either with a simple resistor (limited amplitude range in practice), or an operational amplifier. The current sources and the drive switches can be implemented with various electronic techniques. For more information, the reader should refer himself or herself to the specialized literature [96].

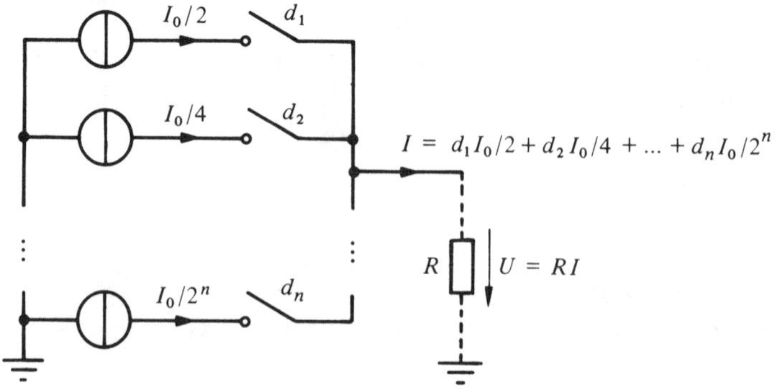

Fig. 10.8

D/A converters have a parallel structure. The $I = d_1 I_0/2 + \ldots + d_n I_0/2^n$ analog output corresponds almost immediately to the simultaneous application of the $d_1 \ldots d_n$ bits to the converter input. The current-voltage conversion imposes some speed limitations (capacitive loads, limited slew-rate of operational amplifiers).

In practice, all the switches do not react at exactly the same time. This yields glitches that must be eliminated by filtering or re-sampling with delayed hold (sample-and-hold: sub-sect. 10.2.4).

10.2 MAXIMAL A/D CONVERSION RATES

10.2.1 Signal variation during the conversion duration

A/D conversion is not an instantaneous operation. It can be relatively fast in flash converters and clearly slower in successive approximation converters.

Let V be the available conversion range. If the quantization law is uniform, this range is divided into $q = 2^n$ intervals of constant width $\Delta = V/2^n$. The conversion procedure gives the value $x_k = k\Delta$ to each sample with amplitude between $k\Delta - \frac{1}{2}\Delta$ and $k\Delta + \frac{1}{2}\Delta$. If the signal varies during the *conversion duration* τ_c, the digital result obtained can be wrong (fig. 10.9). In first approximation, the variation Δx of the analog signal during the duration τ_c is

$$\Delta x \approx \tau_c \left| \frac{dx}{dt} \right| \tag{10.3}$$

A usual accuracy requirement is to limit the variation of Δx to $\frac{1}{2}\Delta$:

$$\Delta x \leqslant \tfrac{1}{2}\Delta \tag{10.4}$$

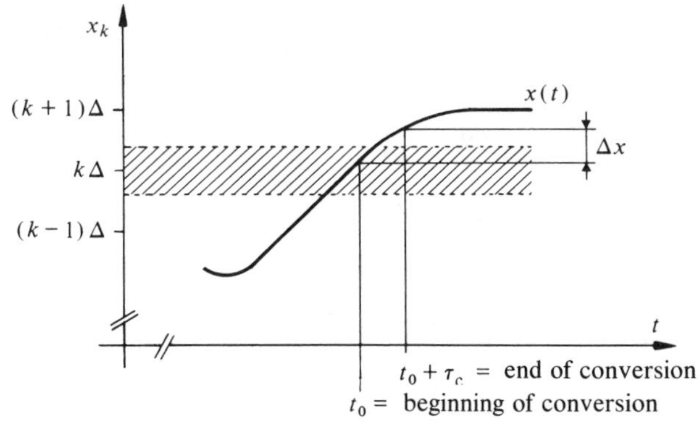

Fig. 10.9

The global error introduced is thus less than or equal to Δ in absolute value. By combining (10.3) and (10.4), we obtain

$$\left|\frac{\mathrm{d}x}{\mathrm{d}t}\right|_{\max} = \frac{\Delta}{2\tau_c} = \frac{V}{2^{n+1} \cdot \tau_c} \tag{10.5}$$

10.2.2 Example: sinewave signal

If we consider a sine wave signal, the peak-to-peak amplitude of which corresponds to the V conversion range of the converter, its maximum slope is

$$\left|\frac{\mathrm{d}x}{\mathrm{d}t}\right|_{\max} = \frac{\mathrm{d}}{\mathrm{d}t}\left[\frac{V}{2}\sin(2\pi ft)\right]_{t=0} = \pi fV \tag{10.6}$$

By combining (10.5) and (10.6), we obtain

$$f_{\max} = \frac{1}{2^{n+1} \cdot \pi\tau_c} \tag{10.7}$$

For example, for a 1 μs conversion by a $n = 8$ bit converter, the acceptable maximum frequency of the sine wave with peak-to-peak value equal to the converter range is only 622 Hz! It is four times weaker for $n = 10$ and ten times weaker for $\tau_c = 10$ μs.

The abacus of fig. 10.10 illustrates relation (10.7)

10.2.3 Example: Gaussian random signal

In the case of a random signal, the slope is also a random variable, and we must think in terms of probability of exceeding a fixed limit.

Consider a Gaussian signal $x(t)$ with a band-limited white spectrum $\Phi_x(f) = \frac{1}{2}\eta\ \mathrm{rect}(f/2B)$ and variance $\sigma_x^2 = \eta B$. The probability of having $|x| > 3\sigma_x$ is smaller than 3‰, so we can choose the conversion range $V = 6\ \sigma_x$ for the converter.

By (8.71), the derivative $\dot{x} = \mathrm{d}x/\mathrm{d}t$ has a spectrum $(2\pi f)^2\Phi_x(f)$; hence, a variance:

$$\sigma_{\dot{x}}^2 = 4\pi^2\eta \int_0^B f^2\mathrm{d}f = 4\pi^2\eta B^3/3 = 4\pi^2B^2\sigma_x^2/3 \tag{10.8}$$

Moreover, $\dot{x} = \mathrm{d}x/\mathrm{d}t$ also has a Gaussian distribution, since the differentiation is a linear operation. Then, the slope exceeds, in absolute value, $3\sigma_{\dot{x}}$ only with a probability lower than 3‰. Choosing this value as the maximum slope

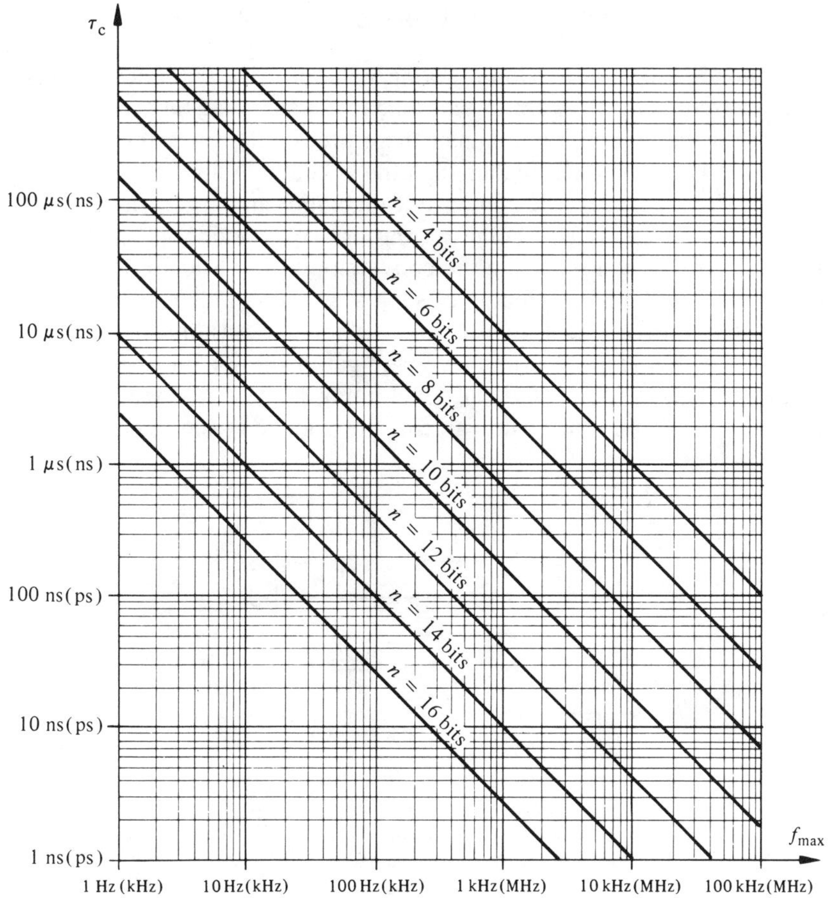

Fig. 10.10

that does not introduce any conversion error, we have

$$\left|\frac{dx}{dt}\right|_{max} = 3\sigma_{\dot{x}} = \frac{6\pi}{\sqrt{3}} B\sigma_x = \frac{V\pi B}{\sqrt{3}} \tag{10.9}$$

Equating this result to (10.5) yields the condition

$$B_{max} = \frac{\sqrt{3}\Delta}{2\pi V \tau_c} = 1.73 \frac{1}{2^{n+1}\pi\tau_c} \tag{10.10}$$

Results (10.7) and (10.10) have then the same order of magnitude. The abacus of fig. 10.10 seems therefore a good indicator of the signal maximum frequency that a converter having a conversion duration τ_c can convert, without other precaution.

10.2.4 Temporary analog storage by sample and hold

A converter with conversion duration τ_c is, in principle, able to work with a sampling frequency $f_e \approx 1/\tau_c$. According to the sampling theorem, the highest frequency of the analog input signal, therefore, could reach the value $B_{max} = 1/(2\tau_c)$. This is a value $2^n\pi$ higher than the acceptable limit frequencies (10.7) or (10.10). Consequently, the converter capabilities are largely underemployed.

This disadvantage is overcome by the use of a device which temporarily stores each sample value during the conversion cycle. Such a device is called *sample and hold*. The functioning of such a circuit is illustrated by fig. 10.11.

When the control signal level is low, switch I is on and the output signal $x_m(t)$ of the sense amplifier (unity voltage gain) follows the input signal $x(t)$. The storage strobe (control signal level going high at t_m) opens the switch. The capacitor C then stores the value of the voltage at the opening (the input

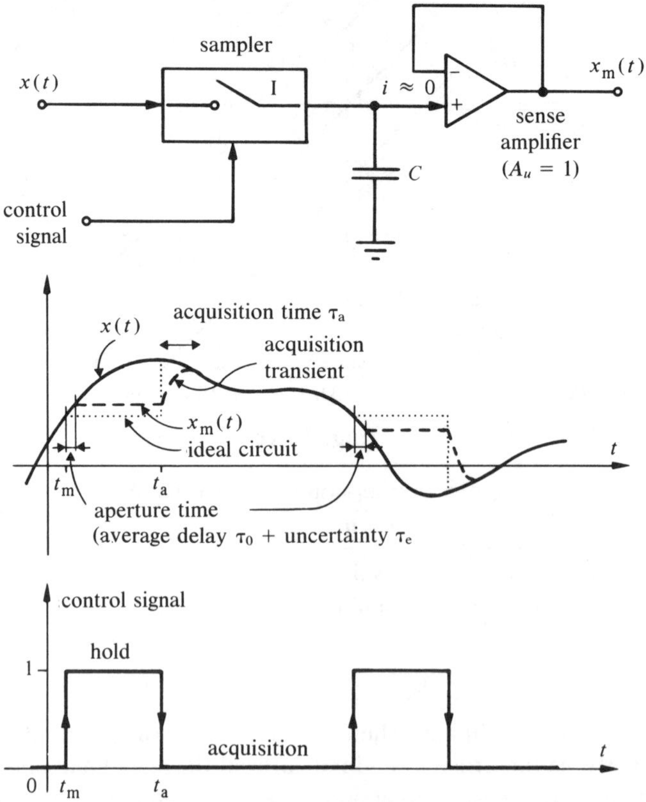

Fig. 10.11

current of the sense amplifier is very weak). This value, available at the sense amplifier output, can be converted without risk of variation by a downstream converter (fig. 10.12).

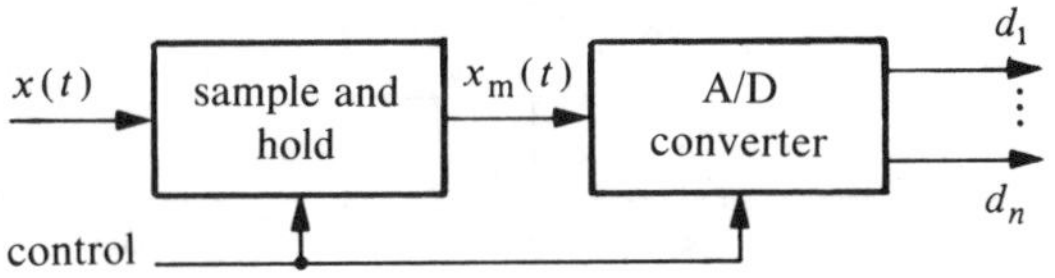

Fig. 10.12

There is, however, a certain delay between the aperture order instant t_m and the effective storage time. This delay, called *aperture time,* is divided into an average delay τ_0 and a variable delay τ_e that leads to some uncertainty concerning the exact sampling instant. This uncertainty is called *aperture uncertainty* or *jitter.* It plays, for the amplitude error of the sample and hold, the same role as the conversion duration τ_c for the converter, according to (10.5).

The maximum sampling rate of a sample and hold depends particularly on the acquisition time τ_a, which is the interval separating the hold end (instant of application of an acquisition enabling command, t_a) and the instant at which the $x_m(t)$ output has reached the $x(t)$ input. This duration is tied to the circuit transient response and depends on the storage capacitor value.

For example, if the $n = 8$ bit converter with a $\tau_c = 1$ μs conversion time of example 10.2.2 is preceded by a sample and hold having an aperture uncertainty lower than 1 ns and an acquisition time $\tau_a = 1$ μs, the sampling rate could be 500 kHz, allowing the sampling of a signal with maximum frequency around 250 kHz.

10.3 QUANTIZATION

10.3.1 General principle

Quantization is a mapping of the infinitely many values of the input signal $x(t)$ in a finite number of values assigned to the output signal $x_q(t)$. The corresponding operator is of an amnesic nonlinear type (sub-sect. 8.4.2).

The mapping (fig. 10.13) is obtained by dividing the total conversion range V of the input signal variations into q adjacent intervals Δ_i with $i = 1, \ldots, q$ and by assigning the value x_i to the output signal when the input signal amplitude belongs to the interval Δ_i. All the input values belonging to the same interval are, therefore, represented by the same quantized level, which gen-

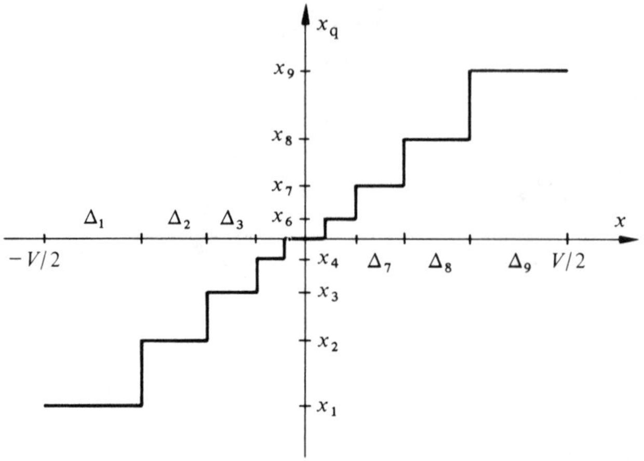

Fig. 10.13

erally corresponds to the interval median value (*rounded quantization*) or to the minimal value (*truncated quantization*).

Such a process naturally introduces a specific distortion, which depends as much as on the signal nature as on the adopted quantization law.

10.3.2 Quantization distortion or quantization noise

The difference

$$n_q(t) = x_q(t) - x(t) \tag{10.11}$$

is called *quantization distortion* or *quantization noise*. This distortion depends on the signal amplitude and can be considered as the output of a nonlinear operator stimulated by the signal $x(t)$. Figure 10.14 illustrates the distortion characteristic generated by the quantization law of fig. 10.13.

When the signal $x(t)$ is random, it is convenient to consider the quantized signal as resulting from the sum of $x(t)$ and of a random quantization noise (fig. 10.15). In many situations, the quantization noise is practically uncorrelated with the input signal.

The quantization noise appears not only during the signal analog-to-digital conversion, but also in the subsequent computation operations (rounding or truncation error).

10.3.3 Quantization noise variance

The quantization noise n_q, being considered as a random process, has a mean value μ_q and a variance σ_q^2. A value $\mu_q \neq 0$ indicates the presence of a

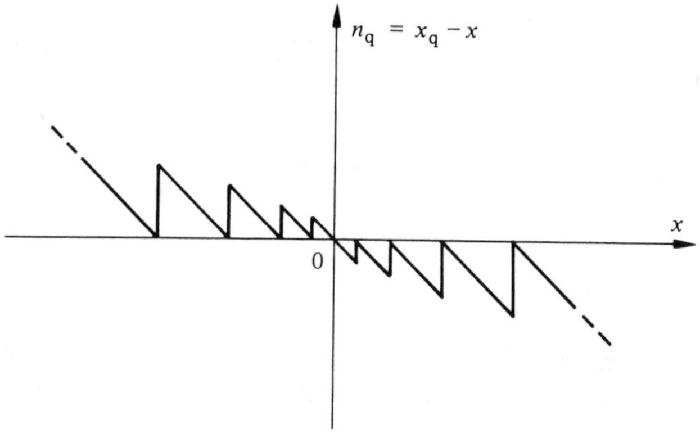

Fig. 10.14

systematic bias. This situation occurs, in particular, in the case of truncated quantization. This bias is equal to zero when two conditions are met: the input signal has an even probability density with respect to the mean value μ_x and the quantization law has an odd symmetry with respect to μ_x. The variance is identical to the fluctuation power (tab. 5.2):

$$\sigma_q^2 = \sum_{i=1}^{q} \int_{\Delta_i} (x - x_i)^2 p(x) \mathrm{d}x \tag{10.12}$$

If the number of intervals q is high, a reasonable approximation can be made by considering $p(x) \approx p(x_i) =$ constant over interval Δ_i:

$$\sigma_q^2 \approx \sum_{i=1}^{q} p(x_i) \int_{\Delta_i} (x - x_i)^2 \mathrm{d}x \tag{10.13}$$

Moreover, if the quantized values x_i correspond to the median value of interval Δ_i (rounded quantization), the quantization noise variance becomes simply

$$\sigma_q^2 \approx \sum_{i=1}^{q} p(x_i) \cdot \int_{x_i-\Delta_i/2}^{x_i+\Delta_i/2} (x - x_i)^2 \mathrm{d}x$$

$$= \sum_{i=1}^{q} p(x_i)\Delta_i^3/12 \tag{10.14}$$

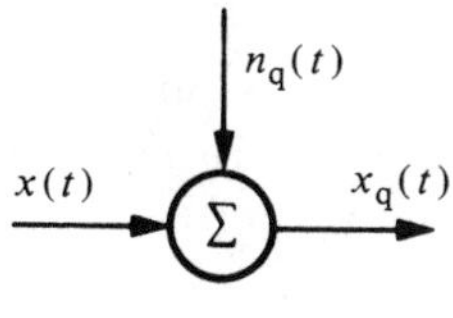

Fig. 10.15

When the probability density $p(x)$ of the input signal is known, it is possible to look for the optimal quantization law [112] which minimizes σ_q^2 for a given number of quantization levels.

In some cases, it is useful to maintain constant the ratio of the input signal power to the quantization noise power. This is the approach used in digital telephony (PCM), which leads to a logarithmic quantization law (vol. XVIII).

10.3.4 Uniform quantization

The most frequently used quantization law is the *uniform law* (sometimes improperly called *linear*) in which the quantization steps Δ_i are constant (fig. 10.16). Thus,

$$\Delta_i = \Delta \qquad \forall i \tag{10.15}$$

Since, according to (14.14), $\Sigma p(x_i)\Delta = 1$, we get by inserting (10.15) in (10.14):

$$\sigma_q^2 \approx \frac{\Delta^2}{12} \tag{10.16}$$

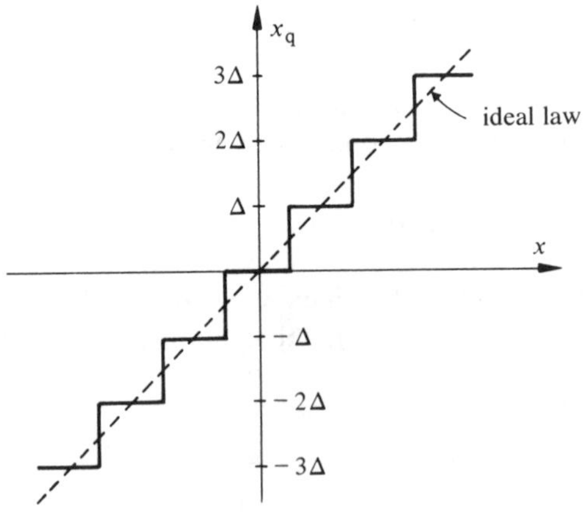

Fig. 10.16

This widely used approximation is, in fact, correct when the quantization noise, distributed between $\pm\Delta/2$ (fig. 10.17), has a uniform probability density. This situation is exactly or almost perfectly realized in many practical situations, as the analysis of sub-sect. 10.3.10 shows.

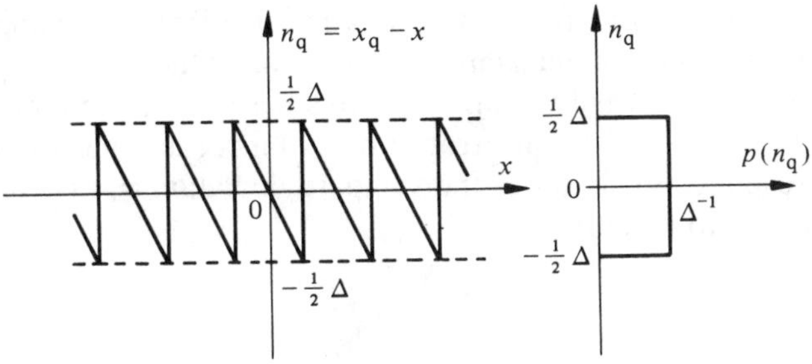

Fig. 10.17

The *signal-to-quantization noise ratio* is defined by

$$\xi_q = \sigma_x^2/\sigma_q^2 \tag{10.17}$$

where the standard deviation σ_x is identical to (tab. 5.2) the rms value of the input signal $x(t)$. Inserting (10.16), we get

$$\xi_q \approx 12(\sigma_x/\Delta)^2 \tag{10.18}$$

and

$$\xi_{qdB} = 10 \log_{10} \xi_q \approx 20 \log_{10}(\sigma_x/\Delta) + 10.8 \qquad \text{dB} \tag{10.19}$$

If the useful range V of the input signal is divided into $q = 2^n$ intervals of width Δ, relation (10.19) becomes, with $V/\sigma_x = a$:

$$\xi_{qdB} \approx 6n + 10.8 - 20 \log_{10} a \qquad \text{dB} \tag{10.20}$$

Thus, ***for an n-bit analog-to-digital converter, the signal to quantization noise ratio, measured in decibels, varies linearly with n with a 6 dB increase for each additional bit.*** In this analysis, we have disregarded the clipping distortion that occurs if the amplitude range of the input signal is larger than the conversion range V.

10.3.5 Examples

Consider first an input signal with ***uniform distribution*** $p(x) = V\ \text{rect}(x/V)$, the variance of which is equal to $\sigma_x^2 = V^2/12$. By (10.18), $\xi_q = (V/\Delta)^2 = q^2 = 2^{2n}$ and

$$\xi_{qdB} \approx 6n \qquad \text{dB} \tag{10.21}$$

This result is often mentioned without specifying the particular conditions under which it is obtained!

In practice, uniformly distributed signals are uncommon. An approximately Gaussian distribution is not unusual. An estimation of the signal to quantization noise ratio can also be easily obtained in this case assuming a quasi-uniform distribution of the quantization error. For instance, assume the conversion range V_1 corresponds to six times the standard deviation σ_x (fig. 10.18). Then, we get, according to (10.20):

$$\xi_{qdB} \approx 6n - 4.76 \qquad \text{dB} \tag{10.22}$$

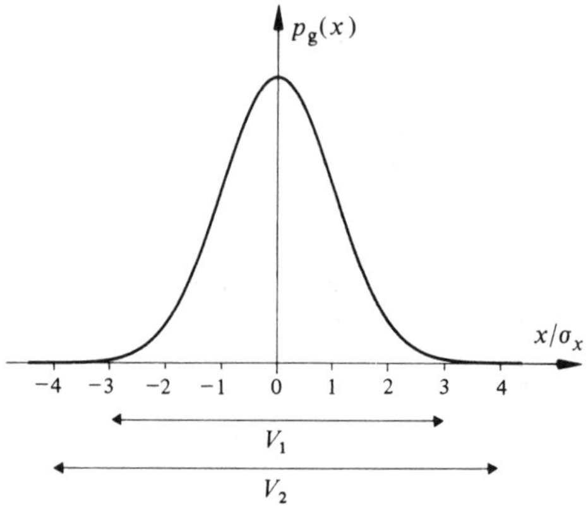

Fig. 10.18

If the conversion range is $V_2 = 8\sigma_x$, we have

$$\xi_{qdB} \approx 6n - 7.27 \qquad \text{dB} \tag{10.23}$$

In the case of a sine wave with a peak-to-peak amplitude V, we have $\sigma_x = V/(2\sqrt{2})$; thus,

$$\xi_{qdB} \approx 6n + 1.77 \qquad \text{dB} \tag{10.24}$$

These results are summarized in fig. 10.19.

A 16-level quantization ($n = 4$ bits) is sufficient for industrial television. Values of $n = 8$, 10, or 12 bits ($q = 256$, 1024, or 4096) are commonplace in signal processing and image processing and in data acquisition in general. The higher values ($n = 14$ to 16, or even 18) are reserved for very particular applications (high accuracy measurements, high fidelity acoustical signals).

10.3.6 Quantized signal distribution

In the case of a rounded uniform quantization for which the number of levels q is, for simplification, considered to be unbounded, the probability

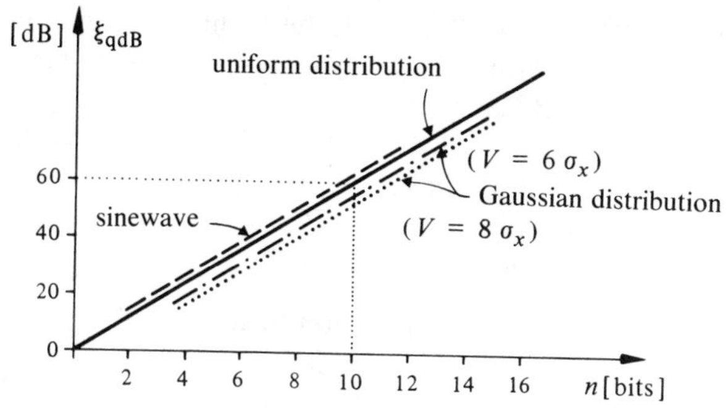

Fig. 10.19

density of the quantized signal is the discrete law (fig. 10.20):

$$p(x_q) = \sum_{k=-\infty}^{\infty} p_k \delta(x_q - k\Delta) \tag{10.25}$$

where (fig. 10.21)

$$p_k = \int_{(k-1/2)\Delta}^{(k+1/2)\Delta} p(x)\mathrm{d}x = \int_{-\infty}^{\infty} p(x) \cdot \mathrm{rect}[(x - k\Delta)/\Delta]\mathrm{d}x \tag{10.26}$$

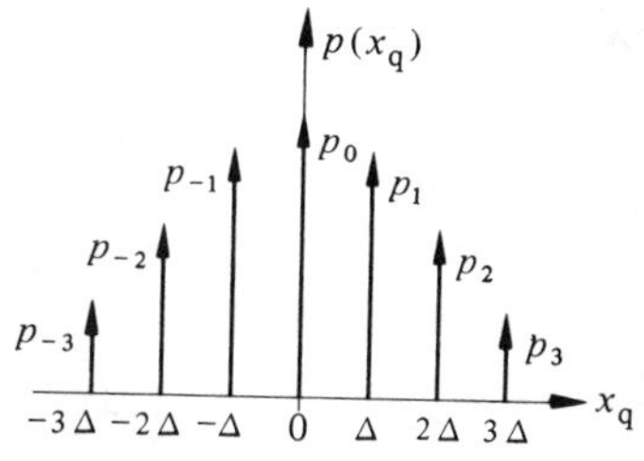

Fig. 10.20

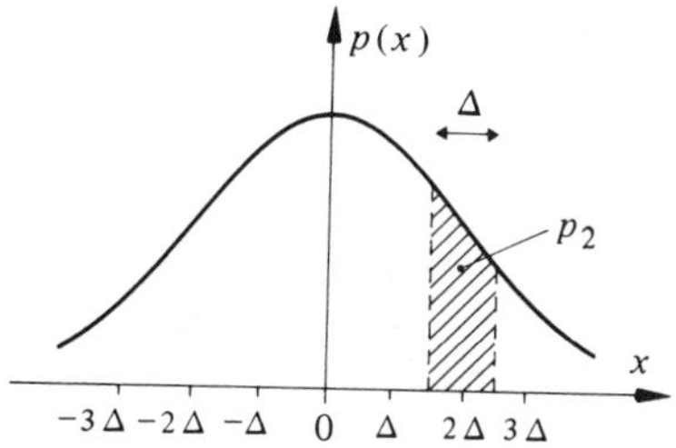

Fig. 10.21

Probabilities p_k are, therefore, equal to the samples in $x = k\Delta$ of a function:

$$\begin{aligned} w(x) &= \int_{-\infty}^{\infty} p(x')\mathrm{rect}[(x - x')/\Delta]\mathrm{d}x' \\ &= p(x) * \mathrm{rect}(x/\Delta) \\ &= \mathrm{Prob}(x - \Delta/2 \leqslant \mathbf{x} \leqslant x + \Delta/2) \end{aligned} \tag{10.27}$$

Results established in chapter 9 with regard to the ideal sampling of a time signal can, therefore, be used in the quantization case. ***The quantization is equivalent to an ideal sampling of a probability law*** $w(x)$.

The characteristic function $\Pi_{xq}(\upsilon)$ of the quantized signal is, by (14.64), the (inverse) Fourier transform of the probability density $p(x_q)$. According to (9.12), (10.25), and (10.26), we have

$$\Pi_{xq}(\upsilon) = \sum_{n=-\infty}^{\infty} \Pi_x(\upsilon - n/\Delta)\text{sinc}(\Delta\upsilon - n)$$
$$= \text{rep}_{\Delta^{-1}}\{\Pi_x(\upsilon)\text{sinc}(\Delta\upsilon)\} \quad (10.28)$$

where $\Pi_x(\upsilon)$ is the characteristic function of the input signal $x(t)$.

Function $\Pi_x(\upsilon)$ sinc$(\Delta\upsilon)$ is, in particular, equal to zero for all the values $\upsilon = k/\Delta$, where k is an integer. Moreover, if the amplitude range of the input signal $x(t)$ is V, the dispersion of $\Pi_x(\upsilon)$ on the υ-axis is, according to (4.185), approximately equal to V^{-1} (fig. 10.22). Since, in practice, $\Delta \ll V$, we have approximately (fig. 10.23)

$$\Pi_{xq}(\upsilon) \approx \text{rep}_{\Delta^{-1}}\{\Pi_x(\upsilon)\} \quad (10.29)$$

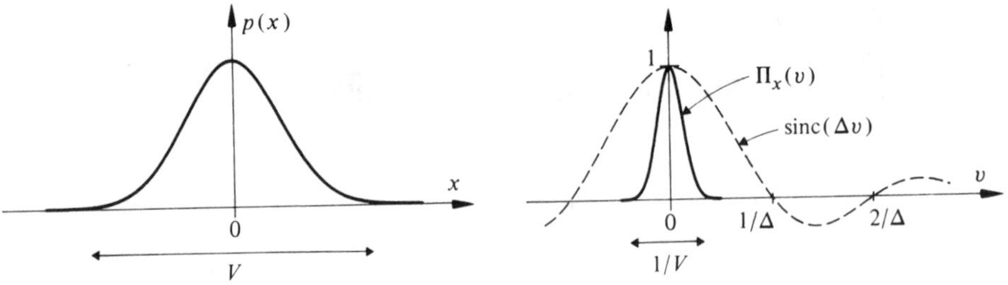

Fig. 10.22

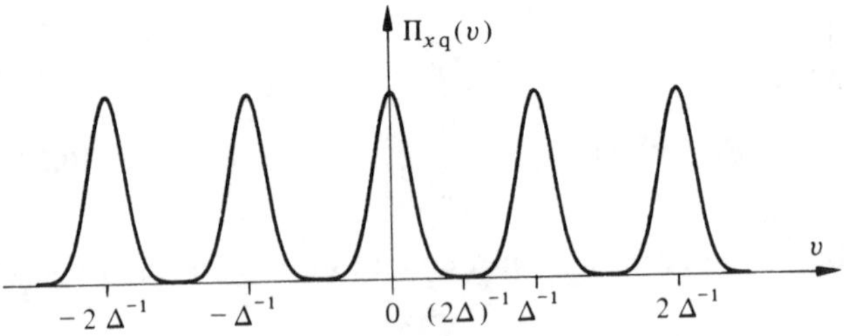

Fig. 10.23

10.3.7 Quantization theorem

By analogy with the sampling theorem of sub-section 9.3.3, we can state the following theorem:

- ***the probability density*** $p(x)$ ***of a continuous variable*** $\mathbf{x}$ ***is fully described by the probability density*** $p(x_q)$ ***of the variable*** $\mathbf{x}$ ***uniformly quantized with a step*** Δ***, if the characteristic function*** $\Pi_x(\upsilon)$ ***is such that***

$$\Pi_x(\upsilon) = 0 \qquad \text{for } |\upsilon| \geqslant (2\Delta)^{-1} \tag{10.30}$$

For the same reasons as in sub-section 9.3.2, when the signal amplitude range is bounded, the characteristic function cannot be equal to zero on a non-zero interval. It tends, however, toward zero when υ tends toward infinity so that (10.30) can be satisfied with good approximation.

10.3.8 Theorem of first-order moments

By (14.68), the moment of the first-order and of degree k : $m_{xk} = \mathrm{E}[\mathbf{x}^k]$ is proportional (factor $j - k$) to the kth derivative of characteristic function $\Pi_x(u)$ evaluated at $u = 2\pi\upsilon = 0$. This derivative can be evaluated without error from $\Pi_{xq}(\upsilon)$, if the translated functions composing it do not overlap in the vicinity of $\upsilon = 0$. This condition is, therefore, less severe than (10.30):

$$\Pi_x(\upsilon) = 0 \qquad \text{for } |\upsilon| > \Delta^{-1} - \epsilon;\ \epsilon > 0 \tag{10.31}$$

The so-called Sheppard relations [64] can then be established, which are reproduced here only for $k = 1$ and $k = 2$:

$$m_{x1} = m_{xq1} \tag{10.32}$$

$$m_{x2} = m_{xq2} - \Delta^2/12 \tag{10.33}$$

10.3.9 Extension to the second order statistics

By a similar analysis and taking into account (14.70), it can be shown that if the second-order characteristic function:

$$\Pi_{xy}(\upsilon,\nu) = 0 \qquad \text{for } |\upsilon| > \Delta^{-1} - \epsilon,\ |\nu| > \Delta^{-1} - \epsilon;\ \epsilon > 0 \tag{10.34}$$

then the second order moments $\mathrm{E}[\mathbf{x}_1^k \mathbf{y}_2^l]$, with $\mathbf{x}_1 = \mathbf{x}(t)$ and $\mathbf{y}_2 = \mathbf{y}(t + \tau)$, can be obtained directly by differentiating the characteristic function $\Pi_{xq}(\upsilon,\nu)$ at the origin $\upsilon = \nu = 0$.

Under these conditions, the following relation between the correlation functions of nonquantized and quantized signals can be derived:

$$R_{xy}(\tau) = \begin{cases} R_{xyq}(0) - \Delta^2/12 & \tau = 0 \\ R_{xyq}(\tau) & \tau \neq 0 \end{cases} \tag{10.35}$$

In the case of autocorrelation, then we have

$$R_x(\tau) = \begin{cases} R_{xq}(0) - \Delta^2/12 & \tau = 0 \\ R_{xq}(\tau) & \tau \neq 0 \end{cases} \tag{10.36}$$

This result indicates that, when condition (10.34) is satisfied, the quantization noise $n_q = x_q - x$ is a zero mean value process with variance $\Delta^2/12$ and uniform distribution, comparable to a white noise uncorrelated with $x(t)$.

10.3.10 Properties of quantization noise

When $x_q = k\Delta$, $n_q = x_q - x$ is distributed according to the conditional law $p_x(n_q - k\Delta)\,\mathrm{rect}(n_q/\Delta)$. The probability density of the rounded quantization noise, assuming a uniform quantization law with an infinite number of levels, is

$$p(n_q) = \sum_{k=-\infty}^{\infty} p_x(n_q - k\Delta)\,\mathrm{rect}(n_q/\Delta) \tag{10.37}$$

Consequently, its characteristic function is, according to (4.15), (4.17), (4.21), and (1.36):

$$\begin{aligned}\Pi_{nq}(\upsilon) &= \left[\Pi_x(\upsilon) \cdot \sum_{k=-\infty}^{\infty} \exp(\,j2\pi\upsilon k\Delta)\right] * \Delta\,\mathrm{sinc}(\Delta\cdot\upsilon) \\ &= \Pi_x(\upsilon)\cdot\delta_{\Delta^{-1}}(\upsilon) * \mathrm{sinc}(\Delta\cdot\upsilon) \\ &= \sum_{k=-\infty}^{\infty} \Pi_x(k\Delta^{-1})\,\mathrm{sinc}(\Delta\cdot\upsilon - k)\end{aligned} \tag{10.38}$$

Components of $\Pi_{nq}(\upsilon)$ for $k = 0$ and $k = 1$ are represented in fig. 10.24.

A necessary and sufficient condition [113] to have a quantization noise with a uniform distribution:

$$p(n_q) = \Delta^{-1}\,\mathrm{rect}(n_q/\Delta) \tag{10.39}$$

is that the characteristic function $\Pi_x(\upsilon)$ of the input signal is equal to zero for $\upsilon = k\Delta^{-1}$, with k being a non-zero integer. In this case, we have, indeed, with $\Pi_x(0) = \int p(x)\mathrm{d}x = 1$:

$$\Pi_{nq}(\upsilon) = \mathrm{sinc}(\Delta\cdot\upsilon) = \mathrm{F}^{-1}\{p(n_q)\} \tag{10.40}$$

This condition is, of course, also satisfied when condition (10.31) is valid.

Therefore, it appears clearly that ***the quantization noise can be considered to be uniformly distributed even in coarse quantization conditions.*** Its mean value is equal to zero, in the case of rounded quantization, and its variance $\sigma_q^2 = \Delta^2/12$ is precisely equal to the right-hand part of (10.16). By a bidimensional

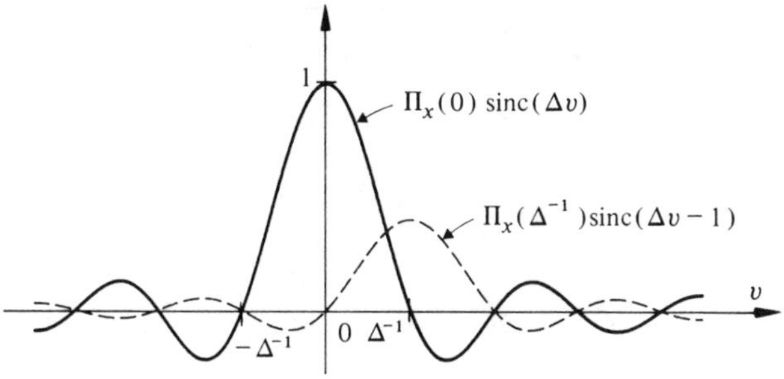

Fig. 10.24

generalization, it can be shown that under the same, not very severe, conditions, the ***quantization error is comparable to a white noise uncorrelated (but not statistically independent) with the input signal*** $x(t)$.

10.3.11 Example: Gaussian signal

Let $x(t)$ be such that

$$p(x) = (2\pi\sigma_x^2)^{-1/2} \exp(-\tfrac{1}{2}x^2/\sigma_x^2) \tag{10.41}$$

with, according to (14.97) and $u = 2\pi\upsilon$,

$$\Pi_x(\upsilon) = \exp(-2\pi^2\sigma_x^2\upsilon^2) \tag{10.42}$$

For $\upsilon \geqslant (2\sigma_x)^{-1}$, $\Pi_x(\upsilon) < 1\%$. We deduce that coarse quantization with a step:

$$\Delta \leqslant \sigma_x \tag{10.43}$$

is already sufficient to ensure a quantization noise having a practically uniform distribution and a white spectrum, weakly correlated with $x(t)$.

In such a situation, the digital measurements of the statistical properties of first and second orders of the input signal can be achieved with an acceptable accuracy with only eight quantization levels ($V = q\Delta = 8\sigma_x$) for example, using a 3-bit A/D converter only.

10.3.12 Uniform quantization with stochastic reference (dither quantization)

The addition, before quantization (fig. 10.25), of an auxiliary random noise $a(t)$ to the useful input signal $s(t)$ is analogous to the presence of a stochastic fluctuation of the origin of the quantization law.

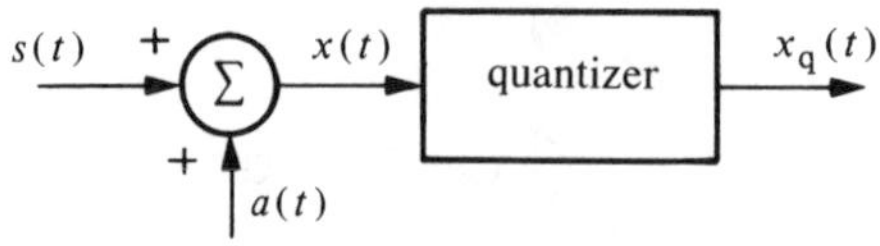

Fig. 10.25

If $a(t)$ and $s(t)$ are ***statistically independent,*** we have, by (5.187), for $x(t) = s(t) + a(t)$:

$$\Pi_x(\upsilon) = \Pi_s(\upsilon) \cdot \Pi_a(\upsilon) \tag{10.44}$$

It is enough that $\Pi_a(\upsilon) = 0$ for $\upsilon = k\Delta^{-1}$, with k a non-zero integer, to have the quantization noise uniformly distributed according to (10.39) with a zero mean value. The quantized signal is

$$x_q(t) = x(t) + n_q(t) = s(t) + a(t) + n_q(t) \tag{10.45}$$

By (14.46), we obtain the mean value identity

$$\mu_{xq} \equiv \mu_s \tag{10.46}$$

if $a(t)$ has also a zero mean value. An obvious distribution of $a(t)$ obeying the stated condition is the uniform law:

$$p(a) = \Delta^{-1} \operatorname{rect}(a/\Delta) \tag{10.47}$$

This approach is used [114] in stochastic computing, in which the variables are represented by the mean values of coarsely quantized random variables (often with $q = 2$), or in order to improve the performances of simplified correlators (sub-sect. 13.2.4).

10.4 BINARY CODING

10.4.1 Binary representation of quantization levels

Each one of the $q = 2^n$ discrete levels defined by the quantization law can be designated by a distinct binary word of n bits. The choice of a solution from among the $q!$ theoretically possible ones is tied to practical considerations concerning the A/D or D/A conversion method and the source or destination of the digital information.

All the codes used in practice are deduced from the natural binary code (tab. 10.26). An important difference is introduced between the coding of strictly unipolar signals and the more frequent case of bipolar signals.

Table 10.26 Principal codes for unipolar and bipolar signals ($n = 4$)

M	$m = \dfrac{M}{2^n}$	binary $d_1\,d_2\,d_3\,d_4$	Gray code	BCD 8421	BCD 2421
15	15/16	1111	1000		
14	14/16	1110	1001		
13	13/16	1101	1011		
12	12/16	1100	1010		
11	11/16	1011	1110	⋮	⋮
10	10/16	1010	1111	0001 0000	0001 0000
9	9/16	1001	1101	1001	1111
8	8/16	1000	1100	1000	1110
7	7/16	0111	0100	0111	0111
6	6/16	0110	0101	0110	0110
5	5/16	0101	0111	0101	0101
4	4/16	0100	0110	0100	0100
3	3/16	0011	0010	0011	0011
2	2/16	0010	0011	0010	0010
1	1/16	0001	0001	0001	0001
0	0	0000	0000	0000	0000

M	$m = \dfrac{M}{2^n}$	offset binary	sign-magnitude	2-complement	1-complement
7	7/16	1 111	0 111	0 111	0 111
6	6/16	1 110	0 110	0 110	0 110
5	5/16	1 101	0 101	0 101	0 101
4	4/16	1 100	0 100	0 100	0 100
3	3/16	1 011	0 011	0 011	0 011
2	2/16	1 010	0 010	0 010	0 010
1	1/16	1 001	0 001	0 001	0 001
(–) 0	(–) 0	1 000	(1) 0 000	0 000	0 000 (1111)
– 1	– 1/16	0 111	1 001	1 111	1 110
– 2	– 2/16	0 110	1 010	1 110	1 101
– 3	– 3/16	0 101	1 011	1 101	1 100
– 4	– 4/16	0 100	1 100	1 100	1 011
– 5	– 5/16	0 011	1 101	1 011	1 010
– 6	– 6/16	0 010	1 110	1 010	1 001
– 7	– 7/16	0 001	1 111	1 001	1 000
		↑	↑	↑	↑

sign bit

10.4.2 Natural binary code

This is the simplest and best known method of representation of a number M in binary form. Standardizing the scale of numbers between zero and one by writing $m = M/2^n$ ($M < 2^n$), the corresponding *natural binary code* is the set of binary digits, or bits, $d_i (i = 1, \ldots, n)$ defined by

$$m = \sum_{i=1}^{n} d_i \cdot 2^{-i} \qquad ; d_i = 0 \text{ or } 1 \tag{10.48}$$

The powers of two are called code *weights*. In the notation used, d_1 is the *most significant bit* = MSB (weight 1/2) and d_n is the *least significant bit* = LSB (weight $1/2^n$).

So, for a uniform quantization law with conversion range V, the level x_{qm} is equal to

$$x_{qm} = mV = M\Delta \tag{10.49}$$

where Δ is the quantization step.

10.4.3 Gray code

The *Gray code* is not a weighted code. It is deduced from the natural binary code by successively examining each bit, from the most significant to the least significant one. For each binary digit of the natural binary code equal to zero, the next most significant digit is unchanged. For each binary digit equal to one, on the contrary, the next most significant digit must be complemented ($0 \rightarrow 1$ and $1 \rightarrow 0$). This transcoding is defined by the relations:

$$d_{1g} = d_{1b} \tag{10.50}$$

$$d_{ig} = d_{ib} \oplus \mathrm{d}_{(i-1)b} \qquad ; i = 2, 3, \ldots, n \tag{10.51}$$

where indices b and g are assigned to the natural binary code and the Gray code, respectively, and the symbol $\oplus$ represents the exclusive-OR logic operation: $1 \oplus 1 = 0 \oplus 0 = 0$, $1 \oplus 0 = 0 \oplus 1 = 1$.

The inverse transcoding is obtained analogously, with (10.51) replaced by

$$d_{ib} = d_{ig} \oplus \mathrm{d}_{(i-1)b} \qquad ; i = 2, 3, \ldots, n \tag{10.52}$$

These relations are easily implemented with logic circuits.

The Gray code has the property such that in order to go from one level to the next, only one bit has to be changed. For this reason, it is widely used in continuous conversion systems (parallel converter, electromechanical shaft encoder, *et cetera*) in order to limit to $\pm\Delta$ the possible reading errors made at the time of a level change.

10.4.4 BCD code

In the *BCD* (*binary coded decimal*) code, each decimal unit (units, tens, hundreds, *et cetera*) of the integer digit M is translated into a binary code word. For this purpose, the weighting system 8-4-2-1 (natural binary code), or sometimes 2-4-2-1, is used. This coding pattern appears mainly in digital voltmeters and other digital display devices, because it allows the use of a simple decoder to translate each group of four bits into a decimal digit. It is also used, for the same reasons, in pocket and desk calculators.

10.4.5 Offset binary code

The *offset binary code* simply corresponds to a shift of the natural binary code of a quantity equal to half the total range. The most significant bit, equal to one for all the positive or zero values and equal to zero for all the negative values, represents the sign of the encoded value. This code is easily obtained from a unipolar converter to which an analog offset voltage is applied.

10.4.6 Sign-magnitude (signed) binary code

In the *sign-magnitude binary code,* the positive and negative levels of the same absolute value differ only by the sign bit. This type of code, not very often used in digital processing, allows us to maintain good accuracy and good linearity around the zero level, while all the other bipolar codes imply a change of all the bits from level $M = 0$ to level $M = -1$. This code can be directly obtained with a unipolar converter and an additional circuit for the polarity inversion.

A transcoding of the offset binary code to the sign-magnitude binary code implies the inversion of the MSB and, if the new MSB is equal to 1, the inversion of all the other bits with the addition (with carry) of a one to the LSB.

10.4.7 2-Complement code

The *2-complement code* is widely used for the digital representation of bipolar signals. It particularly eases the further arithmetical operations (addition and subtraction) of negative numbers, the result automatically taking the sign into account (problem 10.4.9).

If M is a positive number, the negative number $-M$ is represented by the binary code of the 2-complement of M defined by

$$M' = 2^n - M \tag{10.53}$$

The representation of the negative value of a positive number is practically obtained by inverting each bit (1-complement) and adding (with carry) a one to the less significant bit.

We go from the offset binary code to the 2-complement code simply by inverting the sign bit. The successive approximation A/D converters supply a natural binary code (shifted or not, depending on whether the range to be converted is bipolar or unipolar). They also are directly applicable in 2-complement code with a simple inversion of the most significant bit (MSB), which plays the role of a sign bit. Manufacturers generally incorporate a logic inverter in their converters, so that they can simultaneously supply the MSB and the $\overline{\text{MSB}}$ variables.

10.4.8 1-Complement code

For the *1-complement code,* the negative value of a positive number is simply obtained by inverting all the bits. This code has the disadvantage, like the sign-magnitude binary code, of offering two representations of zero. The 1-complement code differs from the 2-complement code by only one unit for negative numbers. It is sometimes used because of its simplicity when systematic quantization step error can be accepted.

10.4.9 Example

If $M = 5$ and $n = 4$, $-M = -5$ is represented in 2-complement code by the natural binary code of $M' = 16 - 5 = 11 \leftrightarrow 1011$. $M = 5$ corresponds in natural binary to 0101, and the 1-complement code of $-M = -5$ is 1010.

If we have two positive numbers, for example $M_1 = 5 \leftrightarrow 0101$ and $M_2 = 2 \leftrightarrow 0010$, the addition and subtraction operations are equivalent in 2-complement code:

- $M_1 + M_2 = 7 \leftrightarrow 0101 + 0010 = 0111$
- $M_1 - M_2 = 3 \leftrightarrow M_1 + M_2' \leftrightarrow 0101 + 1110 = (1)0011$ (the fifth bit is disregarded)
- $M_2 - M_1 = -3 \leftrightarrow M_2 + M_1' \leftrightarrow 0010 + 1011 = 1101$

10.5 DATA ACQUISITION

10.5.1 Multi-channel systems

A data acquisition system (fig.10.27) is a system that generally groups several analog-to-digital conversion channels which interface sensors to a central processor and its memory. The central processor first ensures the system and data management. Depending on its processing capabilities, we

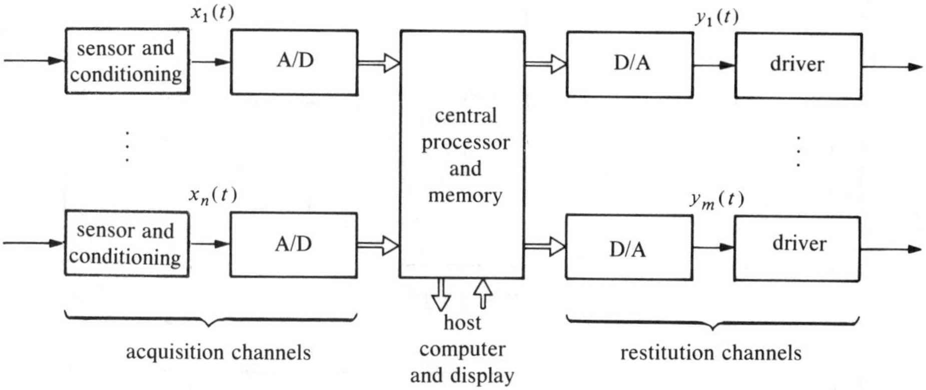

Fig. 10.27

can also assign some other tasks to it: filtering, compensation for sensor nonlinearities, statistical analysis, control, *et cetera*.

The central processor is often connected to a host computer, which takes care of more important processing, and to display devices, *et cetera*. Depending on the needs, the system also incorporates channels to recover the processed information with digital-to-analog converters and drivers.

These systems are mainly used for measurement techniques, industrial control, and the management of important systems (energy production, aerospace equipments, *et cetera*). If one or a few channels are adequate in simple applications, it is not unusual to meet systems incorporating several hundreds, or even several thousands of channels in some cases.

Depending on the needs, the A/D converters are connected to the measurement sensors by means of signal conditioning circuits (amplification, filtering, current-voltage conversion, *et cetera*).

When the number of channels is large, the acquired information is often multiplexed, either at the analog level (fig. 10.28), or at the digital level (fig. 10.29).

The first solution is used when the A/D converter is the most expensive part. A temporary storage on each analog channel by a sample-and-hold circuit is theoretically incorporated in the multiplexer. A decentralized solution that allows us to assign to each sensor a complete preprocessing device including the *ad-hoc* signal conditioning circuits and an A/D converter, which is completed if necessary by a microprocessor, is the current trend offering many advantages (direct preprocessing carried out close to the sensor, digital transmission and multiplexing, better immunity to environmental EM disturbances, *et cetera*).

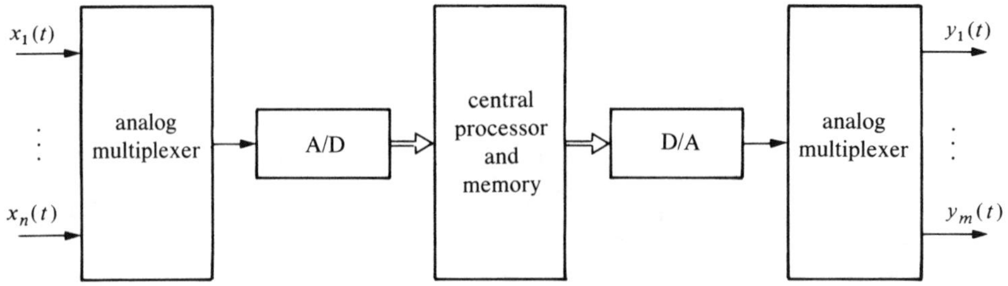

Fig. 10.28

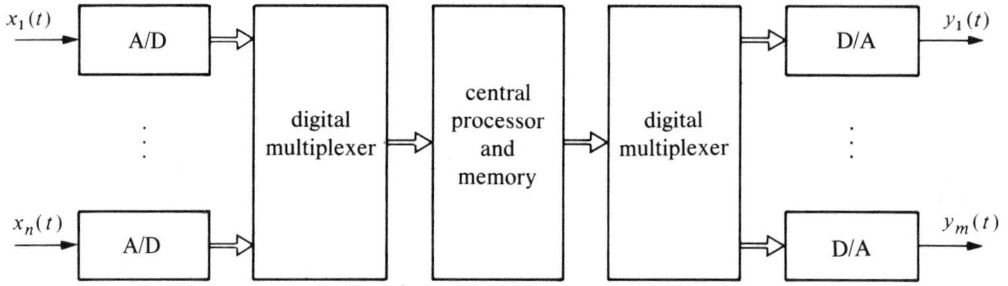

Fig. 10.29

In most systems, the available sampling rates are low, and therefore only suitable for relatively low-frequency signals.

10.5.2 Transient recorder

A transient recorder is a special storage device designed to be an interface between an analog source of transient signals and a convenient display (e.g., oscilloscope, graphic recorder) or analysis equipment. Its principle is illustrated by fig. 10.30.

During the acquisition phase, the input signal is sampled and digitized immediately at a f_{e1} rate compatible with its frequency and time characteristics. An important over-sampling ($f_{e1} \gg 2B_m$) is often used to guarantee a good reconstruction later via a simple zero-order extrapolation (sub-sect. 9.4.4). When the N samples corresponding to the capacity of the internal memory are stored, the device is ready to perform a reconstruction, with a new rate f_{e2}, which is higher or lower than f_{e1} depending on the specific needs. Insofar as the storage device has a cyclic sequential memory structure, the available output signal can be periodic and, therefore, quite suitable for

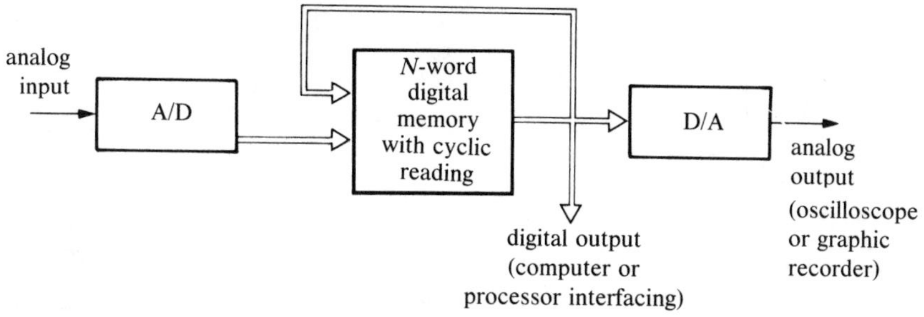

Fig. 10.30

observation on an oscilloscope or spectrum analyzer. Such a time conversion also allows us to record and, later on, conveniently study very slow, short-lived phenomena (e.g., seismic waves, thermal transients) or, when the input converter is fast enough, very short pulses (thunder, *et cetera*).

If T_{e1} is the acquisition sampling step, the signal segment $x_1(t,T_1)$ stored in memory has a duration $T_1 = NT_{e1}$ (fig. 10.31). The acceleration or deceleration factor is

$$a = T_{e1}/T_{e2} = f_{e2}/f_{e1} \tag{10.54}$$

with $a > 1$ in the case of acceleration and $0 < a < 1$ in the case of deceleration. The reproduced signal thus becomes $x_2(t,T_2) = x_1(at,T_1/a)$. The cyclical reading creates a periodic signal $x_3(t) = \text{rep}_{T_3}\{x_2(t,T_2)\}$. From (4.18) and (4.125), the

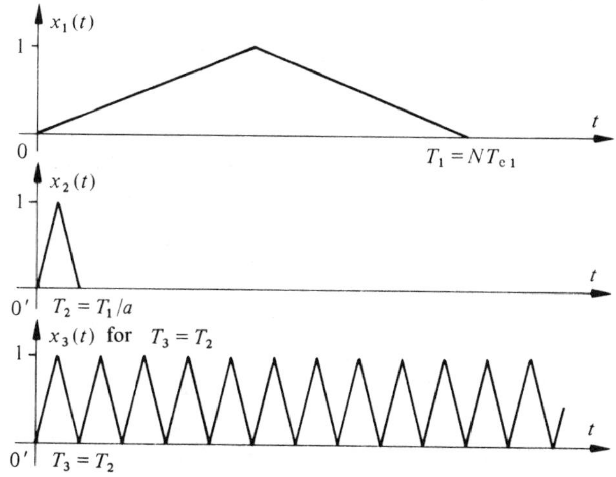

Fig. 10.31

Fourier transforms (fig. 10.32) of $x_2(t,T_2)$ and $x_3(t)$ are deduced from that of the $x_1(t,T_1)$:

$$X_2(f,T_2) = a^{-1}X_1(f/a,T_1) \tag{10.55}$$

$$\begin{aligned} X_3(f) &= T_{e2}^{-1}X_2(f,T_2) \cdot \delta_{1/T_3}(f) \\ &= T_{e1}^{-1}X_1(f/a,T_1)\delta_{1/T_3}(f) \end{aligned} \tag{10.56}$$

The spectral expansion provided by a cyclical memory with an acceleration factor $a > 1$ is used by some spectrum analyzers (sub-sect. 12.3.6).

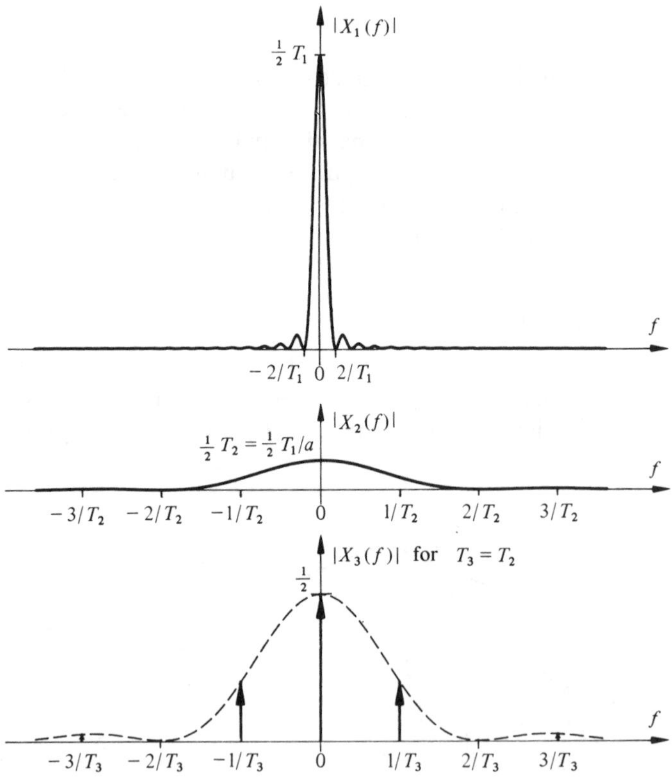

Fig. 10.32

10.5.3 Acquisition errors

In addition to the aliasing distortion (sub-sect. 9.3.1) and the quantization distortion, produced by the sampling and quantization operations, other sources of acquisition errors exist. These are due to the non-ideal characteristics of the A/D and D/A converter circuits. Their detailed description is beyond the scope of this book and can be found in the specialized literature [111]. Their principal effect is to alter the ideal quantization law, which we studied in sub-section 10.3.4, in the following way:

- offset: the average slope of the quantization law does not run through the origin (fig. 10.33);
- gain error: the quantization law has an average slope growing faster or slower than theoretically expected (fig. 10.34);
- nonlinearity: the quantization steps are not equal, and hence the average slope is no longer a straight line;
- differential nonlinearity: excessive rippling of the average quantization law with respect to the theoretical straight line results in missing codes in the A/D conversion or a nonmonotonic slope in the D/A conversion characteristic.

An efficient test procedure for converters is to apply a signal with a known statistical distribution at the input (e.g., a uniform distribution) and to check the output distribution.

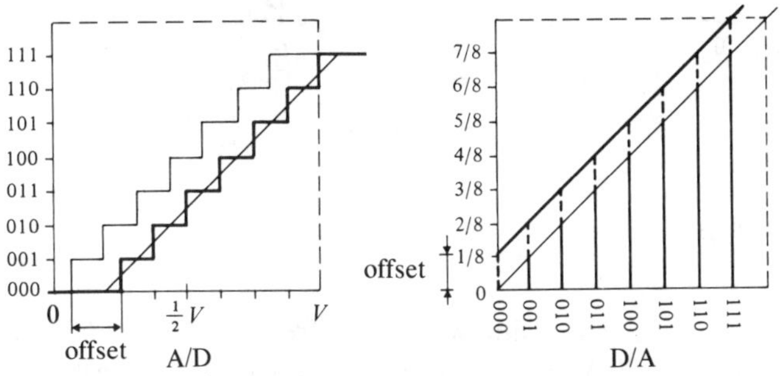

Fig. 10.33

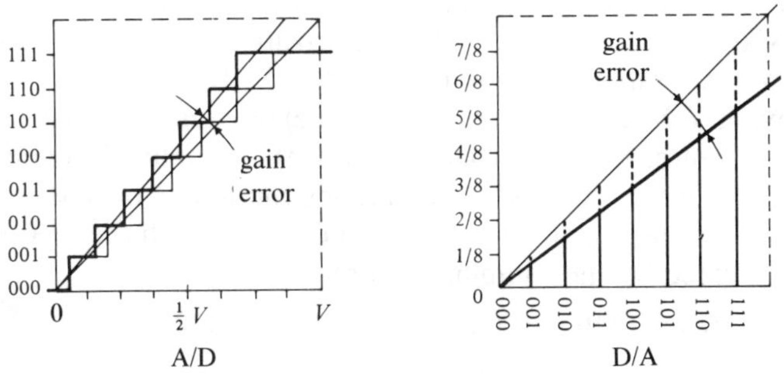

Fig. 10.34

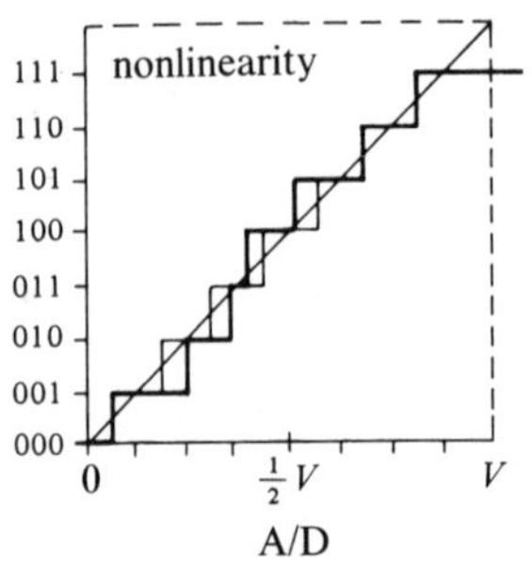

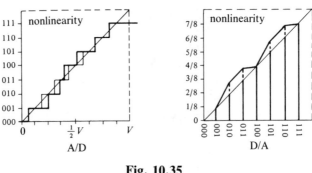

Fig. 10.35

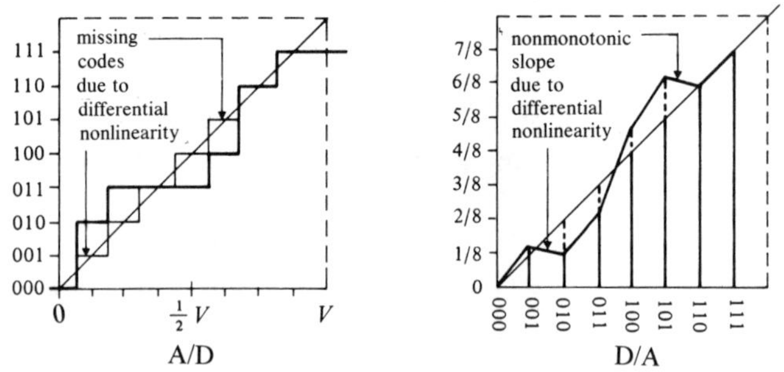

Fig. 10.36

10.6 PROBLEMS

10.6.1 We want to record a sine wave with frequency $f_0 = 500$ Hz and peak-to-peak amplitude of 20 V in natural binary coded digital form:

- What are the main characteristics (conversion range, conversion time, number of bits) that the selected analog-to-digital converter should have, if the signal is sampled without memorization, and if the signal-to-quantization noise ratio has to be larger than 45 dB?
- If we want to be able to restore this signal by digital-to-analog conversion with linear interpolation, what should be the sampling frequency required so that the absolute value of the reconstruction error does not exceed a quantization step? Compare the result obtained with the theoretical lower limit indicated by the sampling theorem.

10.6.2 Verify results (10.21), (10.22), (10.23), and (10.24).

10.6.3 Compute the signal-to-quantization noise ratio ξ_{qdB} obtained in the case of a conversion range V and of an input signal probability density equal to: (a) $p(x) = 2V^{-1}$ rect$(2x/V)$; (b) $p(x) = 2V^{-1}$ tri$(2x/V)$; (c) a zero mean value Gaussian distribution with $V = 10\sigma_x$.

10.6.4 A random signal $x(t)$, having a uniform amplitude distribution between $-V/4$ and $V/4$ is uniformly quantized by a quantization device with 256 discrete levels representing a total range V. Define the signal-to-quantization noise ratio ξ_{qdB} so obtained.

10.6.5 A signal $x(t)$ has a probability density $p(x) = ½a \exp(-a|x|)$. It is uniformly quantized over $q = 2^n$ discrete levels between zero and $\pm 5/a$. Find the lowest value of n from which we obtain a signal-to-quantization noise ratio ξ_{qdB} larger than 30 dB.

10.6.6 Verify (10.32) and (10.33).

10.6.5 A signal $x(t)$ has a probability density $p(x) = ½a \exp(-a|x|)$. It is uniformly quantized over $q = 2^n$ discrete levels between $-5/a$ and $+5/a$. Find the lowest value of n from which we obtain a signal-to-quantization noise ratio ξ_{qdB} larger than 30 dB.

10.6.6 Verify (10.32) and (10.33).

Chapter 11

Modulation and Frequency Translation

11.1 GENERAL PRINCIPLES

11.1 Objectives

Modulation is a process in which a *primary signal,* called the *modulating signal,* modifies an auxiliary signal, called the *carrier signal* or simply *carrier,* to create a *secondary signal,* or *modulated signal,* the characteristics of which are better adapted to the desired conditions for application (fig. 11.1).

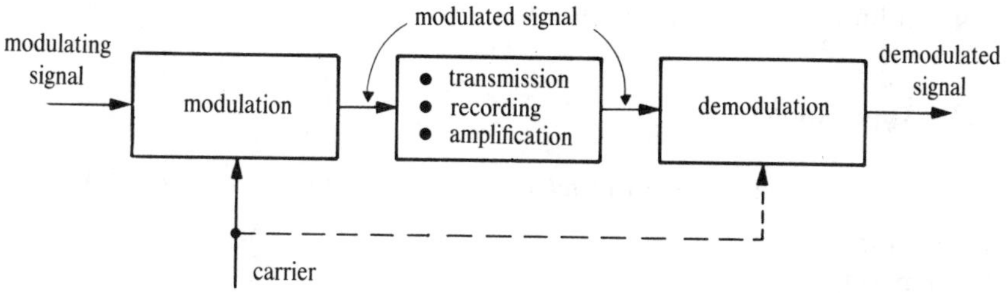

Fig. 11.1

Modulation is, therefore, a means of information representation; it is, in this sense, similar to coding.

Modulation is mainly used to:

- shift a signal spectrum, without information loss, in ***another frequency range*** to match the emission-reception constraints (antenna efficiency and dimensions) and to satisfy the conditions imposed by a transmission channel (propagation characteristics, available bandwidth), or to facilitate some signal processing operations (frequency translation in radio receiver tuning circuits or in swept-frequency spectrum analyzers, for example);
- share the available communication channel between several simultaneously transmitted signals (*frequency multiplexing*: allocation of a different frequency band to each simultaneously transmitted message; *time multiplexing*: sequential transmission of sampled values for each message);

- obtain the amplification and the efficient filtering of weak low frequency signals without having to cope with the $1/f$ background noise described in section 6.4 (lock-in amplifier: sub-sect. 13.2.10);
- record very low frequency signals on magnetic tape (measurement recorders);
- modify the spectrum of the emitted signal in order to improve detection (e.g., radar; see sub-sect. 7.4.8 and sect.13.4) and the noise immunity (angular modulation or pulse modulation), or in order to make the communication channel more confidential and difficult to jam (spread spectrum systems, sub-sect. 13.2.11);
- modify an appropriate parameter to drive or control a machine or an industrial process.

Demodulation is the reverse of modulation: it is the reconstruction of the modulating signal from the modulated signal. Depending on the situation, this reconstruction requires an auxiliary signal identical to, or deduced from, the carrier used in modulation. The possible addition of noise, interference, and distortion limits the reconstruction fidelity.

11.1.2 Applications

Modulation is mainly used in ***telecommunications*** systems (vol. XVIII) for:

- telegraphy
- telephony
- radio broadcasting
- television broadcasting
- data transmission
- remote control and telemetry

Modulation is also used to improve the resolution of ***detection systems*** in:

- radar
- sonar
- remote sensing

Electronic instrumentation and ***metrology*** also use frequency translations to facilitate other signal processing operations.

11.1.3 Methods

Modulation methods are often classified in two categories:

- analog modulation
- digital modulation

Digital modulation is a term used in telecommunications to identify the various methods of representing analog information in a ***coded form.*** The most usual form of this representation, called *pulse code modulation* (PCM), is similar to the digitization techniques described in chapter 10. We will, therefore, mention it only briefly.

Analog modulation obeys the definition given in section 11.1. The carrier signal is either a sine wave (*continuous modulation*), or a periodic sequence of pulses (*sampled modulation*). According to the nature of the carrier signal parameter that is modified by the modulating signal, we have a variety of modulation types.

For modulation of a sine wave carrier:

- amplitude modulation (AM)
- angle modulation: frequency (FM) or phase (ΦM)

For pulse carrier modulation:

- pulse amplitude modulation (PAM)
- pulse duration modulation (PDM)
- pulse position modulation (PPM) or pulse frequency modulation (PFM)

When the modulating signal is digital, or simply quantized, we obtain a *discrete modulation.* This case is encountered in data transmission (vol. XVIII), for which discrete frequency or phase modulation is mainly used.

11.1.4 Modulation of a sine wave carrier: general model

With the representation of a bandpass signal defined in section 7.4, the secondary signal $s(t)$ of a modulator (sub-sect. 8.3.12), the inputs of which are the modulating signal $m(t)$ and the carrier $u_p(t) = \hat{U}\cos(2\pi f_p t + \alpha_p)$ can be written (fig. 11.2), according to (8.100), as

$$s(t) = \text{Re}\{\underline{s}(t) = \underline{r}(t)\exp[\text{j}(2\pi f_p t + \alpha_p)]\}$$
$$= a(t)\cos(2\pi f_p t + \alpha_p) - b(t)\sin(2\pi f_p t + \alpha_p) \qquad (11.1)$$

where

$$\underline{r}(t) = a(t) + \text{j}b(t) = r(t)\exp[\text{j}\Delta\phi(t)] \qquad (11.2)$$

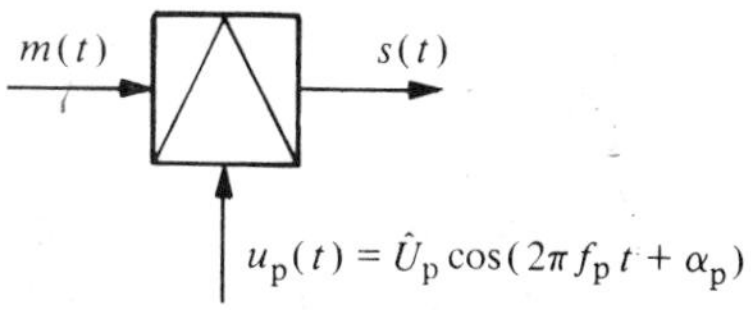

Fig. 11.2

is the ***complex envelope*** associated to the anlytic signal $\underline{\underline{s}}(t)$ that contains all the information brought by the modulating signal. Depending on the modulation methods, the information is carried by the real envelope $r(t)$, the instantaneous phase variation $\Delta\phi(t)$, or a combination of both.

Each modulation type can thus be characterized [99] by a functional relationship:

$$\underset{\sim}{r}(t) = \mathrm{S_m}\{m(t)\} \tag{11.3}$$

where $\mathrm{S_m}$ represents the adequate modulation operator (fig. 11.3). Conversely, the modulating signal recovery can be defined by

$$m(t) = \mathrm{S_d}\{\underset{\sim}{r}(t)\} \tag{11.4}$$

where $\mathrm{S_d}$ represents the necessary demodulation operator (fig. 11.4).

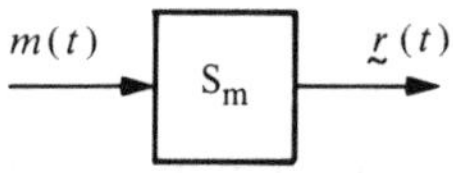

Fig. 11.3

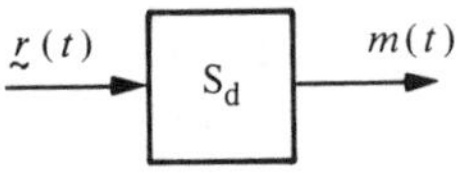

Fig. 11.4

The spectral densities of the modulated signal and the complex envelope are simply related by (7.88):

$$\Phi_s(f) = \tfrac{1}{4}[\Phi_{\underset{\sim}{r}}(-f-f_p) + \Phi_{\underset{\sim}{r}}(f - f_p)] \tag{11.5}$$

11.1.5 Pulse carrier modulation: general model

When the carrier (fig. 11.5) is a periodic sequence of pulses of period T and shape $g(t/\Delta)$, the modulated signal $s(t)$ can be written as

$$s(t) = \sum_{k=-\infty}^{\infty} a_k g\left(\frac{t - kT - \tau_k}{\Delta_k}\right)$$

$$= \sum_{k=-\infty}^{\infty} a_k g(t/\Delta_k) * \delta(t - kT - \tau_k) \tag{11.6}$$

where a_k is an amplitude parameter, Δ_k is a duration parameter, and τ_k is a position parameter, which may vary as a function of the sampled values $m(kT)$ of the modulating signal (fig. 11.6).

No general expression of the spectral density of the modulated signal can be given here. Note, however, that in the particular case (PAM) where $\tau_k = 0$ and $\Delta_k = \Delta$ = constant, the model (11.6) becomes similar to the sampled signal (9.1), with a periodic sampling function $e(t) = \delta_T(t)$. The corresponding spec-

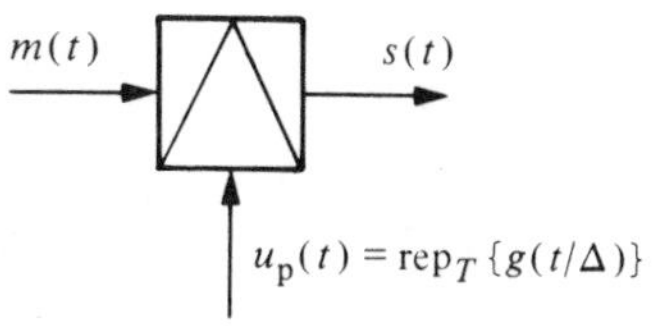

Fig. 11.5

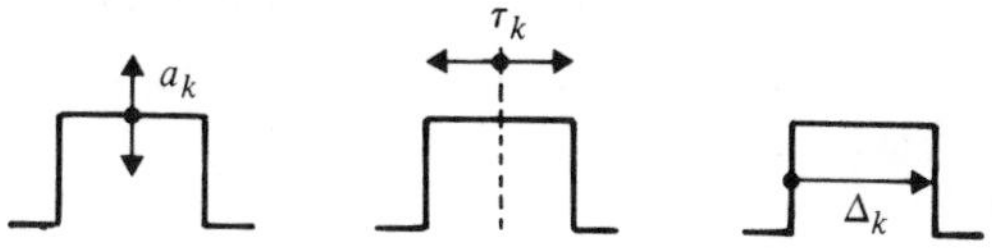

Fig. 11.6

trum, therefore, is directly deduced from (9.4). If $a_k = a$ and $\Delta_k = \Delta$ are constant (PPM, PFM), expression (11.6) is similar to the equation for the output signal of an invariant linear system, with pulse response $ag(t/\Delta)$, stimulated by an aperiodic sequence of delta-functions. Denoting by $\Phi_\delta(f)$ the spectral density of such a sequence, that of the modulated signal becomes, by (4.18) and (8.24): $\Phi_s(f) = a^2\Delta^2|G(\Delta f)|^2 \cdot \Phi_\delta(f)$. When only Δ_k varies (PDM), expression (11.6) is similar to the output signal of a linear parametric operator (sub-sect. 8.3.2) stimulated by a periodic sequence of delta-functions $\delta_T(t)$.

11.2 AMPLITUDE MODULATION

11.2.1 Introduction

Amplitude modulation is the generic name for a family of modulation methods in which the amplitude of a sine wave carrier varies as a function of the modulating signal. These family members are designated by the following conventional terms:

- amplitude modulation with carrier (AM)
- double-sideband suppressed carrier modulation (DSBSC)
- single-sideband modulation (SSB)
- vestigial sideband modulation (VSB)

Amplitude modulation is a ***linear operation*** generally preserving the unilateral spectrum morphology of the modulating signal $m(t)$ by imposing frequency translation. Table 11.7 summarizes the expressions for the complex envelope $\underline{r}(t)$ and the modulated signal spectrum $\Phi_s(f)$ corresponding to each type of modulation.

The fundamental operation implied by any amplitude modulation is the ***multiplication*** (sub-sect. 8.3.4) of the modulating signal—or of a function of this signal—by a sine wave carrier (fig. 11.8). This multiplication is obtained, either directly with an appropriate parametric device (analog multiplier), or indirectly by superimposing the carrier and the modulating signal at the input

Table 11.7

Modulating signal	$m(t) = U_{\mathrm{m}} \cdot m_0(t)$
AM	$\underset{\sim}{r}(t) = \hat{U}_{\mathrm{p}}[1 + m_0(t)]$
DSBSC	$\underset{\sim}{r}(t) = \hat{U}_{\mathrm{p}} \cdot m_0(t)$
SSB	$\underset{\sim}{r}(t) = \dfrac{\hat{U}_{\mathrm{p}}}{2}[m_0(t) + \mathrm{j}\,\check{m}_0(t)]$
VSB	$\underset{\sim}{r}(t) = \dfrac{\hat{U}_{\mathrm{p}}}{2}\{(1) + m_0(t) + \mathrm{j}\,[\check{m}_0(t) * h(t)]\}$

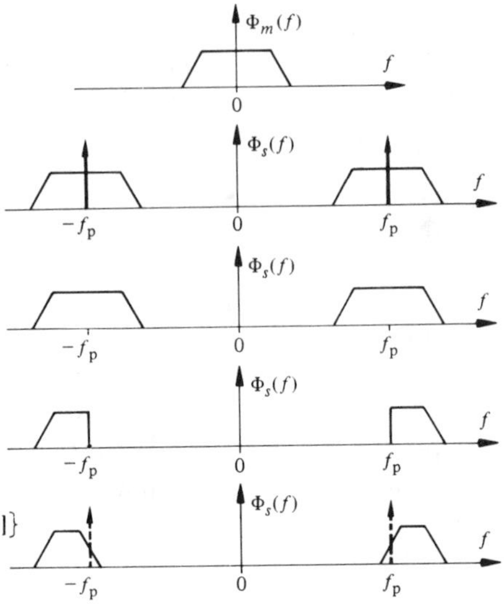

of an amnesic nonlinear operator (sub-sect. 8.4.2) of characteristic $y = g(x)$, or with a nonlinear reactance circuit. In practice, we generally use the nonlinearity of an electronic component (bipolar or field-effect transistor, vacuum tube, nonlinear capacitance of a reverse-biased diode). A postfiltering rejection of the undesired spectral components (fig. 11.9) is added.

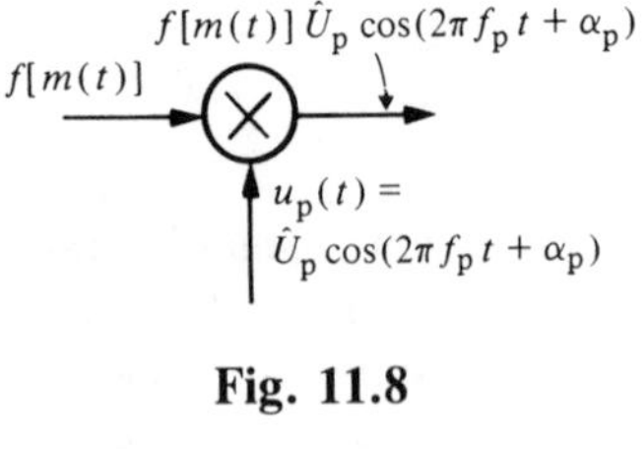

Fig. 11.8

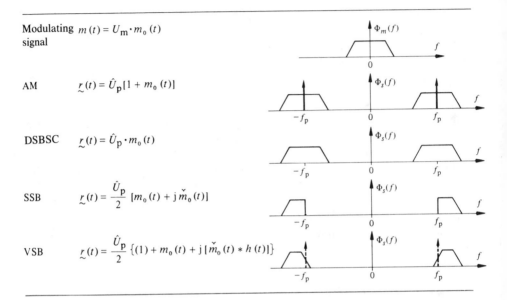

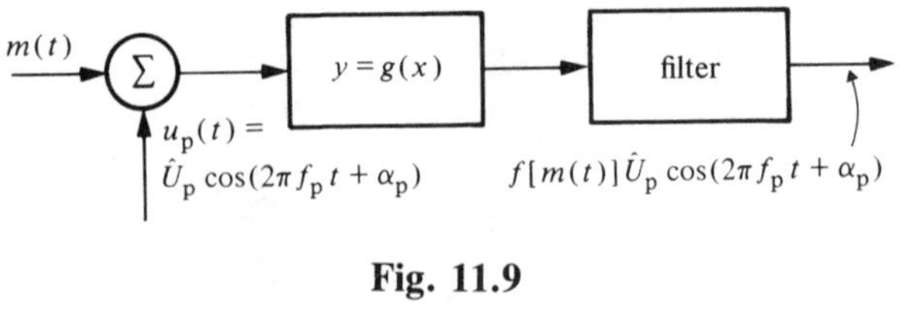

Fig. 11.9

A simple example is a quadratic modulator of characteristic $y = a_1x + a_2x^2$. If $x(t) = m(t) + \hat{U}_{\mathrm{p}} \cos(\omega_{\mathrm{p}}t + \alpha_{\mathrm{p}})$, $y(t) = \{a_1m(t) + a_2m^2(t) + a_2\hat{U}_{\mathrm{p}}^2 \cos^2(\omega_{\mathrm{p}}t + \alpha_{\mathrm{p}})\} + \{a_1\hat{U}_{\mathrm{p}} \cos(\omega_{\mathrm{p}}t + \alpha_{\mathrm{p}})[1 + (2a_2/a_1)m(t)]\}$. The second term enclosed in braces contains the desired modulated signal, the first term has to be eliminated by filtering.

11.2.2 Amplitude modulation with carrier

Amplitude modulation (with carrier), denoted AM, is the most classic form of amplitude modulation. Its principle is illustrated by fig. 11.10.

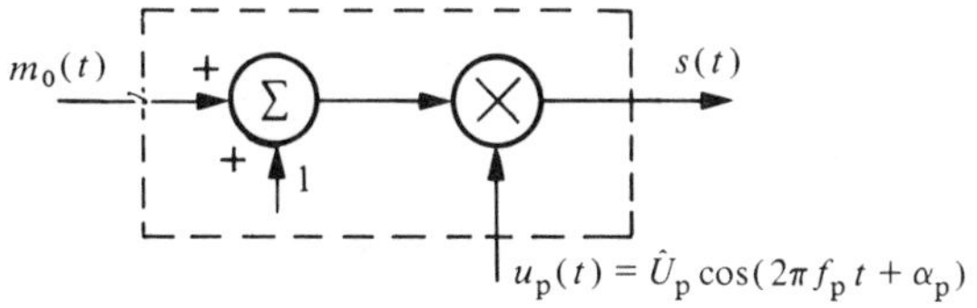

Fig. 11.10

The modulated signal $s(t)$ corresponds to the product of the sine wave carrier $u_p(t)$ by the modulating signal $x(t) = 1 + m_0(t)$, which is assumed to be statistically independent. The notation $m_0(t) = m(t)/U_m$ designates dimensionless signal (obtained by dividing the original signal by a characteristic value U_m). Thus,

$$s(t) = [1 + m_0(t)]\hat{U}_p \cos(2\pi f_p t + \alpha_p) \tag{11.7}$$

In the absence of a modulating signal $[m(t) = 0]$, the signal $s(t)$ is equal to the sole carrier (fig. 11.11).

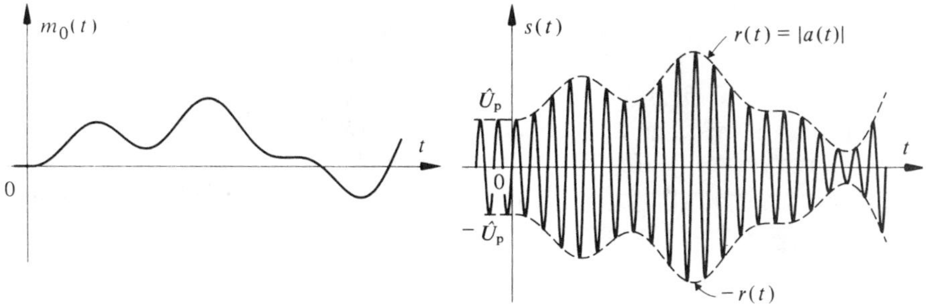

Fig. 11.11

The comparison of (11.7) with the general model (11.1) shows that, in this case,

$$\underline{r}(t) = a(t) = [1 + m_0(t)]\hat{U}_p \tag{11.8}$$

The spectral density of the complex envelope is $\Phi_{\underline{r}}(f) = \hat{U}_p^2[\delta(f) + \Phi_{m0}(f)]$ and the spectrum of the modulated signal (fig. 11.12), according to (11.5), becomes

$$\Phi_s(f) = \frac{\hat{U}_p^2}{4}[\delta(f + f_p) + \delta(f - f_p)] + \frac{\hat{U}_p^2}{4}[\Phi_{m0}(f + f_p) + \Phi_{m0}(f - f_p)] \tag{11.9}$$

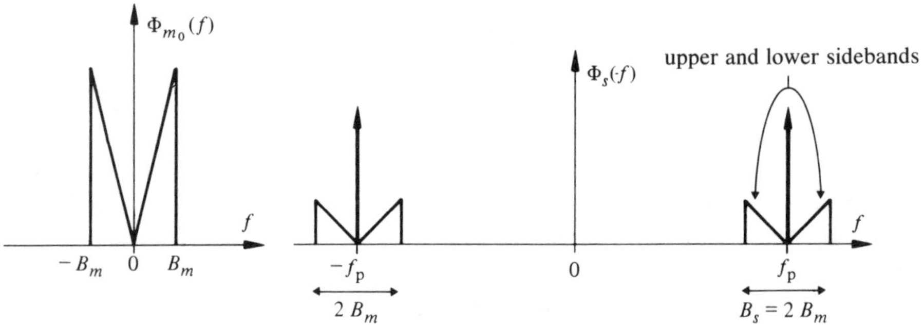

Fig. 11.12

where we have taken into account the even character of $\Phi_{m0}(f)$ and $\delta(f)$. The dimension of the spectral density of the dimensionless signal $m_0(t)$ is in Hz^{-1} and $\Phi_m(f) = U_m^2\Phi_{m0}(f)$.

This amplitude modulation is characterized by a simple translation of the modulating signal's ***bilateral spectrum,*** of a quantity corresponding to the *carrier frequency* $\pm f_p$, with the addition of lines at the same frequencies $\pm f_p$. The total power of this signal is equal to $P_s = (\hat{U}_p^2/2)(1 + P_{m0})$, where $P_{m0} = P_m/U_m^2$ is the power coefficient of the modulating signal.

It is easily shown that if B_m is the modulating signal bandwidth, the bandwidth B_s occupied by the modulated signal is doubled:

$$B_s = 2B_m \tag{11.10}$$

Amplitude modulation (with carrier) has the advantage, as long as

$$a(t) = [1 + m_0(t)]\hat{U}_p \geqslant 0 \tag{11.11}$$

of allowing a simple demodulation by envelope detection (sub-sect. 11.2.7). It is still widely used in radio broadcasting (long, medium, and short waves).

One of its disadvantages is that the modulated signal power is largely concentrated in the lines located at $f = \pm f_p$ which do not carry information.

The ratio of the power in the sidebands to the total power gives a measure of the efficiency of this modulation:

$$\eta_m = P_{m0}/(1 + P_{m0}) \tag{11.12}$$

where $P_{m0} = \int \Phi_{m0}(f)\mathrm{d}f$ is the power coefficient of the modulating signal. For example, in the case of a sine wave $m_0(t)$ with unit amplitude [maximum value still complying with (11.11); hence, allowing an envelope detection], $P_{m0} = ½$ and $\eta_m = 33.3\%$. If we consider a Gaussian random signal $m_0(t)$, with zero mean value, and reaching the limit set by (11.11) for amplitudes equal to $3\sigma_{m0}$, the efficiency decreases down to $\eta_m = 10\%$.

11.2.3 Double-sideband suppressed carrier modulation

Double-sideband suppressed carrier modulation (DSBSC) corresponds to a simple multiplication (fig. 11.3):

$$s(t) = m_0(t) \cdot u_p(t) = m_0(t)\hat{U}_p \cos(2\pi f_p t + \alpha_p) \tag{11.13}$$

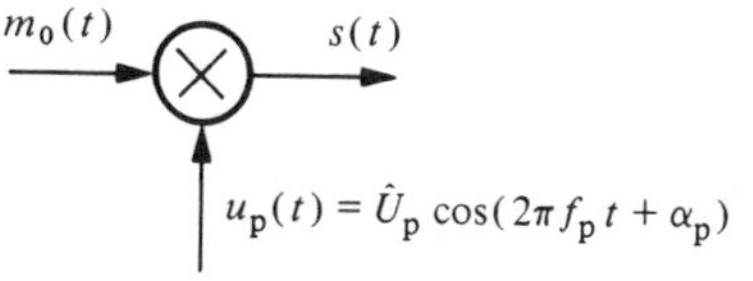

Fig. 11.13

The complex envelope here is directly proportional to the modulating signal:

$$\underline{r}(t) = a(t) = \hat{U}_p m_0(t) \tag{11.14}$$

The modulated signal is, hence, equal to zero in the absence of the modulating signal (fig. 11.14), and its spectrum contains only the translated contributions of the modulating signal bilateral spectrum (fig. 11.15):

$$\Phi_s(f) = \frac{\hat{U}_p^2}{4} [\Phi_{m0}(f + f_p) + \Phi_{m0}(f - f_p)] \tag{11.15}$$

The occupied bandwidth B_s is the same as in the case of amplitude modulation with carrier, but the lines at $f = \pm f_p$ have disappeared. This allows for concentration of the total emitted power in the sidebands, which contain all of the information: the efficiency is consequently 100%.

It is not possible, on the other hand, to demodulate with a simple envelope detection. An isochronous detection (sub-sect. 11.2.6), requiring carrier reconstruction, is needed.

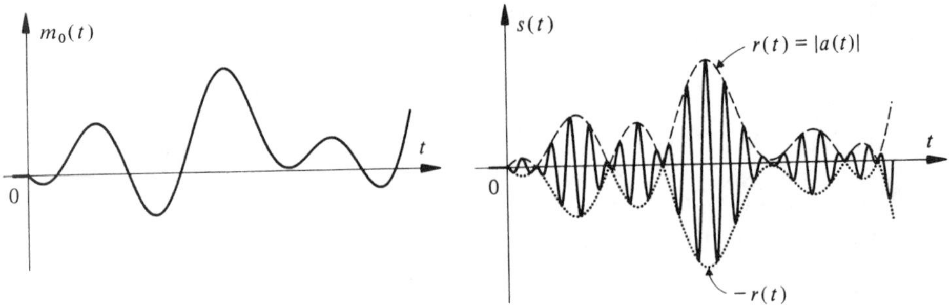

Fig. 11.14

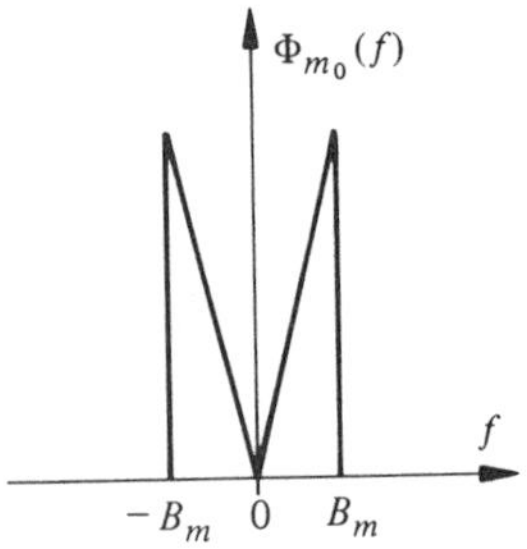

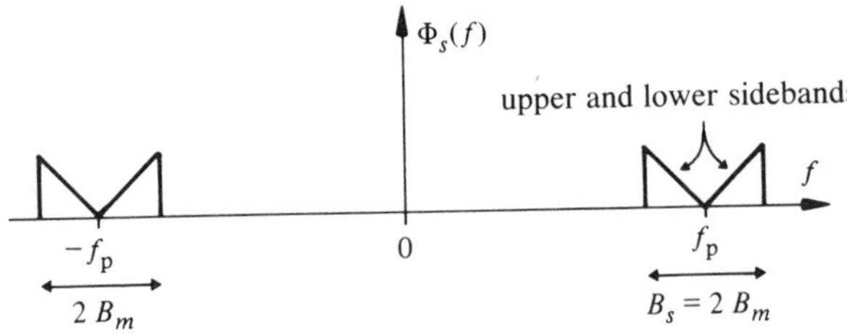

Fig. 11.15

11.2.4 Single-sideband modulation

Since the spectrum is entirely defined by its unilateral form (sub-sect. 4.5.2), all of the modulating signal information is, in fact, represented by a single sideband of the modulated signal spectrum. This statement enables us to contemplate a ***reduction of the occupied bandwidth by a factor of two,*** by getting rid of one of the sidebands. This is the principle of *single-sideband modulation* (SSB). This method is mainly used in telephony in the frequency multiplexing scheme called *carrier system* (vol. XVIII). It also requires an isochronous detection (sub-sect. 11.2.6). Consider the case (fig. 11.16) in which only the upper frequency band of the spectrum is kept. The new spectrum corresponds to the translation by f_p of the unilateral spectrum of the modulating signal and its image.

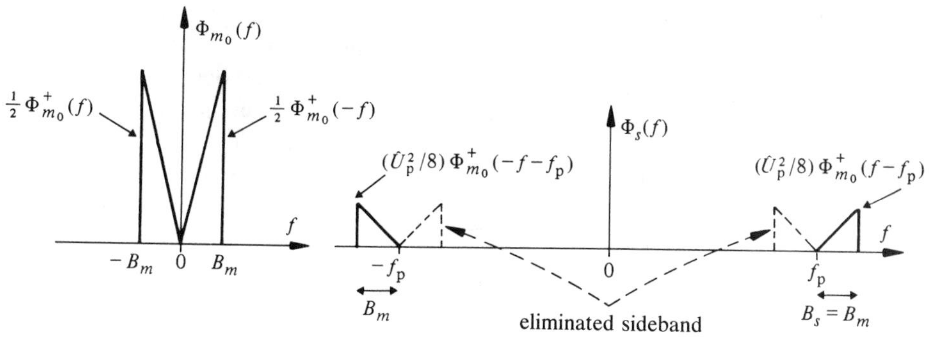

Fig. 11.16

According to (4.162) and (7.41), the modulated signal spectrum becomes

$$\Phi_s(f) = \frac{\hat{U}_p^2}{8}\left[\Phi^+_{m0}(-f - f_p) + \Phi^+_{m0}(f - f_p)\right]$$

$$= \frac{\hat{U}_p^2}{16}\left[\Phi_{\underline{\underline{m0}}}(-f - f_p) + \Phi_{\underline{\underline{m0}}}(f - f_p)\right] \qquad (11.16)$$

Comparing with (11.5), we obtain

$$\Phi_{\underline{r}}(f) = \frac{\hat{U}_p^2}{4}\,\Phi_{\underline{\underline{m0}}}(f) \qquad (11.17)$$

from which we deduce that the complex envelope here is analytic and equal to

$$\underline{r}(t) \equiv \underline{\underline{r}}(t) = \tfrac{1}{2}\hat{U}_p \underline{\underline{m_0}}(t) \qquad (11.18)$$

with in-phase and quadrature components related by the Hilbert transform:

$$a(t) = \tfrac{1}{2}\hat{U}_p m_0(t);\; b(t) = \tfrac{1}{2}\hat{U}_p \check{m}_0(t) \qquad (11.19)$$

The equation for a single-(upper)sideband modulated signal is, thus,

$$s(t) = \frac{\hat{U}_p}{2}\left[m_0(t)\cos(2\pi f_p t + \alpha_p) - \check{m}_0(t)\sin(2\pi f_p t + \alpha_p)\right] \qquad (11.20)$$

The choice of the lower sideband only changes the sign of the second term of (11.20).

From (11.19), we deduce that the single-sideband modulation simultaneously generates an amplitude variation (envelope) and a phase variation of the modulated signal.

SSB modulated signal can theoretically be obtained either by ideal filtering of a DSBSC signal (fig. 11.17), or by starting directly from equation (11.20), i.e., replacing the sideband-eliminating ideal filter with a Hilbert operator (subsect. 8.2.18) and using two carriers in quadrature. This second solution leads to the block diagram of fig. 11.18. Both the ideal sideband filter and the Hilbert operator are theoretical models which are difficult to approximate in practice.

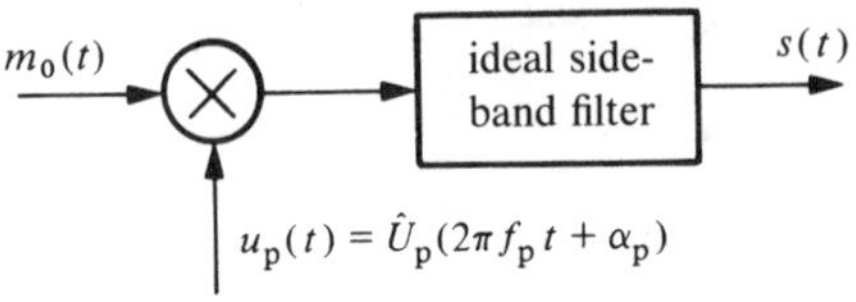

Fig. 11.17

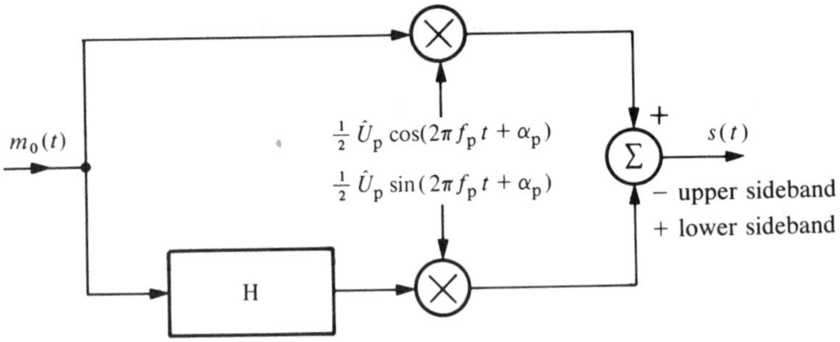

Fig. 11.18

11.2.5 Vestigial sideband modulation

Vestigial sideband (VSB) *modulation* is a trade-off between SSB modulation and AM or DSBSC modulation. By retaining a vestige of the undesired sideband, we reduce the requirements imposed on the sideband filter, and we facilitate the transmission of a modulating signal with significant low-frequency components (e.g., video signal). In television, the occupied bandwidth B_s is practically equal to 1.25 B_m.

The block diagram of fig. 11.19 illustrates the generation principle of a VSB (with carrier) modulated signal. The VSB filter must have a transfer function $G(f)$, exhibiting a localized odd symmetry in the vicinity of the carrier frequency (fig. 11.20). It can be shown (problem 11.6.8) that the complex envelope of the modulated signal becomes in this case,

$$\underline{r}(t) = \frac{\hat{U}_p}{2} \{[1 + m_0(t)] + j[\check{m}_0(t) * h(t)]\} \tag{11.21}$$

where $h(t)$ is the impulse response of a highpass filter.

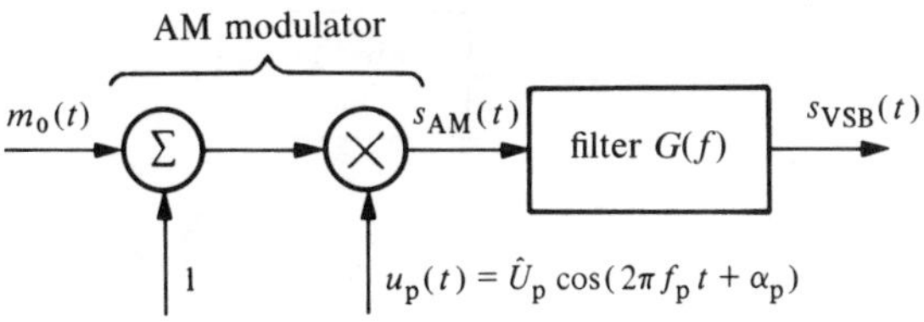

Fig. 11.19

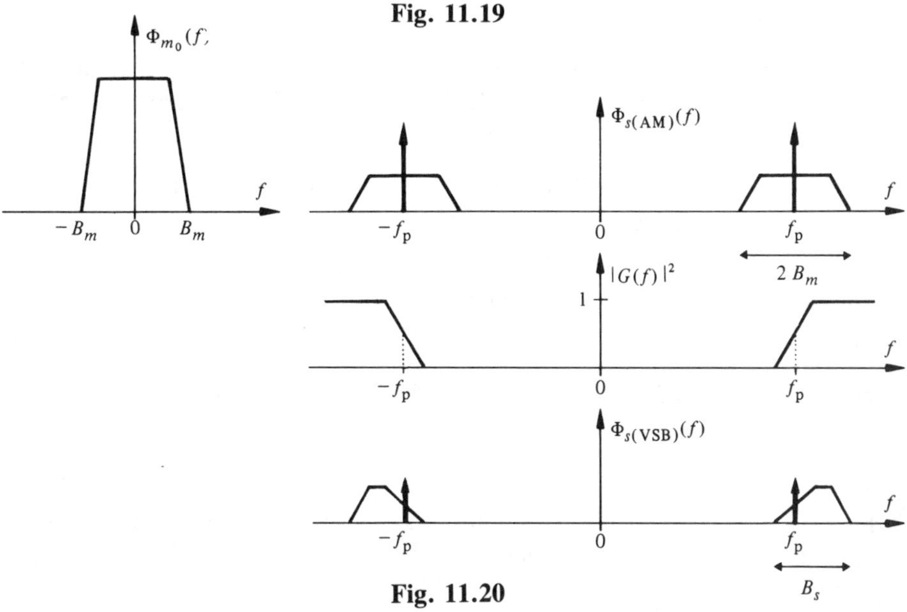

Fig. 11.20

If we start with DSBSC modulation, the demodulation can only be synchronous (sub-sect. 11.2.6). If the carrier is retained, a simple envelope demodulation (sub-sect. 11.2.7), introducing an acceptable level of distortion if $m_0(t) \ll 1$, can be satisfactory (TV).

11.2.6 Demodulation by synchronous detection

Demodulation by *synchronous detection* is analogous to amplitude modulation (fig. 11.21). The modulated signal is multiplied by a periodic auxiliary signal of the same frequency as the carrier, and the resulting undesired components are eliminated with a lowpass filter.

Consider the case in which the auxiliary signal is equal to the carrier, but with phase α_d and unit amplitude:

$$u_d(t) = \cos(2\pi f_p t + \alpha_d) \tag{11.22}$$

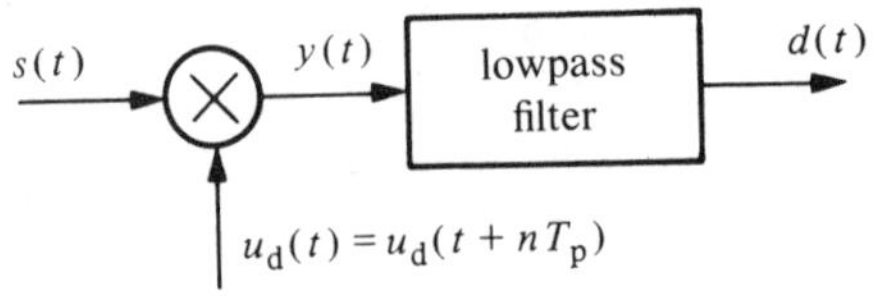

Fig. 11.21

The output signal of the multiplier, the inputs of which are $u_d(t)$ and the modulated signal $s(t)$ as defined by (11.1), thanks to the trigonometric identities of sect. 15.2, becomes

$$y(t) = s(t) \cdot u_d(t) = \tfrac{1}{2}[a(t)\cos(\alpha_p - \alpha_d) - b(t)\sin(\alpha_p - \alpha_d)] + \tfrac{1}{2}[a(t)\cos(4\pi f_p t + \alpha_p + \alpha_d) - b(t)\sin(4\pi f_p t + \alpha_p + \alpha_d)] \quad (11.23)$$

If the (assumed ideal) lowpass filter has a cut-off frequency:

$$f_c < 2f_p - B_m \quad (11.24)$$

where B_m is the spectral bandwidth of the modulating signal $m(t)$, then the output signal is equal to

$$d(t) = \tfrac{1}{2}[a(t)\cos\Delta\alpha - b(t)\sin\Delta\alpha] \quad (11.25)$$

where $\Delta\alpha = \alpha_p - \alpha_d$ is the phase difference between $u_p(t)$ and $u_d(t)$.

The general demodulation operator (11.4), therefore, in the case of the synchronous detection, is characterized by

$$d(t) = S_d\{\underline{r}(t)\} = \tfrac{1}{2}\mathrm{Re}\{\underline{r}(t)\exp(\,j\Delta\alpha)\} \quad (11.26)$$

In particular, $d(t) = \tfrac{1}{2}a(t)$ if the detection is coherent (or isochronous), i.e., if $\Delta\alpha = 0$.

Table 11.22 summarizes the results deduced from (11.26) for the various types of amplitude modulation indicated in table 11.7.

In the case of AM and VSB with carrier modulation, the modulating signal is recovered from $d(t)$ by getting rid of a dc component (highpass filtering). A nonisochronous detection ($\Delta\alpha \neq 0$) exhibits distortion in the SSB and VSB demodulated signal, depending on the Hilbert transform of the modulating signal. This distortion can be important if the modulating signal has sharp transients such as rectangular pulses (see fig. 7.3).

Synchronous detection requires the reconstruction, at the demodulator, of a periodic function synchronous with the carrier signal. This can be achieved either by the additional transmission of an appropriate pilot synchronization signal, or by directly extracting this information from the modulated signal (sub-sect. 13.2.8).

Table 11.22

	$\underset{\sim}{r}(t)$	$d(t)$: synchronous $\Delta\alpha \neq 0$	$d(t)$: isochronous $\Delta\alpha = 0$
AM	$\hat{U}_p[1 + m_0(t)]$	$\frac{\hat{U}_p}{2}[1 + m_0(t)]\cos\Delta\alpha$	$\frac{\hat{U}_p}{2}[1 + m_0(t)]$
DSBSC	$\hat{U}_p m_0(t)$	$\frac{\hat{U}_p}{2} m_0(t)\cos\Delta\alpha$	$\frac{\hat{U}_p}{2} m_0(t)$
SSB	$\frac{\hat{U}_p}{2}[m_0(t) + j\check{m}_0(t)$	$\frac{\hat{U}_p}{4}[m_0(t)\cos\Delta\alpha - \check{m}_0(t)\sin\Delta\alpha]$	$\frac{\hat{U}_p}{4} m_0(t)$
VSB	$\frac{\hat{U}_p}{2}[(1) + m_0(t) + j\check{m}_0(t) * h(t)]$	$\frac{\hat{U}_p}{4}[(1) + m_0(t)]\cos\Delta\alpha - [\check{m}_0(t) * h(t)]\sin\Delta\alpha$	$\frac{\hat{U}_p}{4}[(1) + m_0(t)]$

11.2.7 Demodulation by envelope detection

In some situations, synchronous detection can be replaced by simple envelope detection, which requires no auxiliary reference signal. This is the method favored in AM radio broadcasting receivers. It is used not only in amplitude modulation, but, with some distortion, in SSB and VSB modulation with carrier.

A general expression of the complex envelope valid for these different cases is

$$\underset{\sim}{r}(t) = a(t) + jb(t) = k\hat{U}_p\{[1 + m_0(t)] + j\check{m}_0(t) * h(t)\} \tag{11.27}$$

with $k = 1$ and $b(t) = \hat{U}_p\check{m}_0(t) * h(t) = 0$ in AM, and $k = ½$ and $b(t) = ½\hat{U}_p\check{m}_0(t)$ in SSB $[h(t) = \delta(t)]$.

The real envelope $r(t)$ is the absolute value of the complex envelope (11.2):

$$r(t) = |\underset{\sim}{r}(t)| = \sqrt{a^2(t) + b^2(t)} \tag{11.28}$$

In AM modulation, $b(t) = 0$ and

$$r(t) = |\underset{\sim}{r}(t)| = |a(t)| = k\hat{U}_p|1 + m_0(t)| \tag{11.29}$$

We can see immediately that if condition (11.11) is satisfied, i.e., if

$$m_0(t) \geqslant -1 \tag{11.30}$$

then demodulation by envelope detection, after filtering the dc component, provides a signal proportional to the modulating signal.

In SSB or VSB modulation with carrier, the term $b(t)$ is not equal to zero. However, if

$$|b(t)| \ll |a(t)| \tag{11.31}$$

then result (11.29) is approximately obtained.

The residual distortion due to the quadrature component generally can be neglected if $|m_0(t)| \ll 1$.

An envelope detector is a ***nonlinear*** device. The principle of such a detector is represented in fig. 11.23. It includes an absolute value operator (bipolar rectifier) followed by a lowpass filter, which eliminates the undesired rectified components. This detector is often incorrectly referred to as "linear" in the specialized literature.

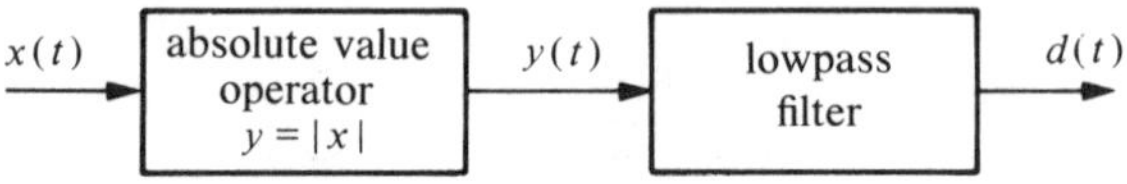

Fig. 11.23

If the input signal is amplitude modulated, then by (11.8) $\underset{\sim}{r}(t) = a(t) = [1 + m_0(t)]\hat{U}_p$ and $x(t) = s(t) = a(t) \cos(2\pi f_p t + \alpha_p)$; thus,

$$y(t) = |s(t)| = |a(t)| \cdot |\cos(2\pi f_p t + \alpha_p)| \tag{11.32}$$

Expanding $|\cos(2\pi f_p t + \alpha_p)|$ in Fourier series, we obtain

$$y(t) = \frac{2}{\pi}|a(t)| \cdot \left[1 + 2\sum_{n=1}^{\infty} \frac{(-1)^n}{4n^2 - 1} \cos\{2n(2\pi f_p t + \alpha_p)\}\right] \tag{11.33}$$

If the cut-off frequency f_c of the (assumed ideal) lowpass filter is such that

$$f_c < 2f_p - B_m \tag{11.34}$$

where B_m represents the maximum frequency of the spectrum of $m(t)$, then the demodulated signal becomes

$$d(t) = \frac{2}{\pi}|a(t)| = \frac{2}{\pi}\hat{U}_p|1 + m_0(t)| \tag{11.35}$$

and is similar to (11.29).

A variant of the envelope detector model is shown in fig. 11.24. It consists of a squaring operator, a lowpass filter, and a square-root operator in cascade. With $x(t) = s(t) = a(t) \cos(2\pi f_p t + \alpha_p)$, the output signal of the squaring operator becomes

$$\begin{aligned} y(t) &= x^2(t) = a^2(t)\cos^2(2\pi f_p t + \alpha_p) \\ &= \tfrac{1}{2}a^2(t) + \tfrac{1}{2}a^2(t)\cos(4\pi f_p t + 2\alpha_p) \end{aligned} \tag{11.36}$$

The signal $a(t) = \hat{U}_p[1 + m_0(t)]$ has its spectrum contained within the frequency range $|f| \leqslant B_m$. According to the results of sub-section 8.4.4, it

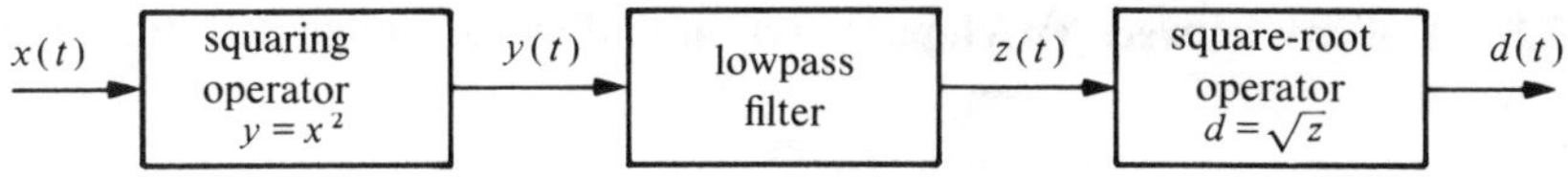

Fig. 11.24

appears that the term $a^2(t)$ has a spectrum within the range $|f| \leqslant 2B_m$. The multiplication of this term by $\cos(4\pi f_p t + 2\alpha_p)$ shifts this spectrum by a quantity $\pm 2f_p$. Hence, if

$$f_p > 2B_m \tag{11.37}$$

we can recover the first term of $y(t)$, according to (11.36), by a lowpass filter—again assumed ideal—with cut-off frequency:

$$f_c < 2(f_p - B_m) \tag{11.38}$$

Thus, we get

$$z(t) = \tfrac{1}{2}a^2(t) = \frac{\hat{U}_p^2}{2}\,|1 + m_0(t)|^2 \tag{11.39}$$

At the output of the square-root operator, we find

$$d(t) = \sqrt{z(t)} = \frac{\hat{U}_p}{\sqrt{2}}\,|1 + m_0(t)| \tag{11.40}$$

which is, once again, similar to (11.29).

If $|m_0(t)| \ll 1$, the output signal of the lowpass filter $z(t) \approx \tfrac{1}{2}\hat{U}_p^2 m_0(t)$ can be directly used to get a good approximation of the modulating signal.

In practice, the envelope detector is approximately implemented as a simple half-wave diode rectifier, with a capacitive load R_1C_1 (fig. 11.25) playing the role of a lowpass filter and a coupling capacitor C_2 eliminating the dc component.

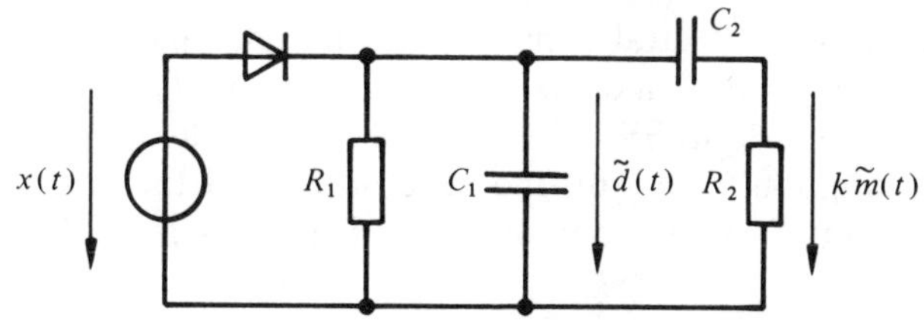

Fig. 11.25

11.2.8 Analogy between envelope detection and synchronous detection of an AM signal

Equation (11.32) which describes the output signal of the absolute value operator of fig. 11.23, if the AM signal envelope satisfies the condition $r(t) = a(t) \geqslant 0$, is equal to the product:

$$\begin{aligned} y(t) &= x(t) \cdot u_d(t) \\ &= a(t) \cos(2\pi f_p t + \alpha_p) \cdot \text{sgn}[\cos(2\pi f_p t + \alpha_p)] \end{aligned} \tag{11.41}$$

where $u_d(t) = \text{sgn}[\cos(2\pi f_p t + \alpha_p)]$ is synchronous and coherent (isochronous: $\Delta\alpha = 0$) with the carrier. This shows the existence of a formal link between synchronous detection, described in sub-section 11.2.6, and envelope detection.

11.2.9 Signal-to-noise ratio in amplitude modulation

With additive noise, the signal $x(t)$ at the demodulator input is

$$x(t) = s(t) + n(t) \tag{11.42}$$

where $s(t)$ is the modulated signal with bandpass spectrum defined by (11.1), and $n(t)$ is an independent noise with bandpass spectrum limited to the same bandwidth [e.g., $x(t)$ is the output signal of a (assumed ideal) filter rejecting all the signals or noise with frequencies outside the useful band]. We can also express it as (7.102):

$$n(t) = a_n(t) \cos(2\pi f_p t + \alpha_p) - b_n(t) \sin(2\pi f_p t + \alpha_p) \tag{11.43}$$

where $a_n(t)$ and $b_n(t)$ are the in-phase and quadrature components which have the same power as $n(t)$. They are Gaussian if $n(t)$ is also Gaussian (sub-sect. 7.4.6).

The signal-to-noise ratio at the demodulator input is

$$\xi_x = P_s / P_n \tag{11.44}$$

where P_s can be evaluated in the different modulation cases by integrating the corresponding power spectral densities. If the noise has a constant spectral density $\eta/2$ in the useful band, its power is equal to $P_n = 2\eta B_m$ in AM and DSBSC, and $P_n = \eta B_m$ in SSB. In the case of demodulation by synchronous detection (assumption: $\Delta\alpha = 0$), the general form of the demodulated signal (after suppression of the dc component in AM), taking (11.43) into account, becomes

$$d(t) = \tfrac{1}{2}[k\hat{U}_p m_0(t) + a_n(t)] \tag{11.45}$$

where $k = 1$ in AM and DSBSC, and $k = \tfrac{1}{2}$ in SSB.

The signal-to-noise ratio after detection is

$$\xi_d = k^2 \hat{U}_p^2 P_{m0}/P_n \tag{11.46}$$

where $P_{m0} = P_m/U_m^2$ is the power coefficient of the modulating signal.

We can measure the demodulation efficiency (noise immunity) with the ratio ξ_d/ξ_x. The results corresponding to the principal amplitude modulation types are summarized in table 11.26. We can see that the DSBSC efficiency is twice that of SSB. This can be explained by the action of the synchronous detector: it coherently adds the bilateral components of the DSBSC signal, while it incoherently adds those of the noise.

Table 11.26

Modulation	B_s	P_s	P_m	ξ_x	ξ_d	ξ_d/ξ_x	ξ_d/ξ_0
AM	$2B_m$	$\frac{1}{2}\hat{U}_p^2(1+P_{m0})$	$2\eta B_m$	$\frac{1}{4}\hat{U}_p^2\frac{1+P_{m0}}{\eta B_m}$	$\frac{1}{2}\hat{U}_p^2\frac{P_{m0}}{\eta B_m}$	$2P_{m0}<\frac{2P_{m0}}{1+P_{m0}}<2$	$P_{m0}<\frac{P_{m0}}{1+P_{m0}}<1$
DSBSC	$2B_m$	$\frac{1}{2}\hat{U}_p^2P_{m0}$	$2\eta B_m$	$\frac{1}{4}\hat{U}_p^2\frac{P_{m0}}{\eta B_m}$	$\frac{1}{2}\hat{U}_p^2\frac{P_{m0}}{\eta B_m}$	2	1
SSB	B_m	$\frac{1}{4}\hat{U}_p^2P_{m0}$	ηB_m	$\frac{1}{4}\hat{U}_p^2\frac{P_{m0}}{\eta B_m}$	$\frac{1}{4}\hat{U}_p^2\frac{P_{m0}}{\eta B_m}$	1	1

The performance analysis in presence of noise of an AM signal demodulation by envelope detection is more difficult because of the nonlinear character of the detectors. In the case of a Gaussian noise $n(t)$, the statistical distribution of the output signal of the envelope detector of fig. 11.23 is the Rice-Nakagami law established in sub-section 7.3.9. It can be used [40] to estimate the signal-to-noise ratio ξ_d obtained after demodulation. For a high signal-to-noise ratio at the input ($\xi_x \gg 1$), the obtained efficiency matches the result indicated in table 11.26 [with $P_{m0} < 1$ to ensure the nondistortion condition $|m_0(t)| < 1$]. This can be explained by the analogy mentioned in the previous sub-section. When $\xi_x \ll 1$, we observe a threshold effect such that the signal-to-noise ratio after detection, ξ_d, becomes proportional to ξ_x^2; this practically forbids the use of envelope detection in these circumstances.

To get an ***overall*** comparison of modulation-demodulation systems, we define a figure of merit ξ_d/ξ_0, the quotient of the signal-to-noise ratio after demodulation ξ_d, and a conventional signal-to-noise ratio:

$$\xi_0 = P_s/\eta B_m \tag{11.47}$$

This would be the signal-to-noise ratio of the direct (without modulation-demodulation) transmission of a modulating signal with a power P_s, disturbed by the same noise with spectral density $\Phi_n(f) = \eta/2$, ideally filtered at reception by a lowpass filter,with a $|G(f)|^2 = \text{rect}(f/2B_m)$ characteristic. This figure of merit, also shown in table 11.26, indicates that the DSBSC and SSB systems, in this respect, are equal, and in any case superior to the AM system.

11.3 FREQUENCY AND PHASE MODULATION

11.3.1 Angle modulation

The results of sub-section 11.2.9 show that in order to combat noise, the only solution, in the case of amplitude modulation, is to increase the modulated signal power. Angle modulation, which varies the phase or frequency of the sine wave carrier in proportion to the modulating signal, offers a way of increasing the noise immunity at the expense of spreading the modulated signal spectrum.

By contrast with amplitude modulation, the angle, sometimes called exponential, modulation is a ***nonlinear*** operation, which directly explains the broadening of the occupied bandwidth.

The modulated signal $s(t)$ is once again represented by equation (11.1). The complex envelope here becomes

$$\underline{r}(t) = \mathrm{S_m}\{m(t)\} = \hat{U}_\mathrm{p} \exp[\mathrm{j}\Delta\phi(t)] \tag{11.48}$$

where the argument $\Delta\phi(t)$ is a function of the modulating signal. Denoting the instantaneous phase of the modulated signal by $\phi_s(t)$, we thus have

$$\begin{aligned} s(t) &= \mathrm{Re}\{\underline{r}(t) \exp[\mathrm{j}(2\pi f_\mathrm{p} t + \alpha_\mathrm{p})]\} \\ &= \hat{U}_\mathrm{p}\mathrm{Re}\{\exp[\mathrm{j}\phi_s(t)]\} \\ &= \hat{U}_\mathrm{p}\cos\phi_s(t) = \hat{U}_\mathrm{p}\cos[2\pi f_\mathrm{p} t + \Delta\phi(t) + \alpha_\mathrm{p}] \\ &= \hat{U}_\mathrm{p}[\cos\Delta\phi(t)\cos(2\pi f_\mathrm{p} t + \alpha_\mathrm{p}) - \sin\Delta\phi(t)\sin(2\pi f_\mathrm{p} t + \alpha_\mathrm{p})] \end{aligned} \tag{11.49}$$

If the argument $\Delta\phi(t)$ of the complex envelope is directly proportional to the modulating signal, we deal with *phase modulation* (denoted by PM, or preferably by ΦM in order to avoid confusion with pulse modulation):

$$\Delta\phi(t) = \beta \cdot m(t) \tag{11.50}$$

If it is the derivative of $\Delta\phi(t)$, or, in other words, the instantaneous angular frequency (7.47), which is proportional to the modulating signal, then we are dealing with *frequency modulation* (FM):

$$\Delta f(t) = \frac{1}{2\pi}\frac{\mathrm{d}\Delta\phi(t)}{\mathrm{d}t} = v \cdot m(t) \tag{11.51}$$

and assuming an initial phase equal to zero:

$$\Delta\phi(t) = 2\pi\nu \int_0^t m(t)\mathrm{d}t \tag{11.52}$$

The β and ν constants are arbitrary factors measured in rad/V and in Hz/V, respectively, if $m(t)$ is a voltage. Figure 11.27 represents an example comparing ΦM and FM signals. Variables $\Delta\phi(t)$ and $\Delta f(t)$ are called *phase deviation* and *frequency deviation,* respectively.

The modulated signal can be produced by an electronic oscillator, the frequency of which is controlled by the modulating signal (VCO = voltage-controlled oscillator). When the carrier frequency f_p must be very stable and accurate, an indirect method with low-level phase modulation (Armstrong method: sub-sect. 11.3.3) followed by frequency multiplication (sub-sect. 11.5.3) is used.

Relations (11.51) and (11.52) show that the phase and frequency modulations relate to each other by a simple ***linear*** operation on the modulating signal: integration or derivation. Thus, we can use a frequency modulator to produce a phase modulated signal if the modulating signal has been previously differentiated.

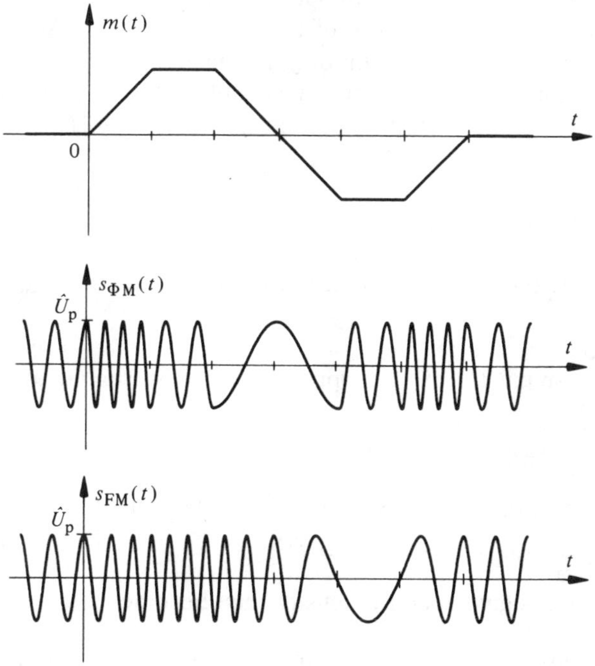

Fig. 11.27

11.3.2 Spectrum of angle modulated signals

The modulated signal (11.49) can be equivalently written as

$$s(t) = \hat{U}_p \mathrm{Re}\{\exp[j(2\pi f_p t + \alpha_p)] \cdot \exp[j\Delta\phi(t)]\}$$
$$= \hat{U}_p \mathrm{Re}\{\exp[j(2\pi f_p t + \alpha_p)] \cdot \left[1 + j\Delta\phi(t) - \frac{\Delta\phi^2(t)}{2!} - j\frac{\Delta\phi^3(t)}{3!} + \cdots\right] \tag{11.53}$$

where the expansion in series of $\exp[j\Delta\phi(t)]$ is emphasized. Because of the non-linear functions of $\Delta\phi(t)$, it is practically impossible to define a general expression for the modulated signal spectrum $\Phi_s(f)$. Some specific results have been established [26, 41, 66] under the assumption of a random, especially Gaussian, modulating signal.

In a general way, the spectral density of the modulated signal is equal to (1.5)

$$\Phi_s(f) = \tfrac{1}{4}[\Phi_{\underset{\sim}{r}}(-f - f_p) + \Phi_{\underset{\sim}{r}}(f - f_p)] \tag{11.54}$$

where $\Phi_{\underset{\sim}{r}}(f)$ is the spectral density (real, but not necessarily symmetrical with respect to f_p) of the complex envelope $\underset{\sim}{r}(t)$. The spectrum $\Phi_{\underset{\sim}{r}}(f)$ theoretically has an infinite bandwidth because of the contributions of $\Delta\phi^n(t)$ terms, with $n \rightarrow \infty$, appearing in the expansion of $\underset{\sim}{r}(t)$. The analysis shows, however, that these contributions rapidly become negligible when $|f|$ increases, so that $\Phi_s(f)$ is really different from zero only in the vicinity of $|f| = f_p$.

11.3.3 Approximation: low-level modulation

The rms value (standard deviation) of the phase $\Delta\phi(t)$ is $\sigma_{\Delta\phi}$. When

$$\sigma_{\Delta\phi} \ll 1 \tag{11.55}$$

the modulated signal (11.49) is approximately represented by the simplified expression:

$$s(t) \approx \hat{U}_p[\cos(2\pi f_p t + \alpha_p) - \Delta\phi(t) \sin(2\pi f_p t + \alpha_p)] \tag{11.56}$$

The complex envelope and its spectrum here are equal to

$$\underset{\sim}{r}(t) \simeq \hat{U}_p[1 + j\Delta\phi(t)] \text{ and } \Phi_{\underset{\sim}{r}}(f) \cong \hat{U}_p^2[\delta(f) + \Phi_{\Delta\phi}(f)].$$

The modulated signal spectral density then becomes

$$\Phi_s(f) \cong \frac{\hat{U}_p^2}{4}[\delta(f + f_p) + \delta(f - f_p)] + \frac{\hat{U}_p^2}{4}[\Phi_{\Delta\phi}(f + f_p) + \Phi_{\Delta\phi}(f - f_p)] \tag{11.57}$$

We can see, by comparing this expression with (11.9), the strong analogy between AM modulation and this low-level angular modulation, often called ***narrowband angle modulation.*** The occupied bandwidth B_s is also equal to twice the bandwidth of the modulating signal (fig. 11.28). Taking (11.50), (11.41), and (8.71) into account, the spectral density $\Phi_{\Delta\phi}(f)$ is related to the modulating signal spectrum by

$$\Phi_{\Delta\phi}(f) = \beta^2 \Phi_m(f) \tag{11.58}$$

in phase modulation, and by

$$\Phi_{\Delta\phi}(f) = \nu^2 \Phi_m(f)/f^2 \tag{11.59}$$

in frequency modulation.

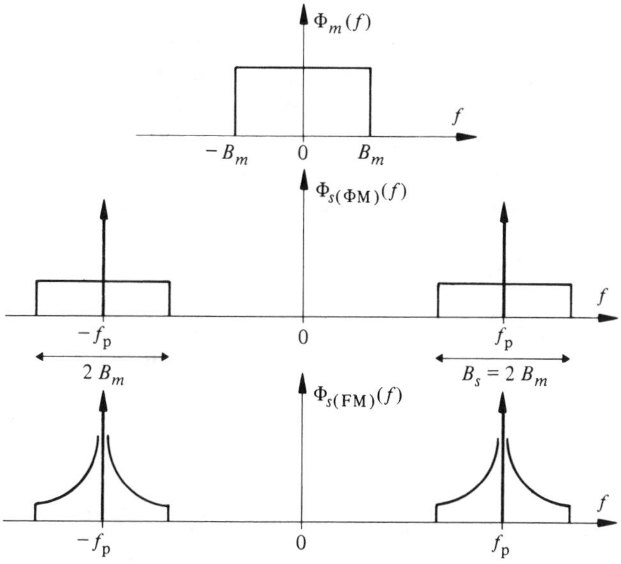

Fig. 11.28

The above mentioned analogy suggests a phase modulation method deduced from equation (11.56). It is the Armstrong method illustrated by the block diagram of fig. 11.29.

11.3.4 Approximation: high-level modulation

If the level of angular modulation is high, i.e., if

$$\sigma_{\Delta\phi} \gg 1 \tag{11.60}$$

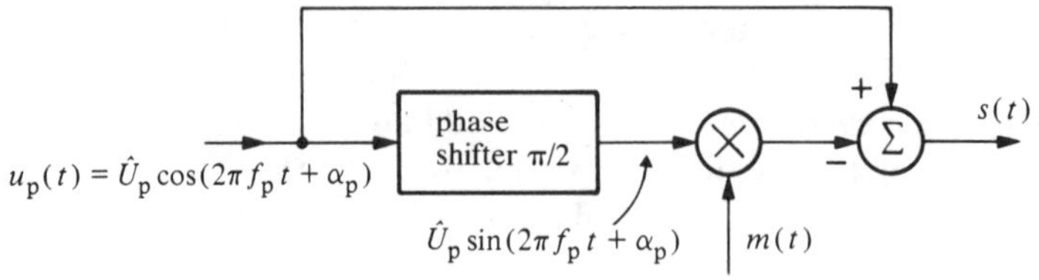

Fig. 11.29

the previous approximation is no longer valid. The modulated signal spectrum is then said to be ***wideband.***

A new approximation can be established [41] when the instantaneous frequency of the modulated signal $s(t)$ varies slowly enough so that the modulated signal looks like a sine wave of frequency $f_i = f_p + \Delta f$ during several periods. This occurs if $\Delta f \gg B_m$. We can then deduce the following: if $s(t)$ is ***frequency modulated*** according to (11.51) by a signal $m(t)$, the probability that the instantaneous frequency f_i is in a range df centered on $f_p + \Delta f$ is directly related to the probability that the amplitude of $m(t)$ is in a range dm centered on $\Delta f/\nu$. Taking (5.39) into account, the modulated signal spectrum is, under these conditions, approximately equal to

$$\Phi_s(f) \approx \frac{\hat{U}_p^2}{4\nu}\left[p_m\left(\frac{-f - f_p}{\nu}\right) + p_m\left(\frac{f - f_p}{\nu}\right)\right] \tag{11.61}$$

where $p_m(m)$ is the ***probability density*** of the modulating signal.

Using the concept of useful bandwidth B_u, introduced in sub-section 7.5.6, we have for $\sigma_{\Delta f} \gg B_m$:

$$B_s = B_u = \alpha\sigma_{\Delta f} = \alpha\nu\sigma_m \tag{11.62}$$

where α is a coefficient depending on the criterion used to define B_u; $\sigma_{\Delta f}$ and σ_m are the respective rms values of the instantaneous frequency and the modulating signal.

For example (fig. 11.30), if $m(t)$ is a Gaussian random signal, the ***spectral density*** $p_m(f/\nu)$ is a Gaussian law. If B_u is defined as the bandwidth containing about 95% of the total modulated signal power $P_s = \int\Phi_s(f)df = \hat{U}_p^2/2$, we have (see sect. 15.8): $\alpha \approx 4$ and $B_s = B_u = 4\sigma_{\Delta f} = 4\nu\sigma_m$.

11.3.5 Example: modulation by a Gaussian random signal

If the modulating signal is Gaussian, $\Delta\phi(t)$ is also Gaussian, the modulation being ΦM or FM, since we go from one to the other by a linear operation (sub-sect. 11.3.1 and 5.7.3). A general expression for the autocorrelation func-

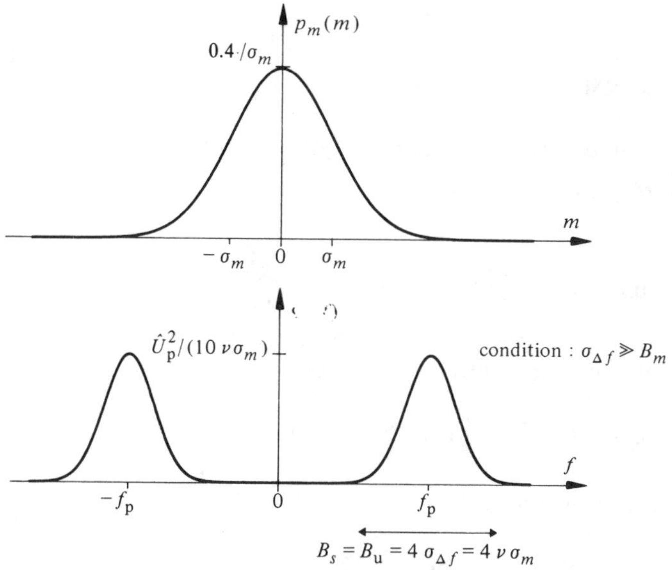

Fig. 11.30

tion $R_s(\tau)$ of the modulated signal can be established in this case. We can deduce from it the power spectral density by Fourier transform.

According to (7.40) and (7.92), the autocorrelation function of the modulated signal is

$$R_s(\tau) = \tfrac{1}{2}\mathrm{Re}\{\underline{\underline{R}}_s(\tau)\} \tag{11.63}$$

$$= \tfrac{1}{2}\mathrm{Re}\{\underline{R}_r(\tau)\exp(\,\mathrm{j}2\pi f_p\tau)\}$$

with, in the case of an angle modulation,

$$\underline{R}_r(\tau) = \mathrm{E}[\underline{\mathbf{r}}^*(t)\underline{\mathbf{r}}(t+\tau)]$$

$$= \hat{U}_p^2\mathrm{E}\{\exp(\,\mathrm{j}[\Delta\boldsymbol{\phi}(t+\tau) - \Delta\boldsymbol{\phi}(t)])\} \tag{11.64}$$

$$= \hat{U}_p^2\Pi_{\Delta\phi}(1, -1)$$

where $\Pi_{\Delta\phi}(1, -1)$ is the value in $u = 1$ and $v = -1$ of the second-order characteristic function $\Pi_{\Delta\phi}(u,v)$ of variables $\mathbf{x} = \Delta\boldsymbol{\phi}(t)$ and $\mathbf{y} = \Delta\boldsymbol{\phi}(t+\tau)$, defined by (14.66).

For a Gaussian process with zero mean value, $R_{\Delta\phi}(\tau) \equiv C_{\Delta\phi}(\tau)$, $R_{\Delta\phi}(0) \equiv C_{\Delta\phi}(0) = \sigma_{\Delta\phi}^2$ and, from (14.104),

$$\Pi_{\Delta\phi}(1, -1) = \exp[R_{\Delta\phi}(\tau) - \sigma_{\Delta\phi}^2] \tag{11.65}$$

Finally,

$$R_s(\tau) = \frac{\hat{U}_p^2}{2} \exp[R_{\Delta\phi}(\tau) - \sigma_{\Delta\phi}^2] \cos(2\pi f_p \tau) \tag{11.66}$$

The modulated signal spectral density then becomes, according to (11.54):

$$\Phi_s(f) = \tfrac{1}{4}[\Phi_{\underset{\sim}{r}}(f + f_p) + \Phi_{\underset{\sim}{r}}(f - f_p)] \tag{11.67}$$

with

$$\Phi_{\underset{\sim}{r}}(f) = \Phi_{\underset{\sim}{r}}(-f) = \mathrm{F}\{R_{\underset{\sim}{r}}(\tau)\} = \mathrm{F}\{\hat{U}_p^2 \exp[R_{\Delta\phi}(\tau) - \sigma_{\Delta\phi}^2]\} \tag{11.68}$$

Unfortunately, this Fourier transform usually cannot be expressed analytically.
Expanding the term $\exp[R_{\Delta\phi}(\tau)]$, we get

$$R_{\underset{\sim}{r}}(\tau) = \hat{U}_p^2 \exp(-\sigma_{\Delta\phi}^2) \cdot \left[1 + \sum_{n=1}^{\infty} R_{\Delta\phi}^n(\tau)/n!\right] \tag{11.69}$$

and by Fourier transform:

$$\Phi_{\underset{\sim}{r}}(f) = \hat{U}_p^2 \exp(-\sigma_{\Delta\phi}^2) \cdot \left[\delta(f) + \sum_{n=1}^{\infty} \frac{1}{n!} \mathop{\boxast}_{i=1}^{n} \Phi_{\Delta\phi}(f)\right] \tag{11.70}$$

where $\boxast$ denotes a multiple convolution product:

$$\mathop{\boxast}_{i=1}^{n} \Phi_{\Delta\phi}(f) = \underbrace{\Phi_{\Delta\phi}(f) * \Phi_{\Delta\phi}(f) * \ldots * \Phi_{\Delta\phi}(f)}_{n \text{ times}} \tag{11.71}$$

Whatever $\Phi_{\Delta\phi}(f)$, the multiple convolution leads to a Gaussian law when $n \rightarrow \infty$, in direct analogy with the central limit theorem (sub-sect. 5.5.3). The higher is $\sigma_{\Delta\phi}$, the larger is the occupied bandwidth B_s.

11.3.6 Example: sinusoidal modulation

The modulated signal spectrum can be exactly defined, assuming that the modulating signal is a pure sine wave.

Let

$$m(t) = \hat{U}_m \sin(2\pi f_m t) \tag{11.72}$$

According to (11.50) and (11.51), we obtain the instantaneous phase and frequency deviations:

$$\begin{aligned} \Delta\phi(t) &= \beta \hat{U}_m \sin(2\pi f_m t) \\ &= \Delta\phi_{max} \cdot \sin(2\pi f_m t) \end{aligned} \tag{11.73}$$

and

$$\Delta f(t) = \beta \hat{U}_m f_m \cos(2\pi f_m t)$$
$$= \Delta f_{\max} \cdot \cos(2\pi f_m t) \tag{11.74}$$

We call the *modulation index,* generally noted δ, the maximum phase deviation $\Delta\phi_{\max}$. Thus,

$$\delta = \Delta\phi_{\max} = \Delta f_{\max}/f_m \tag{11.75}$$

where $\Delta f_{\max}$ represents the maximum frequency deviation.

The complex envelope (11.48) becomes

$$\underset{\sim}{r}(t) = \hat{U}_p \exp[j\delta \sin(2\pi f_m t)] \tag{11.76}$$

It is a periodic complex function, of period $T = 1/f_m$, which can be expanded in Fourier series as a result of the Bessel-Jacobi identity [58]:

$$\exp[j\delta \sin(2\pi f_m t)] = \sum_{n=-\infty}^{\infty} J_n(\delta) \exp(j2\pi f_m t) \tag{11.77}$$

The Fourier coefficients (3.75) of expansion (11.77) are particular values of the Bessel functions of the first kind (sect. 15.7):

$$J_n(\delta) = f_m \int_{-(2f_m)^{-1}}^{(2f_m)^{-1}} \exp\{j[\delta \sin(2\pi f_m t) - n2\pi f_m t]\}dt \tag{11.78}$$

Thus, the complex envelope can be written here as

$$\underset{\sim}{r}(t) = \sum_{n=-\infty}^{\infty} \hat{U}_p J_n(\delta) \exp(j2\pi n f_m t) \tag{11.79}$$

and its spectral density (4.139) becomes with $\underset{\sim}{R}_n = \hat{U}_p J_n(\delta)$,

$$\Phi_{\underset{\sim}{r}}(f) = \sum_{n=-\infty}^{\infty} \hat{U}_p^2 J_n^2(\delta) \cdot \delta(f - nf_m) \tag{11.80}$$

The modulated signal spectral density (fig. 11.31) is then deduced from (11.67), taking into account that here $\Phi_{\underset{\sim}{r}}(-f) = \Phi_{\underset{\sim}{r}}(f)$:

$$\Phi_s(f) = \frac{1}{4}[\Phi_{\underset{\sim}{r}}(f + f_p) + \Phi_{\underset{\sim}{r}}(f - f_p)]$$
$$= \frac{\hat{U}_p^2}{4} \sum_{n=-\infty}^{\infty} J_n^2(\delta) \cdot [\delta(f + f_p - nf_m) + \delta(f - f_p - nf_m)] \tag{11.81}$$

with, as a result of the Parseval theorem (3.76),

$$\sum_{n=-\infty}^{\infty} J_n^2(\delta) = 1 \tag{11.82}$$

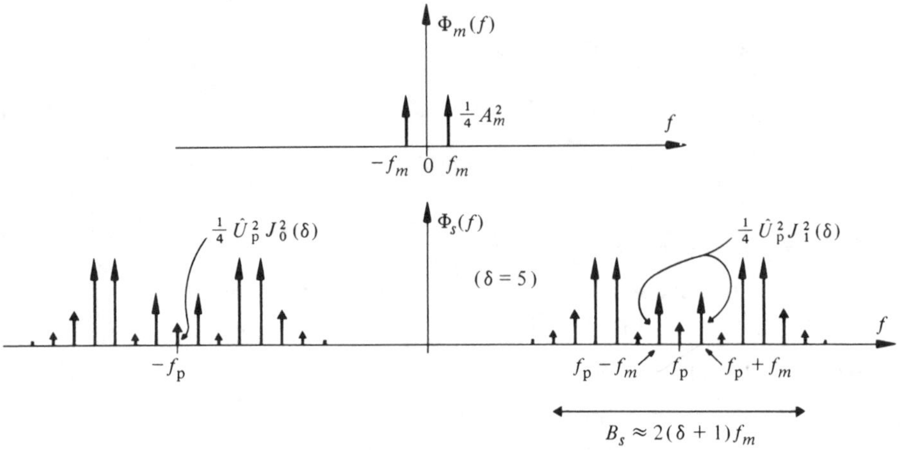

Fig. 11.31

The modulated signal can be expressed as

$$s(t) = \mathrm{Re}\{\underset{\sim}{r}(t) \exp[\mathrm{j}(2\pi f_p t + \alpha_p)]\}$$

$$= \hat{U}_p \sum_{n=-\infty}^{\infty} J_n(\delta) \mathrm{Re}\{\exp[\mathrm{j}(2\pi(f_p + nf_m)t + \alpha_p)]\}$$

$$= \hat{U}_p \sum_{n=-\infty}^{\infty} J_n(\delta) \cos[2\pi(f_p + nf_m)t + \alpha_p] \tag{11.83}$$

Although the series is theoretically infinite, coefficients $J_n(\delta)$ tend rapidly towards zero (fig. 15.2) when n tends toward infinity in compliance with (11.82). The useful bandwidth, therefore, remains limited and can be estimated by the "*empirical Carson's rule*":

$$B_s \approx 2(\Delta f + f_m)$$

$$= 2(\delta + 1)f_m \tag{11.84}$$

At a low level ($\delta \ll 1$), the useful bandwidth is roughly equal to twice the frequency of the modulating signal, the lines at $|f_p|$ and $|f_p \pm f_m|$ being the only ones having non-negligible amplitudes. This conforms to the general result (11.57),with $J_0(\delta) \approx 1$ and $J_1(\delta) = \delta/2$ for $\delta \ll 1$.

In wideband modulation ($\delta \gg 1$), an approximate spectral envelope of $\Phi_s(f)$ is obtained using relation (11.61). The probability density of the signal (11.74), which frequency modulates the carrier, from (5.23) and with $\nu = \Delta f$

$= \delta \cdot f_m$, is

$$\frac{1}{\nu} p_m \left(\frac{f}{\nu}\right) = \frac{1}{\pi\sqrt{(\delta \cdot f_m)^2 - f^2}} ; \quad f \leqslant \delta \cdot f_m \tag{11.85}$$

yielding

$$\Phi_s(f) \approx \frac{\hat{U}_p^2}{4\pi} \left[\frac{1}{\sqrt{(\delta \cdot f_m)^2 - (f + f_p)^2}} + \frac{1}{\sqrt{(\delta \cdot f_m)^2 - (f - f_p)^2}}\right] \tag{11.86}$$

The occupied bandwidth estimated by this model is obviously equal to $B_s = 2\delta \cdot f_m$, in agreement with the Carson's rule for $\delta \gg 1$.

11.3.7 Difference between ΦM and FM modulations

In the case of a sinusoidal phase modulation, the modulation index defined by (11.75) is a constant that does not depend on the modulating signal frequency, f_m. A modification of f_m produces a simple expansion (or contraction) of the spectrum $\Phi_s(f)$ with respect to $|f_p|$, which results in a variation of the useful bandwidth proportional to f_m.

For a sinusoidal frequency modulation, the maximum frequency deviation Δf_{max} is a constant and the modulation index now varies proportionally to the inverse of f_m. Consequently, a change of the modulating signal frequency yields not only a spectral expansion (or contraction), but also a modification of the coefficients $J_n^2(\delta)$. The useful bandwidth remains, in this case, relatively constant according to the Carson rule.

11.3.8 Example: phase modulation by a random binary signal

Discrete angle modulation is widely used for the transmission of digital information (data transmission). If the modulated variable is the phase, we speak of PSK (*phase shift keying*) *modulation,* if it is the frequency, we speak of FSK (*frequency shift keying*) *modulation.*

Consider a binary PSK modulation

$$\Delta\phi(t) = \frac{\pi}{2}[1 + m_0(t)] = \begin{cases} 0 \\ \pi \end{cases} \tag{11.87}$$

where

$$m_0(t) = \pm 1 \tag{11.88}$$

is a random binary signal, carrying digital information, with two states +1 and −1 that are equally likely. The complex envelope becomes simply

$$\underline{r}(t) = \hat{U}_p \exp[j\Delta\phi(t)] \triangleq \hat{U}_p m_0(t) \tag{11.89}$$

and the modulated signal here is equivalent to an amplitude modulated signal:

$$s(t) \triangleq \hat{U}_p m_0(t) \cos(2\pi f_p t + \alpha_p) \tag{11.90}$$

According to (11.15)

$$\Phi_s(f) = \frac{\hat{U}_p^2}{4} [\Phi_{m0}(f + f_p) + \Phi_{m0}(f - f_p)] \tag{11.91}$$

with, in the case of a clocked NRZ binary signal (sub-sect. 5.3.7 and 5.6.4) having a rate $1/T$ bits/s and levels ± 1:

$$\Phi_{m0}(f) = T \text{ sinc}^2(Tf) \tag{11.92}$$

and, in the case of a diphase binary signal (sub-sect. 5.3.8) of same rate:

$$\Phi_{m0}(f) = T[\text{sinc}^2(Tf/2) - \text{sinc}^2(Tf)] \tag{11.93}$$

The useful bandwidth is on the order of $2/T$ in the first case and $4/T$ in the second, as can be deduced from fig. 5.16.

For binary FSK modulation, the exact determination of the modulated signal spectrum is much more complicated [115] and will not be attempted here. Simply note that

$$\Delta f(t) = \Delta f \cdot m_0(t) = \pm \Delta f \tag{11.94}$$

and that the phase variation, which here is a ***continuous function,*** becomes

$$\Delta\phi(t) = 2\pi\Delta f \cdot \int m_0(t)\mathrm{d}t \tag{11.95}$$

If phase discontinuities at each frequency change are allowed (problem 11.6.12), we can model the result as a combination of two sinusoidal OOK (*on-off keying*) *modulated* signals.

11.3.9 Demodulation methods

The basic operation of frequency demodulation is differentiation:

$$d(t) = S_d\{\underline{r}(t)\} = \frac{1}{2\pi}\frac{\mathrm{d}}{\mathrm{d}t} \arg[\underline{r}(t)] = k\frac{1}{2\pi}\frac{\mathrm{d}}{\mathrm{d}t}\Delta\phi(t) \tag{11.96}$$

An indirect phase demodulation is obtained by cascade connection of a frequency demodulator (discriminator) and an integrator circuit.

Among the various methods used to implement the frequency discrimination function, there are

- FM/AM conversion followed by envelope detection (fig. 11.32);
- FM/PFM (sequence of frequency modulated pulses) conversion followed by a time-averager (fig. 11.33);
- phase-locked-loop (PLL) detection (fig. 11.34).

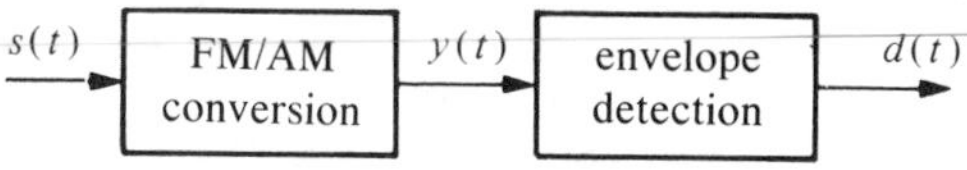

Fig. 11.32

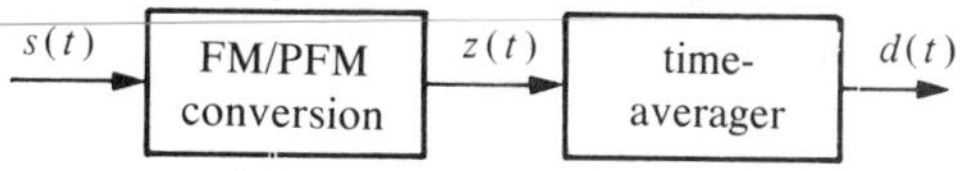

Fig. 11.33

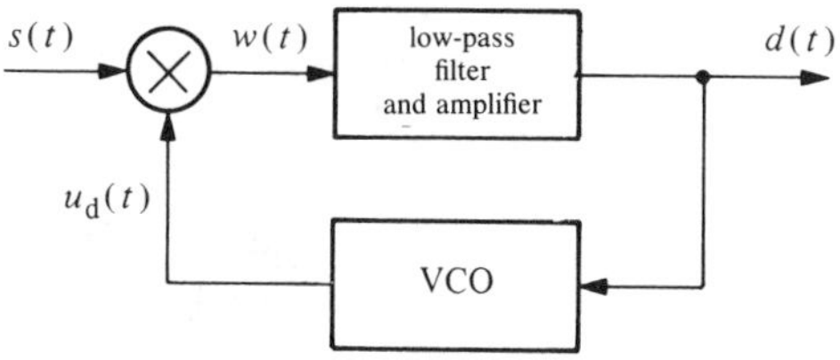

Fig. 11.34

A differentiation operator (sub-sect. 8.2.28) is the ideal model of a FM/AM converter because the module of its harmonic transfer function is proportional to frequency: $|G(f)| = 2\pi f$. If $s(t) = \hat{U}_p \cos[\omega_p t + \Delta\phi(t)]$, the differentiator output is

$$y(t) = \frac{\mathrm{d}}{\mathrm{d}t}\,s(t) = -\hat{U}_p[\omega_p + \frac{\mathrm{d}}{\mathrm{d}t}\,\Delta\phi(t)] \cdot \sin[\omega_p t + \Delta\phi(t)] \tag{11.97}$$

In frequency modulation: $\mathrm{d}\Delta\phi(t)/\mathrm{d}t = 2\pi\nu m(t)$. As long as $\Delta f(t) = \nu m(t) < f_p$, the output of the envelope detector will be (after suppression of the dc component $\hat{U}_p\omega_p$):

$$d(t) = \hat{U}_p 2\pi\nu m(t) = k_1 m(t) \tag{11.98}$$

To improve the immunity to possible amplitude fluctuations in the modulated signal, a clipper circuit (tab. 8.48) is placed ahead of the FM/AM converter.

In practice, any filter with a transfer function modulus that varies as a function of f in the spectral range occupied by $s(t)$ can be used as a FM/AM converter. This is obtained at the expense of some distortion generated by the $|G(f)|$ nonlinearity. The simplest example is a resonant circuit tuned at a frequency close to the carrier frequency. An approximate localized compensation for the nonlinearity of the characteristic $|G(f)|$ is often implemented by combining two opposite nonlinearities [116]. This principle is used in radio-wave receivers (ratio detector).

The FM/PFM conversion is implemented by detecting, with a comparator, the zero crossings of $s(t)$ and by producing at each zero-crossing instant t_n a calibrated pulse $g(t)$. The output signal of the FM/PFM converter is thus a pulse sequence, the density of which varies with the instantaneous frequency of $s(t)$:

$$z(t) = \sum_n g(t - t_n) \tag{11.99}$$

The time-averager (sub-sect. 8.2.19) computes a running average of $z(t)$ on an interval $f_p^{-1} \ll T \ll B_m^{-1}$

$$d(t) = \frac{1}{T} \int_{t-T}^{t} z(t')\mathrm{d}t' \approx d_0 + k_2 \cdot m(t) \tag{11.100}$$

This method yields a very linear demodulation for very wide frequency deviations (instrumentation, telemetry, measurement recorders).

The frequency or phase demodulation by a phase-locked-loop (PLL) [117, 118] is also widely used in telecommunications. Its principle is simple: a voltage-controlled oscillator (VCO) produces a periodic signal $u_d(t)$ synchronized with the incident signal $s(t)$ by the feedback loop. If $s(t)$ is frequency modulated, the control voltage of the VCO varies in proportion to the instantaneous frequency deviation (problem 11.6.11). The detailed analysis of the phase-locked-loop behavior is beyond the scope of this book.

11.3.10 Signal-to-noise ratios in angle modulations

The detailed performance analysis of a ΦM or FM modulated system in the presense of noise is complex because of the nonlinear relationships involved [26, 40, 41]. It is, however, possible to deduce easily some basic results for the usual case where the noise power is much smaller than the carrier power.

Consider simply an unmodulated carrier (it can be shown that the noise plays an identical role) received in the presence of an additive noise $n(t)$ with a bandpass spectrum fixed by the receiver selective filter centered on the carrier frequency f_p. Using the expansion (11.43), the combined signal becomes

$$\begin{aligned} x(t) &= s(t) + n(t) \\ &= [\hat{U}_p + a_n(t)] \cos(2\pi f_p t + \alpha_p) - b_n(t) \sin(2\pi f_p t + \alpha_p) \\ &= r_x(t) \cos(2\pi f_p t + \Delta\phi_n(t) + \alpha_p) \end{aligned} \tag{11.101}$$

with

$$r_x(t) = \{[\hat{U}_p + a_n(t)]^2 + b_n^2(t)\}^{1/2} \tag{11.102}$$

and

$$\Delta\phi_n(t) = \arctan \frac{b_n(t)}{\hat{U}_p + a_n(t)} \tag{11.103}$$

The noise at the demodulator input thus influences both the amplitude and the phase. However, the amplitude fluctuations are eliminated by the clipper circuit incorporated in the demodulator. Consequently the sole disturbance that ought to be considered is the ***phase noise*** $\Delta\phi_n(t)$ defined by (11.103).

Under the assumption of a strong input signal-to-noise ratio $\xi_x = P_s/P_n = \frac{1}{2}(\hat{U}_p/\sigma_n)^2 \gg 1$, relation (11.103) becomes approximately

$$\Delta\phi_n(t) \approx b_n(t)/\hat{U}_p \tag{11.104}$$

The phase noise is thus nearly proportional to the quadrature component of the input noise, and its spectral density is equal to

$$\Phi_{\Delta\phi}(f) \approx \hat{U}_p^{-2}\Phi_{bn}(f) \tag{11.105}$$

In the case of frequency demodulation, we obtain, from (11.96), a ***frequency noise***:

$$\Delta f_n(t) = \frac{1}{2\pi}\frac{\mathrm{d}}{\mathrm{d}t}\Delta\phi_n(t) \tag{11.106}$$

the spectral density of which is, from (8.71),

$$\Phi_{\Delta f}(f) = f^2\Phi_{\Delta\phi}(f) \tag{11.107}$$

The demodulator output lowpass filter eliminates all the frequency components above B_m, the bandwidth of the modulating signal. If the noise $n(t)$ has a uniform spectrum $\Phi_n(f) = \eta/2$ in the considered band B_s centered on f_p, the spectrum of the quadrature component is equal to (sub-sect. 7.4.6): $\Phi_{bn}(f) = \eta\,\mathrm{rect}(f/B_s)$. The phase noise spectrum, therefore, is also uniform and the frequency noise spectrum has a parabolic shape (fig. 11.35).

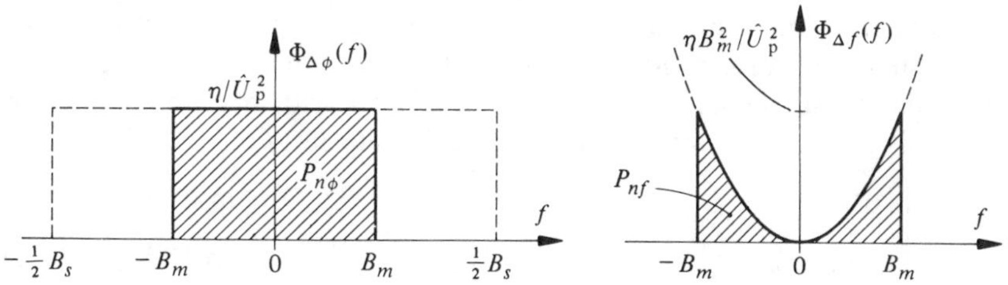

Fig. 11.35

The power (variance) of the phase noise at a ΦM demodulator output is equal to

$$P_{n\phi} = 2\int_0^{B_m} \Phi_{\Delta\phi}(f)\mathrm{d}f = 2\eta B_m/\hat{U}_\mathrm{p}^2 \tag{11.108}$$

The power of the frequency noise observed at the FM demodulator output is equal to

$$P_{nf} = 2\int_0^{B_m} \Phi_{\Delta f}(f)\mathrm{d}f = 2\eta B_m^3/(3\hat{U}_\mathrm{p}^2) \tag{11.109}$$

If B_s is the useful bandwidth occupied by the modulated signal, the signal-to-noise ratio at the demodulator input is equal to

$$\xi_x = P_s/P_n = \tfrac{1}{2}\hat{U}_\mathrm{p}^2/(\eta B_s) \tag{11.110}$$

In the absence of noise, the demodulated signal is proportional to the phase deviation $\Delta\phi(t)$ in ΦM,and to the frequency deviation $\Delta f(t)$ in FM. Denoting the respective variances by $\sigma_{\Delta\phi}^2$ and $\sigma_{\Delta f}^2$, and assuming a proportionality coefficient equal to one, the signal-to-noise ratio after demodulation becomes

$$\xi_d = \begin{cases} \dfrac{\hat{U}_\mathrm{p}^2}{2\eta B_m}\,\sigma_{\Delta\phi}^2 & \text{in } \Phi\text{M} \\[2ex] \dfrac{3}{2}\,\dfrac{\hat{U}_\mathrm{p}^2}{\eta B_m^3}\,\sigma_{\Delta f}^2 & \text{in FM} \end{cases} \tag{11.111}$$

The ξ_d/ξ_x ratio provides, as in 11.2.9, an indication of the demodulator efficiency (noise immunity):

$$\xi_d/\xi_x = \begin{cases} \sigma_{\Delta\phi}^2 B_s/B_m & \text{in } \Phi\text{M} \\ 3\sigma_{\Delta f}^2 B_s/B_m^3 & \text{in FM} \end{cases} \tag{11.112}$$

While this efficiency is bounded in amplitude demodulation (tab. 11.26), we can see that ***an important improvement of the noise immunity is possible in angle modulation at the expense of a spreading of the bandwidth B_s occupied by the modulated signal.*** Indeed, this bandwidth depends directly on $\sigma_{\Delta\phi}$ or $\sigma_{\Delta f}$ (subsect. 11.3.4).

To obtain an overall comparison (tab. 11.36) with amplitude modulation systems, we use the figure of merit ξ_d/ξ_0, where $\xi_0 = \hat{U}_\mathrm{p}^2/(2\eta B_m)$ is the conventional signal-to-noise ratio of the direct transmission (without modulation) of a modulating signal with power $P_s = \hat{U}_\mathrm{p}^2/2$ and with maximum frequency B_m, which is affected by a noise of spectral density $\Phi_n(f) = \eta/2$:

$$\xi_d/\xi_0 = \begin{cases} \sigma_{\Delta\phi}^2 & \text{in } \Phi\text{M} \\ 3\sigma_{\Delta f}^2/B_m^3 & \text{in FM} \end{cases} \tag{11.113}$$

In amplitude modulation, this ratio is always equal or lower than one.

Table 11.36

Modulation	$\underset{\sim}{r}(t)$	B_s/B_m	ξ_d/ξ_0
AM	$\hat{U}_p[1 + m_0(t)]$	2	< 1
DSBSC	$\hat{U}_p m_0(t)$	2	1
SSB	$\frac{\hat{U}_p}{2}[m_0(t) + j\check{m}_0(t)]$	1	1
ΦM	$\hat{U}_p \exp[j\beta m(t)]$	$2(\delta + 1)$	$\delta^2/2$
FM	$\hat{U}_p \exp[j2\pi\nu \int_0^t m(t)\,dt]$	$2(\delta + 1)$	$3\delta^2/2$

11.3.11 Example: sinusoidal angle modulation in the presence of noise

In the case of a sinusoidal modulating signal of frequency f_m, the rms values of the phase or frequency deviations become, from (11.73), (11.74) and (11.75):

$$\sigma_{\Delta\phi} = \delta/\sqrt{2} \tag{11.114}$$

$$\sigma_{\Delta f} = \delta f_m/\sqrt{2} \tag{11.115}$$

Identifying the bandwidth B_m with f_m (which is questionable!), and inserting it in (11.13), we get

$$\xi_d/\xi_0 = \begin{cases} \tfrac{1}{2}\delta^2 & \text{in } \Phi\text{M} \\ \tfrac{3}{2}\delta^2 & \text{in FM} \end{cases} \tag{11.116}$$

This figure of merit is proportional to the square of the modulation index δ. Consequently, it can be much greater than one, if $\delta \gg 1$. The bandwidth, given by Carson's rule (11.84), is then proportional to the modulation index: $B_s \approx 2\delta f_m = 2\delta B_m \gg B_m$.

11.3.12 Observation

The parabolic variation of the frequency noise spectrum (fig. 11.35) suggests an additional means of improving the demodulator efficiency. We can indeed modify the modulating signal power distribution at emission, by accentuating the high frequencies with an appropriate filtering (pre-emphasis). The linear distortion thereby introduced is compensated, after demodulation, by a second filtering (de-emphasis) that strongly reduces the frequency noise, which is mainly distributed over the high frequencies.

This method, which, in fact, creates a mixed FM-ΦM modulation, is used, for instance, in FM broadcasting and in transmissions via radio link (vol. XVIII).

11.4 PULSE MODULATIONS

11.4.1 Introduction

Pulse modulation is associated with the sampling principle described in chapter 9. Depending on the type of pulse modulation, it can offer an improved noise immunity at the expense of a larger occupied bandwidth, as in angular modulation. Nonetheless, its main advantage is to allow the simultaneous transmission of several messages by time multiplexing (sub-sect. 11.1.1). It is also used in some industrial-control applications. Compared to analog pulse modulation (fig. 11.37: PAM, PDM, PPM, PFM), the representation method—already described in chap. 10—known as pulse code modulation (PCM) is gaining more and more interest and applications.

We will limit ourselves here to a very short description of these various modulations. The reader will find more details in vol. XVIII and in the important specialized literature dedicated to telecommunications. References [40, 41] contain elaborate mathematical analysis of spectra and of performances in the presence of noise for these various modulations.

11.4.2 Pulse amplitude modulation (PAM)

In *pulse amplitude modulation* (PAM), the modulated signal corresponds to the general equation (11.6) with $a_k = m(kT)$, $\tau_k = \frac{1}{2}\Delta$ (or any other arbitrary value) and $\Delta_k = \Delta = \text{constant}$. Such a signal is obtained by periodic sampling and holding (sub-sect. 9.2.4), when the pulse is rectangular, and its spectral density is deduced, by analogy, from equation (9.17).

A demodulation (reconstruction) is, hence, achievable by a simple lowpass filtering with equalization.

As is the case with any amplitude modulation, this method offers limited performance with additive noise and, consequently, is not widely used.

11.4.3 Pulse duration modulation (PDM)

In *pulse duration modulation* (PDM, or PWM, *pulsewidth modulation*), the modulated signal parameters (11.6) are $a_k = a = \text{constant}$, $\tau_k = \frac{1}{2}T$ and $\Delta_k = \frac{1}{2}T + \alpha m(kT)$, where α is a proportionality coefficient.

The integral of each pulse varies linearly with the corresponding sample $m(kT)$, as in PAM modulation. This also allows us to obtain demodulation by a simple lowpass filtering (averaging filter). A constant amplitude makes pulse regeneration easier in the presence of additive noise.

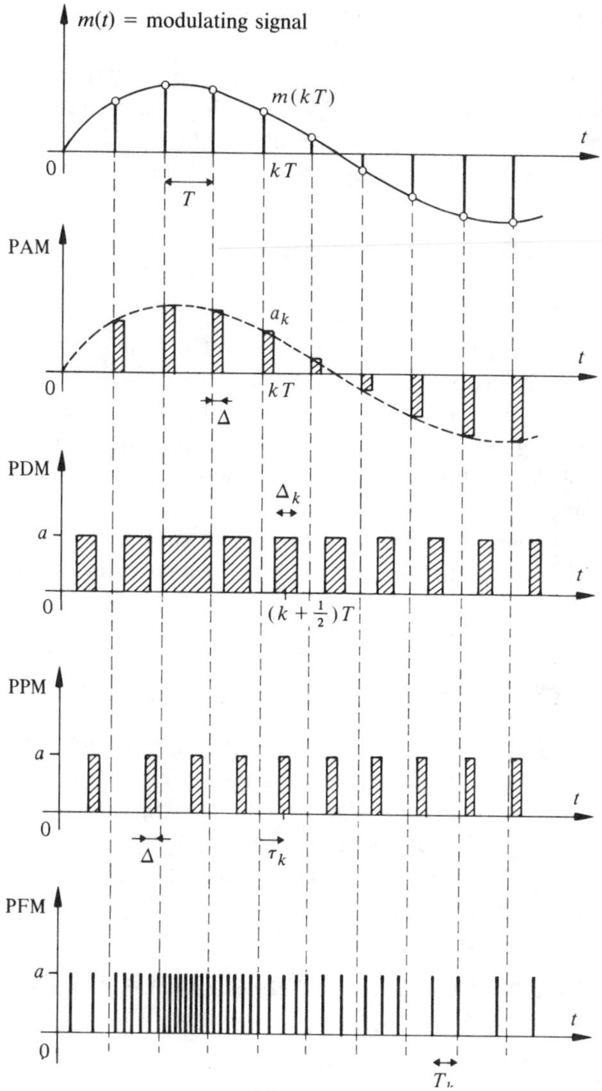

Fig. 11.37

11.4.4 Pulse position modulation (PPM)

In *pulse position modulation* (PPM), the modulated signal parameters are $a_k = a$ = constant, $\tau_k = T + \alpha m(kT)$ and $\Delta_k = \Delta$ = constant.

The invariable pulse format makes this type of modulation attractive in some circumstances: the modulated signal power is constant and independent of the modulating signal; pulse regeneration is easily accomplished in the presence of noise.

This modulation can be deduced from a PDM modulation: the calibrated pulse being generated at the trailing edge of each pulse of the PDM signal.

Conversely, the indirect demodulation of a PPM signal is implemented by a PPM/PDM conversion, followed by a lowpass filtering.

11.4.5 Pulse frequency modulation (PFM)

Pulse frequency modulation (PFM) is similar to PPM modulation. The parameters a_k and Δ_k of (11.6) are also constant in this case, but instead of varying the delay τ_k in proportion to the modulating signal, it is the inverse of the instantaneous period $1/T_k = 1/T + \nu m(t)$ that linearly varies as a function of the modulating signal. This prohibits its use in time multiplexing. PFM modulation is to PPM modulation what FM modulation of a sinusoidal carrier is to ΦM modulation.

Because the number of pulses per unit time varies in proportion to the modulating signal, simple demodulation by a lowpass filter (averager) is also possible.

11.4.6 Digital pulse modulations

As we have already seen, the name *pulse code modulation* (PCM) is given, in telecommunications, to the digital representation described in chap. 10. The names *differential pulse code modulation* (DPCM) and *delta modulation* (ΔM) are given to special digital representation methods of the primary signal derivative. In fact, these are more properly coding than modulation methods, if we define modulation as in sub-section 11.1.1.

11.5 FREQUENCY TRANSLATION AND MULTIPLICATION

11.5.1 Principle of frequency translation

Frequency translation is a spectral translation operation similar to amplitude modulation. In the latter, the spectrum of a low-frequency primary signal is shifted of a quantity equal to the carrier frequency f_p by multiplying the primary signal with a sine wave of frequency f_p. The same principle applied to a primary signal with bandpass spectrum is called frequency translation. It is often implemented with an auxiliary oscillator (heterodyne), the frequency f_0 of which can be varied.

If the signal with bandpass spectrum is

$$s(t) = m_0(t)\hat{U}_p \cos(2\pi f_p t) \tag{11.117}$$

and if the auxiliary signal is

$$u_0(t) = \cos(2\pi f_0 t) \tag{11.118}$$

then the product

$$z(t) = s(t)u_0(t) = \tfrac{1}{2}\hat{U}_p m_0(t)\{\cos[2\pi(f_0 + f_p)t] + \cos[2\pi(f_0 - f_p)t]\} \tag{11.119}$$

has a spectral density (fig. 11.38):

$$\Phi_z(f) = \frac{\hat{U}_p^2}{16}[\Phi_{m0}(f + f_0 + f_p) + \Phi_{m0}(f + f_0 - f_p) + \Phi_{m0}(f - f_0 + f_p) + \Phi_{m0}(f - f_0 - f_p)] \tag{11.120}$$

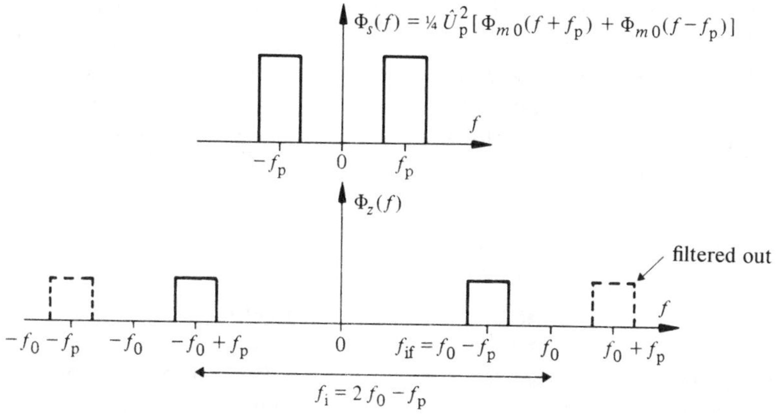

Fig. 11.38

A bandpass postfiltering eliminates the undesired spectral components, for instance, those around $|f_0 + f_p|$.

As in amplitude modulation, the multiplication can be achieved either directly, with an appropriate parametric operator (sub-sect. 8.3.4), or indirectly by superimposing the carrier and the primary signal $s(t)$ at the input of an amnesic nonlinear operator (sub-sect. 8.4.2).

In a receiver with frequency translation (superheterodyne), the frequency of the local oscillator f_0 is chosen in order to keep $f_0 - f_p = f_{if} =$ constant, where f_p is the frequency of the received carrier. This allows us to amplify the received signal with a selective amplifier tuned on the fixed intermediate frequency f_{if} (usual values of f_{if} in radio broadcasting: 455 kHz in amplitude modulation and 10.7 MHz in frequency modulation).

11.5.2 Image-frequency

The signal with bandpass spectrum $s(t)$ defined by (11.117) is not the only one that, after multiplication with (11.118), generates spectral components in the vicinity of the intermediate frequency f_{if}. Indeed, it results from the spectral convolution corresponding to the product (11.119) that such components will also be given by a bandpass spectrum signal $s'(t)$, the spectral bands of which are centered on frequencies $f = |f_i|$ with

$$f_i = 2f_0 \pm f_p \tag{11.121}$$

After multiplication by $u_0(t)$, we have a contribution shifted to $f_{if} = f_0 \pm f_p$.

The frequency f_i, defined by (11.121), is called the *image frequency* of the desired frequency f_p. It is the image (mirror-like symmetry) of f_p with respect to f_0.

Any signal of frequency f_i that is simultaneously present at the receiver input with the desired signal of frequency f_p produces interferences. It is, therefore, necessary to get rid of it, or, at least, to weaken it enough with a filter tuned on frequency f_p.

11.5.3 Frequency multiplication

It is sometimes necessary to create a periodic or narrowband signal, the frequency of which is proportional to the frequency of another signal (fig. 11.39). The narrowband signal can be supplied by a reference generator (e.g., quartz oscillator), or it may correspond to a frequency modulated signal.

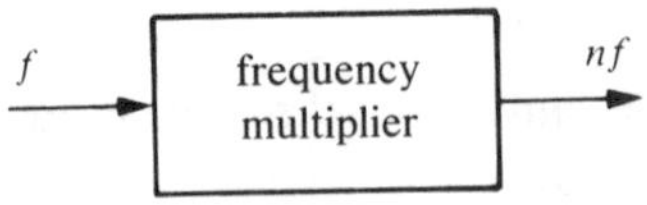

Fig. 11.39

Such a frequency multiplication is used, for instance,

- to generate one or several accurate frequency signals from a single reference (subcarriers, frequency synthesizer);
- to convert a narrowband FM signal (sub-sect. 11.3.3) into a wideband FM signal.

The frequency multiplication can be implemented:

- with a nonlinear circuit (sub-sect. 8.4.2), the characteristic of which generates spectral components at multiples of the original frequency, followed by a filter that eliminates the undesired components;
- with a phase-locked-loop (PLL) containing a logic frequency divider circuit (fig. 11.40).

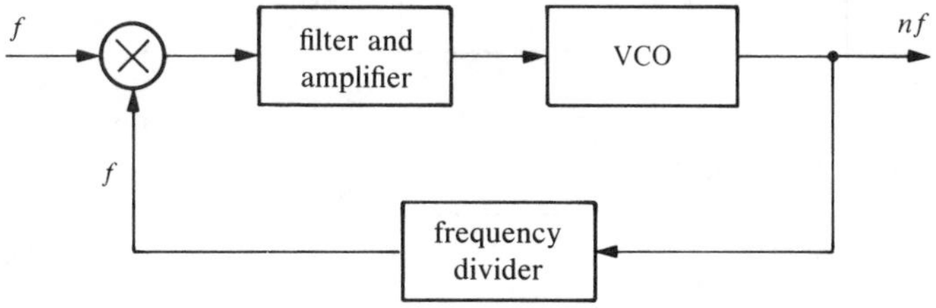

Fig. 11.40

The existence of adequate integrated electronic circuits has made the second method very popular now.

If the original signal is frequency modulated, with carrier frequency f_p and deviation Δf, the new signal has a frequency nf_p and a deviation $n\Delta f$.

11.6 PROBLEMS

11.6.1 Evaluate the efficiency η_m of an amplitude modulation where the normalized modulating signal $m_0(t)$:

- has an autocorrelation function $R_{m0}(\tau) = \sigma_{m0}^2 \exp(-a|\tau|)$ with $a = 1000$ Hz and $\sigma_{m0} = 0.5$;
- is a square wave of unit amplitude;
- is a square wave of amplitude ½.

11.6.2 Express the efficiency ξ_d/ξ_x of an amplitude demodulation by synchronous detection as a function of the modulation factor m defined, in the case of a sinusuoidal AM modulation with carrier, by $m_0(t) = m \cdot \cos(2\pi f_m t)$.

11.6.3 Demonstrate that the modulated signal of a single-sideband system which retains the lower sideband is obtained by changing the sign of the second term of (11.20).

11.6.4 Write the expressions for the modulated signal $s(t)$ and for the spectral density $\Phi_s(f)$ of a single sideband system keeping the (a) upper sideband, or (b) lower sideband, for a sinusoidal modulating signal $m_0(t) = \cos(2\pi f_m t)$.

11.6.5 What kind of modulation is achieved by the double modulator of fig. 11.41, if the Fourier transform of the modulating signal is $M_0(f) = 0$ for $|f| \geqslant f_m$?

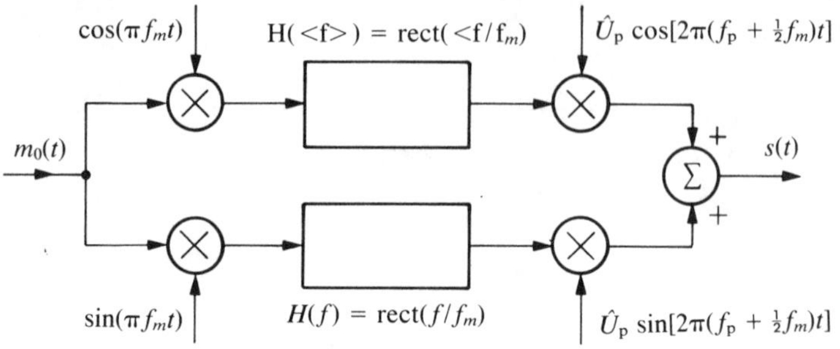

Fig. 11.41

11.6.6 Show that it is possible to transmit two independent signals, $m_1(t)$ and $m_2(t)$, with same bandwidth, via the same amplitude modulated sine wave carrier, if there is at the receiver a double isochronous detection with auxiliary sinusoidal signals $u_{d1}(t)$ and $u_{d2}(t)$ in quadrature (QAM modulation).

11.6.7 Find the condition under which envelope detection can be used for the demodulation of a VSB signal with carrier, modulated by a signal $m(t)$ of power P_m.

11.6.8 Demonstrate (11.21), given the assumption, for the sake of simplicity, that $m_0(t)$ is deterministic.

11.6.9 Compare graphically (11.81) and (11.86) for $\delta = 10$.

11.6.10 Find the approximate expression for the power spectral density of a sine wave $s(t)$, which has been frequency modulated at high level by a random signal $m(t)$, with a uniform distribution between $\pm m$.

11.6.11 Demonstrate that the control voltage $d(t)$ of fig. 11.34 is proportional to the instantaneous frequency of the modulated signal $s(t)$.

11.6.12 Find the power spectral density of a sine wave, which has been frequency modulated (with phase discontinuities) by a random binary signal $m_0(t)$ in NRZ mode, such that $s(t) = \hat{U}_p \cos(2\pi f_1 t + \alpha_1)$ for $m_0(t) = 1$ and $s(t) = \hat{U}_p \cos(2\pi f_2 t + \alpha_2)$ for $m_0(t) = -1$.

11.6.13 Demonstrate that a phase shift keying (PSK) modulation with four independent and equally likely states: 0, $\pi/2$, $-\pi/2$, π, occupies the same useful band as the diphase modulation of sub-sect. 11.3.8.

11.6.14 Find the spectrum of the modulating signal $m(t)$ of a stereophonic FM modulator generated by the system of fig. 11.42, where $m_d(t)$ and $m_g(t)$ represent, respectively, the right and left channels of the stereophonic signal, of bandwidth limited to 15 kHz.

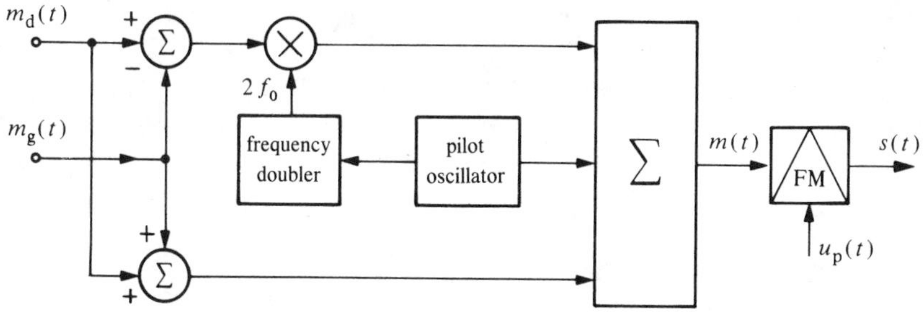

Fig. 11.42

11.6.15 Demonstrate that the block diagram of fig. 11.43 represents a stereophonic FM demodulator adapted to the processing of the modulated signal produced by the modulator of fig. 11.42.

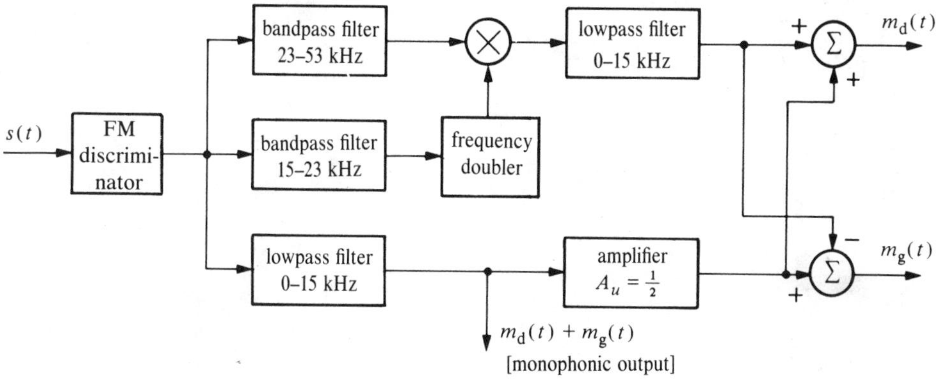

Fig. 11.43

Chapter 12

Experimental Spectrum Analysis

12.1 GENERAL PRINCIPLES

12.1.1 Objectives, definitions, and terminology

Spectrum analysis is the expansion of a time-varying quantity in its frequency components. It is one of the most usual signal processing techniques.

Experimental spectrum analysis is an invaluable research tool in many fields. Various approaches, based on analog, digital, or hybrid methods, are used to achieve such an analysis.

Autonomous measurement instruments performing such an operation are called *spectrum analyzers. Fourier processors* (or *FFT processors*) are dedicated digital computers that mainly play this role. However, they can also perform other tasks: correlation analysis, estimation of coherence functions, measurement of impulse responses or transfer functions, transient signal and shock analysis, *et cetera. Wave Analyzers* are essentially selective voltmeters. With some limitations, they can be used as spectrum analyzers.

Experimental spectrum analysis is also often realized with a computer processing previously acquired data. In this case appropriate software is used to implement the required algorithms.

The aim of this chapter is to describe the general principles of the most common experimental spectrum analysis methods. The relations binding the theoretical models of spectra to the experimental measurements are established. This relationship is illustrated by some application examples.

12.1.2 Main application fields

Many signals are difficult to interpret as time functions. On the other hand, their spectra provide us with valuable information. This information enables us to understand the signal nature, to identify its origin, when possible, and to detect the variations of its characteristics.

In metrology, electronics, automatic control systems, telecommunications, applied physics, chemistry, *et cetera*, spectrum analysis is used to determine the energy frequency distribution of the observed signals, to measure the transfer function of linear systems (filters, sensors, amplifiers, *et cetera*), to reveal the distortions induced by nonlinearities, to estimate the phase noise of oscillators, *et cetera*.

In telecommunications, radar, and sonar techniques, it is moreover a valuable tool to characterize modulated signals and to scrutinize transmission and reception conditions (background noise and other parasitic phenomena, channel separation in frequency multiplexing, *et cetera*).

Linked to a calibrated microphone, the spectrum analyzer is a powerful research instrument in acoustics.

Spectrum analyzers are also used in vibration analysis (structure analysis, rotating machines, transport vehicles, aeronautics, *et cetera*) in which the signals are generally provided by acceleration sensors. For instance, the vibrations produced by cog-wheels, or the noise generated by bearings, can be characterized by some frequency ***signature,*** due to the presence of spectral lines associated with mechanical periodicities.

Finally, the spectrum analysis of many natural (biological, seismic, oceanographic, atmospheric, galactic) signals, or signals stemming from human activities (oil exploration, nuclear explosions, EM interference, *et cetera*) is a compulsory investigative tool.

12.1.3 Fundamental methods of spectrum analysis

The usual techniques of spectrum analysis can be divided into two main groups:

- direct methods (selective filtering, periodogram computation);
- indirect methods (correlogram computation, high resolution parametric methods: autoregressive, maximum entropy, Pisarenko's, *et cetera*).

Direct methods are similar to the principle of the general orthogonal function analyzer depicted in fig. 3.6. The scanned frequency range is broken down into a sequence of contiguous channels in which the local spectral density is evaluated. This decomposition is achieved by the computation of the discrete Fourier transform (DFT) in the *periodogram method* (fig. 12.1), or by means of a set of narrowband filters (fig. 12.2). This set can be implemented with N filters working simultaneously in parallel (multichannel spectrum analyzers; sub-sect. 12.2.1), or with a single filter sequentially sweeping the entire desired

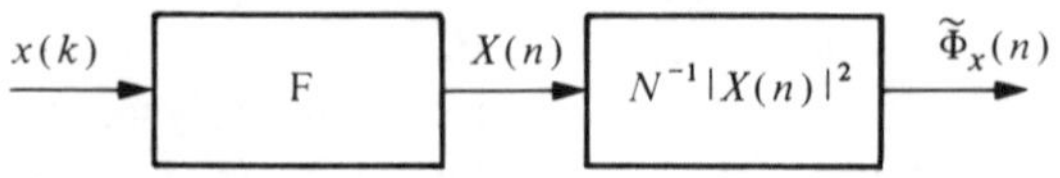

Fig. 12.1 Periodogram method.

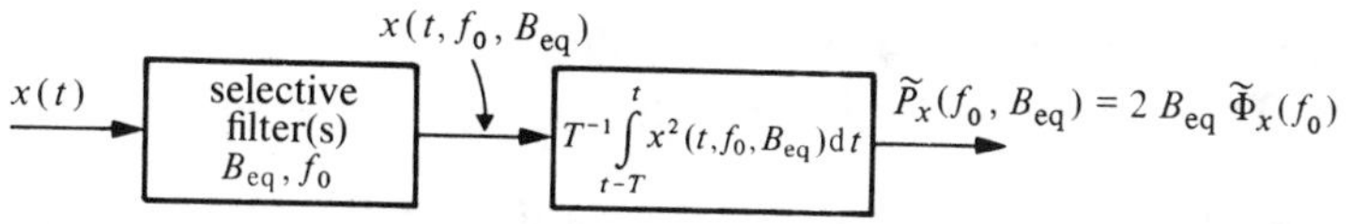

Fig. 12.2 Selective filter method.

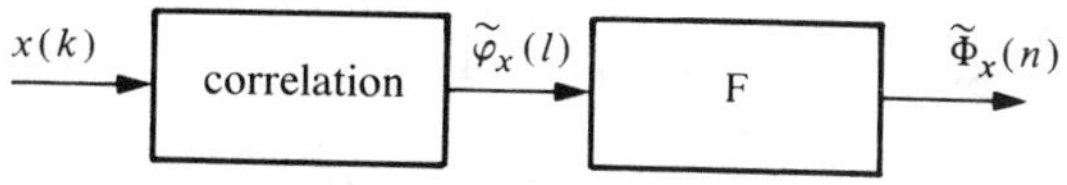

Fig. 12.3 Correlogram method.

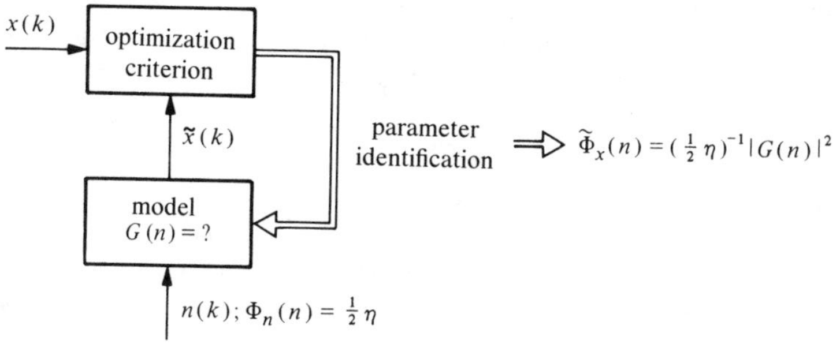

Fig. 12.4 Parametric method.

frequency range (swept-frequency spectrum analyzers; sub-sect. 12.3). Spectrum analyzers based on optical diffraction also belong to this group [67].

Indirect methods use digital algorithms based on either the Wiener-Khinchin theorem (*correlogram method* [31]), or the spectral estimation of a signal model, which must fit the observed data (*parametric methods* [119]). In the first case (fig. 12.3), the autocorrelation function of the (assumed ergodic) signal, and then its Fourier transform, are evaluated. The most typical approach among the parametric methods is the *autoregressive method* (AR, ARMA). It consists of assuming the observed signal as the response of a linear system stimulated by a white noise (fig. 12.4). The spectrum estimation is then replaced, according to (8.24), by the estimation of the transfer function $G(f)$ of the system. The identification of the model parameters is based on the solution of a set of linear equations describing the relations between those parameters and the signal autocorrelation function. The *maximum entropy method* (MEM) is a similar, but nonlinear, approach in which, among all the spectra consistent with the limited available data, the spectrum corresponding to a random signal of maximum entropy is selected. *Pisarenko's method* considers the spectrum as the sum of a discrete spectrum and a white noise spectrum. This approach yields the highest resolution if the assumptions are verified.

Direct methods are rather straightforwardly implemented. They generate very good results as long as the duration of the observed signal is not too short (sub-sect. 12.1.5). When it is too short, parametric methods are usually more efficient [120]. However, their implementation is far more complex and their success depends on the suitability of the selected method to the kind of signal analyzed.

The indirect correlogram method, which was for a long time a favorite choice, has now been superseded by the periodogram method, since fast Fourier transform (FFT) algorithms are available.

Only direct methods, widely used to design spectrum analyzers, are detailed further in this chapter. For more information on indirect methods, the reader should refer himself or herself to the specialized literature mentioned.

12.1.4 Measurement errors

Experimental spectrum analysis is different from theoretical spectrum analysis for many reasons:

- signal observation on a limited duration T (finite number of samples in digital signal processing);
- bandwidth B_{eq} and frequency response of the analysis filters in analog processing;
- frequency sampling in digital processing;
- approximation of the true rms value measurement by an envelope detection;
- distortion generated by the sweeping speed in swept-frequency analyzers;
- statistical fluctuations in random signal analysis;
- violation of the sampling theorem, quantization noise, and computational errors (rounding, truncation) in digital processing;
- internal noise of the acquisition and analysis systems;
- bias induced by an exagerated spectral smoothing in the frequency domain.

The main causes of errors in classical methods are, by far, the finite observation duration T, the equivalent bandwidth B_{eq} of the analysis filter, and the statistical fluctuations. They are inherent to the experimental approach and cannot be reduced at will for practical reasons. Their origins and consequences are explained below.

12.1.5 Resolution, bandwidth, and observation duration

Resolution is the aptitude of an analysis method to discriminate between two contiguous spectral lines of the same amplitude.

In selective filter analyzers (fig. 12.2), this resolution corresponds approximately to the equivalent bandwidth B_{eq}, defined by (8.53), of the filters used. It is necessarily coarse in multichannel analyzers (sub-sect. 12.2.1) where it usually varies with the center frequency f_0 of each filter (B_{eq}/f_0 = constant). The swept-frequency analyzers (sect. 12.3) can achieve very high resolutions, but at the expense of a drastic reduction of the sweeping speed (sub-sect. 12.3.4). This means that the signal must exist and be stationary during the entire analysis duration.

The resolution of analyzers based on the periodogram method, according to fig. 12.1, depends on the necessarily finite observation (or acquisition) duration T of the signal. This duration is equal to the product of the number N of available samples $x(k)$ and of the sampling interval T_e:

$$T = NT_e \tag{12.1}$$

In fact, the samples $x(k)$ are taken from a limited duration signal $x_u(t)$ defined by the product:

$$x_u(t) = x(t) \cdot u(t) \tag{12.2}$$

where $x(t)$ is the signal to be analyzed and $u(t)$ is an adequate dimensionless weighting function—or time window—that is zero outside the observation interval T.

According to (8.88), the spectral densities of $x(t)$, $x_u(t)$, and $u(t)$ are related by the convolution:

$$\mathring{\Phi}_{xu}(f) = \Phi_x(f) * \mathring{\Phi}_u(f) \tag{12.3}$$

The evaluation of the power spectral density of $x(t)$ is obtained through dividing (12.3) by T:

$$\tilde{\Phi}_x(f) = T^{-1}\mathring{\Phi}_{xu}(f) \tag{12.4}$$

The outcome is an approximation of the periodogram (5.108).

If $x(t)$ has two sine wave components of amplitudes A_1 and A_2 and frequencies f_1 and f_2, its theoretical bilateral power spectral density (4.154) contains the lines $\frac{1}{4}A_1^2[\delta(f + f_1) + \delta(f - f_1)] + \frac{1}{4}A_2^2[\delta(f + f_2) + \delta(f - f_2)]$. From (12.4) and (12.3), these lines are replaced by a sum of translated versions of the continuous function $T^{-1}\mathring{\Phi}_u(f)$: $\frac{1}{4}T^{-1}A_1^2[\mathring{\Phi}_u(f + f_1) + \mathring{\Phi}_u(f - f_1)] + \frac{1}{4}T^{-1} A_2^2[\mathring{\Phi}_u(f + f_2) + \mathring{\Phi}_u(f - f_2)]$. The power $A_k^2/2$ of each pair of lines with frequencies $\pm f_k$ is spread over the frequency axis according to a law: $\Phi(f) =$

$T^{-1}[\mathring{\Phi}_u(f + f_k) + \mathring{\Phi}_u(f - f_k)]$. This law can be interpreted as the modulus square of a filter transfer function. By analogy with (8.53), the equivalent bandwidth associated with the time window $u(t)$, under the assumption that $T \ll f_k^{-1}$, becomes

$$B_{eq} = \frac{1}{\Phi_{max}} \int_0^{\infty} \Phi(f)\mathrm{d}f = \frac{1}{\mathring{\Phi}_u(0)} \int_{-\infty}^{\infty} \mathring{\Phi}_u(f)\mathrm{d}f = \frac{\mathring{\varphi}_u(0)}{\mathring{\Phi}_u(0)} \tag{12.5}$$

12.1.6 Illustration

The simplest window is a uniform weighting function (fig. 12.5):

$$u_1(t) = \mathrm{rect}[(t - t_0)/T] \tag{12.6}$$

From (4.64), its spectral density is

$$\mathring{\Phi}_{u1}(f) = T^2\mathrm{sinc}^2(Tf) \tag{12.7}$$

and, from (4.63) and (12.5):

$$B_{eq} = 1/T \tag{12.8}$$

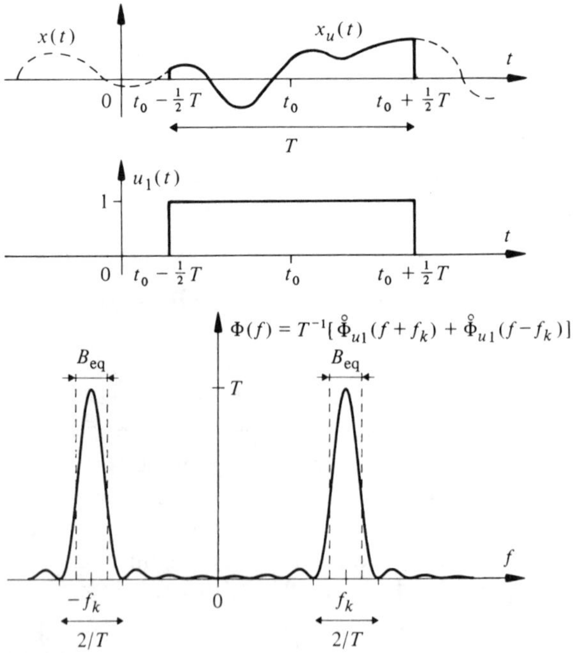

Fig. 12.5

Figure 12.6 shows three examples of the approximative unilateral spectral density $\tilde{\Phi}_x^+(f)$ of a signal $x(t)$ composed of three sinusoidal components of the same normalized power $A_k^2/2 = 1\ \mathrm{V}^2$. The frequency difference between components with indices $k = 2$ and $k = 3$ is $\Delta f = f_3 - f_2 = 1/128$ Hz. Each example corresponds to an observation duration $T = 64$s, $T = 128$s, and $T = 256$s, respectively. The respective equivalent bandwidths are, thus, $B_{eq} = 2\Delta f$, $B_{eq} = \Delta f$, and $B_{eq} = \frac{1}{2}\Delta f$. The resolution limit obviously appears at $B_{eq} \approx \Delta f$.

The discontinuities of the uniform weighting window $u_1(t)$ generate relatively significant sidelobes in the $\mathring{\Phi}_{u1}(f)$ function. These sidelobes can hide the existence of a weak line located close to a strong line. To overcome this problem, other weighting windows with more gradual variation have been

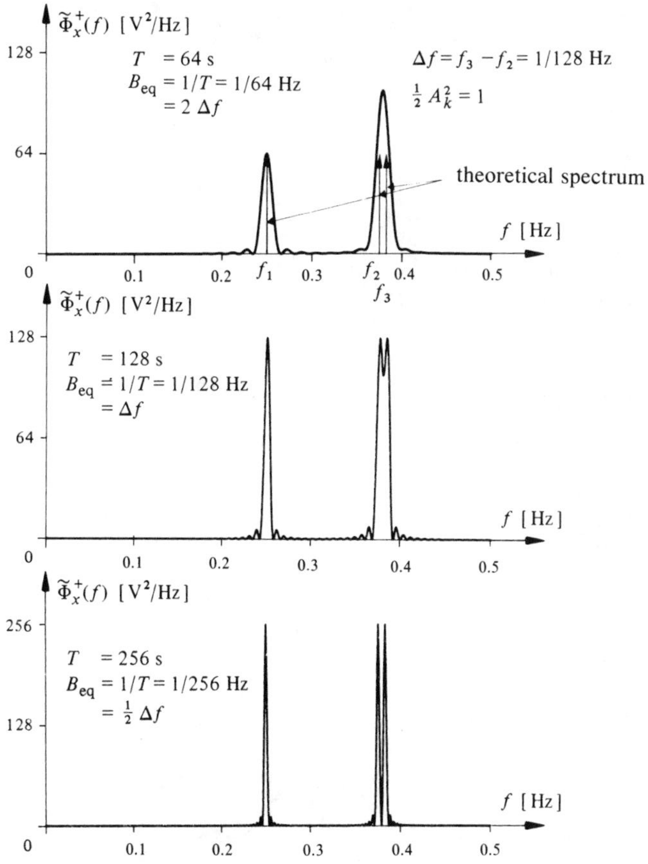

Fig. 12.6

proposed [67, 122, 123]. The properties of some of them are detailed in volume XX. As a general rule, the reduction of the sidelobe is achieved at the expense of an increase in the equivalent bandwidth and, consequently, a reduction of the resolution.

A classical set of weighting functions is generated by the following relation, called the *generalized Hamming window* (fig. 12.7):

$$u_\alpha(t) = [\alpha + (1 - \alpha)\cos(2\pi t/T)] \cdot \mathrm{rect}(t/T) \tag{12.9}$$

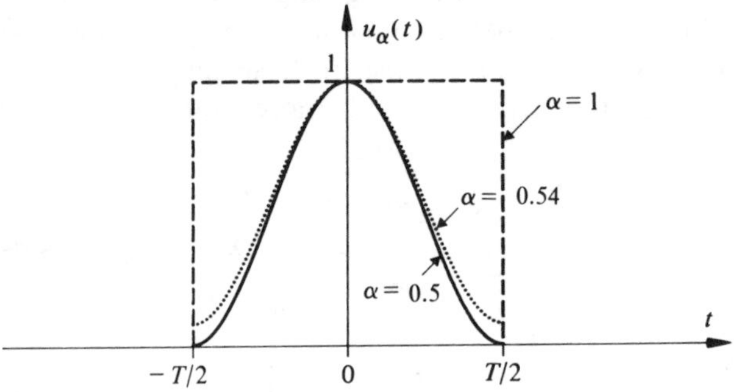

Fig. 12.7

The rectangular function (12.6) corresponds to the particular case $\alpha = 1$. For $\alpha = 0.5$, we get the *Hanning window,* and for $\alpha = 0.54$ we get the *Hamming window.*

We can easily determine (problem 12.5.1) the corresponding spectral density $\overset{\circ}{\Phi}_{u\alpha}(f)$ and the equivalent bandwidth B_{eq} (fig. 12.8):

$$\overset{\circ}{\Phi}_{u\alpha}(f) = T^2|\alpha\,\mathrm{sinc}(Tf) + \tfrac{1}{2}(1 - \alpha)[\mathrm{sinc}(Tf + 1) + \mathrm{sinc}(Tf - 1)]|^2 \tag{12.10}$$

$$B_{eq} = \left[1 + \frac{1}{2}\left(\frac{1-\alpha}{\alpha}\right)^2\right] \cdot T^{-1} \tag{12.11}$$

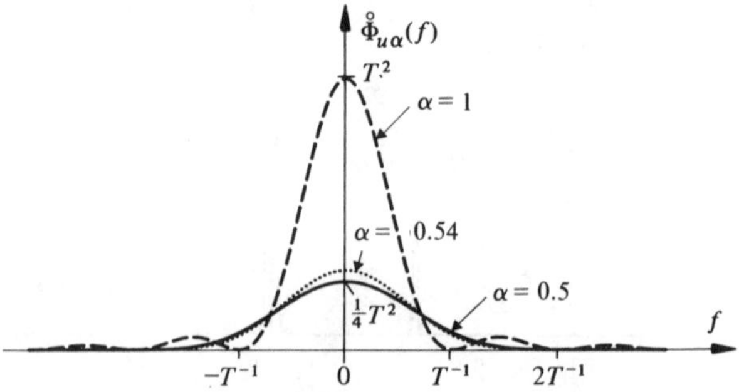

Fig. 12.8

The equivalent bandwidth of the Hanning window, thus, is equal to $1.5T^{-1}$ and the Hamming one equals $1.36T^{-1}$.

12.1.7 Comments

The indirect correlogram method (fig. 12.3) is, in principle, equivalent to the direct periodogram method (fig. 12.1). The resolutions that can be obtained in both cases for the same observation durations are, therefore, similar. If a more severe autocorrelation function weighting is used to reduce the statistical uncertainties, the resolution can be even worse.

Modern high resolution methods (fig. 12.4) replace the zero weighting outside the observation interval by assumptions about the signal nature and its evolution outside this interval [120]. They cannot be universally applied, but, under suitable conditions, they bring about a significant improvement in resolution.

12.1.8 Statistical fluctuations of the spectrum estimation obtained by a selective filter analyzer

The spectrum estimation of a random signal is a particular case of statistical interpretation of measurements. It belongs to the type of estimation methods studied in chapter 13. The statistics of the random measurement error provided by direct methods will be examined here.

Consider first the conceptual block diagram of a selective filter analyzer (fig. 12.2). As a first approximation, it is composed of three operators (fig. 12.9): an ideal filter (sub-sect. 8.2.21) of bandwidth B_{eq} centered on the analysis frequency f_0, a squaring operator (sub-sect. 8.4.16), and an ideal time averager (sub-sect. 8.2.19).

The output signal $z(t)$ corresponds to the estimation $\tilde{P}_x(f_0,B_{eq})$ of the filtered signal power. The estimation of the bilateral power spectral density $\tilde{\Phi}_x(f_0)$ is obtained through dividing it by $2B_{eq}$ [or by B_{eq} if the unilateral spectral density $\tilde{\Phi}_x^+(f_0)$ is looked after].

Taking into account the reduced bandwidth of the analysis filter, it is rational (sub-sect. 8.2.14) to represent the statistics of the filtered signal $x(t,f_0,B_{eq})$ approximately by a Gaussian distribution with zero mean value and variance (i.e., power) σ_x^2. Moreover, it can be assumed that the spectral density $\Phi_x(f)$ is approximately constant on the narrow intervals $f_0 - \frac{1}{2}B_{eq} < |f| < f_0 + \frac{1}{2}B_{eq}$. Thus, $\sigma_x^2 \equiv P_x(f_0,B_{eq}) = 2B_{eq}\Phi_x(f_0)$.

The power estimation performed by the system depicted on fig. 12.9 is linked to the theoretical value $P_x(f_0,B_{eq})$ by relation

$$z(t) \triangleq \tilde{P}_x(f_0,B_{eq}) = P_x(f_0,B_{eq}) + \epsilon(t) \tag{12.12}$$

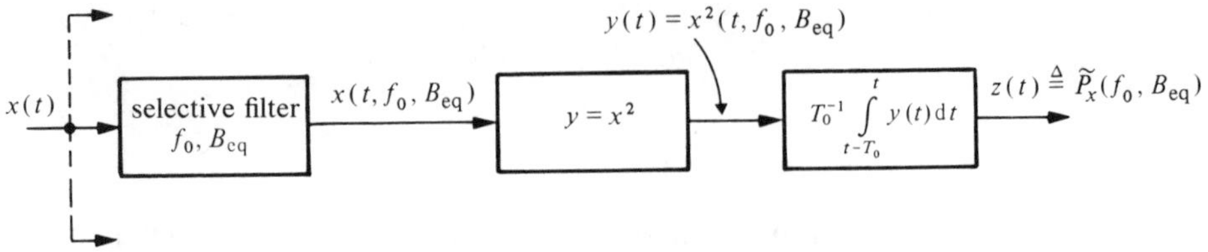

Fig. 12.9

where $\epsilon(t)$ represents the random measurement error for this case. This error has a mean value μ_ϵ and a variance σ_ϵ^2. The mean value is called the bias of the measurement. The mean square error is related to μ_ϵ and σ_ϵ^2 by (14.53):

$$\mathrm{E}[\boldsymbol{\varepsilon}^2] = \mu_\epsilon^2 + \sigma_\epsilon^2 \tag{12.13}$$

The standard deviation $\sigma_\epsilon = \sigma_z$ characterizes the extent of the measurement dispersion with respect to the mean value. It is often used as the principal estimation quality criterion, for instance, to define a kind of measurement signal-to-noise ratio. ***This is valid only if the bias μ_ϵ is zero or negligible.***

From (12.12), it can be deduced that the bias is given by the difference:

$$\mu_\epsilon = P_x(f_0, B_{eq}) - \mu_z \tag{12.14}$$

In the case of the fig. 12.9 system, the estimation (12.12) can be written as

$$z(t) = \frac{1}{T_0}\int_{t-T_0}^{t} x^2(t,f_0,B_{eq})\mathrm{d}t \tag{12.15}$$

This estimator is *unbiased,* since

$$\mu_z = \mathrm{E}[\mathbf{z}] = \frac{1}{T_0}\int_{t-T_0}^{t} \mathrm{E}[\mathbf{x}^2(t,f_0,B_{eq})]\mathrm{d}t = \sigma_x^2\frac{1}{T_0}\int_{t-T_0}^{t}\mathrm{d}t$$
$$= \sigma_x^2 \equiv P_x(f_0,B_{eq}) \tag{12.16}$$

hence, $\mu_\epsilon = 0$.

The variance $\sigma_\epsilon^2 = \sigma_z^2$ can be found using the results of sub-sections 8.2.19 and 8.4.16. The spectrum of the output signal $y(t)$ of the squaring operator is identical to the one represented in fig. 8.54, with the correspondence $\eta B = \sigma_x^2$. According to (8.44):

$$\sigma_\epsilon^2 = \sigma_z^2 = \int_{-\infty}^{\infty} [\Phi_y(f) - \mu_y^2\delta(f)]\,\mathrm{sinc}^2(T_0 f)\mathrm{d}f \tag{12.17}$$

with, in this case,

$$\Phi_y(f) - \mu_y^2\delta(f) = \frac{\sigma_x^4}{B_{eq}}\left[\text{tri}\left(\frac{f}{B_{eq}}\right) + \frac{1}{2}\,\text{tri}\left(\frac{f+f_0}{B_{eq}}\right) + \frac{1}{2}\,\text{tri}\left(\frac{f-f_0}{B_{eq}}\right)\right] \quad (12.18)$$

For $B_{eq}T_0 < 1$ and $f_0 \gg 1/T_0$, the integrand (12.17) is approximately given by

$$[\Phi_y(f) - \mu_y^2\delta(f)]\ \text{sinc}^2(T_0 f) \approx (\sigma_x^4/B_{eq})\ \text{tri}(f/B_{eq}) \quad (12.19)$$

Therefore, the measurement variance is independent of B_{eq} and the observation duration T_0:

$$\sigma_\epsilon^2 \approx \sigma_x^4 \quad (12.20)$$

The statistical distribution of the measurement fluctuations is easily found by noting that, following (7.86), the filtered signal can be expressed as a function of its low frequency in-phase and quadrature components:

$$x(t,f_0,B_{eq}) = a(t)\cos(2\pi f_0 t) - b(t)\sin(2\pi f_0 t) \quad (12.21)$$

Thanks to the assumptions we made (sub-sect. 7.4.6), $a(t)$ and $b(t)$ are independent Gaussian variables, with zero mean value and variance σ_x^2. The output signal of the squaring operator is

$$y(t) = x^2(t,f_0,B_{eq}) = a^2(t)\cos^2(2\pi f_0 t) + b^2(t)\sin^2(2\pi f_0 t) - 2a(t)b(t)\cos(2\pi f_0 t)\sin(2\pi f_0 t) \quad (12.22)$$

For $B_{eq}T_0 < 1$ and $f_0 \gg 1/T_0$, the averager gets rid of the spectral contributions around $2f_0$ and keeps the term:

$$z(t) = \tfrac{1}{2}[a^2(t) + b^2(t)] \quad (12.23)$$

From (14.115), the random variable $z = \tilde{P}_x(f_0,B_{eq})$ is distributed according to a χ_m^2 law with $m = 2$ degrees of freedom (exponential law; fig. 12.10):

$$p_z(z) = \sigma_z^{-1}\exp(-z/\sigma_z) \approx \sigma_x^{-2}\exp(-z/\sigma_x^2) \quad (12.24)$$

for $z \geq 0$, with $\mu_z = \sigma_z = \sigma_x^2$.

Here, the standard deviation is equal to the power to be estimated! The signal-to-noise ratio of the measurement is $\xi = \mu_z^2/\sigma_z^2 = 1$ (0 dB). This situation is unacceptable in practice. To reduce the variance, the product of equivalent bandwidth times integration duration has to be increased.

For $B_{eq}T_0 \gg 1$, relation (12.17) becomes

$$\sigma_\epsilon^2 = \sigma_z^2 \approx \frac{\sigma_x^4}{B_{eq}}\int_{-\infty}^{\infty}\text{sinc}^2(T_0 f)\,df = \frac{\sigma_x^4}{B_{eq}T_0} \quad (12.25)$$

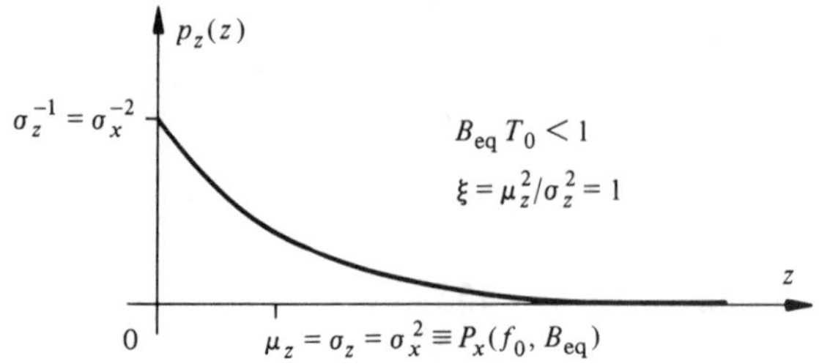

Fig. 12.10

Since $\mu_z = \tilde{P}_x(f_0, B_{eq}) = \sigma_x^2$, the signal-to-noise ratio of the measurement becomes

$$\xi = \mu_z^2/\sigma_z^2 = B_{eq} T_0 \tag{12.26}$$

Because of the strong lowpass filtering imposed by the averager for $T_0 \gg B_{eq}^{-1}$, the estimation $z(t)$ now has an approximately Gaussian distribution (fig. 12.11):

$$p_z(z) \cong \frac{1}{\sqrt{2\pi}\sigma_z} \exp\left[-\frac{(z-\mu_z)^2}{2\sigma_z^2}\right] \tag{12.27}$$

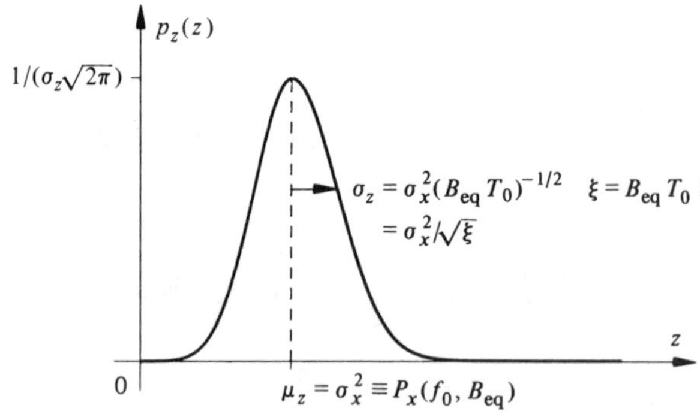

Fig. 12.11

with $\mu_z = \sigma_x^2 = P_x(f_0, B_{eq})$ and $\sigma_z = \sigma_x^2/\sqrt{\xi} = \mu_z/\sqrt{\xi}$. These results have been experimentally confirmed [121]. The table of section 15.8 enables us to interpret (12.27), saying, for instance, that the sought theoretical value μ_z has a 95% probability of being located between $\pm 2\sigma_z$ of the estimated value z (*confidence interval*). For a 95% confidence interval that is smaller than 20% of the sought mean value, a ratio $\xi > 400$ (26 dB) is needed.

12.1.9 Statistical fluctuations of the spectrum estimation obtained by a digital analyzer

Consider next the case of a digital estimation of the periodogram according to the block-diagram of fig. 12.1. Hence,

$$\tilde{\Phi}_x(n) = \frac{1}{N}|X(n)|^2 = \frac{1}{N}[A_x^2(n) + B_x^2(n)] \tag{12.28}$$

where $X(n) = A_x(n) + jB_x(n)$. The periodogram $\tilde{\Phi}_x(n)$ is theoretically again a random variable distributed according to a χ_m^2 law with $m = 2$ degrees of freedom if the analyzed signal is Gaussian. Indeed, the real $A_x(n)$ and imaginary $B_x(n)$ parts of the Fourier transform $X(n)$ of a Gaussian process are independent Gaussian variables (linear transform and orthogonality conditions). Hence, the probability density is similar to (12.24):

$$p_{\Phi}(\tilde{\Phi}_x) = \Phi_x^{-1}(n)\exp[-\tilde{\Phi}_x(n)/\Phi_x(n)] \tag{12.29}$$

for $\tilde{\Phi}_x(n) \geqslant 0$, with $\mu_{\Phi} = \sigma_{\Phi} = \Phi_x(n)$.

The variance of this estimator is once again unacceptable in practice. To improve it, we have two methods:

- compute the arithmetical mean of $\tilde{\Phi}_x(n)$ from K independent estimations (sub-sect. 13.1.24):
- smooth (filter) the sequence of $\tilde{\Phi}_x(n)$ values at the expense of a resolution reduction and the introduction of a possible bias (sub-sect. 12.1.12).

In the first case, the statistical fluctuation variance decreases, according to (13.82), with the number K of independent measurements, and the signal-to-noise ratio ξ is simply equal to K. From (12.8), the minimum resolution is equal to $1/T$, where T is the observation duration needed for a spectrum measurement. The observation (acquisition) total duration needed for K independent measurements, hence, is $T_0 = KT$, thus,

$$\xi = K = B_{eq}T_0 \tag{12.30}$$

which is analogous to (12.26).

The estimation obtained from (14.115) is now distributed according to a χ_m^2 law with $m = 2K = 2B_{eq}T_0$ degrees of freedom (4.113) with mean value m and variance $2m$. It tends toward a Gaussian distribution similar to (12.27) when $m \rightarrow \infty$, thanks to the central limit theorem (sub-sect. 5.5.3).

12.1.10 Comparison

The methods of direct analysis by selective filter or periodogram evaluation, thus, have equal performance. The quality criterion of the measurement is in both cases given by the signal-to-noise ratio: $\xi = B_{eq}T_0$.

For selective filter analyzers, the resolution is set by the equivalent bandwidth B_{eq} of the selected filters. The integration duration to be used is then deduced from the desired signal-to-noise ratio ξ: $T_0 = \xi / B_{eq}$.

In multichannel analyzers, the entire frequency range B_m is scanned in parallel. In the case of a constant resolution, the number of required channels is $N = B_m / B_{eq}$ and the integration duration T_0 also represents the total required observation duration. If the resolution is not constant, the minimum value of T_0 is set by the most selective channel (the signal-to-noise ratio is then better in the other channels).

Swept-frequency analyzers sequentially scan the frequency range B_m. Each partial measurement requires T_0 seconds to reach the desired signal-to-noise ratio, and the total sweeping duration is then equal to $T_b = (B_m / B_{eq}) T_0$. If the analyzer contains a memory with fast cyclical recirculation (sub-sect. 10.5.2 and 12.3.6), the minimum signal observation duration is again T_0; otherwise, it equals T_b.

Analysis based on the discrete Fourier transform (DFT) and the periodogram evaluation is very similar to constant resolution multichannel analysis, as long as the number of discrete values $\tilde{\Phi}_x^+(n)$ with $n = 0, 1, \ldots, N - 1$ of the periodogram is greater than or equal to B_m / B_{eq}. The frequency coverage directly depends on the sampling rate f_e : $B_m = \frac{1}{2} f_e$ (sub-sect. 12.2.3). The signal-to-noise ratio is equal to the number K of independent spectrum estimations normally performed sequentially on the same signal (ergodism). The chosen resolution sets the minimum duration T required for each independent observation (acquisition): $T = B_{eq}^{-1}$ in the case of a uniform time-weighting window. The total observation duration needed is then: $T_0 = KT = \xi / B_{eq}$.

The above discussion shows that, in any case, when the duration of the signal to be analyzed is limited (setting T_0 or T_b), the experimenter must choose an acceptable trade-off between resolution and signal-to-noise ratio.

The experimental parameters so defined are limiting values. The use of imperfect averagers, nonuniform weighting windows (sub-sect. 12.1.6), of non-independent spectrum estimations, or spectrum smoothing procedures (sub-sect. 12.1.12) slightly deteriorate those results.

A usual approximation of the perfect averager (problem 8.5.6 and sub-sect. 13.1.24) is, in analog techniques, the first-order RC lowpass filter (sub-sect. 8.2.24). It can be shown that, in this case, the same signal-to-noise ratio is achieved by substituting 2 RC for T_0 in (12.26). Because of the unit-step response of such a filter, a measurement performed after a settling time equal to $T_0 = 2RC$ has a 14% bias. In practice, it is advisable to adopt a minimum settling time (i.e., observation duration) $T_0' = 4RC$ in order to reduce the bias below 2%.

12.1.11 Examples

Consider the spectrum analysis of a low frequency random signal with a unilateral spectrum covering the 0–20 kHz range.

For a 100 Hz resolution and a measurement signal-to-noise ratio of $\xi = 100$ (20 dB) leading to a 95% confidence interval in the range of 0.4 μ_z, the experimental parameters are:

- for a swept-frequency analyzer: $B_{eq} = 100$ Hz, $T_0 = 1$s and $T_b = 200$s;
- for a multichannel analyzer with a filter bank: $B_{eq} = 100$ Hz, $N = 200$ channels, $T_0 = 1$s;
- for an FFT multichannel analyzer: $f_e = 40$ kHz, $N \geqslant 200$, $T = 10$ ms, $K = 100$, and $T_0 = 1$s.

The reader can easily extend these results to other analysis conditions.

12.1.12 Spectrum smoothing

Spectrum smoothing is an alternative to the mean value computation (by integration or weighted sum) introduced in sub-sect. 12.1.9 to reduce the variance of measurement fluctuations. It is a simple linear filtering (in the frequency domain) of the measured spectral density, which can be expressed as the convolution:

$$\tilde{\Phi}_{xl}(f) = \tilde{\Phi}_x(f) * U(f) \tag{12.31}$$

where $U(f)$ is a spectral window playing a role analogous to the impulse response of a filter. Relation (12.31) is similar to relation (12.3). Hence, we see that:

- the spectrum smoothing is equivalent to the multiplication of the signal autocorrelation function by a weighting function $u(\tau) = F^{-1}\{U(f)\}$:
- it necessarily introduces a resolution reduction ($B_{eq}\nearrow$);
- as in any situation where a large bandwidth B_{eq} is used, the reduction of the estimation error variance is achieved at the expense of an increasing risk of bias ($\mu_\epsilon \neq 0$).

Our main interest in smoothing is that it is rather handy for interactive processing, where the experimenter tentatively proceeds to select the best analysis parameters.

For a more detailed study of this problem, the reader should refer himself or herself to references [65, 122].

In analog techniques, smoothing is implemented by a simple lowpass filtering (RC filter) of the time function $\tilde{\Phi}_x(t)$ generated by the sweeping. A weighted running average is usually computed in digital techniques.

12.1.13 Substitution of envelope detection to the squaring operation

The squaring operation considered in the direct methods of spectrum analysis yields a limited dynamic range of measurements. To improve this, the squaring operator can be replaced by a linear envelope detector (fig. 11.23). This trick, however, introduces a measurable bias between the measurements of purely random and sinusoidal components.

In the case of a sine wave with an rms value σ_x, the mean value of the detected envelope is the peak value $\sqrt{2}\sigma_x$.

In the case of a purely random—Gaussian by assumption—component of variance σ_x^2, the envelope distribution is a Rayleigh law (sub-sect. 7.3.8), the mean value (14.111) of which is $\sqrt{\pi/2}\sigma_x$. The ratio of those two values is $2/\sqrt{\pi} = 1.128$, corresponding to about 1 dB. If the analyzer is calibrated for noise measurements, the level of possible spectral lines, due to the presence of sinusoidal components, will automatically be affected by a positive bias in the range of 1 dB.

12.1.14 Display modes

The results of a spectrum analysis are usually displayed as a X-Y plot (Y = measurement, X = frequency) on a calibrated screen (oscilloscope, video monitor, graphic terminal), or printed on graph paper (graphic plotter).

Depending on the operating principle of the analyzer, the displayed measurement is:

- normalized power (calibration in V^2);
- rms value (calibration in V);
- unilateral power spectral density (calibration in V^2/Hz).
- rms spectral density corresponding to the square root of the unilateral power spectral density (calibration in $V/\sqrt{Hz}$).

Fourier transform analyzers also offer, in principle, the possibility for computing and displaying the phase spectrum (calibration in degrees).

Power spectral measurements are reserved for cases in which the dynamic range is small (~30 dB). A display with rms value or rms spectral density value is usually preferred in order to get improved dynamic range (fig. 12.12).

If the dynamic range is too large, the same measurements can be displayed using a logarithmic scale (fig. 12.13) linearly calibrated in decibels (e.g., dBV scale with 0 dBV = 1 Vrms).

Depending on the analysis method, differences are encountered in the calibration of measurements of purely random or sinusoidal components.

In filter analyzers with bandwidth B_{eq}, we measure and display the normalized power $\tilde{P}_x(f_0,B_{eq})$ or the rms value $\sqrt{\tilde{P}_x(f_0,B_{eq})}$. By computation, we can

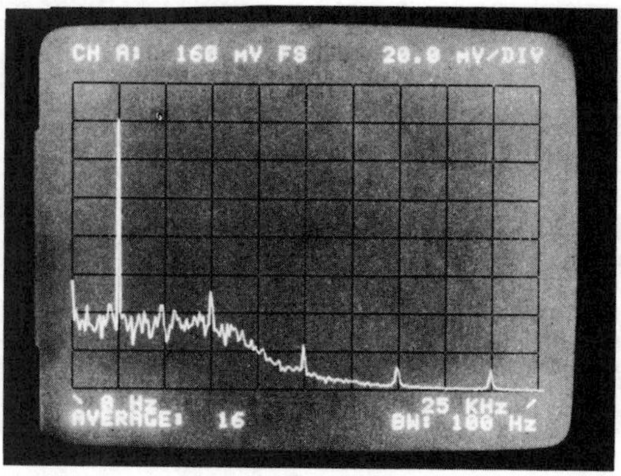

Fig. 12.12 Spectrum of a noisy periodic signal analyzed between 0 and 25 kHz with a 100 Hz resolution; the vertical scale of the display is the rms value measured with a linear scale (20 mV/div).

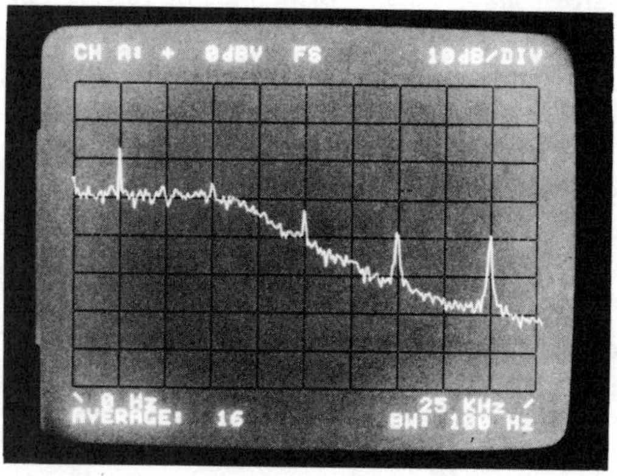

Fig. 12.13 Logarithmic display of the fig. 12.12 spectrum with 10 dB/div on the vertical axis.

deduce from it the unilateral power spectral density $\tilde{\Phi}_x^+(f_0) = \tilde{P}_x(f_0,B_{eq})/B_{eq}$, or the rms spectral density $\sqrt{\tilde{\Phi}_x(f_0)}$.

In the case of a discrete spectrum, the power of each sinusoidal component is independent of the choice of B_{eq}: the level is, hence, constant, irrespective of B_{eq}. On the other hand, in the case of a continuous spectrum, the measured power $\tilde{P}_x(f,B_{eq}) = \tilde{\Phi}_x^+(f_0)B_{eq}$ is proportional to B_{eq}: consequently, the display level varies with a change of the equivalent bandwidth selected (fig. 12.14).

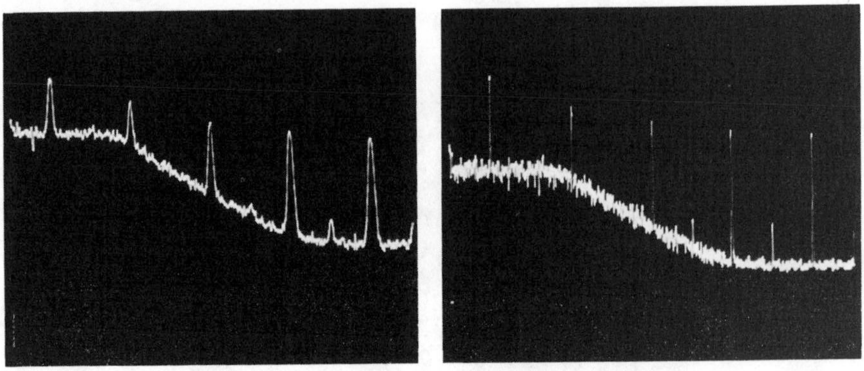

Fig. 12.14 Analog spectrum analysis of a noisy periodic signal for two bandwidths B_{eq}: 300 Hz (left) and 30 Hz (right); the calibration of the frequency scale (horizontal axis) is 5 kHz/div, and the calibration of the amplitude scale (vertical axis) is 10 dB/div.

A Fourier transform analyzer, in principle, directly computes the periodogram $\tilde{\Phi}_x(f,T) = T^{-1}|X(f,T)|^2$, or its square root. As indicated in sub-sect. 12.1.6, the equivalent bandwidth here is proportional to $1/T$.

In the case of a continuous spectrum, the spectral density does not depend on the duration T of the analyzed signal and the display level is independent of the chosen resolution. On the other hand, the Fourier transform $X(f,T)$ of a sinusoidal component perceived through a time window of duration T is proportional to T. The display level of a line spectrum, hence, varies with the selected resolution (fig. 12.15). The measurement converges to the theoretical line spectrum model represented by delta-functions when $T \to \infty$. When the

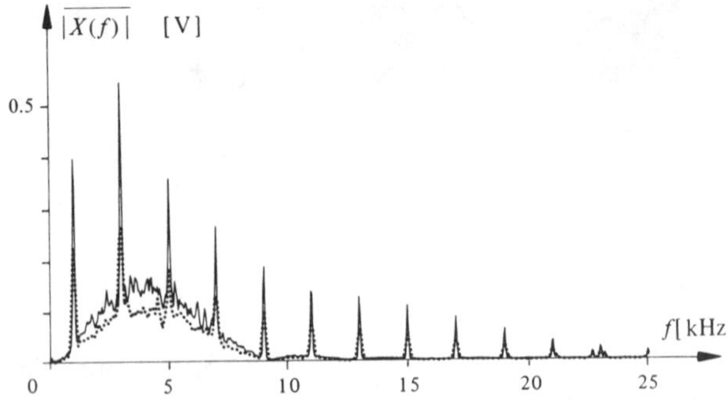

Fig. 12.15 Digital spectrum analysis of a noisy periodic signal for two observation durations $T = NT_e = 1.024$ ms (N = 512, dotted line) and $T = 2.048$ ms (N = 1024, continuous line).

display has a log-scale (dB), the measured lines are simply vertically shifted by $-10 \log B_{eq}$.

A pseudo-3D representation is used in acoustics (fig. 12.16), or in other situations, where we want to emphasize the frequency evolution of a nonstationary signal (fig. 12.17).

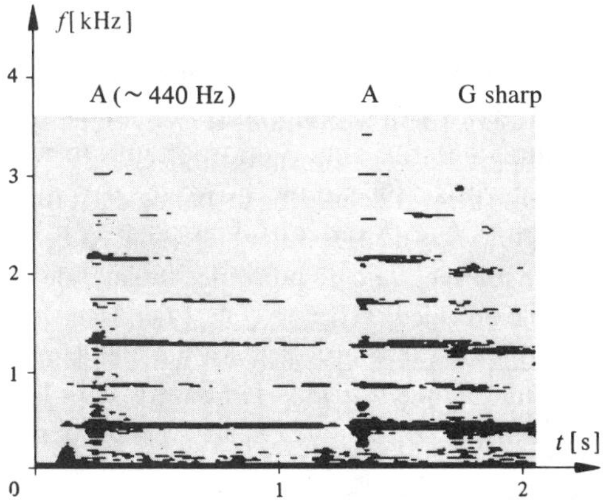

Fig. 12.16 Sonogram of piano notes: graphic representation of the spectrum evolution in function of time (B_{eq} = 20 Hz). The amplitude of the measured spectral components is translated into more or less dark stains; the image so obtained emphasizes the frequency-time structure of the analyzed sounds.

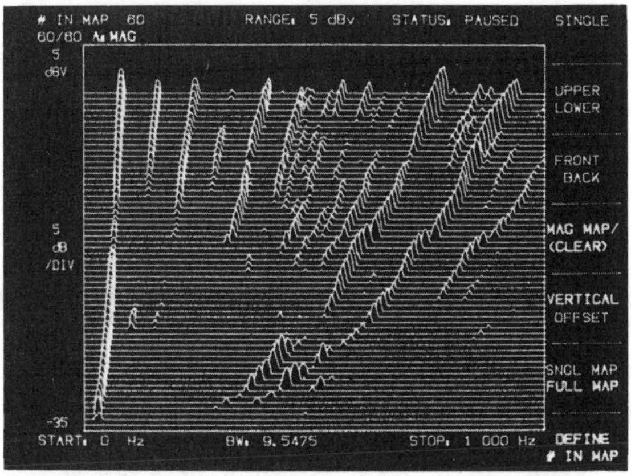

Fig. 12.17 Waterfall representation: the analyzer's internal memory allows the storage and display of a sequence of successive spectral estimations revealing the evolution of frequency components (horizontal axis) in function of time (vertical axis).

12.2 MULTICHANNEL SPECTRUM ANALYZERS

12.2.1 Filter bank analyzers

The simplest principle of the spectrum analyzer is represented on figure 12.18.

The signal to be analyzed is applied simultaneously at the inputs of a set of filters, the bandwidths of which are contiguous, so that the entire required spectral range is covered. In principle, each filter allows measurement of the power—or the rms value—of the signal components in a different bandwidth (channel). A partial superposition stems, however, from the non-ideal nature of the transfer functions of the practical filters used.

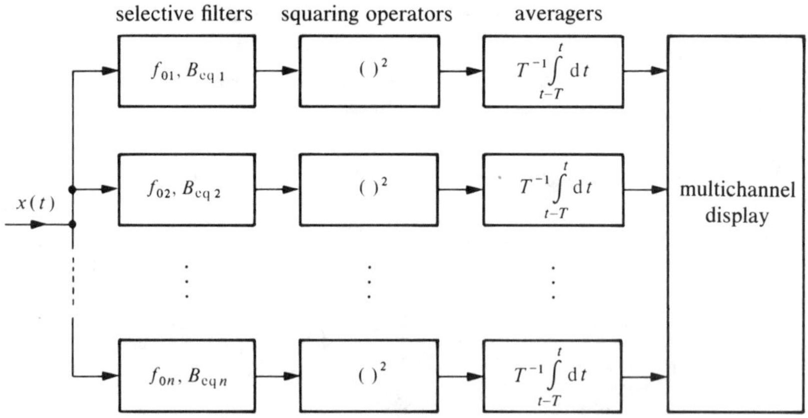

Fig. 12.18

The resolution of such an analyzer is obviously limited by the number (a few tens) of available filters and the extent of the frequency range to be analyzed.

The main advantage of this kind of system is that it enables the continuous supervision of each channel. This is the reason for its use in some applications: nonstationary phenomenon analysis in geophysics; in acoustics: mechanical vibration study, *et cetera*. Its use is mainly limited to low frequency signals.

Filters have long been implemented using classical analog techniques. The current trend is towards the use of integrated devices processing sampled signals (CCD or switched capacitor ICs), or digital simulation of the filter bank.

12.2.2 Variable resolution

A filter bank analyzer has a variable resolution if the equivalent bandwidths are not identical. This allows the monitoring of a rather wide frequency range with a small set of filters. This solution is mainly used in acoustics.

The repartition of the center frequencies f_{0i} and of the equivalent bandwidths B_{eqi} is usually exponential:

$$f_{0(i+1)} = 2^n \cdot f_{0i}; \qquad B_{eq(i+1)} = 2^n \cdot B_{eqi} \tag{12.32}$$

The center frequency is defined here as the geometric mean of the upper and lower cut-off (e.g., at −3 dB) frequencies f_1 and f_2:

$$f_0 = \sqrt{f_1 f_2} \tag{12.33}$$

For $n = 1/3$, we get the standard 1/3 octave analysis (fig. 12.19). By connecting three adjacent filters in parallel (fig. 12.20), we can easily get a full octave ($n = 1$) analysis. The attenuation specifications of 1 octave and 1/3 octave filters (fig. 12.21) are standardized [IEC (International Electrotechnical Commission); ANSI (American National Standards Institute)]. A modular solution to the digital simulation of the 1/3 octave filter is described in [123].

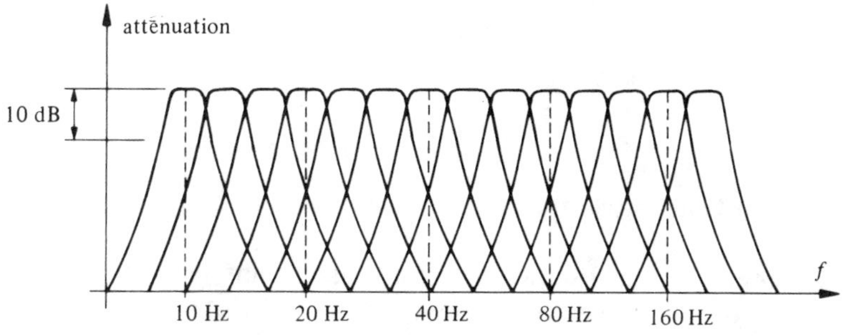

Fig. 12.19

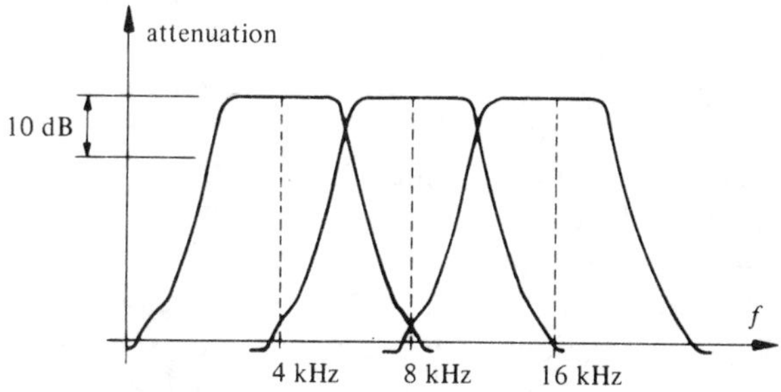

Fig. 12.20

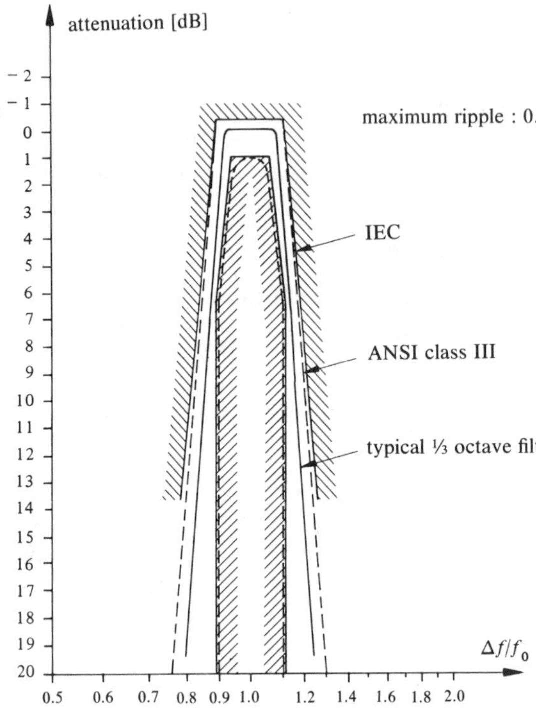

Fig. 12.21

12.2.3 Digital Fourier transform multichannel analyzers

The existence of fast Fourier transform algorithms (sub-sect. 9.3.11) and the evolution of digital electronic devices have led to the implementation of signal analyzers used as multichannel spectrum analyzers (fig. 12.22).

The potential resolution is rather high, since the number of equivalent channels is usually in the range of a few hundreds, or even a few thousands (this is the reason why manufacturers call them high resolution analyzers). The effective resolution depends, however, on the weighting time window and the possible spectrum smoothing function (sub-sect. 12.1.5 and 12.1.12).

The weighted combination of channels enables the synthesis of a 1 octave or ⅓ octave analysis (sub-sect. 12.2.2).

The functioning of such equipments is sequential: an initial acquisition operation (sampling, A/D conversion, storage) of a signal segment is followed by a set of calculus operations: FFT computation, spectrum averaging or smoothing, polar coordinate evaluation (amplitude and phase spectra or power spectrum); and, finally, display, *et cetera*.

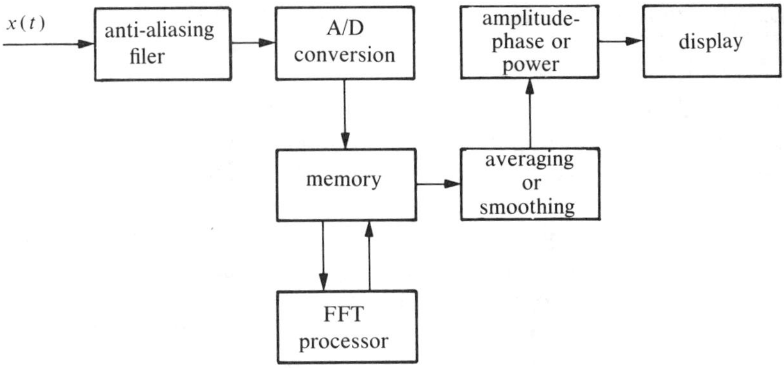

Fig. 12.22

Real-time processing is achieved when the set of operations is executed in a time lap smaller than the acquisition time. Because of the sampling constraints (sect. 9.3) and the computation time, real-time processing is limited to rather low frequency signals. With higher sampling rates, the processing is discontinuous: the signal segments successively recorded and analyzed are no longer contiguous.

The number N of samples stored during the acquisition phase of a signal segment depends on the algorithm used to evaluate the discrete Fourier transform (DFT). Usually, it is a power of 2. The fast algorithms described in volume XX take advantage of the many redundancies and symmetries appearing in the DFT calculus. They reduce the number of basic operations (complex additions with multiplications) from N^2 down to something like $N \mathrm{lb} N$. The achieved savings are about 100 times less operations for $N = 2^{10} = 1024$.

This digital processing has some consequence. If $T_e = 1/f_e$ is the sampling time interval, the observation duration is $T = NT_e$. The computed DFT is represented by a sequence of N complex samples with a frequency interval $\Delta f = 1/T$. The time sampling generates a periodic spectrum with a period f_e. In principle, only the main half-period, covering the 0 to $\frac{1}{2} f_e$ frequency range, is displayed (unilateral spectrum). Thanks to the reciprocity theorem (sub-sect. 9.3.10), this sampled spectrum corresponds to the spectrum of a periodic repetition of a N samples segment.

The experimental analysis conditions, hence, directly depend on the choice of the sampling rate f_e and the number N of samples stored. The rate f_e sets the analyzed frequency range. It must comply with the sampling theorem (9.25) in order to avoid aliasing errors. The number N then defines the lowest analyzed frequency and sets the resolution limit given by (12.8):

$$\Delta f = B_{eq} = 1/T = f_e/N \tag{12.34}$$

Digital premultiplication of the signal by a cosine function followed by lowpass filtering and resampling are sometimes used to get a spectrum shift (frequency translation; sub-sect 11.5.1) and a zoom effect.

By means of a Fourier transform computation performed on one or two signals $x(t)$ and $y(t)$, it is possible to evaluate, from (4.65), (5.168), and (5.178), other important relations: spectral density $\Phi_x(f)$ and cross-spectral density $\Phi_{xy}(f)$, auto- and cross-correlation functions $\phi_x(\tau) = F^{-1}\{\Phi_x(f)\}$ and $\phi_{xy}(\tau) = F^{-1}\{\Phi_{xy}(f)\}$, coherence function $\Gamma_{xy}(f) = |\Phi_{xy}(f)|^2/[\Phi_x(f)\Phi_y(f)]$, transfer function $G(f) = Y(f)/X(f)$, *et cetera*. Digital analyzers usually offer a wider range of test possibilities than conventional spectrum analyzers.

12.2.4 Discrete Fourier transform with linear frequency modulation

Inserting $2nk = n^2 + k^2 - (n - k)^2$ into (9.49) converts the DFT into the following equivalent forms:

$$X(n) = \sum_{k=k_0}^{k_0+N-1} x(k) \exp(-j2\pi nk/N)$$

$$= \exp(-j\pi n^2/N) \times \left\{ \sum_{k=k_0}^{k_0+N-1} x(k) \exp(-j\pi k^2/N) \cdot \exp[j\pi(n - k)^2/N] \right\} \tag{12.35}$$

The second form is called the *chirp z-transform* (CZT). A *chirp* is an expression used in radar to designate a linearly frequency modulated sinusoidal pulse. The sampled and complex expression for such a signal is of the $\exp(j\pi k^2/N)$ kind.

The interpretation of (12.35) is straightforward; the DFT can be evaluated by a system performing the following successive operations:

- premultiplication of the signal $x(k)$ by a linearly frequency modulated complex exponential $\exp(-j\pi k^2/N)$;
- discrete convolution (8.13) of the outcome by a complex exponential $\exp(j\pi k^2/N)$ with linear frequency modulation of reverse sign;
- postmultiplication by a linearly frequency modulated complex exponential $\exp(-j\pi n^2/N)$.

The discrete convolution corresponds to a linear filtering by a transversal filter with complex impulse response $g(k) = \exp(j\pi k^2/N)$, leading to the block diagram of fig. 12.23. When the sole power spectral density is needed, postmultiplication can be deleted.

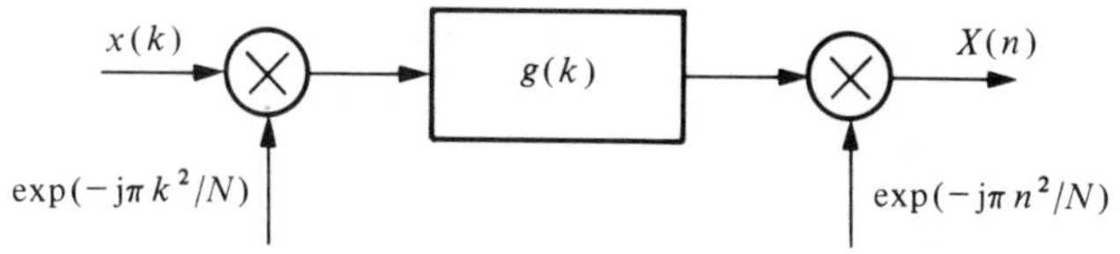

Fig. 12.23

Such a structure is especially well suited for integrated implementation in a charge-coupled device (CCD [124]), or in surface acoustic wave (SAW [125]) technology. These devices are used in specific applications in telecommunications, radar, and instrumentation. Their operating ranges are rather different: from a few kilohertz to a few megahertz for the CCD, and from a few megahertz to a few hundreds of megahertz for the SAW [126].

12.2.5 Optical spectrum analyzer

For coherent light, an optical lens is a bidimensional Fourier transformer [127]. Its principle is illustrated by fig. 1.21 (chap. 1). N values for the spectral density of the transmittance function located in the lens front focal plane can be simultaneously measured by placing in the rear focal plane a photoelectric sensor composed of an array of N photosensors with quadratic response.

The translation of the electrical signal to be analyzed into a variable transmittance image can be achieved by photographic film when real-time processing is not required. However, the interest in this kind of device lies precisely in its naturally parallel operation and its real-time processing capabilities.

Various optoelectronic modulators performing the direct voltage-optical transmittance conversion have been developed [67]. The most promising approach is that of integrated optics, which gathers in a monolithic circuit of a few centimeters (fig. 12.24) a semiconductor laser, two geodesic lenses separated by a transducer (acousto-optical or optoelectronic), diffracting the optical beam at an angle depending on the applied signal frequency, and, finally, a linear array of photoelectric sensors. This kind of analyzer is reserved to high frequency signals ($\gg 1$ MHz).

12.3 SWEPT-FREQUENCY SPECTRUM ANALYZERS

12.3.1 General principle

Swept-frequency analyzers are based on the frequency translation described in section 11.5 to shift the spectrum of the signal to be analyzed. It progres-

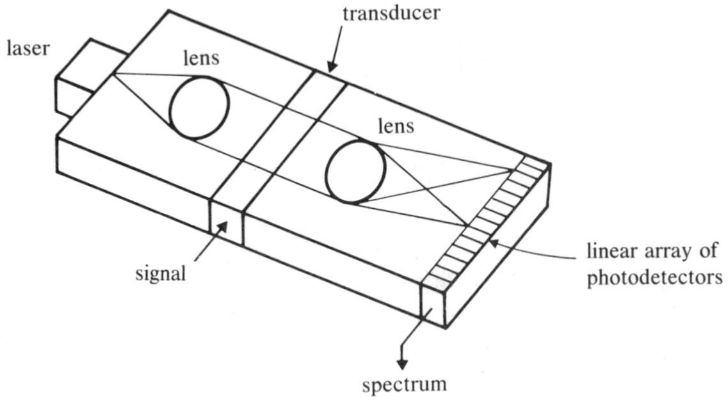

Fig. 12.24

sively scans the signal spectrum with a single selective filter. Figure 12.25 shows the conceptual block diagram.

The input signal $x(t)$ is multiplied by an auxiliary sinusoidal signal (or a square wave) with a frequency f_p. This frequency is controlled by a voltage supplied by the scanning generator and varies between two limits setting the analyzed frequency range. The input signal spectrum is thus frequency-shifted by a quantity f_p. The multiplier output signal is applied to the input of a narrow bandpass filter with a frequency response $G(f)$ centered on a fixed frequency f_0. Its equivalent bandwidth B_{eq} determines the analysis resolution. This kind of analyzer usually offers a choice of bandwidths. The filtered signal is finally applied to a device measuring its average power (or its rms value). The output, according to a linear or logarithmic (dB scale) law, controls the vertical displacement (Y) of the spot on the display screen. The voltage produced by the scanning generator usually has a saw-toothed shape. It simulta-

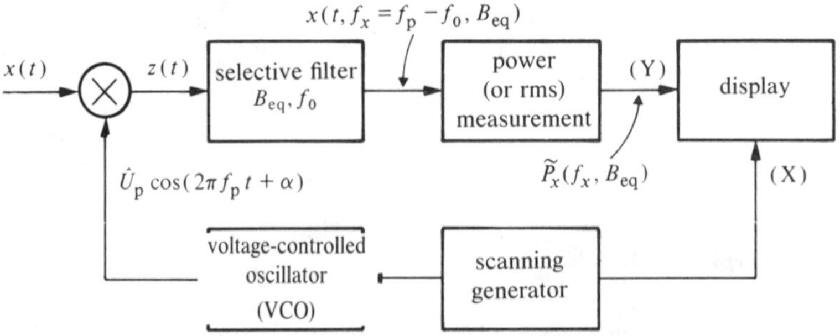

Fig. 12.25

neously controls the oscillator variable frequency f_p and the horizontal scanning (X) of the display screen.

If $\Phi_x(f)$ is the input signal spectrum, the multiplication $z(t) = x(t)\hat{U}_p \cos(2\pi f_p t + \alpha)$, from (5.199) and (4.144), corresponds to

$$\Phi_z(f) = k^2[\Phi_x(f + f_p) + \Phi_x(f - f_p)] \tag{12.36}$$

where $k = \hat{U}_p/2$ is an arbitrary calibration parameter. It is suitable here to equate it to one.

Assuming $f_0 < f_p$, the filter output signal can be denoted $x(t,f_p - f_0,B_{eq})$. It corresponds to the component of the signal $x(t)$, with a unilateral spectrum reduced to a bandwidth B_{eq} centered on frequency $f_x = f_p - f_0$ (fig. 12.26). The measured power is equal to $\tilde{P}_x(f_x,B_{eq})$.

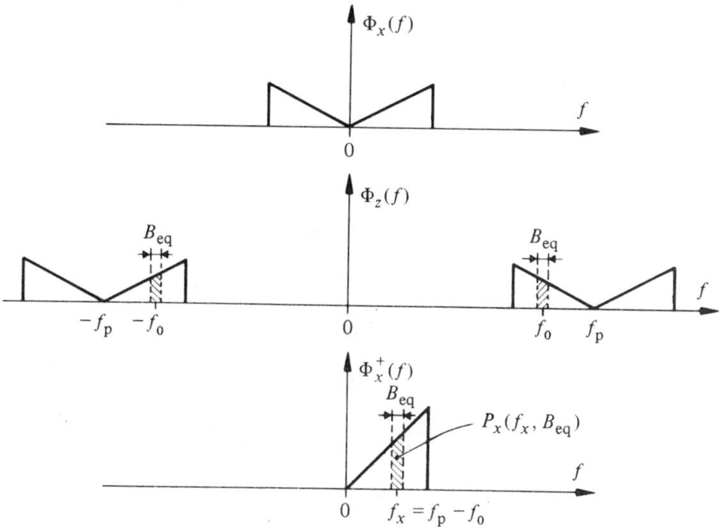

Fig. 12.26

The bilateral spectral density estimation of signal $x(t)$ at frequency $f_x = f_p - f_0$ is obtained by dividing the measured power by $2B_{eq}$:

$$\tilde{\Phi}_x(f_x) = \tfrac{1}{2}\tilde{P}_x(f_x,B_{eq})/B_{eq} \tag{12.37}$$

By progressively shifting the oscillator frequency f_p across the $[f_0,f_0 + f_m]$ range, we sequentially scan the entire spectrum of signal $x(t)$ between $f_x = 0$ and the maximum frequency $f_x = f_m$.

If the oscillator control enables us to bring f_p below f_0, the analyzer also scans the symmetrical part of the bilateral spectrum $\Phi_x(f)$, which is theoretically located on negative frequencies $f_x = f_p - f_0$!

The origin corresponding to $f_x = 0$ is usually marked on the display by a pseudospectral line created (sub-sect. 12.3.3) by the addition of an artificial but calibrated dc component to signal $x(t)$. The possible true dc component of $x(t)$ was previously eliminated by a highpass filter.

12.3.2 Relation between theoretical spectrum and measured spectrum

The measured power $\tilde{P}_x(f_x,B_{eq}) = 2\tilde{\Phi}_x(f)B_{eq}$ is identical to the power that we would get by directly filtering the signal $x(t)$ with a selective filter similar to $G(f)$, but with an equivalent bandwidth B_{eq} centered on frequency f_x.

The power being the integral of the power spectral density, the measurement (12.37) can also be expressed, taking (8.24) into account, by the frequency convolution product:

$$\begin{aligned}\tilde{\Phi}_x(f_x) &= \frac{1}{B_{eq}}\int_{-\infty}^{\infty}\Phi_x(f)|H(f_x-f)|^2\mathrm{d}f\\ &= B_{eq}^{-1}\,\Phi_x(f_x) * \mathring{\Phi}_h(f_x)\end{aligned} \tag{12.38}$$

where $\mathring{\Phi}_h(f) = |H(f)|^2 \triangleq |G(f+f_0)|^2\epsilon(f+f_0)$ is a filter function obtained through shifting by f_0 the modulus square of the transfer function $G(f)$ defined for $f > 0$ (fig. 12.27).

The convolution (12.38) is similar to (12.3), established under the assumption of a multiplication of the signal to be analyzed by a time weighting function $u(t)$. This results in a similar resolution limit. This concept is illustrated in the next sub-section.

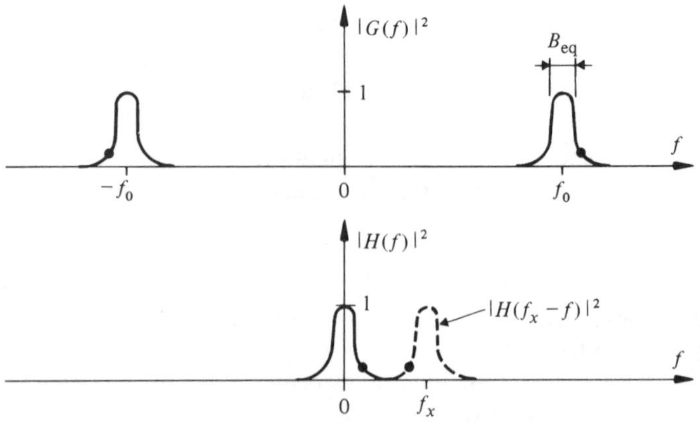

Fig. 12.27

12.3.3 Example: discrete spectrum measurement

Any line of a theoretical spectrum, represented by a delta-function at frequency f_*, is transformed by convolution (12.38) and property (1.48) into a continuous function proportional to $\mathring{\Phi}_h(f - f_*)$.

Spectral lines are only associated with the presence of a dc component or sinusoidal components. In this case, the signal can be written as

$$x(t) = \bar{x} + \sum_k A_k \sin(2\pi f_k t + \alpha_k) \tag{12.39}$$

and its power spectral density, from (4.154), is

$$\Phi_x(f) = \bar{x}^2\delta(f) + \sum_k \frac{1}{4} A_k^2[\delta(f + f_k) + \delta(f - f_k)] \tag{12.40}$$

From (12.38), the measured spectral density becomes (fig. 12.28):

$$\tilde{\Phi}_x(f_x) = B_{\text{eq}}^{-1}\left\{\bar{x}^2\mathring{\Phi}_h(f_x) + \sum_k \frac{1}{4} A_k^2[\mathring{\Phi}_h(f_x + f_k) + \mathring{\Phi}_h(f_x - f_k)]\right\} \tag{12.41}$$

Hence, the resolution is directly determined by the equivalent bandwidth B_{eq} of the analysis filter.

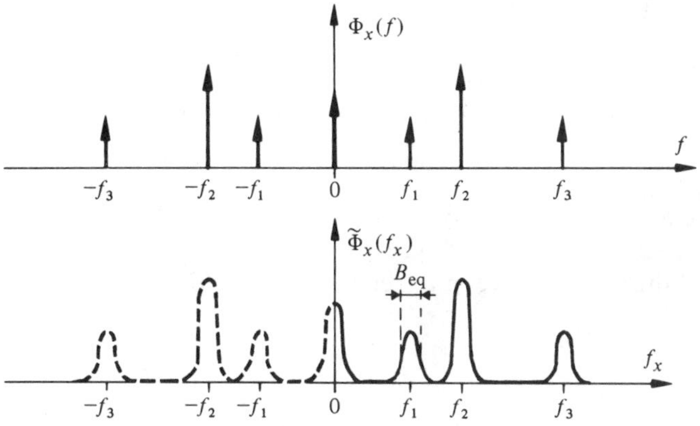

Fig. 12.28

12.3.4 Resolution-scanning speed trade-off

The scanning speed ν_b, measured in Hz/s, is compelled to be lower than or equal to a limit depending on the chosen resolution—hence, on the bandwidth B_{eq} of the analysis filter—and on the measurement duration T of the averager used for the power estimation $\tilde{P}_x(f_x, B_{\text{eq}})$.

For the analyzer to obtain a correct response to any discontinuity in the spectrum $\Phi_x(f)$, the spectral pattern, corresponding to the transposed spectrum $\Phi_z(f)$, must be frequency-shifted by a quantity lower than or equal to B_{eq} during the measurement interval T:

$$\nu_b \leqslant B_{eq}/T \tag{12.42}$$

However, T is also related to the signal-to-noise ratio ξ of the measurement (sub-sect. 12.1.8). For a perfect averager, $T = T_0$ is the total integration duration and, from (12.26), $\xi = B_{eq}T_0$. In the usual practical case of an exponentially weighted averager with time constant τ (e.g., lowpass RC filter), the same signal-to-noise ratio is reached with a time constant $\tau = T_0/2$.

However, the bias introduced by the transient term of the filter time response becomes acceptable (<2%) only for $T \geqslant 4\tau = 2T_0$.

Condition (12.42) can then be directly expressed as a function of the resolution and the desired signal-to-noise ratio by

$$\nu_b \leqslant \alpha B_{eq}^2/\xi \tag{12.43}$$

where $\alpha = 1$ for a perfect averager and $\alpha = ½$ for an exponentially weighted averager.

Another restriction is due to the rise time of the analysis bandpass filter. This rise time is roughly equal to the inverse of the equivalent bandwidth (sub-sect. 8.2.22). Thus, we obtain

$$\nu_b \leqslant B_{eq}/t_m \cong B_{eq}^2 \tag{12.44}$$

Combining (12.43) and (12.44), we end up with the double condition:

$$\nu_b \leqslant \alpha B_{eq}^2/\xi \leqslant B_{eq}^2 \tag{12.45}$$

The first condition is usually the more severe, insofar as the desired signal-to-noise ratio is normally much greater than one.

The total duration to sweep a frequency range B_m is thus given by

$$T_b = B_m/\nu_b \geqslant \xi B_m/(\alpha B_{eq}^2) \tag{12.46}$$

12.3.5 Examples

The spectrum analysis obtained by sweeping a frequency range $B_m = 1$ kHz with a resolution $B_{eq} = 10$ Hz and a measurement signal-to-noise ratio $\xi = 100$ requires a scanning speed $\nu_b \leqslant 1$ Hz/s in the case of a perfect averager, and $\nu_b \leqslant 0.5$ Hz/s in the case of an exponentially weighted averager (bias <2%). The total scanning durations are, respectively, $T_b \geqslant 1000$s $\cong 17$ minutes and $T_b \geqslant 2000$s $\cong 33$ minutes!

Under the same signal-to-noise ratio and bias conditions, the scanning of a 100 MHz range with a 1 MHz resolution requires a scanning speed $v_b \leqslant 5$ GHz/s with an exponentially weighted averager, and, therefore, a total scanning duration of $T_b \geqslant 20$ ms.

Thus, with an equal relative resolution B_{eq}/B_m, the swept-frequency spectrum analyzer is better suited to the high frequency range than to the low frequency range.

12.3.6 Recirculation procedure

To overcome the scanning speed limitation imposed on analyzers designed for low frequency signals, the principle of digital storage and recirculation described in sub-section 10.5.2 can be applied ahead of the analyzer itself (fig. 12.29).

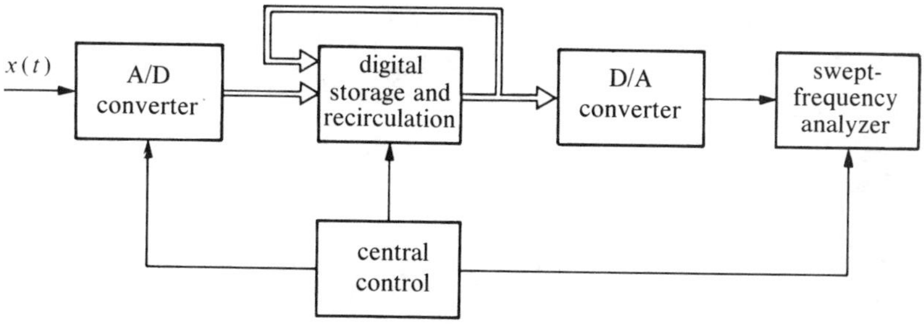

Fig. 12.29

A segment $x_1(t, T_1)$ of duration T_1 of the input signal to be analyzed is first digitized and stored. Its cyclical reading, speeded up by a factor $a > 1$, generates a periodic signal $x_3(t) = \text{rep}_{T_3}\{x_2(t, T_2)\}$, the Fourier transform of which is, from (4.18) and (10.56):

$$X_3(f) = T_{e1}^{-1} X_1(f/a, T_1)\delta_{1/T_3}(f) \tag{12.47}$$

where T_{e1} is the acquisition sampling interval and $T_2 = T_1/a$.

By analogy with (4.139), the power spectral density becomes

$$\Phi_{x3}(f) = T_{e1}^{-2} \mathring{\Phi}_{x1}(f/a, T_1)\delta_{1/T_3}(f) \tag{12.48}$$

The recirculation process generates a discrete spectrum, the ***spectral envelope*** of which (sub-sect. 4.4.9) is proportional to the spectrum of $x_1(t, T_1)$ expanded by a factor a.

By applying this accelerated signal to a swept-frequency analyzer using a selective filter of bandwidth B_{eq}, we get for the input signal $x_1(t)$ an *equivalent resolution* B_{eq}/a, as long as the condition $T_1 \gg a/B_{eq}$ is satisfied, according to (12.8). It is, thus, possible to keep a good resolution while increasing the scanning speed by a factor a^2. The total scanning duration is then reduced by a factor a, since the swept-frequency range has been expanded by a factor a.

Acceleration factors in the range of 10^2 to 10^5 are commonly used, depending on the chosen frequency range.

This kind of memory device also allows for the swept-frequency analysis of transient signals.

12.4 APPLICATION EXAMPLES

12.4.1 Analysis of a periodic signal

The periodic signal of fig. 12.30 is analyzed by an FFT digital analyzer (fig. 12.31) and a swept-frequency analyzer (fig. 12.32) with the indicated parameters.

12.4.2 Analysis of a transient signal

Digital FFT analysis of the transient signal of fig. 12.33 is represented in fig. 12.34, with the given experimental parameters.

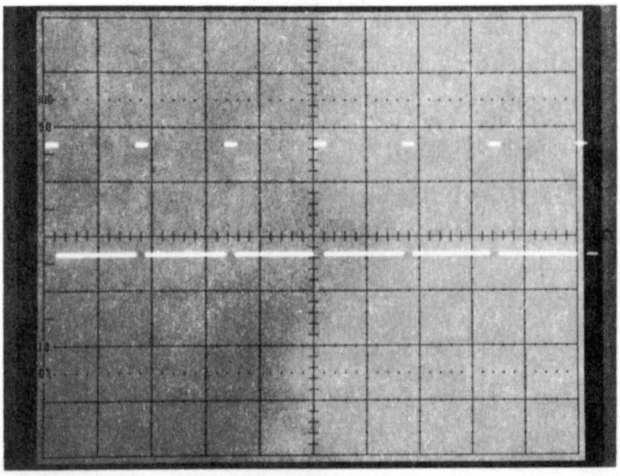

Fig. 12.30 Oscillogram of a periodic sequence of rectangular pulses with zero mean value, 0.2 ms duration, and 600 Hz repetition frequency (horizontal scale: 1 ms/div; vertical scale: 2 V/div).

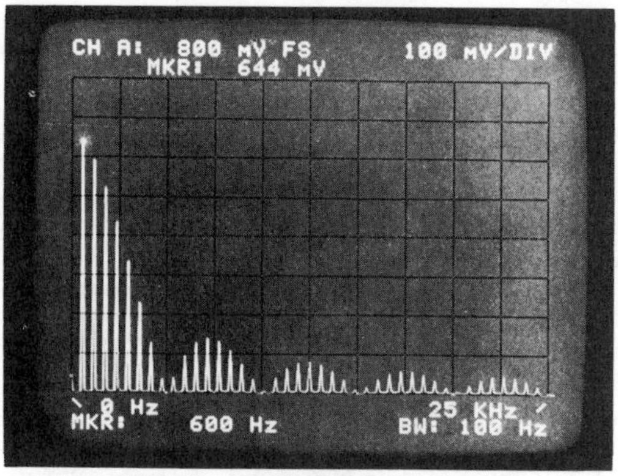

Fig. 12.31 Digital spectrum analysis of the signal of fig. 12.30 over a frequency range: 0 to 25 kHz with B_{eq} = 100 Hz; the bright marker (MKR) gives the fundamental amplitude and frequency.

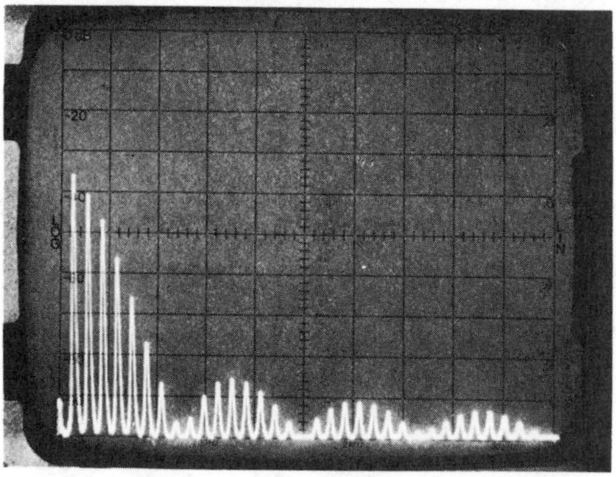

Fig. 12.32 Analog swept-frequency spectrum analysis of the signal of fig. 12.30 with a 100 Hz equivalent bandwidth filter (vertical scale: 0.1 V/div; horizontal scale: 2.5 kHz/div).

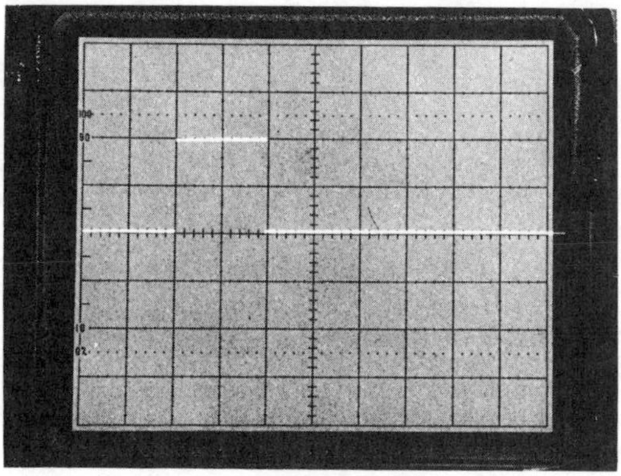

Fig. 12.33 Oscillogram of a 4 V rectangular pulse with a 0.2 ms duration.

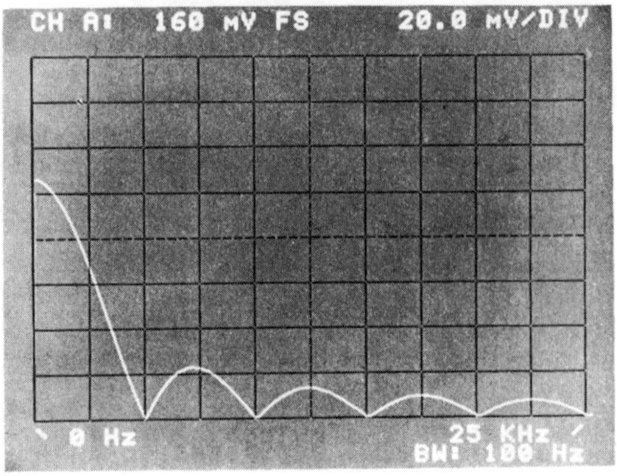

Fig. 12.34 Amplitude spectrum of the signal of fig. 12.33 measured between 0 and 25 kHz with an observation duration $T = 10$ ms ($B_{eq} = 100$ Hz).

12.4.3 Analysis of a random signal

Figure 12.36 reproduces two digitally evaluated periodograms of the signal displayed in fig. 12.35. Swept-frequency analysis of the same signal is reproduced in fig. 12.37.

12.4.4 Analysis of vibration signals

Two examples of vibration signals of a water-turbine are represented in figure 12.38. The presence of spectral lines in the signal obtained when subjecting the turbine to a shock (percussion) reveals the existence of various mechanical resonances also present in the vibration spectrum of the rotating turbine.

12.4.5 Modulated signals

Figures 12.39, 12.40, and 12.41 display the spectrum of a sine wave, an amplitude-modulated sine wave, and a frequency-modulated sine wave by a sinusoidal modulating signal.

12.4.6 Sampled signals

The spectrum of a sine wave of frequency f_0, sampled and held at a rate $f_e = 5f_0$ is shown in fig. 12.42.

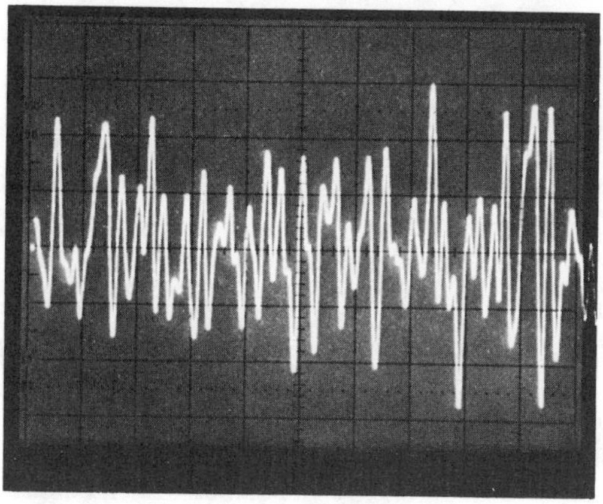

Fig. 12.35 Oscillogram of a random signal (horizontal scale: 1 ms/div; vertical scale: 1 V/div).

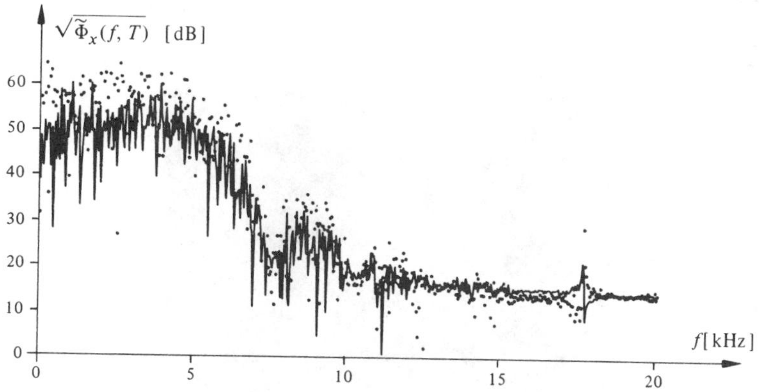

Fig. 12.36 Superposition of two periodograms (dotted and broken lines) evaluated on different segments of the signal of fig. 12.35 and given in rms dB value (vertical logarithmic scale).

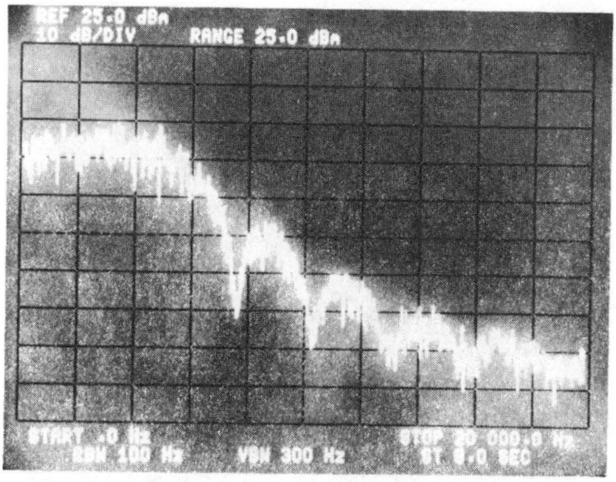

Fig. 12.37 Swept-frequency analysis on the 0 to 20 kHz range with a 100 Hz filter of the signal of fig. 12.35.

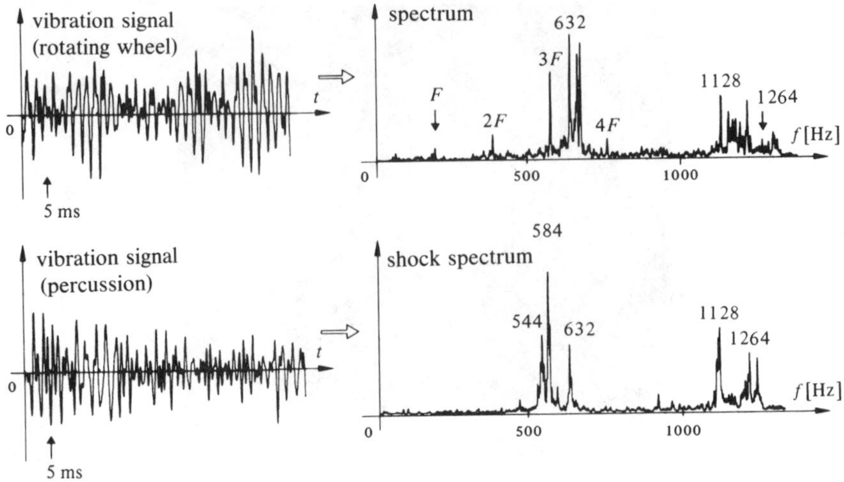

Fig. 12.38 Signals and vibration spectra of a Pelton turbine with 23 buckets; frequency F is the product of the number of buckets and the wheel rotation frequency (8⅓ rps).

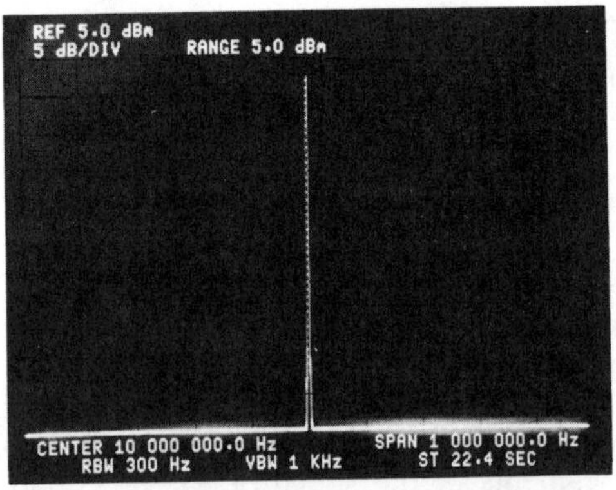

Fig. 12.39 Spectrum of a 10 MHz sine wave measured with a 300 Hz resolution.

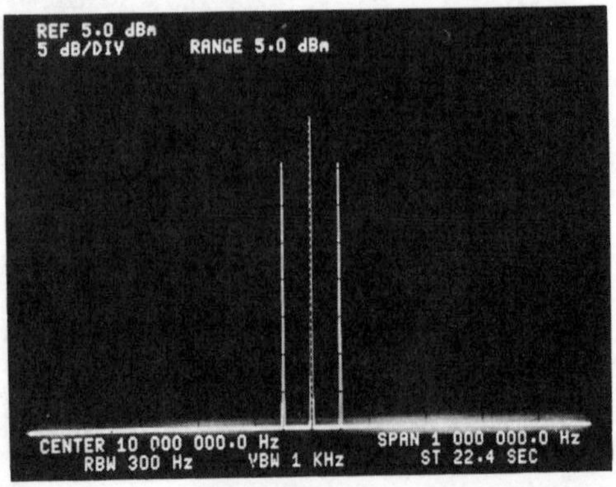

Fig. 12.40 Spectrum of a 10 MHz sine wave amplitude modulated by a 50 kHz sine wave.

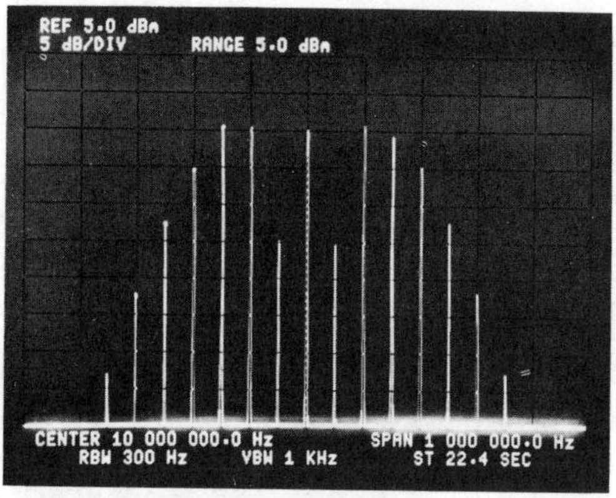

Fig. 12.41 Spectrum of a 10 MHz sine wave frequency modulated by a 50 kHz sine wave (modulation index: $\delta \approx 6$).

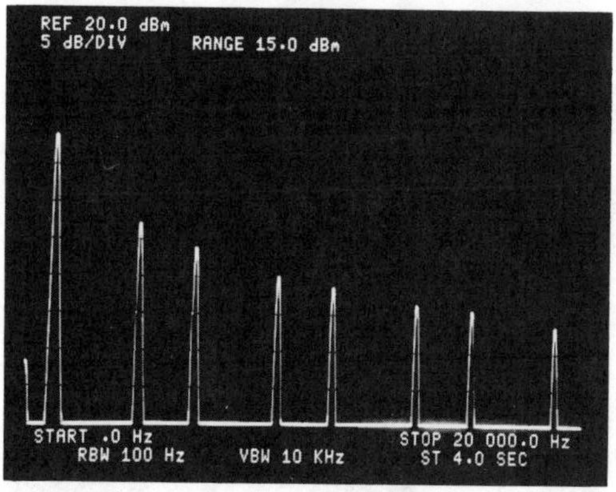

Fig. 12.42 Spectrum of a 1 kHz sine wave, sampled and held at a 5 kHz rate.

12.4.7 Measurement of the purity of an oscillator or the nonlinear distortion of an amplifier

Spectrum analysis enables us to measure easily the harmonic distortion of an oscillator output signal (fig. 12.43).

Analyzing the output signal of an amplifier stimulated by an input signal composed of two sine waves with different frequencies (fig. 12.44) exhibits its nonlinear distortion as revealed by crossmodulation products (sub-sect. 8.4.6).

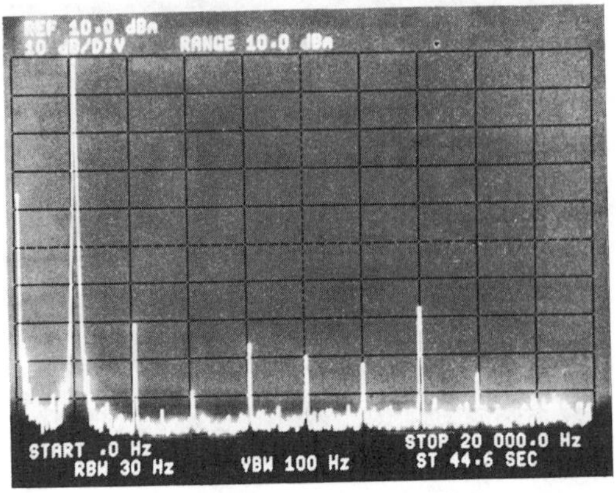

Fig. 12.43 Spectrum of the signal produced by an oscillator operating at 2 kHz.

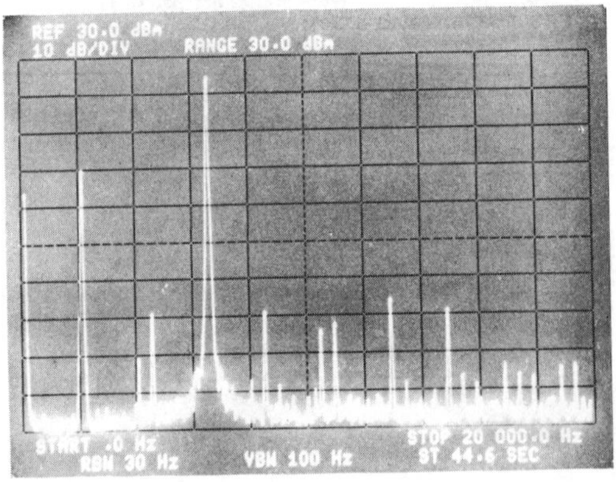

Fig. 12.44 Spectrum analysis of the output signal of an amplifier stimulated by the sum of a 2 kHz and a 6.5 kHz sine wave; the various secondary lines are due to harmonic and crossmodulation distortions.

12.4.8 Frequency and phase stability

The frequency or phase instabilities of an oscillator can be revealed by spectrum analysis. In the same way, we can measure the fluctuations of the winding speed of tape recorders by analyzing the spectrum of a previously recorded sine wave (fig. 12.45).

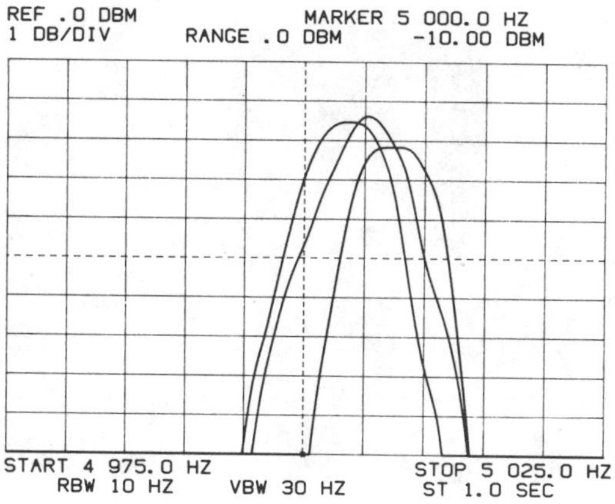

Fig. 12.45 Frequency fluctuations of a recorded sine wave revealed by repeated spectrum analysis.

12.4.9 Electrical power supply network

The impurely sinusoidal character of the electrical power supply network (Europe: 220 V, 50 Hz) is shown in fig. 12.46.

12.4.10 Video signal

Spectrum analysis allows us to measure the bandwidth occupied by a television signal (fig. 12.47).

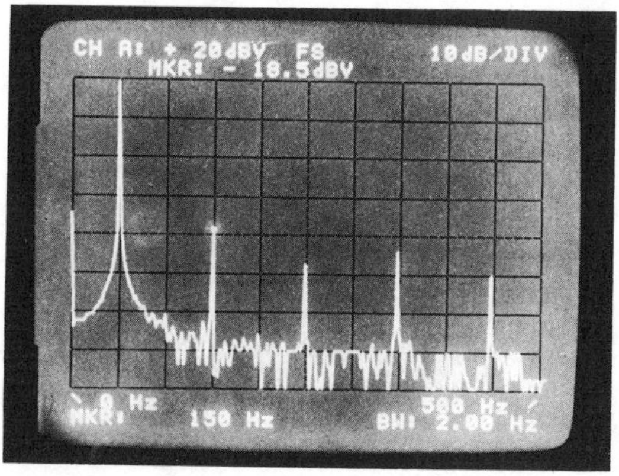

Fig. 12.46 Spectrum analysis of a 50 Hz power network voltage; the bright marker (MKR) shows the relative amplitude of the third frequency harmonic.

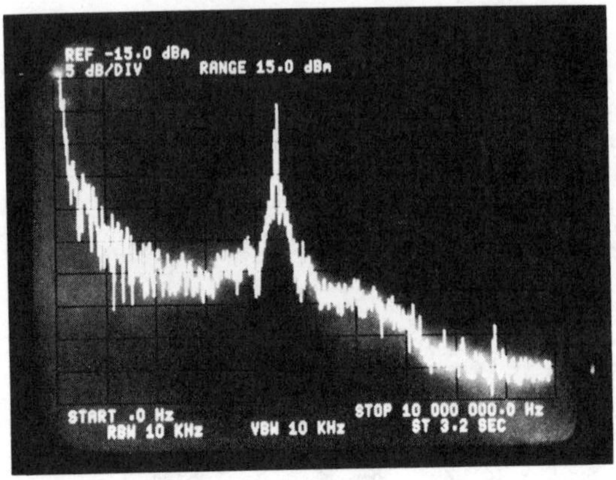

Fig. 12.47 Spectrum analysis of the video signal produced by a TV camera in the 0–10 MHz frequency range.

12.4.11 Frequency response of a filter or an amplifier

The swept-frequency analyzer, or its so-called wave analyzer version, is well suited for tracing the amplitude frequency response of linear systems such as filters (fig. 12.48) or amplifiers (fig. 12.49). A large dynamic range of measurement (70 to 100 dB) is obtained by stimulating the linear system with a sine wave controlled by the VCO in order to get a frequency strictly centered on the analysis bandwidth, as a result of appropriate frequency translation. Very narrow bandwidths can then be used to reduce the residual background noise to a minimum level.

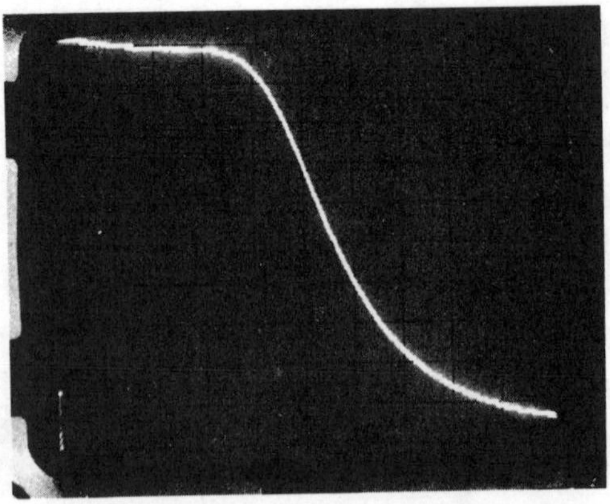

Fig. 12.48 Frequency response of a lowpass filter with cut-off frequency $f_c = 10$ kHz analyzed over a 0–20 kHz linear frequency range with a vertical linear amplitude scale.

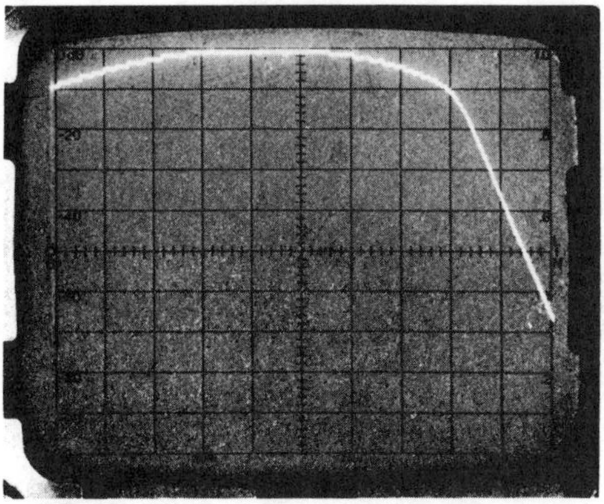

Fig. 12.49 Frequency response of an amplifier analyzed with a vertical logarithmic scale (10 dB/div) and a horizontal logarithmic frequency scale (20 Hz, 200 Hz, 2 kHz, 20 kHz marks at the bottom of the scope).

Digital analyzers offer the possibility for simultaneous measurement of the amplitude and phase responses (fig. 12.50), but usually with a reduced dynamic range. The implemented method is, for instance, the one described in subsect. 8.2.10, or by the application of relation (8.25).

12.4.12 Detection of parasitic oscillations

A poorly designed amplifier often produces parasitic oscillations appearing at frequencies much greater than the normal operating range. Such oscillations are easily revealed and measured by spectrum analysis (fig. 12.51 and 12.52).

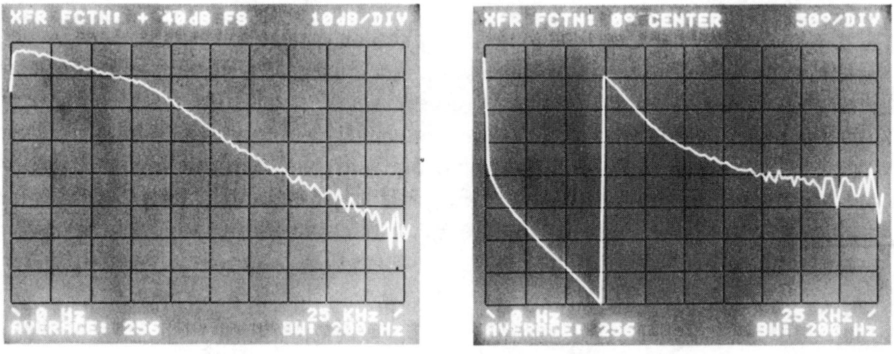

Fig. 12.50 Amplitude (left) and phase (right) responses of an amplifier.

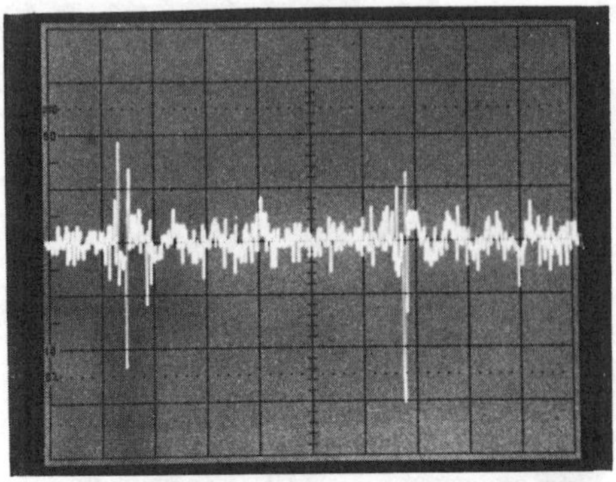

Fig. 12.51 Oscillogram of the output voltage fluctuations of a dc voltage regulator (vertical scales: 2 mV/div; horizontal scale: 5 μs/div).

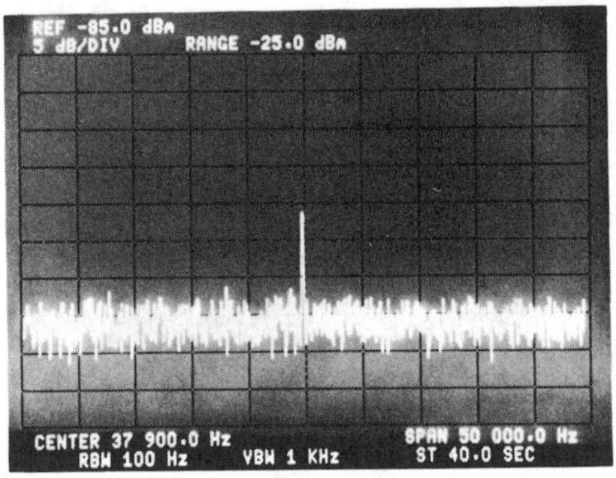

Fig. 12.52 Spectrum analysis of the signal of fig. 12.51 revealing an oscillation at 37.9 kHz.

12.4.13 Cepstral analysis

The *cepstrum* of a signal is a time function defined as either the Fourier transform of the logarithms of its power spectral density, or the modulus square of this transform. This function is convenient for exhibiting the spectral periodicities. It is used, for instance, in speech processing (formant extraction) and in the analysis of vibration signal of cog-wheel devices (gear boxes, *et cetera*). Digital analyzers are sometimes designed to perform cepstral analysis (fig. 12.53). The cepstrum has some similarities with the autocorrelation function.

12.5 PROBLEMS

12.5.1 Verify the results (12.10) and (12.11).

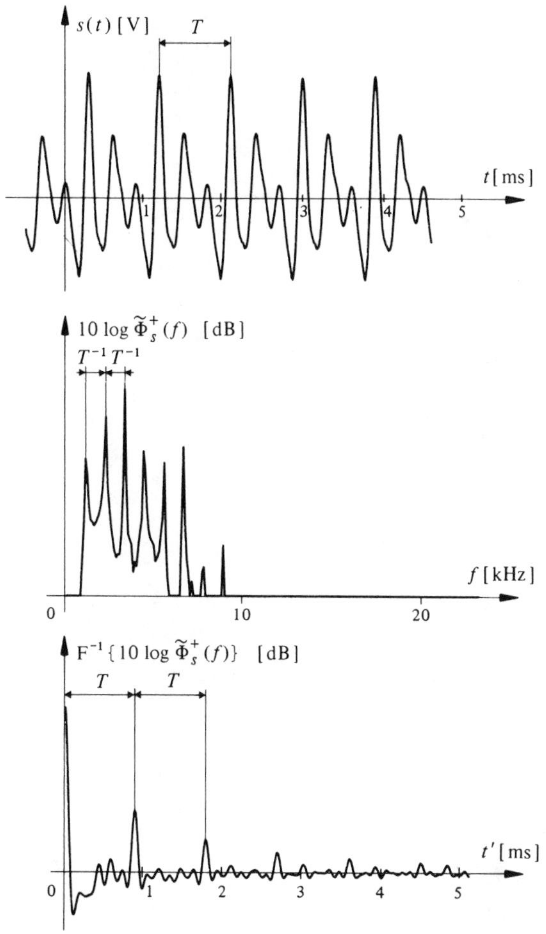

Fig. 12.53 Signal, logarithmic spectral density, and corresponding cepstrum.

12.5.2 Find the main experimental parameters for the direct spectrum analysis of a random signal in the 0–5 MHz frequency range, if a 10 kHz resolution and a 95% confidence interval, corresponding to 0.1 μ_z, are required.

12.5.3 Demonstrate for which reasons a selective filter spectrum analyzer provides, in the case of a discrete spectrum, a measurement of the line amplitude independent of the resolution, and, in the case of a continuous spectrum, a result that is dependent on the resolution. Also demonstrate why this is just the reverse of what happens with a digital analyzer evaluating a periodogram.

[illegible]

[illegible]

[illegible]

Chapter 13

Detection and Estimation

13.1 PARAMETER ESTIMATION

13.1.1 Introduction

The observation and interpretation of signals contaminated by noise are among the major tasks of signal processing.

The interpretation is performed with the objective of making an optimal decision, i.e., to select, according to a given criterion, the best solution among a (continuous or discrete) set of possible solutions. It consists of:

- the measurement of one or more signal parameters, which can be deterministic but unknown, or random, such as mean value, power or energy, amplitude, frequency, phase, position, duration, *et cetera*. This is an estimation problem;
- the recovery of a signal contaminated by noise. This is an optimal filtering problem which is a special case of estimation;
- the prediction of the future behavior of a signal based on the knowledge of its past or the interpolation of an intermediate value between two known values. This is another special case of estimation;
- the detection of the presence or absence of a signal and its classification in a known category. This is a problem of identification or pattern recognition that requires a prior operation of parameter estimation.

The core of the estimation problem is to minimize a function of the generated error. The core of the detection problem is to minimize the probability of wrong decisions.

The statistical theory of estimation and decision [34 and 128–133] offers various methods that allow us to find the optimal solution to these problems in a given context. Such a solution does not always yield an achievable system or a system of an acceptable level of complexity. Nonetheless, knowledge of the optimal solution provides the system designer with a guideline to set a trade-off for a simple but operational suboptimal system.

13.1.2 Application examples

Experimental spectrum analysis, presented in chapter 12, is a typical problem, in which statistical estimation is important.

Radar or sonar signal processing is a classic example, in which the detection and estimation of parameters are implemented. The decision is based on the observation of a possible echo of a pulse previously emitted by the antenna in a given direction. This decision directly controls the presence or absence of a spot on the display screen. The measure of the time delay between pulse emission and echo reception provides the range information. The Doppler frequency shift between the echo and the emitted signal allows us to deduce the target radial velocity.

Long-distance—especially space—radio-wave telecommunications require high performance sophisticated detection systems.

Optimal filtering of the position measurements of a satellite—or any other space object—is used to get an accurate determination of its orbit.

By attempting to predict the short-term evolution of a signal from the observation of its previous states, we can in some instance drastically reduce the amount of information needed in order to store it into memories (e.g., speech synthesis), or to transmit it over long distances (communication systems with limited data rate or limited bandwidth).

The recovery of signals embedded in background noise appears in many research fields of physics, astrophysics, and biomedical engineering.

Oil exploration tries to determine geological structures with methods based on the detection and interpretation of sonic echoes perceived by an array of sensors after the detonation of an explosive charge or some other mechanical stimulation.

The automatic identification of printed characters or simplified graphic symbols (i.e., bar-codes) is used in inventory management of commercial or industrial products, mail sorting, *et cetera*.

Object recognition and the estimation of their relative positions is a central problem of robotics.

Medical diagnostic proposals can be suggested through the use of classification performed on some physiological signal features (electrocardiograms, electroencephalograms, *et cetera*).

The estimation of acoustical voice parameters is one of the key problems of speech recognition.

13.1.3 Deterministic evaluation of parameters.

Let us first consider the determination of nonrandom—but unknown—parameters of a noiseless signal. We know that the signal belongs to a given family of signals described by a finite number N of parameters varying in known ranges.

Such a family generates a signal space (sect. 3.1) of N dimensions, where each vector represents a given signal. The determination of N parameters (i.e., coordinates) requires $M \geqslant N$ measurements.

13.1.4 Examples

The angular velocity and the phase of a cosine signal of known amplitude $x(t) = \cos(\omega_0 t + \alpha)$ with $0 \leqslant \omega_0 \leqslant \omega_{max}$ and $0 \leqslant \alpha \leqslant \alpha_{max}$ is to be determined. If $\alpha_{max} < \pi$ and $\omega_0 t + \alpha \leqslant \pi$, the two measurements $x(0) = \cos\alpha$ and $x(t_1) = \cos(\omega_0 t_1 + \alpha)$ are enough to determine ω_0 and α. When this is not the case, many solutions are possible and a number $M > 2$ of measurements is needed to find these two parameters.

As a general rule, thanks to the sampling theorem, we know that any signal of finite spectrum with bandwidth B and duration T, is theoretically completely defined by $N = 2BT$ samples (sub-sect. 9.3.9). Considering that band-limited signals do not exist (sub-sect. 9.3.2) and taking into account the practical interpolation considerations (sect. 9.4), a satisfactory reconstruction of the signal indeed requires $M > N$ samples.

13.1.5 Statistical estimation of parameters

In signal processing, the purpose of statistical estimation is to evaluate as accurately as possible one or many parameters of a signal $s(t)$ from an observation disturbed by noise, $x(t)$.

A classical approach is to take into account assumed statistical models of the signal, its parameters, and the noise. We then seek a processing strategy that minimizes a given distance $d(a,\tilde{a})$ between the estimate $\tilde{a}$ and the estimated quantity a. A usual distance is the mean square error $E[(\mathbf{a} - \tilde{\mathbf{a}})^2]$ similar to (3.3).

When the available *a priori* knowledge is small, we look for more robust procedures, leading to performances close to the optimum, while tolerating rather wide deviations of the real statistics with respect to the assumed nominal model (fig. 13.1). The theoretical solutions are, in such cases, often very close to those inspired by mere common sense!

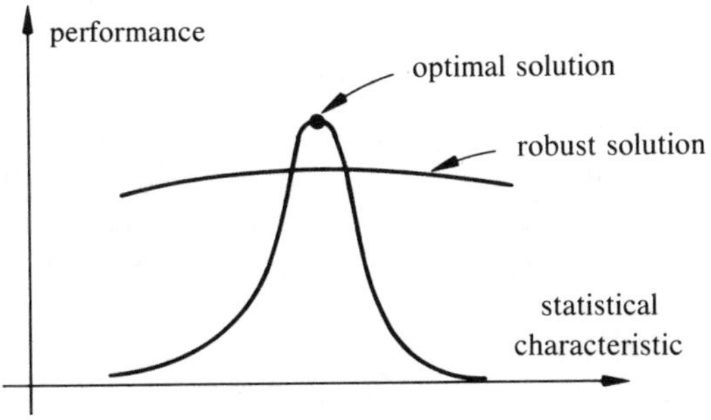

Fig. 13.1

Representation in a signal space, as mentioned in sub-section 13.1.3, is still valid. The set of estimated parameters defines an estimation vector $\tilde{\boldsymbol{a}} = (\tilde{a}_1, \tilde{a}_2, \ldots, \tilde{a}_m)$. However, the set of the possible estimation results, because of noise, generates a ***cluster of solutions***, in which is located the point representing the sought vector $\boldsymbol{a}$ (fig. 13.2).

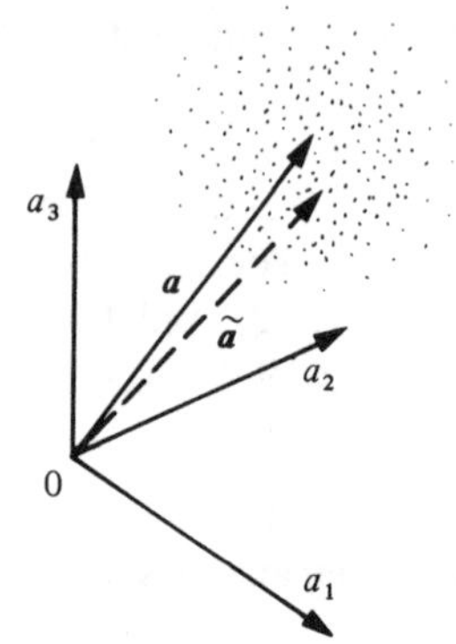

Fig. 13.2

We can deduce from the mathematical estimation formula, the structure of the estimator: analog or digital processor that deduces $\tilde{\boldsymbol{a}}$ from the observation of $x(t)$.

Classical estimation theory assumes knowledge of a probability distribution $p(x;\boldsymbol{a})$ of the observed signal $x(t;\boldsymbol{a})$, depending on M unknown parameters represented by vector $\boldsymbol{a} = (a_1, a_2, \ldots, a_M)$. In presence of an additive noise $n(t)$, the observed signal becomes

$$x(t;\boldsymbol{a}) = s(t,\boldsymbol{a}) + n(t) \tag{13.1}$$

where $s(t;\boldsymbol{a})$ is the useful signal. From a vector of N samples (measurements) $\boldsymbol{x} = (x_1, x_2, \ldots, x_N)$, or from continuous observation of $x(t)$ on an interval T, the aim of the estimation is to assign a value to the M components of vector $\tilde{\boldsymbol{a}}$.

Various strategies can be used for this purpose, depending on the available *a priori* information. The maximum likelihood method (sub-sect. 13.1.6) is applied when the parameters' statistics are unknown or when they are just not random (as is frequently the case). The minimum risk method (sub-sect. 13.1.7) requires knowledge of the probability distributions of the parameters to be estimated, and depends on the choice of an appropriate weighting function for the error. Linear estimation (sub-sect. 13.1.17) considers estimators that can be implemented with linear operators (sect. 8.2).

13.1.6 Maximum likelihood method

If $\boldsymbol{a}$ is a vector of parameters that are either nonrandom but unknown, or random but with unknown statistics, the estimation must be based on the sole *a priori* knowledge of the conditional probability density of the observation vector $\boldsymbol{x}$, depending on the parameter vector $\boldsymbol{a}$ and on the noise statistics: $p(\boldsymbol{x}|\boldsymbol{a})$. This distribution is called the *likelihood function*.

The maximum likelihood estimation consists merely of selecting for $\tilde{\boldsymbol{a}}$ the vector, noted $\tilde{\boldsymbol{a}}_{ml}$, that maximizes $p(\boldsymbol{x}|\boldsymbol{a})$ (fig. 13.3).

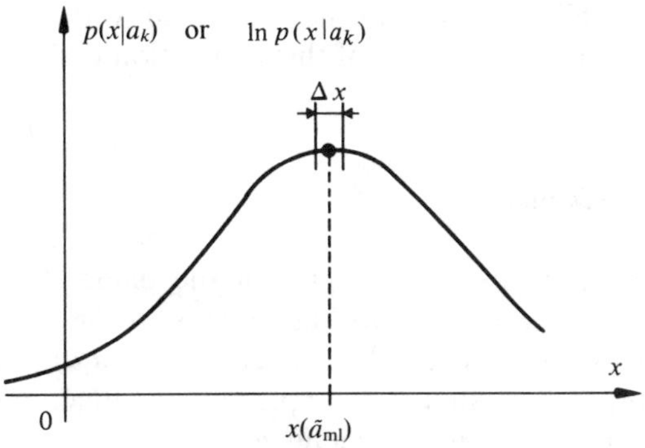

Fig. 13.3

Usually, it is found by solving the set of *likelihood equations*:

$$\left.\frac{\partial p(\boldsymbol{x}|a_k)}{\partial a_k}\right|_{a_k=\tilde{a}_{ml}} = 0 \text{ with } k = 1, 2, \ldots, M \tag{13.2}$$

Since the logarithm is a monotonic function, it is sometimes more convenient (Gaussian noise, *et cetera*) to differentiate the function $\ln p(\boldsymbol{x}|a_k)$.

The maximum likelihood estimation simply takes for the $\boldsymbol{a}_{ml}$ estimation the value that corresponds to the most frequent observation $\boldsymbol{x}$ (measured on a small interval $\Delta\boldsymbol{x}$).

13.1.7 Example

Let $x(t)$ be a Gaussian random signal with unknown mean value $a = \mu_x$. Its conditional probability distribution is simply

$$p(x|a) = (2\pi\sigma^2)^{-1/2} \exp[-\tfrac{1}{2}(x - a)^2/\sigma^2] \tag{13.3}$$

If N samples, statistically independent (i.e., uncorrelated), of this signal are taken, they form, from (14.36), a vector $\boldsymbol{x} = (x_1, x_2, \ldots, x_N)$ with an Nth-order conditional probability distribution:

$$p(\boldsymbol{x}|a) = \prod_{i=1}^{N} p(x_i|a) = (2\pi\sigma^2)^{-N/2} \exp\left[-\frac{1}{2\sigma^2}\sum_{i=1}^{N}(x_i - a)^2\right] \qquad (13.4)$$

Taking the natural logarithm of (13.4) and setting its derivative to zero with respect to the unknown parameter a, we get

$$\tilde{a}_{\text{ml}} = \frac{1}{N}\sum_{i-1}^{N} x_i \qquad (13.5)$$

The maximum likelihood estimate here is equal to the arithmetic mean of the samples. The variance behavior of this estimation is studied in sub-section 13.1.24.

13.1.8 Minimum risk method

When the probability distribution $p(\boldsymbol{a})$ of the parameter vector a is also known, the *Bayes estimation method* associates to the error $(\boldsymbol{a} - \tilde{\boldsymbol{a}})$ a cost $c(\boldsymbol{a},\tilde{\boldsymbol{a}})$ and minimizes the average cost, called *risk*, $d_r(\boldsymbol{a},\tilde{a}) = \mathrm{E}[c(\boldsymbol{a},\tilde{\boldsymbol{a}})]$.

Figure 13.4 shows three familiar examples of cost functions associated with the estimation error of a single parameter a:

- quadratic criterion:

$$c(a,\tilde{a}) = (a - \tilde{a})^2 \qquad (13.6)$$

- absolute value criterion:

$$c(a,\tilde{a}) = |a - \tilde{a}| \qquad (13.7)$$

- uniform cost criterion outside a dead zone Δ:

$$c(a,\tilde{a}) = 1 - rect[(a - \tilde{a})/\Delta] \qquad (13.8)$$

The quadratic criterion is the most frequently used.

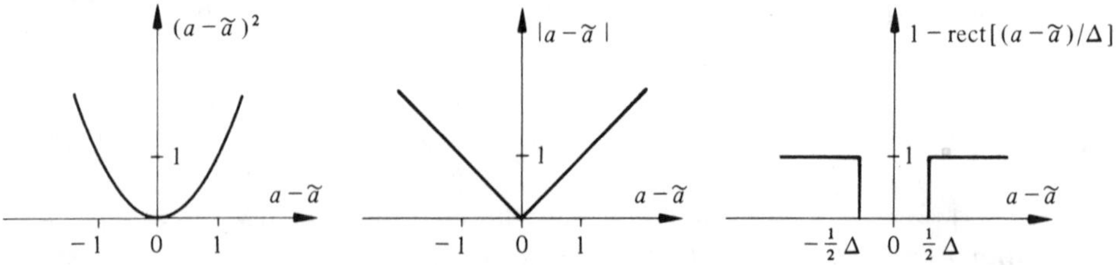

Fig. 13.4

The risk associated with the estimation $\tilde{\boldsymbol{a}}$, based on observation $\boldsymbol{x}$, in fact, expresses a weighted distance $d_r(\boldsymbol{a},\tilde{\boldsymbol{a}})$, similar to those introduced in subsection 3.1.4, and defined by

$$d_r(\boldsymbol{a},\tilde{\boldsymbol{a}}) = \mathrm{E}[c(\boldsymbol{a},\tilde{\boldsymbol{a}})] = \int_{-\infty}^{\infty}\int_{-\infty}^{\infty} c(\boldsymbol{a},\tilde{\boldsymbol{a}})p(\boldsymbol{x},\boldsymbol{a})\mathrm{d}\boldsymbol{x}\mathrm{d}\boldsymbol{a} \tag{13.9}$$

However, according to (14.32), the joint probability distribution $p(\boldsymbol{x},\boldsymbol{a}) = p(\boldsymbol{a}|\boldsymbol{x})p(\boldsymbol{x})$, thus,

$$d_r(\boldsymbol{a},\tilde{\boldsymbol{a}}) = \int_{-\infty}^{\infty} \underbrace{\left[\int_{-\infty}^{\infty} c(\boldsymbol{a},\tilde{\boldsymbol{a}})p(\boldsymbol{a}|\boldsymbol{x})\mathrm{d}\boldsymbol{a}\right]}_{I} p(\boldsymbol{x})\mathrm{d}\boldsymbol{x} \tag{13.10}$$

Minimizing $d_r(\boldsymbol{a},\tilde{\boldsymbol{a}})$ is equivalent to minimizing the conditional risk—or distance—represented by the I integral:

$$I = \int_{-\infty}^{\infty} c(\boldsymbol{a},\tilde{\boldsymbol{a}})p(\boldsymbol{a}|\boldsymbol{x})\, \mathrm{d}\boldsymbol{a} \tag{13.11}$$

since $p(\boldsymbol{x})$ is a non-negative function.

In the case of the quadratic criterion, $c(\boldsymbol{a},\tilde{\boldsymbol{a}}) = \Sigma_i(a_i - \tilde{a}_i)^2$ and the conditional risk express the *mean square error*. The optimal estimate $\tilde{\boldsymbol{a}}_{\mathrm{ms}}$ is obtained in this case by setting the derivative of (13.11) with respect to each parameter a_i equal to zero. Also, taking into account the condition:

$$\int_{-\infty}^{\infty} p(\boldsymbol{a}|\boldsymbol{x})\mathrm{d}\boldsymbol{a} = 1 \tag{13.12}$$

the *minimum mean square error estimate* is

$$\tilde{\boldsymbol{a}}_{\mathrm{ms}} = \int_{-\infty}^{\infty} \boldsymbol{a}p(\boldsymbol{a}|\boldsymbol{x})\, \mathrm{d}\boldsymbol{a} \tag{13.13}$$

This is, by definition, the ***mean value*** of the *a posteriori* conditional probability distribution $p(\boldsymbol{a}|\boldsymbol{x})$, called the *conditional mean*.

Similarly, it can be shown that the absolute value criterion leads to an estimate $\tilde{\boldsymbol{a}}_{\mathrm{abs}}$ corresponding to the ***median*** of $p(\boldsymbol{a}|\boldsymbol{x})$, i.e., to the value $\tilde{\boldsymbol{a}}_{\mathrm{abs}}$ that splits the probability distribution into two areas, the integrals of which are equal to $\frac{1}{2}$.

For the uniform cost, the minimum conditional risk is obtained, when Δ tends toward zero, with the *maximum a posteriori estimate* $\tilde{\boldsymbol{a}}_{\mathrm{map}}$ defined by the abscissa of the ***maximum***—or *mode*— of $p(\boldsymbol{a}|\boldsymbol{x})$.

These three estimates are represented in fig. 13.5 in the case of a single parameter a, with a law conditional to the observation vector: $p(a|\boldsymbol{x})$.

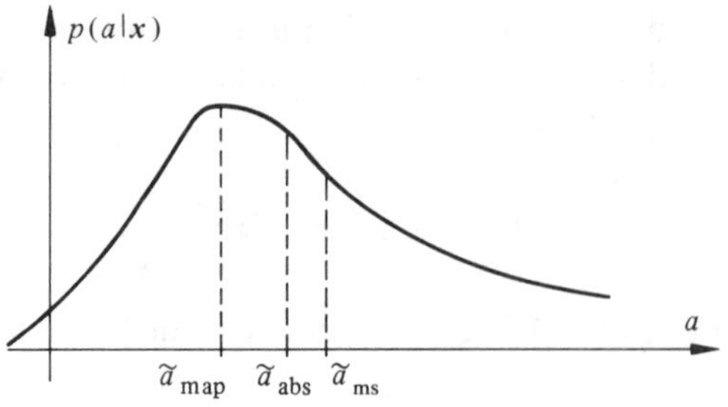

Fig. 13.5

13.1.9 Properties

The maximum *a posteriori* estimate $\tilde{\boldsymbol{a}}_{\text{map}}$ is often easier to determine than $\tilde{\boldsymbol{a}}_{\text{abs}}$ or $\tilde{\boldsymbol{a}}_{\text{ms}}$. ***In the case of symmetrical unimodal distributions, like the Gauss law, these three estimates are identical.***

From (14.32) and (14.33), the *a posteriori* probability distribution $p(\boldsymbol{a}|\boldsymbol{x})$ can be written as

$$p(\boldsymbol{a}|\boldsymbol{x}) = p(\boldsymbol{x}|\boldsymbol{a})p(\boldsymbol{a})/p(\boldsymbol{x}) \tag{13.14}$$

where $p(\boldsymbol{x})$ is, in fact, a known value when the observation $\boldsymbol{x}$ has been performed. The estimations obtained by the minimum risk method depend hence only on the product of probability distributions $p(\boldsymbol{x}|\boldsymbol{a})p(\boldsymbol{a}) = p(\boldsymbol{x},\boldsymbol{a})$. Thus, the maximum likelihood estimate $\tilde{\boldsymbol{a}}_{\text{ml}}$ defined in sub-sect. 13.1.6 coincides with the maximum a posteriori estimate $\tilde{\boldsymbol{a}}_{\text{map}}$ if the probability distribution $p(\boldsymbol{a})$ is uniform. When such is not the case, these two estimates are often similar. This sometimes justifies the use of the simplified estimate $\tilde{\boldsymbol{a}}_{\text{ml}}$, even in situations where the *a posteriori* statistics could be established.

Moreover, the $\tilde{\boldsymbol{a}}_{\text{ml}}$ estimation is rather ***robust,*** since it is not influenced by the variations of the statistics of the parameter to be evaluated.

Another useful property of the maximum likelihood estimate (not necessarily common to other estimates) is that it is invariant to any reversible transformation, i.e., the same result is found by estimating $\boldsymbol{a}$ or a monotonic function $f(\boldsymbol{a})$ with simply $f_{\text{ml}} = \text{f}(\tilde{\boldsymbol{a}}_{\text{ml}})$.

The performances of an estimate are essentially characterized by its bias (systematical error) and its variance (dispersion of results). The *bias of an estimate* is the difference between the estimate mean value and the true mean of the estimated value:

$$\boldsymbol{b} = \text{E}[\tilde{\boldsymbol{a}}] - \text{E}[\boldsymbol{a}] \tag{13.15}$$

with $\text{E}[\boldsymbol{a}] \equiv \boldsymbol{a}$ for nonrandom parameters. It is usually desirable to have a zero bias.

The variance of an unbiased estimate has a lower bound (*Cramer-Rao inequality* [134]). An estimation complying with this bound is said to be *efficient*.

The estimate is *consistent* if it tends toward the true value $\boldsymbol{a}$ when the number of observations N—or the observation duration T in the continuous case—tends toward infinity.

It can be shown that for rather weak restrictive conditions the maximum likelihood estimate $\tilde{\boldsymbol{a}}_{ml}$ of nonrandom parameters is consistent, asymptotically efficient, and, additionally, asymptotically Gaussian [134]. It can be easily demonstrated that, in the case of the estimation of random parameters, if there is an efficient estimate, it is unique and given by $\tilde{\boldsymbol{a}}_{map} \equiv \tilde{\boldsymbol{a}}_{ms}$.

13.1.10 Example

Consider the following situation: a random signal $a(t)$ with a Rayleigh distribution (14.110) and variance σ_a^2 is observed in the presence of an additive noise $n(t)$ with zero mean value and variance σ_n^2. We try to estimate a from a single sample of $x(t) = a(t) + n(t)$.

The *a priori* conditional probability distribution $p(x|a)$ is simply the noise probability distribution shifted by a:

$$p(x|a) = (2\pi\sigma^2)^{-1/2} \exp[-\tfrac{1}{2}(x - a)^2/\sigma_n^2] \tag{13.16}$$

The maximum likelihood estimate here is simply

$$\tilde{a}_{ml} = x \tag{13.17}$$

By (13.14), the *a posteriori* conditional probability distribution becomes

$$p(a|x) = Ka \exp[-\tfrac{1}{2}(\alpha/\sigma_n)^2(a - x/\alpha)^2] \tag{13.18}$$

where $K = 2\pi\sigma_n^2\sigma_a^2/p(x)$ is a constant, when x is known, which can be evaluated using condition (13.12) and $\alpha = 1 + \sigma_n^2/\sigma_a^2$.

The maximum *a posteriori* estimate is obtained by setting the derivative of (13.18) with respect to a equal to zero:

$$\tilde{a}_{map} = x(1 + \sqrt{1 + 4\alpha\sigma_n^2/x^2})/(2\alpha) \tag{13.19}$$

This is a nonlinear function of the observation converging towards (13.17) when $\sigma_n^2 \to 0$.

13.1.11 Example

Let us estimate the random parameter $\mathbf{a} \geqslant 0$ of an observation x with an exponential distribution (14.90):

$$p(x|a) = a \exp(-ax)\epsilon(x) \tag{13.20}$$

knowing that **a** has a similar distribution:

$$p(a) = \lambda \exp(-\lambda a)\epsilon(a) \tag{13.21}$$

with $\lambda \geqslant 0$.

The above example [135] can describe a traffic situation on a road or telecommunication network: **a** represents the average rate of vehicles or messages crossing a network node, and this average rate also fluctuates from node to node, according to the law (13.21) with a mean value $E[\mathbf{a}] = \lambda^{-1}$. The observation x here is the interval between the crossing instants of two successive vehicles or messages (Poisson point process, see sect. 5.8).

The *a posteriori* distribution takes the form (problem 13.5.3):

$$p(a|x) = (x + \lambda)^2\, a \exp[-a(x + \lambda)]\epsilon(a) \tag{13.22}$$

The mean value of this distribution is the minimum mean square error estimate:

$$\tilde{a}_{\text{ms}} = \int_0^\infty a p(a|x) \mathrm{d}a = 2(x + \lambda)^{-1} \tag{13.23}$$

The abscissa of the $p(a|x)$ maximum is the maximum *a posteriori* estimate $\tilde{a}_{\text{map}}$ obtained by solving equation $\partial p(a|x)/\partial a = 0$ for $a = \tilde{a}_{\text{map}}$:

$$\tilde{a}_{\text{map}} = (x + \lambda)^{-1} \tag{13.24}$$

The median, representing estimate $\tilde{a}_{\text{abs}}$, is the solution of the transcendental equation:

$$[1 - (x + \lambda)\, \tilde{a}_{\text{abs}}] \exp[-\tilde{a}_{\text{abs}}(x + \lambda)] + \tfrac{1}{2} = 0 \tag{13.25}$$

which is found by solving the integral:

$$\int_0^{\tilde{a}_{\text{abs}}} p(a|x) \mathrm{d}a = \tfrac{1}{2} \tag{13.26}$$

13.1.12 Example

Consider the estimation of an unknown mean value a in presence of a random Gaussian fluctuation $n(t)$, with zero mean value and variance σ_n^2. The estimation is performed with a set $\boldsymbol{x} = (x_1, x_2, \ldots, x_N)$ of N independent samples:

$$x_i = a + n_i \tag{13.27}$$

The conditional probability distribution $p(\boldsymbol{x}|a)$ is the same as (13.4), with σ_n^2 replacing σ^2, and the maximum likelihood estimate $\tilde{a}_{\text{ml}}$ is identical to result (13.5), since this estimation does not take into account $p(a)$. This is an unbiased estimate with variance

$$\sigma_{\text{ml}}^2 = \sigma_n^2/N \tag{13.28}$$

tending towards zero when N approaches infinity (sub-sect. 13.1.24).

Now, assume that a is a Gaussian random variable with mean value μ_a and variance σ_a^2. Applying (13.14), we get, after some algebraic calculus [34], the distribution:

$$p(a|\boldsymbol{x}) = (2\pi\sigma^2)^{-1/2} \exp[-\tfrac{1}{2}(a - \mu)^2/\sigma^2] \tag{13.29}$$

with

$$\sigma^2 = \frac{\sigma_a^2\sigma_n^2/N}{\sigma_a^2 + \sigma_n^2/N}\ \alpha\sigma_n^2/N \tag{13.30}$$

and

$$\mu = \frac{\sigma_a^2\bar{x} + \mu_a\sigma_n^2/N}{\sigma_a^2 + \sigma_n^2/N} = \alpha\bar{x} + \beta \tag{13.31}$$

where $\bar{x}$ is the arithmetic mean corresponding to the maximum likelihood estimate

$$\tilde{a}_{\text{ml}} = \bar{x} = \frac{1}{N}\sum_{i=1}^{N} x_i \tag{13.32}$$

and where $\alpha = \sigma_a^2/(\sigma_a^2 + \sigma_n^2/N) < 1$ and $\beta = (1 + N\sigma_a^2/\sigma_n^2)^{-1}\mu_a$. Since the abscissa of the maximum, the median, and the mean value are identical in a Gaussian law (sub-sect. 13.1.9), we have

$$\tilde{a}_{\text{ms}} = \tilde{a}_{\text{map}} = \tilde{a}_{\text{abs}} = \mu \neq \bar{x} = \tilde{a}_{\text{ml}} \tag{13.33}$$

The optimal Bayesian estimate is also linear, but it is biased. It tends toward the unbiased arithmetic mean $\bar{x} = \tilde{a}_{\text{ml}}$ when σ_n^2/N tends toward zero, i.e., especially when N tends toward infinity. When σ_a^2 tends toward zero, the optimal estimate tends toward μ_a, the *a priori* mean value, and is not very dependent on the observation.

We check that the *a posteriori* estimate is optimal, since its variance, taking (13.28) and (13.31) into account, is

$$\sigma_\mu^2 = \text{Var}[\alpha\bar{\mathbf{x}}] = \alpha^2\text{Var}[\bar{\mathbf{x}}] = \alpha^2\sigma_{\text{ml}}^2 = \alpha^2\sigma_n^2/N \tag{13.34}$$

and since $\alpha < 1$.

13.1.13 Example: maximum likelihood estimation of the amplitude of a known signal. Discrete case

Let $x(t) = a \cdot s(t) + n\,(t)$, where $s(t)$ is a known signal (sine wave, pulse, *et cetera*). The parameter a is random and expresses an amplitude modulation, a variable attenuation due to propagation (e.g., fading phenomenon; sub-sect. 7.3.11), or reflection (radar, sonar) conditions, *et cetera*. As previously, $n(t)$ is a Gaussian noise with zero mean value and variance σ_n^2.

The estimation is again based on a set of N independent samples $\boldsymbol{x} = (x_1, x_2, \ldots, x_N)$. Each sample $x_i = as_i + n_i$ has a conditional distribution:

$$p(x_i|a) = (2\pi\sigma_n^2)^{-1/2} \exp[-\tfrac{1}{2}(x_i - as_i)^2/\sigma_n^2] \tag{13.35}$$

The vector $\boldsymbol{x}$ of the N independent observations has the conditional distribution:

$$p(\boldsymbol{x}|a) = \prod_{i=1}^{N} p(x_i|a) = (2\pi\sigma_n^2)^{-N/2} \exp\left[-\frac{1}{2\sigma_n^2}\sum_{i=1}^{N}(x_i - as_i)^2\right] \tag{13.36}$$

The maximum likelihood estimate is given by the value $a = \tilde{a}_{\text{ml}}$ satisfying the equation $\partial p(\boldsymbol{x}|a)/\partial a = 0$:

$$\tilde{a}_{\text{ml}} = \sum_{i=1}^{N} s_i x_i \Big/ \sum_{i=1}^{N} s_i^2 \tag{13.37}$$

The numerator is merely the inner product (3.10) of the vectors $\boldsymbol{x}$ and $\boldsymbol{s}$ (alias correlation), and the denominator is a scale factor, according to (3.11), equal to the square of the norm (alias energy) of $\boldsymbol{s}$. The corresponding estimator is of the matched filter kind (sect. 13.4). In the particular case where $s_i = 1\ \forall i$, we can identify (13.37) with (13.32).

13.1.14 Maximum likelihood estimation of parameters of a known signal. Continuous case

Let the signal observed during interval T be

$$x(t) = s(t,a) + n(t); \qquad 0 \leqslant t \leqslant T \tag{13.38}$$

where $s(t,a)$ is a known signal depending on a parameter a (amplitude, frequency, phase, position, *et cetera*) and $n(t)$ is a Gaussian noise with autocorrelation function $R_n(\tau)$.

The *a priori* distribution of parameters a seldom being known, we will only consider its maximum likelihood estimation.

The logarithm of the $p(x|a)/p_0(x)$ ratio is another form of likelihood function, where $p_0(x)$ denotes the distribution that the observation $x(t)$ would have in the absence of a signal $s(t,a)$. It can be written [34] as

$$l(a) = \ln[p(x|a)/p_0(x)] = \int_0^T x(t)g(t,a)\mathrm{d}t - \tfrac{1}{2}\int_0^T s(t,a)g(t,a)\mathrm{d}t \tag{13.39}$$

where $g(t,a)$ is the solution of the integral equation

$$s(t,a) = \int_0^T R_n(t-\tau)g(\tau,a)\mathrm{d}\tau; \quad 0 \leq t \leq T \tag{13.40}$$

Comparing (13.40) and (8.23), we see that, if $T \to \infty$, the function $g(t,a)$ tends toward the impulse response of a linear filter such that $s(t,a)$ represents the crosscorrelation between input and output signals when the input is $n(t)$. This result is similar to (13.64).

In the case of a Gaussian white noise with a bilateral power spectral density $\Phi_n(f) = \frac{1}{2}\eta$ and with autocorrelation function $R_n(\tau) = \frac{1}{2}\eta\delta(\tau)$, the solution of (13.40) is $g(t,a) = (2/\eta)s(t,a)$, hence,

$$l(a) = \frac{2}{\eta}\int_0^T x(t)s(t,a)\mathrm{d}t - \frac{1}{\eta}\int_0^T s^2(t,a)\mathrm{d}t \tag{13.41}$$

The maximum likelihood estimate $\tilde{a}_{\mathrm{ml}}$ is the value of a maximizing (13.41) and satisfying the equation:

$$\int_0^T [x(t) - s(t,a)]\frac{\partial s(t,a)}{\partial a}\mathrm{d}t\bigg|_{a=\tilde{a}_{\mathrm{ml}}} = 0 \tag{13.42}$$

In other cases, the solution of the integral equation (13.40) is tricky, or even impossible [34, 131].

13.1.15 Example: correlation receiver

Let $s(t,a) = as(t)$ as in sub-sect. 13.1.13. Then $g(t,a) = ag(t)$, where $g(t)$ is the solution of the equation:

$$s(t) = \int_0^T R_n(t-\tau)g(\tau)\mathrm{d}\tau; \quad 0 \leq t \leq T \tag{13.43}$$

The likelihood function (13.39) becomes

$$l(a) = a\int_0^T g(t)x(t)\mathrm{d}t - \tfrac{1}{2}a^2\int_0^T g(t)s(t)\mathrm{d}t \tag{13.44}$$

and is maximum for

$$\tilde{a}_{\mathrm{ml}} = \int_0^T g(t)x(t)\mathrm{d}t \bigg/ \int_0^T g(t)s(t)\mathrm{d}t \tag{13.45}$$

In case of ***Gaussian white noise*** with spectral density $\Phi_n(f) = \frac{1}{2}\eta$, $g(t) = (2/\eta)s(t)$ and the estimation becomes

$$\tilde{a}_{\mathrm{ml}} = \int_0^T s(t)x(t)\mathrm{d}t \bigg/ \int_0^T s^2(t)\mathrm{d}t \tag{13.46}$$

This relation can be compared with the discrete case (13.37).

For the sake of simplicity, normalizing the energy of signal $s(t)$ in interval T, it becomes

$$\tilde{a}_{\text{ml}} = \int_0^T s(t)x(t)\text{d}t = <s,x> \tag{13.47}$$

The block diagram of such an estimator (fig. 13.6) is directly deduced from (13.47). It performs a ***correlation*** of the observed signal with a stored copy of the known signal $s(t)$. For this reason it is called a *correlation receiver.* An equivalent implementation of the same estimator is the matched filter (sect. 13.4).

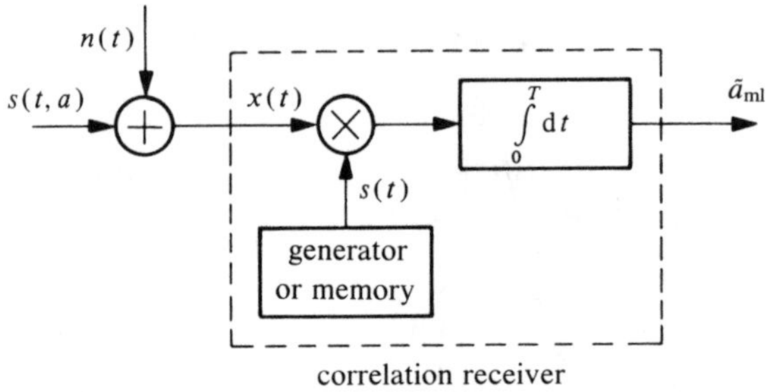

Fig. 13.6

13.1.16 Example: estimation of the phase of a sine wave

If the unknown parameter is the phase, the useful signal is $s(t,a) = A \sin(\omega_0 t + a)$, where A and ω_0 are known. The estimate $\tilde{a}_{\text{ml}}$ of the phase in presence of a Gaussian white noise is obtained by solving (13.42):

$$\int_0^T [x(t) - A \sin(\omega_0 t + \tilde{a}_{\text{ml}})] \cos(\omega_0 t + \tilde{a}_{\text{ml}})\text{d}t = 0 \tag{13.48}$$

For $\omega_0 T = k\pi$ or $\omega_0 T \gg 1$, the second integral of the (13.48) expansion vanishes, and the estimation is given by

$$\int_0^T x(t) \cos(\omega_0 t + \tilde{a}_{\text{ml}})\text{d}t = 0 \tag{13.49}$$

Taking advantage of the trigonometric relations of section 15.2, the estimate becomes

$$\tilde{a}_{\text{ml}} = \arctan \left\{ \frac{\int_0^T x(t) \cos(\omega_0 t)\text{d}t}{\int_0^T x(t) \sin(\omega_0 t)\text{d}t} \right\} \tag{13.50}$$

The evaluation of these integrals again corresponds to a correlation (fig. 13.7).

The integral (13.49) suggests, however, a much more interesting solution: the phase locked loop (fig. 13.8), which has already been mentioned in the discussion of phase and frequency demodulation (sub-sect. 11.3.9). The average voltage of the error $\bar{\epsilon}$ continuously drives a voltage-controlled oscillator (VCO), so that it introduces an automatic phase correction $\tilde{a}_{ml}$, maintaining $\bar{\epsilon}$ close to zero. Other examples, by far more complicated, are studied in the specialized literature mentioned.

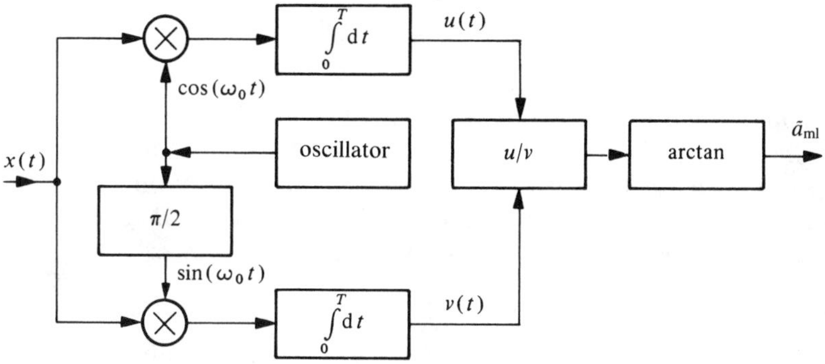

Fig. 13.7

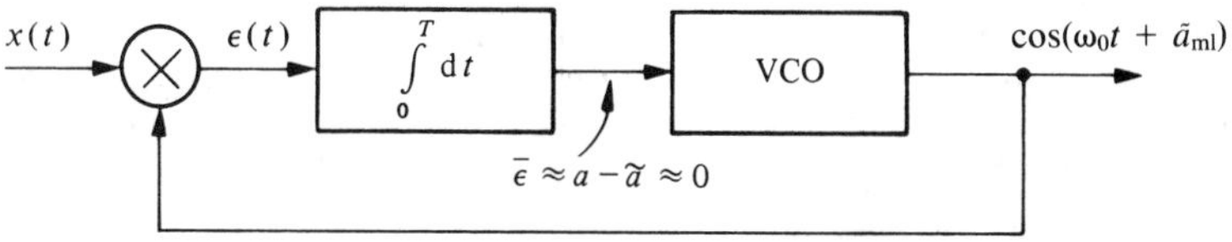

Fig. 13.8

13.1.17 Linear estimation

The estimation criteria previously mentioned require, at least, knowledge of a statistical model of the observed signal; hence, often leading to nonlinear estimators.

With a *linear estimation*:

$$\tilde{a}_\ell = S_\ell\{x\} \tag{13.51}$$

where S_ℓ represents a discrete or continous linear operator (sect. 8.2), the necessary *a priori* information is reduced to the sole autocorrelation function of observation x and crosscorrelation function of x with the estimated quantity a.

The linear estimator (fig. 13.9) can often be considered as a filter that is digital in the discrete case and analog in the continuous case, whose impulse response g must satisfy (13.51).

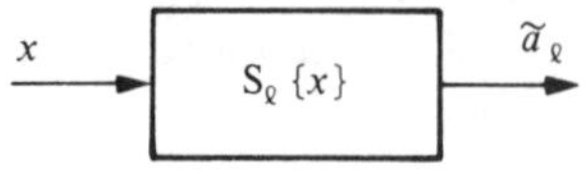

Fig. 13.9

By analogy with the projection theorem (3.26), which defines the optimal linear approximation conditions of a deterministic signal in the least square sense, the optimal linear estimate $\tilde{a}_{\ell o}$ that minimizes the mean square error must satisfy the orthogonality principle [24]. The latter indicates that ***the optimal estimation error*** $a - \tilde{a}_{\ell o}$ ***and the observation*** x ***are orthogonal,*** i.e., their statistical crosscorrelation (sect. 5.4) must be zero:

$$\mathrm{E}[\mathbf{a} - \tilde{\mathbf{a}}_{\ell o})^*\mathbf{x}] = \mathrm{E}[\mathbf{x}^*(\mathbf{a} - \tilde{\mathbf{a}}_{\ell o})] = 0 \tag{13.52}$$

The minimal mean square error achieved thusly, by analogy with (3.29), is then given by

$$\mathrm{E}[(\mathbf{a} - \tilde{\mathbf{a}}_{\ell o})^2] = \mathrm{E}[\mathbf{a}^*(\mathbf{a} - \tilde{\mathbf{a}}_{\ell o})] = \mathrm{E}[|\mathbf{a}|^2] - \mathrm{E}[|\tilde{\mathbf{a}}_{\ell o}|^2]$$

Expanding (13.52), the optimal linear estimate is finally defined by the solution of the equation:

$$\mathrm{E}[\mathbf{x}^*\tilde{\mathbf{a}}_{\ell o}] = \mathrm{E}[\mathbf{x}^*\mathbf{a}] \tag{13.53}$$

It can be shown [24] that, ***in the case where the observation has Gaussian statistics, there is no nonlinear estimator that can yield a smaller mean square error than the optimal linear estimator.***

13.1.18 Application to digital estimation

Consider a discrete linear estimator that has to interpret N samples of a stationary random signal $x(t)$:

$$\tilde{a}_{\ell o} = \mathrm{S}_\ell\{x\} = \sum_{j=1}^{N} g_j x_j \tag{13.54}$$

The optimal coefficients g_j are given by the solution of (13.53), which reduces to the solution of N equations with N unknowns:

$$\sum_{j=1}^{N} g_j R_x(i - j) = R_{xa}(i); \qquad i = 1, 2 \ldots, N \tag{13.55}$$

where $R_x(i - j) = \mathrm{E}[\mathbf{x}_i^* \mathbf{x}_j]$ are the elements (sub-sect. 5.2.8) of the autocorrelation matrix R_x of the observation vector $\boldsymbol{x} = (x_1, x_2, \ldots, x_N)$ and $R_{xa}(i) = \mathrm{E}[\mathbf{x}_i^* \mathbf{a}]$ are the crosscorrelation of the x_i with the estimated quantity a.

Rearranging these crosscorrelation values in a column vector $\boldsymbol{R}_{xa}$ and the g_j coefficients in a column vector $\boldsymbol{g} = (g_1, g_2, \ldots, g_N)$, the equation set (13.55) becomes simply

$$R_x \boldsymbol{g} = \boldsymbol{R}_{xa} \tag{13.56}$$

the solution of which is, with R_x^{-1}, the inverse matrix of R_x:

$$\boldsymbol{g} = R_x^{-1} \boldsymbol{R}_{xa} \tag{13.57}$$

13.1.19 Example: simple observation of a signal in presence of noise

Let $x(t) = s(t) + n(t)$, where $s(t) = a(t)$ is the signal to be estimated and $n(t)$ is a noise with zero mean value uncorrelated with $s(t)$. We take the simple estimation $\tilde{a}_\ell = \tilde{s}(t) = gx(t)$. From (13.57), the optimal coefficient g corresponds to

$$g_o = R_{xa}(0)/R_x(0) = R_s(0)/[R_s(0) + R_n(0)] \tag{13.58}$$

13.1.20 Example: linear prediction

Let $x(t)$ be a random signal known at instant t. We want to predict its value $a(t) = x(t + T)$ at instant $t + T$ by a simple linear prediction $\tilde{a}_\ell = \tilde{x}(t + T) = gx(t)$. By (13.57), the optimal coefficient g of the prediction system (fig. 13.10) is

$$g_o = R_x(T)/R_x(0) \tag{13.59}$$

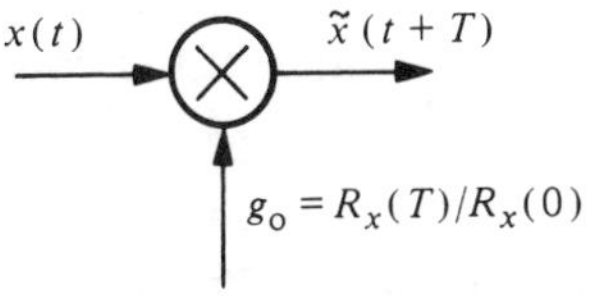

Fig. 13.10

In a more general way, the estimated value of the $\tilde{x}(t + T)$ signal is predicted by a linear combination of the N previous values $x(t - iT)$ with $i = 0, 1, \ldots,$ N $- 1$:

$$\tilde{a}_\ell = \tilde{x}(t + T) = \sum_{i=0}^{N-1} g_i x(t - iT) \tag{13.60}$$

The structure of the estimator (fig. 13.11) is that of a transversal digital filter, whose g_i coefficients are given by the solution of (13.57), which becomes

$$\boldsymbol{g} = R_x^{-1}\boldsymbol{R}_x \tag{13.61}$$

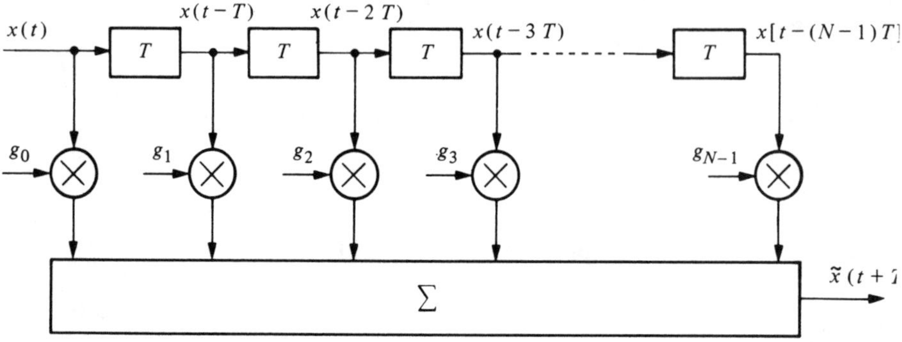

Fig. 13.11

where $\boldsymbol{R}_x = \{R_x(0), R_x(T), R_x(2T), \ldots, R_x[(N-1)T]\}$ and R_x^{-1} is the inverse of the correlation matrix:

$$R_x = \begin{bmatrix} R_x(0) & R_x(T) & \cdots & R_x[(N-1)T] \\ R_x(T) & \ddots & & \vdots \\ \vdots & & \ddots & R_x(T) \\ R_x[(N-1)T] & \cdots & R_x(T) & R_x(0) \end{bmatrix} \tag{13.62}$$

Linear prediction is used, for instance, in speech processing (vocoder encoding, synthesis, automatic recognition), in spectrum analysis (parametric methods), in dynamic linear system identification (automatic control), and to correct the imperfect characteristics of the communication channels in digital data transmission (equalizer).

13.1.21 Application: optimal linear filtering

If estimation $\tilde{a}_\ell$ is represented by the output signal of a linear filter with impulse response $g(t)$, excited by the observation $x(t)$ (here assumed to be stationary), we have, from (8.12):

$$\tilde{a}_\ell(t) = \mathrm{S}_\ell\{x(t)\} = x(t) * g(t) \tag{13.63}$$

Inserting (13.63) into (13.53) leads to the convolution product:

$$R_x(\tau) * g_o(\tau) = R_{xa}(\tau) \quad \mathbf{(13.64)}$$

By Fourier transform, the transfer function of the optimal (usually noncausal) filter is

$$G_o(f) = \Phi_{xa}(f)/\Phi_x(f) \quad (13.65)$$

If $x(t) = s(t) + n(t)$, with $s(t)$ being the signal to be extracted from the independent noise $n(t)$ with zero mean value, the solution (13.65) becomes

$$G_o(f) = \Phi_s(f)/[\Phi_s(f) + \Phi_n(f)] \quad (13.66)$$

In the very specific case where the observation $x(t)$ is a white noise with autocorrelation function $\frac{1}{2}\eta\delta(\tau)$, the optimal impulse response, derived from (13.64), according to (1.47), becomes

$$g_o(t) = 2\eta^{-1}R_{xa}(t) \quad (13.67)$$

Non-real-time processing does not necessarily require a causal impulse response. On the contrary, real-time processing and, especially, analog processing require it. In this case, a good causal approximation of the optimal filter can often be obtained by introducing an appropriate delay t_0 and by considering the function:

$$\tilde{g}_o(t) = g_o(t - t_0)\epsilon(t) \quad (13.68)$$

where $\epsilon(t)$ is, as usual, the unit-step function.

13.1.22 Wiener filter

The *Wiener filter* is associated with the solution of equation (13.64) under the additional causality condition: $g_o(t) = 0$ for $t < 0$.

For instance, if $x(t)$ is a white noise with spectral density $\frac{1}{2}\eta$, the optimal causal impulse response becomes simply

$$g_o(t) = 2\eta^{-1}R_{xa}(t)\epsilon(t) \quad (13.69)$$

A more general solution is obtained by expanding the Wiener filter in a cascade (fig. 13.12) of two filters with impulse responses and transfer functions $g_1(t) \leftrightarrow G_1(f)$ and$g_2(t) \leftrightarrow G_2(f)$. The first filter transforms the $x(t)$ observation, with spectral density $\Phi_x(f)$, into a white noise $z(t)$ with spectral density $\Phi_z(f) = 1$. From (8.24), we immediately get

$$|G_1(f)|^2 = G_1^*(f)G_1(f) = \Phi_x^{-1}(f) \quad (13.70)$$

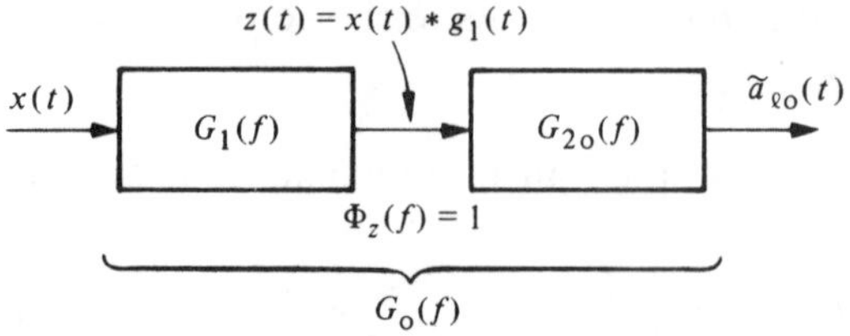

Fig. 13.12

The second filter is optimized for the transformed observation $z(t)$ by writing, according to (13.69):

$$g_{2o}(t) = R_{za}(t)\epsilon(t) \tag{13.71}$$

Denoting by $[\Phi_{za}(f)]_+$ the Fourier transform of (13.71), we get, with $R_{za}(\tau) = g_1(-\tau) * x(-\tau) * a(\tau) = g_1(-\tau) * R_{xa}(\tau)$:

$$G_{2o}(f) = [\Phi_{za}(f)]_+ = [G_1^*(f)\Phi_{xa}(f)]_+ \tag{13.72}$$

Finally, by writing $\Phi_x(f)$ as a product of two complex conjugate functions (always possible if this spectral density is represented by a rational function: its non-negative, real, and even character implying that its roots are complex conjugates or real of even order):

$$\Phi_x(f) = \Psi_x(f)\Psi_x^*(f) \tag{13.73}$$

so that, by identification with (13.70), the transfer function:

$$G_1(f) = \Psi_x^{-1}(f) \tag{13.74}$$

is also that of a causal filter (roots of $f > 0$), the Wiener filter is defined by

$$G_o(f) = G_1(f)G_{2o}(f) = \frac{1}{\Psi_x(f)}\left[\frac{\Phi_{xa}(f)}{\Psi_x^*(f)}\right]_+ \tag{13.75}$$

13.1.23 Kalman filter

If we discard the stationarity restriction, relation (13.64) is replaced by the more general integral equation (*Wiener-Hopf equation*):

$$\int R_x(t_1,u)g_o(t_2,u)\mathrm{d}u = R_{xa}(t_1,t_2) \tag{13.76}$$

The name *Kalman filter* [136] is given to a kind of solution of the Wiener-Hopf equation in which the discrete or continuous estimation is recurrent.

In a simple case, we have, for instance, a digital estimate:

$$\tilde{a}(k) = \tilde{a}(k-1) + \beta(k)[x(k) - \tilde{a}(k-1)] \tag{13.77}$$

The estimation of the kth sample is equal to the previous estimation completed with a corrective term dependent on the difference between the current estimation and the previous estimation. The coefficient $\beta(k)$ is, if necessary, adapted after each step.

The Kalman filter is a generalization of the Wiener filter and is mainly applied in nonstationary signal processing. Its study is beyond the scope of this book. It is principally used to identify evolving systems permanently, to predict the effective values of imposed parameters in automatic control equipment, and to estimate the orbits of space vehicles.

13.1.24 Estimation of mean values

The estimation of a mean value $a = \mu_x$ plays a key role in signal processing. It is encountered in many applications: spectrum analysis, correlation, power measurement, *et cetera*.

The usual linear estimator is the perfect time-averager described in sub-section 8.2.19. It evaluates the running average:

$$\tilde{a}_\ell = \bar{x}(t,T) = \frac{1}{T}\int_{t-T}^{t} x(t)\mathrm{d}t \tag{13.78}$$

in its continuous form. In its digital form, it computes the arithmetic mean:

$$\tilde{a}_\ell = \bar{x}_k = \frac{1}{N}\sum_{i=k-N+1}^{k} x_i = \bar{x}_{k-1} + (x_k - x_{k-N})/N \tag{13.79}$$

This is an unbiased, consistent, and asymptotically efficient estimator. On the other hand, it is not optimal in the sense of sub-section 13.1.17, as shown by example 13.1.12. In this case (Gaussian statistics), it performs a maximum likelihood estimation: $\tilde{a}_\ell = \tilde{a}_{\mathrm{ml}}$. The variance of the estimation (13.78) depends on the integration duration T and is given by (8.44):

$$\sigma_{\bar{x}}^2 = \frac{1}{T}\int_{-\infty}^{\infty} \mathrm{tri}(\tau/T)C_x(\tau)\mathrm{d}\tau \tag{13.80}$$

where $C_x(\tau) = R_x(\tau) - \mu_x^2$ is the autocovariance function of the observation (assumed here to be stationary and ergodic).

The signal-to-noise ratio before estimation is $\xi_x = a^2/\sigma_x^2$. After estimation, we have $\xi_{\bar{x}} = a^2/\sigma_{\bar{x}}^2$ and the signal-to-noise ratio improvement achieved simply corresponds to the variance ratio.

By analogy, for the discrete estimation based on N samples $x_i = x(iT_e)$ with $i = 0, \ldots, N-1$, $T = NT_e$, $\tau = iT_e$, and taking (9.14) into account, we get

$$\sigma_{\bar{x}}^2 = \frac{1}{N}\sum_{i=-(N-1)}^{N-1} \mathrm{tri}(i/N)C_x(iT_e) = \frac{\sigma_x^2}{N} + \frac{2}{N}\sum_{i=1}^{N-1}(N-i)C_x(iT_e) \tag{13.81}$$

If the samples are ***uncorrelated***: $C_x(iT_e) = 0$ for $i \neq 0$ and

$$\sigma_{\bar{x}}^2 = \sigma_x^2/N \tag{13.82}$$

In this case, the signal-to-noise ratio improvement is proportional to the number N of available samples. This is a classical statistical result.

In a general way, we see from (8.37) that any causal linear system, whose unit-step response $\gamma(t)$ complies with condition:

$$\gamma(\infty) = \int_0^\infty g(t)\mathrm{d}t = 1 \tag{13.83}$$

can be used as (imperfect) mean value estimator. In practice, since the measurement is performed after a non-infinite settling time T_0, this value has to be chosen such that $\gamma(T_0) \approx 1$. Otherwise, it introduces a significant bias.

The most familiar example is the exponential averager defined by

$$g(t) = \alpha^{-1} \exp(-t/\alpha)\epsilon(t) \tag{13.84}$$

corresponding to a first-order lowpass filter with time constant α. The variance of the estimation is given, according to (8.38), by

$$\sigma_{\bar{x}}^2 = \frac{1}{\alpha} \int_0^\infty C_x(\tau) \exp(-\tau/\alpha)\mathrm{d}\alpha \tag{13.85}$$

Such an estimator is usually implemented in practice with the RC network of sub-section 8.2.24 in analog equipment. In order to display easily the progressively computated results, digital equipment often simulates the RC network by means of an approximative recurrent equation, such as

$$\tilde{a}_\ell = \bar{x}_k = \bar{x}_{k-1} + (x_k - \bar{x}_{k-1})/K \approx \frac{1}{K} \sum_{i=k-N+1}^{k} x_i \exp[-(k-i)/K] \tag{13.86}$$

where K is a fixed integer > 1 (usually a power of 2 to simplify the division). The equivalent time constant is proportional to K.

13.1.25 Compared performances

In the case of white noise, we have the autocovariance function $C_x(\tau) = \frac{1}{2}$ $\eta\delta(\tau)$ and $\sigma_{\bar{x}}^2 = \frac{1}{2}\eta/T$ for the perfect averager and $\sigma_{\bar{x}}^2 = \frac{1}{4}\eta/\alpha$ for the first-order lowpass filter with time constant α. The same signal-to-noise ratio performances are achieved if $T = 2\alpha$.

Because of its unit-step response $\gamma(t) = 1 - \exp(-t/\alpha)$, the first-order filter needs a settling time in the range of 4α or 5α to ensure a reasonably unbiased estimation. With the perfect averager, this settling time is limited to T.

13.1.26 Application: signal averaging

The following principle for recovering a highly disturbed signal, which recurs with a known chronology, is used in various fields, especially physics and biomedical engineering. It is known as *signal averaging*.

Let $\{s(t - t_i)\}$, $i = 1, 2, \ldots, N$ be a set of signals with identical shapes and finite duration T, occuring as responses to stimuli applied at known instants t_i (fig. 13.13). If the available observation is

$$x(t) = \sum_{i=1}^{N} s(t - t_i) + n(t) \tag{13.87}$$

where $n(t)$ is a noise with zero mean value and a standard deviation (rms value) much greater than the amplitudes of signals $s(t - t_i)$, then the latter are totally masked by the noise and no direct analysis can be used.

An estimation of the shape of the $s(t)$ response to the stimulus is always possible by taking advantage of the recurring (whether periodic or not) character of the phenomenon, as long as $t_{i+1} - t_i \geqslant T$. A sample is taken every $t_i + kT_e$ with $T_e \ll T$. For a given k, we have

$$x_i \triangleq s(kT_e) + n_i \tag{13.88}$$

The instantaneous value $s(kT_e)$, thus, represents the mean value of the x_i samples. The estimation (13.79) here gives

$$\bar{x}(kT_e) = \frac{1}{N} \sum_{i=1}^{N} x_i = s(kT_e) + \bar{n} \tag{13.89}$$

where $\bar{n}$ is the arithmetic mean of the N noise samples alone. If these samples are independent (i.e., uncorrelated), $\bar{n}$ becomes Gaussian thanks to the central limit theorem (sub-sect. 5.3.3) and its variance, from (13.82), is

$$\sigma_{\bar{n}}^2 = \sigma_n^2/N \tag{13.90}$$

The signal-to-noise ratio after N additions is thus improved by a factor N. For instance, for $N = 1000$, we get a 30 dB improvement. Theoretically, there is no limitation on the achievable improvement as long as $N \rightarrow \infty$. The practical limitations are due to $N \ll \infty$, the inaccurate knowledge of instants t_i (synchronization signal jitter), and the possible modifications of the $s(t - t_i)$ signals.

Performing this mean value estimation on a set of values $k = 0, 1, \ldots, K$, we recover the signal $s(t)$ on interval $[0, KT_e]$.

This reconstruction can be processed in ***parallel*** if, for each $s(t - t_i)$ signal, a set of samples $s(kT_e - t_i)$ with $k = 0, 1, \ldots, K$ is recorded.

It is also possible to use a ***sequential*** processing mode: this is the solution implemented in the analog instruments called *boxcar integrators*. Real sampling (sub-sect. 9.2.2) of observation $x(t)$ is performed at instants $t_i + \tau$ and

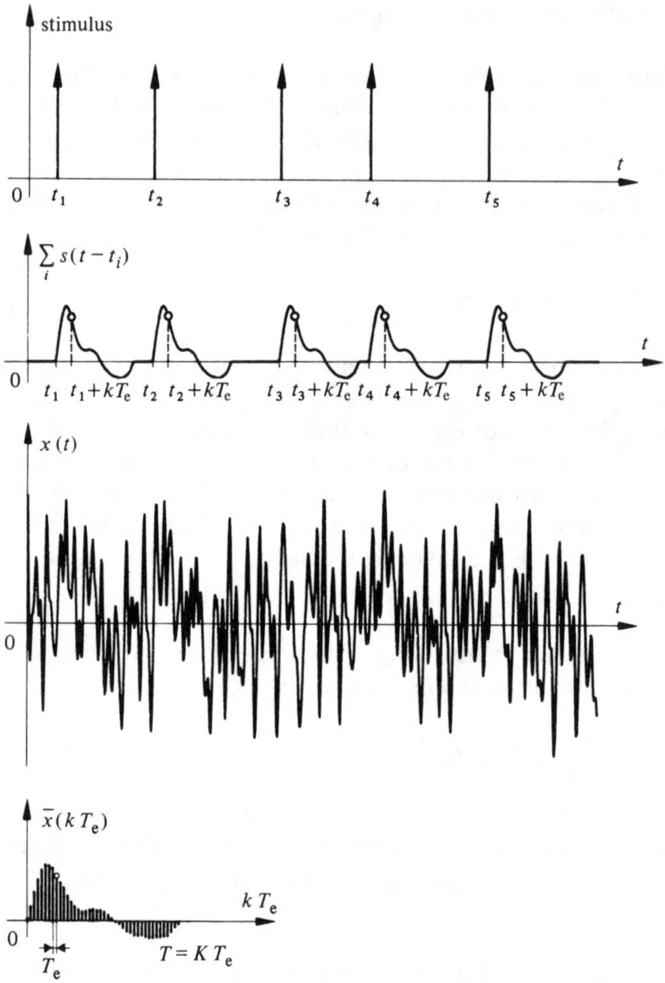

Fig. 13.13

the sequence of amplitude modulated pulses so obtained feeds an averaging filter. By slowly varying the delay τ, the averager output reconstructs $\tilde{s}(\tau)$. It can be shown (problem 13.5.14) that with a T-periodic stimulation, the averager's behavior is similar to that of a very selective comb-filter (i.e., the frequency response is non-zero only near the harmonics $nf_1 = n/T$ of the signal to be recovered).

13.2 SIGNAL COMPARISON

13.2.1 Correlation techniques

The importance of correlation—or of the inner product of two signals—in detection and estimation problems has already been mentioned in the previous section. The formal relationship between the inner product and the Euclidian distance has been established in sub-section 3.1.13.

Various correlation techniques are used [38, 67, 137]. They sometimes help to reveal the existence of an expected signal and to measure some of its parameters. More often, they are a means of comparing two signals in order to display their degree of likeness, or to evaluate their relative delay. This allows us, for instance [138], to ensure synchronization, to estimate moving speed (fig. 13.14), to locate sources of noise or vibration (fig. 13.15), and to measure various acoustical (fig. 13.16) or biophysical (fig. 13.17) characteristics.

As in spectrum analysis, various approaches for estimating the correlation function are available. The direct solution (fig. 13.18) is deduced from the equation:

$$\tilde{\varphi}_{xy}(t,\tau) = \frac{1}{T}\int_{t-T}^{t} x(t' - \tau)y(t')\mathrm{d}t' \tag{13.91}$$

or from its discrete (digital) version:

$$\tilde{\varphi}_{xy}(m,k) = \frac{1}{K}\sum_{l=m-K+1}^{m} x(l - k)y(l) \tag{13.92}$$

The indirect digital approach (fig. 13.19) is deduced from (4.65) and takes advantage of the fast Fourier transform (FFT) algorithms.

13.2.2 Estimation error

For a given delay τ, the measured value $\tilde{\varphi}_{xy}(\tau)$ is a particular realization of a random variable if the signals $x(t)$ and $y(t)$ are random (and assumed here to be stationary and ergodic). Its mean value:

$$\mathrm{E}[\tilde{\varphi}_{xy}(\tau)] = \frac{1}{T}\int_{0}^{T} \mathrm{E}[\mathbf{x}(t - \tau)\mathbf{y}(t)]\mathrm{d}t = R_{xy}(\tau) = \varphi_{xy}(\tau) \tag{13.93}$$

is equal to the theoretical limit value; hence, the estimation is unbiased.

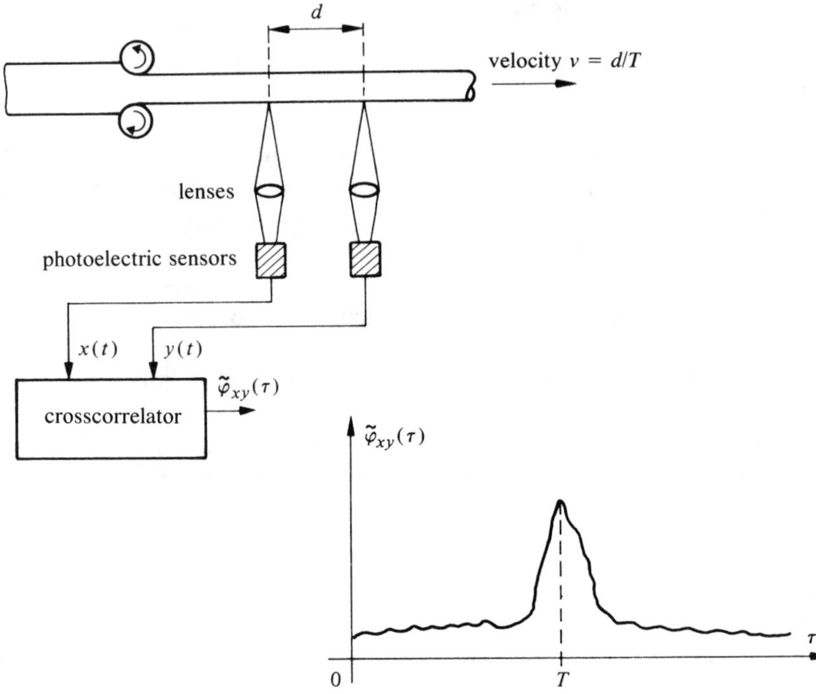

Fig. 13.14 A contactless measurement of a moving surface velocity (e.g., of a rolling mill product) can be obtained by determining the position of the peak of the crosscorrelation function of the signals of two photoelectric sensors. The latter receive the light, modulated by surface ruggedness, reflected in two different points of the moving material and located at a distance d from one another.

Its variance is expressed by (13.80):

$$\sigma_{\tilde{\varphi}}^2 = \frac{1}{T} \int_{-\infty}^{\infty} C_z(u) \operatorname{tri}(u/T) \mathrm{d}u \tag{13.94}$$

where $C_z(u)$ is the covariance function of the product $z(t) = x(t - u)y(t)$.

If the integration duration T is much greater than the correlation equivalent duration (7.144) of $z(t)$:

$$\sigma_{\tilde{\varphi}}^2 \approx \frac{1}{T} \int_{-\infty}^{\infty} C_z(u) \mathrm{d}u = \frac{C_z(0)}{B_\tau T} \tag{13.95}$$

where $C_z(0)$ is the autocovariance at origin of $z(t) = x(t)y(t)$ and $B_\tau = D_\tau^{-1}$ is the approximate bandwidth of the signal $z(t)$.

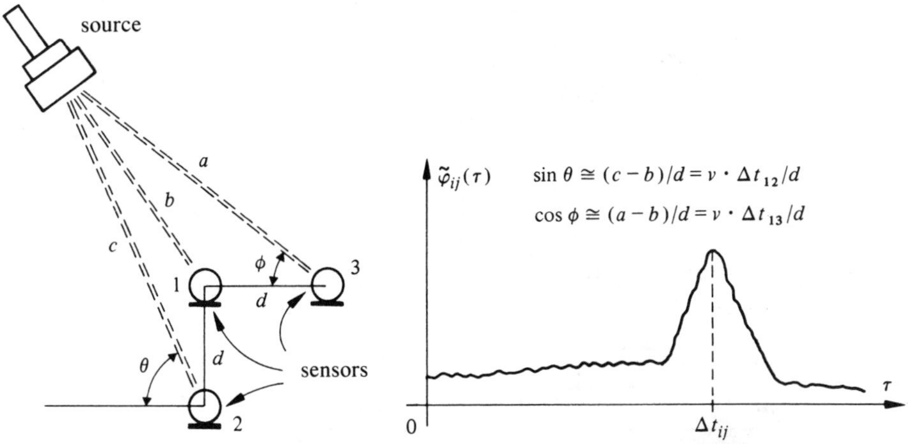

Fig. 13.15 Two consecutive measurements are enough to determine the angles θ and φ defining the location of a vibration source with respect to three sensors located in the same plane on the corners of a square of side d, knowing the vibration propagation velocity v. The difference Δt_{ij} of the propagation times between the source and the sensors 1 and 2, on one hand, and 1 and 3, on the other hand, can be estimated with the crosscorrelation.

13.2.3 Example: estimation of the autocorrelation of a Gaussian signal with zero mean value

Consider the estimation of the value at origin $\tilde{\varphi}_x(0)$ of the autocorrelation function of a Gaussian signal $x(t)$ with zero mean value. We have here $z(t) = x^2(t)$ and $R_x(\tau) \equiv C_x(\tau)$.

From (8.148), $R_z(\tau) = 2R_x^2(\tau) + R_x^2(0)$ and $C_z(\tau) = 2R_x^2(\tau)$, thus,

$$\sigma^2_{\tilde{\varphi}(0)} = \frac{2}{T}\int_{-\infty}^{\infty} R_x^2(u)\,\mathrm{tri}(u/T)\mathrm{d}u \tag{13.96}$$

If $x(t)$ is a band-limited white noise with spectral density $\Phi_x(f) = \frac{1}{2}\eta\,\mathrm{rect}(\frac{1}{2}f/B)$, $R_x^2(u) = \eta^2B^2\,\mathrm{sinc}^2(2Bu)$ and, for $T \gg B^{-1}$

$$\sigma^2_{\tilde{\varphi}(0)} \approx \frac{R_x^2(0)}{BT} \tag{13.97}$$

with $R_x(0) = \eta B \equiv \varphi_x(0)$. The measurement signal-to-noise ratio is

$$\xi = \varphi_x^2(0)/\sigma^2_{\tilde{\varphi}(0)} = BT \tag{13.98}$$

This emphasizes, once again, the role of the product BT in the estimation of mean values.

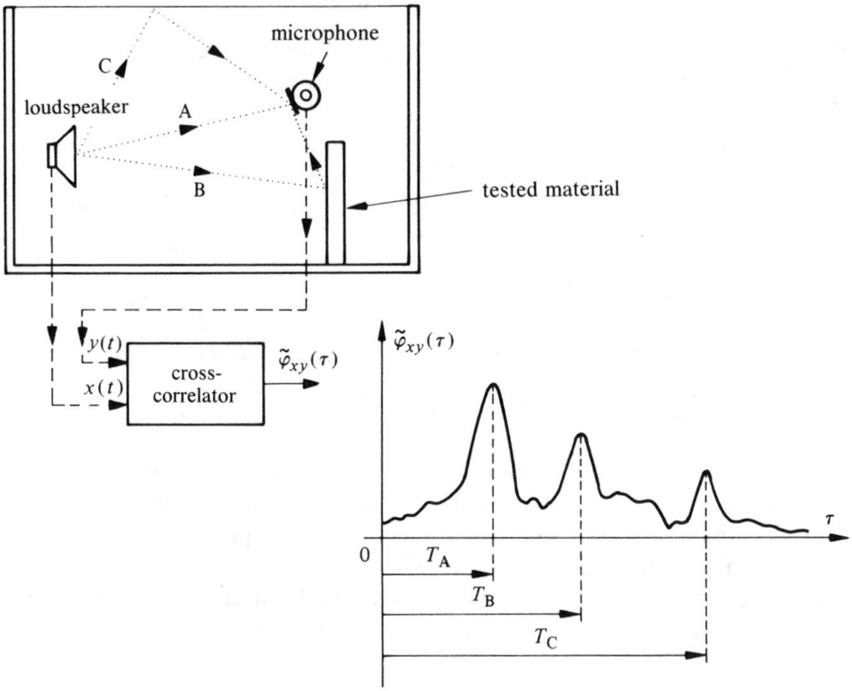

Fig. 13.16 Acoustical measurement example: the crosscorrelation peaks of the loudspeaker input signal and of the microphone output signal reveal the various propagation paths and allow measurement of their absorption coefficients.

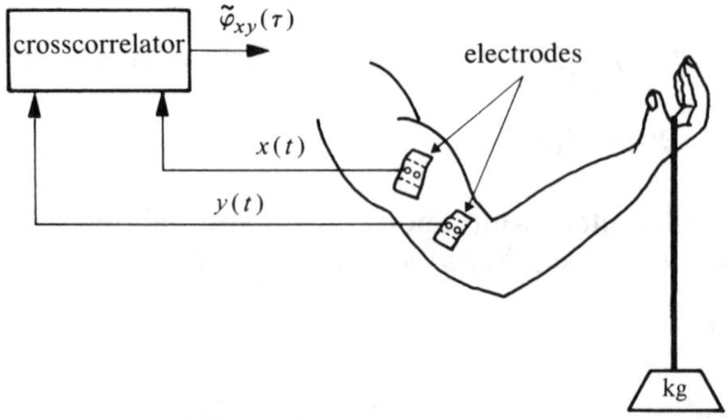

Fig. 13.17 The crosscorrelation of signals sent by two electrodes placed on a muscle allows detection of whether adjacent muscle cells work in a coherent manner or not. This crosscorrelation provides useful information on the condition of musculature.

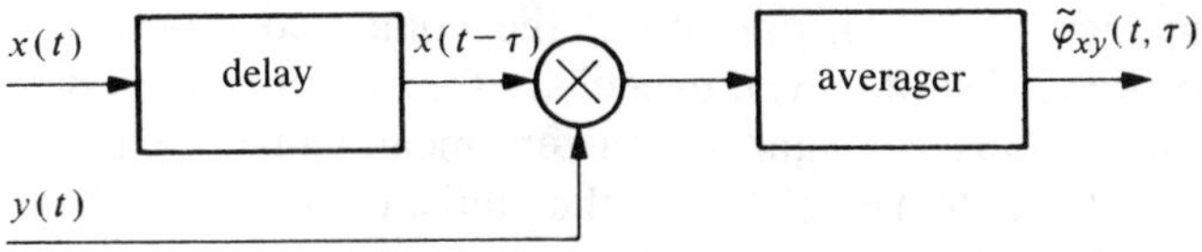

Fig. 13.18

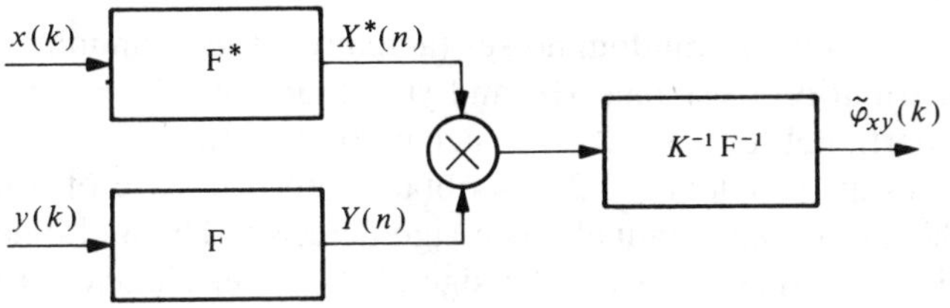

Fig. 13.19

13.2.4 Simplified correlators

Sometimes, it may be sufficient to evaluate the correlation by means of simplified structures taking advantage of the large tolerance of this function to the signal's nonlinear deformations.

The simpliest solution (fig.13.20) consists of considering the correlation of the sign functions of the signals to be compared:

$$\tilde{\varphi}_{uv}(t,\tau) = \frac{1}{T}\int_{t-T}^{t} \operatorname{sgn}[x(t' - \tau)] \operatorname{sgn}[y(t')]\mathrm{d}t' \tag{13.99}$$

with $u(t) = \operatorname{sgn}[x(t)]$ and $v(t) = \operatorname{sgn}[y(t)]$.

The two-level quantization (1 bit) facilitates the implementation of the delay operator and the multiplier (exclusive-OR operator). Despite the introduction

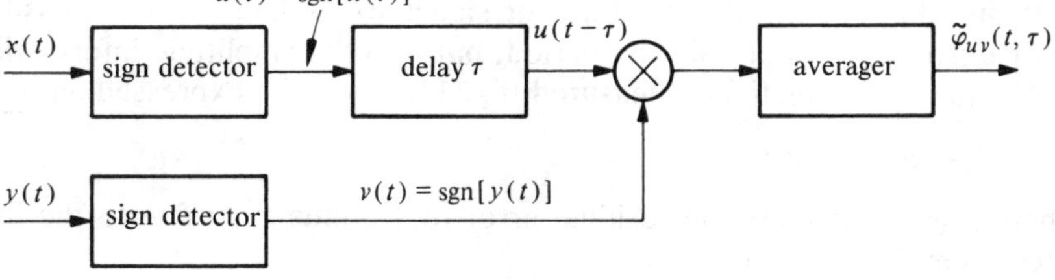

Fig. 13.20

of nonlinear sign detection operators, the computed crosscorrelation $\tilde{\varphi}_{uv}(\tau)$ keeps some properties of $\varphi_{xy}(\tau)$, especially the position of its peak value.

In the case of Gaussian signals with zero mean value, it can be shown (sub-sect. 8.4.15 and problem 8.5.28) that the statistical correlation of $u(t)$ and $v(t)$ is linked to that of $x(t)$ and $y(t)$ by the simple relation

$$R_{uv}(\tau) = \frac{2}{\pi} \arcsin \frac{R_{xy}(\tau)}{\sigma_x \sigma_y} \tag{13.100}$$

By adding independent random noises (actually, pseudorandom noises) with a uniform distribution to signals $x(t)$ and $y(t)$, functions $R_{uv}(\tau)$ and $R_{xy}(\tau)$ can be made proportional for a very large signal group [67].

A similar result (problem 8.5.29) is obtained for the case of a hybrid correlator in which only one channel has a sign detector. The multiplication here simply consists of either inverting the sign of the other channel or not.

13.2.5 Detection of periodic signals by autocorrelation

Consider the detection of the presence and the estimation of the period T_0 of an unknown periodic signal $s(t) = s(t + mT_0)$ embedded in an independent background noise $n(t)$ with zero mean value. The observed signal is $x(t) = s(t) + n(t)$. The autocorrelation function of a T_0-periodic signal is also T_0-periodic (sub-sect. 4.4.12). On the other hand, the autocorrelation of the background noise (that can be assumed filtered) is a function tending toward zero when τ tends toward infinity. According to (5.191) and assuming the considered signals are ergodic, we have

$$\varphi_x(\tau) = \varphi_s(\tau) + \varphi_n(\tau) \tag{13.101}$$

For τ much larger than the equivalent correlation duration (sub-sect. 7.5.13) of the noise $D_\tau = 1/B_\tau$, we get

$$\varphi_x(\tau \gg D_\tau) \approx \varphi_s(\tau) \tag{13.102}$$

If, in addition, the general form of signal $s(t)$ is known, it is possible to deduce from $\varphi_x(\tau)$, not only the period, but also the amplitude information.

The function effectively measured (fig. 13.21) can be expressed as

$$\tilde{\varphi}_x(\tau) = \varphi_x(\tau) + \varphi_\epsilon(\tau) \tag{13.103}$$

where $\varphi_\epsilon(\tau)$ represents the residual error (noise plus bias) due to the limited integration duration with

$$\lim_{T\to\infty} \varphi_\epsilon(\tau) = 0 \tag{13.104}$$

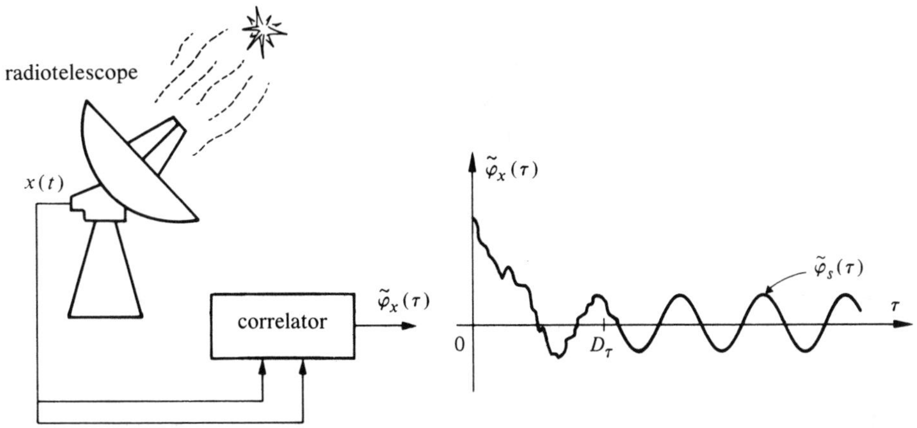

Fig. 13.21 Illustration of the detection of weak cosmic periodic signals in astrophysics.

13.2.6 Signal detection by crosscorrelation

Consider two disturbed signals containing delayed versions of the same useful signal $s(t)$:

$$x(t) = s(t - t_1) + n_1(t) \tag{13.105}$$

and

$$y(t) = s(t - t_2) + n_2(t) \tag{13.106}$$

where $n_1(t)$ and $n_2(t)$ are independent noises with zero mean values.

Always under the ergodic assumption, the theoretical crosscorrelation function of $x(t)$ and $y(t)$ is

$$\varphi_{xy}(\tau) = \varphi_s(\tau - t_2 + t_1) \tag{13.107}$$

and its maximum is positioned in $\tau = t_2 - t_1$ (fig. 13.22).

The measured function can again be expressed as

$$\tilde{\varphi}_{xy}(\tau) = \varphi_{xy}(\tau) + \varphi_\epsilon(\tau) \tag{13.108}$$

where $\varphi_\epsilon(\tau)$ takes into account the residual errors due to the limited integration duration.

Detection by crosscorrelation is similar to the matched filtering described in section 13.4.

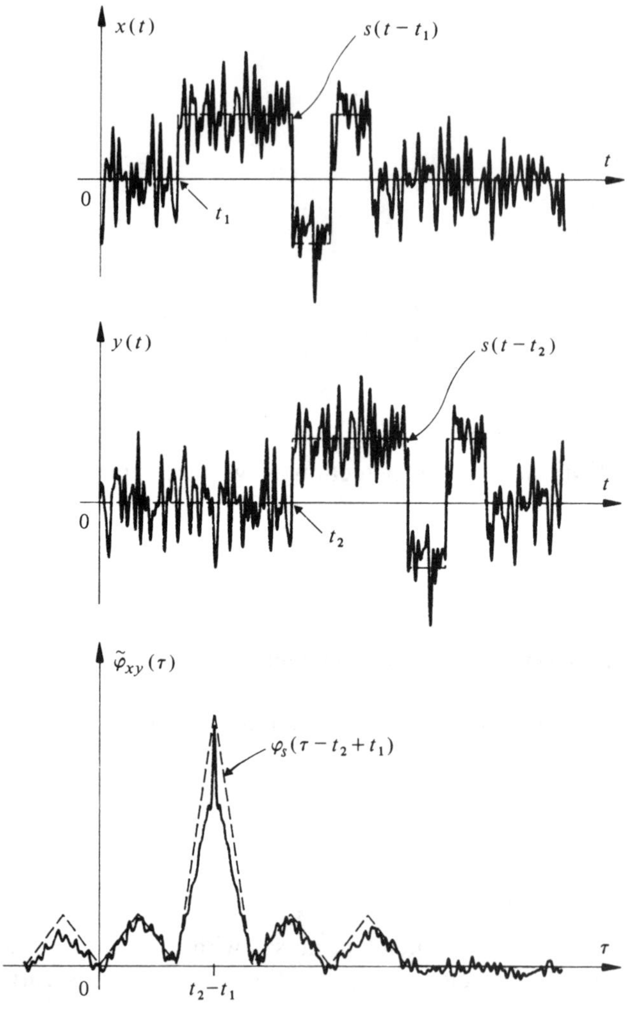

Fig. 13.22

13.2.7 Application: measurement of the delay between two random signals

The relative phase between two sine waves is measured by comparing their zero-crossings. The relative shift between two delayed versions of a random signal is much more difficult to evaluate.

The most efficient method is that of crosscorrelation (fig. 13.23). The sought delay corresponds to the abscissa of the peak value of (13.107). The accuracy

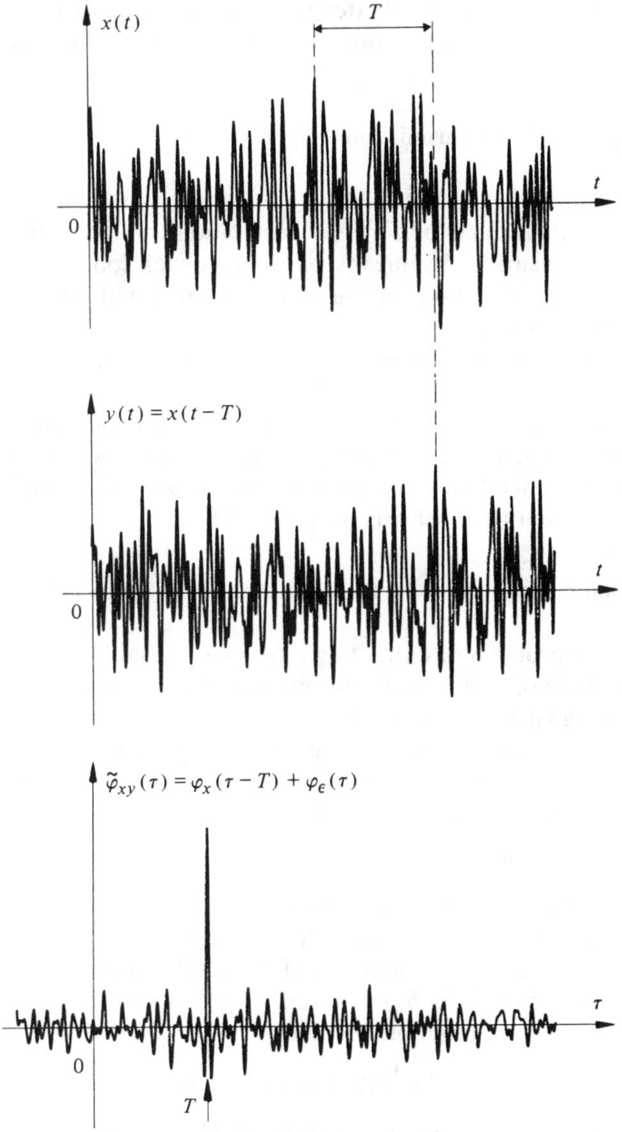

Fig. 13.23

of this measurement is all the better when the random signal autocorrelation function is close to a delta-function, i.e., when its spectrum has a wide bandwidth.

13.2.8 Application: synchronization

The phase locked loop (fig. 13.8) is a classical synchronization means for period signals. It, in fact, performs a correlation (multiplication followed by an integration) ensuring, by means of the feedback loop, the maximum orthogonality (zero correlation) of the received signal with the signal generated by the controlled oscillator.

With a very low signal-to-noise ratio, the initial acquisition lock is difficult to achieve.

The crosscorrelation of an incoming signal (with an autocorrelation similar to a delta-function), with its image generated by the receiver, is an efficient means of synchronization. The pseudorandom periodic signal described in section 5.10 is often used for this purpose [73].

13.2.9 Periodic signal recovery by synchronous crosscorrelation

Consider the problem of extracting a periodic signal $s(t)$ out of an independent background noise $n(t)$, with zero mean value, when a perfectly synchronous auxiliary signal $u(t)$ is available.

The signal to be extracted can be expressed in the equivalent forms $s(t) = s(t + mT_0) = s(t, T_0)*\delta_{T_0}(t)$. The observed signal is $x(t) = s(t) + n(t)$. The measured crosscorrelation (fig. 13.24) is

$$\tilde{\varphi}_{xu}(\tau) = \varphi_{xu}(\tau) + \varphi_{\epsilon}(\tau) \tag{13.109}$$

where $\varphi_{\epsilon}(\tau)$ represents the residual error.

If $u(t)$ is a periodic sequence of very short pulses with same period T_0 and very large amplitude—represented here for the sake of simplicity by a periodic sequence of delta-functions $\delta_{T_0}(t)$—from (1.57), (4.98), and (4.138), the theoretical crosscorrelation function becomes

$$\varphi_{xu}(\tau) = \delta_{T_0}(-\tau)\overset{*}{\ast}x(\tau) = T_0^{-1}\delta_{T_0}(\tau)*s(\tau,T_0) \equiv T_0^{-1}s(\tau) \tag{13.110}$$

In the frequency domain, this crosscorrelation corresponds to a comb-filtering: very selective filtering at every discrete frequency $f_n = n/T_0$. By (4.14), (4.123), and (4.126), we get

$$\Phi_{xu}(f) = T_0^{-2}S(f,T_0)\delta_{1/T_0}(f) = T_0^{-1}\sum_{n=-\infty}^{\infty} S_n\delta(f - n/T_0) \tag{13.111}$$

$$= T_0^{-1}S(f)$$

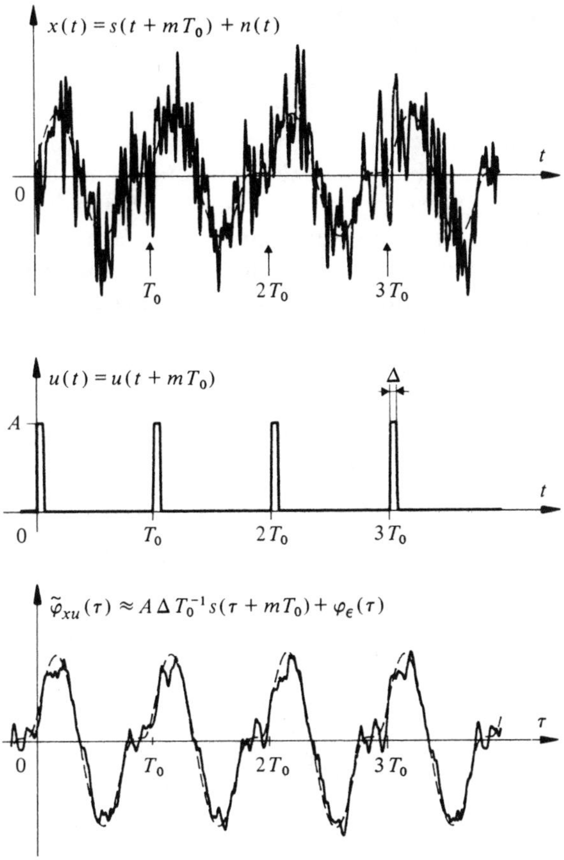

Fig. 13.24

Theoretically, such a filtering extracts out of the observed signal $x(t)$ only the components S_n that are present at the discrete frequencies $f_n = n/T_0$, i.e., the various harmonics of the periodic signal $s(t)$. The noise $n(t)$ is totally eliminated.

In practice (problem 13.5.15), the real periodic filtering function corresponds to the convolution of the periodic sequence $\delta_{1/T_0}(f)$ with the transfer function $H(f)$ of a lowpass filter, the bandwidth of which is non-zero. The latter depends on the integration duration T and the non-ideal shape of the periodic pulses of the auxiliary signal $u(t)$.

This recovery technique is equivalent to the signal averaging technique described in sub-sect. 13.1.26 in the case of periodic signals.

13.2.10 Particular case: lock-in amplifier

A *lock-in amplifier* is a very selective kind of voltmeter (fig. 13.25) based on detection by crosscorrelation for measuring a dc or very slowly varying signal embedded in an independent noise. It transposes the signal to be measured into a more favorable frequency range (avoiding the spectrum range where the $1/f$ amplifier noise is prevalent), and provides very efficient filtering.

This method can be applied when the signal to be measured, denoted here by $m(t)$, can be amplitude modulated by means of a known periodic auxiliary function $u_p(t)$ (electrical, acoustical, optical, mechanical, *et cetera* stimulating signal). The modulated signal is $s(t) = m(t)u_p(t)$.

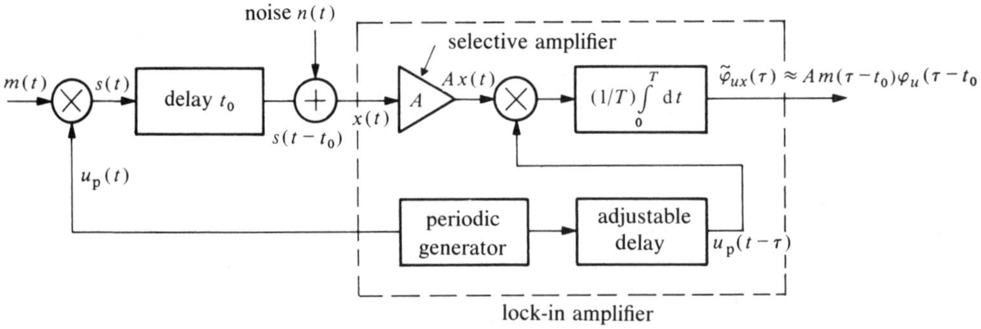

Fig. 13.25

The signal observed at the amplifier input is $x(t) = s(t - t_0) + n(t)$, where $n(t)$ takes into account both the noise perceived by the observation of $m(t)$ and the amplifier internal noise. The delay t_0 is an unknown parameter dependent on the signal propagation conditions.

The lock-in amplifier has a gain A at the repetition frequency $1/T$ of the auxiliary signal $u_p(t)$. Additionally, it performs the product of the amplified signal $Ax(t)$ with the auxiliary signal $u_p(t - \tau)$ and averages the result. The parameter τ is adjustable.

For a given τ, the measure corresponds to the correlation:

$$\tilde{\varphi}_{ux}(\tau) = \frac{1}{T}\int_0^T Ax(t)u_p(t - \tau)\mathrm{d}t \tag{13.112}$$

the asymptotic value of which is

$$\begin{aligned}\varphi_{ux}(\tau) &= \lim_{T\to\infty} \tilde{\varphi}_{ux}(\tau) = Au_p(-\tau)\bar{*}[m(\tau - t_0)u_p(\tau - t_0) + n(\tau)] \\ &\approx Am(\tau - t_0)\varphi_u(\tau - t_0)\end{aligned} \tag{13.113}$$

because, by assumption, $m(t)$ varies very slowly with respect to $u_p(t)$ and the residual term $\varphi_{un}(\tau)$ vanishes if the noise $n(t)$ is independent of $u_p(t)$ and if one of the two signals has a zero mean value.

Since the autocorrelation function $\varphi_u(\tau)$ has a maximum at origin which corresponds to the power P_u, the measure has a maximum at $\tau = t_0$. Thus, both $m(t)$ and the delay t_0 can be estimated.

13.2.11 Spread spectrum communication systems

A typical example of the application of correlation to ensure information transmission despite strong disturbances is used in spread-spectrum systems. Spread-spectrum techniques are mainly used in military electronic countermeasures (prevention of deliberate jamming). However, they can also be used in civilian applications because of their natural interference immunity and their capability of ensuring secure (confidential) communications with selective addressing.

The basic principle of such a system is depicted in fig. 13.26.

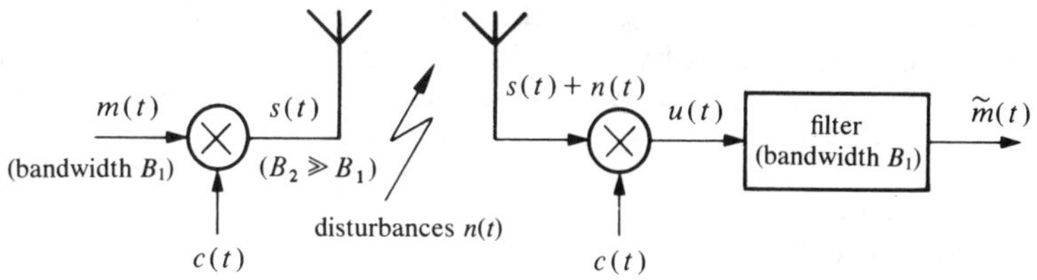

Fig. 13.26

The power of the emitted secondary signal $s(t)$ is spread over a very wide bandwidth B_2 by multiplying the message signal $m(t)$ and an independent auxiliary signal $c(t)$ with a very wide bandwidth, which is used as a coding key. The multiplication of two independent signals implies the convolution of their spectra:

$$s(t) = m(t)c(t) \leftrightarrow \Phi_s(f) = \Phi_m(f)*\Phi_c(f) \tag{13.114}$$

If $m(t)$ has a rather narrow bandwidth B_1 and $c(t)$ has a very wide one, their product has approximately the same spectrum as $c(t)$.

At the receiver, the sum of the emitted signal $s(t)$ and of the disturbances $n(t)$ is also multiplied by the same auxiliary signal $c(t)$, which implies that adequate synchronization is available:

$$u(t) = [s(t) + n(t)]c(t) = m(t)c^2(t) + n(t)c(t) \tag{13.115}$$

If $c(t)$ is, for instance, a pseudorandom binary sequence (± 1 level), the result corresponds to the message signal $m(t)$ accompanied by a noise term $n(t)c(t)$ with a very wide band that is essentially eliminated by the final filtering.

Only receivers with the same coding key can reconstruct the information. Different users can thus communicate on the same channel bandwidth without mutual interference by employing different coding keys.

The efficiency of the system increases with the bandwidth ratio B_2/B_1.

13.3 ELEMENTS OF DECISION-THEORY

13.3.1 Decision process

In signal processing, for pattern recognition (sub-sect. 13.4.8), a decision process usually arises in the following way: on the basis of a set of observations (parameter estimation, features extraction), the analyzed signal [pattern] is attributed to a given class of possible signals [patterns].

Such a decision process can be based on the observation of a single parameter (amplitude, frequency, phase), or on a set of values of the same parameter (set of N samples, *et cetera*). A general model consists of representing, as in section 13.1, the observations by a parameter vector $\tilde{\boldsymbol{a}} = (\tilde{a}_1, \tilde{a}_2, \ldots, \tilde{a}_n)$.

If only two classes are considered, it is a *binary decision*: choice between the two terms of an alternative. It is, for instance,

- the presence or absence of a radar echo signal;
- the choice between the logic states 0 and 1 represented by a telecommunication signal in digital data transmission.

In other instances, the classification operation implies a large number of choices: diagnostic proposals based on the interpretation of a biological signal, deduction of the probable causes of a system misfunctioning after the analysis of the appropriate sensor signals, identification of alphanumerical characters, selection of mechanical parts observed by a television camera, labelling of geographical entities in remote sensing, *et cetera*. The number of parameters to be measured (spectral coefficients, statistical moments, correlation factors, *et cetera*), or of features to be determined (number of intersections, of line or curve segments, of angles of a contour, *et cetera*), can be high. This is a problem of *decision with multiple hypothesis*.

The observation set is called an *event*. A classical formulation of the decision process consists of representing each event by a point in a vector space called *observation space* or *feature space*. The dimension of this space is less than or equal to the number of observations performed; the latter case corresponding to independent observations.

Each signal [pattern] class to be identified is represented by a typical combination of parameter values forming a vector $\boldsymbol{a}$: hence, it is defined by a specific point with, in principle, known coordinates.

Because of noise, measurement inaccuracies, pattern distortion, *et cetera*, each event is translated into a point located at a variable distance from the typical point. The set of possible events associated with a given class thus forms a ***cluster of points*** surrounding the typical point (fig. 13.27).

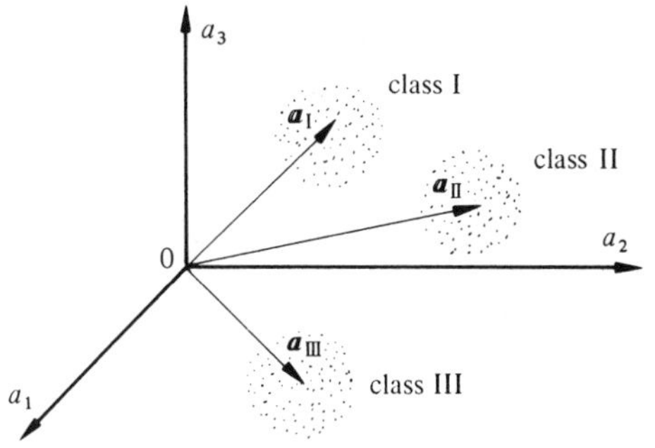

Fig. 13.27

The (identification) decision rule is obtained by dividing the observation space into as many distinct regions as there are classes, in order to minimize the risks of wrong decisions. Hence, a fundamental theoretical problem is to define a partitioning mode (choice of the region boundaries).

A high probability of correct decisions can only be obtained if the clusters of points are clearly far apart. In order to comply with this condition, we must, on one hand, look to limit the dispersion of points around the typical value, i.e., to reduce noise and other uncertainty factors. On the other hand, we must also pay special attention to the choice of the parameters used, in order to separate as much as possible the clusters from one another (maximization of the distance between classes), and define an appropriate distance estimation.

13.3.2 Usual distances

In the frequent case of a unidimensional observation (sample or group of samples of a signal), the usual distance is a simple difference of normalized mean values (unit variance) that can be interpreted as the square root of the observation signal-to-noise ratio:

$$d(x,y) = (\mu_x - \mu_y)/\sigma_n = \sqrt{\xi} \tag{13.116}$$

When the observation space is multidimensional, a weighted version of the Euclidian distance (3.3) is often used:

$$d(\mathbf{x},\mathbf{y}) = \left\{ \sum_{i=1}^{n} w_i(x_i - y_i)^2 \right\}^{1/2} \tag{13.117}$$

where w_i is the weighting factor of the ith observed characteristics.

A generalization of this approach is the *Mahalanobis distance* [140], which weights the influence of the dependence between characteristics:

$$d(\mathbf{x},\mathbf{y}) = \left\{ \sum_{i=1}^{n} \sum_{j=1}^{n} w_{ij}(x_i - y_i)(x_j - y_j) \right\}^{1/2} \tag{13.118}$$

The weighting factor is usually the element of index ij of the inverse of the covariance matrix C_{xy}.

When the attributes are binary (0 and 1 symbols, *et cetera*), we frequently use the Hamming distance (3.7).

13.3.3 Illustration

Consider the identification of a binary message A or B represented by a sequence of three bits (attributes). Under the influence of additive noise, each bit of the message can be misinterpreted at the receiver. The observation space, thus, is made up of $2^3 = 8$ different words (fig. 13.28). The Hamming distance between these words has a maximum equal to 3: for instance, between 000 and 111.

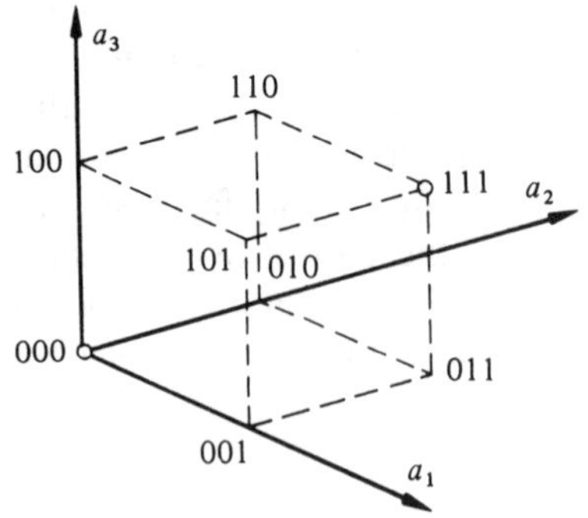

Fig. 13.28

The optimal partitioning of this space consists of attributing to information A one of the words of this set, for instance 000, and all those that have a Hamming distance equal to one: 001,010,100. Information B is then represented by the word that is the farthest from 000, i.e., 111, completed by all the words having a distance equal to one: 110, 101, 011.

A decision error can only occur if at least two bits of a word are misinterpreted.

13.3.4 Binary decision

The classical binary decision problem can be formulated in the following way.

A source of information produces two possible signals s_0 and s_1, with the respective probabilities $p_0 = \text{Prob}(s_0)$ and $p_1 = \text{Prob}(s_1)$. These signals reach the observer in a deteriorated form (for instance, $x = s_i + n$) because of their contamination by various random disturbances. Knowing the binary nature of the source, the observer can set two *hypotheses*, denoted H_0 and H_1, about the emitted signal identity. On the basis of his or her (continuous or discrete) observation of the received signal x, or, more generally, of a set of observations $\boldsymbol{x}$, the observer has to choose the most likely hypothesis. For this, he or she must apply a *decision criterion.*

The choice between those two hypotheses can lead to the four following situations:

- choice of H_0 when H_0 is true;
- choice of H_1 when H_1 is true;
- choice of H_0 when H_1 is true;
- choice of H_1 when H_0 is true.

The first two cases correspond to correct decisions; the other two cases correspond to wrong decisions. The decision criterion sets a strategy designed to minimize the risk of taking a wrong decision. It can be interpreted as a rule allowing division of the observation space O in two mutually exclusive regions O_0 and O_1. If the observation is an element of O_0, or of O_1, the decision is taken to admit hypothesis H_0, or, otherwise, hypothesis H_1.

The three main criteria are the Bayesian criterion, the minimax criterion, and the Neyman-Pearson criterion.

13.3.5 Bayesian criterion

The Bayesian criterion can be applied when we have an *a priori* knowledge of the probabilities p_0 and p_1 of the occurrence of the two signals s_0 and s_1. The criterion consists of determining the regions O_0 and O_1 of the decision space in order to minimize the total error probability. This probability, denoted p_ϵ, is equal to the sum of the joint probability of accepting the hypothesis H_1

when signal s_0 is emitted, and of the joint probability of accepting the hypothesis H_0 when s_1 is emitted. According to (14.6):

$$\begin{aligned} p_\epsilon &= \text{Prob}\,(H_1, s_0) + \text{Prob}\,(H_0, s_1) \\ &= p_0\,\text{Prob}\,(H_1|s_0) + p_1\,\text{Prob}\,(H_0|s_1) \end{aligned} \tag{13.119}$$

The wrong decisions are taken if the observation $\boldsymbol{x}$ belongs to the O_0 region when $s_i = s_1$ or to the O_1 region when $s_i = s_0$. The corresponding conditional probabilities are consequently obtained by integrating the conditional probability distributions $p(\boldsymbol{x}|s_0)$ and $p(\boldsymbol{x}|s_1)$ on the $\boldsymbol{x}$ domain respectively belonging to O_1 or O_0:

$$\begin{aligned} \text{Prob}\,(H_1|s_0) &= \int_{O_1} p(\boldsymbol{x}|s_0)\,\mathrm{d}\boldsymbol{x} = 1 - \int_{O_0} p(\boldsymbol{x}|s_0)\,\mathrm{d}\boldsymbol{x} \\ \text{Prob}\,(H_0|s_1) &= \int_{O_0} p(\boldsymbol{x}|s_1)\,\mathrm{d}\boldsymbol{x} = 1 - \int_{O_1} p(\boldsymbol{x}|s_1)\,\mathrm{d}\boldsymbol{x} \end{aligned} \tag{13.120}$$

Inserting (13.120) into (13.119) in order to regroup integrals belonging to the same integration domain, the total error probability becomes

$$\begin{aligned} p_\epsilon &= p_0 + \int_{O_0} [p_1 p(\boldsymbol{x}|s_1) - p_0 p(\boldsymbol{x}|s_0)]\mathrm{d}\boldsymbol{x} \\ &= p_1 + \int_{O_1} [p_0 p(\boldsymbol{x}|s_0) - p_1 p(\boldsymbol{x}|s_1)\mathrm{d}\boldsymbol{x} \end{aligned} \tag{13.121}$$

This error probability is obviously minimal if the regions O_0 and O_1 are chosen in order to make the result of the respective integrals as negative as possible. A probability and a probability distribution being by definition positive, this condition is achieved if O_0 is defined as the region of the observation space for which

$$p_0 p(\boldsymbol{x}|s_0) > p_1 p(\boldsymbol{x}|s_1) \tag{13.122}$$

and if O_1 is defined with

$$p_1 p(\boldsymbol{x}|s_1) > p_0 p(\boldsymbol{x}|s_0) \tag{13.123}$$

Figure 13.29 illustrates the optimal partitioning obtained in the case of unimodal Gaussian probability distributions with a unidimensional observation.

The Bayesian criterion can be summarized as follows

$$\frac{p(\boldsymbol{x}|s_1)}{p(\boldsymbol{x}|s_0)} \underset{H_0}{\overset{H_1}{\gtrless}} \frac{p_0}{p_1} \tag{13.124}$$

and stated as follows:

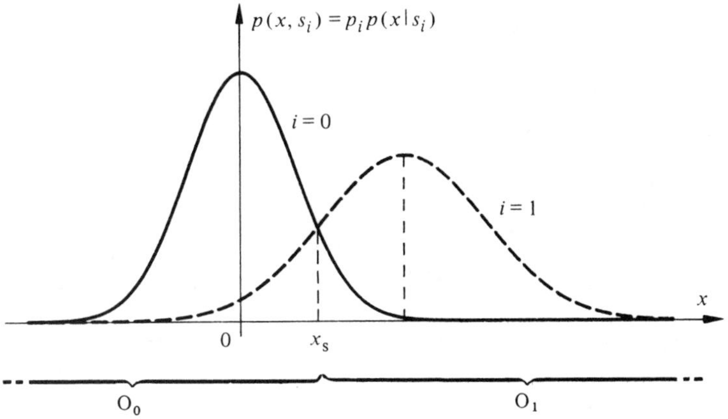

Fig. 13.29

- if observation $\boldsymbol{x}$ is in the O_0 region determined by (13.122), then the H_0 hypothesis is accepted;
- if observation $\boldsymbol{x}$ is in the O_1 region determined by (13.123), then the H_1 hypothesis is accepted.

13.3.6 Likelihood ratio

The ratio

$$\Lambda(\boldsymbol{x}) = \frac{p(\boldsymbol{x}|s_1)}{p(\boldsymbol{x}|s_0)} \tag{13.125}$$

is called the likelihood ratio.

An equivalent form of relation (13.24) is obtained by taking the logarithm of the two terms. The logarithm being a monotonically increasing function, we get for the Bayesian criterion:

$$\ln \Lambda(\boldsymbol{x}) \underset{H_0}{\overset{H_1}{\gtrless}} \ln \left(\frac{p_0}{p_1}\right) \tag{13.126}$$

13.3.7 Terminology

The conditional probabilities occuring in relations (13.120) are often referred to, in the specialized literature, as (fig. 13.30):

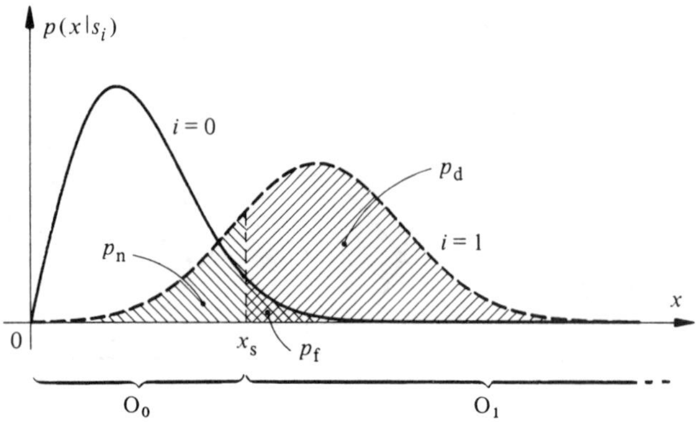

Fig. 13.30

- *false alarm probability*

$$p_f = \text{Prob}(H_1|s_0) = \int_{O_1} p(\boldsymbol{x}|s_0)d\boldsymbol{x} \tag{13.127}$$

- *detection probability*

$$p_d = \text{Prob}(H_1|s_1) = \int_{O_1} p(\boldsymbol{x}|s_1)d\boldsymbol{x} \tag{13.128}$$

- *nondetection probability*

$$p_n = \text{Prob}(H_0|s_1) = 1 - p_d = \int_{O_0} p(\boldsymbol{x}|s_1)d\boldsymbol{x} \tag{13.129}$$

and, thanks to (13.116), with the total error probability

$$p_\epsilon = p_0 p_f + p_1 p_n \tag{13.130}$$

This terminology is taken from the jargon used in radar where hypotheses H_1 and H_0, respectively, correspond to the presence or absence of a target in the exploration beam path. In pattern recognition, the terms *false positive* decision and *false negative* decision are preferred over false alarm and non-detection.

In statistics, the error made by rejecting hypothesis H_0 when it is true is called *error of the first kind*. The probability of making an error of the first kind is denoted α. The error made by accepting H_0 when it is wrong is called *error of the second kind*. The probability of making an error of the second kind is denoted β. Thus, we have: $p_f = \alpha$ and $p_n = \beta$.

13.3.8 Cost coefficients and average risk

To emphasize the relative importance given to each of the four situations inferred by the choice of one of the two hypotheses, we can modify the decision rule by introducing weighting coefficients. These coefficients, called *cost coefficients,* are usually denoted c_{ij}, where the first index represents the correct hypothesis, and the second represents the decision. They can be seen as elements of a *cost matrix*:

$$c = \begin{bmatrix} c_{00} & c_{01} \\ c_{10} & c_{11} \end{bmatrix} \tag{13.131}$$

We define then an *average risk*:

$$R = \sum_{i=0}^{1} \sum_{j=0}^{1} c_{ij} p_i \int_{O_j} p(\mathbf{x}|s_i)\mathrm{d}\mathbf{x} \tag{13.132}$$

The decision rule, according to the Bayesian criterion, is obtained by minimizing the average risk. Hence, we obtain

$$\Lambda(\mathbf{x}) \underset{H_0}{\overset{H_1}{\gtrless}} \frac{p_0}{p_1} \frac{c_{01} - c_{00}}{c_{10} - c_{11}} \tag{13.133}$$

with, in some instances, $c_{00} = c_{11} = 0$ (differentiated weighting of the nondetection and false alarm risks).

13.3.9 Example: single observation

Consider a voltage source delivering a binary random signal $s(t)$ taking either a value A (hypothesis H_1) with a probability p_1, or the value zero (hypothesis H_0) with a probability p_0. Before it reaches the detection system (observer), this signal is corrupted by an additive Gaussian noise $n(t)$ with zero mean value and variance σ_n^2. The source output voltage and the noise are two statistically independent phenomena. The normalized distance between the two clusters of observation points can be simply defined, according to (13.116), by $d(x_{H0}, x_{H1}) = (\mu_{xH1} - \mu_{xH0})/\sigma_n = A/\sigma_n$.

At reception, the observer must decide, on the basis of a sample x taken from the signal received at a given instant t, which value of the emitted signal corresponds to the observed value. To take his or her decision, the observer applies the criterion (13.126). Because of the unimodal character of the Gaussian probability distribution, the test here becomes a simple comparison of the sample with a reference voltage: the optimal decision threshold minimizing the total error probability.

Under the H_0 hypothesis, the observed signal is $x_{H0}(t) = n(t)$ with a conditional probability distribution:

$$p(x|s_0) = (2\pi\sigma_n^2)^{-1/2} \exp(-\tfrac{1}{2}x^2/\sigma_n^2) \tag{13.134}$$

Under the H_1 hypothesis, we have $x_{H1}(t) = A + n(t)$ with

$$p(x|s_1) = (2\pi\sigma_n^2)^{-1/2} \exp[-\tfrac{1}{2}(x - A)^2/\sigma_n^2] \tag{13.135}$$

Inserting these densities into (13.126) and solving with respect to observation x, the optimal test becomes

$$x \underset{H_0}{\overset{H_1}{\gtrless}} \frac{\sigma_n^2}{A} \ln(p_0/p_1) + \tfrac{1}{2}A \tag{13.136}$$

and the decision threshold is the value x_s corresponding to the equality of the two members of (13.136).

The probabilities of false alarm, nondetection, detection, and total error are deduced from (13.127), (13.128), (13.129), and (13.130) with, for this case: $O_1 = [x_s, +\infty]$ and $O_0 = [-\infty, x_s]$.

For example, for $p_0 = p_1 = \frac{1}{2}$, $\sigma_n = 2$ V and $A = 6$ V, we get $x_s = A/2 = 3$ V, and according to the normal distribution table of section 15.8: $p_f = p_n = p_\epsilon = 6.68 \times 10^{-2}$.

On the other hand, in the case of unequal probabilities of emission $p_0 = 0.9$ and $p_1 = 0.1$ with same signal level A and same variance σ_n^2, we have: $x_s = 4.46$ V, $p_f \cong 1.3 \times 10^{-2}$, $p_n \cong 2.2 \times 10^{-1}$, and $p_\epsilon = 3.36 \times 10^{-2}$. Keeping, in this case, a nonoptimal decision threshold at $A/2$ would give a higher total error probability equal to 6.68×10^{-2}.

13.3.10 Example: multiple independent observations

Under the same signal and noise conditions, consider not a single sample of the received signal, but a set of N samples $x_1, x_2, \ldots, x_N$, the noise contributions of which are statistically independent. These N samples form a vector $\boldsymbol{x}$ in an observation space with N dimensions, which, under hypothesis H_0 or H_1, has the conditional probability distributions deduced from (5.33):

$$p(\boldsymbol{x}|s_0) = \prod_{i=0}^{N} (2\pi\sigma_n^2)^{-1/2} \exp(-\tfrac{1}{2}x_i^2/\sigma_n^2) \tag{13.137}$$

and

$$p(\boldsymbol{x}|s_1) = \prod_{i=0}^{N} (2\pi\sigma_n^2)^{-1/2} \exp[-\tfrac{1}{2}(x_i - A)^2/\sigma_n^2) \tag{13.138}$$

The test (13.126) becomes then:

$$\bar{x} = \frac{1}{N} \sum_{i=1}^{N} x_i \underset{H_0}{\overset{H_1}{\gtrless}} \frac{\sigma_n^2}{NA} \ln(p_0/p_1) + \tfrac{1}{2}A \tag{13.139}$$

where the experimental mean value $\bar{x}$ represents statistics sufficient for taking the decision. The noise variance is then reduced by a factor N according to (13.28).

13.3.11 Minimax criterion

The fundamental assumption of the Bayesian criterion is the knowledge of the *a priori* probabilities p_0 and p_1. This assumption enables us to determine the test that minimizes the total error (or average risk) probability. However, in many problems, the source statistics (p_0 and p_1) are unknown or nonstationary. If the threshold is arbitrarily assigned, the total error probability is not necessarily minimum. We can, on the other hand, determine the most unfavorable case as a function of the source statistics, leading to the largest average risk and try to minimize it. Of course, this implies the choice of cost coefficients. The Bayesian criterion used to ***minimize*** the ***maximum*** average risk is called *minimax criterion,* and the corresponding risk is noted R_{minimax}.

In this kind of decision, we usually introduce a differentiated cost to account for the relative importance of the wrong decision cases, without introducing a relative weighting for the correct decisions. The expression for the average risk (13.132) with $c_{00} = c_{11} = 0$ can thus be written, with $p_0 = 1 - p_1$, as

$$R = c_{01}p_0p_f + c_{10}p_1p_n = c_{01}p_f + p_1(c_{10}p_n - c_{01}p_f) \tag{13.140}$$

For a given decision threshold x_s and given conditional probability distributions $p(\mathbf{x}|s_0)$ and $p(\mathbf{x}|s_1)$, the probabilities p_f and p_n are determined. Consequently, the average risk R is a linear function of p_1 (fig. 13.31), with slope $c_{10}p_n - c_{01}p_f$. This straight line $R(p_1)$ is tangent to the locus $R_{\min}(p_1)$ of the minimal average risks evaluated according to the Bayesian criterion for each value of p_1.

The minimization of the maximum risk, whatever p_1, is obtained by choosing for threshold x_s the one that cancels the $R(p_1)$ slope, i.e., that satisfies the *minimax equation*:

$$c_{10}p_n - c_{01}p_f = 0 \tag{13.141}$$

In the special case where $c_{01} = c_{10} = 1$ (errors of the first and second kinds of the same cost), the minimax test comes down to choosing x_s such that the conditional error probabilities p_f and p_n are equal.

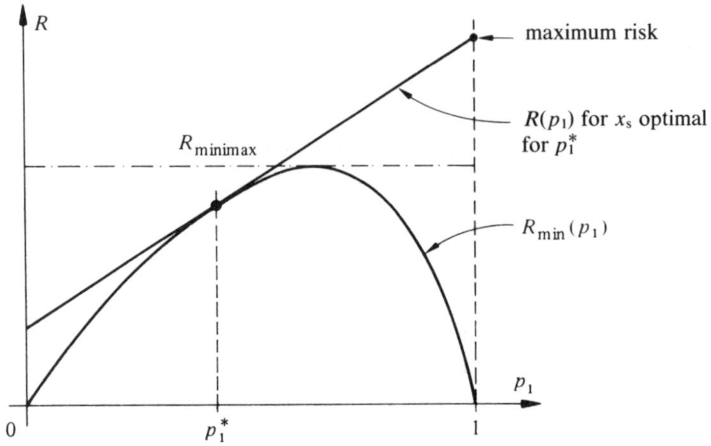

Fig. 13.31

13.3.12 Neyman-Pearson criterion

If the *a priori* probabilities p_0 and p_1 are unknown and if it is difficult to assign realistic values to the risk coefficients, we can try to establish a decision criterion using only the conditional probabilities p_f and $p_d = 1 - p_n$. In principle, we want to make p_f and p_n simultaneously as small as possible. Unfortunately, the decrease of one usually induces an increase of the other. We then make use of the *Neyman-Pearson criterion,* which consists of assigning to the false alarm probability an arbitrary acceptable value α_a and, under this constraint, trying, as far as possible, to maximize the detection probability p_d (or, equivalently, to minimize p_n).

In general, this minimization under constraint problem can be solved using an extreme-value optimization method (Lagrange multiplier λ). The likelihood test (13.126) associated with the Bayesian criterion is then replaced by the test: $\ln \Lambda(x) \gtrless \ln \lambda(\alpha_a)$. The threshold $\lambda(\alpha_a)$ is determined by the relation: $p_f = \text{Prob}[\Lambda(x) > \lambda|H_0] = \alpha_a$.

In the simplest case (single decision threshold), $\lambda = x_s$ is directly determined, for a given α_a, by

$$p_f = \alpha_a = \int_{x_s}^{\infty} p(x|s_0)\mathrm{d}x \tag{13.142}$$

The nondetection probability is then simply given by

$$p_n = \int_{-\infty}^{x_s} p(x|s_1)\mathrm{d}x \tag{13.143}$$

13.3.13 Example: radar detection

Consider a radar signal, made of possible echoes with shape A rect (t/T) cos $(2\pi f_0 t)$, observed in the presence of an additive Gaussian noise with zero mean value and variance σ_n^2 by a receiver containing an envelope detector (fig. 11.23) followed by a comparator. The latter recognizes the presence or absence of echo by comparing a sample value of the envelope $r(t)$ with a reference threshold V.

In the absence of echo, the envelope $r(t)$ of the noise alone, according to (7.49), has a Rayleigh distribution and the false alarm probability is

$$p_{\mathrm{f}} = \int_V^{\infty} (r/\sigma_n^2) \exp(-\tfrac{1}{2}r^2/\sigma_n^2)\mathrm{d}r = \exp(-\tfrac{1}{2}V^2/\sigma_n^2) \tag{13.144}$$

from which, for a given value $p_{\mathrm{f}} = \alpha_{\mathrm{a}}$, we derive

$$V = \sigma_n \sqrt{-2 \ln \alpha_{\mathrm{a}}} \tag{13.145}$$

In the presence of echo, the envelope statistics form the Rice-Nakagami distribution (7.57). Then, the detection probability is

$$p_{\mathrm{d}} = \int_V^{\infty} \frac{r}{\sigma_n^2} \exp\left(-\frac{r^2 + A^2}{2\sigma_n^2}\right) I_0\left(\frac{rA}{\sigma_n^2}\right) \mathrm{d}r = Q\left(\frac{A}{\sigma_n}, \frac{V}{\sigma_n}\right) \tag{13.146}$$

The integral

$$Q(a,b) = \int_b^{\infty} \exp[-\tfrac{1}{2}(a^2 + u^2)]I_0(au)u \, \mathrm{d}u \tag{13.147}$$

is known as *Marcum's Q-function* and can be digitally evaluated [141] by the recurrence

$$Q(a,b) = 1 - \sum_{n=0}^{\infty} g_n k_n \tag{13.148}$$

with

$$g_n = g_{n-1} - (\tfrac{1}{2}b^2/n!) \exp(-\tfrac{1}{2}b^2) \; ; \; k_n = \tfrac{1}{2}a^2 k_{n-1}/n \tag{13.149}$$

A graph of Marcum's Q-function is given in section 15.9.

13.3.14 Comparison of the binary decision criteria

The characteristics of the various decision criteria mentioned are summarized in table 13.32.

Table 13.32

Criterion	*A Priori* Knowledge	Minimized Figure of Merit	Test
Bayes	• p_0 and p_1 • $p(x\|s_i)$ with $i = 0.1$	Total error probability $p_\epsilon = p_0 p_f + p_1 p_n$	$\Lambda(x) = \dfrac{p(x\|s_1)}{p(x\|s_0)} \underset{H_0}{\overset{H_1}{\gtrless}} \dfrac{p_0}{p_1}$
Bayes with weighted errors	• p_0 and p_1 • $p(x\|s_i)$ with $i = 0.1$ • $c = \begin{bmatrix} 0 & c_{01} \\ c_{10} & 0 \end{bmatrix}$	Average risk: $R = c_{01} p_0 p_f + c_{10} p_1 p_n$	$\Lambda(x) \underset{H_0}{\overset{H_1}{\gtrless}} \dfrac{p_0 c_{01}}{p_1 c_{10}}$
Bayes with weighted decisions	• p_1 and p_1 • $p(x\|s_i)$ with $i = 0.1$ • $c = \begin{bmatrix} c_{00} & c_{01} \\ c_{10} & c_{11} \end{bmatrix}$	Average risk: $R = \sum_{i=0} \sum_{j=0} c_{ij} p_i \int_{Oj} p(x\|s_i)\,dx$	$\Lambda(x) \underset{H_0}{\overset{H_1}{\gtrless}} \dfrac{p_0(c_{01} - c_{00})}{p_1(c_{10} - c_{11})}$
Minimax	• $p(x\|s_i)$ with $i = 0.1$ • $c = \begin{bmatrix} 0 & c_{01} \\ c_{10} & 0 \end{bmatrix}$	Maximum average risk: R_{minimax}	Threshold deduced from $c_{10} p_n = c_{01} p_f$
Neyman-Pearson	• $p(x\|s_i)$ with $i = 0.1$ • $p_f = \alpha_a$	Nondetection probability: p_n	$\Lambda(x) \underset{H_0}{\overset{H_1}{\gtrless}} \lambda(\alpha_a)$ with λ deduced from $p_f = \alpha_a$

13.3.15 Receiver operating characteristics

The application of decision criteria to concrete cases emphasizes an important parameter: the distance $d = d(x_{H_0}, x_{H_1})$ between the mean values of distributions $p(x|s_0)$ and $p(x|s_1)$. In example 13.3.9, this distance is simply equal to the signal amplitude divided by the noise rms value (standard deviation). What we call the *operating characteristics* of a receiver consist of the graphical representation of the detection probability p_d as a function of the false alarm probability p_f for constant values of the distance d. For a Gaussian noise, we get the characteristics presented in fig. 13.33.

It is also possible to represent the detection probability p_d as a function of the distance d for p_f constant. In the case of a Gaussian noise, representation with a Gaussian y-axis and a linear x-axis, leads to the straight-line pattern shown in fig. 13.34.

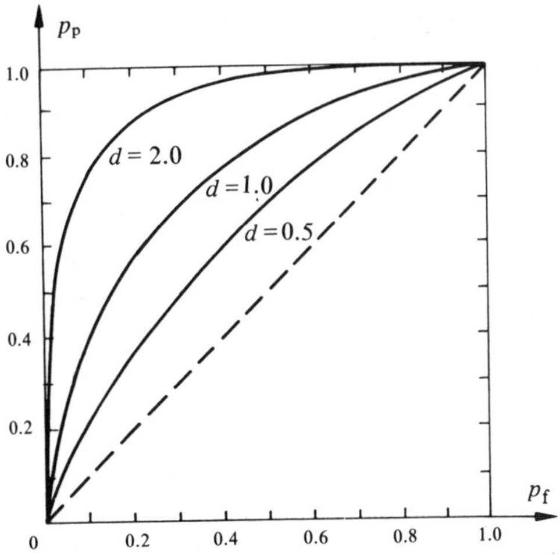

Fig. 13.33

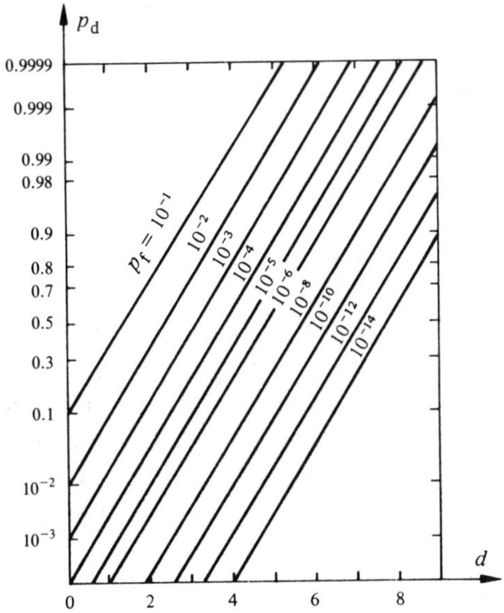

Fig. 13.34

13.4 DETECTION OF KNOWN SIGNALS

13.4.1 Optimal detection of a known signal in the presence of noise: matched filtering

A *matched filter* is a linear system (sub-sect. 8.2.27) optimizing the detection of a known signal $s(t)$ in the presence of an independent additive noise. ***It maximizes the signal-to-noise ratio at the decision instant.***

In the case of a ***white noise***, the matched-filter impulse response is simply deduced from the signal $s(t)$ by the relation:

$$g_a(t) = ks^*(t_0 - t) \tag{13.150}$$

where k is an arbitrary constant (depending on the filter gain) and t_0 is a delay parameter, usually corresponding to the signal duration T.

The output signal of the filter stimulated by the sum $x(t) = s(t) + n(t)$ then becomes

$$y(t) = x(t) * g_a(t) = k\mathring{\varphi}_s(t - t_0) + k\varphi_{sn}(t - t_0) \tag{13.151}$$

The matched filter itself behaves like a ***correlator*** *vis-à-vis* the signal to be detected. The signal-to-noise ratio is optimum at $t = t_0$ and depends only on the signal energy and the noise spectral density.

13.4.2 Demonstration

Consider the processing system shown in fig. 13.35.

Let $x(t) = s_1(t) + n_1(t)$ and $y(t) = s_2(t) + n_2(t)$. The useful input signal $s_1(t)$ is known. Its model can be real or complex (envelope). The input noise $n_1(t)$ is a stationary random process with zero mean value and with an assumed constant spectrum (white noise): $\Phi_{n1}(f) = \frac{1}{2}\eta$. Let $\xi(t_0)$ be the ratio of the signal instantaneous power at $t = t_0$ to the noise average power (variance) at the filter output:

$$\xi(t_0) = |s_2(t_0)|^2/P_{n2} \tag{13.152}$$

Of course, we will choose t_0 as the instant where $|s_2(t)|$ is maximum.

The power spectral density of the filter output noise, following (8.24), is given by

$$\Phi_{n2}(f) = \Phi_{n1}(f)|G(f)|^2 \tag{13.153}$$

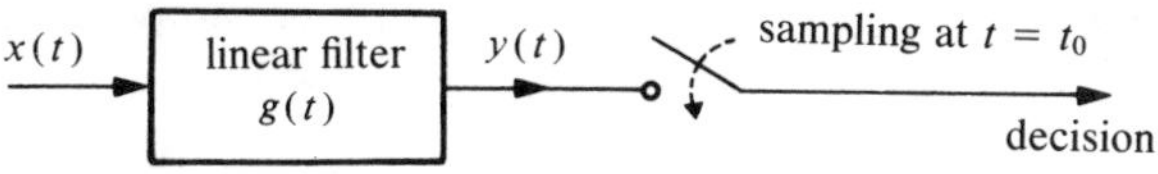

Fig. 13.35

where $G(f) = \mathrm{F}\{g(t)\}$ is the filter transfer function. Thus, the average power of the filter output noise is

$$P_{n2} = \int_{-\infty}^{\infty} \Phi_{n2}(f)\mathrm{d}f = \tfrac{1}{2}\eta \int_{-\infty}^{\infty} |G(f)|^2 \mathrm{d}f = \tfrac{1}{2}\eta\, \mathring{\varphi}_g(0) \tag{13.154}$$

where $\mathring{\varphi}_g(\tau)$ is the autocorrelation function of the impulse response $g(t)$.

The useful signal of the filter output is given by

$$s_2(t) = g(t) * s_1(t) = \int_{-\infty}^{\infty} G(f)S_1(f) \exp(\,\mathrm{j}2\pi ft)\mathrm{d}f \tag{13.155}$$

We finally get

$$\xi(t_0) = \left| \int_{-\infty}^{\infty} G(f)S_1(f) \exp(\,\mathrm{j}2\pi ft_0)\mathrm{d}f \right|^2 \Big/ \left(\tfrac{1}{2}\eta \int_{-\infty}^{\infty} |G(f)|^2 \mathrm{d}f\right) \tag{13.156}$$

The optimal filter that maximizes $\xi(t_0)$ is easily obtained using the Schwarz inequality (3.21), yielding here

$$\left| \int_{-\infty}^{\infty} G(f)S_1(f) \exp(\,\mathrm{j}2\pi ft_0)\mathrm{d}f \right|^2 \leqslant \int_{-\infty}^{\infty} |G(f)|^2 \mathrm{d}f \cdot \int_{-\infty}^{\infty} |S_1(f)|^2 \mathrm{d}f \tag{13.157}$$

Equality is satisfied, following (3.22), only if

$$G(f) = G_a(f) = kS_1^*(f) \exp(-\mathrm{j}2\pi ft_0) \tag{13.158}$$

where k is an arbitrary complex constant.

Thus, we get the inequality:

$$\xi(t_0) \leqslant (\tfrac{1}{2}\eta)^{-1} \int_{-\infty}^{\infty} |S_1(f)|^2 \mathrm{d}f = 2W_s/\eta \tag{13.159}$$

for $|S_1(f) \exp(\,\mathrm{j}2\pi ft)|^2 \equiv |S_1(f)|^2$ and $W_s = \int |S_1(f)|^2 \mathrm{d}f$ is the incoming signal energy.

The maximum value of $\xi(t_0)$ is hence reached when the filter is defined by (13.158), or, in other words, while taking (4.54) into account, when its impulse response is given by

$$g_a(t) = \mathrm{F}^{-1}\{G_a(f)\} = k \int_{-\infty}^{\infty} S_1^*(f) \exp[\mathrm{j}2\pi f(t - t_0)]\mathrm{d}f$$

$$= ks_1^*(t_0 - t) \tag{13.160}$$

When $s_1(t)$ is a real function: $s_1^*(t) = s_1(t)$, and we simply have

$$g_a(t) = ks_1(t_0 - t) \tag{13.161}$$

In order to implement this filter, it must be causal, i.e., $g(t) = 0$ for $t < 0$. Thus, if $s_1(t)$ is a pulse of duration T, we must have $t_0 \geqslant T$ to comply with the causality condition. In practice, the general rule is $t_0 = T$.

The useful output signal is given by convolution:

$$s_2(t) = s_1(t) * g_a(t) = ks_1(t) * s_1^*(t_0 - t) = k\mathring{\varphi}_{s1}(t - t_0) \tag{13.162}$$

and can be identified, from (4.51), with a t_0-delayed version of the autocorrelation function of the incoming signal $s_1(t)$. Its peak value occurs, according to (4.49), at $t = t_0$. ***The matched filter is hence equivalent to a crosscorrelator comparing the noisy input signal*** $x(t)$ ***to an image of the signal to be detected*** $s_1(t)$.

13.4.3 Instantaneous signal-to-noise ratio and time resolution

As emphasized by the second member of inequality (13.159), the optimal signal-to-noise ratio at instant $t = t_0$ ***depends only on the energy of the useful input signal*** and on the white noise power spectral density:

$$\xi_{opt}(t_0) = 2W_s/\eta \tag{13.163}$$

In the presence of the same noise, different filters matched to different signals having the same energy will exhibit equivalent behavior in terms of signal-to-noise ratio at the decision instant.

This is no longer true for the time resolution that expresses the filter's capability of discriminating between two close pulses.

Without matched filtering, this resolution is almost equal to the incoming pulse duration T. This duration cannot be reduced without simultaneously reducing the energy W_s, for the average power W_s/T delivered by the signal emitter is naturally limited in practice.

After a matched filtering, the resolution is determined by the signal ***correlation duration*** D_τ, as suggested by (13.162). We know, from (7.144), that this duration is inversely proportional to the approximate bandwidth B_τ. The matched filtering, thus, offers the possibility of a substantial resolution gain, provided that it processes wideband signals (narrow autocorrelation function); hence, with a ***high BT product.***

The most frequently used signals of this kind are the frequency modulated sine-wave pulse, best known as *chirp* signal [27, 90], and the pseudorandom sequences (sect. 5.10). Other sequences with good correlation properties are described in [142].

Figure 13.36 illustrates the instantaneous signal-to-noise ratio and time-resolution performances obtained by matched filtering of two signals with the same energy W_s and same duration T embedded in an independent white noise: a simple rectangular pulse and a chirp signal.

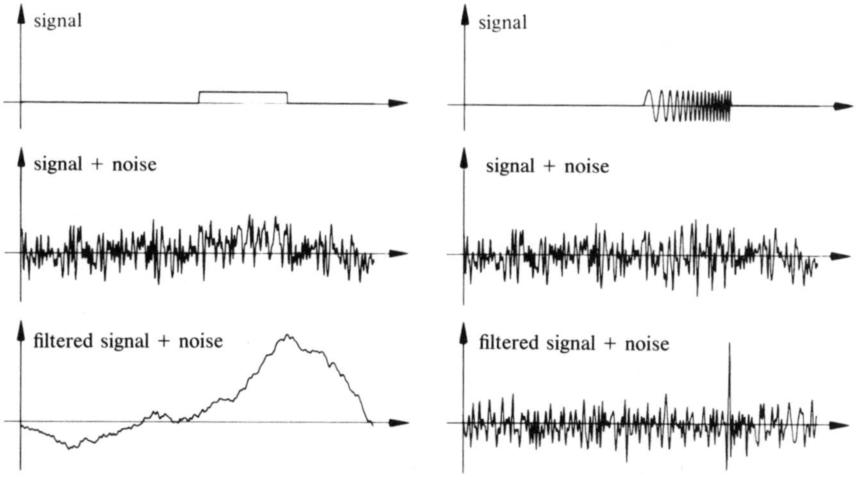

Fig. 13.36

13.4.4 Matched filtering in presence of a colored noise

If the noise is not white, the matched filter transfer function $G_a(f)$ can be obtained by cascading a first filter of transfer function $G_1(f)$ that transforms the colored noise into a white noise, with a second filter of transfer function $G_2(f)$ matched to the useful output signal of the first one (fig. 13.37).

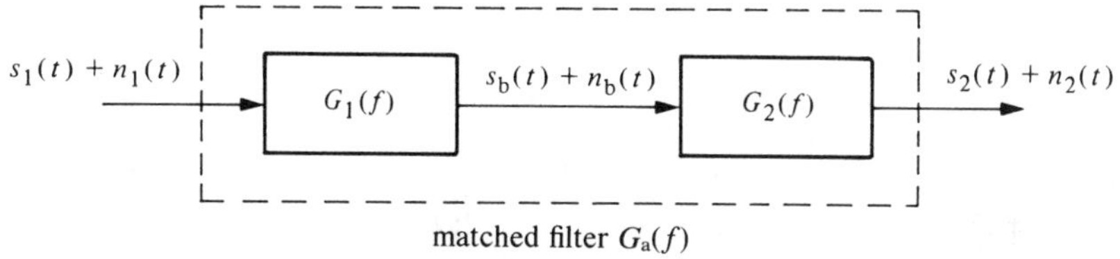

Fig. 13.37

With $\Phi_{n1}(f)$, the power spectral density of the incoming colored noise $n_1(t)$ and with $\Phi_{nb}(f) = \frac{1}{2}\eta$, the power spectral density of the white noise at the first filter output, its transfer function, according to (8.24), must satisfy the relation:

$$|G_1(f)|^2 = \tfrac{1}{2}\eta/\Phi_{n1}(f) \tag{13.164}$$

Following (13.158), the second filter has a transfer function:

$$G_2(f) = kS_b^* \exp(-j2\pi f t_0) \tag{13.165}$$

where $S_b(f)$ is the Fourier transform of the useful output signal of the first filter: $s_b(t) = s_1(t) * g_1(t)$. Thus, from (8.14), we have

$$S_b(f) = S_1(f)G_1(f) \tag{13.166}$$

The matched filter total transfer function, thus, according to (8.33) is

$$\begin{aligned} G_a(f) &= G_1(f)G_2(f) = k|G_1(f)|^2 S_1^*(f) \exp(-j2\pi f t_0) \\ &= k' S_1^*(f) \exp(-j2\pi f t_0)/\Phi_{n1}(f) \end{aligned} \tag{13.167}$$

with $k' = \tfrac{1}{2}\eta k$.

13.4.5 Suboptimal filters

The optimal matched filter defined by (13.150) or (13.167) is not always easily implementable. Simpler suboptimal filters can be used at the expense of a slight reduction of the instantaneous signal-to-noise ratio (13.152).

Resolution performance depends mainly on the match of the filter's phase response to the signal phase spectrum [143].

The relative efficiency of the suboptimal filter is given by the quotient of its instantaneous signal-to-noise ratio to that of the optimal filter. In the case of white noise, according to (13.152), (13.154), and (13.163), we have

$$\xi(t_0)/\xi_{opt} = |s_2(t_0)|^2 W_s^{-1} \Big/ \int_{-\infty}^{\infty} |G(f)|^2 \mathrm{d}f = |s_2(t_0)|^2/[W_s \mathring{\varphi}_g(0)] \tag{13.168}$$

with t_0 chosen so that $s_2(t_0)$ is maximum.

13.4.6 Example

Let $s_1(t) = A \operatorname{rect}[(t - \tfrac{1}{2}T)/T]$ with $W_s = A^2T$. By (13.150), the impulse response of the optimal matched filter is, with $k = 1$: $g_a(t) = A \operatorname{rect}[(t_0 - t - \tfrac{1}{2}T)/T] = A \operatorname{rect}[(t - \tfrac{1}{2}T)/T]$ thanks to the $s_1(t)$ symmetry and choosing $t_0 = T$. The optimal matched filter is hence an integration operator during time T (sub-sect. 8.2.19).

We know that an approximate implementation of an integrator is a first-order lowpass RC filter (sub-sect. 8.2.24) with transfer function $G(f) = [1 + j(f/f_c)]^{-1}$, where $f_c^{-1} = 2\pi RC$, and impulse response $g(t) = (RC)^{-1} \exp[-t/RC] \cdot \epsilon(t)$. The response $s_2(t)$ of the filter to the stimulation $s_1(t)$ for

$0 \leqslant t \leqslant T$ is the unit-step response $s_2(t) = A\{1 - \exp[-t/(RC)] \cdot \epsilon(t)\}$ that is maximum for $t_0 = T$. The value at the origin is $\mathring{\varphi}_g(0) = (2RC)^{-1}$. By insertion into (13.168), the relative efficiency becomes

$$\xi(t_0 = T)/\xi_{\text{opt}} = 2RC\{1 - \exp[-T/(RC)]\}^2/T \tag{13.169}$$

that is maximum for $T/(RC) = 1.25$ and then equals 0.816. The corresponding relative reduction of the signal-to-noise ratio is 0.88 dB.

In the case of a sine wave pulse with rectangular envelope, an identical result is obtained for an *RLC* resonant network by substituting for the low-pass filter time constant *RC* in (13.169) by $(\pi B)^{-1}$, where *B* is the −3 dB bandwidth of the resonant network.

Other approximations are possible (problem 13.5.24).

13.4.7 Applications: correlation or matched-filter detector

The matched-filter theory shows that an optimal technique for identification of known signals in the presence of independent noise is the correlation. It can be implemented either by means of correlators (fig. 13.38)—even simplified ones (sub-sect. 13.2.4)—or by means of matched filters as shown in fig. 13.39.

13.4.8 Introduction to pattern recognition

Pattern recognition is a decision process. Its purpose is the identification, despite a usually imperfect observation, of a uni-, bi-, or tridimensional signal (acoustic, seismic, biological phenomena, written or printed characters, finger prints, industrial objects, *et cetera*). Its potential applications in robotics,

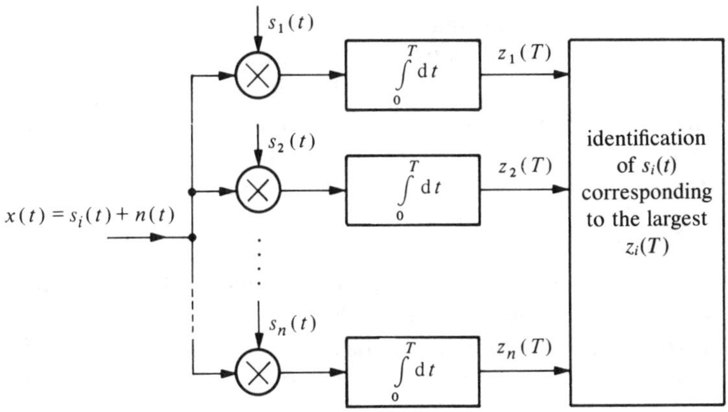

Fig. 13.38

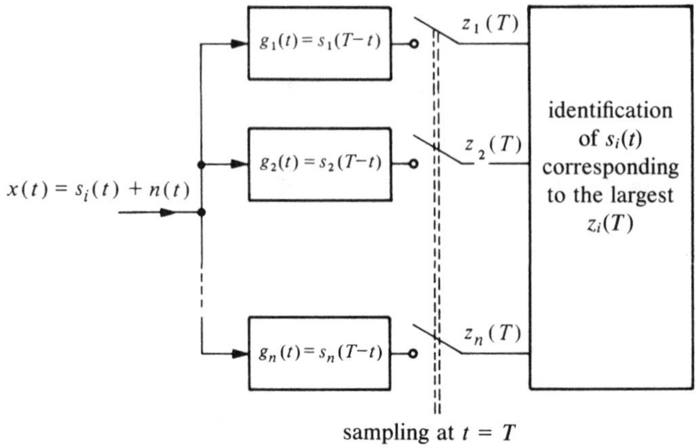

Fig. 13.39

automatic inspection machines, or patient monitoring systems are very important. It is based on the theoretical principles of estimation, comparison, and decision described in this chapter.

Pattern recognition methods can be subdivided into three categories [144–148]:

- *template matching methods* in which the pattern to be identified is compared with a set of templates that are the prototypes of each known class;
- *vector space methods* (often referred to as *decision-theoretic approaches*) in which each pattern corresponds to a point in the observation space;
- *syntactic methods* based on a structural description of the patterns.

The template matching methods are the easiest to implement. They encompass the matched filtering described in this section and can be considered as a special case of the vector space or syntactic methods. A measurement of similarity (e.g., correlation) between the pattern to be identified and each template is computed. The criterion for deciding whether it belongs to a given class is that of greatest similarity. The principle of a correlative pattern recognition system is similar to the ones of figs. 13.38 and 13.39.

In vector space methods (fig. 13.40), each observed pattern is represented in the observation space (sub-sect. 13.3.1) by a *description vector* $\boldsymbol{x} = (x_1, x_2, \ldots, x_n)$. Each component x_k is a measured particular feature (e.g., variance, Fourier coefficients, area, perimeter, momentum, *et cetera*). Thus, to each pattern class, there corresponds a more or less compact cluster of possible results. The identification is based on a test to determine if a vector $\boldsymbol{x}$ belongs to a given cluster (segmentation of the observation space). There are many decision criteria. A frequent approach is the use of *discriminant functions* defining the class boundaries. To M classes $\Omega_1, \ldots, \Omega_M$ of patterns

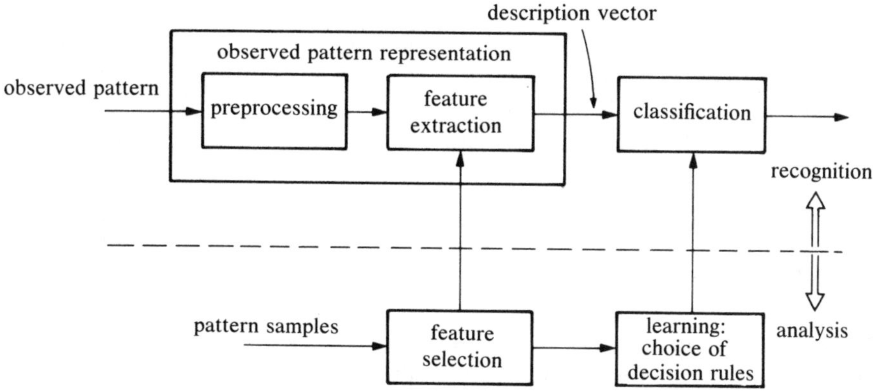

Fig. 13.40 Block diagram of a vector space pattern recognition system.

to be identified there are M associated functions $\Delta_1(\boldsymbol{x}), \ldots, \Delta_M(\boldsymbol{x})$, such that, for $i \neq j$, $\Delta_i(\boldsymbol{x}_*) > \Delta_j(\boldsymbol{x}_*)$ for the larger number of patterns $\boldsymbol{x}_*$ of a given class Ω_i, in order to minimize the wrong classification risk. This problem can be analytically tackled if the measurement error statistics are known (a simple example is the Bayesian criterion in binary decision described in sub-sect. 13.3.5). In practice, the decision rule optimization is made during an initial learning phase performed on an appropriately selected population of representative patterns. The search for an optimal description mode of a set of patterns to be identified, so as to allow us to regroup them in clusters as far apart as possible, is known as the *clustering* problem.

The syntactic approach (fig. 13.41) is based on language theory. A pattern is described as a specially structured combination of *primitives*: identifiable

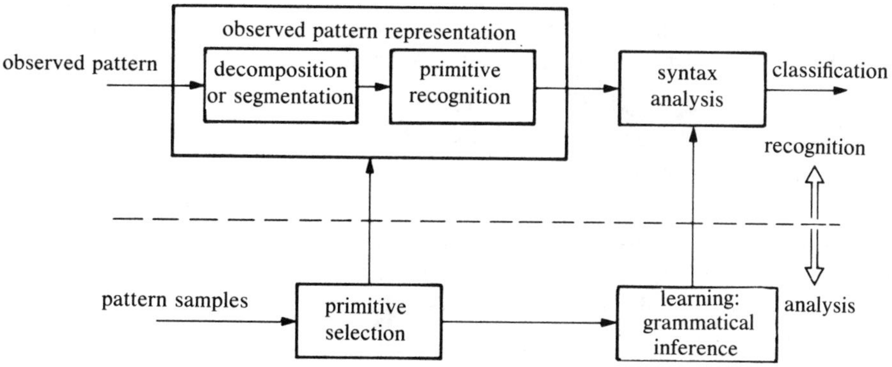

Fig. 13.41 Block diagram of a syntactic pattern recognition system.

pattern elements, the concatenation rules of which define a grammar. Recognition implies an initial identification of the primitives followed by a syntactic analysis. A tree diagram of the primitive concatenation rules is often used. Classification can then be achieved through comparing the path followed in the tree by the observed pattern with typical paths defined during an initial learning phase.

13.5 PROBLEMS

13.5.1 Let a signal $x(t)$ be linearly drifting with time: $x(t) = at + b + n(t)$, where $n(t)$ is a zero mean value Gaussian noise with variance σ_n^2. Find the equations of parameters $\tilde{a}$ and $\tilde{b}$ estimated with the maximum likelihood method by means of a set of m measurements (samples) x_1 to x_m, if the noise samples are independent.

13.5.2 Find the equations of the maximum likelihood estimates of the variance of a Gaussian process $\boldsymbol{x}(t)$ with zero mean value:

- if N independent samples x_1 to x_N are available;
- if a continuous observation $x(t)$ of duration T is available.

13.5.3 Demonstrate relation (13.22) and find the corresponding estimations $\tilde{a}_{ms}$, $\tilde{a}_{map}$, and $\tilde{a}_{abs}$.

13.5.4 Let $x(t) = s(t - a) + n(t)$, where $s(t)$ is a known signal and $n(t)$ is a Gaussian noise. A continuous observation of $x(t)$ with a duration $T \gg a$ is available. Find and interpret the integral equation, the solution of which corresponds to the maximum likelihood estimation $\tilde{a}_{ml}$ of delay a.

13.5.5 Evaluate the mean square error made in the case of the linear prediction of the example 13.1.20.

13.5.6 Consider $z_i = a + n_i$, $i = 1, \ldots, N$ with $\mathrm{E}[\mathbf{n}_i] = 0$, $\mathrm{E}[\mathbf{n}_i\mathbf{n}_j] = 0$ and $\mathrm{E}[\mathbf{n}_i^2] = \sigma_n^2$. Show that the optimal linear estimation $\tilde{a}_{\ell o}$ of the parameter a does not correspond to the z_i arithmetic mean and is biased.

13.5.7 Consider a signal $x(t)$, the values of which at instants $t = 0$ and $t = T$ are known. Find the equation of the optimal linear estimation of $x(t)$ for $0 < t < T$ (linear interpolation).

13.5.8 Let $x(t) = s(t) + n(t)$, where $n(t)$ is a noise independent of the signal $s(t)$. Show that the causal Wiener filter has a transfer function $G_0(f) = 1 - [\Phi_n(f)/\Psi^*(f)]_+/\Psi(f)$, where $|\Psi(f)|^2 = \Phi_x(f)$. Show that if $\Phi_n(f) = \frac{1}{2}\eta$ (white noise) and if the signal power is finite, $G_0(f) = 1 - \sqrt{\frac{1}{2}\eta}/\Psi(f)$.

13.5.9 Let $x(t) = s(t) + n(t)$ with $\Phi_s(f) = \alpha/[\beta^2 + (2\pi f)^2]$ and $\Phi_n(f) = \frac{1}{2}\eta$. Under the assumption that $n(t)$ is independent of $s(t)$, demonstrate that the causal Wiener filter estimating $s(t)$ is a simple first-order lowpass filter.

13.5.10 Find the expression of the variance of the mean value estimation of the binary signal with random transition (sub-sect. 5.8.4) performed by a time averager integrating during a duration T.

13.5.11 Consider a signal $x(t)$, the power spectral density of which is $\Phi_x(f) = \mu_x^2\delta(f) + B^{-1}\operatorname{sinc}^2(f/B)$. This signal is integrated during a time T in order to estimate its mean value μ_x. Find the variance of the estimate $\bar{x}$ in function of the BT product and evaluate the minimum integration time needed to get a measurement with a signal-to-noise ratio better than 30 dB, when the signal parameter B is 1 kHz and the signal-to-noise ratio before integration is 0 dB.

13.5.12 Consider the same signal $x(t)$ as in problem 13.5.11 for which we want to estimate the mean value through discrete periodic samples. Find the minimum total sampling duration T allowing an improvement of the signal-to-noise ratio by the same amount as in 13.5.11:

- when the sampling period is such that the samples are uncorrelated (zero covariance);
- when the sampling period is equal to $1/(2B)$.

13.5.13 Control the results of sub-section 13.1.25.

13.5.14 Show that an averager with periodic sampling is similar to a comb-filter (periodic transfer function).

13.5.15 Compare the ideal theoretical result (13.110) with the one obtained by correlating the observed signal $x(t)$ with a sequence of real pulses with shape $g(t)$ and the same period T as the signal to be recovered.

13.5.16 A very weak light flux with constant intensity must be measured by means of a photomultiplier tube followed by a low-current amplifier (electrometer). The output signal of the amplifier has a mean value $U_0 = 10$ mV proportional to the incident light flux to which is added the amplifier background noise $n(t)$. The power spectral density of this noise is: $\Phi_n(f) = \frac{1}{2}\eta/(1 + f^2/f_c^2)$ with $f_c = 10$ Hz and $\eta = 1\ \mathrm{V^2/Hz}$.

Find the signal-to-noise ratio (ratio of the signal power U_0 to the background noise power), if the output voltage of the amplifier is filtered by an ideal lowpass filter totally eliminating the frequency components higher than 0.1 Hz.

Find the signal-to-noise ratio obtained by periodically interrupting the light flux with a periodic shutter implemented with a four-blade chopper turning at 250 rps (we assume that the shutter open and closed times are equal) and by connecting the amplifiers in cascade with a synchronous detector followed by the same ideal lowpass filter. The synchronous detector must be considered as a circuit performing the multiplication of the amplifier output voltage with a reference square wave, isochronous with the shutter, and having amplitude ± 1.

13.5.17 Find the likelihood ratio corresponding to the envelope detection of a noisy signal $x(t) = s_i(t) + n(t)$, where $s_i(t) = A_i \cos(\omega_0 t)$ with $A_0 = 0$ and $A_1 = A$, in the case where the observation is made of N independent samples and if $p_0 = \text{Prob}(s_0) = \frac{1}{2}$. Find the block diagram of the optimal detector.

13.5.18 A detection system (presence or absence of useful signal) is disturbed by an additive Gaussian noise with zero mean value and variance σ_n^2. Find the value of the decision threshold setting the false alarm probability at $p_f = \alpha_a = 0.05$.

13.5.19 Consider a random binary source producing the signals $s_0 = 0$ with a probability p_0 and $s_1 = A$ with a probability p_1. The transmission channel is disturbed by a zero mean value noise $n(t)$ with a probability distribution $p_n(n) = \exp(-2|n|)$. Find the decision threshold as a function of parameters A, p_0, and p_1, that minimizes the total error probability (or average risk). Application: $p_0 = 0.8$; $p_1 = 0.2$; $A = 3$ V.

13.5.20 Consider a binary source producing the signals $s_0 = 0$ and $s_1 = +A$. Moreover, the receiver noise statistics varies with the presence or absence of signal: $p_{n0}(n|s_0) = \exp(-n)\epsilon(n)$, $p_{n1}(n|s_1) = (2\pi\sigma_n^2)^{-1/2}\exp(-\frac{1}{2}n^2/\sigma_{n1}^2)$. Assuming that the cost coefficients associated with the wrong decision cases are equal to one (and with $c_{00} = c_{11} = 0$), find the decision threshold that minimizes the maximum error probability. Application: $\sigma_{n1} \simeq 11.1$ V and $A = 6\sigma_{n1}$.

13.5.21 Let $s(t) = A\cos(2\pi f_0 t)$ be a radar signal detected in presence of a Gaussian white noise $n(t)$ with power spectral density $\Phi_n(f) = \eta/2$. In the receiver, this signal goes first through an ideal bandpass filter with bandwidth B centered on f_0, then through an envelope linear detector. Assigning $p_f = \alpha_a$ to the maximum false alarm probability, find the optimal threshold at the detector output, according to the Neyman-Pearson criterion, and find the detection probability using Marcum's Q-function table given in section 15.9. Application: $B = 400$ Hz $\gg f_0$, $\eta = 0.01$ V^2/Hz; $A = 16$ V; $p_f = \alpha_a = 10^{-9}$.

13.5.22 Consider a source generating equally likely levels $+V$ and $-V$ (antipolar signal) to which a zero mean value Gaussian noise with variance σ_n^2 is added. What should be, in function of σ_n^2, the power of the emitted signal to get at the receiver a total error probability of 10^{-4}, assuming the two wrong decision cases are equally likely. Use the operating characteristics of the receiver (sub-sect. 13.3.15).

13.5.23 Find the autocorrelation functions of the binary signals with finite energy $W_s = A^2T$ of fig. 13.42 and compare their time resolution performances.

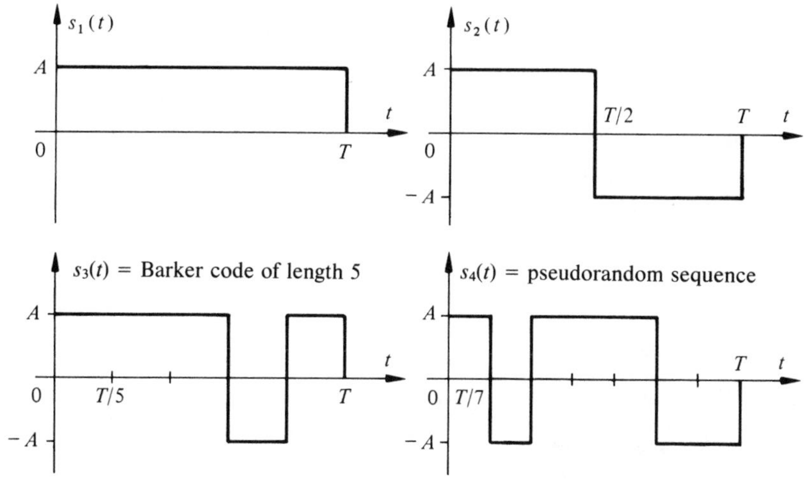

Fig. 13.42

13.5.24 The transfer function of a filter composed of RC filters in cascade can be given by the approximate expression $G(f) = \exp[-\frac{1}{2}(2\pi f)^2/\sigma^2]$ $\exp(-j2\pi f \tau_0)$. This filter is called a Gaussian filter. What is the relative efficiency of such a filter for the detection of a rectangular pulse with amplitude A and duration T? What should be the value of the optimal sampling instant t_0?

Chapter 14

Review of Probability Theory

14.1 BASIC DEFINITIONS

14.1.1 Probability concept

A *statistical experiment* consists of observing events, the outcomes of which are governed by random laws. The set Z of all the possible outcomes ζ comprises the *sample space*.

An *event* A is any subset of Z in which we have a particular interest. If in N trials the event A occurs M times, the M/N ratio measures the *relative frequency* of the outcome of A.

The *probability* of the occurrence of A is defined as the limit for $N \to \infty$ of the relative frequency of A:

$$\mathrm{Prob(A)} = \lim_{N\to\infty} M/N \quad \text{with } 0 \leq \mathrm{Prob(A)} \leq 1 \tag{14.1}$$

14.1.2 Mutually exclusive events

If two events A and B are mutually exclusive, the probability of the occurrence of event C = A ∪ B (A ***or*** B in set theory) is given by

$$\mathrm{Prob(C)} = \mathrm{Prob(A \cup B)} = \mathrm{Prob(A)} + \mathrm{Prob(B)} \tag{14.2}$$

If the fundamental set Z is subdivided into n mutually exclusive subsets $A_i(i = 1, \ldots, n)$ such that $A_jA_k = \varnothing$ for j ≠ k, where ∅ is the empty set, we have

$$\sum_{i=1}^{n} \mathrm{Prob}(\mathrm{A}_i) = \mathrm{Prob(Z)} = 1 \tag{14.3}$$

14.1.3 Joint probability

If D = A ∪ B with A ∩ B ≠ ∅ (non-disjointed sets), we have

$$\mathrm{Prob(D)} = \mathrm{Prob(A \cup B)} = \mathrm{Prob(A)} + \mathrm{Prob(B)} - \mathrm{Prob(A,B)} \tag{14.4}$$

The term Prob(A,B) is the *joint probability* measuring the chance of occurrence of the event $A \cap B$ (A ***and*** B in set theory). If $A \cap B = \emptyset$, Prob(A,B) = 0.

14.1.4 Conditional probabilities

Consider two events A and B that are not mutually exclusive. During an experiment repeated N times, the number of occurrences of events A,B, and $A \cap B$ is given by $M(\mathrm{A})$, $M(\mathrm{B})$, and $M(\mathrm{A,B})$, respectively.

The ratio $M(\mathrm{A,B})/M(\mathrm{B})$ expresses the relative frequency of the occurrence of A ***when*** B ***is realized.***

Dividing the numerator and the denominator by N and taking the limit, for $N \to \infty$, we get the *conditional probability of* A *knowing that* B *has occurred*:

$$\mathrm{Prob}(\mathrm{A}|\mathrm{B}) = \mathrm{Prob}(\mathrm{A,B})/\mathrm{Prob}(\mathrm{B}) \tag{14.5}$$

The joint probability of two events A and B can thus be expressed by the product of simple and conditional probabilities:

$$\mathrm{Prob}(\mathrm{A,B}) = \mathrm{Prob}(\mathrm{A}|\mathrm{B})\,\mathrm{Prob}(\mathrm{B}) = \mathrm{Prob}(\mathrm{B}|\mathrm{A})\,\mathrm{Prob}(\mathrm{A}) \tag{14.6}$$

By extension, the joint probability associated with n events is given by the following general relation (multiplication rule):

$$\mathrm{Prob}(\mathrm{A}_1, \mathrm{A}_2, \ldots, \mathrm{A}_n) = \mathrm{Prob}(\mathrm{A}_1)\,\mathrm{Prob}(\mathrm{A}_2|\mathrm{A}_1)\,\mathrm{Prob}(\mathrm{A}_3|\mathrm{A}_1\mathrm{A}_2) \ldots \mathrm{Prob}(\mathrm{A}_n|\mathrm{A}_1\mathrm{A}_2 \ldots \mathrm{A}_{n-1}) \tag{14.7}$$

14.1.5 Independent events

Two events A and B are statistically independent if:

$$\mathrm{Prob}(\mathrm{A}|\mathrm{B}) = \mathrm{Prob}(\mathrm{A});\ \mathrm{Prob}(\mathrm{B}|\mathrm{A}) = \mathrm{Prob}(\mathrm{B}) \tag{14.8}$$

or, in other words, if:

$$\mathrm{Prob}(\mathrm{A},\ \mathrm{B}) = \mathrm{Prob}(\mathrm{A})\,\mathrm{Prob}(\mathrm{B}) \tag{14.9}$$

14.2 RANDOM VARIABLES

14.2.1 Definitions

A *random variable* is a real quantity, the value of which depends on chance. This dependence is expressed by a probability law, usually called *distribution*.

The distribution of a random variable **x** can be defined either by its *cumulative distribution* $F(x)$, by its *probability density* $p(x)$, or by its *characteristic function* $\Pi_x(u)$.

14.2.2 Cumulative distribution

The *cumulative distribution* expresses the probability that the variable **x** is less than or equal to a given value x:

$$F(x) = \text{Prob}(\mathbf{x} \leqslant x) \tag{14.10}$$

The cumulative distribution is a non-decreasing function of x with the extremal values:

$$F(-\infty) = 0 \tag{14.11}$$

and

$$F(+\infty) = 1 \tag{14.12}$$

14.2.3 Probability density

The *probability density* is, by definition, the derivative of the cumulative distribution:

$$p(x) = \mathrm{d}F(x)/\mathrm{d}x \tag{14.13}$$

From (14.10) and (14.12), we deduce that $p(x) \geqslant 0$ for any x, and that

$$\int_{-\infty}^{\infty} p(x)\mathrm{d}x = F(\infty) = 1 \tag{14.14}$$

We go, therefore, from the probability density to a probability value by integration. The probability to observe the variable **x** between two limits a and b can thus be expressed in the equivalent forms:

$$\text{Prob}(a \leqslant \mathbf{x} \leqslant b) = F(b) - F(a) = \int_a^b p(x)\mathrm{d}x \tag{14.15}$$

14.2.4 Discrete random variable (DRV)

A random variable is said to be ***discrete*** when it can only take a finite, or countable, number of distinct values.

If we denote the different distinct values taken by variable **x** as x_i and the corresponding probabilities as $\text{Prob}(x_i)$, the cumulative distribution of a *DRV* can be expressed by

$$F(x) = \sum_{x_i \leqslant x} \text{Prob}(x_i)\, \epsilon_-(x - x_i) \tag{14.16}$$

where

$$\epsilon_-(x) = \begin{cases} 0 \text{ for } x < 0 \\ 1 \text{ for } x \geqslant 0 \end{cases} \tag{14.17}$$

is the unit-step defined in sub-section 1.3.3, but with unity value at the origin.

The probability density of a *DRV* is then expressed by

$$p(x) = \sum_i \text{Prob}(x_i)\, \delta_-(x - x_i) \tag{14.18}$$

where $\delta_-(x) = \text{d}\epsilon_-(x)/\text{d}x$ is a delta-function, such as the integral:

$$\int_{-\infty}^{0} \delta_-(x)\text{d}x = 1 \tag{14.19}$$

and not $\frac{1}{2}$ as usually allowed.

An example of cumulative distribution and of probability density of a *DRV* is represented in fig. 14.1.

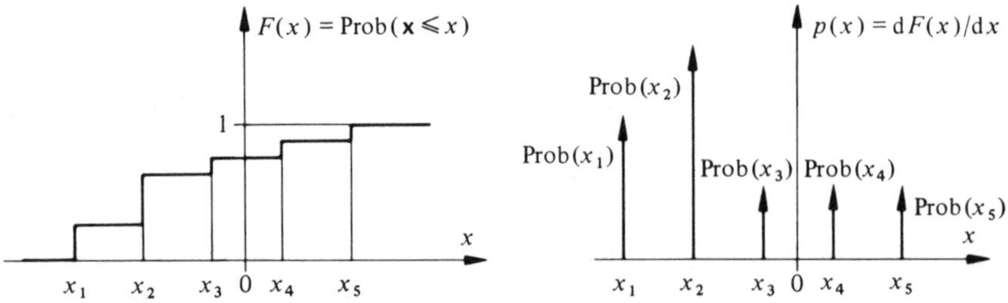

Fig. 14.1

14.2.5 Continuous random variable (CRV)

A random variable is said to be ***continuous*** when it can take any value on a given interval.

An example of a cumulative distribution and a probability distribution of a *CRV* is represented in fig. 14.2.

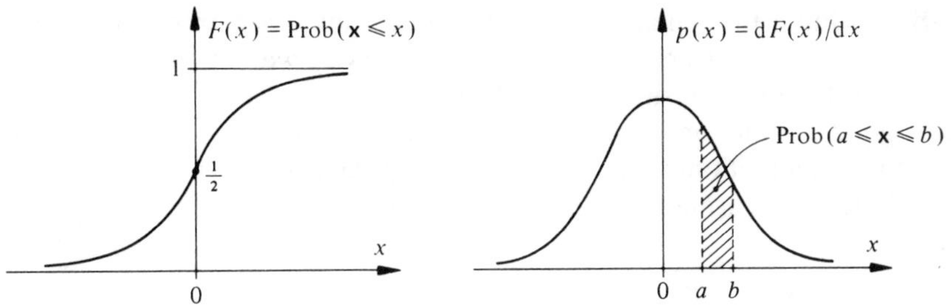

Fig. 14.2

14.2.6 Estimation of a probability distribution: histogram

The distribution of a random variable cannot be determined experimentally. We can, nevertheless, measure the relative frequency $M(x_i, \Delta x)/N$ of the occurence of variable $\mathbf{x}$ in an amplitude interval $[x_i - \frac{1}{2}\Delta x, x_i + \frac{1}{2}\Delta x]$, of width Δx, centered on a particular value x_i. $M(x_i, \Delta x) = M(x_i - \frac{1}{2}\Delta x < \mathbf{x} < x_i + \frac{1}{2}\Delta x)$ denotes the number of favorable events counted on a total of N trials.

The probability of finding $\mathbf{x}$ in this interval is, by (14.1), the value reached when $N \rightarrow \infty$:

$$\text{Prob}(x_i - \Delta x/2 < \mathbf{x} < x_i + \Delta x/2) = \lim_{N\rightarrow\infty} \frac{M(x_i,\Delta x)}{N} \tag{14.20}$$

A local estimation (fig. 14.3) of the probability density is obtained by dividing the measured relative frequency by the width of the interval used:

$$\tilde{p}(x_i,\Delta x) = \frac{M(x_i,\Delta x)}{N \cdot \Delta x} \tag{14.21}$$

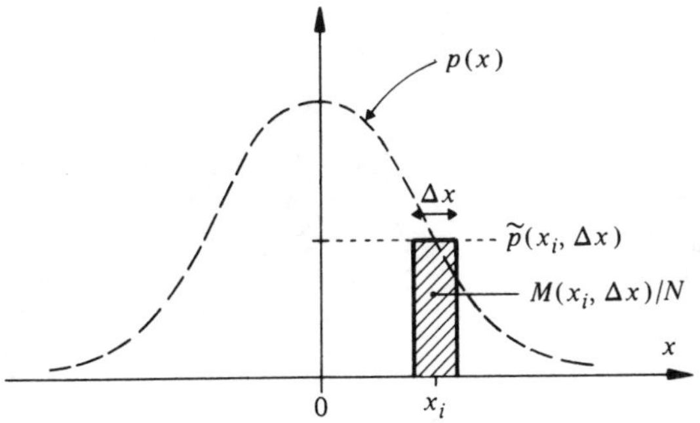

Fig. 14.3

Subdividing the whole range of possible values of **x** into a set of adjacent intervals and measuring (14.21) for each interval, we obtain a total estimation of the probability density called a *histogram*.

A good estimation [65] needs both a large number of trials N (to reduce statistical fluctuations) and a small interval Δx (to limit the estimation error, called bias, appearing in the absence of a local symmetry of $p(x)$ with respect to x_i).

14.2.7 Bidimensional random variables

Let **x** and **y** be two random variables. The pair (**x**,**y**) is a *bidimensional random variable*, the distribution of which can be defined either by its *joint cumulative distribution*:

$$F(x,y) = \text{Prob}(\mathbf{x} \leqslant x \text{ and } \mathbf{y} \leqslant y) \tag{14.22}$$

or by its *joint probability density* (fig. 14.4):

$$p(x,y) = \frac{\partial^2 F(x,y)}{\partial x \, \partial y} \tag{14.23}$$

with

$$F_{xy}(\infty,\infty) = \int_{-\infty}^{\infty} \int_{-\infty}^{\infty} p(x,y) \, \mathrm{d}x \, \mathrm{d}y = 1 \tag{14.24}$$

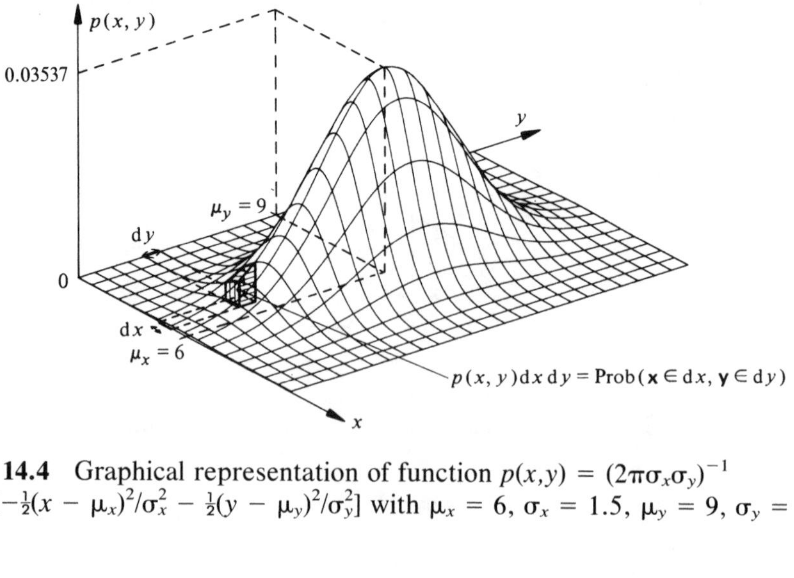

Fig. 14.4 Graphical representation of function $p(x,y) = (2\pi\sigma_x\sigma_y)^{-1} \exp[-\frac{1}{2}(x - \mu_x)^2/\sigma_x^2 - \frac{1}{2}(y - \mu_y)^2/\sigma_y^2]$ with $\mu_x = 6$, $\sigma_x = 1.5$, $\mu_y = 9$, $\sigma_y = 3$.

If **x** and **y** are discrete variables, they can take the values $x_1, x_2, \ldots, x_m$ and $y_1, y_2, \ldots, y_n$, respectively. The bidimensional variable can thus take $m \cdot n$ pairs of values $(x_1, y_1), \ldots, (x_m, y_n)$. It is convenient to represent the set of events (x_i, y_j) and the corresponding joint probabilities $\text{Prob}(x_i, y_j)$ as two $m \times n$ matrices:

$$[\mathbf{x},\mathbf{y}] = \begin{bmatrix} (x_1,y_1) & (x_1,y_2) & \ldots & (x_1,y_n) \\ \vdots & & & \vdots \\ (x_m,y_1) & (x_m,y_2) & \ldots & (x_m,y_n) \end{bmatrix} \tag{14.25}$$

$$[\text{Prob}(\mathbf{x},\mathbf{y})] = \begin{bmatrix} \text{Prob}(x_1,y_1) & \ldots & \text{Prob}(x_1,y_n) \\ \vdots & & \vdots \\ \text{Prob}(x_m,y_1) & \ldots & \text{Prob}(x_m,y_n) \end{bmatrix} \tag{14.26}$$

The joint probability density here becomes an array of delta-functions (fig. 14.5):

$$p(x,y) = \sum_i \sum_i \text{Prob}(x_i,y_j)\, \delta(x - x_i\, ,\, y - y_j) \tag{14.27}$$

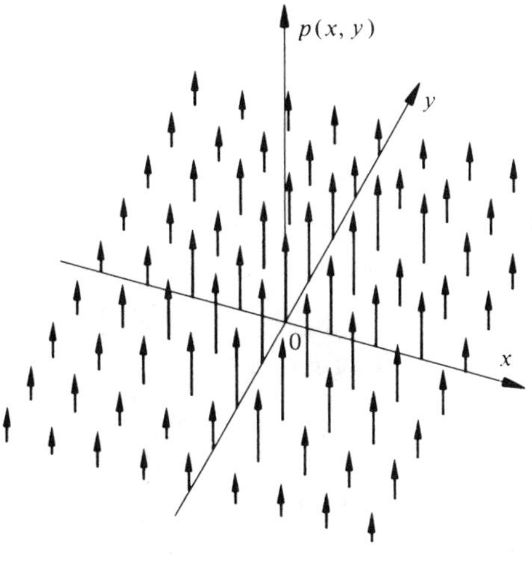

Fig. 14.5

The *marginal probability distributions* define the statistics of one of the variables, independently of the other. They can be derived from the bidimensional distribution by the relations:

- *CRV*

$$p_x(x) = \int_{-\infty}^{\infty} p(x,y)\mathrm{d}y \tag{14.28}$$

$$p_y(y) = \int_{-\infty}^{\infty} p(x,y)\mathrm{d}x \tag{14.29}$$

- *DRV*

$$\text{Prob}(x_i) = \sum_{j=1}^{n} \text{Prob}(x_i,y_j) \tag{14.30}$$

$$\text{Prob}(y_i) = \sum_{i=1}^{m} \text{Prob}(x_i,y_j) \tag{14.31}$$

$\text{Prob}(x_i)$ is equal to the sum of all the elements of the ith row of the joint probability matrix and $\text{Prob}(y_j)$ is equal to the sum of all the elements of the jth column of this matrix.

The corresponding *conditional probability distributions* are defined by relations similar to (14.5):

- *CRV*

$$p(x|y) = p(x,y)/p_y(y) \tag{14.32}$$

$$p(y|x) = p(x,y)/p_x(x) \tag{14.33}$$

- *DRV*

$$\text{Prob}(x_i|y_j) = \text{Prob}(x_i,y_j)/\text{Prob}(y_j) \tag{14.34}$$

$$\text{Prob}(y_j|x_i) = \text{Prob}(x_i,y_j)/\text{Prob}(x_i) \tag{14.35}$$

Thus, all the marginal, joint, and conditional probability distributions of a bidimensional variable can be determined from knowledge of:

- the joint probability density alone;
- or a marginal probability density and a conditional probability density.

In the discrete case, this corresponds to knowledge of:

- the joint probability matrix;
- or a marginal matrix and a conditional matrix.

14.2.8 Statistically independent variables

The statistical independence condition of two continuous random variables is that the joint probability density is equal to the product of the marginal probability densities:

$$p(x,y) = p_x(x) \cdot p_y(y) \tag{14.36}$$

In the discrete case, this condition is expressed in terms of probability, according to (14.9), as

$$\text{Prob}(x_i,y_j) = \text{Prob}(x_i) \cdot \text{Prob}(y_j) \quad \forall i \text{ and } j \tag{14.37}$$

14.2.9 Example (CRV)

Let a bidimensional distribution be defined by (fig. 14.6):

$$p(x,y) = \begin{cases} 2 \text{ for } 0 \leqslant x \leqslant 1 \text{ and } 0 \leqslant y \leqslant x \\ 0 \text{ otherwise} \end{cases}$$

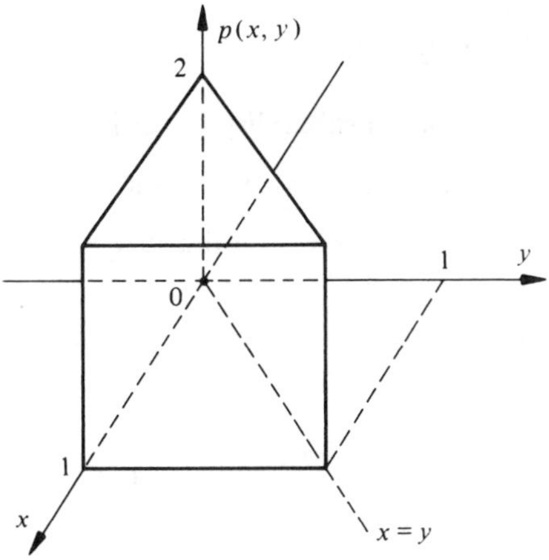

Fig. 14.6

- Calculus of the marginal probability densities (fig. 14.7)

$$p_x(x) = \begin{cases} \int_{-\infty}^{\infty} p(x,y)\mathrm{d}y = \int_{0}^{y=x} 2 \text{ d}x = 2x \text{ ; } 0 \leqslant x \leqslant 1 \\ 0 \text{ otherwise} \end{cases}$$

$$p_y(y) = \begin{cases} \int_{-\infty}^{\infty} p_x(x,y)\mathrm{d}x = \int_{x=y}^{1} 2 \text{ d}x = 2(1-y) \text{ ; } 0 \leqslant y \leqslant 1 \\ 0 \text{ otherwise} \end{cases}$$

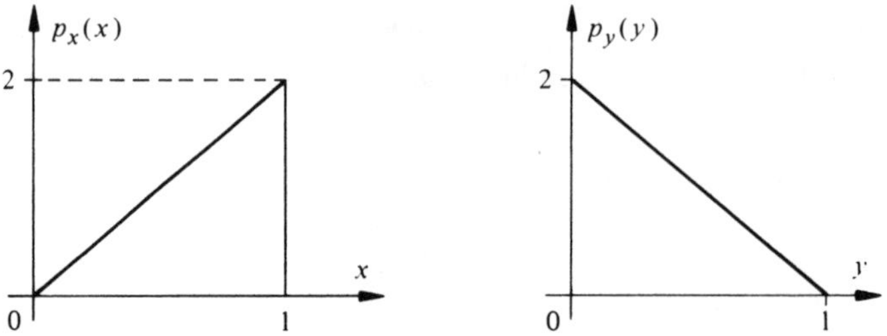

Fig. 14.7

Verifications:

$$\int_0^1 p(x)\mathrm{d}x = \int_0^1 2x\mathrm{d}x = x^2\Big|_0^1 = 1$$

$$\int_0^1 p_y(y)\mathrm{d}y = \int_0^1 2(1 - y)\mathrm{d}y = (2y - y^2)\Big|_0^1 = 1$$

- Calculus of the conditional probability densities (fig. 14.8)

$$p(x|y) = p(x,y)/p_y(y) = 1/(1 - y);\ 0 \leqslant y \leqslant 1 \text{ and } y \leqslant x \leq 1$$

$$p(x|y) = p(x,y)/p_x(x) = 1/x;\ 0 \leqslant x \leqslant 1 \text{ and } 0 \leqslant y \leq x$$

with

$$\int_y^1 p(x|y)\mathrm{d}x = 1\ ;\ \int_0^x p(y|x)\mathrm{d}y = 1$$

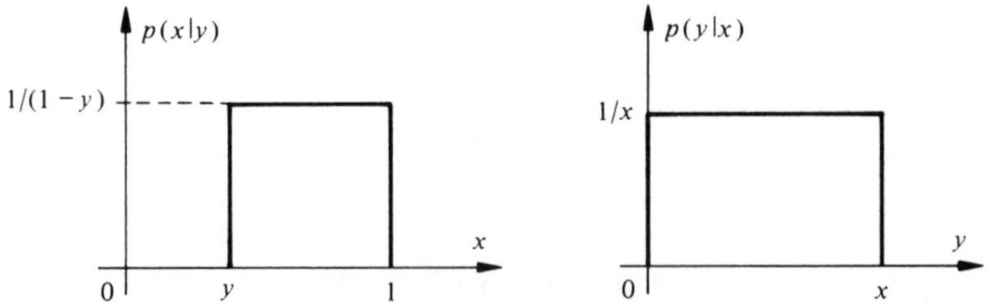

Fig. 14.8

Variables **x** and **y** are not statistically independent, because

$$p(x,y) \neq p_x(x) \cdot p_y(y)$$

or

$$p(x|y) \neq p_x(x) \text{ and } p(y|x) \neq p_y(y)$$

14.2.10 Example (DRV)

Let

$$[\text{Prob}(\mathbf{x},\mathbf{y})] = \begin{bmatrix} 1/9 & 1/3 & 0 \\ 2/9 & 1/9 & 2/9 \end{bmatrix}$$

Then, we have

$\text{Prob}(x_1) = 1/3 + 1/9 = 4/9$, $\text{Prob}(x_2) = 5/9$,

$\Sigma \text{ Prob } (x_i) = 1$ (verification);

$\text{Prob}(y_1) = 3/9$, $\text{Prob}(y_2) = 4/9$, $\text{Prob}(y_3) = 2/9$,

$\Sigma \text{ Prob } (y_j) = 1$ (verification);

$$[\text{Prob}(\mathbf{x}|\mathbf{y})] = \begin{bmatrix} 1/3 & 3/4 & 0 \\ 2/3 & 1/4 & 1 \end{bmatrix}$$

$$[\text{Prob}(\mathbf{y}|\mathbf{x})] = \begin{bmatrix} 1/4 & 3/4 & 0 \\ 2/5 & 1/5 & 2/5 \end{bmatrix}$$

The two variables are not independent since

$\text{Prob}(x_i,y_j) \neq \text{Prob}(x_i) \cdot \text{Prob}(y_i)$

or, respectively, since

$\text{Prob}(x_i|y_j) \neq \text{Prob}(x_i)$ and $\text{Prob}(y_j|x_i) \neq \text{Prob}(y_j)$

14.3 STATISTICAL MEANS AND MOMENTS

14.3.1 Mathematical expectation of a random variable function

Consider a discrete random variable **x**, with n states, N realizations of which are observed. The state x_1 occurs N_1 times, the state x_2 occurs N_2 times, *et cetera,* with $N_1 + N_2 + \ldots + N_n = N$.

The *experimental average value* of **x** is

$$\bar{\mathbf{x}} = (x_1N_1 + x_2N_2 + \ldots + x_nN_n)/N \tag{14.38}$$

The *mathematical expectation of* **x** is the *statistical mean value,* i.e., the theoretical limit value reached by x when N tends toward infinity. It is noted μ_x, and, taking into account (14.1), is equal to

$$\mu_x = \text{E}[\mathbf{x}] = \lim_{N\to\infty} \bar{\mathbf{x}} = \sum_{i=1}^{n} x_i \text{ Prob}(x_i) \tag{14.39}$$

By analogy and in a general way, we have for a continuous or discrete random variable $\mathbf{x}$:

$$\mu_x = \mathrm{E}[\mathbf{x}] = \int_{-\infty}^{\infty} x\, p(x)\, \mathrm{d}x \tag{14.40}$$

The mean value is the centroid of the probability density.

Let $f(\mathbf{x})$ be a function of a random variable $\mathbf{x}$. Its statistical mean value is similarly defined by

$$\mathrm{E}[f(\mathbf{x})] = \int_{-\infty}^{\infty} f(x)\, p(x)\, \mathrm{d}x \tag{14.41}$$

In the case of a discrete variable, this expression, after integration, becomes

$$\mathrm{E}[f(\mathbf{x})] = \sum_{i=1}^{n} f(x_i)\, \mathrm{Prob}(x_i) \tag{14.42}$$

In the case of a bidimensional random variable $(\mathbf{x}_1,\mathbf{x}_2)$, the statistical mean value of a function $f(\mathbf{x}_1,\mathbf{x}_2)$ is given by

$$\mathrm{E}[f(\mathbf{x}_1,\mathbf{x}_2)] = \int_{-\infty}^{\infty}\int_{-\infty}^{\infty} f(x_1,x_2)\, p(x_1,x_2)\, \mathrm{d}x_1\, \mathrm{d}x_2 \tag{14.43}$$

which becomes, in the discrete case,

$$\mathrm{E}[f(\mathbf{x}_1,\mathbf{x}_2)] = \sum_i \sum_j f(x_i,y_j)\, \mathrm{Prob}(x_i,y_j) \tag{14.44}$$

14.3.2 Mean value of the sum of random variables

Let $\mathbf{z} = f(\mathbf{x},\mathbf{y}) = \mathbf{x} + \mathbf{y}$. From (14.43), (14.28), and (14.29):

$$\mu_z = \mathrm{E}[\mathbf{z}] = \int_{-\infty}^{\infty}\int_{-\infty}^{\infty} (x + y)p(x,y)\, \mathrm{d}x\, \mathrm{d}y$$

$$= \int_{-\infty}^{\infty} x\, p_x(x)\, \mathrm{d}x + \int_{-\infty}^{\infty} y\, p_y(y)\, \mathrm{d}y = \mu_x + \mu_y \tag{14.45}$$

In a general way, ***the mathematical expectation of a sum of*** (dependent or independent) ***random variables is equal to the sum of their mathematical expectations***:

$$\mathrm{E}\left[\sum_{i=1}^{n} \mathbf{x}_i\right] = \sum_{i=1}^{n} \mathrm{E}[\mathbf{x}_i] \tag{14.46}$$

14.3.3 Statistical moments

The mathematical expectation of the *n*th power of a random variable is called its *statistical moment* of first-order and *n*th degree:

$$m_{xn} = \mathrm{E}[\mathbf{x}^n] = \int_{-\infty}^{\infty} x^n \, p(x) \, \mathrm{d}x \tag{14.47}$$

The mean value μ_x, therefore, corresponds to the moment of first degree.

14.3.4 Mean square value

The *mean square value* of a random variable is its moment of second degree:

$$m_{x2} = \mathrm{E}[\mathbf{x}^2] = \int_{-\infty}^{\infty} x^2 \, p(x) \, \mathrm{d}x \tag{14.48}$$

In the case of a discrete variable, (14.48) becomes

$$m_{x2} = \sum_i x_i^2 \, \mathrm{Prob}(x_i) \tag{14.49}$$

14.3.5 Central moments

The mathematical expectation of the *n*th power of the difference between a random variable and its mean value is called its *central moment* of first-order and *n*th degree:

$$m_{x-\mu,n} = \mathrm{E}[(\mathbf{x} - \mu_x)^n] = \int_{-\infty}^{\infty} (x - \mu_x)^n p(x) \, \mathrm{d}x \tag{14.50}$$

14.3.6 Variance and standard deviation

The *variance* of a random variable is its central moment of second degree. Since it is a very important characteristic, it is noted σ_x^2, or sometimes Var($\mathbf{x}$):

$$\sigma_x^2 = m_{x-\mu,2} = \mathrm{E}[(\mathbf{x} - \mu_x)^2] = \int_{-\infty}^{\infty} (x - \mu_x)^2 \, p(x) \, \mathrm{d}x \tag{14.51}$$

In the case of a discrete variable, (14.51) becomes

$$\sigma_x^2 = \sum_i (x_i - \mu_x)^2 \, \mathrm{Prob}(x_i) \tag{14.52}$$

The square root σ_x of the variance is called the *standard deviation*. This is a measure of the dispersion of values of $\mathbf{x}$ around the mean value μ_x (fig. 14.9).

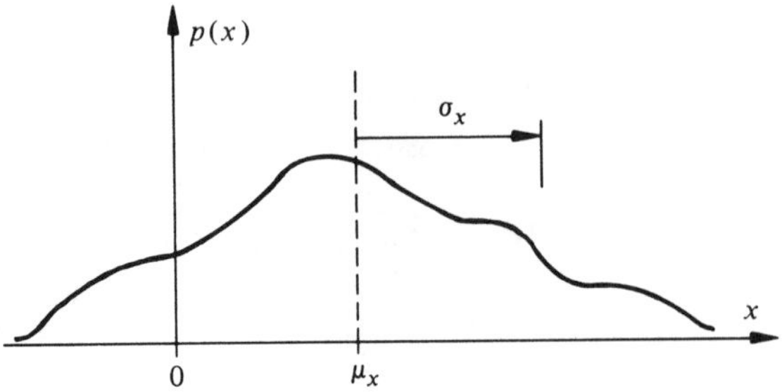

Fig. 14.9

Expanding (14.51), we get the following relation between the mean value, the variance, and the mean square value:

$$m_{x2} = \sigma_x^2 + \mu_x^2 \qquad (14.53)$$

Therefore, the variance can be identified with the mean square value for any variable with zero mean value.

14.3.7 Chebyshev's inequality

The following inequality can be established [24]:

$$\text{Prob}(\mu_x - \epsilon < \mathbf{x} < \mu_x + \epsilon) \geqslant 1 - \sigma_x^2/\epsilon^2 \qquad (14.54)$$

Thus, regardless of the shape of the probability density $p(x)$, the probability that $\mathbf{x}$ takes a value belonging to interval $[\mu_x - \epsilon, \mu_x + \epsilon]$ is almost one, if $\sigma_x \ll \epsilon$.

14.3.8 Correlation and covariance

In the case of a pair of random variables $(\mathbf{x},\mathbf{y})$, we can also define the following second-order moments:

$$R_{xy} = \text{E}[\mathbf{x}\ \mathbf{y}] = \begin{cases} \displaystyle\int_{-\infty}^{\infty}\int_{-\infty}^{\infty} xy\ p(x,y)\ \text{d}x\ \text{d}y & \text{(CRV)} \\ \displaystyle\sum_i \sum_j x_i y_j\ \text{Prob}(x_i,y_j) & \text{(DRV)} \end{cases} \qquad (14.55)$$

$$C_{xy} = \text{E}[(\mathbf{x} - \mu_x)(\mathbf{y} - \mu_y)] = \begin{cases} \displaystyle\int_{-\infty}^{\infty}\int_{-\infty}^{\infty} (x - \mu_x)(y - \mu_x)p(x,y)\ \text{d}x\ \text{d}y & \text{(CRV)} \\ \displaystyle\sum_i \sum_j (x_i - \mu_x)(y_j - \mu_x)\ \text{Prob}(x_i,y_i) & \text{(DRV)} \end{cases}$$

(14.56)

that are related by

$$R_{xy} = C_{xy} + \mu_x\mu_y \tag{14.57}$$

The moment R_{xy} measures the *statistical correlation* between **x** and **y**. The central moment C_{xy}, also noted Cov(**x**,**y**), is called the *covariance* of the pair (**x**,**y**).

The correlation is, therefore, identical to the covariance when at least one of the variables has a zero mean value.

We easily verify, by (14.36) or(14.37), (14.6), and (14.57), that if variables **x** and **y** are ***independent***:

$$C_{xy} = 0 \text{ and } R_{xy} = \mu_x\mu_y \tag{14.58}$$

The reciprocal is not necessarily true!

14.3.9 Correlation coefficient

The normalized covariance is called *correlation coefficient*:

$$\rho_{xy} = \frac{C_{xy}}{\sigma_x\sigma_y} \text{ with } |\rho_{xy}| \leqslant 1 \tag{14.59}$$

The variables **x** and **y** are said to be ***uncorrelated*** if $\rho_{xy} = 0$. They are totally correlated if $|\rho_{xy}| = 1$. The variables **x** and **y** then have a linear relationship. The correlation coefficient is a measurement of the similarity of **x** and **y**.

14.3.10 Variance of a sum of two random variables

If **z** = **x** + **y**, the variance σ_z^2, taking into account (14.53), (14.46), (14.55), and (14.57), is given by

$$\begin{aligned}\sigma_z^2 &= m_{z2} - \mu_z^2 = \mathrm{E}[(\mathbf{x} + \mathbf{y})^2] - \mathrm{E}^2[\mathbf{x} + \mathbf{y}] \\ &= m_{x2} + 2R_{xy} + m_{y2} - \mu_x^2 - \mu_y^2 - 2\mu_x\mu_y \\ &= \sigma_x^2 + \sigma_y^2 + 2C_{xy}\end{aligned} \tag{14.60}$$

If the variables **x** and **y** are independent: $C_{xy} = 0$, and

$$\sigma_z^2 = \sigma_x^2 + \sigma_y^2 \tag{14.61}$$

In a general way, the variance of a sum of independent variables $\mathbf{x}_1, \ldots, \mathbf{x}_n$ ***is equal to the variance sum***:

$$\sigma_z^2 = \sum_{i=1}^{n} \sigma_{xi}^2 \tag{14.62}$$

if $\mathbf{z} = \Sigma\mathbf{x}_i$ and if the variables $\mathbf{x}_i$ are independent.

14.3.11 Characteristic function

The *characteristic function* of a random variable $\mathbf{x}$ is, by definition, the mathematical expectation of the function $\exp(\mathrm{j}u\mathbf{x})$:

$$\Pi_x(u) = \mathrm{E}[\exp(\mathrm{j}u\mathbf{x})] = \int_{-\infty}^{\infty} p(x)\exp(\mathrm{j}ux)\,\mathrm{d}x \tag{14.63}$$

with $\Pi_x(0) = 1$ from (14.14).

Writing $u = 2\pi\upsilon$, we see that the characteristic function corresponds to the ***inverse Fourier transform*** (4.2) of the probability density:

$$\Pi_x(\upsilon) = \mathrm{F}^{-1}\{p(x)\} \tag{14.64}$$

and, conversely,

$$p(x) = \mathrm{F}\{\Pi_x(\upsilon)\} \tag{14.65}$$

In the case of a bidimensional variable $(\mathbf{x},\mathbf{y})$, we have

$$\Pi_{xy}(u,v) = \mathrm{E}\{\exp[\mathrm{j}(u\mathbf{x} + v\mathbf{y})]\} = \int_{-\infty}^{\infty}\int_{-\infty}^{\infty} p(x,y)\exp[\mathrm{j}(ux + vy)]\mathrm{d}x\,\mathrm{d}y \tag{14.66}$$

which corresponds to the bidimensional inverse Fourier transform of the joint probability density.

With the expansion in series of $\exp(\mathrm{j}u\mathbf{x})$, we can write the characteristic function (14.63), if all the moments of $\mathbf{x}$ are finite, as

$$\Pi_x(u) = 1 + \sum_{k=1}^{\infty} \frac{(\mathrm{j}u)^k}{k!}\mathrm{E}[\mathbf{x}^k] \tag{14.67}$$

If $\Pi_x(u)$ can be differentiated at origin, we can thus get the different statistical moments from the relation:

$$m_{xk} = \mathrm{E}[\mathbf{x}^k] = \mathrm{j}^{-k}\left.\frac{\mathrm{d}^k\Pi_x(u)}{\mathrm{d}u^k}\right|_{u=0} \tag{14.68}$$

In the bidimensional case, if the moments exist, we have

$$\Pi_{xy}(u,v) = \sum_{k=0}^{\infty}\sum_{l=0}^{\infty}\frac{\mathrm{j}^{(k+l)}u^k v^l}{k!l!}\mathrm{E}[\mathbf{x}^k\mathbf{y}^l]$$

$$= 1 + \mathrm{j}(u\mu_x + v\mu_y) - \frac{1}{2}(u^2 m_{x2} + v^2 m_{y2}) - uvR_{xy} + \tag{14.69}$$

and

$$\mathrm{E}[\mathbf{x}^k\mathbf{y}^l] = \mathrm{j}^{-(k+l)}\left.\frac{\partial^{k+l}\Pi_{xy}(u,v)}{\partial u^k\partial v^l}\right|_{u=v=0} \tag{14.70}$$

Therefore, the correlation (14.55) is equal to the derivative at the origin for the particular case $k = l = 1$

$$R_{xy} = -\left.\frac{\partial^2 \Pi_{xy}(u,v)}{\partial u \partial v}\right|_{u=v=0} \tag{14.71}$$

14.4 IMPORTANT DISTRIBUTIONS

14.4.1 Binary variable

If $\mathbf{x} \in \{a,b\}$ with $\text{Prob}(\mathbf{x} = a) = p$ and $\text{Prob}(\mathbf{x} = b) = 1 - p$, the probability density of $\mathbf{x}$ is (fig. 14.10):

$$p(x) = p\delta(x - a) + (1 - p)\delta(x - b) \tag{14.72}$$

with

$$\mu_x = ap + b(1 - p) \tag{14.73}$$

$$\sigma_x^2 = p(1 - p)(b - a)^2 \tag{14.74}$$

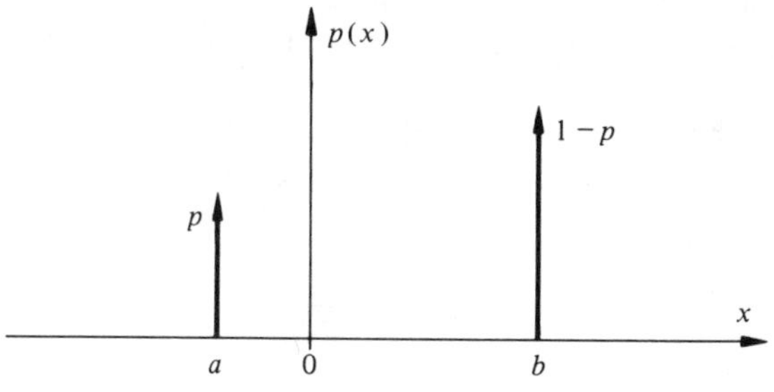

Fig. 14.10

14.4.2 Binomial distribution

The *binomial distribution* is the statistical law of the discrete random variable obtained by counting the number of occurrences of an event during n independent trials. If p is the probability of the event occurring, the probability of obtaining exactly k occurrences in n trials is

$$\text{Prob}(k,n) = \binom{n}{k} p^k (1 - p)^{n-k} \tag{14.75}$$

with the notation

$$\binom{n}{k} = \frac{n!}{k!(n-k)!} \tag{14.76}$$

The corresponding probability density can be written as

$$p(x) = \sum_{k=0}^{n} \text{Prob}(k,n)\delta(x-k) \tag{14.77}$$

with

$$\mu_x = np \tag{14.78}$$

$$\sigma_x^2 = np(1-p) \tag{14.79}$$

The binomial law allows, for example, determination of the probability of having more than k symbols 1 in a binary word of n bits.

14.4.3 Poisson distribution

The *Poisson distribution* characterizes many random-point processes, the occurrence instants of which are random (sub-sect. 5.8.1).

Consider a random sequence of independent events that can occur at any instant with the same chance. The average number of events per time unit is a constant λ.

Let $\text{Prob}(N,\tau)$ be the probability of counting exactly N events during a time interval τ. On an infinitesimal interval $\mathrm{d}\tau$, we have

$$\text{Prob}(1,\mathrm{d}\tau) = \lambda \mathrm{d}\tau \tag{14.80}$$

and

$$\text{Prob}(0,\mathrm{d}\tau) = 1 - \lambda \mathrm{d}\tau \tag{14.81}$$

because $\text{Prob}(N > 1, \mathrm{d}\tau) \cong 0$.

On an interval $\tau + \mathrm{d}\tau$, we have

$$\text{Prob}(N, \tau + \mathrm{d}\tau) = \text{Prob}(N, \tau)\,\text{Prob}(0, \mathrm{d}\tau) + \text{Prob}(N-1, \tau)\,\text{Prob}(1, \mathrm{d}\tau) \tag{14.82}$$

and

$$\text{Prob}(0, \tau + \mathrm{d}\tau) = \text{Prob}(0, \tau)\,\text{Prob}(0, \mathrm{d}\tau) \tag{14.83}$$

We derive, from these relations, the differential equations:

$$\begin{aligned} \mathrm{d}\,\text{Prob}(N, \tau)/\mathrm{d}\tau &= [\text{Prob}(N, \tau + \mathrm{d}\tau) - \text{Prob}(N, \tau)]/\mathrm{d}\tau \\ &= \lambda[\text{Prob}(N-1, \tau) - \text{Prob}(N, \tau)] \end{aligned} \tag{14.84}$$

and

$$d\,\mathrm{Prob}(0, \tau)/d\tau = [\mathrm{Prob}(0, \tau + d\tau) - \mathrm{Prob}(0, \tau)]/d\tau$$

$$= -\lambda\,\mathrm{Prob}(0, \tau) \tag{14.85}$$

The solution of this last equation gives, with the initial condition $\mathrm{Prob}(0, 0) = 1$:

$$\mathrm{Prob}(0, \tau) = \exp(-\lambda\tau) \tag{14.86}$$

We can then solve the first differential equation for $N = 1$, with the initial condition $\mathrm{Prob}(1, 0) = 0$, leading to

$$\mathrm{Prob}(1, \tau) = \lambda\tau \exp(-\lambda\tau) \tag{14.87}$$

Proceeding by recurrence with N increasing, we finally get (fig. 14.11):

$$\mathrm{Prob}(N, \tau) = \frac{(\lambda\tau)^N}{N!} \exp(-\lambda\tau) = \frac{\mu^N}{N!} \exp(-\mu) \tag{14.88}$$

by admitting as a parameter the mean value: $\mu = \lambda\tau$. The variance can be identified here with the mean value: $\sigma^2 = \mu = \lambda\tau$.

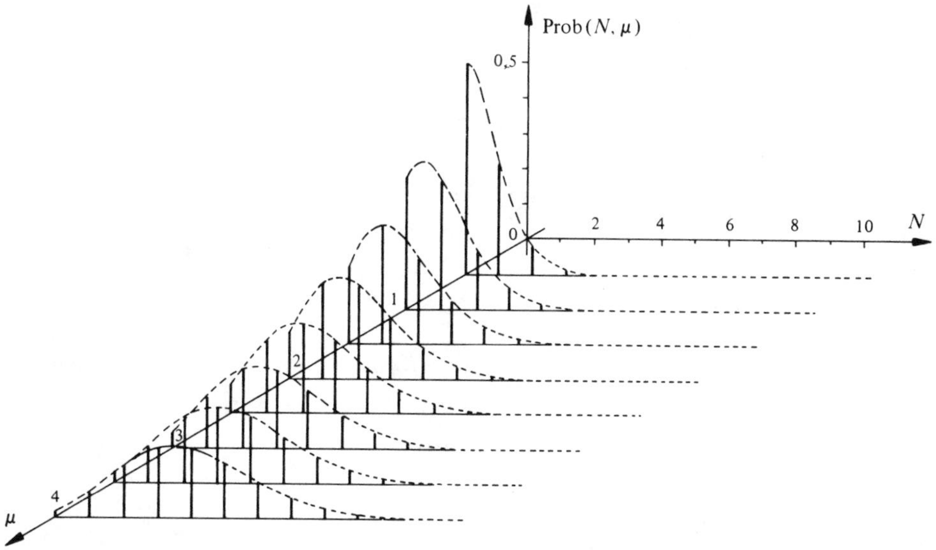

Fig. 14.11 Graphic representation of the Poisson distribution: $\mathrm{Prob}(N;\mu) = \mu^N \exp(-\mu)/N!$

Let $\mathbf{z}$ be the random variable representing the time interval separating two consecutive events. This situation corresponds to the absence of events on the interval z followed by an event on the succeeding infinitesimal interval dz. Thus, from (14.80) and (14.86):

$$p(z)\,\mathrm{d}z = \mathrm{Prob}(0, z)\,\mathrm{Prob}(1, \mathrm{d}z) = \exp(-\lambda z)\cdot\lambda\mathrm{d}z \tag{14.89}$$

hence,

$$p(z) = \lambda\exp(-\lambda z);\ z \geqslant 0 \tag{\textbf{14.90}}$$

with

$$\mu_z = \lambda^{-1} \tag{14.91}$$

and

$$\sigma_z^2 = \lambda^{-2} = \mu_z^2 \tag{14.92}$$

This is an *exponential distribution* (fig. 14.12).

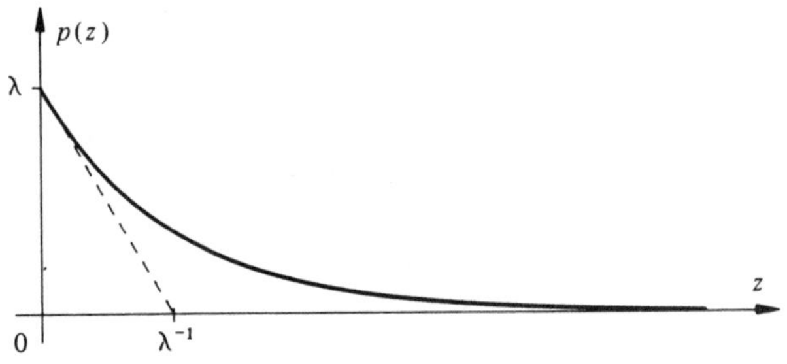

Fig. 14.12

14.4.4 Uniform distribution

A continuous random variable has a *uniform distribution* if its probability density has the general form (fig. 14.13):

$$p(x) = (b-a)^{-1}\,\mathrm{rect}\,\{[x - \tfrac{1}{2}(a+b)]/(b-a)\} \tag{14.93}$$

with

$$\mu_x = \tfrac{1}{2}(a+b) \tag{14.94}$$

$$\sigma_x^2 = (b-a)^2/12 \tag{14.95}$$

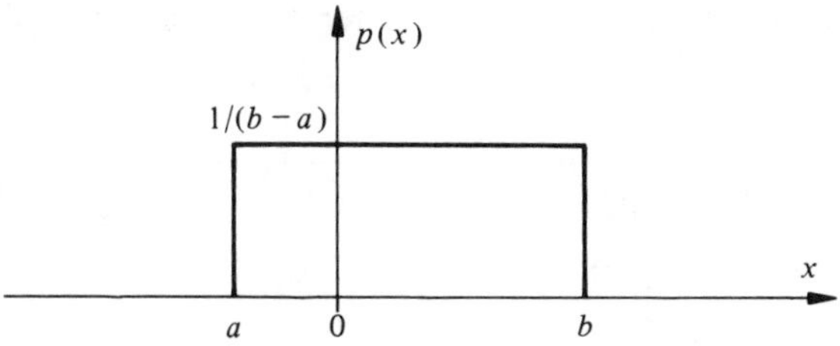

Fig. 14.13

14.4.5 Normal or Gaussian distribution

A continuous random variable with mean value μ_x and variance σ_x^2 has a *normal* (or *Gaussian*) *distribution* if its probability density has the general form (fig. 14.14):

$$p(x) = \frac{1}{\sqrt{2\pi}\sigma_x} \exp\left[-\frac{(x-\mu_x)^2}{2\sigma_x^2}\right] \tag{14.96}$$

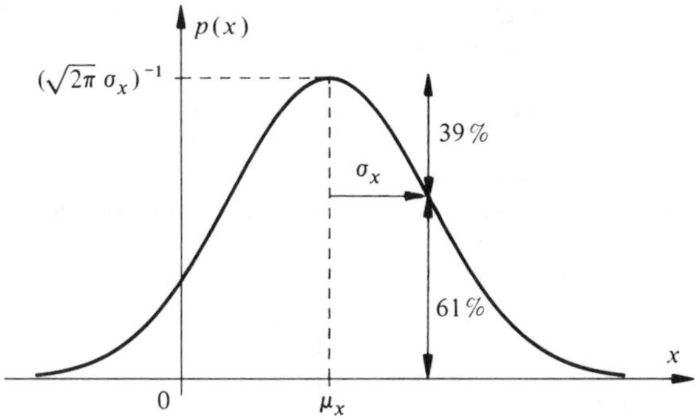

Fig. 14.14

Its characteristic function is

$$\Pi(u) = \exp(\,j\mu_x u - \tfrac{1}{2}\sigma_x^2 u^2) \tag{14.97}$$

In the case of a bidimensional Gaussian random variable (**x**,**y**), **x** and **y** both have a marginal distribution similar to (14.96). The bidimensional distribution can be characterized by the following generalized joint probability density:

$$p(x,y) = \frac{1}{2\pi\sigma_x\sigma_y\sqrt{1-\rho_{xy}^2}} \exp\left\{\frac{-1}{2(1-\rho_{xy}^2)}\left[\frac{(x-\mu_x)^2}{\sigma^2} - \frac{2\rho_{xy}(x-\mu_x)(y-\mu_y)}{\sigma_x\sigma_y} + \frac{(y-\mu_y)^2}{\sigma_y^2}\right]\right\} \tag{14.98}$$

where ρ_{xy} is the correlation coefficient (14.59) of **x** and **y**. ***The uncorrelation*** ($\rho_{xy} = 0$) ***of* x *and* y *induces their statistical independence.***

The conditional probability densities are determined by (14.32), (14.33), (14.96), and (14.98). They are Gaussian laws. If $\mu_x = \mu_y = 0$, we obtain, for instance,

$$p(x|y) = \frac{1}{\sigma_{x|y}\sqrt{2\pi}} \exp\left[-\frac{(x-\mu_{x|y})^2}{2\sigma_{x|y}^2}\right] \tag{14.99}$$

defining the conditional mean value and the conditional variance by

$$\mu_{x|y} = \sigma_{xy}(\sigma_x/\sigma_y)y = (C_{xy}/\sigma_y^2)y \tag{14.100}$$

$$\sigma_{x|y}^2 = \sigma_x^2(1-\rho_{xy}^2) \tag{14.101}$$

The *multidimensional Gaussian joint probability density* can be expressed, using matrix notation, in the following way:

$$p(x) = (2\pi)^{-n/2}|C_x|^{-1/2} \exp[-\tfrac{1}{2}(\boldsymbol{x}-\boldsymbol{\mu}_x)C_x^{-1}(\boldsymbol{x}-\boldsymbol{\mu}_x)^{\mathrm{T}}] \tag{14.102}$$

where $\mathbf{x} = (\mathbf{x}_1, \mathbf{x}_2, \ldots, \mathbf{x}_n)$ is an n-dimensional Gaussian random vector, $(\boldsymbol{x} - \boldsymbol{\mu}_x)$ is a row vector, $(\boldsymbol{x} - \boldsymbol{\mu}_x)^{\mathrm{T}}$ is the same transposed vector and C_x is the covariance matrix:

$$C_x = \begin{bmatrix} \sigma_{x1}^2 & C_{x1x2} \cdots\cdots & C_{x1xn} \\ \vdots & \sigma_{x2}^2 \cdots & \vdots \\ C_{xnx1} & \cdots\cdots\cdots & \sigma_{xn}^2 \end{bmatrix} \tag{14.103}$$

If the variables $\mathbf{x}_i$ are independent, $C_{xixj} = 0$ and the covariance matrix is diagonal.

The multidimensional Gaussian characteristic function is

$$\Pi_x(\boldsymbol{u}) = \exp[j\boldsymbol{\mu}_x^{\mathrm{T}}\boldsymbol{u} - \tfrac{1}{2}\boldsymbol{u}C_x\boldsymbol{u}^{\mathrm{T}}] \tag{14.104}$$

Because of its great importance, numerical tables of the Gaussian distribution are given in the specialized literature in normalized form (*standardized normal distribution*), obtained by introducing the ***standardized central variable***

$$z = (x - \mu_x)/\sigma_x \tag{14.105}$$

giving:

$$p(z) = \frac{1}{\sqrt{2\pi}} \exp\left[-\frac{z^2}{2}\right] \tag{14.106}$$

With the notation introduced in sub-section 1.3.16, we have the equivalence:

$$p(z) = \frac{1}{\sqrt{2\pi}} \mathrm{ig}\left(\frac{z}{\sqrt{2\pi}}\right) \tag{14.107}$$

Usually given in tables are (see sect. 15.8): the standardized probability density $p(z)$, the corresponding cumulative distribution $F(z)$, and the complementary cumulative distribution $F_c(z) = 1 - F(z)$. We also find the *error function* and its complement defined by (with $z = \sqrt{2u}$):

$$\mathrm{erf}(u) = \frac{2}{\sqrt{\pi}} \int_0^u \exp(-\alpha^2)\, d\alpha = 2F(z) - 1 \tag{14.108}$$

$$\mathrm{erfc}(u) = \frac{2}{\sqrt{\pi}} \int_u^\infty \exp(-\alpha^2)\, d\alpha = 2F_c(z) \tag{14.109}$$

14.4.6 Rayleigh distribution

A continuous random variable has a *Rayleigh distribution* if its probability distribution has the general form (fig. 14.15):

$$p(x) = (x/\sigma^2) \exp(-\tfrac{1}{2}x^2/\sigma^2)\epsilon(x) \tag{14.110}$$

with

$$\mu_x = \sigma\sqrt{\pi/2} \tag{14.111}$$

$$\sigma_x^2 = \sigma^2(2 - \pi/2) \tag{14.112}$$

14.4.7 χ^2 distribution

The general form of the χ_m^2 *distribution with m degrees of freedom* is given by

$$p(\chi_m^2) = \frac{1}{2^{m/2}\Gamma(m/2)} \chi_m^{2(m/2-1)} \exp(-\chi_m^2/2) \tag{14.113}$$

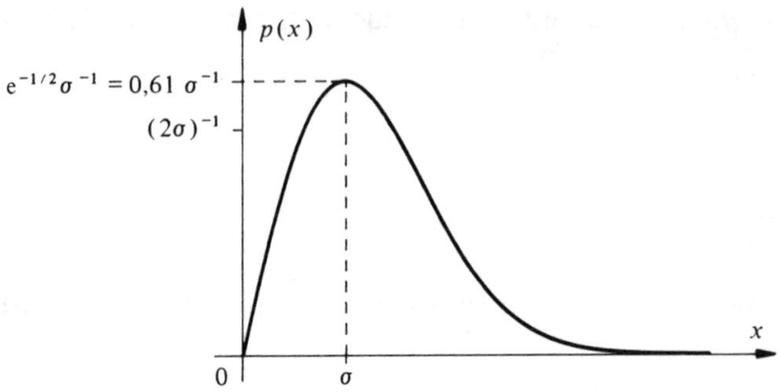

Fig. 14.15

where

$$\Gamma(n) = \int_0^{\infty} \alpha^{n-1} \exp(\alpha) \, d\alpha \tag{14.114}$$

with the particular values $\Gamma(1) = 1$, $\Gamma(\frac{1}{2}) = \sqrt{\pi}$, $\Gamma(2) = 1$, and $\Gamma(n + 1) = n\Gamma(n)$. Especially, if n is an integer: $\Gamma(n + 1) = n!$

The mean value is equal to the number of degrees of freedom: $\mu_{x^2} = m$, the variance is equal to twice this number: $\sigma^2_{x^2} = 2m$.

This distribution is used to define the statistics of a variable:

$$\chi^2_m = \sum_{i=1}^{m} \mathbf{x}_i^2 \tag{14.115}$$

where the $\mathbf{x}_i$ are independent Gaussian random variables, with zero mean value and variance unity.

For $m \to \infty$, the χ^2_m law asymptotically tends toward a Gaussian law.

The variable $\chi_2 = (\mathbf{x}_1^2 + \mathbf{x}_2^2)^{1/2}$ has a Rayleigh distribution (14.110).

The sum of two χ^2 variables, respectively having a and b degrees of freedom, is a new χ^2 variable with $(a + b)$ degrees of freedom.

Chapter 15

Formulas and Reference Tables

15.1 SUMMARY OF MAIN FORMULAS

15.1.1 Foreword

A selection of the main formulas used in this book is reproduced here. The formulas are grouped by topic and are identified by their original number.

Section 15.2 is devoted to the usual trigonometric identities. The relations relative to the Fourier transform and its properties form the subject of section 15.3.

An illustrated table of the main Fourier transforms is reproduced in section 15.4.

15.1.2 Delta-function properties

$$x(t_0) = \int_{-\infty}^{\infty} x(t)\delta(t - t_0)\mathrm{d}t \tag{1.36}$$

$$x(t) * \delta(t) = x(t) \tag{1.47}$$

$$x(t) * \delta(t - t_0) = x(t - t_0) \tag{1.48}$$

$$x(t - t_1) * \delta(t - t_2) = x(t - t_1 - t_2) \tag{1.49}$$

$$\delta(t - t_1) * \delta(t - t_2) = \delta(t - t_1 - t_2) \tag{1.50}$$

$$\delta(at) = |a|^{-1}\delta(t) \tag{1.51}$$

15.1.3 Time average, normalized energy and power of real signals

$$\overline{\mathrm{x}} = \lim_{T\to\infty} \frac{1}{T} \int_{-T/2}^{T/2} x(t)\mathrm{d}t \tag{1.30}$$

$$\overline{\mathrm{x}^n} = \lim_{T\to\infty} \frac{1}{T} \int_{-T/2}^{T/2} x^n(t)\mathrm{d}t \tag{5.17}$$

$$W_x = \int_{-\infty}^{\infty} x^2(t)\mathrm{d}t \tag{2.11}$$

$$P_x = \lim_{T\to\infty} \frac{1}{T} \int_{-T/2}^{T/2} x^2(t)\mathrm{d}t \tag{2.12}$$

15.1.4 Vector representation in the L^2 space

- Euclidian distance, norm and inner product

$$d(x,y) = \|x - y\| = \left[\int_{t_1}^{t_2} |x(t) - y(t)|^2 \mathrm{d}t\right]^{1/2} \tag{3.9}$$

$$\|x\| = \left[\int_{t_1}^{t_2} |x(t)|^2 \mathrm{d}t\right]^{1/2} \tag{3.8}$$

$$<x,y^*> = \int_{t_1}^{t_2} x(t)y^*(t)\mathrm{d}t \tag{3.12}$$

- Schwarz inequality

$$|<x,y^*>|^2 \leqslant <x,x^*> \cdot <y,y^*> \tag{3.20}$$

$$\left|\int_{t_1}^{t_2} x(t)y^*(t)\mathrm{d}t\right|^2 \leqslant \int_{t_1}^{t_2} |x(t)|^2\mathrm{d}t \cdot \int_{t_1}^{t_2} |y(t)|^2\mathrm{d}t \tag{3.21}$$

- Orthogonality condition

$$<x,y^*> = \int_{t_1}^{t_2} x(t)y^*(t)\mathrm{d}t = 0 \tag{3.15}$$

- Orthogonal expansion

$$x(t) = \sum_{k=1}^{\infty} \alpha_k\psi_k(t) \tag{3.50}$$

$$\alpha_k = \frac{1}{\lambda_k} <x,\psi_k^*> = \frac{1}{\lambda_k}\int_{t_1}^{t_2} x(t)\psi_k^*(t)\mathrm{d}t \tag{3.43}$$

$$\lambda_k = <\psi_k,\psi_k^*> = \|\psi_k\|^2 = \int_{t_1}^{t_2} |\psi_k(t)|^2\mathrm{d}t \tag{3.44}$$

$$\int_{t_1}^{t_2} |x(t)|^2\mathrm{d}t = \sum_{k=1}^{\infty} |\alpha_k|^2 \cdot \lambda_k \tag{3.49}$$

15.1.5 Stationary random signals

- Statistical mean values

$$\mathrm{E}[f(\mathbf{x})] = \int_{-\infty}^{\infty} f(x)p(x)\mathrm{d}x \tag{14.41}$$

$$E[\mathbf{x}^n] = \int_{-\infty}^{\infty} x^n p(x)\mathrm{d}x \tag{5.18}$$

$$\mu_x = E[\mathbf{x}] = \int_{-\infty}^{\infty} xp(x)\mathrm{d}x \tag{14.40}$$

$$m_{x2} = E[\mathbf{x}^2] = \int_{-\infty}^{\infty} x^2 p(x)\mathrm{d}x \tag{14.48}$$

$$\sigma_x^2 = m_{x-\mu,2} = E[(\mathbf{x} - \mu)^2] = \int_{-\infty}^{\infty} (x - \mu)^2 p(x)\mathrm{d}x \tag{14.51}$$

$$m_{x2} = \sigma_x^2 + \mu_x^2 \tag{14.53}$$

- Probability densities

$$p_x(x) = \int_{-\infty}^{\infty} p(x,y)\mathrm{d}y \tag{14.28}$$

$$p_y(y) = \int_{-\infty}^{\infty} p(x,y)\mathrm{d}x \tag{14.29}$$

$$p(x|y) = p(x,y)/p_y(y) \tag{14.32}$$

$$p(y|x) = p(x,y)/p_x(x) \tag{14.33}$$

- Real signal correlation and covariance functions

$$R_x(\tau) = E[\mathbf{x}(t)\mathbf{x}(t + \tau)] = \int_{-\infty}^{\infty} x_1 x_2 p(x_1,x_2;\ \tau)\mathrm{d}x_1\mathrm{d}x_2 \tag{5.55}$$

$$\varphi_x(\tau) = \overline{x(t)x(t + \tau)} = \lim_{T\to\infty} \frac{1}{T} \int_{-T/2}^{T/2} x(t)x(t + \tau)dt \tag{5.58}$$

$$C_x(\tau) = E\{[\mathbf{x}(t) - \mu_x][\mathbf{x}(t + \tau) - \mu_x]\} = R_x(\tau) - \mu_x^2 \tag{5.63}$$

$$R_x(\tau) = R_x(-\tau) \tag{5.65}$$

$$R_x(0) = \sigma_x^2 + \mu_x^2 \tag{5.68}$$

$$R_{xy}(\tau) = E[\mathbf{x}_1\ \mathbf{y}_2] = \int_{-\infty}^{\infty}\int_{-\infty}^{\infty} x_1 y_2 p(x_1,y_2;\ \tau)\mathrm{d}x_1\mathrm{d}y_2 \tag{5.161}$$

$$R_{xy}(\tau) = R_{yx}(-\tau) \tag{5.165}$$

$$C_{xy}(\tau) = E[(\mathbf{x}_1 - \mu_x)(\mathbf{y}_2 - \mu_y)] = R_{xy}(\tau) - \mu_x\mu_y \tag{5.169}$$

$$\rho_{xy}(\tau) = C_{xy}(\tau)/(\sigma_x\sigma_y) \tag{5.171}$$

$$|\rho_{xy}(\tau)| \leqslant 1 \tag{5.174}$$

• Independent signals

$$p_{xy}(x,y) = p_x(x)p_y(y) \tag{5.33}$$

$$C_{xy}(\tau) = C_{yx}(\tau) = \rho_{xy}(\tau) = 0 \tag{5.175}$$

• Sum $z(t) = x(t) + y(t)$ of independent signals

$$p_z(z) = \int_{-\infty}^{\infty} p_x(x)p_y(z - x)\mathrm{d}x = p_x(z) * p_y(z) \tag{5.182}$$

$$\left.\begin{aligned} R_z(\tau) &= R_x(\tau) + R_y(\tau) + 2\mu_x\mu_y \\ \Phi_z(f) &= \Phi_x(f) + \Phi_y(f) + 2\mu_x\mu_y\delta(f) \end{aligned}\right\} \quad \text{adapted from (5.189) and (5.190)}$$

• Product $z(t) = x(t)y(t)$ of independent signals

$$R_z(\tau) = R_x(\tau) \cdot R_y(\tau) \tag{5.198}$$

$$\Phi_z(f) = \Phi_x(f) * \Phi_y(f) \tag{5.199}$$

15.1.6 Main statistical distributions

• Binomial

$$\text{Prob}(k,n) = \binom{n}{k} p^k(1 - p)^{n-k} \tag{14.75}$$

• Poisson

$$\text{Prob}(N,\tau) = \frac{(\lambda\tau)^N}{N!} \exp(-\lambda\tau) = \frac{\mu^N}{N!} \exp(-\mu) \tag{14.88}$$

• Gauss

$$p(x) = \frac{1}{\sqrt{2\pi}\sigma_x} \exp\left[-\frac{(x - \mu_x)^2}{2\sigma_x^2}\right] \tag{14.96}$$

$$p(x,y) = \frac{1}{2\pi\sigma_x\sigma_y\sqrt{1 - \rho_{xy}^2}} \exp\left\{\frac{-1}{2(1 - \rho_{xy}^2)}\left[\frac{(x - \mu_x)^2}{\sigma_x^2} - \frac{2\rho_{xy}(x - \mu_x)(y - \mu_y)}{\sigma_x\sigma_y} + \frac{(y - \mu_y)^2}{\sigma_y^2}\right]\right\} \tag{14.98}$$

• Rayleigh

$$p(x) = (x/\sigma^2) \exp(-\tfrac{1}{2}x^2/\sigma^2)\epsilon(x) \tag{14.110}$$

15.1.7 Background noise

- White noise

$$\Phi_x(f) = \tfrac{1}{2}\eta \text{ for } |f| < \infty \qquad (5.151)$$

$$R_x(\tau) = \mathrm{F}^{-1}\{\tfrac{1}{2}\eta\} = \tfrac{1}{2}\eta\delta(\tau) \qquad (5.152)$$

- Thermal noise

$$P_{\text{th}} = kTB \quad \text{W} \qquad (6.4)$$

$$\sigma_{u0}^2 = 4kTRB \quad \text{V}^2 \qquad (6.13)$$

- Shot noise

$$\sigma_{ig}^2 = 2eI_0B \quad \text{A}^2 \qquad (6.33)$$

- Noise factor of a multi-stage linear system

$$F = F_1 + \frac{F_2 - 1}{G_1} + \frac{F_3 - 1}{G_1G_2} + \ldots + \frac{F_m - 1}{\prod_{i=1}^{m-1} G_i} \qquad (6.58)$$

15.1.8 Analytic signal and complex envelope

$$\underline{\underline{x}}(t) = x(t) + \mathrm{j}\check{x}(t) \qquad (7.5)$$

$$\check{x}(t) = \frac{1}{\pi t} * x(t) = \frac{1}{\pi}\int_{-\infty}^{\infty} \frac{x(\tau)}{t - \tau}\,\mathrm{d}\tau \qquad (7.9)$$

$$x(t) = -\frac{1}{\pi t} * \check{x}(t) = -\frac{1}{\pi}\int_{-\infty}^{\infty} \frac{\check{x}(\tau)}{t - \tau}\,\mathrm{d}\tau \qquad (7.11)$$

$$R_{\underline{\underline{x}}}(\tau) = 2[R_x(\tau) + \mathrm{j}\check{R}_x(\tau)] \qquad (7.40)$$

$$\Phi_{\underline{\underline{x}}}(f) = 4\epsilon(f)\Phi_x(f) = 2\Phi_x^+(f) \qquad (7.41)$$

$$\underline{\underline{x}}(t) = r_x(t) \exp[\mathrm{j}\phi_x(t)] \qquad (7.44)$$

$$\underset{\sim}{r}(t) = \underline{\underline{x}}(t) \exp(-\mathrm{j}\omega_0 t) \qquad (7.80)$$

$$\underset{\sim}{r}(t) = a(t) + \mathrm{j}b(t) = r(t) \exp[\mathrm{j}\alpha(t)] \qquad (7.81)$$

$$\begin{aligned} x(t) &= \mathrm{Re}\{\underset{\sim}{r}(t) \exp(\mathrm{j}\omega_0 t)\} \\ &= r(t) \cos[\omega_0 t + \alpha(t)] \\ &= a(t) \cos(\omega_0 t) - b(t) \sin(\omega_0 t) \end{aligned} \qquad (7.86)$$

$$\Phi_x(f) = \tfrac{1}{4}[\Phi_{\underset{\sim}{r}}(-f - f_0) + \Phi_{\underset{\sim}{r}}(f - f_0)] \qquad (7.88)$$

$$\Phi_{\underset{\sim}{r}}(f) = \Phi_{\underline{\underline{x}}}(f + f_0) = 2\Phi_x^+(f + f_0) = 2[\Phi_a(f) + \mathrm{j}\Phi_{ab}(f)] \qquad (7.89)$$

15.1.9 Linear system with impulse response $g(t)$

$$y(t) = x(t) * g(t) = \int_{-\infty}^{\infty} x(\tau)g(t-\tau)\mathrm{d}\tau \tag{8.12}$$

$$Y(f) = X(f) \cdot G(f) \tag{8.14}$$

$$\varphi_y(\tau) = \varphi_x(\tau) * \mathring{\varphi}_g(\tau) \tag{8.22}$$

$$\varphi_{xy}(\tau) = \varphi_x(\tau) * g(\tau) \tag{8.23}$$

$$\Phi_y(f) = \Phi_x(f) \cdot |G(f)|^2 \tag{8.24}$$

$$\Phi_{xy}(f) = \Phi_x(f) \cdot G(f) \tag{8.25}$$

$$\mu_y = \mu_x \int_{-\infty}^{\infty} g(t)\mathrm{d}t = \mu_x G(0) \tag{8.37}$$

$$\sigma_y^2 = \int_{-\infty}^{\infty} [\Phi_x(f) - \mu_x^2\delta(f)] \cdot |G(f)|^2 \mathrm{d}f$$

$$= \int_{-\infty}^{\infty} C_x(\tau)\mathring{\varphi}_g(\tau)\mathrm{d}\tau \tag{8.38}$$

15.1.10 Sampling and quantization

- Sampling theorem for signals with low-pass spectrum

$$f_e = 1/T_e \geqslant 2f_{\max} \tag{9.25}$$

$$x(t) = \sum_{k=-\infty}^{\infty} x(kT_e)\,\mathrm{sinc}[(t/T_e) - k] \tag{9.29}$$

- Dimension of a signal of duration T

$$N = 2BT \tag{9.38}$$

- Discrete Fourier transform

$$x(k) = N^{-1} \sum_{n=-N/2}^{N/2-1} X(n)W_N^{nk} \leftrightarrow X(n) = \sum_{k=k_0}^{k_0+N-1} x(k)W_N^{-nk} \tag{9.49}$$

$$W_N \triangleq \exp(\,\mathrm{j}2\pi/N) \tag{9.48}$$

- Uniform quantization noise (step Δ)

$$\sigma_q^2 \approx \frac{\Delta^2}{12} \tag{10.16}$$

15.2 MAIN TRIGONOMETRIC IDENTITIES

- Product of trigonometric functions

$\sin\alpha \sin\beta = \frac{1}{2}\cos(\alpha - \beta) - \frac{1}{2}\cos(\alpha + \beta)$

$\cos\alpha \cos\beta = \frac{1}{2}\cos(\alpha - \beta) + \frac{1}{2}\cos(\alpha + \beta)$

$\sin\alpha \cos\beta = \frac{1}{2}\sin(\alpha + \beta) + \frac{1}{2}\sin(\alpha - \beta)$

$\cos\alpha \sin\beta = \frac{1}{2}\sin(\alpha + \beta) - \frac{1}{2}\sin(\alpha - \beta)$

- Sum and difference of trigonometric functions

$\sin\alpha + \sin\beta = 2\sin[(\alpha + \beta)/2]\cos[(\alpha - \beta)/2]$

$\sin\alpha - \sin\beta = 2\cos[(\alpha + \beta)/2]\sin[(\alpha - \beta)/2]$

$\cos\alpha + \cos\beta = 2\cos[(\alpha + \beta)/2]\cos[(\alpha - \beta)/2]$

$\cos\alpha - \cos\beta = -2\sin[(\alpha + \beta)/2]\sin[(\alpha - \beta)/2]$

- Sum and difference of angles

$\sin(\alpha + \beta) = \sin\alpha\cos\beta + \cos\alpha\sin\beta$

$\sin(\alpha - \beta) = \sin\alpha\cos\beta - \cos\alpha\sin\beta$

$\cos(\alpha + \beta) = \cos\alpha\cos\beta - \sin\alpha\sin\beta$

$\cos(\alpha - \beta) = \cos\alpha\cos\beta + \sin\alpha\sin\beta$

- Square of trigonometric functions

$\sin^2\alpha = (1 - \cos 2\alpha)/2$

$\cos^2\alpha = (1 + \cos 2\alpha)/2$

- Exponential relations

$\exp(j\alpha) = \cos\alpha + j\sin\alpha \qquad j = \sqrt{-1}$

$$\sin\alpha = \frac{\exp(j\alpha) - \exp(-j\alpha)}{2j}$$

$$\cos\alpha = \frac{\exp(j\alpha) + \exp(-j\alpha)}{2}$$

- Series expansion

$\sin\alpha = \alpha - \alpha^3/3! + \alpha^5/5! - \alpha^7/7! + \ldots$

$\cos\alpha = 1 - \alpha^2/2! + \alpha^4/4! - \alpha^6/6! + \ldots$

$\exp\alpha = 1 + \alpha + \alpha^2/2! + \alpha^3/3! + \ldots$

15.3 PROPERTIES OF THE FOURIER TRANSFORM AND ASSOCIATED RELATIONS

15.3.1 Finite energy signals

Property	Time domain	Frequency domain
Direct transformation	$x(t)$	$X(f) = \int_{-\infty}^{+\infty} x(t)\, e^{-j2\pi ft}\, dt$
Inverse transformation	$x(t) = \int_{-\infty}^{+\infty} X(f)\, e^{j2\pi ft}\, df$	$X(f)$
Complex conjugation	$y(t) = x^*(t)$	$Y(f) = X^*(-f)$
Time reversal	$y(t) = x(-t)$ complex real	$Y(f) = X(-f)$ $Y(f) = X(-f) \hat{=} X^*(f)$
Multiplication of the variable by a positive constant > 0	$y(t) = x(at)$ $a > 1$ acceleration $0 < a < 1$ deceleration	$Y(f) = \frac{1}{\lvert a \rvert} X\left(\frac{f}{a}\right)$
Time translation (time-shift theorem)	$y(t) = x(t - t_0)$	$Y(f) = e^{-j2\pi f t_0} \cdot X(f)$
Frequency translation	$y(t) = e^{j2\pi f_0 t} \cdot x(t)$	$Y(f) = X(f - f_0)$
Modulation	$y(t) = x(t) \cdot \cos 2\pi f_0 t$	$Y(f) = \frac{1}{2}[X(f + f_0) + X(f - f_0)]$
Differentiation	$y(t) = \frac{d^n}{dt^n} x(t)$	$Y(f) = (j2\pi f)^n X(f)$
Integration	$y(t) = \int_{-\infty}^{t} x(u)\, du$	$Y(f) = \frac{1}{j2\pi f} X(f) + \frac{1}{2} X(0)\, \delta(f)$ with $X(0) = \int_{-\infty}^{+\infty} x(t)\, dt$
Symmetry	$y(t) = X(t)$	$Y(f) = x(-f)$
Superposition	$z(t) = ax(t) + by(t)$	$Z(f) = aX(f) + bY(f)$
Multiplication	$z(t) = x(t) \cdot y(t)$	$Z(f) = X(f) * Y(f) =$ $\int_{-\infty}^{+\infty} X(f_0)\, Y(f - f_0)\, df_0$

Property	Time domain	Frequency domain
Convolution	$z(t) = x(t) * y(t) = \int_{-\infty}^{+\infty} x(\tau)\, y(t-\tau)\, \mathrm{d}\tau$	$Z(f) = X(f) \cdot Y(f)$
Correlation	$\mathring{\varphi}_{xy}(\tau) = \int_{-\infty}^{+\infty} x^*(t)\, y(t+\tau)\, \mathrm{d}t = x^*(-\tau) * y(\tau)$	$\mathring{\Phi}_{xy}(f) = X^*(f)\, Y(f)$
Product theorem	$\mathring{\varphi}_{xy}(0) = \int_{-\infty}^{+\infty} x^*(t)\, y(t)\, \mathrm{d}t$	$\mathring{\varphi}_{xy}(0) = \int_{-\infty}^{\infty} X^*(f)\, Y(f)\, \mathrm{d}f$
Autocorrelation and energy spectral density	$\mathring{\varphi}_x(\tau) = \int_{-\infty}^{+\infty} x^*(t)\, x(t+\tau)\, \mathrm{d}t = x^*(-\tau) * x(\tau)$	$\mathring{\Phi}_x(f) = X^*(f)\, X(f) = \lvert X(f)\rvert^2$
Parseval identity	$W_x = \mathring{\varphi}_x(0) = \int_{-\infty}^{+\infty} \lvert x(t)\rvert^2\, \mathrm{d}t$	$W_x = \int_{-\infty}^{+\infty} \mathring{\Phi}_x(f)\, \mathrm{d}f = \int_{-\infty}^{+\infty} \lvert X(f)\rvert^2\, \mathrm{d}f$

15.3.2 Finite average power signals

Property	Time domain	Frequency domain
Correlation	$\varphi_{xy}(\tau) = \lim_{T\to\infty} \frac{1}{T} \int_{-T/2}^{T/2} x^*(t)\, y(t+\tau)\, \mathrm{d}t$	$\Phi_{xy}(f) = \mathrm{F}\,[\varphi_{xy}(\tau)]$
Autocorrelation and power spectral density	$\varphi_x(\tau) = \lim_{T\to\infty} \frac{1}{T} \int_{-T/2}^{T/2} x^*(t)\, x(t+\tau)\, \mathrm{d}t$	$\Phi_x(f) = \mathrm{F}\,\{\varphi_x(\tau)\}$
Parseval identity	$P_x = \varphi_x(0) = \lim_{T\to\infty} \frac{1}{T} \int_{-T/2}^{T/2} \lvert x(t)\rvert^2\, \mathrm{d}t$	$P_x = \int_{-\infty}^{\infty} \Phi_x(f)\, \mathrm{d}f$

15.3.3 Particular case of periodic signals

Property	Time domain	Frequency domain
Transformation	$x(t) = x(t+T)$ $= \text{rep}_T[x(t,T)] =$ $= x(t,T) * \delta_T(t)$ $= \sum_{n=-\infty}^{\infty} X_n \, e^{jn2\pi f_1 t}$ with $f_1 = \frac{1}{T}$	$X(f) = \sum_{n=-\infty}^{\infty} X_n \delta(f - nf_1)$ $\doteq X(f,T) \frac{1}{T} \delta_{1/T}(f)$ with $X_n = \frac{1}{T} \int_{t_0}^{t_0+T} x(t) \, e^{-jn2\pi f_1 t} \, dt$ $= \frac{1}{T} X(f_n = nf_1, T)$
Multiplication of the variable by a constant	$y(t) = x(at)$	$Y(f) = \sum_{n=-\infty}^{\infty} X_n \delta(f - anf_1)$
Time translation (time-shift theorem)	$y(t) = x(t - \tau)$	$Y_n = X_n \, e^{-jn2\pi f_1 \tau}$
Differentiation	$y(t) = \frac{d^n}{dt^n} x(t)$	$Y_n = (jn2\pi f_1)^n X_n$
Integration	$y(t) = \int_{-\infty}^{t} x(u) \, du$	$Y_n = \frac{X_n}{jn2\pi f_1} \quad (n \neq 0)$ if $\bar{x} = X_0 = \frac{1}{T} \int_{t_0}^{t_0+T} x(t) \, dt = 0$
Convolution	$z(t) = x(t) \,\bar{*}\, y(t)$ $= \frac{1}{T} \int_{t_0}^{t_0+T} x(\tau) y(t-\tau) \, d\tau$ with $x(t) = x(t+T)$ et $y(t) = y(t+T)$	$Z_n = X_n \cdot Y_n$
Correlation	$\varphi_{xy}(\tau) = \frac{1}{T} \int_{t_0}^{t_0+T} x^*(t) y(t+\tau) \, dt$ $= x^*(-\tau) \,\bar{*}\, y(\tau)$ with $x(t) = x(t+T)$ et $y(t) = y(t+T)$	$\Phi_{xy}(f) =$ $\sum_{n=-\infty}^{\infty} X_n^* Y_n \delta(f - nf_1)$ with $f_1 = \frac{1}{T}$

15.3.3 Particular case of periodic signals *(continued)*

Property	Time domain	Frequency domain
Autocorrelation and power spectral density	$\varphi_x(\tau) = \frac{1}{T}\int_{t_0}^{t_0+T} x^*(t)\,x(t+\tau)\,\mathrm{d}t$ $= x^*(-\tau) \,\overline{*}\, x(\tau)$ $= \mathrm{rep}_T[T^{-1}\mathring{\varphi}_x(\tau,T)]$ with $x(t) = x(t+T)$; $y(t) = y(t+T)$ and $\mathring{\varphi}_x(\tau,T) = x^*(-\tau,T) * x(\tau,T)$	$\Phi_x(f) = \sum_{n=-\infty}^{\infty} \|X_n\|^2\,\delta(f-nf_1)$ $= \frac{1}{T^2}\,\mathring{\Phi}_x(f,T)\,\delta_{1/T}(f)$ with $\mathring{\Phi}_x(f,T) = \|X(f,T)\|^2$
Parseval identity	$P_x = \varphi_x(0) = \frac{1}{T}\int_{t_0}^{t_0+T} \|x(t)\|^2\,\mathrm{d}t$	$P_x = \int_{-\infty}^{+\infty} \Phi_x(f)\,\mathrm{d}f$ $= \sum_{n=-\infty}^{\infty} \|X_n\|^2$

15.3.4 Stationary random signals

Property	Time domain	Frequency domain
Correlation	$R_{xy}(\tau) = \mathrm{E}[\mathbf{x}^*(t)\,\mathbf{y}(t+\tau)]$	$\Phi_{xy}(f) = \mathrm{F}\{R_{xy}(\tau)\}$
Wiener-Khinchin theorem	$R_x(\tau) = \mathrm{E}[\mathbf{x}^*(t)\,\mathbf{x}(t+\tau)]$ $\equiv \varphi_x(\tau)$ if $\mathbf{x}(t)$ ergodic	$\Phi_x(f) = \mathrm{F}\{R_x(\tau)\}$ $= \lim_{T\to\infty} \mathrm{E}[\Phi_{xi}(f,T)]$ where $\Phi_{xi}(f,T) = T^{-1}\|X_i(f,T)\|^2$ (periodogram)
Parseval identity	$P_x = R_x(0) = \mathrm{E}[\mathbf{x}^2] = \sigma_x^2 + \mu_x^2$ $= \int_{-\infty}^{\infty} x^2 p(x)\,\mathrm{d}x$	$P_x = \int_{-\infty}^{\infty} \Phi_x(f)\,\mathrm{d}f$

15.4 GRAPHIC TABLE OF FOURIER TRANSFORMS

$x(t)$		$X(f)$	$\lvert X(f)\rvert$
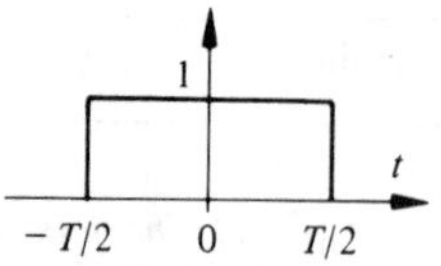 rectangular pulse	$\mathrm{rect}\,(t/T)$	$T\dfrac{\sin(\pi f T)}{\pi f T} = T\,\mathrm{sinc}\,(fT)$	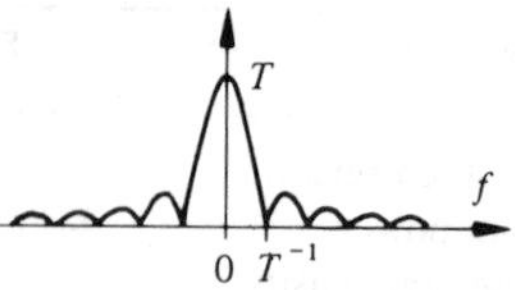
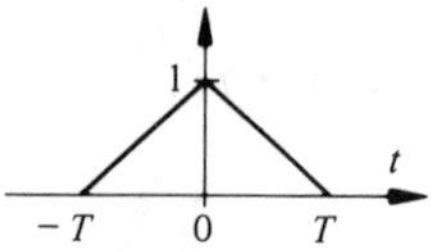 triangular pulse	$\mathrm{tri}\,(t/T)$	$= T\,\mathrm{sinc}^2\,(fT)$	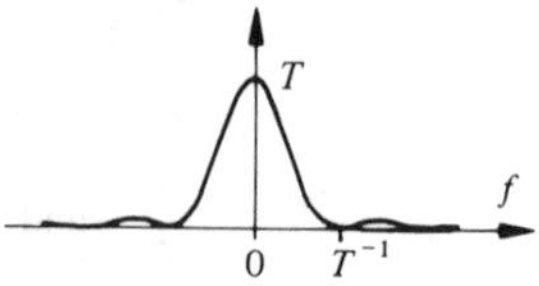
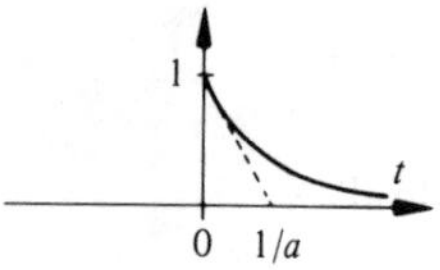 exponential pulse	$\mathrm{e}^{-at}\epsilon\,(t)$	$\dfrac{1}{a + \mathrm{j}2\pi f}$	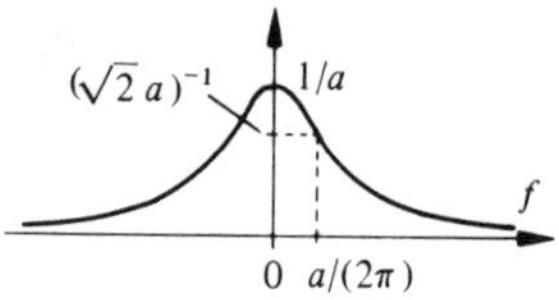
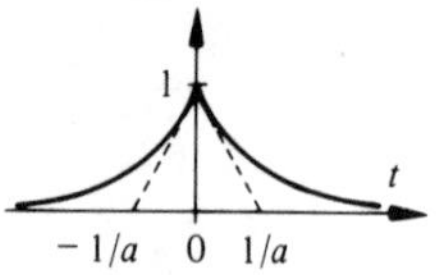 double exponential	$\mathrm{e}^{-a\lvert t\rvert}$	$\dfrac{2a}{a^2 + (2\pi f)^2}$	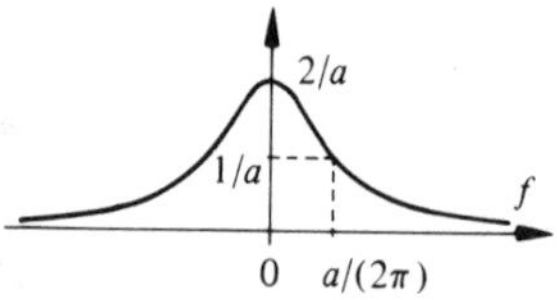
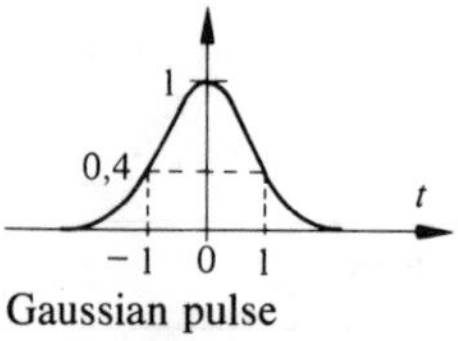 Gaussian pulse	$\mathrm{ig}\,(t) = \mathrm{e}^{-\pi t^2}$	$\mathrm{ig}\,(f) = \mathrm{e}^{-\pi f^2}$	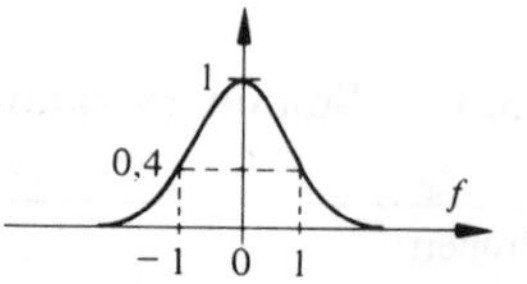
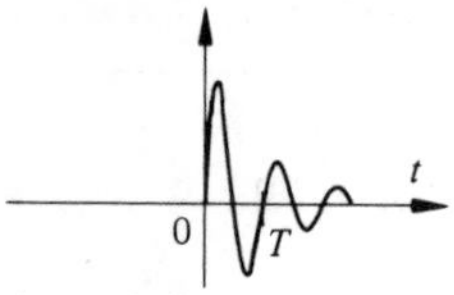 damped sine wave	$\mathrm{e}^{-at}\sin\,(2\pi f_0\,t)\,\epsilon\,(t)$	$\dfrac{2\pi f_0}{(a + \mathrm{j}2\pi f)^2 + (2\pi f_0)^2}$	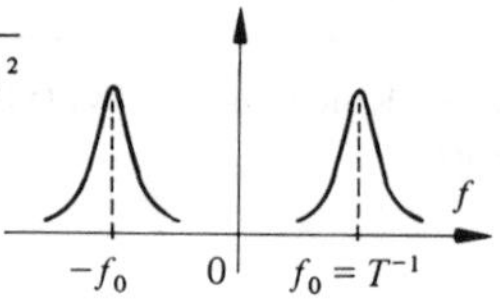
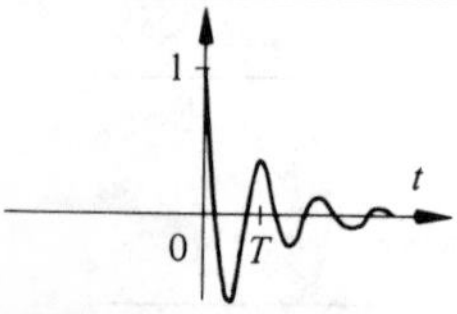 damped cosine wave	$\mathrm{e}^{-at}\cos\,(2\pi f_0\,t)\,\epsilon\,(t)$	$\dfrac{a + \mathrm{j}2\pi f}{(a + \mathrm{j}2\pi f)^2 + (2\pi f_0)^2}$	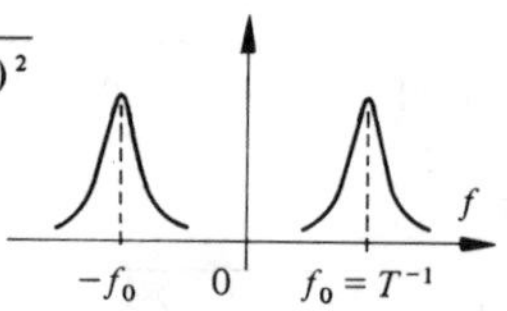

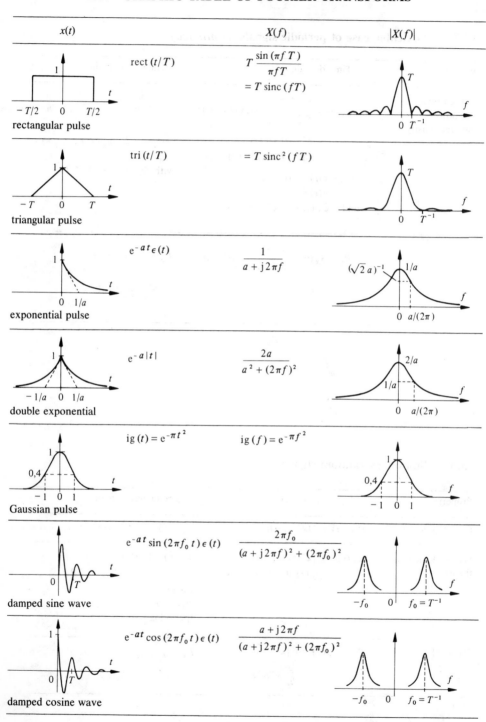

$x(t)$	$X(f)$	$\lvert X(f)\rvert$
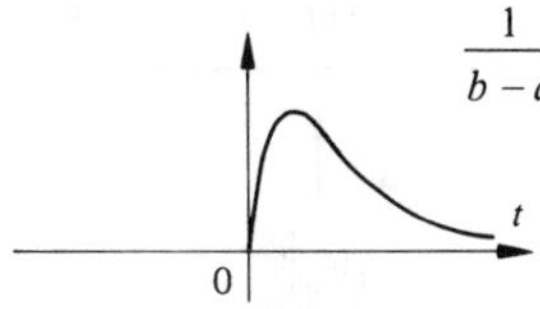 $\frac{1}{b-a}\left(e^{-at}-e^{-bt}\right)\epsilon(t)$	$\frac{1}{(a+j2\pi f)(b+j2\pi f)}$	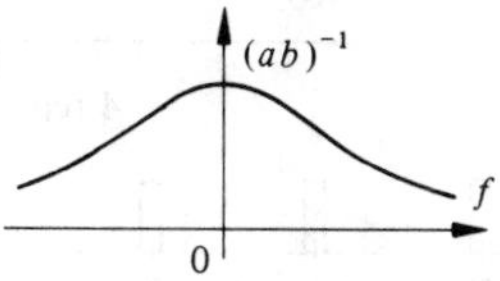
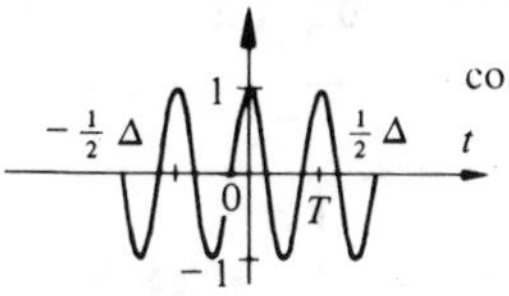 cosine pulse $\cos(2\pi f_0 t)\cdot \mathrm{rect}(t/\Delta)$	$\frac{\Delta}{2}\{\mathrm{sinc}[\Delta(f+f_0)] + \mathrm{sinc}[\Delta(f-f_0)]\}$	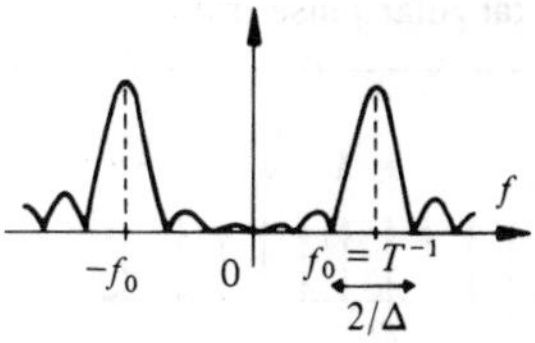
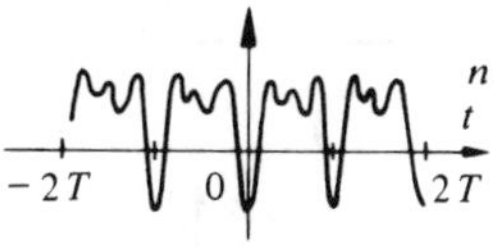 periodic signal $\sum_{n=-\infty}^{+\infty} X_n e^{-j2\pi \frac{nt}{T}}$	$\sum_{n=-\infty}^{+\infty} X_n \delta\left(f-\frac{n}{T}\right)$	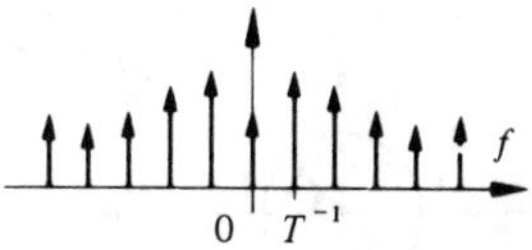
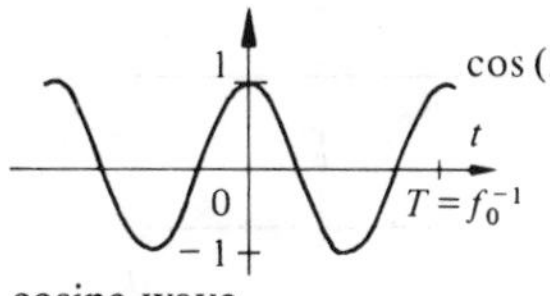 cosine wave $\cos(2\pi f_0 t)$	$\frac{1}{2}[\delta(f+f_0) + \delta(f-f_0)]$	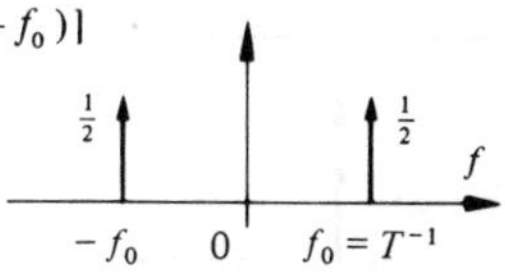
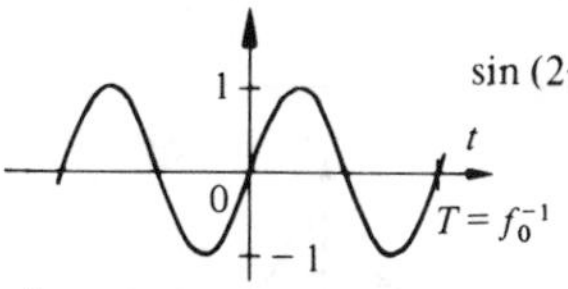 sine wave $\sin(2\pi f_0 t)$	$\frac{j}{2}[\delta(f+f_0) - \delta(f-f_0)]$	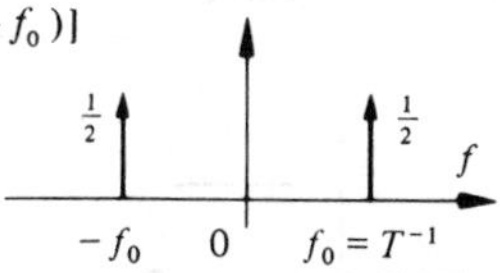
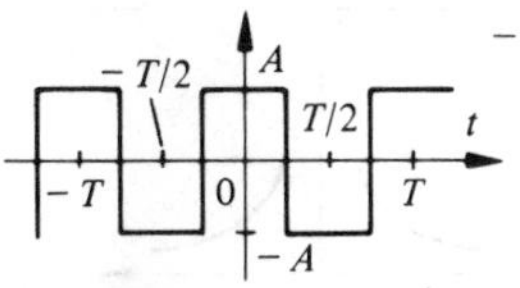 square wave $A\,\mathrm{rep}_T[2\,\mathrm{rect}(2t/T) - \mathrm{rect}(t/T)]$	$\sum_n X_n \delta(f-n/T)$ with $X_n = A\,\mathrm{sinc}(n/2)$ $= \begin{cases} \left\lvert \frac{2A}{\pi n} \right\rvert & \text{for } n = \pm 1, \pm 5, \ldots \\ -\left\lvert \frac{2A}{\pi n} \right\rvert & \text{for } n = \pm 1, \pm 5, \ldots \end{cases}$ $X_n = 0$ for n even or zero	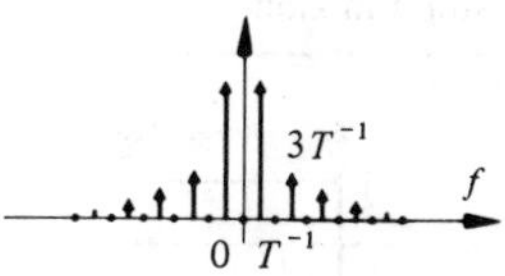

15.4 GRAPHIC TABLE OF FOURIER TRANSFORMS *(continued)*

$x(t)$		$X(f)$	$\|X(f)\|$
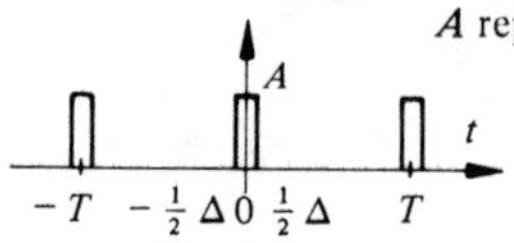 rectangular pulse train	$A \text{ rep}_T[\text{rect}(t/\Delta)]$	$\sum_n X_n \delta(f - n/T)$ with $X_n = \frac{A\Delta}{T} \text{sinc}(n\Delta/T)$	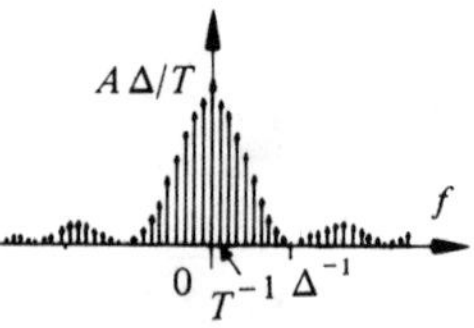
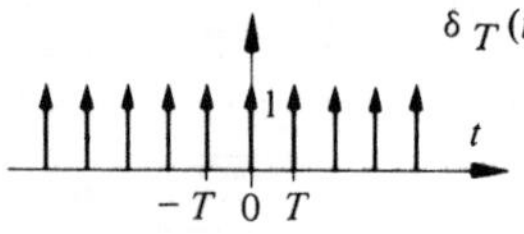 comb function	$\delta_T(t) = \sum_{n=-\infty}^{+\infty} \delta(t - nT)$	$\frac{1}{T} \sum_{n=-\infty}^{+\infty} \delta\left(f - \frac{n}{T}\right) = \frac{1}{T}\delta_{1/T}(f)$	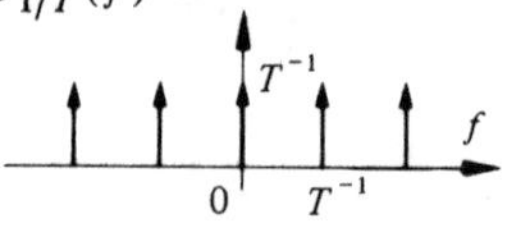
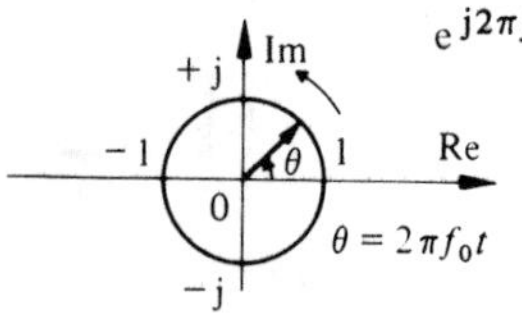	$e^{j2\pi f_0 t}$	$\delta(f - f_0)$	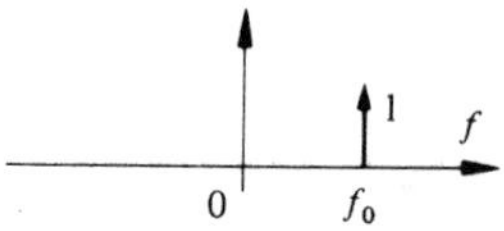
 delta-function (Dirac)	$\delta(t)$	1	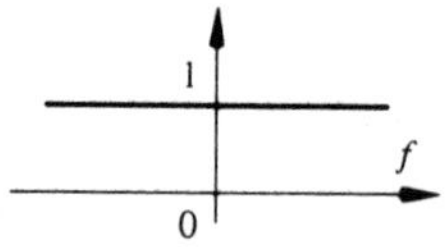
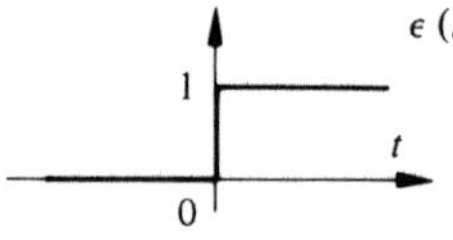 unit-step function	$\epsilon(t)$	$\frac{1}{2}\delta(f) + \frac{1}{j2\pi f}$	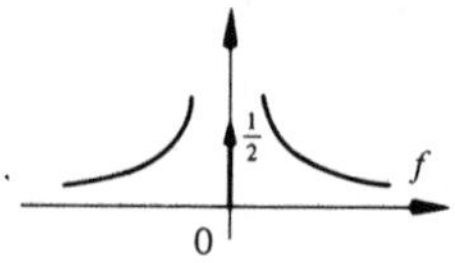
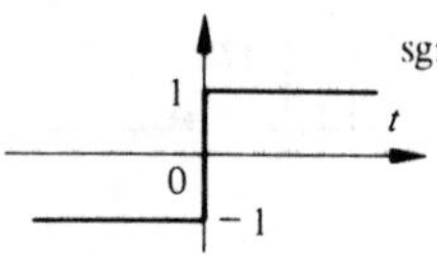 signum function	$\text{sgn}(t) = \frac{t}{\|t\|}$	$\frac{1}{j\pi f}$	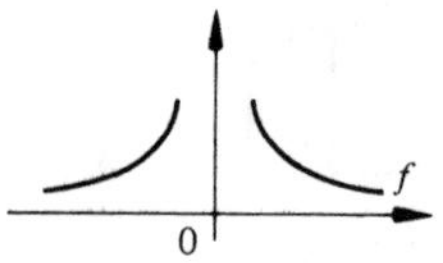
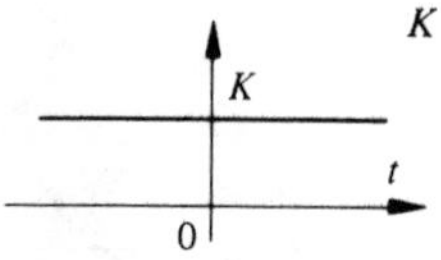 constant	K	$K\delta(f)$	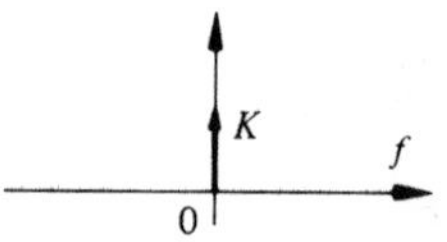

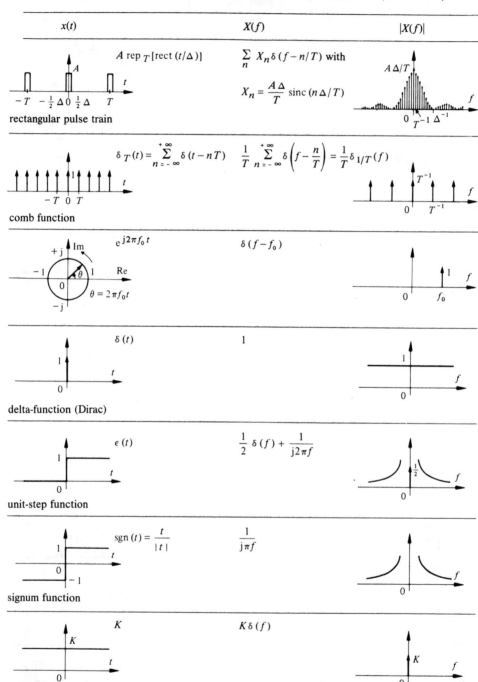

15.5 TIME, FREQUENCY, AND STATISTICAL DESCRIPTIONS OF TYPICAL SIGNALS

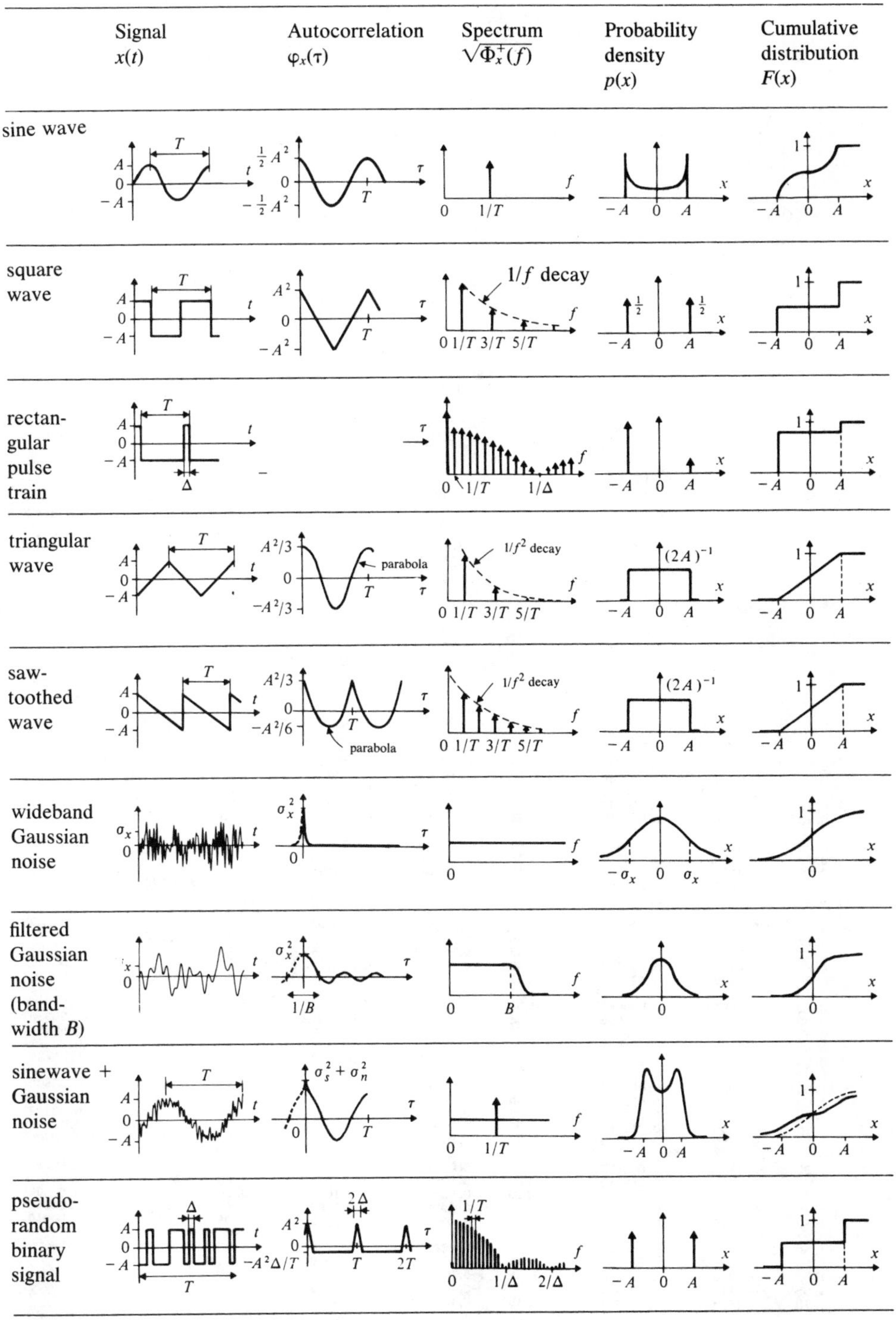

15.6 GRAPHS AND TABLES OF FUNCTIONS sinc(α) AND sinc²(α)

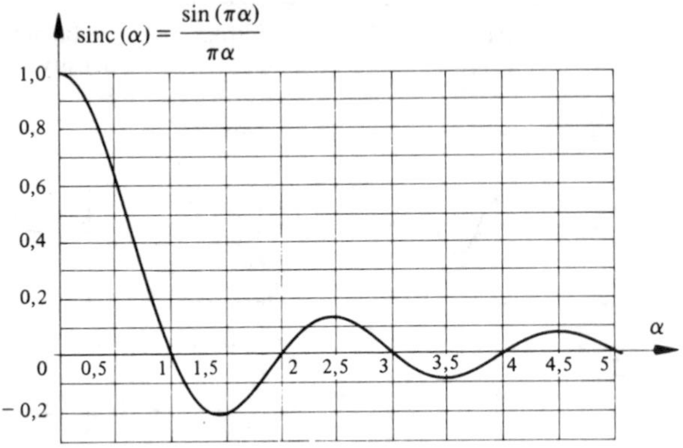

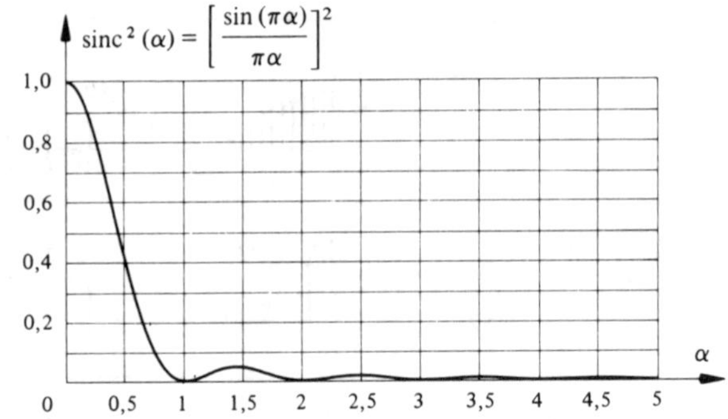

α	sinc (α)	sinc² (α)	α	sinc (α)	sinc² (α)	α	sinc (α)	sinc² (α)
0, 00	1, 000000	1, 000000	1, 75	−0, 128617	0, 016542	3, 50	−0, 090946	0, 008271
0, 05	0, 995893	0, 991802	1, 80	−0, 103943	0, 010804	3, 55	−0, 088561	0, 007843
0, 10	0, 983632	0, 967531	1, 85	−0, 078113	0, 006102	3, 60	−0, 084092	0, 007071
0, 15	0, 963398	0, 928135	1, 90	−0, 051770	0, 002680	3, 65	−0, 077703	0, 006038
0, 20	0, 935489	0, 875140	1, 95	−0, 025536	0, 000652	3, 70	−0, 069599	0, 004844
0, 25	0, 900316	0, 810569	2, 00	0, 000000	0, 000000	3, 75	−0, 060021	0, 003603
0, 30	0, 858394	0, 736840	2, 05	0, 024290	0, 000590	3, 80	−0, 049236	0, 002424
0, 35	0, 810332	0, 656638	2, 10	0, 046840	0, 002194	3, 85	−0, 037535	0, 001409
0, 40	0, 756827	0, 572787	2, 15	0, 067214	0, 004518	3, 90	−0, 025221	0, 000636
0, 45	0, 698647	0, 488107	2, 20	0, 085044	0, 007233	3, 95	−0, 012606	0, 000159
0, 50	0, 636620	0, 405285	2, 25	0, 100035	0, 010007	4, 00	0, 000000	0, 000000
0, 55	0, 571620	0, 326749	2, 30	0, 111964	0, 012536	4, 05	0, 012295	0, 000151
0, 60	0, 504551	0, 254572	2, 35	0, 120688	0, 014566	4, 10	0, 023991	0, 000576
0, 65	0, 436333	0, 190386	2, 40	0, 126138	0, 015911	4, 15	0, 034822	0, 001213
0, 70	0, 367883	0, 135338	2, 45	0, 128323	0, 016467	4, 20	0, 044547	0, 001984
0, 75	0, 300105	0, 090063	2, 50	0, 127324	0, 016211	4, 25	0, 052960	0, 002805
0, 80	0, 233872	0, 054696	2, 55	0, 123291	0, 015201	4, 30	0, 059888	0, 003587
0, 85	0, 170011	0, 028904	2, 60	0, 116435	0, 013557	4, 35	0, 065199	0, 004251
0, 90	0, 109292	0, 011945	2, 65	0, 107025	0, 011454	4, 40	0, 068802	0, 004734
0, 95	0, 052415	0, 002747	2, 70	0, 095377	0, 009097	4, 45	0, 070650	0, 004991
1, 00	−0, 000000	0, 000000	2, 75	0, 081847	0, 006699	4, 50	0, 070736	0, 005004
1, 05	−0, 047423	0, 002249	2, 80	0, 066821	0, 004465	4, 55	0, 069097	0, 004774
1, 10	−0, 089421	0, 007996	2, 85	0, 050705	0, 002571	4, 60	0, 065811	0, 004331
1, 15	−0, 125661	0, 015791	2, 90	0, 033918	0, 001150	4, 65	0, 060993	0, 003720
1, 20	−0, 155915	0, 024309	2, 95	0, 016880	0, 000285	4, 70	0, 054791	0, 003002
1, 25	−0, 180063	0, 032423	3, 00	−0, 000000	0, 000000	4, 75	0, 047385	0, 002245
1, 30	−0, 198091	0, 039240	3, 05	−0, 016326	0, 000267	4, 80	0, 038979	0, 001519
1, 35	−0, 210086	0, 044136	3, 10	−0, 031730	0, 001007	4, 85	0, 029796	0, 000888
1, 40	−0, 216236	0, 046758	3, 15	−0, 045876	0, 002105	4, 90	0, 020074	0, 000403
1, 45	−0, 216821	0, 047011	3, 20	−0, 058468	0, 003419	4, 95	0, 010060	0, 000101
1, 50	−0, 212207	0, 045032	3, 25	−0, 069255	0, 004796	5, 00	−0, 000000	0, 000000
1, 55	−0, 202833	0, 041141	3, 30	−0, 078036	0, 006090			
1, 60	−0, 189207	0, 035799	3, 35	−0, 084662	0, 007168			
1, 65	−0, 171889	0, 029546	3, 40	−0, 089038	0, 007928			
1, 70	−0, 151481	0, 022947	3, 45	−0, 091128	0, 008304			

15.7 BESSEL FUNCTIONS OF THE FIRST KIND

15.7.1 General relations

The Bessel function of the first kind and nth (integer) order $J_n(x)$ is one of the particular solutions of the differential Bessel equation [88]:

$$x^2 \frac{d^2y}{dx^2} + x \frac{dy}{dx} + (x^2 - n^2)y = 0 \tag{15.1}$$

The $J_n(x)$ integral representation is given by

$$J_n(x) = \frac{1}{\pi} \int_0^{\pi} \cos(x \sin \theta - n\theta)d\theta$$

$$= \frac{1}{2\pi} \int_{-\pi}^{\pi} \exp(jx \sin \theta - n\theta)d\theta \tag{15.2}$$

The $J_n(x)$ series expansion is given by

$$J_n(x) = \sum_{k=0}^{\infty} \frac{(-1)^k}{k!(n + k)!} \left(\frac{x}{2}\right)^{n+2k} \tag{15.3}$$

We also have the relation:

$$J_{-n}(x) = (-1)^n J_n(x) \tag{15.4}$$

The generating functions and the associated series are

$$\cos(x \sin \theta) = J_0(x) + 2 \sum_{k=1}^{\infty} J_{2k}(x) \cos(2k\theta) \tag{15.5}$$

$$\sin(x \sin \theta) = 2 \sum_{k=0}^{\infty} J_{2k+1}(x) \sin[(2k + 1)\theta] \tag{15.6}$$

$$\exp(jx \sin \theta) = \sum_{n=-\infty}^{\infty} J_n(x) \exp(jn\theta) \tag{15.7}$$

and, by writing $\theta = 0$:

$$\sum_{n=-\infty}^{\infty} J_n(x) = 1 \tag{15.8}$$

The $J_n(x)$ functions for n between 0 and 4 are represented in fig. 15.1. Figure 15.2 gives the amplitudes obtained as a function of the order n for some constant values of the argument.

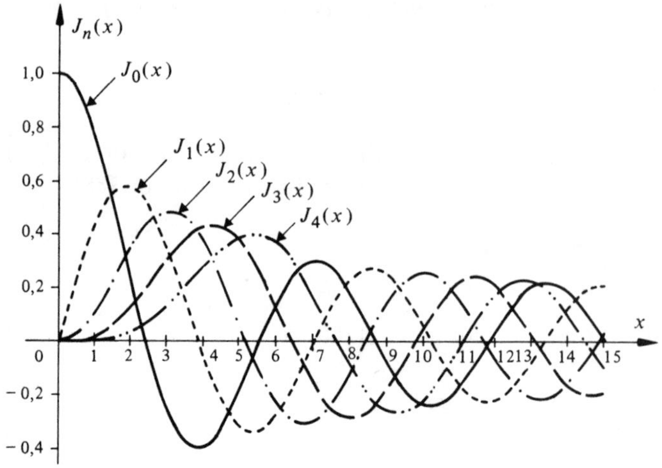

Fig. 15.1

15.8 NORMAL DISTRIBUTION

15.8.1 Standardization of the Gaussian distribution

The probability density

$$p(x) = \frac{1}{\sqrt{2\pi}\sigma_x} \exp\left[-\frac{(x-\mu_x)^2}{2\sigma_x^2}\right] \tag{15.9}$$

and its integral $F(x)$ are normalized by the change of variable

$$z = (x - \mu_x)/\sigma_x \tag{15.10}$$

We then obtain the ***standardized normal probability density*** (Fig. 15.3):

$$p(z) = \frac{1}{\sqrt{2\pi}} \exp\left(-\frac{1}{2}z^2\right) \tag{15.11}$$

The corresponding cumulative distribution is

$$F(z) = \int_{-\infty}^{z} p(z')\mathrm{d}z' \tag{15.12}$$

and the complementary cumulative distribution (fig. 15.4) is equal to

$$F_c(z) = 1 - F(z)$$

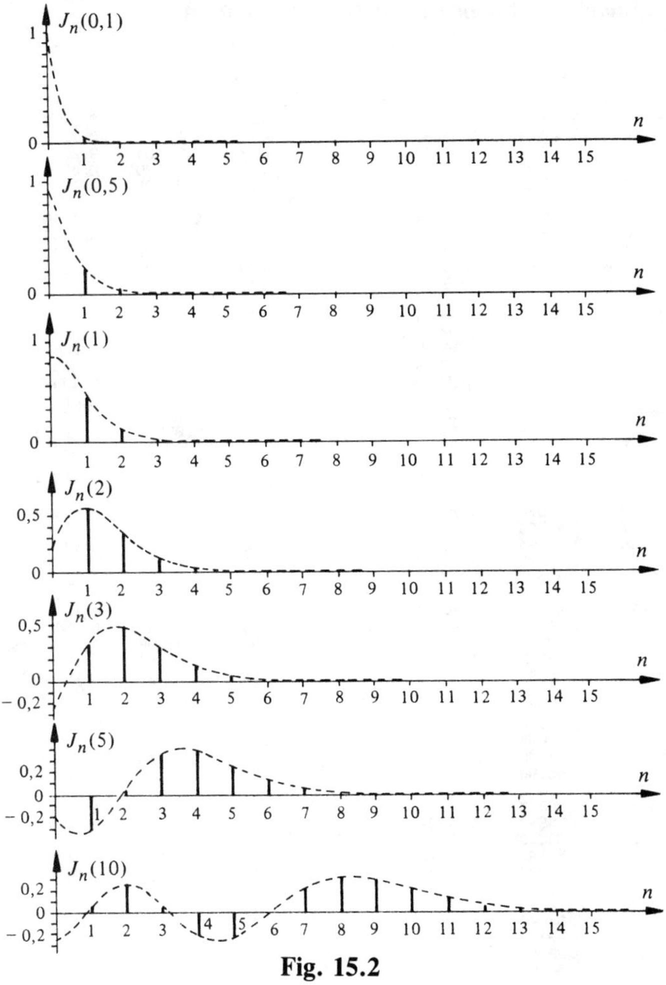

Fig. 15.2

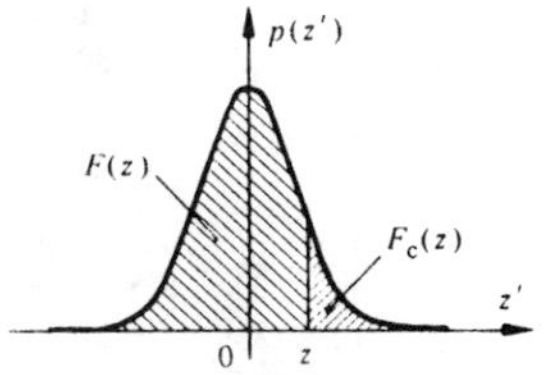

Fig. 15.3

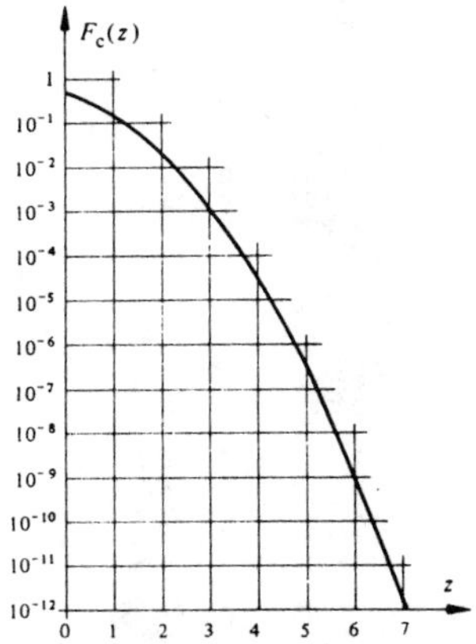

Fig. 15.4

15.8.2 Numerical table of the normal distribution

z	$F(z)$	$F_c(z)=1-F(z)$	$p(z)$	z	$F(z)$	$F_c(z)=1-F(z)$	$p(z)$
0.00	0.50000	0.50000	0.39894	0.70	0.75804	0.24196	0.31225
0.01	0.50399	0.49601	0.39892	0.71	0.76115	0.23885	0.31006
0.02	0.50798	0.49202	0.39886	0.72	0.76424	0.23576	0.30785
0.03	0.51197	0.48803	0.39876	0.73	0.76730	0.23270	0.30563
0.04	0.51595	0.48405	0.39862	0.74	0.77035	0.22965	0.30339
0.05	0.51994	C.48006	0.39844	0.75	0.77337	0.22663	0.30114
0.06	0.52392	0.47608	0.39822	0.76	0.77637	0.22363	0.29887
0.07	0.52790	0.47210	0.39797	0.77	0.77935	0.22065	0.29659
0.08	0.53188	0.46812	0.39767	0.78	0.78230	0.21770	0.29431
0.09	0.53586	0.46414	0.39733	0.79	0.78524	0.21476	0.29200
0.10	0.53983	0.46017	0.39695	0.80	0.78814	0.21186	0.28969
0.11	0.54380	0.45620	0.39654	0.81	0.79103	0.20897	0.28737
0.12	0.54776	0.45224	0.39608	0.82	0.79389	0.20611	0.28504
0.13	0.55172	0.44828	0.39559	0.83	0.79673	0.20327	0.28269
0.14	0.55567	0.44433	0.39505	0.84	0.79955	0.20045	0.28034
0.15	0.55962	0.44038	0.39448	0.85	0.80234	0.19766	0.27798
0.16	0.56356	0.43644	0.39387	0.86	0.80511	0.19489	0.27562
0.17	0.56749	0.43251	0.39322	0.87	0.80785	0.19215	0.27324
0.18	0.57142	0.42858	0.39253	0.88	0.81057	0.18943	0.27086
0.19	0.57535	0.42465	0.39181	0.89	0.81327	0.18673	0.26848
0.20	0.57926	0.42074	0.39104	0.90	0.81594	0.18406	0.26609
0.21	0.58317	0.41683	0.39024	0.91	0.81859	0.18141	0.26369
0.22	0.58706	0.41294	0.38940	0.92	0.82121	0.17879	0.26129
0.23	0.59095	0.40905	0.38853	0.93	0.82381	0.17619	0.25888
0.24	0.59483	0.40517	0.38762	0.94	0.82639	0.17361	0.25647
0.25	0.59871	0.40129	0.38667	0.95	0.82894	0.17106	0.25406
0.26	0.60257	0.39743	0.38568	0.96	0.83147	0.16853	0.25164
0.27	0.60642	0.39358	0.38466	0.97	0.83398	0.16602	0.24923
0.28	0.61026	0.38974	0.38361	0.98	0.83646	0.16354	0.24681
0.29	0.61409	0.38591	0.38251	0.99	0.83891	0.16109	0.24439
0.30	0.61791	0.38209	0.38139	1.00	0.84134	0.15866	0.24197
0.31	0.62172	0.37828	0.38023	1.01	0.84375	0.15625	0.23955
0.32	0.62552	0.37448	0.37903	1.02	0.84614	0.15386	0.23713
0.33	0.62930	0.37070	0.37780	1.03	0.84849	0.15151	0.23471
0.34	0.63307	0.36693	0.37654	1.04	0.85083	0.14917	0.23230
0.35	0.63683	0.36317	0.37524	1.05	0.85314	0.14686	0.22988
0.36	0.64058	0.35942	0.37391	1.06	0.85543	0.14457	0.22747
0.37	0.64431	0.35569	0.37255	1.07	0.85769	0.14231	0.22506
0.38	0.64803	0.35197	0.37115	1.08	0.85993	0.14007	0.22265
0.39	0.65173	0.34827	0.36973	1.09	0.86214	0.13786	0.22025
0.40	0.65542	0.34458	0.36827	1.10	0.86433	0.13567	0.21785
0.41	0.65910	0.34090	0.36678	1.11	0.86650	0.13350	0.21546
0.42	0.66276	0.33724	0.36526	1.12	0.86864	0.13136	0.21307
0.43	0.66640	0.33360	0.36371	1.13	0.87076	0.12924	0.21069
0.44	0.67003	0.32997	0.36213	1.14	0.87286	0.12714	0.20831
0.45	0.67364	0.32636	0.36053	1.15	0.87493	0.12507	0.20594
0.46	0.67724	0.32276	0.35889	1.16	0.87698	0.12302	0.20357
0.47	0.68082	0.31918	0.35723	1.17	0.87900	0.12100	0.20121
0.48	0.68439	0.31561	0.35553	1.18	0.88100	0.11900	0.19886
0.49	0.68793	0.31207	0.35381	1.19	0.88298	0.11702	0.19652
0.50	0.69146	0.30854	0.35207	1.20	0.88493	0.11507	0.19419
0.51	0.69497	0.30503	0.35029	1.21	0.88686	0.11314	0.19186
0.52	0.69847	0.30153	0.34849	1.22	0.88877	0.11123	0.18954
0.53	0.70194	0.29806	0.34667	1.23	0.89065	0.10935	0.18724
0.54	0.70540	0.29460	0.34482	1.24	0.89251	0.10749	0.18494
0.55	0.70884	0.29116	0.34294	1.25	0.89435	0.10565	0.18265
0.56	0.71226	0.28774	0.34105	1.26	0.89617	0.10383	0.18037
0.57	0.71566	0.28434	0.33912	1.27	0.89796	0.10204	0.17810
0.58	0.71904	0.28096	0.33718	1.28	0.89973	0.10027	0.17585
0.59	0.72240	0.27760	0.33521	1.29	0.90147	0.09853	0.17360
0.60	0.72575	0.27425	0.33322	1.30	0.90320	0.09680	0.17137
0.61	0.72907	0.27093	0.33121	1.31	0.90490	0.09510	0.16915
0.62	0.73237	0.26763	0.32918	1.32	0.90658	0.09342	0.16694
0.63	0.73565	0.26435	0.32713	1.33	0.90824	0.09176	0.16474
0.64	0.73891	0.26109	0.32506	1.34	0.90988	0.09012	0.16256
0.65	0.74215	0.25785	0.32297	1.35	0.91149	0.08851	0.16038
0.66	0.74537	0.25463	0.32086	1.36	0.91309	0.08691	0.15822
0.67	0.74857	0.25143	0.31874	1.37	0.91466	0.08534	0.15608
0.68	0.75175	0.24825	0.31659	1.38	0.91621	0.08379	0.15395
0.69	0.75490	0.24510	0.31443	1.39	0.91774	0.08226	0.15183

15.8.2 Numerical table of the normal distribution *(continued)*

z	$F(z)$	$F_c(z)=1-F(z)$	$p(z)$	z	$F(z)$	$F_c(z)=1-F(z)$	$p(z)$
1.40	0.91924	0.08076	0.14973	2.10	0.98214	0.01786	0.04398
1.41	0.92073	0.07927	0.14764	2.11	0.98257	0.01743	0.04307
1.42	0.92220	0.07780	0.14556	2.12	0.98300	0.01700	0.04217
1.43	0.92364	0.07636	0.14350	2.13	0.98341	0.01659	0.04128
1.44	0.92507	0.07493	0.14146	2.14	0.98382	0.01618	0.04041
1.45	0.92647	0.07353	0.13943	2.15	0.98422	0.01578	0.03955
1.46	0.92785	0.07215	0.13742	2.16	0.98461	0.01539	0.03871
1.47	0.92922	0.07078	0.13542	2.17	0.98500	0.01500	0.03788
1.48	0.93056	0.06944	0.13344	2.18	0.98537	0.01463	0.03706
1.49	0.93189	0.06811	0.13147	2.19	0.98574	0.01426	0.03626
1.50	0.93319	0.06681	0.12952	2.20	0.98610	0.01390	0.03547
1.51	0.93448	0.06552	0.12758	2.21	0.98645	0.01355	0.03470
1.52	0.93574	0.06426	0.12566	2.22	0.98679	0.01321	0.03394
1.53	0.93699	0.06301	0.12376	2.23	0.98713	0.01287	0.03319
1.54	0.93822	0.06178	0.12188	2.24	0.98745	0.01255	0.03246
1.55	0.93943	0.06057	0.12001	2.25	0.98778	0.01222	0.03174
1.56	0.94062	0.05938	0.11816	2.26	0.98809	0.01191	0.03103
1.57	0.94179	0.05821	0.11632	2.27	0.98840	0.01160	0.03034
1.58	0.94295	0.05705	0.11450	2.28	0.98870	0.01130	0.02965
1.59	0.94408	0.05592	0.11270	2.29	0.98899	0.01101	0.02898
1.60	0.94520	0.05480	0.11092	2.30	0.98928	0.01072	0.02833
1.61	0.94630	0.05370	0.10915	2.31	0.98956	0.01044	0.02768
1.62	0.94738	0.05262	0.10741	2.32	0.98983	0.01017	0.02705
1.63	0.94845	0.05155	0.10567	2.33	0.99010	0.00990	0.02643
1.64	0.94950	0.05050	0.10396	2.34	0.99036	0.00964	0.02582
1.65	0.95053	0.04947	0.10226	2.35	0.99061	0.00939	0.02522
1.66	0.95154	0.04846	0.10059	2.36	0.99086	0.00914	0.02463
1.67	0.95254	0.04746	0.09893	2.37	0.99111	0.00889	0.02406
1.68	0.95352	0.04648	0.09728	2.38	0.99134	0.00866	0.02349
1.69	0.95449	0.04551	0.09566	2.39	0.99158	0.00842	0.02294
1.70	0.95543	0.04457	0.09405	2.40	0.99180	0.00820	0.02239
1.71	0.95637	0.04363	0.09246	2.41	0.99202	0.00798	0.02186
1.72	0.95728	0.04272	0.09089	2.42	0.99224	0.00776	0.02134
1.73	0.95818	0.04182	0.08933	2.43	0.99245	0.00755	0.02083
1.74	0.95907	0.04093	0.08780	2.44	0.99266	0.00734	0.02033
1.75	0.95994	0.04006	0.08628	2.45	0.99286	0.00714	0.01984
1.76	0.96080	0.03920	0.08478	2.46	0.99305	0.00695	0.01936
1.77	0.96164	0.03836	0.08329	2.47	0.99324	0.00676	0.01888
1.78	0.96246	0.03754	0.08183	2.48	0.99343	0.00657	0.01842
1.79	0.96327	0.03673	0.08038	2.49	0.99361	0.00639	0.01797
1.80	0.96407	0.03593	0.07895	2.50	0.99379	0.00621	0.01753
1.81	0.96485	0.03515	0.07754	2.51	0.99396	0.00604	0.01709
1.82	0.96562	0.03438	0.07614	2.52	0.99413	0.00587	0.01667
1.83	0.96638	0.03362	0.07477	2.53	0.99430	0.00570	0.01625
1.84	0.96712	0.03288	0.07341	2.54	0.99446	0.00554	0.01585
1.85	0.96784	0.03216	0.07206	2.55	0.99461	0.00539	0.01545
1.86	0.96856	0.03144	0.07074	2.56	0.99477	0.00523	0.01506
1.87	0.96926	0.03074	0.06943	2.57	0.99492	0.00508	0.01468
1.88	0.96995	0.03005	0.06814	2.58	0.99506	0.00494	0.01431
1.89	0.97062	0.02938	0.06687	2.59	0.99520	0.00480	0.01394
1.90	0.97128	0.02872	0.06562	2.60	0.99534	0.00466	0.01358
1.91	0.97193	0.02807	0.06438	2.61	0.99547	0.00453	0.01323
1.92	0.97257	0.02743	0.06316	2.62	0.99560	0.00440	0.01289
1.93	0.97320	0.02680	0.06195	2.63	0.99573	0.00427	0.01256
1.94	0.97381	0.02619	0.06077	2.64	0.99585	0.00415	0.01223
1.95	0.97441	0.02559	0.05959	2.65	0.99598	0.00402	0.01191
1.96	0.97500	0.02500	0.05844	2.66	0.99609	0.00391	0.01160
1.97	0.97558	0.02442	0.05730	2.67	0.99621	0.00379	0.01130
1.98	0.97615	0.02385	0.05618	2.68	0.99632	0.00368	0.01100
1.99	0.97670	0.02330	0.05508	2.69	0.99643	0.00357	0.01071
2.00	0.97725	0.02275	0.05399	2.70	0.99653	0.00347	0.01042
2.01	0.97778	0.02222	0.05292	2.71	0.99664	0.00336	0.01014
2.02	0.97831	0.02169	0.05186	2.72	0.99674	0.00326	0.00987
2.03	0.97882	0.02118	0.05082	2.73	0.99683	0.00317	0.00961
2.04	0.97932	0.02068	0.04980	2.74	0.99693	0.00307	0.00935
2.05	0.97982	0.02018	0.04879	2.75	0.99702	0.00298	0.00909
2.06	0.98030	0.01970	0.04780	2.76	0.99711	0.00289	0.00885
2.07	0.98077	0.01923	0.04682	2.77	0.99720	0.00280	0.00861
2.08	0.98124	0.01876	0.04586	2.78	0.99728	0.00272	0.00837
2.09	0.98169	0.01831	0.04491	2.79	0.99736	0.00264	0.00814

15.8.2 Numerical table of the normal distribution *(continued)*

z	$F(z)$	$F_c(z)=1-F(z)$	$p(z)$	z	$F(z)$	$F_c(z)=1-F(z)$	$p(z)$
2.80	0.99744	0.00256	0.00792	3.50	0.9997674	2.33E-004	8.73E-004
2.81	0.99752	0.00248	0.00770	3.51	0.9997759	2.24E-004	8.43E-004
2.82	0.99760	0.00240	0.00748	3.52	0.9997842	2.16E-004	8.14E-004
2.83	0.99767	0.00233	0.00727	3.53	0.9997922	2.08E-004	7.85E-004
2.84	0.99774	0.00226	0.00707	3.54	0.9997999	2.00E-004	7.58E-004
2.85	0.99781	0.00219	0.00687	3.55	0.9998074	1.93E-004	7.32E-004
2.86	0.99788	0.00212	0.00668	3.56	0.9998146	1.85E-004	7.06E-004
2.87	0.99795	0.00205	0.00649	3.57	0.9998215	1.78E-004	6.81E-004
2.88	0.99801	0.00199	0.00631	3.58	0.9998282	1.72E-004	6.57E-004
2.89	0.99807	0.00193	0.00613	3.59	0.9998347	1.65E-004	6.34E-004
2.90	0.99813	0.00187	0.00595	3.60	0.9998409	1.59E-004	6.12E-004
2.91	0.99819	0.00181	0.00578	3.61	0.9998469	1.53E-004	5.90E-004
2.92	0.99825	0.00175	0.00562	3.62	0.9998527	1.47E-004	5.69E-004
2.93	0.99831	0.00169	0.00545	3.63	0.9998583	1.42E-004	5.49E-004
2.94	0.99836	0.00164	0.00530	3.64	0.9998637	1.36E-004	5.29E-004
2.95	0.99841	0.00159	0.00514	3.65	0.9998689	1.31E-004	5.10E-004
2.96	0.99846	0.00154	0.00499	3.66	0.9998739	1.26E-004	4.92E-004
2.97	0.99851	0.00149	0.00485	3.67	0.9998787	1.21E-004	4.74E-004
2.98	0.99856	0.00144	0.00470	3.68	0.9998834	1.17E-004	4.57E-004
2.99	0.99861	0.00139	0.00457	3.69	0.9998879	1.12E-004	4.41E-004
3.00	0.99865	0.00135	0.00443	3.70	0.9998922	1.08E-004	4.25E-004
3.01	0.99869	0.00131	0.00430	3.71	0.9998964	1.04E-004	4.09E-004
3.02	0.99874	0.00126	0.00417	3.72	0.9999004	9.96E-005	3.94E-004
3.03	0.99878	0.00122	0.00405	3.73	0.9999043	9.57E-005	3.80E-004
3.04	0.99882	0.00118	0.00393	3.74	0.9999080	9.20E-005	3.66E-004
3.05	0.99886	0.00114	0.00381	3.75	0.9999116	8.84E-005	3.53E-004
3.06	0.99889	0.00111	0.00370	3.76	0.9999150	8.50E-005	3.40E-004
3.07	0.99893	0.00107	0.00358	3.77	0.9999184	8.16E-005	3.27E-004
3.08	0.99896	0.00104	0.00348	3.78	0.9999216	7.84E-005	3.15E-004
3.09	0.99900	0.00100	0.00337	3.79	0.9999247	7.53E-005	3.03E-004
3.10	0.99903	0.00097	0.00327	3.80	0.9999277	7.23E-005	2.92E-004
3.11	0.99906	0.00094	0.00317	3.81	0.9999305	6.95E-005	2.81E-004
3.12	0.99910	0.00090	0.00307	3.82	0.9999333	6.67E-005	2.71E-004
3.13	0.99913	0.00087	0.00298	3.83	0.9999359	6.41E-005	2.60E-004
3.14	0.99916	0.00084	0.00288	3.84	0.9999385	6.15E-005	2.51E-004
3.15	0.99918	0.00082	0.00279	3.85	0.9999409	5.91E-005	2.41E-004
3.16	0.99921	0.00079	0.00271	3.86	0.9999433	5.67E-005	2.32E-004
3.17	0.99924	0.00076	0.00262	3.87	0.9999456	5.44E-005	2.23E-004
3.18	0.99926	0.00074	0.00254	3.88	0.9999478	5.22E-005	2.15E-004
3.19	0.99929	0.00071	0.00246	3.89	0.9999499	5.01E-005	2.07E-004
3.20	0.99931	0.00069	0.00238	3.90	0.9999519	4.81E-005	1.99E-004
3.21	0.99934	0.00066	0.00231	3.91	0.9999539	4.61E-005	1.91E-004
3.22	0.99936	0.00064	0.00224	3.92	0.9999557	4.43E-005	1.84E-004
3.23	0.99938	0.00062	0.00216	3.93	0.9999575	4.25E-005	1.77E-004
3.24	0.99940	0.00060	0.00210	3.94	0.9999593	4.07E-005	1.70E-004
3.25	0.99942	0.00058	0.00203	3.95	0.9999609	3.91E-005	1.63E-004
3.26	0.99944	0.00056	0.00196	3.96	0.9999625	3.75E-005	1.57E-004
3.27	0.99946	0.00054	0.00190	3.97	0.9999641	3.59E-005	1.51E-004
3.28	0.99948	0.00052	0.00184	3.98	0.9999655	3.45E-005	1.45E-004
3.29	0.99950	0.00050	0.00178	3.99	0.9999670	3.30E-005	1.39E-004
3.30	0.99952	0.00048	0.00172	4.00	0.9999683	3.17E-005	1.34E-004
3.31	0.99953	0.00047	0.00167	4.05	0.9999744	2.56E-005	1.09E-004
3.32	0.99955	0.00045	0.00161	4.10	0.9999793	2.07E-005	8.93E-005
3.33	0.99957	0.00043	0.00156	4.15	0.9999834	1.66E-005	7.26E-005
3.34	0.99958	0.00042	0.00151	4.20	0.9999867	1.33E-005	5.89E-005
3.35	0.99960	0.00040	0.00146	4.25	0.9999893	1.07E-005	4.77E-005
3.36	0.99961	0.00039	0.00141	4.30	0.9999915	8.54E-006	3.85E-005
3.37	0.99962	0.00038	0.00136	4.35	0.9999932	6.81E-006	3.10E-005
3.38	0.99964	0.00036	0.00132	4.40	0.9999946	5.41E-006	2.49E-005
3.39	0.99965	0.00035	0.00127	4.45	0.9999957	4.29E-006	2.00E-005
3.40	0.99966	0.00034	0.00123	4.50	0.9999966	3.40E-006	1.60E-005
3.41	0.99968	0.00032	0.00119	4.55	0.9999973	2.68E-006	1.27E-005
3.42	0.99969	0.00031	0.00115	4.60	0.9999979	2.11E-006	1.01E-005
3.43	0.99970	0.00030	0.00111	4.65	0.9999983	1.66E-006	8.05E-006
3.44	0.99971	0.00029	0.00107	4.70	0.9999987	1.30E-006	6.37E-006
3.45	0.99972	0.00028	0.00104	4.75	0.9999990	1.02E-006	5.03E-006
3.46	0.99973	0.00027	0.00100	4.80	0.9999992	7.93E-007	3.96E-006
3.47	0.99974	0.00026	0.00097	4.85	0.9999994	6.17E-007	3.11E-006
3.48	0.99975	0.00025	0.00094	4.90	0.9999995	4.79E-007	2.44E-006
3.49	0.99976	0.00024	0.00090	4.95	0.9999996	3.71E-007	1.91E-006

15.8.2 Numerical table of the normal distribution *(continued)*

z	$F(z)$	$F_c(z) = 1 - F(z)$	$p(z)$	z	$F(z)$	$F_c(z) = 1 - F(z)$	$p(z)$
5,00	0,9999997	2,87E-007	1,49E-006	6,00	1,0000000	9,87E-010	6,08E-009
5,05	0,9999998	2,21E-007	1,16E-006	6,10	1,0000000	5,30E-010	3,32E-009
5,10	0,9999998	1,70E-007	8,97E-007	6,20	1,0000000	2,82E-010	1,79E-009
5,15	0,9999999	1,30E-007	6,94E-007	6,30	1,0000000	1,49E-010	9,60E-010
5,20	0,9999999	9,96E-008	5,36E-007	6,40	1,0000000	7,77E-011	5,09E-010
5,25	0,9999999	7,60E-008	4,13E-007	6,50	1,0000000	4,02E-011	2,67E-010
5,30	0,9999999	5,79E-008	3,17E-007	6,60	1,0000000	2,06E-011	1,39E-010
5,35	1,0000000	4,40E-008	2,43E-007	6,70	1,0000000	1,04E-011	7,13E-011
5,40	1,0000000	3,33E-008	1,86E-007	6,80	1,0000000	5,23E-012	3,63E-011
5,45	1,0000000	2,52E-008	1,42E-007	6,90	1,0000000	2,60E-012	1,83E-011
5,50	1,0000000	1,90E-008	1,08E-007	7,00	1,0000000	1,28E-012	9,13E-012
5,60	1,0000000	1,07E-008	6,18E-008	8,00	1,0000000	6,13E-016	5,05E-015
5,70	1,0000000	5,99E-009	3,51E-008	9,00	1,0000000	1,13E-019	1,03E-018
5,80	1,0000000	3,32E-009	1,98E-008	10,00	1,0000000	7,62E-024	7,69E-023
5,90	1,0000000	1,82E-009	1,10E-008				

15.9 MARCUM'S Q-FUNCTION

15.9.1 Introduction

The Marcum function $Q(a,b)$ is defined by

$$Q(a,b) = \int_b^\infty \exp\left(-\frac{a^2+x^2}{2}\right) I_0(ax)x\,\mathrm{d}x$$

$$= 1 - \int_0^b \exp\left(-\frac{a^2+x^2}{2}\right) I_0(ax)x\,\mathrm{d}x \qquad (15.13)$$

It can be numerically evaluated by an iterative method based on the series expansion of the modified Bessel function of zero order [141]:

$$I_0(x) = \sum_{n=0}^{\infty} \left(\frac{x}{2}\right)^{2n} \frac{1}{(n!)^2} \qquad (15.14)$$

We then obtain

$$Q(a,b) = 1 - \sum_{n=0}^{\infty} g_n k_n \qquad (15.15)$$

with

$$g_n = g_{n-1} - \frac{1}{n!}\left(\frac{b^2}{2}\right) \exp\left[-\frac{b^2}{2}\right] \qquad (15.16)$$

and

$$k_n = \frac{a^2}{2}\,\frac{k_{n-1}}{n} \qquad (15.17)$$

15.9.2 Main properties and asymptotic relation

$$Q(\sqrt{2a}, \sqrt{2b}) = \int_b^\infty \exp(-a - y)\, I_0(2\sqrt{ay})\mathrm{d}y \tag{15.18}$$

$$Q(0,b) = \exp(-b^2/2) \tag{15.19}$$

$$Q(a,0) = 1 \tag{15.20}$$

$$Q(a,b) \cong \frac{1}{2}\,\mathrm{erfc}\left(\frac{b-a}{\sqrt{2}}\right) \text{ for } b \gg 1 \text{ and } b \gg b - a \tag{15.21}$$

15.9.3 Graphs

Graphs of functions $Q(a,b)$ and $1 - Q(a,b)$ are reproduced in figs. 15.5 and 15.6.

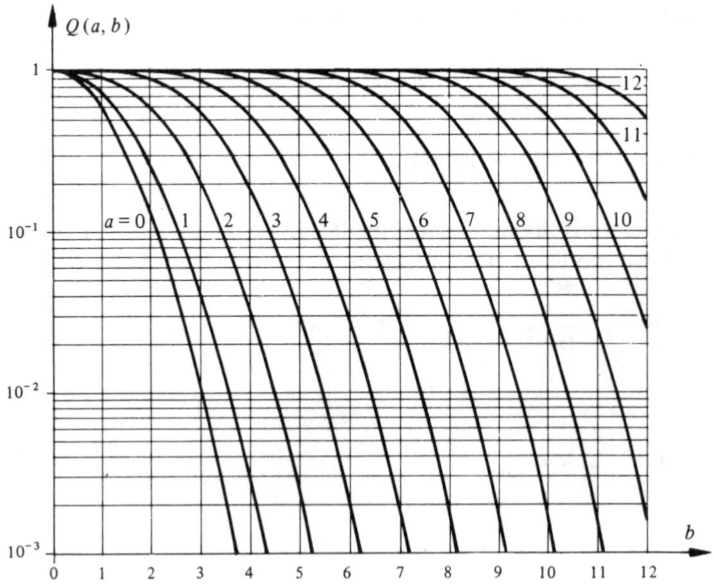

Fig. 15.5

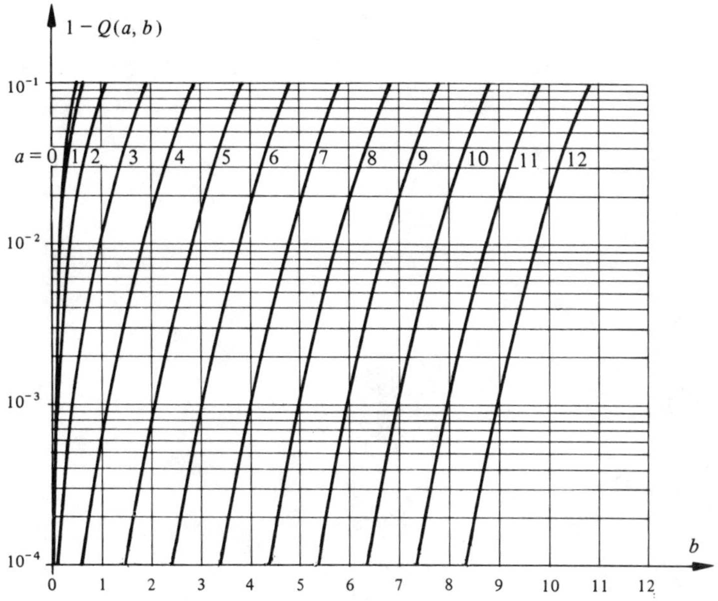
1 − Q(a, b)
10⁻¹
10⁻²
10⁻³
10⁻⁴
a = 0
1
2
3
4
5
6
7
8
9
10
11
12
b
0
1
2
3
4
5
6
7
8
9
10
11
12

Fig. 15.6

Solutions to the Problems

CHAPTER 1

1.4.1 $A\Delta$; $\bar{x} = \bar{y} = A\Delta/T$.

1.4.2 $x(t) = \frac{1}{2}A\,[\operatorname{sgn}(t-t_0) - \operatorname{sgn}(t-t_0-T)]$.

1.4.3

$$\begin{aligned} z_1(t) = \ & AB\,[\delta(t+t_0) + \delta(t-t_0)] + \tfrac{1}{2}AB\,[\delta(t+t_0+t_1) + \delta(t-t_0+t_1) \\ & + \delta(t+t_0-t_1) + \delta(t-t_0-t_1)] \\ z_2(t) = \ & A\,\mathrm{rep}_T\,[\cos(\pi t/T)\,\mathrm{rect}(t/T)] = A\,|\cos(\pi t/T)|. \end{aligned}$$

1.4.5

$$\bar{x}(t, T_1) = (A/\pi)(T_0/T_1)\sin[(\pi T_1/T_0)\sin[(2\pi t/T_0) - \pi T_1/T_0]$$

and

$$\bar{x}(t, T_1 = T_0/2) = -(2A/\pi)\cos(2\pi t/T_0)$$

and

$$\bar{x}(t, T_1 = kT_0) = 0.$$

1.4.6

$$z(t) = x(t) * y(t) = \sum_{k=0}^{\infty} c_k\,\delta(t-kT)$$

with

$$c_0 = 1,\ c_1 = \exp(-1) - 1 = -0{,}63$$

and

$$c_k = \exp(-k) - \exp(-k+1) + \exp(-k+2) \text{ for } k \geqslant 2.$$

1.4.7 $\bar{x}(T_1 = 2T) = A/2$; $W_x(T_1 = 2T) = 2A^2T/3$, $P_x(T_1 = 2T) = A^2/3$.

CHAPTER 2

2.6.1 $P = 0$, $W = A^2T$; $P = A^2/2$, $W = \infty$; $P = A^2/4$, $W = \infty$; $P = \frac{1}{2}$, $W = \infty$; $P = \infty$, $W = \infty$; $P = 0$, $W = A^2/2a$; $P = \infty$, $W = \infty$; $P = 0$, $W = 2A^2T/3$.

2.6.2 $P(T) = \frac{1}{2}A^2[1 - \mathrm{sinc}(2T/T_0)]$; $P(T) = P = A^2/2$ for $T = kT_0/2$, with k integer $\neq 0$.

2.6.3 $x_p(t) = -A \sin\alpha \cos \omega t;\ \ x_i(t) = A \cos\alpha \sin \omega t.$

2.6.4 Show that $\overline{x}_i = \overline{x} - \overline{x} = 0$.

CHAPTER 3

3.5.1 $d_1(x_k, x_l) = A\sqrt{T}\ \ \forall k, l;\ \ d_1^2(x_k, x_l) = A^2 T = \|x_k\|^2 + \|x_l\|^2$ for $k \neq l$ = sum of energies of each signal; x_k and x_l are orthogonal for $k \neq l$.

3.5.2 $d_1(x_i, x_i) = d_2(x_i, x_i) = 0$ for $i = 1, 2, 3$; $d_1(x_1, x_3) = d_1(x_2, x_3) =$
$= \sqrt{2/3}A; d_1(x_1, x_2) = 2A/\sqrt{3};\ \ \frac{1}{2} d_2(x_1, x_2) = d_2(x_1, x_3) = d_2(x_2, x_3) = A/2.$

3.5.3 $<x, y^*> = -T/2 \sin \Delta\theta;\ \ <x, y^*> = 0$ for $\Delta\theta = 0 \pm k2\pi$.

3.5.4

$$\tilde{x}(t) = -1{,}917\, e^{-t} + 10{,}914\, e^{-2t} - 9{,}472\, e^{-3t}$$

$$\|e\|^2 = \|x\|^2 - \|\tilde{x}\|^2 = \int_{-\infty}^{\infty} x^2(t)\, dt - \sum_{k=1}^{3} \sum_{l=1}^{3} \alpha_k \alpha_l \lambda_{kl} = 0{,}152.$$

3.5.5

$$\tilde{x}(t) = (2/\pi)[\text{rect}(t - \tfrac{1}{2}) - \text{rect}(t + \tfrac{1}{2})];\ \|e\|^2 = 1 - 8/\pi^2 \cong 0{,}18943.$$

3.5.6 Any odd function on the interval cannot be represented by a linear combination of the $\psi_k(t)$.

3.5.7 Expand $x(t)$ and $y(t)$ using the set $\{\psi_k(t)\}$ and take into account their orthogonality property.

3.5.10

$$\psi_4(t) = -2\sqrt{2}(4e^{-t} - 30e^{-2t} + 60e^{-3t} - 35e^{-4t})$$
$$\psi_5(t) = \tfrac{1}{2}\sqrt{10}(10e^{-t} - 120e^{-2t} + 420e^{-3t} - 560e^{-4t} + 252e^{-5t}).$$

3.511

$$\psi_1(t) = \nu_1(t)/\sqrt{12}$$
$$\psi_2(t) = (1/\sqrt{8})[\nu_2(t) + \sqrt{3}\psi_1(t)] = (1/\sqrt{2})\,\text{rect}[(t-2)/2];$$
$$\psi_k(t) = 0 \text{ for } k = 3 \text{ and } 4;$$
$$\nu_1(t) = \sqrt{12}\,\psi_1(t);$$
$$\nu_2(t) = -\sqrt{3}\,\psi_1(t) + \sqrt{8}\,\psi_2(t);$$
$$\nu_3(t) = \sqrt{3}\,\psi_1(t) - \sqrt{2}\,\psi_2(t);$$
$$\nu_4(t) = -\sqrt{3}\,\psi_1(t) - \sqrt{8}\,\psi_2(t).$$

3.5.12 Case a): non-zero α_k only for $k = 0$: $\tilde{x}(t) = \alpha_0 = 1/2$; $\xi = 4$ and $\xi_{dB} =$ 6dB; no other value of m allows a better approximation.
Case b): $\alpha_0 = 1/2$, $\alpha_2 = -1/4$, $\alpha_6 = -1/8$, $\alpha_{14} = -1/16$, $\alpha_1 = \alpha_3 = \alpha_5 = \alpha_7 = \alpha_9 = \alpha_{11} = \alpha_{13} = \alpha_4 = \alpha_{12} = \alpha_8 = 0$; for $m_{min} = 14$ (but 4 coeff. $\neq 0$): $\xi_{dB} = 24$ dB; each added non-zero coefficient yields a 6 dB improvement.

3.5.13 Real $x(t)$: $X_{-k} = X_k^*$, with $X_0 = A/2$ and $X_k = \mathrm{j}(A/2\pi k)$ for $k \neq 0$.

CHAPTER 4

4.6.1 Expand (4.10) and examine each term.

4.6.2 As (4.6.1).

4.6.3 Use (4.76).

4.6.4 Use (4.24) and (4.14).

4.6.5 Use (4.67) with $\mathrm{rect}(t) * \mathrm{tri}(t) \leftrightarrow \mathrm{sinc}^3(f)$ and $\mathrm{tri}(t) * \mathrm{tri}(t) \leftrightarrow \mathrm{sinc}^4(f)$.

4.6.6 Convolution evaluated by indirect method with (4.14), (4.18), and (4.24): $v(t) = \pi \sin(t)/t$.

4.6.7

$$X(f) = \tfrac{1}{2}AT[\mathrm{sinc}(fT) - \tfrac{1}{2}\mathrm{sinc}^2(fT/2)];$$
$$Y(f) = [\mathrm{j}A/(2\pi f)][\cos(\pi fT) - \mathrm{sinc}(fT)].$$

4.6.8 Differentiate expression (4.1).

4.6.9

$$\mathrm{d}x/\mathrm{d}t = -2\pi t \exp(-\pi t^2) \leftrightarrow \mathrm{j}2\pi f X(f) = -\mathrm{j}\,\mathrm{d}X/\mathrm{d}f$$

hence

$$X(f) = \text{solution of equation } \mathrm{d}X/X = -2\pi f \mathrm{d}f.$$

4.6.10 By (4.14), (4.68), and(4.81):

$$x(\tau) * y(\tau) = \mathring{\varphi}_{xy}(\tau) = T_1 T_2 [T_1^2 + T_2^2]^{-1/2} \cdot \mathrm{ig}[t(T_1^2 + T_2^2)^{-1/2}].$$

4.6.11 $|X(f)| = 2AT\,|\,\mathrm{sinc}(2Tf) - \mathrm{sinc}(Tf)\cos(7\pi Tf)|$; $\mathring{\Phi}_x(f) = |X(f)|^2$.
The autocorrelation $\mathring{\varphi}_x(\tau) = \mathring{\varphi}_x(-\tau)$ can be determined graphically, it is represented in fig. 4.31.

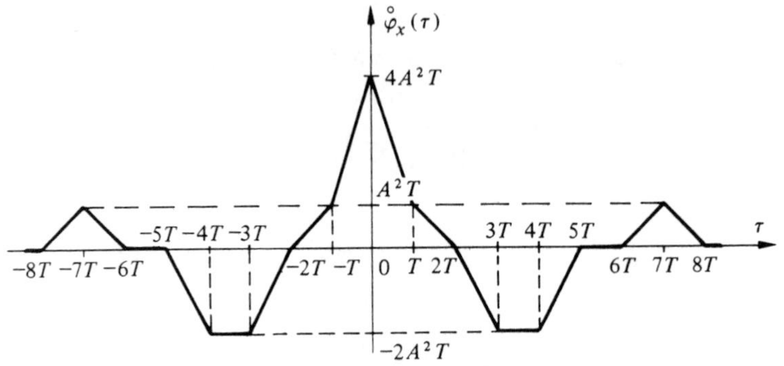

Fig. 4.31

4.6.12 $X(f) = \frac{1}{2}AT[\mathrm{sinc}^2(Tf + 1) - \mathrm{sinc}^2(Tf - 1)]\exp[\mathrm{j}(\pi/2 - \pi fT)]$. Graphs: see fig. 4.32.

4.6.13 $\mathring{\varphi}_{xy}(\tau)=0$ for $|\tau|>T$; $\mathring{\varphi}_{xy}(\tau)=AB(\tau^2/T-2\tau+T)$ for $\frac{1}{2}T<|\tau|<T$; $\mathring{\varphi}_{xy}(\tau)=AB(\frac{1}{2}T-\tau^2/T)$ for $0<|\tau|<\frac{1}{2}T$.

4.6.14 $X(f)=(\mathrm{j}2\pi f)^{-1}\,\mathrm{sinc}(af)+\frac{1}{2}\delta(f)$ from (4.87).

4.6.15

$$\varphi_x(\tau) = (A^2\Delta/T)\,\mathrm{rep}_T\,\{2\,\mathrm{tri}(\tau/\Delta)-\mathrm{tri}[(\tau+T/2)/\Delta]-\mathrm{tri}[(\tau-T/2)/\Delta]\};$$
$$\Phi_x(f) = \Sigma_n(2A\Delta/T)^2\,\mathrm{sinc}^2(n\Delta/T)\,\delta(f-n/T) \quad \text{for } n \text{ odd.}$$

4.6.16

$$x(t) = A\,\mathrm{rep}_T[\cos(2\pi f_0 t)\cdot\mathrm{rect}(t/\Delta)];$$
$$\cos(2\pi f_0 t)\cdot\mathrm{rect}(t/\Delta)\leftrightarrow\tfrac{1}{2}\Delta\,\{\mathrm{sinc}[\Delta(f+f_0)]+\mathrm{sinc}[\Delta(f-f_0)]\};$$
$$\Phi_x(f) \cong (\tfrac{1}{2}A\Delta/T)^2\,\{\mathrm{sinc}^2[\Delta(f+f_0)]+\mathrm{sinc}^2[\Delta(f-f_0)]\}\,\delta_{1/T}(f) \quad \text{for } f \gg 1/\Delta.$$

4.6.17 By the Parseval identity (4.142) applied to a periodic sequence of rectangular pulses $\mathrm{rep}_T[\mathrm{rect}(t/\Delta)]$.

4.6.18

$$\varphi_{xy}(\tau) = 0;\ \varphi_z(\tau) = \tfrac{1}{2}A^2\cos(2\pi f_0\tau)+(A^2/8)\cos(2f_0\tau);$$
$$\Phi_z(f) = \tfrac{1}{4}A^2\,\{\delta(f+f_0)+\delta(f-f_0)+\tfrac{1}{4}[\delta(f+f_0/\pi)+\delta(f-f_0/\pi]\};$$
$$P_z = 5A^2/8.$$

4.6.19 By (4.67): $< z_k, z_l^* > = \mathrm{sinc}(k-l)$ that is equal to 1 for $k = l$ and zero for $k \neq l$.

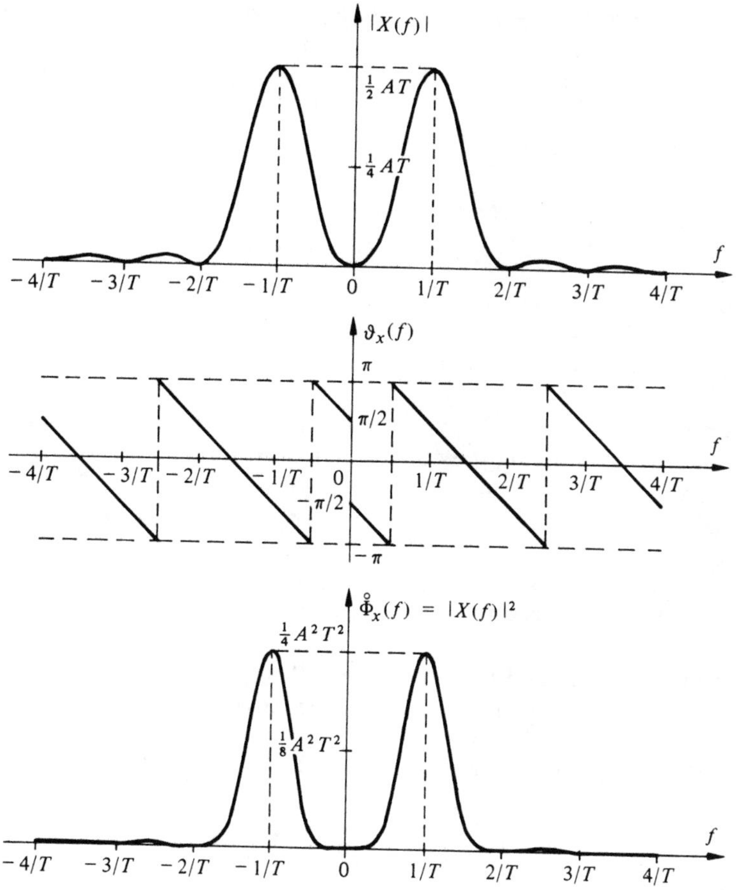

Fig. 4.32

CHAPTER 5

5.11.1 $\mu_x = 2$, $\sigma_x^2 = 1/3$: $\mu_y = 2/3$, $\sigma_y^2 = 2/9$.

5.11.2 Prob $(|z| < 1{,}5) = 11/12$.

5.11.4 Prob $(x \geqslant A/2 = F_x(x = -A/2) = 0{,}33$ by (5.22).

5.11.5 $\mu_x = \Sigma_i \text{Prob}(x_i) \cdot x_i = 3/16$; $P_x = \Sigma_i \text{Prob}(x_i) \cdot x_i^2 = 73/32$.

5.11.6 **x** and **y** are dependent because $p_{xy}(x,y) \neq p_x(x)p_y(y) = x(\sqrt{2} - y)$; solution similar to example 14.2.9.

5.11.7 Show that $p_{xy}(x,y) = p_x(x)p_y(y)$ with $p_x(x) = 2/a$ and $p_y(y) = 1/(2b)$.

5.11.8 $\mu_y = A\mu_x$; $\sigma_y^2 = A^2\sigma_x^2$.

5.11.9 $p_y(y) = (|a|/y^2)\, p_x(a/y)$.

5.11.11

$$p_y(y) = p_0\delta(y) + p_x(y)\,\epsilon(y) \text{ with } p_0 = \int_{-\infty}^{0} p_x(x)\,\mathrm{d}x.$$

5.11.12 Instantaneous power $y(t) = x^2(t)$: $p_y(y) = 1/(2A\sqrt{y})$ and $F_y(y) = \sqrt{y}/A$ with $0 < y < A^2$; $\text{Prob}(y > A^2/2) = 1 - F_y(A^2/2) = 0.293$.

5.11.13 Compare $p_y(y) = (\sigma_x^2 2\pi a y)^{-1/2}\exp[-y/(2a\sigma_x^2)]$ for $y > 0$ with (14.113) for $m = 1$ and by considering the correspondence $p_y(y)\mathrm{d}y = p(\chi_1^2)\mathrm{d}\chi_1^2$.

5.11.14 $F_r(r) = 1 - \exp[-r^2/(2\sigma^2)]$; $\text{Prob}(r > 3\sigma) = \exp(-4{,}5) = 1.11 \cdot 10^{-2}$.

5.11.15 $\mathrm{E}[\mathbf{x}] = 0$; $R_x(t_1, t_2) = R_x(\tau) = \sigma_y^2 \cdot \cos\omega\tau$ but $\mathrm{E}[\mathbf{x}^3(t)] = \mathrm{E}[\mathbf{y}^3](\sin^3\omega t + \cos^3\omega t)$ is a function of t if $\mathrm{E}[\mathbf{y}^3] \neq 0$.

5.11.16 $\mu_y(t) = \mu_x\cos\omega t$ and $R_y(t, t+\tau) = \frac{1}{2}R_x(\tau)[\cos\omega\tau + \cos(2\omega t + \omega\tau)]$ is a function of t; $R_z(t, t+\tau) = R_z(\tau) = \frac{1}{2}R_x(\tau)\cos\omega\tau$.

5.11.17 $a = R_x(T)/R_x(0)$ for $P(e = x - \tilde{x}) = R_e(0)$ minimum.

5.11.18 $R_x(\tau) = [\text{tri}(\tau/T) + 1]/a^2$.

5.11.19 Particular case of example 5.2.7 with $\mu_x = 0$ and $P_x \equiv \sigma_x^2 = 1/2$: $\Phi_x(f) = (T/2)\,\text{sinc}^2(Tf)$.

5.11.20 $|R_x(\tau)| = |\int \Phi_x(f)\exp(\mathrm{j}2\pi f t)\,\mathrm{d}f| \leqslant \int |\Phi_x(f)\exp(\mathrm{j}2\pi f t)|\mathrm{d}f = \int \Phi_x(f)\,\mathrm{d}f$.

5.11.21 Particular case of example 5.2.7 with $\mu_x = A/3$ et $P_x \equiv \sigma_x^2 = 8A^2/9$: $\Phi_x(f) = A^2(8T/9)\,\text{sinc}^2(Tf) + (A^2/9)\,\delta(f)$.

5.11.22 $\Phi_y(f) = (\Delta^2/T)\,\sigma_x^2\,\text{sinc}^2(\Delta f) + (\Delta/T)^2\mu_x^2 \cdot \Sigma_{n=-\infty}^{\infty}\text{sinc}^2(n\Delta/T)\,\delta(f - n/T)$; general formula, of which (5.132) is only a particular case for $\Delta = T$ and which corresponds also to the result of example 4.4.17 when $\sigma_x^2 = 0$.

5.11.24

$$\Phi(f) = (A/4)^2\,[T\,\text{sinc}^2(Tf/2) + \delta(f)] + (A/2\pi)^2\,\Sigma_{n=-\infty}^{\infty}(2n+1)^{-2} \cdot \delta(f - [2n+1]/T).$$

5.11.25 Introducing (5.152) into (5.102) with $\sigma_k^2 = \eta/2$.

5.11.26 Write $\mathbf{z} = \mathbf{x} + \mathbf{y}$ and $\mathbf{w} = \mathbf{x}$ and apply (5.35).

5.11.27 $p(z) = [1/(3\sqrt{\eta B \pi})]\ \{2 \exp[-(z-A)^2/\eta B] + \exp[-(z+A)^2/\eta B]\}$; $\mu_z = \mu_x = A/3$ because $\mu_y = 0$; $\sigma_z^2 = \sigma_x^2 + \sigma_y^2 = 8A^2/9 + \eta B/2$; $\rho_{xy} = 0$ since $x(t)$ and $y(t)$ are independent.

5.11.28 Prob $(z > 3) = 1/3$.

5.11.29 Write $\mathbf{z} = \mathbf{x} \cdot \mathbf{y}$ and $\mathbf{w} = \mathbf{x}$ and apply (5.35).

5.11.30 $p_z(z) = \ln z - 2 \ln a$ for $a^2 \leqslant z < a(a+1)$ and $p_z(z) = 2 \ln(a+1) - \ln z$ for $a(a+1) \leqslant z \leqslant (a+1)^2$.

5.11.31 By (5.198), (5.199), (4.138), (4.123), (5.92), and (5.132): $R_z(\tau) = \sigma_x^2 T^{-1} \delta(\tau) + \mu_x^2 T^{-1} \delta_T(\tau)$; $\Phi_z(f) = \sigma_x^2/T + (\mu_x/T)^2\, \delta_{1/T}(f)$.

5.11.32 Prob $(2 \leqslant x \leqslant 3) = 0.0433$; Prob $(x > 3) = 0.0485$.

5.11.33 $\tau = kT\ \forall k$ integer $\neq 0$.

5.11.34 By (5.180), (5.181), (5.202), and(5.207).

5.11.35 See sub-sect. 14.4.5: $p(x|y) = p(x, y)/p(y) = (\sigma_{x|y}\sqrt{2\pi})^{-1} \cdot \exp[-(x - \mu_{x|y})^2/2\sigma_{x|y}^2]$ with $\mu_{x|y} = y\rho\, \sigma_x/\sigma_y$ and $\sigma_{x|y}^2 = \sigma_x^2(1-\rho^2)$.

5.11.36 For $x(t)$ and $y(t)$ independent: $R_{xy}(\tau) = C_{xy}(\tau) = 0$; for $x(t) = y(t)$: $R_z(\tau) = R_x^2(0) + 2R_x^2(\tau)$ and $\Phi_z(f) = R_x^2(0)\,\delta(f) + 2\Phi_x(f) * \Phi_x(f)$.

5.11.37 $R_x(\tau) = A^2 \exp(-2\lambda|\tau|)$; $\Phi_x(f) = A^2 \lambda/(\lambda^2 + \pi^2 f^2)$.

CHAPTER 6

6.8.1 $P_1 = 12{,}4 \cdot 10^{-18}$ W $= 12{,}4$ aW; $P_2 = 20 \cdot 10^{-15}$ W $= 20$ fW.

6.8.2 $R_{eq} = 524\ \Omega$: $\sigma_u = 0{,}92\ \mu$V.

6.8.3 $U = 0.6$V and $I = 11.4$ mA; $\sigma_i^2 = \sigma_{ig}^2 + \sigma_{io}^2 = 3{,}66 \cdot 10^{-15}$ A^2: $\sigma_u \cong \sigma_i \cdot g_d^{-1} = 0{,}13\,\mu$V.

6.8.4 $U_R = 2kT/e = 50$ mV.

CHAPTER 7

7.6.1 $-A\cos(\omega t+\alpha)$.

7.6.2 By (7.8), (7.7), and (4.17): $\check{x}(t) = 2\pi B^2 t\ \mathrm{sinc}^2(Bt)$ and $\underline{\underline{x}}(t) = 2B\ \mathrm{sinc}(Bt)\cdot\exp(\mathrm{j}\pi Bt)$.

7.6.3 $P_{\underline{\underline{x}}} = R_{\underline{\underline{x}}}(0) = R_x(0) + R_{\check{x}}(0) = 2P_x$ because $R_{x\check{x}}(0) = 0$ and $R_x(\tau) = R_{\check{x}}(\tau)$.

7.6.4 By (4.129): $\varphi_{xy}(\tau) = (A^2/2)\sin(\omega_0\tau)$ and using the result of problem 7.6.1: $\check{\varphi}_{xy}(\tau) = -(A^2/2)\cos(\omega_0\tau)$.

7.6.5 $\mathrm{Prob}\,(r_n < V_0 = 2\sigma_n) = 1 - \mathrm{e}^{-2} = 0.865$.

7.6.6 1) $f_0 = \frac{1}{2}(f_1 + f_2)$; 2) $f_0 = f_2$.

7.6.7 By (2.25), (4.15), (4.89), and (7.9).

7.6.8 Show that $<\underline{\underline{x}},\underline{\underline{y}}^*> = <\underset{\sim}{r}_x, \underset{\sim}{r}_y^*>$ and that $<x,y> = <\check{x},\check{y}> = \mathrm{Re}\,<\underline{\underline{x}},\underline{\underline{y}}^*>$.

7.6.9 Solve (7.111) with $\underset{\sim}{r}_1(t) = \mathrm{rect}[(t - T/2)/T]$ and $\underset{\sim}{r}_2(t)$ given by (7.122).

7.6.10 Introduce (7.111) into (7.145) and consider (7.146), (1.48), and (7.115).

7.6.11 By (7.144):

CHAPTER 8

8.5.1 $\varphi_y(\tau) = \varphi_x(\tau) * \mathring{\varphi}_g(\tau)$ with $\mathring{\varphi}_g(\tau) = \frac{1}{2}\omega_c\exp(-\omega_c|\tau|)$; $B_{eq} = \omega_c/4$; $R_y(\tau) = (\eta\omega_c/4)\exp(-\omega_c|\tau|)$ and $\Phi_y(f) = (\eta/2)/[1 + (f/f_c)^2]$; $P_y = \eta\omega_c/4$.

8.5.2 Show that $y(t) = \Sigma_i\Sigma_j x_i(t) * g_j(t) = x(t) * g(t)$ for $x(t) = \Sigma_i x_i(t)$, $g(t) = \Sigma_j g_j(t)$ and $j = 1, \ldots, n$.

8.5.3 $G(f) = [1 + \mathrm{j}f/f_c]^{-1} - \exp(-\mathrm{j}2\pi f t_0)$ with $f_c = (2\pi RC)^{-1}$; for $t_0 = 0$: $G(f) = -\mathrm{j}(f/f_c)[1 + \mathrm{j}f/f_c]^{-1}$; first-order highpass filter with cut-off frequency f_c.

8.5.4 Write $G(f) = \mathrm{j}2\pi f G_2(f)$, hence, by (4.13) and table 15.4: $g(t) = \mathrm{d}g_2(t)/\mathrm{d}t = [2\pi G_0 f_2/(f_2 - f_1)][f_2\exp(-2\pi f_2 t) - f_1\exp(2\pi f_1 t)]\cdot\epsilon(t)$; $P_y \cong 3142\ \mathrm{V}^2$; $B_{-3\mathrm{dB}} \cong 20$ kHz and $B_{eq} \cong 31.42$ kHz.

8.5.5 $P_{sy}/P_{ny} = (2\pi BA^2T^2/\eta)\,[2\pi BT + \exp(-2\pi BT) - 1]^{-1}$; $T \cong 0.995$ s.

8.5.6 $P_{sy}/P_{ny} = A^2\,(1 + 2\pi B\,RC)/(\eta\pi B)$; $RC = 0.5$ s.

8.5.7 $t_\phi = (2\pi f_0)^{-1} \arctan(f_0/f_c)$ and $t_g = (2\pi f_c)^{-1}\,[1 + (f_0/f_c)^2]^{-1}$; for $f_0 = f_c$: $t_\phi = (8f_c)^{-1}$ and $t_g = (4\pi f_c)^{-1}$.

8.5.8 Introduce (8.24) and (8.25) into (5.178).

8.5.9 By (5.222), (8.88), (8.24), and (1.35): $\Phi_n(f) = [\lambda^2\Phi_\alpha(f) + \lambda R_\alpha(0)]\cdot|G(f)|^2$.

8.5.10 $\alpha(t) = e$ and $g(t) = \delta(t)$.

8.5.11 $\Phi_n(f) = I_0\,\delta(f) + I_0\,e\,\mathrm{sinc}^4(\Delta f)$.

8.5.12 Express $Y(f)$ considering (8.74), $x(\tau) = \mathrm{F}^{-1}[X(\nu)]$ and (8.76).

8.5.13 $\Phi_y(f) = (\Delta/T)^2\,\Sigma_n\,\mathrm{sinc}^2(n\Delta/T)\,\Phi_x(f - n/T)$.

8.5.14 $\mathring{\Phi}_y(f) = \Phi_x(f) * T^2\,\mathrm{sinc}^2(Tf) = T^2\,\Sigma_n|X_n|^2\,\mathrm{sinc}^2[T(f - n/T_x)]$.

8.5.15 $|Y(f)| = \frac{1}{2}|A_1|[|U_i(f+f_1)| + |U_i(f-f_1)|] + \frac{1}{2}|A_2|[|U_i(f+f_2)| + |U_i(f-f_2)|]$ with $U_1(f) = T\,\mathrm{sinc}(Tf)$, $U_2(f) = T\,\mathrm{sinc}^2(Tf)$, $U_3(f) = T\cdot\mathrm{ig}(Tf)$.

8.5.16 $Y(f) = \Sigma_n(A/2)\,[\mathrm{sinc}(n - \frac{1}{2})] + \mathrm{sinc}(n + \frac{1}{2})]\,\delta(f - 2n/T)$.

8.5.17

$$R_y(\tau) = \tfrac{1}{2}A^2\,[(\sigma_x^2/2RC)\exp(-|\tau|/RC) + \mu_x^2]\cos(2\pi f_0\tau);$$
$$\Phi_y(f) = (A^2/4)\,[\sigma_x^2\,\{1 + [2\pi(f+f_0)RC]^2\}^{-1} + \sigma_x^2\,\{1 + [2\pi(f-f_0)RC]^2\}^{-1} + \mu_x^2\,[\delta(f+f_0) + \delta(f-f_0)];$$
$$P_y = R_y(0) = A^2\,\sigma_x^2/(4RC) + A^2\mu_x^2/2.$$

8.5.18 By (8.41) and (7.8): $y(t) = x(t)\cos(2\pi f_0 t) - \check{x}(t)\sin(2\pi f_0 t)$; by (8.103), (7.31) and (7.33): $R_y(\tau) = R_x(\tau)\cos(2\pi f_0\tau) + R_x(\tau)\sin(2\pi f_0\tau)$ and by (7.35): $\Phi_y(f) = \frac{1}{2}[\Phi_x^+(f + f_0) + \Phi_x^+(-f - f_0)]$.

8.5.19 $y(t) = A^2\,\mathrm{ig}(\sqrt{2}\,t/T)$; $Y(f) = (A^2T/\sqrt{2})\,\mathrm{ig}(Tf/\sqrt{2})$; $\mathring{\Phi}_y(f) = (A^4T^2/2)\,\mathrm{ig}(Tf)$.

8.5.20

$$Y_a(f) = \frac{A_1^2 + A_2^2}{2}X(f) + \frac{A_1^2}{4}[X(f+2f_1) + X(f-2f_1)] + \frac{A_2^2}{4}[X(f+2f_2) +$$

$$+ X(f-2f_2)] + \frac{A_1 A_2}{2} [X(f+f_2-f_1) + X(f-f_2+f_1) + \\ + X(f+f_1+f_2) + X(f-f_1-f_2)].$$

with

$$X(f) = \frac{A_1}{2} [\delta(f+f_1) + \delta(f-f_1)] + \frac{A_2}{2} [\delta(f+f_2) + \delta(f-f_2)]$$

$$Y_b(f) = \begin{cases} A^3(3B^2 - f^2) \text{ for } |f| \leqslant B \\ A^3(9B^2/2 - 3Bf + f^2/2) \text{ for } B \leqslant |f| \leqslant 3B \text{ and zero elsewhere.} \end{cases}$$

8.5.21 $d_2 = c\hat{U}/(2b) = 8\%$.

8.5.22 $d_a \cong 8.3 \cdot 10^{-4}$; $d_b \cong 3.3 \cdot 10^{-3}$.

8.5.23

$\mu_{y1} = b^3/32$, $m_{y12} = b^6/448$, $\sigma_{y1}^2 = 9b^6/7168$;
$\mu_{y2} = 1 + b^2/12 + b^3/32$, $m_{y22} = 1 + b^2/6 + b^3/16 + b^4/80 + b^5/96 + b^6/448$,
$\sigma_{y2}^2 = b^4/180 + b^5/192 + 9b^6/7168$.

8.5.25 $R_y(\tau) = 2\pi^{-1} \{R_x(\tau) \arcsin [R_x(\tau)/R_x(0)] + [R_x^2(0) - R_x^2(\tau)]^{1/2}\}$.

8.5.27 $R_y(\tau) = [1 + 3R_x(0)]^2 R_x(\tau) + 6R_x^3(\tau)$.

8.5.28 $R_{y1y2}(\tau) = 2\pi^{-1} \arcsin [R_{x1x2}(\tau)/(\sigma_{x1}\sigma_{x2})]$.

8.5.29 $R_{y1y2}(\tau) = (2/\pi)^{1/2} A \sigma_{x2}^{-1} R_{x1x2}(\tau)$.

CHAPTER 9

9.5.1 $y(t) = x(t) \cdot e(t) = \text{rep}_{T_e} \{\text{rect}(t/T_f)\}$ and $|Y(f)| = |(T_f/T_e) \Sigma_n \text{sinc}(nT_f f_0) \cdot X(f - nf_0)|$; $z(t) \approx 2\pi^{-1} \exp(-a|t|) \cdot \cos(2\pi f_0 t)$.

9.5.2 $f_e = 6$ kHz introduces an aliasing distortion; an ideal reconstruction would imply a minimum sampling rate of $f_e' = 8$ kHz.

9.5.3 No, we should have $f_e = (2\Delta t)^{-1}$.

9.5.4 $f_e = 2B$ with $G_{2i}(f) = \text{rect}(f/2B)$; $f_e' \geqslant 5B$.

9.5.5 Evaluate P_r using the approximation $|G(f)|^2 \cong (f/f_c)^{2n}$ for $f > f_c$.

9.5.6 $f_e = 1272\, f_c\, (n = 1)$, $f_e = 13{,}4\, f_c\, (n = 2)$, $f_e = 5{,}7\, f_c\, (n = 3)$, $f_e = 4{,}1\, f_c\, (n = 4)$.

9.5.8 $\Phi_y(f) \approx 255\ \Sigma_{i=1}^{2}\ \mathrm{sinc}^4\,(f_i\Delta)\,[\delta\,(f + f_i) + \delta\,(f - f_i)]$.

9.5.9 $\mu_n = 2\,\mathrm{V}$; $\sigma_n^2 = 16\,\mathrm{V}^2$; $\Phi_n(f) = 4\delta\,(f) + 16 \cdot 10^{-3}\ \mathrm{tri}\,(10^{-3} f)$; $f_e = 2$ kHz; $f'_e = (kT)^{-1} = k^{-1}$ kHz with $k = 1, 2, 3, \ldots$

CHAPTER 10

10.6.1 $V = 20\,\mathrm{V}$; $\tau_c = 1.25\ \mu\mathrm{s}$; $n = 8$ bits; $f_e \geqslant 25 f_0 = 12.5\ \mathrm{kHz} \gg 2 f_0$.

10.6.3 a) $6n - 6$ dB; b) $6n - 3$ dB; c) $6n - 9{,}2$ dB.

10.6.4 $\xi_{\mathrm{qdB}} \approx 42$ dB.

10.6.5 $\xi_{\mathrm{qdB}} \approx 32.8$ dB for $n = 7$.

10.6.6 By differentiation at $\upsilon = 0$, according to (14.68), of $\Pi_x(\upsilon)$ and $\Pi_{xq}(\upsilon) = \Pi_x(\upsilon)\mathrm{sinc}(\Delta\upsilon)$ with $\Pi_x(0) = 1$, $\mathrm{d}\,(\Pi \cdot F)\ \mathrm{d}\alpha = F\mathrm{d}\Pi/\mathrm{d}\alpha + \Pi F/\mathrm{d}\alpha$ and $\mathrm{d}^2\,(\Pi \cdot F)\ \mathrm{d}\alpha^2 = F\mathrm{d}^2\Pi/\mathrm{d}\alpha^2 + 2\mathrm{d}\Pi/\mathrm{d}\alpha \cdot \mathrm{d}F/\mathrm{d}\alpha + \Pi\mathrm{d}^2F/\mathrm{d}\alpha^2$ where $F(\alpha) = \mathrm{sinc}(\alpha) = 1 - (\pi\alpha)^2/3! + \ldots$ by (1.59).

CHAPTER 11

11.6.1 $\eta_m = 20\%$; $\eta_m = 50\%$; $\eta_m = 20\%$.

11.6.2 $\xi_d/\xi_x = 2m^2/(2 + m^2)$.

11.6.4

a) $\underset{\sim}{r}(t) = \hat{U}_p \exp(\mathrm{j}2\pi f_m t)$, $s(t) = (\hat{U}_p/2)\cos[2\pi(f_p + f_m)t + \alpha_p]$,
$\Phi_s(f) = (\hat{U}_p^2/16)[\delta(f + f_p + f_m) + \delta(f - f_p - f_m)]$;

b) $\underset{\sim}{r}(t) = \hat{U}_p \exp(-\mathrm{j}2\pi f_m t)$, $s(t) = (U_p/2)\cos[2\pi(f_p - f_m)t + \alpha_p]$,
$\Phi_s(f) = (\hat{U}_p/16)[\delta(f + f_p - f_m) + \delta(f - f_p + f_m)]$.

11.6.5 SSB with upper sideband.

11.6.6 $\underset{\sim}{r}(t) = m_1(t) + \mathrm{j}m_2(t)$.

11.6.7 $\hat{U}_p \gg \sigma_m = \sqrt{P_m}$.

11.6.8 $S(f) = (\hat{U}_p/2)\{M_0(f+f_p)[1 - H(f+f_p)\,\mathrm{sgn}\,(f+f_p)] + M_0(f-f_p)\cdot[1 + H(f-f_p)\,\mathrm{sgn}\,(f-f_p)]\}$ where $H(f)$ is the transfer function of a highpass filter (fig. 11.44).

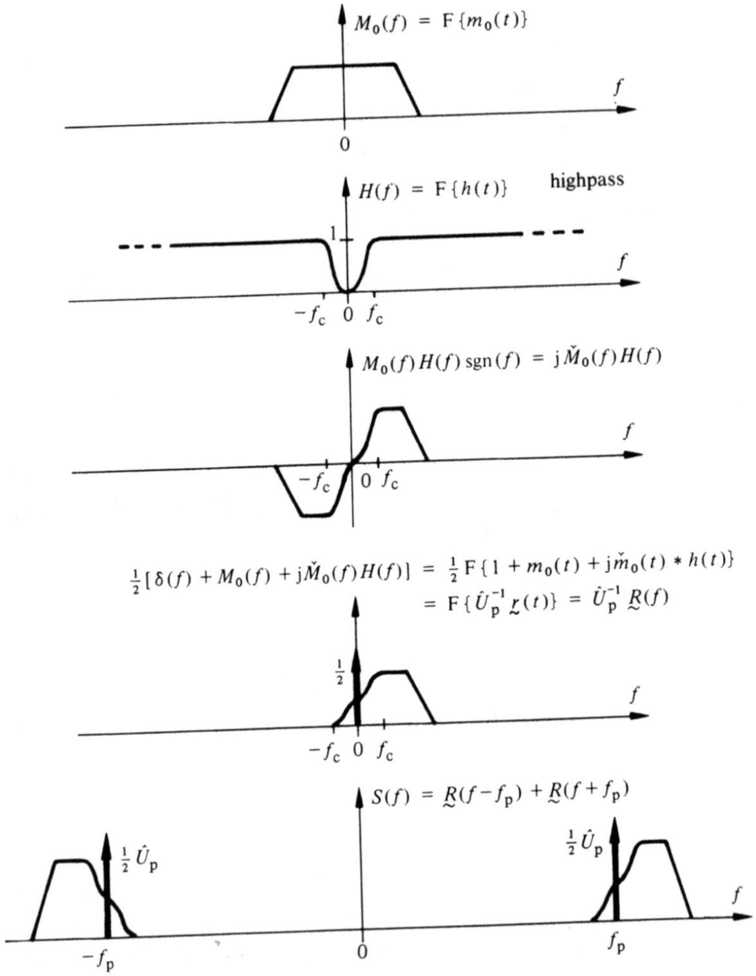

Fig. 11.44

11.6.9 Use fig. 15.2 for $\delta = 10$ and fig. 5.4 for $A = \delta f_m$

11.6.10 $\Phi_s(f) \approx \hat{U}_p^2 (8\nu\hat{m})^{-1}\{\mathrm{rect}\,[(f+f_p)(2\nu\hat{m})^{-1} + \mathrm{rect}\,[(f-f_p)(2\nu\hat{m})^{-1}]\}$.

11.6.12

$$s(t) = \tfrac{1}{2}[m_0(t)+1]\,\hat{U}_p \cos(2\pi f_1 t+\alpha_1) - \tfrac{1}{2}[m_0(t)-1]\,\hat{U}_p \cos(2\pi f_2 t+\alpha_2);$$
$$\Phi_s(f) = (\hat{U}_p^2/16)\,[\Phi_{m0}(f+f_1)+\Phi_{m0}(f-f_1)+\delta(f+f_1)+\delta(f-f_1)+$$
$$+\Phi_{m0}(f+f_2)+\Phi_{m0}(f-f_2)+\delta(f+f_2)+\delta(f-f_2)]$$ with, by (5.132),
$$\Phi_{m0}(f) = T\,\mathrm{sinc}^2(Tf).$$

11.6.13 Similar situation to the one of example 5.2.7 with complex envelope $\underline{r}(t) = \hat{U}_p \exp[jm_0(t)\pi/2]$, where $m_0(t)$ takes, during each interval of duration T, one of the independent states 0, 1, 2, or 3: $\Phi_{\underline{r}}(f) = \hat{U}_p^2 T \,\mathrm{sinc}^2(Tf)$; $B_u \cong 2/T$.

11.6.14 $\Phi_m(f) = \Phi_{m_d+m_g}(f) + \frac{1}{4}[\delta(f+f_0)+\delta(f-f_0)] + \frac{1}{4}[\Phi_{m_d-m_g}(f+2f_0)+\Phi_{m_d-m_g}(f-2f_0)]$.

CHAPTER 12

12.5.1 By (4.55), (4.26), (4.78), (4.117), (4.57), and (7.144).

12.5.2 $\xi \approx 1600$(32 dB); for a swept-frequency analyzer: resolution B_{eq} = 10 kHz, integration duration T_0 = 0.16 s, sweeping duration T_b = 80 s. For a bank-of-filters multichannel analyzer: B_{eq} = 10 kHz, N = 500 channels, T_0 = 0.16 s. For a digital analyzer: B_{eq} = 10 kHz, f_e = 10 MHz, $N \geqslant 500$, T = 100μs, K = 1600, T_0 = 0.16 s.

12.5.3 In analog techniques, the displayed quantity is the power $P_x(f_0, B_{eq}) \cong \Phi^+(f_0)B_{eq}$ or the corresponding rms value measured on a bandwidth B_{eq}; the displayed level depends therefore on B_{eq} in the case of a line spectrum, because the power of each component is concentrated on a discrete frequency.

In digital techniques, the analyzed signal is the product of the signal observed by a window $u(t)$ of duration T: the computed periodogram is $\tilde{\Phi}_x(f) = T^{-1}\Phi_x(f) * \mathring{\Phi}_u(f)$, according to (12.3) and (12.4), where $\mathring{\Phi}_u(f)$ is proportional to T^2 and its integral is proportional to T; this convolution with a continuous spectral density corresponds approximately to $\Phi_x(f)$, independently of the choice of T. On the contrary, considering (1.48), the result is proportional to T in the case of a line spectrum.

CHAPTER 13

13.5.1 Let $x_i = at_i + b + n_i$: $p(\mathbf{x}|a,b) = \Pi_i p(n_i = x_i - at_i - b)$ with $p(n)$ a Gaussian function with zero mean value. By resolving $\partial \ln p(\mathbf{x}|a,b)/\partial a = 0$ and $\partial \ln p(\mathbf{x}|a,b)/\partial b = 0$: $\tilde{a}_{ml} = (\Sigma_i x_i t_i - m\overline{x}\overline{t})/(\Sigma_i t_i^2 - m\overline{t}^2)$ and $\tilde{b}_{ml} = \overline{x} - a\overline{t}$ with $\overline{x} = m^{-1}\Sigma_i^m x_i$ and $\overline{t} = m^{-1}\Sigma_i^m t_i$.

13.5.2 $\tilde{a}_{mv} = N^{-1} \Sigma_i^N x_i^2$; $\tilde{a}_{mv} = T^{-1} \int_0^T x^2(t)\,dt$.

13.5.3 $p(a|x) = p(x|a)p(a)/p(x)$ with $p(x) = \lambda/(x+\lambda)^2$ given by condition $\int_0^\infty p(a|x)da = 1$.

13.5.4 $\tilde{a}_{ml}$ is the solution of the equation:

$$\int_0^T x(t) \frac{\partial s(t-a)}{\partial a} \, dt \approx 0$$

which corresponds to the correlation of observation $x(t)$ with the derivative of the known signal.

13.5.5 $E[(a - \tilde{a}_\ell)^2] = R_y(0) - R_y^2(T)/R_y(0)$.

13.5.6 $\tilde{a}_{\ell 0} = (N + a^2/\sigma_n^2)^{-1} \Sigma_i^N z_i$ and $\sigma_y^2 = \sigma_n^2/(N + \sigma_n^2/a^2)$.

13.5.7 $\tilde{a}_{\ell 0} = \tilde{x}(t) = g_1 x(0) + g_2 x(T)$ with $g_1 = [R_x(t)R_x(0) - R_x(T-t)R_x(T)]/[R_x^2(0) - R_x^2(T)]$ and $g_2 = [R_x(T-t)R_x(0) - R_x(t)R_x(T)]/[R_x^2(0) - R_x^2(T)]$.

13.5.8 By (13.75) with here $\Phi_{xa}(f) = \Phi_s(f) = |\Psi_x(f)|^2 - \Phi_n(f)$; factor $\Phi_x(f) = |\Psi_x(f)|^2$ under the condition $\Phi_n(f) = \eta/2$.

13.5.9 $G_0(f) = A[B + j2\pi f]^{-1}$ with $A = \alpha[(\eta/2)(\alpha + \eta/2)^{1/2} + \beta]^{-1}$ and $B = (2\alpha/\eta + \beta^2)^{1/2}$.

13.5.10 By (13.80) with $C_x(\tau) = (A^2/4)\exp(-2\lambda|\tau|)$: $\sigma_{\bar{x}}^2 = (A^2/8)[\exp(-2\lambda T) + 2\lambda T) - 1]/(\lambda T)^2$.

13.5.11 By (13.80): $\sigma_{\bar{x}}^2 = (BT)^{-1} - (BT)^{-2}/3 \approx (BT)^{-1}$ for $T \gg B^{-1}$; $T_{min} \approx 1$ s.

13.5.12 $C_x(\tau) = 0$ for $T_e \geqslant B^{-1}$ and $N = 1000$, according to (13.82), hence, $T = NT_e \geqslant 1$ s; for $T_e = (2B)^{-1} = 0.5$ ms, by (13.81): $N_{min} = 2000$ and $T = NT_e = 1$ s.

13.5.14 By analogy with (13.110) and(13.111).

13.5.15 $\varphi_{xu}(\tau) = T^{-1} g(-\tau) * s(\tau)$.

13.5.16 $\xi_1 = 10^{-3}$ (−30 dB); $\xi_2 = 3.13$ (5 dB).

13.5.17 For N independent samples $p_r(r|s_i) = \Pi_{j=1}^{N} p_r(r_j|s_i)$ with $p_r(r_j|s_i)$ given by (7.57) and we draw from (13.126): $\Sigma_j \ln I_0(r_j A/\sigma_n^2) \gtrless \frac{1}{2}A^2\sigma_n^2$; the block diagram of the corresponding optimum detector is given in fig. 13.43.

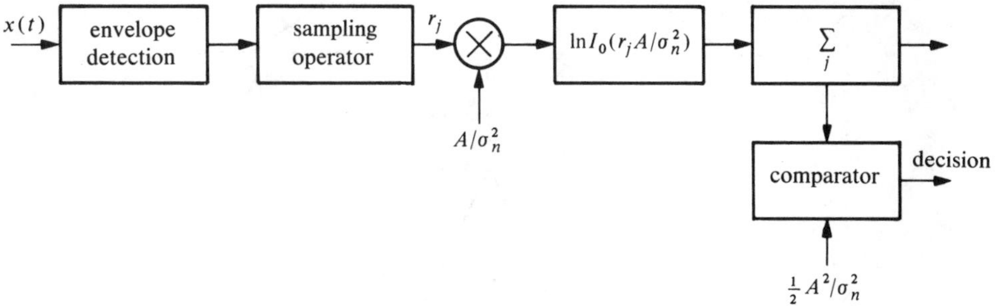

Fig. 13.43

13.5.18 From table 15.8, we draw $x_s = 1.65\sigma_n$.

13.5.19 $x_s = A/2 + (1/4)\ln(p_0/p_1) = 1{,}847$ V; $p_\epsilon \approx 2 \cdot 10^{-2}$.

13.5.20 $x_s \cong 13{,}9$ V; $p_\epsilon = p_f = p_n \cong 10^{-6}$.

13.5.21 By (13.145): $V = (-2\eta B \ln p_f)^{1/2} = 12.88$ V and $p_d = Q(A/\sigma_n;\ V/\sigma_n) = Q(8;\ 6.44) \cong 0.95$.

13.5.22 $V^2 = 14\sigma_n^2$.

13.5.23 See fig. 13.44.

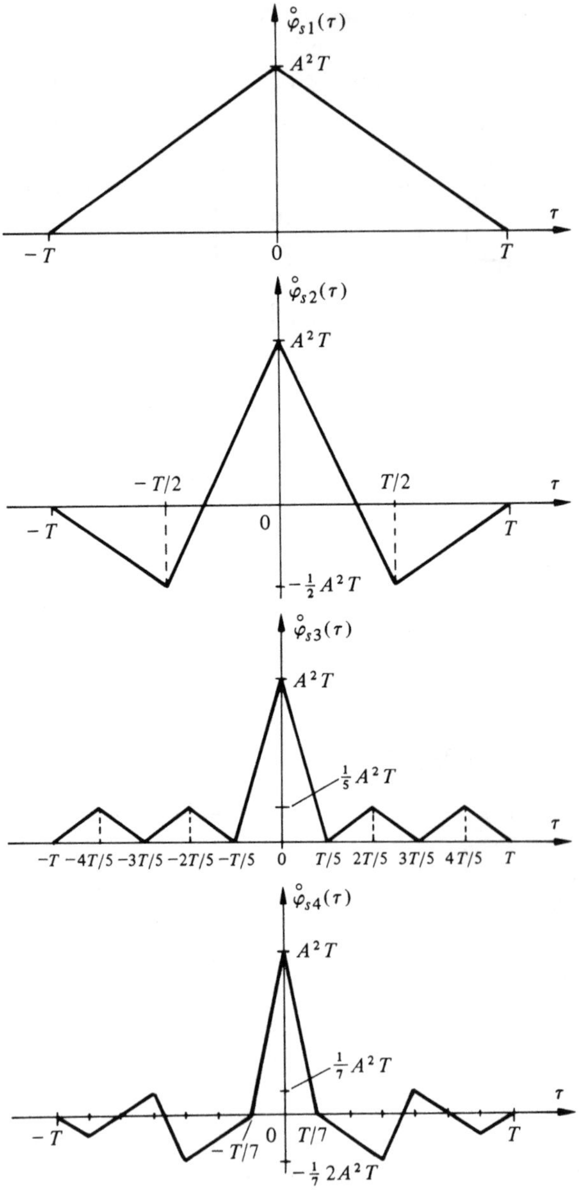

Fig. 13.44

13.5.24 $\xi(T)/\xi_{\text{opt}} = 0.89$; $t_0 = \tau_0 + T/2$.

Bibliography

[1] J. R. Pierce, *Symbols, Signals and Noise: the Nature and Process of Communication*, Harper & Brothers, New York, 1961.

[2] N. Wiener, Generalized Harmonic Analysis, *Acta Mathematica*, 55, 1930, pp. 117–258.

[3] N. Wiener, *The Fourier Integral and Certain of its Applications*, Cambridge Press, New York, 1933.

[4] A. Y. Khintchine, Korrelationstheorie der Stationären Stockastichen Processe, *Math. Annalen*, 109, 1934, pp. 604–615.

[5] C. E. Shannon, The Mathematical Theory of Communication, *Bell Syst. Tech. J.*, 27, July 1948, pp. 379–423 and Oct. 1948, pp. 623–656.

[6] C. E. Shannon, Communication in the presence of noise, *Proc. IRE*, 37, Jan. 1949, pp. 10–21.

[7] N. Wiener, *Cybernetics: Control and Communication in Animal and Machine*, M.I.T. Press, Cambridge, Mass., 1948.

[8] N. Wiener, *The Extrapolation, Interpolation and Smoothing of Stationary Time Series*, M.I.T. Press, Cambridge, Mass., 1949.

[9] K. Küpfmüller, *Die Systemtheorie der Elektrischen Nachrichtenübertragung*, Hirzel Verlag, Stuttgart, 1949.

[10] D. Gabor, Theory of Communication, *J. Inst. Electr. Eng.*, 93, Part III, Nov. 1946, pp. 429–457.

[11] P.M. Woodward, *Probability and Information Theory with Application to Radar*, Artech House, Dedham, Mass., 1980 (1st ed., 1953).

[12] A. N. Kolmogorov, Interpolation und Extrapolation von stationären Zufälligen Folgen, *Acad. Sci., USSR Ser. Math.*, 5, 1941, pp. 3–14.

[13] W. A. Kotelnikov, Über die Kanalkapazität des Athers und der Drahtverbindungen in der elektrischen Nachrichtentechnik, *Tag. Ber. d. 1. Allunionskonf. Nach. Tech.*, Moscow, 1933.

[14] S. O. Rice, Mathematical Analysis of Random Noise, *Bell Syst. Techn. J.*, 23, 1944, pp. 282–332 and 24, 1945, pp. 46–156.

[15] S. Goldman, *Frequency Analysis, Modulation and Noise*, McGraw-Hill, New York, 1948.

[16] J. L. Lawson, G. E. Uhlenbeck, *Threshold Signals*, M.I.T. Rad. Lab. Series, 24, McGraw-Hill, New York, 1950.

[17] J. Ville, Théorie et applications de la notion de signal analytique, *Câbles et transmissions*, 1, Jan. 1948, pp. 61–74.

[18] A. Blanc-Lapierre, R. Fortet, *Théorie des fonctions aléatoires*, Masson, Paris, 1953.

[19] L. Brillouin, *La science et la théorie de l'information*, Masson, Paris, 1959.

[20] J. M. Wozencraft, I. M. Jacobs, *Principles of Communication Engineering*, John Wiley, New York, 1965.

[21] Y. W. Lee, *Statistical Theory of Communication*, John Wiley, New York, 1960.

[22] A. Papoulis, *The Fourier Integral and its Applications*, McGraw-Hill, New York, 1962.

[23] R. Bracewell, *The Fourier Transform and its Applications*, McGraw-Hill, New York, 1965.

[24] A. Papoulis, *Probability, Random Variables and Stochastic Processes*, McGraw-Hill, New York, 1965 (1st ed.), 1984 (2nd ed.).

[25] W. B. Davenport, W. L. Root, *Introduction to the Random Signals and Noise*, McGraw-Hill, New York, 1958.

[26] D. Middleton, *Introduction to Statistical Communication Theory*, McGraw-Hill, New York, 1960.

[27] R. S. Berkowitz (Editor), *Modern Radar, Analysis, Evaluation and System Design*, John Wiley, New York, 1965.

[28] B. Levine, *Théorie des processus aléatoires et son application à la radio-technique*, Sovetskoe radio, Moscow, 1957, (traduit du russe).

[29] E. Hülser, H. Holzwarth, *Theorie und Technik der Pulsmodulation*, Springer Verlag, Berlin, 1957.

[30] J. S. Bendat, *Principles and Applications of Random Noise Theory*, John Wiley, New York, 1958.

[31] R. B. Blackman, J. W. Tukey, *The Measurement of Power Spectra*, Dover, New York, 1958.

[32] M. Schwartz, *Information Transmission, Modulation and Noise*, McGraw-Hill, 1959.

[33] S. J. Mason, H. J. Zimmermann, *Electronic Circuits, Signals and Systems*, John Wiley, New York, 1960.

[34] C. W. Helstrom, *Statistical Theory of Signal Detection*, Pergamon Press, Oxford, 1960 (1st ed.), 1968 (2nd ed.).

[35] J. C. Handcock, *An Introduction to the Principle of Communication Theory*, McGraw-Hill, New York, 1961.

[36] F. M. Reza, *An Introduction to Information Theory*, McGraw-Hill, New York, 1961.

[37] L. A. Wainstein, V. D. Zubakov, *Extraction of Signals from Noise*, Prentice-Hall, London, 1962.

[38] F. H. Lange, *Korrelationselektronik*, Verlag Technik, Berlin, 1962.

[39] W. W. Harman, *Principles of the Statistical Theory of Communication*, McGraw-Hill, New York, 1963.

[40] P. F. PANTER, *Modulation, Noise and Spectral Analysis*, McGraw-Hill, 1965.

[41] H. E. ROWE, *Signals and Noise in Communication Systems*, Van Nostrand, Princeton, 1965.

[42] B. P. LATHI, *Signals, Systems and Communication*, John Wiley, New York, 1965.

[43] P. A. LYNN, *The Analysis and Processing of Signals*, Macmillan, London, 1982 (2nd ed.).

[44] B. GOLD, C. RADER, *Digital Processing of Signals*, McGraw-Hill, New York, 1969.

[45] A. V. OPPENHEIM, R. W. SHAFER, *Digital Signal Processing*, Prentice-Hall, Englewood Cliffs, 1975.

[46] L. R. RABINER, B. GOLD, *Theory and Application of Digital Signal Processing*, Prentice-Hall, Englewood Cliffs, 1975.

[47] R. BOITE, H. LEICH, *Les filtres numériques*, Masson, Paris, 1980.

[48] M. BELLANGER, *Traitement numérique du signal*, Masson, Paris, 1980.

[49] L. E. FRANKS, *Signal Theory*, Prentice-Hall, Englewood Cliffs, 1969.

[50] B. PICINBONO, *Eléments de théorie du signal*, Dunod Université, Paris, 1977.

[51] R. G. GALLAGER, *Information Theory and Reliable Communication*, John Wiley, New York, 1968.

[52] G. CULLMANN, *Codes détecteurs et correcteurs d'erreurs*, Dunod, Paris, 1967.

[53] A. SPĂTARU, *Théorie de la transmission de l'information*, Masson, 1970–1973.

[54] S. LIN, *An Introduction to Error-correcting Codes*, Prentice-Hall, Englewood Cliffs, 1970.

[55] H. STARK, F. B. TUTEUR, *Modern Electrical Communications, Theory and Systems*, Prentice-Hall, Englewood Cliffs, 1979.

[56] H. D. LUEKE, *Signal-übertragung*, Springer-Verlag, Berlin, 1975.

[57] E. ROUBINE, *Introduction à la théorie de la communication*, (Vol. I: *Signaux non aléatoires*, Vol. II: *Signaux aléatoires*), Masson, Paris, 1970.

[58] J. BASS, *Cours de mathématiques*, Masson, Paris, 1968–1971, (3 volumes).

[59] A. J. VITERBI, J. K. OMURA, *Principles of Digital Communication and Coding*, McGraw-Hill, New York, 1979.

[60] C. CARDOT, Définition analytique simple des fonctions de Walsh et application à la détermination exacte de leurs propriétés spectrales, *Annales Télécomm.*, 72, Jan-Feb. 1972, pp. 31–47.

[61] H. F. HARMUTH, *Transmission of Information by Orthogonal Functions*, Springer-Verlag, Berlin, 1969.

[62] K. G. BEAUCHAMP, *Walsh Functions and their Applications*, Academic Press, London, 1975.
[63] N. AHMED, K. R. RAO, *Orthogonal Transforms for Digital Signal Processing*, Springer-Verlag, Berlin, 1975.
[64] J. STERN, J. DE BARBEYRAC, R. POGGI, *Méthodes pratiques d'études des fonctions aléatoires*, Dunod, Paris, 1967.
[65] J. S. BENDAT, A. G. PIERSOL, *Random Data: Analysis and Measurement Procedures*, Wiley-Interscience, New York, 1971.
[66] J. DUPRAZ, *Théorie de la communication (signaux, bruits et modulations)*, Eyrolles, Paris, 1973.
[67] J. MAX et al., *Méthodes et techniques de traitement du signal et applications aux mesures physiques*, Masson, Paris, 1981 (3e édition, 2 volumes).
[68] H. J. LARSON, B. O. SHUBERT, *Probabilistic Models in Engineering Sciences*, Vol. I and II, John Wiley, New York, 1979.
[69] A. BLANC-LAPIERRE, P. PICINBONO, *Fonctions aléatoires*, Masson, Paris, 1981.
[70] P. BECKMANN, *Probability in Communication Engineering*, Harcourt, Brace and World, New York, 1967.
[71] G. CULLMANN, *Initiation aux chaînes de Markov—Méthodes et applications*, Masson, Paris, 1975.
[72] A. V. BALAKRISHNAN et al., *Communication Theory*, McGraw-Hill, New York, 1968.
[73] W. C. LINDSAY, M. K. SIMON, *Telecommunication Systems Engineering*, Prentice-Hall, Englewood Cliffs, 1973.
[74] G. A. KORN, *Random-Process Simulation and Measurements*, McGraw-Hill, New York, 1966.
[75] S. W. GOLOMB (ed.), *Digital Communications with Space Applications*, Prentice-Hall, Englewood Cliffs, 1964.
[76] S. W. GOLOMB, *Shift Register Sequences*, Holden-Day Inc., San Francisco, 1967.
[77] *Reference Data for Radio Engineers*, Howard W. Sams and Co., New York, 1975, (6th ed.).
[78] F. N. H. ROBINSON, *Noise and fluctuations in electronic devices and circuits*, Clarendon Press, Oxford, 1974.
[79] M. SCHWARTZ, *Information Transmission, Modulation and Noise*, McGraw-Hill, New York, 1970, (2nd ed.).
[80] A. VAN DER ZIEL, Noise in solid-state devices and lasers, *Proc. IEEE*, 58, Aug. 1970, pp. 1178–1206, (reproduced in [86]).
[81] M. S. KESHNER, 1/*f* noise, *Proc. IEEE*, 70, March 1982, pp. 212–218.
[82] A. VAN DER ZIEL, *Noise Measurements*, John Wiley, New York, 1976.

[83] C. D. Motchenbacher, F. Fitchen, *Low Noise Electronic Design*, John Wiley, New York, 1973.

[84] Y. Netzer, The Design of low-noise Amplifiers, *Proc. IEEE*, 69, June 1981, pp. 728–741.

[85] M. S. Gupia, Applications of Electrical Noise, *Proc. IEEE*, 63, July 1975, p. 996–1010, (reproduced in [86]).

[86] M. S. Gupia, (ed.), *Electrical Noise: Fundamentals and Sources*, IEEE Press, New York, 1977.

[87] S. O. Rice, Envelope of narrow-band signals, *Proc. IEEE*, 70, July 1982, pp. 692–699.

[88] M. Abramowitz, I. A. Stegun, (ed.), *Handbook of Mathematical Functions*, Applied Mathematics Series no. 55, National Bureau of Standards, U.S. Government Printing Office, Washington, D.C., 1964.

[89] A. Papoulis, Random modulation: a review, *IEEE Trans. ASSP*, 31, February 1983, pp. 96–105.

[90] W. S. Burdic, *Radar Signal Analysis*, Prentice-Hall, Englewood Cliffs, 1968.

[91] R. Deutsch, *System Analysis Techniques*, Prentice-Hall, Englewood Cliffs, 1969.

[92] M. Carpentier, *Radars, Bases modernes*, Masson, Paris, 1981 (4e éd.).

[93] B. Escudie, Représentation en temps et fréquence des signaux d'énergie finie: analyse et observation des signaux, *Annales Télécomm.*, 35, March-April 1979, pp. 101–111.

[94] I. S. Gradshteyn, I. M. Ryzhik, *Table of integrals, series and products*, Academic Press, New York, 1965.

[95] A. F. Arbel, *Analog signal processing and instrumentation*. Cambridge University Press, Cambridge, 1980.

[96] U. Tietze, C. Schenk, *Advanced Electronic Circuits*, Springer-Verlag, Berlin, 1978.

[97] S. K. Mitra, *An Introduction to Digital and Analog Integrated Circuits and Applications*, Harper and Row, New York, 1980.

[98] A. C. Salazar (ed.), *Digital Signal Computers and Processors*, IEEE Press, New York, 1977.

[99] F. Kühne, Modulationssysteme mit Sinusträger, *Archiv f. Electr. Übertrag.*, 24, (1970), pp. 139–150 and 25 (1971), pp. 117–128.

[100] R. Deutsch, *Nonlinear Transformations of Random Processes*, Prentice-Hall, Englewood Cliffs, N.J., 1962.

[101] A. H. Haddad (ed.), Non Linear Systems: *Processing of Random Signals—Classical Analysis*, Dowden, Hutchinson and Ross, Stroudsburg, 1975.

[102] W. J. Rugh, *Nonlinear System Theory*, The John Hopkins University Press, Baltimore, 1981.

[103] S. J. Mason, H. J. Zimmermann, *Electronic Circuits, Signals and Systems*, John Wiley, New York, 1960.

[104] E. D. Sunde, *Communication Systems Engineering Theory*, John Wiley, New York, 1969.

[105] S. A. Tretter, *Introduction to Discrete-Time Signal Processing*, John Wiley, New York, 1976.

[106] J. A. Betts, *Signal Processing, Modulation and Noise*, The English University Press, London, 1970.

[107] H. J. Nussbaumer, *Fast Fourier Transform and Convolution Algorithms*, Springer-Verlag, Berlin, 1981.

[108] K. Arbenz, A. Wolhauser, *Analyse numérique*, Presses polytechniques romandes, Lausanne, 1980.

[109] J. R. Ragazzini, G. F. Franklin, *Sampled-Data Control Systems*, McGraw-Hill, New York, 1958.

[110] B. Loriferne, *La conversion analogique-numérique/numérique-analogique*, Eyrolles, Paris, 1976.

[111] B. M. Gordon, Linear Electronic Analog/Digital Conversion Architectures, their Origins, Parameters, Limitations and Applications, *IEEE Trans. Circuits & Systems*, CAS-25, 7, July 1978, pp. 391–418.

[112] J. Max, Quantizing for minimum distorsion, *IRE Trans. Inform. Theory*, IT-6, March 1960, pp. 7–12.

[113] A. B. Sripad, D. L. Snyder, A necessary and sufficient condition for quantization errors to be uniform and white, *IEEE Trans. Acoustics, Speech and Signal Proc.*, ASSP-25, Oct. 1977, pp. 442–448.

[114] F. Castanie, Signal processing by random reference quantizing, *Signal Processing*, 1, January 1979, pp. 27–43.

[115] W. R. Bennett, J. R. Davey, *Data Transmission*, McGraw-Hill, New York, 1965.

[116] P. M. Chirlian, *Analysis and Design of Integrated Electronic Circuits*, Harper & Row, New York, 1981.

[117] F. M. Gardner, *Phaselock Techniques*, John Wiley, New York, 1979.

[118] A. J. Viterbi, *Principles of Coherent Communication*, McGraw-Hill, New York, 1966.

[119] D. G. Childers (ed.), *Modern Spectrum Analysis*, IEEE Press, New York, 1978.

[120] S. M. Kay, S. L. Marple, Spectrum analysis—A modern perspective, *Proc. IEEE*, 69, 11, Nov. 1981, pp. 1380–1419.

[121] B. Bajić, Distribution of results of analysis by mean of spectrum analyser with ideal bandpass filter and true integrator, *Signal Processing*, 5, 1983, pp. 47–60.

[122] G. M. JENKINS, D. G. WATTS, *Spectral Analysis and its Applications*, Holden-Day, San Francisco, 1968.

[123] R. K. OTNES, L. ENOCHSON, *Digital Time Series Analysis*, John Wiley, New York, 1972.

[124] R. W. BRODERSEN, C. R. HEWES, D. D. BUSS, A 500-stage CCD transversal filter for spectral analysis, *IEEE J. Solid-State Circuits*, SC-11, 1, February 1976, pp. 75–84.

[125] J. D. MAINES, E. G. S. PAIGE, Surface-acoustic-wave devices for signal processing applications, *Proc. IEEE*, 64, 5, May 1976, pp. 639–651.

[126] B. V. MARKEVITCH, Spectral analysis: a comparison of various methods, *Proc. Int. Specialist Seminar on the Impact of New Technologies in Signal Processing*, Aviemore, Scotland, 20–24 September 1976, IEE Conf. Publication No. 144, pp. 96–103.

[127] W. T. CATHEY, *Optical Information Processing and Holography*, John Wiley, New York, 1974.

[128] J. L. MESA, D. L. COHN, *Decision and Estimation Theory*, McGraw-Hill, Tokyo, 1978.

[129] M. D. SRINATH, P. K. RAJASEKARAN, *An Introduction to Statistical Signal Processing with Applications*, John Wiley, New York, 1979.

[130] H. L. VAN TREES, *Detection Estimation and Modulation Theory*, John Wiley, New York, 1968–1971, (3 vol.).

[131] A. D. WAHLEN, *Detection of Signals in Noise*, Academic Press, New York, 1971.

[132] P.-Y. ARQUÈS, *Décisions en traitement du signal*, Masson, Paris, 1979.

[133] J. LIFERMANN, *Les principes du traitement statistique du signal*, Masson, Paris, 1981.

[134] H. CRAMER, *Mathematical Methods of Statistics*, Princeton University Press, Princeton, 1946.

[135] H. SCHWARTZ, L. SHAW, *Signal Processing: Discrete Spectral Analysis, Detection and Estimation*, McGraw-Hill, New York, 1975.

[136] B. D. O. ANDERSON, J. B. MOORE, *Optimal Filtering*, Prentice-Hall, Englewood Cliffs, 1979.

[137] J. S. BENDAT, A. G. PIERSOL, *Engineering Applications of Correlation and Spectral Analysis*, John Wiley, New York, 1980.

[138] Correlation in Action, *Hewlett-Packard Journal*, Nov. 1969.

[139] R. C. DIXON, *Spread Spectrum Systems*, John Wiley, New York, 1976.

[140] C. H. CHEN (ed.), *Digital Waveform Processing and Recognition*, CRC Press Inc., Boca Raton, Florida, 1982.

[141] L. E. BRENNAN, I. S. REED, A recursive method of computing the Q-function, *IEEE Trans. Inf. Theory*, IT-11, April 1965, pp. 312–313.

[142] D. V. SARAVATE, M. B., PURSLEY, Crosscorrelation properties of pseudorandom and related sequences. *Proc. IEEE*, 68, 5, May 1980, pp. 593–619.

[143] T. H. BIRDSALL, On understanding the matched filter in the frequency domain, *IEEE Trans. on Education*, E-19, Nov. 1976, pp. 168–169.

[144] K. FUKUNAGA, *Introduction to Statistical Pattern Recognition*, Academic Press, New York, 1972.

[145] R. O. DUDA, P. E. HART, *Pattern Classification and Scene Analysis*, John Wiley, New York, 1973.

[146] J. R. ULLMAN, *Pattern Recognition Techniques*, Butterworth & Co., London, 1973.

[147] J. T. TOU, R. C. GONZALES, *Pattern Recognition Principles*, Addison-Wesley, Reading, 1974.

[148] R. C. GONZALES, M. G. THOMASON, *Syntactic Pattern Recognition: An Introduction*, Addison-Wesley, Reading, 1978.

[149] H. URKOWITZ, *Signal Theory and Random Processes*, Artech House, Dedham, 1983.

Select Bibliography

The Traité d'Électricité, listed below by volume number, is published by the Presses Polytechniques Romandes (Lausanne, Switzerland) in collaboration with the École Polytechnique Fédérale de Lausanne. The title of each volume is given with the year of publication in parenthesis. English translations by Artech House are denoted by an asterisk with the year of publication in parenthesis.

Vol.	Author	Title
I	Frédéric de Coulon & Marcel Jufer	Introduction à l'électrotechnique (1981).
II	Philippe Robert	Materiaux de l'électrotechnique (1979).
III	Fred Gardiol	Electromagnétisme (1979).
IV	René Boite & Jacques Neirynck	Theorie des reseaux de Kirchhoff (1983).
V	Daniel Mange	Analyse et synthèse des systèmes logiques (1979). *Analysis and Synthesis of Logic Systems (1986).
VI	Frédéric de Coulon	Theorie et traitement des signaux (1984). *Signal Theory and Processing (1986).
VII	Jean-Daniel Chatelain	Dispositifs à semiconducteur (1979).
VIII	Jean-Daniel Chatelain & Roger Dessoulavy	Electronique (1982).
IX	Marcel Jufer	Transducteurs électromécaniques (1979).
X	Jean Chatelain	Machines électriques (1983).
XI	Jacques Zahnd	Machines séquentielles (1980).
XII	Michel Aguet & Jean-Jacques Morf	Energie électrique (1981).
XIII	Fred Gardiol	Hyperfréquences (1981). *Introduction to Microwaves (1984).
XIV	Jean-Daniel Nicoud	Calculatrices (1983).
XV	Hansruedi Bühler	Electronique de puissance (1981).
XVI	Hansruedi Bühler	Electronique de réglage et de commande (1979).
XVII	Philippe Robert	Mesures (1985).
XVIII	Pierre-Gérard Fontolliet	Systèmes de télécommunications (1983). *Telecommunication systems (1986).

XIX	Martin Hasler & Jacques Neirynck	Filtres électriques (1981). *Electric Filters (1986).
XX	Murat Kunt	Traitement numérique des signaux (1980). *Digital Signal Processing (1986).
XXI	Mario Rossi	Electroacoustique (1984).
XXII	Michel Aguet & Mircea Ianovici	Haute tension (1982).

Glossary

Symbol	Designation
a	Parameter
$a=(a_1,a_2, \ldots)$	Parameters vector
$\tilde{a}$	Estimate of parameters a
$\tilde{a}_{abs}$	Absolute value estimate
$\tilde{a}_{mse}$	Minimum mean square error estimate
$\tilde{a}_l$	Linear estimate
$\tilde{a}_{lo}$	Optimal linear estimate
$\tilde{a}_{map}$	Maximum a posteriori estimate
$\tilde{a}_{ml}$	Maximum likelihood estimate
$a(t)$	Real part of the complex envelope $\underset{\sim}{r}(t)$
A	Amplitude
b	Bias
$b(t)$	Imaginary part of the complex envelope $\underset{\sim}{r}(t)$
B	Frequency bandwidth
B_e	Efficient bandwidth
B_{eq}	Noise equivalent bandwidth
B_u	Useful bandwidth
$B_{-3\mathrm{dB}}$	-3 dB bandwidth
B_τ	Approximative bandwidth
c_{ij}	Cost coefficient
$c(a,\tilde{a})$	Cost function
C	Electrical capacitance
C_x	Covariance matrix
C_{xy}	Covariance of random variables $\mathbf{x}$ and $\mathbf{y}$
$C_x(\tau)$	Autocovariance function of the random process $\mathbf{x}(t)$
$C_{xy}(\tau)$	Crosscovariance function of $\mathbf{x}(t)$ and $\mathbf{y}(t)$
d	Distortion
$d(t)$	Demodulated signal
$d(x,y)$	Distance between x and y
D	Duration
D_e	Effective duration
D_u	Useful duration
D_τ	Correlation duration
e	Electron charge
$e(t)$	Sampling function

$e(t)$	Approximation or estimation error
$\|e\|^2$	Mean square error
E	Mathematical expectation
f	Frequency
f_c	Cut-off frequency
f_e	Sampling frequency
f_i	Image frequency
f_p	Carrier frequency
$f_i(t)$	Instantaneous frequency
F	Fourier transform
F_v	Noise factor
F	Frequency resolution
$F(x)$	Cumulative distribution function of the random variable x
$F_c(x)=1-F(x)$	Complementary cumulative distribution function
$g(t),h(t)$	Impulse response of a linear system
$G(f),H(f)$	Transfer function of a linear system
h	Planck's constant
H	Hilbert transform
H_0,H_1	Hypotheses associated with a binary decision
H_N	$N \times N$ Hadamard matrix
$i(t)$	Electrical current (instantaneous value)
$I_0(x)$	Modified Bessel function of zero order
$j=\sqrt{-1}$	Imaginary unit
J	Jacobian
$J_n(x)$	Zero-order Bessel function of the first kind
k	Bolzmann constant
K	Constant
lb	Binary logarithm (base 2)
ln	Natural logarithm (base e)
log	Decimal logarithm (base 10)
$l(a)$	Likelihood function
L^2	Vector space of finite energy signals
m	Amplitude modulation index
$m_{xn}=E[\mathbf{x}^n]$	Statistical moment of degree n of **x**
$m_{x-\mu,n}=[(\mathbf{x}-\mu)^n]$	Statistical central moment of degree n of **x**
$m_{x1}=\mu_x$	Mean value of **x**
m_{x2}	Mean square value of **x**
$m_{x-\mu,2}=\sigma_x^2$	Variance of **x**
$m(t)$	Modulating signal
$n(t)$	Noise
$n_q(t)$	Quantization noise (distortion)

N	Number of samples, components, *et cetera*
O	Observation space
O_0, O_1	Subregions of O
P	Probability (shorthand notation)
P_d	Detection probability
P_f	False alarm probability
P_n	Non-detection probability
P_ϵ	Error probability
P_t	Instantaneous power
$P_x(x)$	Probability density of the random variable **x**
$P_{xy}(x,y)$	Joint probability density of **x** and **y**
$P(x/y)$	Conditional probability density of x knowing y
P_x	Normalized power of $x(t)$
$P_{th}=kTB$	Thermal noise power
Prob(A)	Probability of event A
Prob(A,B)	Joint probability of A and B
Prob(A|B)	Conditional probability of A knowing B
q	Total number of quantization steps
$Q(a,b)$	Marcum function
$r(t)$	Ramp function
$r_x(t)$	Real envelope of $x(t)$
$\underset{\sim}{r}x(t)$	Complex envelope of $x(t)$
R	Electrical resistance
R	Average risk
R_x	Correlation matrix
$R_{xy}=E[\mathbf{x}^*\mathbf{y}]$	Correlation of random variables **x** and **y**
$R_x(\tau)$	Statistical autocorrelation function of $\mathbf{x}(t)$
$R_{xy}(\tau)$	Statistical crosscorrelation function of $\mathbf{x}(t)$ and $\mathbf{y}(t)$
s_k	Sequence of the Walsh function wal$(k,t/T)$
$s(t)$	Modulated signal
$s(t)$	Useful signal
S	Functional operator
t	Time
t_g	Group delay
t_m	Rise time
t_ϕ	Phase delay
T	Absolute temperature
T_a	Absolute ambient temperature
T_d	Time resolution power
T_e	Sampling interval
T_{eb}	Noise equivalent temperature
$u(t)$	Electrical voltage (instantaneous value)

$u(t)$	Auxiliary signal, weighting factor
$u_p(t)$	Carrier
$\hat{U}$	Peak value of voltage $u(t)$
$\underline{\hat{U}}=\hat{U}\exp(\text{j}\alpha)$	Phasor of sinewave voltage $u(t)$
v_b	Sweeping speed
V	Decision threshold
W_x	Normalized energy of signal $x(t)$
$W_N=\exp(\text{j}2\pi/N)$	Nth root of unit
$x_k=x(t_k)$	Sample of $x(t)$ taken at $t=t$
$\{x_k\}$	Sample sequence
$\mathbf{x},\mathbf{y},\mathbf{z},\ldots$	Random variables
$\|x\|$	Norm of $x(t)$
$\bar{x},\bar{x}(t)$	Time average value of $x(t)$ or $\{x_k\}$
$\bar{x}(T)$	Average value over interval T
$x(t,T)$	Running average value over interval T
$\mathbf{x},\mathbf{x}^T$	Line vector (transposed)
$\mathbf{x}=(\mathbf{x}_1,\mathbf{x}_2,\ldots)$	Random vector
$x(t),y(t),z(t),\ldots$	Real or complex signals
$x^*(t)$	Complex conjugate of $x(t)$
$\tilde{x}(t)$	Estimation of $x(t)$
$\mathbf{x}(t)\triangleq\mathbf{x}(t,\zeta)$	Random process
$x(t,T)\leftrightarrow X(f,T)$	Section of $x(t)$ of T length and its Fourier transform
$\check{x}(t)\leftrightarrow\check{X}(f)$	Hilbert transform of $x(t)$ and its Fourier transform
$\underline{\underline{X}}(t)\leftrightarrow\underline{\underline{X}}(f)$	Analytic signal of $x(t)$ and its Fourier transform
$x_e(t)$	Sampled signal
$x_i(t)\leftrightarrow X_i(f)$	Odd part of $x(t)$ and its Fourier transform
$x_p(t)\leftrightarrow X_p(f)$	Even part of $x(t)$ and its Fourier transform
X_n	Fourier coefficient
$X(f)=\text{F}\{x(t)\}$	Fourier transform of $x(t)$
$\lvert X(f)\rvert$	Amplitude spectrum of $x(t)$
$X^+(f)=2\epsilon(f)X(f)$	Unilateral representation of $X(f)$
$X(z)$	z-transform of $x(t)$
$\underline{Z}(f)$	Complex impedance
α,β,θ,ϕ	Phase
α_k	Series expansion coefficient
$\gamma_e=\langle x,\psi_l^*\rangle$	Inner product of $x(t)$ and $\psi_k(t)$
$\gamma(t)$	Unit step response of a linear system
Γ	Column vector of γ_l
$\Gamma(f)$	Coherence function
δ	Angle modulation index (FM,ΦM)
$\delta(t)$	Delta function
$\delta_T(t)$	Sequence of delta function with period T

Δ	Quantization step
$\Delta t, \Delta x$.	Small interval
$\Delta f(t)$	Frequency deviation (FM)
$\Delta\phi(t)$	Phase deviation (ΦM)
ϵ	Measurement error
$\epsilon(t)$	Unit step function
ζ	Result of a random experiment
η	Unilateral power spectral density of the white noise
η_m	Modulation efficiency (AM)
ϑ_a	Ambient temperature in °C
$\vartheta_x(f)=\arg X(f)$	Phase spectrum of signal $x(t)$
$\vartheta_g(f)=\arg G(f)$	Phase response of a linear system
λ	Wavelength
$\lambda_{ke}=<\psi_k,\psi_l^*>$	Inner product of base functions
$\lambda_k=\|\psi_k\|^2$	Square of the $\psi_k(t)$ norm
Λ	Matrix of λ_{kl}
$\Lambda(x)$	Likelihood ratio
$\mu_x=E[\mathbf{x}]$	Statistical mean value of variable x
ν	Doppler shift
ξ	Signal-to-noise ratio
ξ_0	Reference signal-to-noise ratio
$\xi_{dB}=10\log\xi$	Signal-to-noise ratio in decibels
$\Pi_x(u)$	Characteristic function of random variable **x**
$\Pi_{xy}(u,v)$	Joint characteristic function of the **x,y** pair
ρ_{xy}	Correlation coefficient of **x** and **y**
$\rho_x(\tau)$	Normalized autocovariance function
$\rho_{xy}(\tau)$	Normalized crosscovariance function
σ_x	Standard deviation of variable x
$\sigma_x^2=\mathrm{Var}(x)$	Variance of x
τ	Delay parameter, time
$\phi_x(t)$	Instantaneous phase of $x(t)$
$\varphi_x(\tau)$	Time autocorrelation function of a finite average power signal $x(t)$
$\mathring{\varphi}_x(\tau)$	Time autocorrelation function of a finite energy signal $x(t)$
$\mathring{\varphi}_x(\tau,T)$	Autocorrelation function of $x(t,T)$
$\varphi_{xy}(\tau), \mathring{\varphi}_{xy}(\tau)$	Time crosscorrelation function of $x(t)$ and $y(t)$
$\mathring{\varphi}_{xy}(\tau,T)$	Crosscorrelation function of $x(t,T)$ and $y(t,T)$
$\Phi_x(f)=F\{\varphi_x(\tau)\}$	Power spectral density of $x(t)$
$\mathring{\Phi}_x(f)=F\{\mathring{\varphi}_x(\tau)\}$	Energy spectral density of $x(t)$
$\mathring{\Phi}(f,T)$	Energy spectral density of $x(t,T)$
$\mathring{\Phi}_x^+(f)=2\epsilon(f)\mathring{\Phi}_x(f)$	Unilateral energy spectral density of $x(t)$

$\Phi_x^+(f)=2\epsilon(f)\Phi_x(f)$	Unilateral power spectral density of $x(t,T)$
$\Phi_{xy}(f)=F\{\varphi_{xy}(\tau)\}$	Power cross-spectral density of $x(t)$ and $y(t)$
$\mathring{\Phi}_{xy}(f)=F\{\mathring{\varphi}_{xy}(\tau)\}$	Energy cross-spectral density of $x(t)$ and $y(t)$
$\Phi_{xy}(f,T)$	Energy cross-spectral density of $x(t,T)$ and $y(t,T)$
$\Phi_{th}(f)$	Thermal noise power spectral density
$\chi(\tau,\nu)$	Ambiguity function
χ_m^2	Random variable with m degrees of freedom
$\{\psi_k(t)\}$	Set of independent functions, forming the base of a signal space
$\omega=2\pi f$	Angular velocity
$\omega_i(t)$	Instantaneous angular velocity

ABBREVIATIONS

A/D	Analog-digital conversion
AM	Amplitude modulation
AM-C	AM on carrier
CRV	Continuous random variable
D/A	Digital-analog conversion
DFT	Discrete Fourier transform
DPCM	Differential PCM
DRV	Discrete random variable
FFT	Fast Fourier transform
FM	Frequency modulation
FSK	Frequency shift keying
NRZ	Non-return to zero
PAM	Pulse AM
PCM	Pulse code modulation
PDM	Pulse duration modulation
PFM	Pulse FM
PLL	Phase locked loop
PPM	Pulse position modulation
PSK	Phase shift keying
SSB	Single sideband
VCO	Voltage controlled modulation
VSB	Vestigial sideband
ΔM	Delta modulation
ΦM	Phase modulation

SPECIAL SYMBOLS

$*$	Convolution product
$\overline{*}$	Mean value of convolution product
$**$	Convolution product of bidimensional functions
$\mathrm{H}\!\!\!\ast$	Multiple convolution product
$< >$	Scalar product
$\otimes$	Kronecker product
$\oplus$	Modulo-2 addition
$\mathrm{cal}(s_k,t/T)$	Even Walsh function of sequence s_k for k even
$\mathrm{ig}(t)$	Gaussian pulse
$\mathrm{rad}(i,t/T)$	Rademacher function
$\mathrm{rect}(t)$	Rectangular function
$\mathrm{rep}T[\ \ ]$	Repetition operator of period T
$\mathrm{sal}(s_k,t/T)$	Odd Walsh function of sequence s_k for k odd
$\mathrm{sgn}(t)$	Sign function
$\mathrm{sinc}(\alpha)$	Cardinal sine function
$\mathrm{Si}(u)$	Integral sine function
$\mathrm{tri}(t)$	Triangular function
$\mathrm{wal}(k,t/T)$	Walsh function

Index